International Handbooks of Quality-of-Life

Chair of the Editorial Board
Graciela Tonon, Universidad Nacional de Lomas de Zamora and Universidad de Palermo, Argentina

Editorial Board
Alex Michalos, University of Northern British Columbia, Canada
Rhonda Phillips, Purdue University, USA
Don Rahtz, College of William & Mary, USA
Dave Webb, University of Western Australia, Australia
Wolfgang Glatzer, Goethe University, Germany
Dong Jin Lee, Yonsei University, Korea
Laura Camfield, University of East Anglia, UK

Aims and Scope

The International Handbooks of Quality-of-Life Research offer extensive bibliographic resources. They present literature reviews of the many sub-disciplines and areas of study within the growing field of quality of life research. Handbooks in the series focus on capturing and reviewing the quality of life research literature in specific life domains, on specific populations, or in relation to specific disciplines or sectors of industry. In addition, the Handbooks cover measures of quality of life and well-being, providing annotated bibliographies of well-established measures, methods, and scales.

More information about this series at http://www.springer.com/series/8365

Ghozlane Fleury-Bahi • Enric Pol
Oscar Navarro
Editors

Handbook of Environmental Psychology and Quality of Life Research

Library
Quest University Canada
3200 University Boulevard
Squamish, BC V8B 0N8

Springer

Editors
Ghozlane Fleury-Bahi
Faculty of Psychology
University of Nantes
Nantes, France

Enric Pol
Department of Social Psychology
University of Barcelona
Barcelona, Spain

Oscar Navarro
Faculty of Psychology
University of Nantes
Nantes, France

ISSN 2468-7227　　　　　ISSN 2468-7235　(electronic)
International Handbooks of Quality-of-Life
ISBN 978-3-319-31414-3　　ISBN 978-3-319-31416-7　(eBook)
DOI 10.1007/978-3-319-31416-7

Library of Congress Control Number: 2016949049

© Springer International Publishing Switzerland 2017
This work is subject to copyright. All rights are reserved by the Publisher, whether the whole or part of the material is concerned, specifically the rights of translation, reprinting, reuse of illustrations, recitation, broadcasting, reproduction on microfilms or in any other physical way, and transmission or information storage and retrieval, electronic adaptation, computer software, or by similar or dissimilar methodology now known or hereafter developed.
The use of general descriptive names, registered names, trademarks, service marks, etc. in this publication does not imply, even in the absence of a specific statement, that such names are exempt from the relevant protective laws and regulations and therefore free for general use.
The publisher, the authors and the editors are safe to assume that the advice and information in this book are believed to be true and accurate at the date of publication. Neither the publisher nor the authors or the editors give a warranty, express or implied, with respect to the material contained herein or for any errors or omissions that may have been made.

Printed on acid-free paper

This Springer imprint is published by Springer Nature
The registered company is Springer International Publishing AG Switzerland

Preface

Although what we now call environmental psychology had a number of important precursors and prophets in other countries before it had a name, in Germany (Willy Hellpach), Japan (Tetsuro Watsuji), and Canada (Robert Sommer), among others, it was formally founded as a discipline in the United States in the late 1960s. For quite a number of years, most research and writing emanated from that country. However, slowly but surely, that "pebble tossed in the pond" has rippled outward, so much so that one might argue that more research in environmental psychology is now conducted in other countries.

This handbook is a significant and very welcome sign of the internationalization of environmental psychology. Consider that the first handbook of environmental psychology, edited by Daniel Stokols and Irwin Altman in 1987, drew upon the expertise of 66 authors from 11 countries, and 43 of the authors were in the United States. In comparison, the present handbook's editors called upon the expertise of 73 authors from 14 countries, and 71 of the authors are from countries other than the United States. This is one clear indication of the healthy expansion of environmental psychology on the global stage. One might note that this expansion will be complete when some future handbook includes the growing and valuable contributions of authors from the Middle East, Asia, and Africa. However, for now, we can celebrate the progress to date in this comprehensive volume.

This volume is a part of a handbook series centered on quality of life. Its unique contribution is to focus on how environmental psychology understands the notion of quality of life and contributes to its definition. Indeed, quality of life has been a major theme for environmental psychology at many, although certainly not all, points in its history. The challenge for this volume was to review and to rethink the various dominant themes at this moment in history, given that the QoL approach serves as the broad palette of this volume. I infer that "quality of life" was meant in this project to mean life in the neighbourhood, the city, the society, and the planet. These "ripples in the pond" do indeed represent, in general, a widening circle compared to the typical (but certainly not universal) focus of environmental psychology during its early days on more proximate settings, such as the interiors of buildings. Indeed, one mid-1960s name for the field was "architectural psychology."

The widening of the metaphorical ripples on the pond also symbolizes the increased scope of the articles in the handbook to include greater links with neighboring subdisciplines, such as social and health psychology, as well as with other disciplines, including sociology, geography, architecture, anthropology, urban studies, and engineering.

Another value of this handbook lies in its combination of traditional and innovative topics. The reader will find chapters on such core original topics as environmental and urban design, schools, children, workplaces, residential satisfaction, hospitals, risk, and stress, as well as on newer but now well-established topics such as sustainability, nature, restoration, and place attachment. But the reader will also find chapters that have had little space, to my knowledge, in reviews of the field and handbooks: green exercise, global challenges, micropolitics, identity dynamics, cultural practices, and spatial and social inequality.

Overall, the editors have provided you, the reader, with a truly innovative mix of traditional and cutting-edge chapters, crafted by the most international group of authors ever seen in survey of the field. Savor it!

University of Victoria, Victoria, BC, Canada   Robert Gifford
March 1, 2016

Acknowledgements

We would like to thank all the contributors of this Handbook. We would also like to address a special thanks to the authors who accepted, at our request, the challenge of writing a chapter together.

Contents

1 **Introduction: Environmental Psychology and Quality of Life** . . 1
   Ghozlane Fleury-Bahi, Enric Pol, and Oscar Navarro

Part I People-Environment Relations and QoL – *Environmental Quality and Well-Being*

2 **Quality of Life and Sustainability: The End of Quality at Any Price**. 11
   Enric Pol, Angela Castrechini, and Giuseppe Carrus

3 **Some Cues for a Positive Environmental Psychology Agenda**. 41
   Sergi Valera and Tomeu Vidal

4 **Linking People-Environment Research and Design. What Is Missing?**. 65
   Carole Després and Denise Piché

5 **Place Attachment, Sense of Belonging and the Micro-Politics of Place Satisfaction**. 85
   Andrés Di Masso, John Dixon, and Bernardo Hernández

Part II People-Environment Relations and QoL – *Restorative Environments*

6 **Self, Nature and Well-Being: Sense of Connectedness and Environmental Identity for Quality of Life**. 107
   Pablo Olivos and Susan Clayton

7 **Restorative Environments and Health**. 127
   Silvia Collado, Henk Staats, José Antonio Corraliza, and Terry Hartig

8 **Green Exercise, Health and Well-Being**. 149
   Ana Loureiro and Susana Veloso

Part III People-Environment Relations and QoL – *Ecological Behavior*

9 Sustainable Behavior and Quality of Life 173
Cesar Tapia-Fonllem, Victor Corral-Verdugo, and Blanca Fraijo-Sing

10 Self-Determined, Enduring, Ecologically Sustainable Ways of Life: Attitude as a Measure of Individuals' Intrinsic Motivation 185
Florian G. Kaiser, Alexandra Kibbe, and Oliver Arnold

11 Commitment and Pro-Environmental Behaviors: Favoring Positive Human-Environment Interactions to Improve Quality of Life 197
Christophe Demarque and Fabien Girandola

12 Pro-environmentalism, Identity Dynamics and Environmental Quality of Life 211
Marie-Line Félonneau and Elsa Causse

13 Can Engagement in Environmentally-Friendly Behavior Increase Well-Being? 229
Leonie Venhoeven, Linda Steg, and Jan Willem Bolderdijk

Part IV Well-Being and Daily Environments – *Urban Environments*

14 Urban Design and Quality of Life 241
Ombretta Romice, Kevin Thwaites, Sergio Porta, Mark Greaves, Gordon Barbour, and Paola Pasino

15 The City as an Environment for Urban Experiences and the Learning of Cultural Practices 275
Pablo Páramo

16 Adjustment to Geographical Space and Psychological Well-Being 291
Thierry Ramadier

Part V Well-Being and Daily Environments – *Residential Environments*

17 Residential Satisfaction and Quality of Life 311
Juan Ignacio Aragonés, María Amérigo, and Raquel Pérez-López

18 Spatial Inequalities, Geographically-Based Discrimination and Environmental Quality of Life 329
Ghozlane Fleury-Bahi and André Ndobo

19 Children in Cities: The Delicate Issue of Well-Being and Quality of Urban Life 345
Sandrine Depeau

20 Everyday Environments and Quality of Life: Positive School and Neighborhood Environments Influence the Health and Well-Being of Adolescents............................ 369
Taciano L. Milfont and Simon J. Denny

Part VI Well-Being and Daily Environments – *Work and Institutional Environments*

21 The Effect of Workplace Design on Quality of Life at Work... 387
Jacqueline C. Vischer and Mariam Wifi

22 Comfort at Work: An Indicator of Quality of Life at Work... 401
Liliane Rioux

23 Quality of the Hospital Experience: Impact of the Physical Environment... 421
Ann Sloan Devlin and Cláudia Campos Andrade

24 Healthy Residential Environments for the Elderly........... 441
Ferdinando Fornara and Sara Manca

Part VII Quality of Life and Environmental Threats – *Environmental Stressors and Risks*

25 Environmental Stress................................... 469
Birgitta Gatersleben and Isabelle Griffin

26 Living in an "At Risk" Environment: The Example of "Costal Risks"... 487
Élisabeth Michel-Guillou and Catherine Meur-Ferec

27 Social Inequality and Environmental Risk Perception....... 503
Oscar Navarro

28 Living in Industrial Areas: Social Impacts, Adaptation and Mitigation... 519
Maria Luísa Lima and Sibila Marques

Part VIII Quality of Life and Environmental Threats – *Global Change, Energy and Emerging Risks*

29 Emerging Risks and Quality of Life: Towards New Dimensions of Well-Being?............................... 531
Dorothée Marchand, Karine Weiss, and Bouchra Zouhri

30 Energy Issues: Psychological Aspects..................... 543
Rafaella Lenoir-Improta, Patrick Devine-Wright, José Q. Pinheiro, and Petra Schweizer-Ries

31 Global Challenges for Environmental Psychology: The Place of Labor and Production................................ 559
David Uzzell, Nora Räthzel, Ricardo García-Mira, and Adina Dumitru

About the Editors

Ghozlane Fleury-Bahi After gaining a PhD in Psychology at the University Paris Descartes, Ghozlane Fleury-Bahi became a Lecturer at the University of Nantes in 2000. Since 2010, she is Full Professor of Social and Environmental Psychology at the University of Nantes. Her research is based on the study of the psychosocial processes involved in the relationship to the environment, and more precisely on the evaluation of environmental risks, environmental health and the environmental determinants of quality of life.

Enric Pol received his PhD in Psychology from the University of Barcelona in 1986. In 1987, he was promoted to "Profesor Titular" position and since 2003, he is Full Professor on Social and Environmental Psychology at the University of Barcelona in 2003. Since 1987 he has been teaching Applied Social Psychology, with special emphasis on Quality of Life and wellbeing. Since 1988 he is running a professional oriented Master Program on "Environmental Intervention and Management: People and Society". As Director of the Department of Social Psychology (2008–2012) and director of Research Group on Environmental Psychology first (since 1987), and on Social, Environmental and Organizational Psychology (since 2005), he has conducted research on Quality of Life in Barcelona and in Catalonia, and he published a large number of papers on peer-reviewed journals, and books and chapters.

Oscar Navarro is Associate Professor in Social Psychology at University of Nantes in France. He is member of Psychology Laboratory of Pays de la Loire (LPPL) and is currently working on analysis of psychosocial and environmental factors, which can influence the evaluation and management of risks. He is Principal Investigator for the project CLIMATRisk (sense of vulnerability and adaptation to face effects of climate change strategies: the case of perception of coastal risks), which is funded by the ANR (the French national agency for research).

Introduction: Environmental Psychology and Quality of Life

Ghozlane Fleury-Bahi, Enric Pol, and Oscar Navarro

When examining human quality of life, it is essential to take into account the intrinsic quality of different living spaces, for example, housing, neighborhoods, schools, workplaces or, on a larger scale, the planet. This issue of the links between quality of life and the environment is becoming increasingly significant with, at a local level, problems resulting from different types of annoyances, such as pollution and noise, while, at a global level, there is the central question of climate change with its harmful consequences for humans and the planet. The problems caused by pollution are extremely important; however, the lack of basic human needs, such as water, food, shelter and safety, is of greater concern, in that it has an even more drastic impact on the quality of life.

Quality of life is an integrative concept, situated at the intersection of human and social sciences and health; it combines the notion of happiness, philosophical in origin, with that of subjective well-being originating in psychology, and those of physical and mental health emanating from the medical sciences. As Sirgy (2012 p. 5–9) summarizes, we need to distinguish two main approaches of philosophers, which have significant implications for the psychology of quality of life. On one hand, in the tradition of Hobbes, Locke, Bentham, Mill and Rousseau, there is the Hedonic vision that considers people are motivated to enhance their personal freedom, self-preservation, and self-enhancement. This approach focuses on the integrity of the individual and his/her own judgment about what makes him/her happy. This is mainly "contentment", an emotional dimension of well-being. On the other hand, there is the Eudaimonic tradition, which some authors translate as flourishing, well-being, success, or the opportunity to lead a purposeful and meaningful life. This approach is rooted in the Aristotelian concept of a good life, prudence, reason, and justice. People wish to fulfill their potential, contribute to society, and achieve the highest standards of morality. It is congruent with the Christian tradition, as represented by St Thomas Aquinas, with Confucianism, and other religious visions. This approach thus focuses on personal, social, organizational, and societal outcomes, like health, achievement and work, together with social relationships, prosocial behavior, trust, and future happiness. This distinction provides important clarification for scientific research and the decisions of policy makers. As Sirgy concludes, "*happiness maximization is not enough. We need to broaden our happiness*

G. Fleury-Bahi (✉) • O. Navarro
Faculty of Psychology, University of Nantes, Nantes, France
e-mail: ghozlane.fleury@univ-nantes.fr; oscar.navarro@univ-nantes.fr

E. Pol
Department of Social Psychology, University of Barcelona, Barcelona, Spain
e-mail: epol@ub.edu

research from the individual level and do more research at the societal level. We should take into account that happiness is a cultural value that is more embraced in Western than in Eastern cultures. We should broaden our perspective of QOL to deal with both subjective as well as objective aspects of QOL" (2012 p. 567).

From a conceptual viewpoint, there is often confusion between these different ideas. Indeed, in the field of environmental studies, the notions of quality of life, environmental satisfaction and well-being are frequently muddled (VanKamp et al. 2003). Nevertheless, it is possible to distinguish between the different concepts of quality of life. Objective concepts associate material living conditions with an absence of physical illness. On the other hand, subjective concepts perceive quality of life mainly in terms of satisfaction with life and subjective well-being. For example, the World Health Organization (WHO), defines it as *"the perception that an individual has of his/her place in life, in the context of the culture and the value system in which he/she lives, in relation to his/her objectives, expectations, norms, and preoccupations"* (WHO 1984). Lastly, there are integrative concepts that associate objective indicators of living conditions with the self-evaluation of a number of components of a psychological nature (satisfaction, subjective well-being, and happiness). For example, according to Szalai (1980), quality of life refers to the satisfying character of life; it includes well-being and satisfaction with life and is determined not only by exogenous or objective facts and factors, but also by endogenous or subjective factors, which refer to the judgment of these facts and factors, life in general and oneself. Thus, quality of life is defined as both a physical and a psychological state, which provides humans with the feeling of being satisfied with a given environment. It is generally considered to be the result of the interaction of several factors (health-related, social, economic, and environmental), which influence the human and social development of individuals and society. In the same vein, Sen introduced the concept of "capability" in the 1980s, by proposing to view the concept of well-being as dissociated from the utilitarian approach dominating the modern economy. This approach suggests understanding well-being as a group of freedoms underlying the good development of life, i.e. people's real opportunities to do and be according to their values (Nussbaum and Sen 1993). The "Capability Approach" thus refers to the maximization and fair distribution of freedom, that is to say, of the well-being of individuals. *"The capabilities determine what people can do to achieve their potential (i.e. to function at their best), a long and healthy life, access to knowledge, a decent standard of living"* (Sirgy 2012 p. 531).

Most authors agree that quality of life corresponds to a complex system embracing several domains. Thus, it is usual to differentiate between its physical, psychological and social dimensions. An environmental aspect is often added to these three classic components. In fact, certain concepts of the quality of life highlight the role played by the environment in general quality of life (Mitchell 2000; Shafer et al. 2000). Consequently, the WHO suggests including in the environmental dimension of the quality of life those annoyances linked to pollution, noise, and climate as well as certain features of places of residence (healthcare services, leisure facilities, and transport). In the same way, Mitchell (2000) suggests distinguishing six quality of life domains: health, safety, personal development, community development, natural resources, goods and services, and the physical environment. Generally speaking, this environmental component of the quality of life refers to a combination of material factors. These are objective indicators of quality of life, and correspond to goods, services and different attributes provided by the social, physical and economic environment that are characteristic of a living place (Rogerson 1995). However, beyond the objective nature of environmental attributes, it is important to be able to identify the judgment accorded to this reality. In other words, it is equally relevant in this context to consider the evaluation of these objective environmental conditions, their symbolic value, as well as the level of satisfaction or dissatisfaction they produce (Fleury-Bahi et al. 2013).

Beyond the level of satisfaction engendered by the different attributes of the living environment, the issue of environmental pollution arises, together with the impact of potentially harmful environmental conditions on the quality of life. Phenomena such as industrialization, urbanization, population density, (or, on the contrary, shrinking populations), and climate change are clearly at the heart of this debate. Ecological crises and environmental catastrophes have created new challenges such as climate change refugees, energy poverty, various types of pollution, and their impact on the well-being of populations. Moreover, the question of the environmental determinants of quality of life arises at all spatial scales; from the level of housing and the neighborhood, through public and institutional spaces to the global environment. Over several decades, environmental problems of an anthropic nature have generated scientific, political and social interest, mainly in relation to their potential impact on quality of life and people's physical and psychological health, an interest confirmed by it being the subject chosen by the World Health Organization for its World Health Day in 2008. Clearly, the quality and integrity of the environment and natural ecosystems are indispensable to the health and quality of life of human communities. Psychology has tackled this issue for many years, especially through the research carried out in the field of environmental psychology.

1.1 Environmental Psychology and Quality of Life

The concept of quality of life has provided a major challenge for environmental psychology since its beginnings. Whether one believes that environmental psychology started with the early works of Hellpach (firstly in collaboration with Wundt, and subsequently alone) and the meetings between Bauhaus and Gestalt, or one prefers to await the contributions of Terence Lee, David Canter, Harold Proshanky, and those who founded "the golden age of architectural psychology" in the 1960s and 1970s, the question of the quality of life has always been present. Nevertheless, the problem is that the term "quality of life" was explicitly used rather late in environmental psychology, even though its conceptual content has been continuously and permanently present from the very beginnings of the discipline, as demonstrated in Chap. 2 of this Handbook. Reviewing the studies carried out on the quality of life in environmental psychology presents a real challenge, due to the many different viewpoints and theoretical and epistemological visions of environmental psychology, as well as the various conceptualizations and measures of quality of life. In this Handbook, we wished to include all tendencies. We also wished to focus on the links with other specialties in psychology, especially social and health psychology, together with other disciplines such as geography, architecture, sociology, anthropology, urbanism, engineering and law. We begin with the premise that each context generates experiences that favor aspects of the ecosystem that are the most pertinent for the given situation, i.e. natural, cultural, social, economic and psychological aspects. These affect human behavior, which is, by definition, social. In fact, experiences, expectations, hopes, frustrations, and satisfactions depend on the context, in the same way as do the neurological, biological, and physiological functions of individuals. This does not mean, however, that there are no general tendencies that can be standardized, thus enabling us to predict tendencies and laws of functioning, despite the limitations due to the constant and permanent changes in the physical, social and technological context, especially at a time of major globalization. In certain periods of great enthusiasm, which also included much naivety, some researchers tried to establish causal relationships that were too simple, thereby frustrating the expectations of certain social actors. This was especially serious at the end of the 1970s and the beginning of the 1980s, leading some authors to speak of the "crisis of architectural psychology" (Pol 2007). The expectation of architecture was that environmental psychology would give guidance about the way in which space should be improved to enhance the quality of life of residents/users but this was frustrated by too simplistic and rather unsatisfactory answers. Besides, the supposedly implicit environmental and/or architectural determinism encouraged the search for "magic remedies" to improve living conditions with a certain standardization of shapes, colors, textures, spaces and connections. The failure of this model opened the way for other approaches, which reintroduced the symbolic, experiential, cultural and social dimensions at an individual level, while also highlighting the psychosocial processes involved. It was this issue that led Stokols (1987) to declare that space is never purely physical, but socio-physical.

Nevertheless, it is not enough to design a pleasant space for a better life. For the last three decades, the need to raise awareness of sustainable development has become widespread, particularly after the Brundtland report of 1987 (although the discourse of sustainability seemed to weaken after the 2008 crisis). This report focused on the individual's responsibility for his/her environmental performance, as well as the environmental responsibility linked to decisions made by technicians (including architects and planners), economic managers and policy-makers. Environmental psychology has witnessed an increase in research on attitudes and behaviors, with very varied, and sometimes contradictory, theoretical and epistemological approaches. The issue of sustainable development has also led to new constraints in terms of quality norms in the fields of architecture and planning, with living spaces becoming either facilitators or restraints of sustainable behaviors. The question arises, therefore, as to how far sustainability has become a new constraint that limits the contribution of the environment to quality of life or, to what extent sustainability plays, could play, or has played a proactive social role, thereby increasing citizens' responsibility, participation, and commitment, the perception of social support, and cooperation; in a word, empowerment. As some of the literature on quality of life demonstrates, (see Chap. 2), beyond the conditions needed for a decent life, the feeling of being understood, welcomed, and integrated into a social group is of major importance, and increases the capacity for control. Have the move toward sustainability and the organization of towns and territories helped in this direction? Moreover, production processes, whether in agriculture or industry, are responding to new challenges, currently focused on prevention and, where necessary, on adaptation to climate change. However, does this enable a reduction in risks and favor the prediction of potential risk impacts? We can also question the differences at this level between the various regions of the world, particularly between the north and the south, east or west, together with their risk impacts and their capabilities of managing them.

1.2 Chapter Topics

The originality of this Handbook, compared to other works already published in the field of environmental psychology, lies in the fact that it focuses on the links between environment and quality of life. We have sought to show, through the various contributions, how environmental psychology understands the notion of quality of life and contributes to its definition, and how this concept, which is central to this disciplinary field, is at the heart of substantial research carried out over recent years in the different cultural areas of Europe, as well as in North and South America. The different chapters assembled here show how fundamental it is to understand the links between environment and quality of life from the angle of the psychological and psychosocial processes that are at work. These contributions also plead in favor of a better awareness of these dimensions in the context of interventions in the field (information and prevention campaigns, renovation projects, construction, etc.) and within public policies in general.

The Handbook is organized in three main parts, composed in total of 8 subsections and 30 chapters. The first part deals with **People-Environment Relationships and Quality of Life**. After this introductory chapter, the five chapters that make up the first subsection, **Environmental Quality and Well-being**, describe the state of the art of the relationship between the individual, the environment and quality of life. Thus, in Chap. 2, *Quality of Life and Sustainability: The End of Quality at Any Price*, Pol, Castrechini and Carrus look at the evolution of the concepts of quality of life, well-being and happiness. The authors revisit Sprowl's concepts of shrinking cities, ecocities, walkability and sustainable mobility, and the socioeconomic dynamics of globalization, and how these have an impact on the social interactions, living opportunities and well-being of citizens. In Chap. 3, *Some Cues for a Positive Environmental Psychology Agenda*, Valera and Vidal start from the premise that environmental psychology has classically focused on the negative aspects (stress, environmental risk, etc.) of the environmental experience; however, there is research on adaptive processes that allow individuals to attain psychological and social well-being. Based on these concepts and research traditions, the authors reflect on the future of the discipline. In Chap. 4, *Linking Research and Design: What's Missing?*, Desprès and Piché analyze the current state and the future of collaboration between architecture and environmental psychology in terms of working together to improve living conditions and thus quality of life. This first subsection ends with Chap. 5 by Di Masso, Dixon, and Hernandez: *Place Attachment, Sense of Belonging and the Micropolitics of Place Satisfaction*, in which the authors, from different epistemological backgrounds, revisit research using both a qualitative and a quantitative approach to one of the most emblematic research topics in environmental psychology: the link between place, appropriation and sense of place.

The second subsection deals with the question of **restorative environments**. In Chap. 6, *Self, Nature and Well-being: a Sense of Connectedness and Environmental Identity for Quality of Life*, Olivos and Clayton reflect on the relationship between connectedness with nature, self, and well-being, beginning with the premise that the environment could be a contributory factor in the development of a positive identity. In Chap. 7, *Restorative Environments and Health*, Collado, Staats, Corraliza and Hartig present the main concepts and theories positing that the design of certain environments (residential, school, work) can be restorative, that is to say, they can favor health and well-being. This subsection concludes with Chap. 8 by Loureiro and Veloso, *Green Exercise, Health and Well-being*. These authors show that physical activity in a natural setting can have additional positive effects on physical and psychological well-being.

The third subsection examines the issue of **Ecological Behavior**, a topic that questions the relationships between people and the environment on a global scale. Five chapters make up this subsection. It begins with Chap. 9, *Sustainable Behavior and Quality of Life*, in which Tapia-Fonllem, Corral-Verdugo and Fraijo-Sing consider that sustainable development aims to improve quality of life by ensuring the satisfaction of human needs and the protection of the natural environment, as well as having positive psychological consequences for individuals who subscribe to this practice. In Chap. 10, *Self-determined, Enduring, Ecologically Sustainable Ways of Life*, Kaizer, Kibbe and Arnold propose a model that questions the link between sustainable behavior and intrinsic motivation related to the protection of the environment. In the following chapter, *Commitment and Pro-environmental Behavior: Favoring Positive Human-Environment Interactions to Improve Quality of Life*, Demarque and Girandola focus on the effect of commitment and persuasive communication on the adoption of pro-environmental behaviors such as sorting waste, recycling and energy-saving, perceived positively and contributing to quality of life. Chapter 12, *Pro-environmentalism, Identity Dynamics and Environmental Quality of Life*, links pro-environmental attitudes and behaviors with the psychosocial identity of the individual, through the theoretical lens of quality of life. Félonneau and Causse underline here the central role that the environment plays in the constitution of identity as well as its impact on the consolidation of pro-environmental attitudes, through the normative influence and expectations of groups of belonging. This subsection ends with Chap. 13: *Can Engagement in Environmentally-friendly Behavior Increase Well-being?* Here, Venhoeven, Steg and Bolderdijk discuss thoroughly and critically the relationship between behavior that respects the environment and well-being, by putting forward hypotheses demonstrating that this relationship can be both positive and negative.

The second part, entitled **Well-being and Daily Environments**, is also made up of three subsections. Here, the work and theories in environmental psychology applied to quality of life in relation to specific living spaces are highlighted. In the first subsection, **Urban Environments**, Romice, Thwaites, Portz, Greaves and Pasino, in Chap. 14: *Urban Design and Quality of Life*, analyze those aspects of the design of cities that might have an effect on quality of life. They discuss the fact that, although direct causal relationships between physical space and well-being are often difficult to establish, the former plays a fundamental role in the establishment of the latter, a role that must not be ignored. In Chap. 15, the second in this subsection, *The City as an Environment for Urban Experiences and the Learning of Cultural Practices*, Paramo suggests designing the city not only as a spatial phenomenon, but also as a form of socially constructed cultural expression. This subsection ends with Chap. 16: *Adjustments of Acts and Representations to Urban Space and Psychological Well-being*. Here, Ramadier draws on concepts of social readability of space and socio-cognitive accessibility to places in order to understand better the adjustments made by individuals to attain a level of well-being within an urban space.

The subsection **Residential Environments** focuses on quality of life in the residential environment. The chapter by Aragones, Amérigo and Peréz, *Residential Satisfaction and Quality of Life*, describes the long and emblematic tradition of research on the theme of residential satisfaction. In the chapter *Spatial Inequalities, Geographically-based Discrimination and Quality of Life*, Fleury-Bahi and Ndobo put into perspective the question of residential quality of life with problems of residential discrimination and socio-spatial segregation. This subsection ends with two chapters that analyze the everyday environments of children and adolescents. Depeau, in *Children in Residential Contexts*, describes the studies carried out in environmental psychology and social geography on children's use of residential space. Milfont and Denny deal with the use of school and neighborhood spaces by adolescents in the chapter *Everyday Environments and Quality of Life. Positive School and Neighborhood Environments Influence the Health and Well-being of Adolescents*.

This section on everyday environments ends with a subsection on **Work and Institutional Environments**. These spaces give rise to serious questions for society concerning their quality and adaptability to the needs of users. The first two chapters present the classic theoretical models in environmental psychology used to study quality of life in the workplace. The chapter *The Effect of Workplace Design on Quality of Life and Work* by Vischer and Wifi is complemented by that of Rioux on *Comfort at Work: an Indicator of Quality of Life at Work*. The following two chapters deal with quality of life in institutional contexts, particularly health institutions. Thus, Devlin and Andrade analyze the existing literature in environmental psychology on the *Quality of the Hospital Experience: Impact of the Physical Environment* to understand the role of the latter on the quality of life of patients and their families. Fornara and Manca, in the chapter *Healthy Residential Environments for the Elderly*, do the same with institutions specifically dedicated to the elderly, a key issue in an aging world (at least in the North-Western part of the world).

The third and final part of this handbook, **Quality of Life and Environmental Threats**, focuses on the potentially harmful effects of certain environmental features on quality of life. Thus, the first subsection, **Environmental Stressors and Risks**, examines the work on environmental risks and annoyances. The chapter *Environmental Stress* by Gatersleben reviews the research on environmental stress, one of the most emblematic research traditions in environmental psychology. The following three chapters deal with one of the topics most currently studied in environmental psychology: environmental risk evaluation by inhabitants and populations more or less exposed to risk. Thus, the social representations of natural risks are analyzed in the chapter: *Living in an "At Risk" Environment: the Example of Coastal Risk* by Michel-Guillou and Meur-Ferec. The effect of social and environmental inequalities on the perception of risk is examined by Navarro in the chapter *Social*

Inequality and Environmental Risk Perception. Finally, the perception of industrial risk and the adaptation of populations to these living conditions are analyzed by Lima and Marques in Chap. 28: *Living in Industrial Areas: Social Impacts, Adaptation and Mitigation.*

This last part of the Handbook ends with a subsection focusing on the planetary environment, a change of scale that enables the authors to reflect on **Global Change, Energy and Emerging Risks**. In Chap. 29, *Emergent Risks and Quality of Life*, Marchand, Weiss and Zouhri question these new risks called "emergent", which are a source of anxiety for society and represent a new challenge for decision-makers and users. Next, the chapter *Energy Issues: Psychological Aspects* by Lenoir, Devine-Wright, Pinheiro and Schweizer-Ries deals with a currently very sensitive topic, related to the control of energy, new energy sources and their impact on global quality of life.

Finally, Uzzell, Garcia-Mira and Dumitru discuss *Global Challenges for Environmental Psychology*. These authors highlight the contribution that this discipline makes, or should make, to the promotion of necessary changes by societies in response to substantial environmental problems and their impact on the quality of life of communities. A number of crucial questions for the discipline are raised here, not only about globalization, North-South relationships and dependences in terms of environmental issues, but also about implications for the development of Environmental Psychology as a discipline.

1.3 Perspectives

Addressing the concept of quality of life enables an interdisciplinary approach to be adopted, which is especially rich in the field of environmental research. The very nature of the concept of quality of life, situated at the junction of many disciplinary fields, and used in many human and social science disciplines, as well as architecture, technology, and medicine, with meanings that are not always equivalent, undeniably encourages this collaboration. For example, how can we examine the quality of life of communities at risk of sea-submersion without working in tandem with geographers specializing in this natural risk? How can we study the capacities of adaptation of populations exposed to potentially harmful pollution without a dual health-environment approach, working alongside researchers specializing in environmental health? How can we understand better how industry approaches sustainable development without taking into account the organizational, economic and political context? These are just some of many examples that reinforce the value of an interdisciplinary approach. However, the difficulties inherent in the complexity of such an approach are well known. It is clearly not enough to assemble several disciplines around the same scientific topic in order to lead them to construct a common issue. In this type of approach, the main error to avoid is the simple juxtaposition of theories, concepts and methods from different disciplines. We must go beyond this juxtaposition toward a combination of knowledge emanating from each discipline involved, to reach an epistemological exchange and productive methodology. Examples of collaboration of this type are at the heart of some of the chapters published in this Handbook (for example, Pol, Castrechini and Carus, DiMasso, Dixon and Hernandez, Collado et al., Olivos and Clayton, Michel-Guillou and Meur-Ferec, Lenoire et al.). This is one of the main challenges which, in our opinion, needs to be addressed in the future, especially in the context of climate change, its impact on human populations and the profound disruption that it may have on future generations.

The second challenge, in our opinion, is more political as it concerns the public authorities. The recognition of environmental questions through the lens of quality of life and the adoption of an approach focused on psychological and psychosocial processes highlight a need. Indeed, beyond the objective indicators of environmental quality, which are determined by measurements, it is crucial to study the evaluation of the situation. This subjective dimension has already been taken into account over many years in the field of urban studies in relation to renovation and construction, since it enables solutions to be implemented that meet users' expectations. However, it is rarely considered when it comes to environmental risks, whether they are natural, like those generated by climate change, or risks generated by exposure to potentially harmful environmental pollution. In fact, beyond the objective indicators of environmental quality, which are determined by measurements and used to define tolerance thresholds, it is essential to look at the evaluation of the situation and the risk factors. This evaluation, beyond objective exposure, partly predicts well-being and perceived quality of life. It also helps in understanding the adaptation capacity of individuals and communities. Therefore, it would be appropriate, at the level of intervention actions carried out in the field of risk prevention and management or in urban policies, to include the subjective dimension in public actions in a more pertinent way so that the determinants of well-being and quality of life could be better identified. Dealing with existing strategies specific to communities is a real epistemological turning point in risk management and promotion of quality of life, since it requires the expert to have an in-depth knowledge of the local reality. However, let us not be naive. Research shows that psychosocial processes can be linked to a positive environmental change; on the other hand, they can be used to create situations in which the individual and society are rendered more vulnerable and manipulable, for individual not collective interests. This places the issue in the political dimension of psychology, quality of life, and environmental psychology, made up of personal and social choices, in our roles as citizens as well as scientists, and in turn refers to the conclusions of Uzzell and colleagues, in the final chapter, on the challenge of an environmental psychology that is both global and diverse.

Acknowledgments We are grateful to all the authors who have contributed to this unprecedented collective reflection. Thank you to all our colleagues, who, coming from distant backgrounds, countries and universities, have risen to the challenge of working together, sometimes in teams belonging to diametrically opposed traditions and approaches, but always with the aim of developing complementary visions.

References

Fleury-Bahi, G., Marcouyeux, A., Préau, M., & Annabi-Attia, T. (2013). Development and validation of an environmental quality of life scale: Study of a French sample. *Social Indicators Research, 113*(3), 903–913.

Mitchell, G. (2000). Indicators as tools to guide progress on the sustainable development pathway. In R. J. Lawrence (Ed.), *Sustaining human settlement: A challenge for the new millennium* (pp. 55–104). Newcastle: Urban International Press.

Nussbaum, M., & Sen, A. (Eds.). (1993). *The quality of life*. Oxford: Clarendon.

Pol, E. (2007). Blueprints for a history of environmental psychology (II): From architectural psychology to the challenge of sustainability. *Medio Ambiente Comportamiento Humano, 8*(1y2), 1–28.

Rogerson, R. (1995). Environmental and health-related quality of life: Conceptual and methodological studies. *Social Science and Medicine, 411*, 1373–1382.

Shafer, C. S., Koo Lee, B., & Turner, S. (2000). A tale of three greenway trails: User perceptions related to quality of life. *Landscape and Urban Planning, 49*, 163–178.

Sirgy, M. J. (2012). *The psychology of quality of life: Hedonic well-being, life satisfaction, and eudaimonia* (Social indicators research series 50). New York: Springer.

Stokols, D. (1987). Conceptual strategies of environmental psychology. In D. Stokols & I. Altman (Eds.), *Handbook of environmental psychology* (Vol. 1, pp. 41–70). New York: Wiley.

Szalai, A. (1980). The meaning of comparative research on the quality of life. In A. Szalai & F. M. Andrews (Eds.), *The quality of life* (pp. 7–21). London: Sage.

VanKamp, R., Leidelmeijer, K., Marsman, G., & DeHollander, A. (2003). Urban environmental quality and human well-being: Toward a conceptual framework and demarcation of concepts; a literature study. *Landscape and Urban Planning, 65*, 5–18.

WHO. (1984). *Report of the working group on concepts and principles of health promotion*. Copenhagen: World Health Organization.

Part I
People-Environment Relations and QoL – *Environmental Quality and Well-Being*

2

Quality of Life and Sustainability: The End of Quality at Any Price

Enric Pol, Angela Castrechini, and Giuseppe Carrus

Have human beings not been happy until the twenty-first century? Can't we be happy without living in a rich country or neighborhood? These are questions that come to mind when reviewing much of the literature on QoL and happiness, which reveals strong positive correlations between income and happiness. Obviously, there are some errors in this perspective – inaccuracies and deficiencies. It seems reasonable to assume that having the desired or necessary resources contributes to happiness, but it seems unreasonable to suppose that humans were never happy in the past because they were never so well off. The complexity of the matter, with the answer depending on the conceptualizations and the factors taken into consideration in its measurement, can shed light on these questions (rather than providing a clear answer). But how do the physical environment and the interaction of people with their environment influence their well-being? This chapter focuses on these issues.

E. Pol (✉) • A. Castrechini
University of Barcelona, Barcelona, Spain
e-mail: epol@ub.edu; acastrechini@ub.edu

G. Carrus
Roma Tre University, Rome, Italy
e-mail: carrus.giuseppe@gmail.com

2.1 QoL and Environmental Psychology: Antecedents and Evolution

2.1.1 First Steps

Environmental psychology, going right back to its origins with Hellpach (1911, 1924) and Muchow and Muchow (1935/1998)[1] and its coetaneous and close antecedents, such as Simmel (1903) and Tönnies (1887), has always been concerned with citizens' living conditions and QoL, though without using these labels. Those were times when socio-political analysis denounced the precarious, unhealthy living conditions and inhumane lifestyles of the industrial cities (e.g. Engels 1845), and when the new architecture (e.g. Bauhaus, founded in 1919) attempted to find viable, economical solutions. Living conditions were especially important to Environmental Psychology between the 1960s and late 1980s. During this period, the term "Psychology of Architecture" prevailed in Europe, while in North America the preferred expression was "Environmental Design" (Pol 2006, 2007).

Rapoport (1969), at the first EDRA conference,[2] spoke of "environmental quality", and at

---

[1] See Mey and Günther (2015)
[2] EDRA is the mainly North American Environmental Design Research Association, IAPC is the International Architectural Psychology Conference, mainly European,

the first IAPC. Stringer (1969, p. 8) described the psychology of architecture as a way "to help an individual or a social group, (...) (to construct) a more humane and orderly existence". For Canter (1969), environmental psychologists study the individual's satisfaction with the environment while helping architects to produce buildings that benefit people and that can be well used by the occupants in the way the architect really intended (p. 11). At the EDRA conference in Los Angeles (Mitchell 1972), a section on "Indicators of Environmental Quality" appeared for the first time, while in 1985 the effects of "Environmental Change" were depicted strictly as "Social Change", induced by changes in the design of the built environment, extolling the virtues of participatory design and community involvement in some aspects (Klein et al. 1985).

In 1974, under the influence of the National Environmental Policy Act of 1969, the main theme of the EDRA-5 conference (in Milwaukee, USA) was "Social Impact Assessment". At this conference, Wolf (1974, p. 2) affirmed that "social impact assessment" was only a new way of describing what was actually a traditional concern. However, environmental psychology would focus only on the more urban aspects, and systematic research into and the application of more "ambient environmental" aspects did not emerge until the 1990s and the beginning of this century. Bechtel and Churchman's handbook (2002) is an example of consolidation in this direction.

Conversely, Lee (1969) made three admonitions: that "(people)... have an extraordinary capacity for adapting to what they have been given and putting a cheerful face on it..."; that "...the notion that there is a large floating population of people who will move into the optimum architectural environment as soon is provided, is surely false" (p. 20); and that it must be taken into account that people always apply the "principle of least effort" (p. 24).

The Danish architect Jan Gehl (1971), in his contribution to the 1970 IAPC in Kingston (UK), stated that what makes a place attractive are the activities that take place there. He considers that "...We have to compensate the (...) loss of social experiences in the new housing and urban areas (...), and (we must ensure that) primarily the weak and deprived groups in society, have access to a varied environment....." (p. 63).

On the European circuit, more than 15 conferences were held between Dalandui (UK) in 1969 and Eindhoven (NL) in 1998. Until 1979, the theme was Architectural Psychology, and from 1982 onward, People-Environment Studies, the result of the creation of the IAPS (see footnote 2). Only one of them, held in Barcelona in 1982, explicitly refers to QoL in its title.

For Proshansky et al. (1970), although the physical characteristics condition the quality of an environment, it is the psychological and social determinants underlying the activities and social relationships associated with that environment that actually define the quality. It must be kept in mind that people always try to arrange their environment in such a way that it maximizes their freedom of choice (Proshansky et al. 1970).

This places the "multifunctionality" of spaces (possibilities and compatibility of uses and resources in the same place, such as a square or a park, to guarantee freedom of choice) at the center of the debate. However, an excessive number of options can generate distress owing to informative/cognitive overload, conflicts between different types of behavior and users, and so on.

The issues under debate during this first step were not tackled explicitly in terms of QoL but rather as aspects conceptually integrated therein: satisfaction; freedom of choice; the creation of physical conditions that encourage desirable social interactions; capacity for agency but not excessive effort and willingness to change; intervention to prevent the deprivation of the most vulnerable members of society, and so on. Environmental quality is always sought by means of improvements in design, through intuition and creativity, and through research (see Chap. 4 in this book). On the other hand, all this has financial, environmental and social costs that were not called into question in those times.

We would like to emphasize this point, in view of the apparent contradiction presented by the

---

which in 1981 become the current IAPS, International Association of People-Environment Studies.

absence of the ecological component in "environmental" psychology during this period as well as the absence of objections to the economic and environmental limits/costs of the proposals made by architectural psychology, even though some of the first research carried out at the MIT for the Club of Rome had already been published and disseminated: *The Limits to Growth* by Meadows, Randers and Behrens (1972). Moreover, Paul Ehrlich had already published his disturbing book *Population Bomb* and Kenneth Boulding (1966) had expounded his famous and effective metaphor of the cowboy and spaceman economies.

In 1968, the UNESCO Conference on the Biosphere was held in Paris, and in 1972, the important Conference on the Human Environment took place in Stockholm, where the UNEP was approved. In Stockholm, Levi and Anderson (1974) made a suggestive proposal on QoL. They presented it as "a measure consisting of physical, mental and social well-being, as perceived by each individual and group, and happiness, satisfaction and reward (...)" (Levi and Anderson 1974/1980, p. 6).

Moreover, the British journal *The Ecologist* had already published its *Blueprint for Survival* (Goldsmith et al. 1972) while the Nobel laureate Paul A. Samuelson suggested that development should not be measured only by GNP (gross national product) but also in terms of NEW (net economic welfare).[3] NEW calculations showed that we are not as wealthy as might be deduced from GNP statistics. This idea would end up permeating social psychology studies on QoL in the late 1970s and 1980s.

2.1.2 The Explicit Incorporation of Sustainability Into Environmental Psychology

As of the late 1980s, the number of papers that referred to sustainability presented at IAPS and EDRA conferences and those of similar organizations (the Asian MERA,[4] for example) increased very significantly. Some focused on ways of analyzing and conceiving housing, public space and the city; others were part of the growing body of research on citizens' behavior and responsible environmental behavior.

At EDRA 19 (1988), the name of the conference was explicit: "People's Needs/Planet Management: Paths to Co-Existence". The following conference (1989) opted for "Changing Paradigms", while at EDRA 23 (1992) the theme was "Equitable and Sustainable Habitats". This is the last of the unequivocal titles, although they continued to appear in work submitted to the EDRA conferences. The term "sustainability" was not included in the title of an IAPS conference until Paris (2000): "Metropolis: Cities, Social Life and Sustainable Development-16 IAPS". Since then, "QoL" and "sustainability" have appeared in the titles of most of the recent IAPS conferences and have figured strongly at the center of the IAPS "discourse". A simple search in the IAPS digital library[5] returns 646 entries for the term "QoL", 487 for "sustainability" and 1,420 for "sustainable". A preliminary content analysis of the titles of the first 100 items listed under the search term "QoL" shows a representation of the concept of QoL strongly linked to the issue of green space and restorative environments. Two other core elements in the representation of QoL are related to residential space (and associated issues, such as housing and home settings) and urban settings in general, considering the issues of life cycle and life expectancy (i.e. QoL among children and the elderly). This point suggests a need to consider the concept of QoL in relative rather than absolute terms, and also as dynamic and time-dependent. What may be a good life for some may not be for others, especially if we take into account the asymmetry and inequality of the social structure. This approach

---

[3]NEW is equal to GNP minus the social costs and harm caused to the environment attributable to the acquisition of GNP.

[4]Man-Environment Research Association (MERA), founded in 1982 in Osaka, Japan.

[5]http://iaps.scix.net/.

is necessary to avoid the risk of environmental (and architectural) determinism.

Community life, social cohesion, empowerment and participation are seen as prerequisites for the efficacy of any environmental intervention intended to improve citizens' QoL. There is certainly not much quality in modern life without participation, citizenship and agency. At the same time, it can be difficult to promote and sustain the participatory process if the potential participants do not perceive any possibility of improvement in their QoL. In other words, alienation (as opposed to inclusion) and low QoL may reinforce a negative spiral in terms of individual psychological distress (e.g., Evans and Lepore 1993). In conclusion, promoting QoL and social equality may be possible through the appropriate design and management of more inclusive and restorative daily life settings (Stokols 1996).

When discussing the relationships between QoL and sustainability, Vlek (2002), in his keynote speech at the IAPS Conference in A Coruna, established an explicit link between the different QoL components (and the relative importance people assign to them) and the classic paradigm of common dilemmas. When reviewing the possible links and tensions between these two elements (QoL and sustainable lifestyles), Vlek also underscored how QoL could be framed and investigated as an overarching goal of human behavior in relation to the environment, particularly as a driving force for sustainable behavioral change in the wider public.

Though still surprising, it is not by chance that we find only two search results for the associated term "QoL and sustainability" in the IAPS digital library. This suggests a need for critical analysis of the consequences and outcomes of the pursuit of these goals in contemporary human society. What are the environmental costs of improving the quality of contemporary, mainly urban, life? Is there a price to be paid in terms of QoL if we accelerate the transition towards more environmentally sustainable, low carbon lifestyles?

According to Uzzell and Moser (2006), QoL "remains poorly defined, so much so that it has almost come to mean whatever we want it to mean" (p. 1). For these authors, a sustainable QoL is only achieved when, on the one hand, people interact with the environment in a respectful way, and, on the other hand, when that environment does not hinder or threaten what the individual regards as "QoL". This renders possible the capacity of individuals to satisfy their needs. They also emphasize the role of the cultural context: "QoL depends not only on the physical and social "quality" of the environment but is also a result of the way people interact with their environment. Ways of life are ideologically and culturally dependent and individual needs express themselves within that framework" (p. 3).

This leads us to a second aspect that has become a key part of the contributions of environmental psychology in recent decades: the environmentally responsible behavior of citizens as a condition for sustainability.

Stern[6] and colleagues (Stern 1992; Gardner and Stern 1996) suggest that the main contribution of psychology in detaining, slowing or responding to global environmental change is to understand the human causes of this deterioration and present strategies having a bearing on people's behavior; in other words, learning about proximal causes not only related to organizations, social structures, technology, the means of production and politico-economic decisions but, above all, to the attitudes and values associated with each of them.

There is abundant literature on this subject, with contributions from many different cultures and parts of the world (e.g. Bonnes and Bonaiuto 2002; Bonnes et al. 2006; Clayton and Myers 2009; García-Mira and Vega 2009; Clayton 2012; Corral-Verdugo et al. 2013, 2015; Schmuk and Shultz 2002; Weiss and Girandola 2010).

2.2 QoL and Happiness in QoL Literature

We will now review how quality of life is conceptualized by the literature on QoL and examine the role given to the physical environment and living conditions.

---

[6]Paul C. Stern is a Member of the National Research Council's Committee on Global Change Research in the USA.

2.2.1 Concepts and Evolution

The concepts of well-being and quality of life have their antecedents in those such as "standard of living" and "happiness" (Diener 1984; Casas 1996). However, while "standard of living" could be measured by economic indicators (e.g. Pigou 1932), "happiness" was considered an ethereal, philosophical, religious and moral concept, not scientifically measurable. Paradoxically, happiness has become the final link in measurement and dominates research 30 years later.

In 1954, the UN proposed a series of indicators of living standards that included, among others, the dimensions of health, nutrition, working conditions, housing, free time and human rights. In the 1970s, the OECD contributed other important indicators. Altogether, these were the antecedents of the recent World Happiness Reports.

Sirgy, in *The Psychology of Quality of Life* (2012), cites Jeremy Bentham (1789/1969) as an antecedent. Bentham defined happiness as "a state of being that people experience as a result of action by oneself or another." (Sirgy 2012, p. 5). Bentham's idea of QoL and happiness is essentially "contentment", which Sirgy sees as the basis of the hedonic dimension, understood as emotional well-being. In the tradition of Aristotle, he defines the eudemonic vision as a full life, a life with significance, which translates as flourishing, with a sense of well-being, successful, an opportunity to lead a purposeful and meaningful life. The hedonic and eudemonic visions constitute one of the most important dichotomies in the current debate on QoL.

Land et al. (2011), in their *Handbook of Social Indicators and Quality of Life Research*, refer to Bauer (1966), with his *Social Indicators*, as the origin of the so-called social indicators movement. By way of antecedents, he cites authors from the 1930s, such as William F. Ogburn, at the University of Chicago, and Howard W. Odum, at the University of North Carolina, who in collaboration with Margaret Jarman Hagood developed the first well-being index by studying agricultural families (Ferriss 2004).

The link between QoL and health goes back to the 1940s, with Ogburn (1943) and Stouffer (1949), but the key year was when the World Health Organization (WHO, 1948) recognized the importance of the concept of QoL, understood as well-being, in its famous definition of health as "physical, mental and social well-being and not merely the absence of disease." This recognizes the possibility of defining disease as a social fact, as a deviation from the social expectations of the person, or derived from socio-physical conditions, in line with the observations of Talcon Parsons (1958). Dunn (1959) speaks of "wellness" as mental, physical and social balance, which depends on the capabilities of the person and the opportunities offered by the environment. This leads us to some of the critical interpretations that denounce that globalization increases the intentional medicalization of social problems and social inequalities, rather than treating them as deficiencies in social, economic and environmental conditions (Talarn 2007).

The scope of the concept of QoL can be extended to many areas, including the world of work and the physical conditions of the workplace, which has been one of the traditional objects of environmental psychology.[7] Maslow (1954), Argyris (1957), and McGregor (1960) among others, addressed the quality of work life (QWL), influenced by and following up on the impact of the Hawthorne studies and the human relations movement.

Bauer (1966) suggested the need to differentiate between objective and subjective indicators. For this author, the domains that needed objective indicators were participation, employment, leisure, health, income and consumption. Those requiring subjective indicators were perceived sense of belonging, participation and affection, status, respect and power, self-realization, creativity, security, freedom and incentives.

Campbell and Converse (1972); Campbell, Converse and Rodgers (1976); and Andrews and Withey (1976) also worked in this area, and in 1974 Alex Michalos began the publication of *Social Indicators Research*. That same year, the OECD stimulated the issuance of national

---

[7] See, for example, the recent synthesis by Rioux et al. (2013).

social reports based on social indicators. In 1985, Michalos formulated the multiple discrepancies theory, based on the comparison of experienced reality and the expectations of the person in different domains and moments of their life. In the 1980s, Veenhoven set up the world database of happiness. Later, the so-called positive branch of psychology would gain momentum (see Chap. 3 of this book).

Andrews and Withey (1976) found that QoL studies, when making global or abstract assessments, tend to give a positive score, especially when the questions are related to aspects of self-image, identity and self. In contrast, they give lower scores when separable aspects, such as specific habitat conditions, functioning of services, support and care received, are valued. Here, the mechanisms of cognitive dissonance described by Festinger (1957), among others, intervene. To ensure consistency in their overall assessment of quality of life, people distort some aspects. We can afford to be critical if it does not affect our self-image and identity, i.e. if we are dealing with issues that are attributable to external, unrelated factors. All the mechanisms of the so-called "locus of control" come into play in this case (in the sense of Rotter 1954; Montero 1994; Javaloy 2007).

Glatzer and Mohr (1987) warn of the surprisingly low correlations between objective living conditions and subjective well-being. They highlight some factors that may provide the key to understanding current contradictions between data in different studies (to be discussed later). They consider that people value their personal improvements by comparison with their relevant group(s) of reference, not as an autonomous whole. Furthermore, expression of dissatisfaction is culturally learned, and thus to some extent independent of real experience. They consider that individuals are under social pressure to suppress feelings of dissatisfaction, and that expectations will generally be adapted to the circumstances. On the other hand, those living under favorable conditions tend to be more open to new standards of valuation and are therefore more inclined to express criticism and dissatisfaction.

Contributions by other authors show that identification with the place and ideal self-mage also play a role in low correlations (Buttimer 1972), as do expectations, level of aspiration, perceived equity, needs and values (Marans and Rodgers 1975). Attitudinal and contextual variables regarding the community, neighbors, family and friends modulate the valuations of objective characteristics of the place and the available resources (Galster and Hesser 1981; Cutter 1982; Canter and Rees 1982).

These views have been deliberately placed at the center of the theoretical and empirical development of positive psychology and research on human happiness (e.g. Seligman and Csikszentmihalyi 2000) (see Chap. 3). However, as Corral-Verdugo and colleagues (2015) have recently pointed out, positive psychology has rather surprisingly underestimated the role of environmental factors as basic determinants of QoL and happiness, in comparison with individual factors such as traits, skills, behaviors and emotions (see, for example, Lyubomirsky et al. 2005). However, as noted by Stokols (1987), in environmental psychology, the environment, as an object of environmental psychology, is neither physical nor social; it is always socio-physical. This aspect is presented as one of the typical shortcomings of research on QoL, well-being and happiness, probably due to the difficulty and complexity of their measurement, which will be discussed below.

2.2.2 The Happiness-Income Paradox

The assessment of QoL has followed a process of progressive subjectivization. It has shifted from the use of economic indicators of living standards as the only measure to the consideration of experiential dimensions taken as a whole. This opens up an important debate, beginning with the interpretation of the data, which is clearly reflected in the approaches of some authors who present the most iconic standpoints: Richard A. Easterlin, who expounded the happiness-income paradox; Ruut Veenhoven, who affirms the existence of a strong correlation between wealth and increased happiness; and Ed Diener's laboratory team and the group behind the United Nations World Happiness Report, with Peter R.G. Layard, among others, who take a more nuanced and multifactorial stance.

Richard Easterlin (1974), an economist at the University of Southern California, proposed the happiness-income paradox in 1974. Briefly, it states that gains in material well-being have little effect on satisfaction with life after a certain QoL is achieved. Max-Neef (1995), winner of the *Right Livelihood Award* in 1983, takes a similar approach, but with nuances. In his "threshold hypothesis", he suggests that in any society there seems to be a period in which economic growth generates an improvement in QoL, but after crossing a threshold (which remains uncertain, not clearly defined), economic growth generates deterioration in QoL. He also suggests that the "paradigm of accumulation" must be replaced by the "paradigm of solidarity" (an idea we return to below).

Later, in 2010, after heavy criticism from Veenhoven and his advocates, Easterlin, in his work *The Happiness-Income Paradox Revisited* (Easterlin et al. 2010), concluded that happiness and social well-being (SWB) are positively related but, over the long term – here, usually a minimum period of 10 years – the relationship is nil. The main reason is the escalation of material aspirations with economic growth, reflecting the impact of social comparison and hedonic adaptation. In addition, they found no evidence to suggest that poorer countries are somehow exempt from escalating material aspirations as income rises. They concluded that more research is needed. For us, this can be interpreted as a need to explore the role played by the balance of personal, social and environmental aspects in QoL, as discussed below.

In an informative review, comparing data from various sources, but especially from the Pew Research Center in Washington,[8] Stokes (2007) concludes that income alone is not enough to explain why people in some societies are more satisfied than others with their lives. He found strong differences between regions. For instance, he detects an improvement in Kenyans and South Africans' sense of their economic progress, but there is no real change in how satisfied they are with their lives. Factors such as religious and ethnic strife and concern about inequality may contribute to this disconnection between economic well-being and happiness. He concludes that people in poor countries are often more optimistic than those in rich societies. Furthermore, despite the apparent correlation between income growth and happiness in Latin America and Eastern Europe, other researchers warn against overly simple interpretations of the importance of income gains in well-being.

2.2.3 Does Money Lead to Happiness?

For Veenhoven,[9] there is a strong correlation between wealth and happiness. In the 1980s, he set up the World Database of Happiness[10] and he was also the founding editor of the Journal of Happiness Studies, first published in 2000. For years, he has been involved in a fierce controversy with Easterlin over the happiness-income paradox.[11]

Veenhoven highlights the differences in his *quantitative meta-analysis* method of so-called *narrative research synthesis*, which is often

---

[8] http://www.pewresearch.org/.

[9] Ruut Veenhoven, Professor at the Department of Sociology of the Erasmus University Rotterdam, Holland, was one of the pioneers in the study of happiness.

[10] The World Database of Happiness is an archive of "research findings" (not directly research data) on happiness, meant to facilitate meta-analysis, doing "research synthesis" on that subject. http://worlddatabaseofhappiness.eur.nl.

[11] The discussion can be followed in Hagerty and Veenhoven (2003), Easterlin (2005), Veenhoven and Hagerty (2005), Easterlin et al. (2010), and Veenhoven and Vergunst (2013).

theory-driven and screens the available data for evidence for or against a particular theory (Veenhoven 2011).

Veenhoven defines happiness as the subjective enjoyment of one's life taken as a whole. He understands happiness as synonymous with "life satisfaction" and "subjective well-being". He asserts that happiness has two "components", the hedonic component related to affectivity, and a "cognitive" component called contentment. According to his approach, average happiness is high in modern societies and tending to rise, and differs little across social categories such as wealth and gender. He thinks that people are happier in individualistic societies, such as in Denmark, than in collectivistic societies such as in Japan (Veenhoven 1999; Verne 2009). People do not live happier in welfare states than in equally rich nations where the "nanny state" is less open-handed. Happiness inequality does not appear to be less in welfare states either (Veenhoven 2000a, b). This contradicts several communitarian social-science-based approaches. He explicitly mentions his disagreement with Putman's theory (2000), which states that modern society falls short in terms of social cohesion, confidence in governments, likelihood of social and political participation and involvement in community projects; with fewer close friends and confidants and more time spent watching television or new media, and so on, resulting at the end in less happiness and lower perceived QoL.

The problem lies not only in the definition of happiness but also in its measurement, because the former does not always tie in with the latter. In this respect, both Veenhoven and those who hold contrary or differently nuanced views are in agreement. In any case, as he also says, most human beings are or tend to be happy. The conceptualization, the experiential processes involved and the indicators used in the measurement turn out to be the key.

2.2.4 The World Happiness Reports

The series World Happiness Reports (WHR 2012, 2013, 2015) sponsored by the UN Sustainable Development Solutions Network were intended to be a contribution to the World Sustainable Development Goals for the 2015–2030 period. In the second report, De Neve, Diener and Cody (2014, p. 66), in line with Lyubomirsky et al. (2005) and Myers (2000), state that supportive relationships boost subjective well-being, and high subjective well-being leads in turn to better social relationships. Happier people have more and healthier friendships and family relationships (Diener and Seligman 2002).

The World Happiness Report makes few references to specific living conditions such as housing, the city and so on. Instead, it looks at the deterioration in environmental conditions of sustainability as a global threat, while paying little attention to the direct, effective implications for people's everyday lives. The term "housing conditions and environmental quality" only appears in the chapter on indicators proposed by the OECD. From the standpoint of environmental psychology, this once again illustrates the deficit in theoretical conceptualization, as already pointed out by Uzzell and Moser (2006). But what does the UN World Happiness Report say? Its data do not seem to tally with the positive correlations defended by Veenhoven, and provide nuances at the very least.

Looking at the graph in the UN World Happiness Report (Fig. 2.1) comparing the 2005–2007 and 2010–2012 periods, we note a twofold tendency. One group of regions tends to score higher, and another group, lower. The countries in the improving regions clearly fall in the group of so-called emerging countries in Latin America, Asia and Africa; those scoring lower than before are mostly what may be regarded as the developed and richer nations in North America and Europe, and Africa in some cases. Depending on whether we look at the upper or the lower circle, we find arguments either in favor of Easterlin's happiness-income paradox or in favor of Veenhoven's correlation between increasing wealth and increasing happiness.

In any case, it should be taken into account that the crisis that began in 2007 and 2008 emerged between the two sets of WHR measurements,

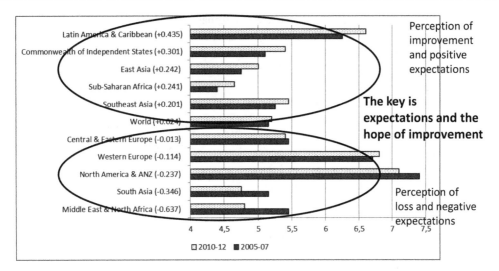

Fig. 2.1 Expectations and hope of improvement (By the authors. Based on the World Happiness Report, Fig. 2.4 Comparing World and Regional Happiness Level: 2005–07 and 2010–12 (WHR 2013, p. 25))

changing the rules and visions of the future. Thus, the factors that seem to explain this twofold trend are positive expectations and hope for improvement in the upper circle, and the perception of loss and negative expectations in the lower circle.

If we shift our focus and look at the case of developments in Spain, we find a similar explanation on a local scale. At the beginning of the transition from Francoism to democracy, income was significantly lower than now (GDP per capita in 1980 was 4,227 euros, less than a fifth of GDP per capita in 2014, 22,780 euros), but the hopes and expectations for improvement were running very high (as they always do at the end of a dictatorship or authoritarian regime). Currently, GDP per capita is more than five times higher, despite the high percentage of unemployed (over 24 % in 2014), but the WHR data show that Spain is the country with the sixth highest drop in its score, only behind Egypt, Greece, Myanmar, Jamaica and Botswana (Helliwell and Wang WHR 2014, p. 29). Hope, disappointment, positive expectations, frustrations, the possibilities of personal fulfillment and human development are aspects that seem to explain much of this loss. The quality of interpersonal relationships, freedom of choice, the locus of interpersonal control, trust in others, cooperation, and, above all, perceived social support from peers (Javaloy 2007, in a study of young Spaniards) are factors that enable us to understand why, despite everything, the country does not explode. At this point, only the multidimensionality of QoL and happiness has become clear so far, with the emergence of significant empirical and conceptual biases that do not allow (and it is unlikely they ever will) a unified vision of the subject. The contextual dimensions are and always will be decisive.

Nonetheless, in this context, there are clear, objective factors – though sometimes we are not sufficiently aware of them – which limit freedom of action and, thus, the very conceptualization and operationalization of QoL and happiness. These are: (1) the planet's finite natural resources, which cannot support the current rate of increasing consumption, i.e. its sustainability; (2) economic wealth, not only as an absolute value but also as the preponderance of enrichment for the sake of enrichment (i.e. accumulation), or enrichment as a basis for covering people's physical, psychological and social needs; (3) basic survival needs (What are they? Are they different in each case?), above which well-being takes priority through social relationships and the fulfillment of expectations.

This places the concept of QoL at the center of a necessary debate – not always sufficiently present in environmental psychology – where two axes of tension can be identified, capable of providing conceptualizations of QoL, forms of analysis, and radically different forms of intervention:

(a) QoL understood as enrichment or capacity for accumulation and the maintenance of this privileged position. This seems congruent with some of Veenhoven's results, while it is the perspective that should be rejected according to Max-Neef. In any case, sustainability is not possible under this parameter because of the finite resources, about which the annual reports on the state of the planet constantly remind us.[12]

(b) QoL understood as a human, social and environmental balance. In this second vision, the need for and importance of social support networks (formal or informal, physical or virtual, as a discussion topic) becomes clear. This seems to fit with Levi and Anderson's old thesis (1974) and a great deal of later literature (including some papers by Veenhoven), which affirm that beyond basic survival needs, human and social balance is more important for the well-being of the individual than an increase in possessions. In this case, sustainability would be possible. But is the evolution of habitat and social (and socioeconomic) structure allowing and facilitating this balance?

The Brundtland Report (1987) and subsequent associated documents, for example *Caring for the Earth, A Strategy for Sustainable Living* (IUCN et al. 1991), place the emphasis on increasing QoL without exceeding the carrying capacity of the planet. But what helps to build social balance, security and good prospects? The answer lies in the quality of the environment as a facilitator and not as a hindrance to positive social dynamics; confidence in personal fulfillment; reinforcement on seeing the positive and tangible results of personal action; the appropriation of space and capacity for agency in a space considered personal; perceived informal social support, and so on. Here, environmental psychology can make and has made a positive contribution.

2.3 Social Dynamics of the City, Sustainability and QoL

The urban form, housing conditions and power relationships that arise from the urban structure have been recurring topics in both urban planning and the behavioral sciences since their inception (Pol 2009).

Living conditions lay in the background of the hygiene movements of the nineteenth and twentieth centuries; Haussmann's renovation of Paris in (Haussmann,1852); extensions with new urban forms, as in the Cerdà Plan in Barcelona in (Cerdà, 1859); Howard's alternative Garden City in (Howard, 1898); and the modern movement in architecture as of 1919, with the Bauhaus of Walter Gropius, Mies van der Rohe and their collaborators, Le Corbusier, the CIAM, Niemeyer and other advocates. This same theme of living conditions was the subject of the work that triggered the nascent social sciences mentioned previously.

The generalization of zoning to prevent annoying and harmful interactions and the appearance of the private car as a means of transport and economic growth dramatically transformed the city and lifestyles. Originally well meant and intended for positive social change, and having provided a large part of the population with "more" decent housing, their social effects (unforeseen by some, and openly premeditated on the part of advocates of the liberal and neoliberal models, such as David Harvey 2012) turned out to be highly problematic for citizens' well-being and effective QoL.

For Jane Jacobs, in her emblematic and radical book from 1961 *The Death and Life of Great American Cities*, functional segregation was a problem. Cities are complex systems resulting from the unplanned actions of individuals and small groups, in which "local knowledge" and diversity are key factors. The complexity and sub-

---

[12]http://www.worldwatch.org/bookstore/state-of-the-world.

tlety of urban dynamics have to be understood. Neighborhoods where various uses (homes, offices, commerce, leisure, etc.) coexist in the same space are safer and more vital that those with little diversity. Functional separation has led to the shopping malls, a form of monopoly that kills social life in the city, impoverishing it economically and making it dependent.

The population explosion of the 1960s, the concentration of population in urban areas with a dominant model of urban sprawl, and the propensity of all human beings to prefer being among their "equals", strengthened the tendency towards social segregation (socio-economic, cultural and racial) in the city, besides the aforementioned functional segregation. The mobility model based on individual transport (often without any planned possible alternatives) involves a maze of urban freeways. This, together with the dominant socio-economic dynamics, generated the fragmentation of urban spaces into isolated lots or tracts of land not easily integrated into a whole, and created conditions favoring the emergence of ghettos with varying degrees of marginality.

Sennett (1970) described a growing sense of personal insecurity, a fear of the unknown and the unfamiliar, a need for mythologized references, and ultimately a lack of maturity that he termed the "adolescent syndrome" in American society. This is the result of concentration in homogeneous suburban developments (sometimes enclosures with high degrees of protection and surveillance), which leads to the disappearance of spontaneous social interaction (any contact not strictly between equals and familiars), breaking the habit of "contact with a difference" and impairing people's development into mature adults. It also leads to a very marked increase in the perception of insecurity and fear of crime. For safety, people must move in a private vehicle, as a protective shield. Any "unprotected" urban environment is perceived as representing a high risk to personal safety. Montaner (2006) speaks of "urban neo-feudalism"). Moreover, given the distances, the diversity of routes and low population density, public transport becomes less effective and efficient and is ultimately rendered meaningless.

In addition, the model of urban sprawl, together with social and functional segregation, has led to the abandonment of public spaces as a venue for interclass and intercultural interaction, which have tended to become spaces of marginality with a high perception of insecurity (Sorkin 1992; Sartori 2001; Lofland 2007; Pol 2009).

Another phenomenon that occurs in modern cities is what has been called "theming" (Montaner 2006; Montaner and Muxí 2015). This is the result of the confluence of historic center rehabilitation programs and a preponderance of tourism and "urban marketing". Historic downtown areas undergo a process of supposed restoration, which is actually a reinvention of their history and conversion into "theme parks" (Sorkin 1992) and has little to do with the social and cultural life that generated them. Theming generates unforeseen – or deliberately disregarded – social and environmental costs. Speculation, gentrification of the district, a certain architectural "virtuosity" and the progressive substitution of commerce and subsistence activities typical of everyday life leaves these neighborhoods barely apt for everyday life.

By contrast, more than 1000 million people currently live in slum neighborhoods and shantytowns, and this figure is expected to rise to 1390 million in 2020 (Bloom and Khanna2007). In 2014, the UN announced that the world population in urban areas had exceeded 54 % of the total: a population in search of more and better opportunities for subsistence and personal development (UN-Habitat, 2014). Paradoxically, in many cases this means a transition from a "humble" life (with the resources of the countryside providing a certain non-monetized type of subsistence) to a "miserable" life, living in self-constructed suburbs or old, degraded and unhealthy neighborhoods in the great metropolises, in a context where everything is monetized and the difficulty is having enough money. To this we must add the frustration of the expectations built up before emigrating, whether people are attracted by images of an affluent society seen in the mass media, or are fleeing tribal, political or religious conflicts. All this, far from facilitating "empowerment", promotes what Seligman (1975) labeled "learned helplessness".

2.3.1 Shrinking Cities

Along with the accelerated growth of populations in cities, there is also the opposite phenomenon: cities that record significant losses of population in a relatively short time, leading to the degradation of urban spaces. These are the so-called "shrinking cities". It is estimated that more than 450 cities with populations of over 100,000 have lost 10 % or more of their inhabitants since 1950 (Stohr 2004).

In Europe, the shrinking city phenomenon took on special relevance with the process of political and economic liberalization in the old Eastern Bloc, which saw the end of a great deal of economic and productive activity, and mass migrations as a result. Ivanovo in Russia, Leipzig in the former East Germany, and cities in Latvia, Romania and other eastern European countries are emblematic cases (Richardson and Nam 2014). Nevertheless, this did not take place only in the East but also in Manchester and Liverpool in the UK, with the deindustrialization of the 1970s and 1980s, especially during the Thatcher years (Schett 2011), and in other European cities too.

In the USA, the phenomenon occurs in two forms with different causes: on the one hand, owing to the effect of the global economic crisis in 2008, which aggravated or accelerated the closure or the offshoring of "classic" economic activities; on the other hand, because of significant changes in the city model. One of the most remarkable aspects is what has been termed "white flight" (Sugrue 1996; Rappaport 2003); the flight of the well-off white population to wealthy, homogeneous, suburban districts. This has led to notable social stratification in cities, with the abandonment and degradation of downtown and historic areas and the questioning of public spaces (Sorkin 1992; Sartori 2001; Lofland 2007; Pol 2009). Streets and public spaces in downtown areas become marginalized, impoverished ghettos that coexist alongside the opulence of the administrative sectors and private services, but with no mixing. Cincinnati, Chicago and Detroit (Martelle 2012; Pallagst 2009) are notable examples. This favors the emergence of what Boddy (1992) called the "analogous city" or "dualized city".

In many cases, the dualized city generates a structure overlying the degraded one, literally "parallel" to it. This has happened in downtown areas, with large buildings connected to each other underground or by elevated walkways where the functions of public space are played out. At first, this measure was justified as protection against the harsh weather conditions. In practice, it has generated a privatized space controlled by the "right of admission", which is encouraged as a way to ensure levels of security not found in the marginalized streets. The only "safe" access to these places is by private vehicle, parking directly in the destination building. Public transport is abandoned. Individualism is reinforced.

To all this Hollander (2010) adds the effect of what has been called "smart decline", when city planners want to reduce the number of buildings, the density, and the human agglomeration, in order to improve the citizens' QoL, but do not take into account peoples' functional, social and emotional needs.

The shrinking of the Halle-Leipzig region in Germany is one of the most studied cases (Bontje 2004; Banzhaf et al. 2007). The plan focused on attracting new activities with a cultural orientation (promoting the university, for example). A program of demolition of obsolete buildings and their replacement with parks and squares was implemented. The idea was to achieve more spacious, attractive, quality urban "places".

Programs with similar objectives were implemented, for example, in Manchester, in the Northern Quarter, a central working-class neighborhood and commercial area dating from the nineteenth century, devastated by the business trends of the indoor shopping mall in the early 1970s (Brown et al. 2004), or in Barcelona, with the establishment of universities, museums, and the opening up of new public spaces – not exempt from controversy (Di Masso et al. 2011) – as a way of regenerating historic neighborhoods in decline (the Raval, Poble Nou, etc.), seeking what Bohigas described as "functionalizing the center, monumentalizing the outskirts" (Bohigas 1985; Maragall et al. 2004).

Another option used for the redevelopment of downtown areas is the creation of Green Retirement Cities (see Chap. 24). This consists of regenerating housing and public spaces and introducing shops, services and local facilities considered essential for senior citizens and people with reduced mobility (Cohen 2007; McGreal 2010; Nefs et al. 2013; Boston Redevelopment Authority 2015).

2.3.2 The Sustainable City

What is meant by a "sustainable city"? As we said at the beginning, not much consideration was originally given to economic costs and environmental constraints in the "new urban planning proposals". It was in 1987, with the Brundtland Report, that the parameters of quality begin to focus on sustainability as one of the essential prerequisites.

The term "sustainable city" refers to "the potential of an urban agglomeration to ensure the environmentally benign development of a city through focused environmental and energy initiatives which stimulate a balance between economic progress, social equity and environmental quality." (Capello et al. 1999, p. 5). The sustainable city is one that allows growing complexity, improving the likelihood of contact between different elements of the urban system, without increasing the consumption of energy and resources (CCCB 1998, p. 96). In other words, a sustainable environment means finding an urban form and type of social interaction that makes it possible to maintain a reasonable level of social diversity and biodiversity, safeguarding the health of the inhabitants, the quality of the air, water and soil, to ensure the development of human well-being, while preserving the flora and fauna. This requires a process of urban design and management involving decision-making that serves not only the interests of the present generations but also the future ones.

Some authors use the term "urban vitality" to refer to social dynamism and the need to maintain elements of social and productive diversity in multifunctional spaces (Rueda 1995). This concept, close to Schilling and Logan's "right-sizing" (2008), considers that a certain number of people (neither too many nor too few) generates an optimal level of interaction necessary for the development of social life and the fulfillment of recreational and relational needs. Moreover, the very presence of people and activity generates a sense of security not found in empty spaces. The perception of public safety is related to residential satisfaction (Valera and Guardia 2014) and age and gender (Carro et al. 2010).

Downton (2009) describes the characteristics needed as prerequisites for eco-cities, with certain echoes of Jacobs (1961). He carries out a very thorough review of theoretical assumptions, authors and experiences of eco-cities (or in his case eco-neighborhoods) in Australia, North America, Europe and China. He situates the theoretical background to eco-cities in Christopher Alexander et al. (1975, 1977), whose studies have become a touchstone for responsive, humanistic, organic design, offering a more humane alternative to mechanistic modernism. Through Alexander, Downton establishes a link, albeit marking a distance, with the *Smart Growth and New Urbanism* of Leon Krier (1984, 1998) and Nikos Salingrados (2001, 2005). It is no coincidence that the first International Ecological City Conference (Eco-city 1) was hosted in Berkeley by Urban Ecology in 1990, under the leadership of Richard Register (1987, 1990).

In relation to urban design that directly affects people and social groups, following Downton, the eco-city model proposes the following items for consideration: layout of dwellings approved by future users, child-friendliness, large natural play areas, private and public spaces; promotion of a sense of neighborhood and identification with the living environment through urban spaces; passive solar orientation; each block with a clearly defined central area; reduced water consumption; healthy, recyclable and durable materials; sound-proofing; primarily high density and low-rise apartments (rather than houses) with one to three stories and balconies, flexible floor plans and a varied architecture in a generally unified urban structure.

Another core challenge is to create settlement structures that are suitable for a mix of different uses and a variety of functions, including the cultural and economic infrastructure. He lays emphasis on proximity and strongly recommends that a shop front should be at or near the intended site of any project, to avoid the shopping mall culture and the disadvantages of a segregated city, which exclude a significant part of the community. He gives special importance to the process of trying to ensure a sense of ownership, understanding and input in shared/communal areas.

For Downton, eco-city planning should be applied not only to the city (in the strictest sense of the term), but also to its whole sphere of metropolitan and regional influence. The eco-city approach reinforces the centrality of social dynamics. However, as Alexander had already pointed out, "it is not only the result which is important, but the process too" (Alexander 1964, p. 133).

One aspect of the sustainability of cities – but exceeding its territorial scope – is the ecological footprint. This is defined as the territory required to obtain the necessary resources to maintain a lifestyle and compensate for emissions and/or the greenhouse effect, whether in a country or a city. As Dodman (2009) says, cities are often blamed for high levels of greenhouse gas emissions. The United Nations Center for Human Settlements report (UN-HABITAT) (UN 2007) attributes 75 % of global energy consumption and 80 % of greenhouse gas emissions to cities. However, an analysis of emission inventories shows that – in most cases – per capita emissions from cities are lower than the average for the countries in which they are located.

In 2006, the ecological footprint of North American cities was between 4 and 5 ha per person, while the European average in cities was 2.8 ha per person (Dodman 2009). The Living Planet Report (2012), with data from 2008, which discusses continents and countries (not cities), situated the US ecological footprint at between 5 and 8 ha per person, while the European average was 3–5 ha per person, in Latin America and China, 2–3, in India, less than 1, and in Africa, between 1 and 4.

Dodman shows that the main causes of emissions are different in each context. In North America, he highlights the contribution of private transport, due to the long distances of daily trips, but contrasts the case of New York, where emissions are relatively low. The density of this city and the extensive public transport system means that levels here are much lower than those nationally (PlaNYC 2007). In the case of Latin American cities (Rio de Janeiro and São Paulo in Brazil), emissions from solid waste are much higher than in many other cities. In Asian cities, the main sources of emissions are related to mobility patterns, the design and distribution of houses, the organization of food and water systems, and individual lifestyle choices. In China, in 1999, industrial activities were responsible for 80 % of Shanghai's emissions and 65 % of Beijing's.

This analysis reveals certain factors that directly or indirectly affect welfare, QoL and sustainability: denser cities seem to consume less energy, require less individual transport and can be equipped with better and more efficient public transportation. Unnecessary mobility and the excessive use of air and maritime transport are part of the human contribution to climate change (with the former responsible for over 3.5 % of human-induced change according to IPPC reports), as is excess waste production. All this is closely linked not only to the urban form and lifestyle but also to the dominant values that end up subtly reformulating these lifestyles, without the citizens themselves being aware of the process.

In any case, lifestyles (including forms of nutrition and waste management), urban technology (also linkable to smart cities), mobility systems, and the user-friendliness of urban spaces, inviting or impeding walkability, appear as key factors in the necessary reduction of the ecological footprint.

2.4 Sustainable Mobility and Walkability

Modern society is characterized by high mobility. The average use of motorized transport per person in the European Union grew by 7 % between 2000 and 2008 (Golisnka and Hajduk 2012), with multiple consequences for the environment. This increase is due to the globalization of economic activities and lifestyles, and urban sprawl, which does not encourage walking as a form of travel. This situation has prompted a need for studies on walkability.

2.4.1 Sustainable Mobility

The European Commission (2006) defines sustainable mobility as a system that meets a society's economic, social and environmental needs whilst minimizing its undesirable impacts on the economy, society and environment. However, Steg and Gifford (2005) note that there is no sufficiently clear, widely accepted definition. Some systems of indicators have been developed, but few include social and behavioral dimensions, due to deficiencies in conceptualization and the complexity of measurement.

Sustainable mobility has two sides to it: on the one hand, the application of new technologies to transport systems (electric and hybrid vehicles, and fossil-fuel vehicles with low emissions, etc.) and on the other, the citizens' choice of private or public transport. Since its beginnings, environmental psychology has explored this second aspect, albeit timidly and taking very varied approaches.

In an early paper published by Proshansky et al. (1970), the only reference to mobility is a chapter by Webber. This author extols the advantages of highways and the private car as the great metropolitan transport solution. He analyzes the psychological factors that influence the choice of private or public transport. He presents the immediate payment of a ticket as a deterrent to the use of public transport, while the deferred payment in the case of the car appears to be a psychological advantage even though the costs are higher. For the author, the need to reduce traffic congestion takes priority over environmental considerations, which he ignores, and he cites the freeway system in Los Angeles as an example to imitate. All this lies at the opposite extreme to the current predominant view.

Everett and Watson (1987), in their review of the state of affairs, oriented the contribution of psychology towards improving mobility management. They defined four possible areas of interest for psychology: (1) physiological, perceptual and ergonomic aspects; (2) personal and contextual characteristics as predictors of the choice of mode of transport (including values and attitudes); (3) the potentialities of learning and behavior modification theories; (4) and the specifics of each stage of life, such as childhood, old age and situations of disability.

Subjective psychological and physiological well-being appears a determining factor in the choice of transport mode. For Everett and Watson (1987), comfort depends on the ergonomics of the vehicle's design and safety measures. Another aspect underscored is the stress generated by commuting (Cassidy 1997), which has an impact in the workplace (Stokols and Novaco 1981). However, time spent traveling is not always necessarily given a negative assessment. Redmond and Mokhtarian (2001) note that a certain amount of traveling time is sought positively, as an interstitial period/space, offering relaxation and serving to accommodate the change in activity.

For Stokols (1976), crowding implies overstimulation, the impossibility of exercising control over the space and unwanted visual contact with other travelers. This appears to be one of the most disliked aspects of public transport. Ride quality is another, highlighted by Golob et al. (1972) and Nicolaidis (1975). Davis and Levine (1967) stress the importance of comfort levels while waiting.

McCormick (1976), in 1976, highlighted the passenger's perception and image of comfort as factors influencing the use of public transport.

However, as pointed out by Seoane et al. (2000), non-users tend to let exclusively negative stereotypes guide their assessment, while regular users tend to make more positive assessments. Then, it appear as essential to develop a common vision sympathetic to public transport, which for Everett and Watson (1987) can act as an intangible boost in its favor.

Rosenbloom (2001) notes that the ageing of the western population may result in increased air pollution because the habits/mobility needs of this sector, which is dependent on other people (children, carers, etc.), tend to multiply the amount of travel. It should be remembered that the elderly are discouraged from using public transport by problems of accessibility, the anxiety caused by jolty driving, and fear of crime (Patterson and Ralston 1983).

In the 1980s, a line of mobility analysis with a gender perspective emerged. The excluding, discriminatory effect of transport systems on women (Turner and Grieco 2000; Michelson 1983; Moore 1982) was denounced. For their part, Matthies et al. (2002) found that German women had stronger ecological values and tended to use private vehicles less.

While studies on mobility have been based mainly on the analysis of attitudes and behavior incentivization, Aarts and his team have emphasized the role played by previous mobility habits in the choice of transport. When there is an established habit, attitudes may be irrelevant as a guide to behavior, and the role of cognitive components in the choice of behavior is minimized. Furthermore, automatic behavior is difficult to prevent or control (Aarts et al. 1998; Ouellette and Woods 1998; Aarts and Dijksterhuis 2000).

As of the 1990s, there was a notable increase in environmental psychology research that took into account the dimension of energy efficiency, emission reduction and citizens' responsibility when choosing mobility systems. Reducing the use of private vehicles, cutting down on unnecessary travel, walking, riding a bike, opting for public transport or more energy-efficient and less polluting vehicles (hybrid and electric vehicles, etc.) are the recurring themes.

The image of private transport is clearly associated with individual QoL. However, as noted by Steg and Gifford, this conflicts with collective QoL if one takes into account sustainability, posing a typical example of a social dilemma. To achieve a sustainable transport system, the use of private vehicles should be reduced, and the safety, accessibility and efficiency of public transport should be improved. However, from an individual point of view, the private car is more attractive because of the associated advantages: independence, flexibility, comfort, speed, perceived security and privacy, and it has become a status symbol and means of self-expression. Therefore, the choice of sustainable transport can be perceived as detrimental to individual QoL (Steg and Gifford 2005).

Golinska and Hajduk (2012) identify the main obstacles to sustainable mobility as: incompatible infrastructures, abusive charges and prices, rising congestion, poor safety, changes in mobility patterns, dependence on fossil fuels and growing CO_2 emissions from transport. To overcome these, they propose a wide range of measures: promotion of walking and cycling, better public transport networks, incentives to use public transport, attractive timetables, improved comfort, easy access, reliability of services, smart intermodal ticketing, among others.

However, as seen above, any discussion of sustainable mobility means more than just fostering transport. In recent decades, hybrid and electric vehicles (EVs) have been promoted as a viable technology to reduce dependence on fossil fuels and the resulting greenhouse emissions associated with conventional vehicles. However, the acceptance of new technologies applied to transport comes up against distrust of the new proposals: they are considered alien and unproved (Egbue and Long 2012), unsafe and of uncertain social status. There is a perception that they offer a lower level of performance than conventional cars (Burgess et al. 2013), reduced range, lower maximum speed, less powerful acceleration, and inconvenient battery-charging requirements (Bunch et al. 1993; Lane 2011).

According to Burgess et al. (2013), the symbolic meanings of EVs are currently in a state of fluctuation. Their study revealed direct experience of EVs to be a crucial factor in shifting the public view from skepticism to acceptance and approval. Other studies insist that activating environmental values motivates consumers to act in keeping with their pro-environmental values, and this includes their decisions on mobility (Verplanken and Holland 2002). However, some studies have shown that the influence of value activation depends on how central the respective value is to a person's self (Verplanken and Holland 2002; Verplanken et al. 2008).

Actions should not only depend on the personal decisions of each citizen. Creating conditions favorable to modes of transport that foster sustainability must be accompanied by measures that discourage unsustainable mobility. For example, the creation of pedestrian zones and reduction of traffic thoroughfares (and the adaptation of streets to make them public-friendly spaces that encourage walking), the creation of cycle lanes, and giving preference to public transport vehicles are successfully tested measures that should accompany attempts to raise public awareness. In Barcelona, for instance, the goal over the last 30 years has been to achieve gradually a 50–50 balance between pedestrian zones and traffic thoroughfares (Bohigas 1985; Barcelona City Council 1987, 2012; FAD 2009). The quality of the urban environment and social qualities of the city are configured as key elements, as shown by studies on walkability.

2.4.2 Walkability

Access to outdoor green areas and the possibility of being physically active in the residential setting is an important component of urban QoL. These ideas have been extensively developed in environmental psychology by studies on restorative environments (e.g. Hartig 2004), as well as on residential satisfaction and residential environmental quality (e.g. Amérigo and Aragonés 1997; Bonaiuto 2004) (see Chaps. 6, 7, 13, 14, and 17 in this book). These studies have explored the following basic questions: Is residential satisfaction related to actual or perceived residential quality? Can we identify specific features of residential settings that are more strictly associated with satisfaction and, in turn, with predictive QoL? A recent paper by Bonaiuto and Alves (2012) discusses some of the basic features characterizing healthy, inclusive residential areas, and focuses in particular on the notions of "environmental fitness" and "affordances", developed by environmental psychology to explain the relationships between residential environmental quality, human activity in the context of everyday life, and actual or perceived QoL.

On the subject of affordances, the concept of "walkability" has emerged recently in various domains of research linked to QoL (e.g., public health, preventive medicine, environmental psychology), providing an interesting angle when addressing these issues (e.g., Brown et al. 2007; Saelens et al. 2003). Different characteristics of the physical environment have been identified as determinants of the walkability of residential settings, related to factors associated with health and well-being (for example, being more physically active). These can therefore contribute to a broader vision of QoL. According to Brown and Werner (2012), physical activity in everyday life, such as walking, should be regarded as a powerful tool for promoting public health, especially to cope with the so-called obesity epidemic rampant in many industrialized countries. Brown and Werner underline the need for ecological and transactional approaches, which share the basic assumption that people and their settings are mutually defining. This vision suggests that the interplay between macro (e.g. street forms and urban density) and micro factors (such as pleasantness of the settings and perceived safety) is crucial to promoting changes in favor of a sustainable lifestyle. Research on macro factors has focused on the so-called "*3Ds*" model: density, diversity, and design (Cervero and Kockelman 1997); this was later expanded into the "*5Ds*", including distance and destination (Cervero et al. 2009). Density, street connectivity and proximity to shops are typical macro factors that emerge as predictors of walkability according to different meta-analyses (e.g., Ewing and Cervero 2010; Saelens and Handy 2008). Among the micro factors, neighborhood aesthetics (pleasantness, green spaces, few signs of antisocial behavior) and perceived safety are strongly associated with walkability, as well as social control and social support (Brown and Werner 2012). An important distinction can be made between different types of walkability: transport walkability (i.e. walking to work or for other practical purposes), and leisure walkability (i.e. walking for enjoyment). The former is more affected by macro factors such as density and proximity, while the latter is more associated with micro factors such as the presence of sidewalks, pleasant surroundings and safety.

To assess and evaluate walkability and compare differential contexts, comprehensive walkability indexes have been drawn up (e.g., Frank et al. 2006). An interesting proposal was made by the Health and Community Design Lab at the School of Population and Public Health, Faculty of Medicine, University of British Columbia (Canada; see http://health-design.spph.ubc.ca/tools/walkability-index/).

Here, a walkability index was created by putting together four components: *Residential density* (i.e. the number of residential units per acre within a neighborhood); *Commercial density* (i.e. the amount of space designated for commercial use within a neighborhood); *Land use mix* (i.e. the degree of mixing of different types of land use in a specific area, such as residential, commercial, entertainment, and office development); *Street connectivity* (i.e. the number of street intersections in a neighborhood).

The relative ease of providing walker-friendly structures is the key to the effective promotion of physical activity in modern cities and, at the same time, it increases the sustainability of the current mobility system. Intervention strategies that combine acting on both the macro and the micro aspects of walkability might then be particularly effective for transforming the residential context into a supportive environment capable of promoting a sustainable QoL with relatively diffused and sustained benefits at a reasonably low cost.

2.5 From Empowerment to Learned Helplessness. Let's Not Be Naive

We began this chapter by asking ourselves whether human beings have notwere ever been happy till before the twenty-first century, and what bearing this has on QoL and environmental conditions. We have seen theoretical proposals and very contradictory data readings. At the end of section three, we concluded that two frameworks of conceptualization can be identified: (a) QoL as accumulation, and (b) QoL as a balance. We now ask ourselves which of these two concepts is favored by socio-economic developments and the evolution of the current habitat.

We have seen how some authors have found sufficient evidence to affirm that human beings tend to be happy and assess their situation positively. Moreover, as Veenhoven insists, average happiness is high in modern societies and tending to rise. This finding contradicts longstanding pessimism about modernization (Cummins 2000; Veenhoven 2005; Veenhoven and Hagerty 2005; Inglehart et al. 2008). Nevertheless, as Easterlin (1974, 2005) and Easterlin et al. warn (2010), in the happiness-income paradox this improvement only lasts for a certain amount of time. What happens after that is what Max-Neef (1995) describes in his "threshold hypothesis": having crossed a certain threshold, economic growth becomes detrimental to QoL.

Back in the 1980s, Glatzer and Mohr (1987) found few correlations between objective living conditions and subjective well-being; perhaps because, as Lee (1969) had pointed out previously, people tend to put on a brave face even in times of adversity, for reasons of self-concept or, as shown by Festinger (1957), due to the reduction of cognitive dissonance. In this respect, the fact that people always tend to have a positive identity seems to have an influence (Tajfel 1978, 1981; Hogg and Abrams 1988), unless there are forces or situations that impede it. However, according to Levi and Anderson (1974/1980), beyond minimum survival needs, QoL is more dependent on the way expectations and realities come together than on the capacity for accumulation, a process that Michalos (1985) tried to systematize in his multiple discrepancies theory. Perceived social support appears as a fundamental element (Diener and Seligman 2002; Javaloy 2007, among others). Social support, the capacity for agency, freedom of choice and the locus of control are key aspects of QoL, but also empowerment (Zimmerman 1995, 2000; Hur 2006; Christens 2012, 2013). This upholds the idea of QoL as a balance.

Psychosocial and communitarian literature, urban planning and environmental psychology, like the speeches of politicians in recent decades (when they want to be "politically correct"), are all full of calls to generate social cohesion and empowerment in groups and communities. But what are the psychological and social mechanisms linked to empowerment? Does the relationship of the individual with his/her sociophysical environment play a role?

The feeling of being part of a network of relationships of mutual support, which can be trusted and as a result of which the individual does not experience feelings of loneliness, anxiety and anguish, is part of what characterizes the community (Sarason 1974; Sánchez 1991; Garcia et al. 1994; Nelson and Prilleltensky 2010; Christens 2013). It also helps to develop a common identity and a sense of being part of a larger, more stable and reliable social structure (McMillan and Chavis 1986).

As Montero (1994) points out when discussing the Latin American tradition, it is necessary to combat self-fulfilling prophecies of personal failure. When the sense of community weakens, people take a fatalistic attitude to their living conditions, as if they were predestined. Finding solutions tends to be left to others. An inability to control life (Rotter 1954), a sense of "learned helplessness" (Seligman 1975), is internalized. On the other hand, the community has characteristics that are both its own and diverse at the same time, which are given in a physical space and form part of the structural elements. These structural elements must be conceived as parts embedded in everyday life, and this interdependence is what gives meaning and significance to the environment, while contributing to the social identity of the person, and being the scenario for the development of functional aspects, in an eminently dynamic relationship (Montero 1994).

A positive assessment is often made of the contribution of the ICTs and the new social networks they have produced. Nevertheless, we must ask to what extent the new virtual social networks complement or supplement (occupy) the role of the proximity "physical" social groups (community associations, clubs and so on). Do they facilitate the emergence of new proactive social movements or generate dependence on the networks and effective social isolation? To what extent can we rely on the accuracy of the unnuanced messages/short sentences circulating in the networks, knowing who forwarded but not who formulated them or what their real intentions are? For the time being, these questions seem to have no clear answers.

Psychologists and social scientists have always intended their contributions to be positive and "utilitarian" (in the sense of Mill, in order to achieve the common good and happiness of the whole community). But an excessively naive vision seems to prevail: the idea that their knowledge will always be used to improve things, improve the social reality. We like to think this is the case, but history tells us it is not.

Sassen showed that the processes of precarious employment, late retirement, cuts in social benefits and subsidies, privatization of services considered fundamental and emblematic, and ultimately, lower QoL, are not random (1988), in cities of a globalized society, which expels a part of its citizens, who lose control and sovereignty (1996, 2014). Sennett (1999) described the dominant social dynamics that generate uncertainty, the fear that the positive cycle will come to an end (as has happened with the 2008 crisis) and how this results in social containment, with an individual and social effect that leads to the "corrosion of character". Bauman (1998, 2000), independently of the crisis, described the new "normality" as a liquid society, an individualized society seeking a new kind of security (2001a, b, 2003), in which economic growth increasingly favors a smaller number of people (2011; 2013). Meanwhile, Putnam (2000) warned against isolation, rather than individualism, with the metaphor that we are "bowling alone". Hessel (2011a, b) invited us to feel "indignant" and become "committed", and Bardi (2011), among others, reminded us that there are limits to growth (environmental and social), that we are forgetting this and must reassess them. However, "limits to growth" do not seem to be in line with the current dominant thinking and action. Furthermore, a long list of thinkers (mostly senior or very senior) warns us with very varying emphasis of the risk – or the reality – of what Seligman (1975), from a psychological standpoint, described as "learned helplessness".

The dynamics of society in recent decades appear to have had the same adverse effect as the electric shocks Seligman gave his dogs when they wanted to open the door of their cage until, despite the door being open, the dogs did not leave. The formal discourse of social cohesion and empowerment is the open door, but the subtle shocks produced by negative experiences (related to economics, employment, identity and its references, including the individualizing urban form) leave the population with the sensation that there is no way out and so it conforms to the prevailing precariousness. There is sufficient evidence to suspect that despite the apparent "good intentions" in society in recent decades, and especially since the 2007–2008 crisis, a subtle (sometimes) or explicit (often) process occurs that creates conditions where those elements thought to generate empowerment (and hence cohesion, strength and the ability to resist, defend oneself and decide for oneself) gradually dissolve away to be replaced by a state of learned helplessness. Taking a clear example (reductionist and demagogic, but real) to illustrate this point, in many parts of Europe, people now work for 10 or 20 % of their previous salary, and they are supposed to be grateful because they have a job (even if a precarious, part-time, "mini job" and so on, and not just in the hardest hit countries in Southern Europe).

We find a loss of self-confidence, intentional uprooting (for those who are interested in promoting helplessness, it is important to ensure that people decrease their attachments to place to "prevent" resistance to the so-called "work mobility") with the loss of physical social networks (replaced by virtual ones), which makes people more vulnerable, a reduction in effective informal social support (street life, spontaneous associative fabric, clubs, etc.), and the discrediting of formalized social support (the welfare state). These functions are then subsumed and offered by large companies, which first perhaps offers a more "efficient" service, but over which the former members (now clients) have no control, and the service or activity loses the "virtue" of a social backbone.

Civil society (associative dynamics of all kinds, greater and different from the "private enterprise" with which it is sometimes equated) was born and grew to provide a direct response to the needs of the people, its associate members. It provided a key structure and support for propinquity. When it appeared to have been consolidated as a "useful" model for well-being, economic globalization changed the scenario.

In Fig. 2.2, three scenarios are represented, based on the interactions between some of these components.

The first diagram shows the components of a society serving the people's well-being, before the "rush" of globalization. The second diagram reflects what may be considered an advanced stage of the new global society, and the third diagram, the situation already existing in some cases or on the way to becoming our society.

Two types of capital are reflected in the diagrams:

(a) Local capital, with an identity, nearby or identified with a place, ready to favor investments that, apart from generating private profit, also meet the needs of the immediate social environment, to which it is assumed to be sensitive. It includes small, medium and large enterprises, provided that they operate in this way. It also includes small local banks and savings banks (in their original version of providers of mutual aid).

(b) Global capital appears on the outermost part of the diagram. It is "anonymous" capital (although it may belong to retirees' investment funds, for example, managed by investment companies) whose purpose is maximum profit, rootless, with no qualms about relocating to obtain the

Fig. 2.2 Stages of socio-economic globalization

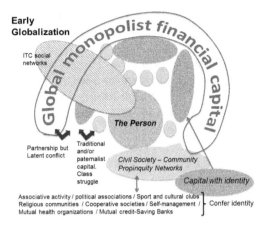

1st stage of socio-economic globalization

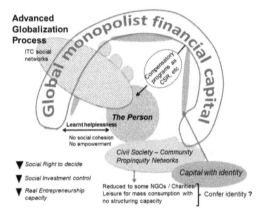

2nd stage. Advanced socio-economic globalization

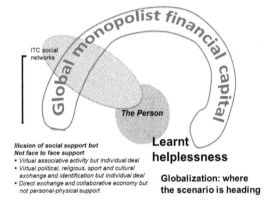

3rd stage of socio-economic globalization

greatest possible yield. It is linked to large multinationals and large global banks.

In between, the uneven presence of small balloons can be observed, which are intended to represent civil society, the associative fabric, which gives the individual a sense of belonging and identity and provides mutual support (formal and informal).

In the center, the person is represented, who may or not be "protected", helped and supported by civil society. The diagram is crossed by a cloud representing the Internet and online social networks. These cut across all levels, but with different (and sometimes contradictory) implications for each of the entities we have mentioned.

In the first stage, capital with an identity coexists and competes with global capital. It is torn between fighting for its independent maintenance, forming partnerships with global capital and its dynamics, or disappearing. Global capital tries to control or absorb local capital. It is a silent, barely visible combat, and dilemmatic. However, it is the key to understanding some of the conflicts between economic powers, and between them and state structures, whose effective political power is subtly undermined.

Civil society, the community and the networks for propinquity display a creative activity (intended for self-protection and for setting up their own services) that offers an opportunity for personal fulfillment and also confers social identity, perception of (community) support and empowerment.

In the second stage, with the advance of the globalization of the economy, local capital with an identity has dwindled considerably and majority rule is exercised by global capital. Control over investment and the right to decide have been lost, difficulties in finding support for entrepreneurial initiatives are increasing, because the decisions are taken in faraway places.

The typical structures of civil society have practically disappeared. There are hardly any non-governmental organizations, which are occasionally used as a palliative for some serious imbalances produced by the system (e.g. food banks). The organization and enjoyment of leisure is no longer a social initiative that generates cohesion. It is merely mass consumption without any structuring capacity and with questionable capacity to confer social identity. We are closer to learned helplessness than empowerment.

Companies are forced to develop compensatory programs (e.g. Corporate Social Responsibility programs) to alleviate some problems of their workers (dysfunctional problems affecting production, although presented in the guise of "social support"). No longer protected by the almost defunct civil society, the person gradually builds up a sense of learned helplessness. To some extent, virtual social networks act as a palliative for the social needs of the person, but suffer from the limitation or the absence of face-to-face contact.

In the third stage, which seems to be where we are heading, we find helpless people without any social support, in a situation of very high geographic mobility because of the precarious, temporary and low-paid jobs that they have no choice but to accept, and that hinders the creation – re-creation – of formal and informal support networks.

Nonetheless, this frightening and depressing scenario, which seems to be coming closer, is highly unlikely. History teaches us that after the most miserable periods, after periods of exclusion and marginalization, human beings recover and reinvent themselves. In the history of the current first-world powers, times of marginalization, destructuration and social decomposition are not very distant. Yet, all social fabrics tend to regenerate and "re"-structure themselves, even if outside the established order.

If we are at stage 2 or heading towards stage 3, it may be due to an accumulation of coincidences or circumstances that have eroded the fundamental environmental conditions and psychosocial processes needed for well-being and QoL. However, it may also be due to the intentional misuse of these same advances contributed by the social sciences to the field of the functioning of personal and social mechanisms. It would be naive to think that they have only been used "for good". Neutralizing certain personal and social trends can help to increase the volume and speed of economic accumulation, albeit at the expense of social and personal balance.

The capacity for recovery and reinvention and the resilience of human beings and society inexorably lead to experimentation with new organizational models and lifestyles. The psychological and life processes will surely differ in form, but not in the basic needs, whose satisfaction we must guarantee.

The dominant form and organization of the environment, habitat and city in recent decades seems to be encouraging an evolution in the direction that favors learned helplessness. Nonetheless, the need for urban models (for both social and sustainability reasons) that favor social cohesion and empowerment is actively proposed in critical reflections and by the new movements. Faced with the reality of new technologies and resources used in a smart way, these urban forms should not be – and cannot be – a "return" to the romantic parochialism (Lofland 2007) of the past. We must invent new urban forms grounded in respect for sustainability (even if it is only for "selfish" survival motivations), which guarantee those psychological and psychosocial processes essential for the individual to achieve greater well-being, happiness and QoL.

References

Aarts, H., & Dijksterhuis, A. (2000). The automatic activation of goal-directed behaviour: The case of travel habit. *Environmental Psychology, 20*, 75–82.
Aarts, H., Verplanken, B., & Van Knippenberg, A. (1998). Predicting behaviour from action in the past: Repeated decision making or a matter of habit? *Journal of Applied Social Psychology, 28*(15), 1355–1374.
Alexander, C. (1964). *Notes on the synthesis of form*. Cambridge: Harvard University Press.
Alexander, C. (1975). *The oregon experiment. Center for environmental structure*. New York: Oxford University Press.
Alexander, C. (1977). *A pattern language: Towns, buildings, construction*. New York: Oxford University Press.
Amérigo, M., & Aragones, J. I. (1997). A theoretical and methodological approach to the study of residential satisfaction. *Journal of Environmental Psychology, 17*, 47–57.

Andrews, F. M., & Withey, S. B. (1976). *Social indicators of well-being: America's perception of life quality*. New York: Plenum Press.
Argyris, C. (1957). *Personality and organization: The conflict between system and the individual*. Oxford: Harpers.
Banzhaf, E., Kindler, A., & Haase, D. (2007). Monitoring, mapping and modelling urban decline: A multiscale approach for Leipzig, Germany. In *EARSeL eProceedings 6, 2/2007* (pp. 101–114). https://www.ufz.de/export/data/1/25597_06_2_banzhaf1.pdf. Accessed 13 July 2013.
Barcelona City Council. (1987). *Programa d'Actuació Municipal 1988–1991* [Municipal Action Plan 1988–1991]. Technical Programming office. Barcelona City Council. http://prod-mobilitat.s3.amazonaws.com/PMU2013-2018IntroDiagnosiEscenaris_llarg_0.pdf. Accessed 9 Mar 2015.
Barcelona City Council. (2012). *Plan de Movilidad Urbana de Barcelona 2013–2018* [Barcelona's Mobility Plan 2013–2018]. http://prod-mobilitat.s3.amazonaws.com/PMU2013-2018IntroDiagnosiEscenaris_llarg_0.pdf. Accessed 9 Mar 2015.
Bardi, U. (2011). *The limits to growth revisited*. New York: Springer.
Bauer, R. A. (1966). *Social indicators*. Cambridge: MIT Press.
Bauman, Z. (1998). *Globalization: The human consequences*. New York: Columbia University Press.
Bauman, Z. (2000). *Liquid modernity*. Cambridge: Polity Press.
Bauman, Z. (2001a). *The individualized society*. Cambridge: Polity Press. http://ca.wikipedia.org/wiki/Especial:Fonts_bibliogr%C3%A0fiques/0745625061.
Bauman, Z. (2001b). *Community. Seeking safety in an insecure world*. Cambridge: Polity Press.
Bauman, Z. (2003). *City of fears, city of hopes*. London: Goldsmith's College.
Bauman, Z. (2011). *Collateral damage: Social inequalities in a global age*. Cambridge: Polity Press.
Bauman, Z. (2013). *Modernity and ambivalence*. New York: Wiley.
Bechtel, R. B., & Churchman, A. (Eds.). (2002). *Handbook of environmental psychology*. New York: Wiley.
Bentham, J. (1789/1969). *Introduction to principles of morals and legislation*. New York: Pegasus.
Bloom, D. E., & Khanna, T. (2007). The urban revolution. *Finance and Development, 44*(3), 9–14.
Boddy, T. (1992). Underground and overhead: Building the analogous city. In M. Sorkin (Ed.), *Variations on a theme park: The new American city and the end of public space* (pp. 123–154). New York: Hill and Wang.
Bohigas, O. (1985). *Reconstrucció de Barcelona* [Reconstruction of Barcelona]. Barcelona: Edicions 62. (Spanish version published by Servicio de Publicaciones, Secretaría General Técnica, Ministerio de Obras Públicas y Urbanismo, Madrid, 1986).
Bonaiuto, M. (2004). Residential satisfaction and perceived urban quality. In C. Spielberger (Ed.),

Encyclopedia of applied psychology (pp. 267–272). New York: Academic/Elsevier.

Bonaiuto, M., & Alves, S. (2012). Residential places and neighbourhoods: Toward healthy life, social integration and reputable residence. In S. D. Clayton (Ed.), *The Oxford handbook of environmental and conservation psychology* (pp. 221–247). New York: Oxford University Press.

Bonnes, M., & Bonaiuto, M. (2002). Environmental psychology: From spatial-physical environment to sustainable development. In R. Bechtel & A. Churchman (Eds.), *Handbook of environmental psychology* (pp. 28–54). New York: Wiley.

Bonnes, M., Carrus, G., & Passafaro, P. (2006). *Psicologia ambientale, sostenibilità e comportamenti ecologici* [Environmental psychology, sustainability and ecological behaviors]. Rome: Carocci Editore.

Bontje, M. (2004). Facing the challenge of shrinking cities in East Germany: The case of Leipzig. *Geojournal, 61*(1), 13–21.

Boston Redevelopment Authority. (2015). *Programs*. http://www.bostonredevelopmentauthority.org.

Boulding, K. (1966). The economics of the coming spaceship earth. In H. Jarrett (Ed.), *Environmental quality in a growing economy* (pp. 3–14). Baltimore: Resources for the Future/Johns Hopkins University Press.

Brown, B. B., & Werner, C. M. (2012). Healthy physical activity and eating: Environmental supports for health. In S. D. Clayton (Ed.), *The Oxford handbook of environmental and conservation psychology* (pp. 459–484). New York: Oxford University Press.

Brown, A., O'Connor, J., & Cohen, S. (2004). Local music policies within a global music industry: Cultural quarters in Manchester and Sheffield. In Philipp Misselwitz (Coord.), *Manchester/Liverpool working papers* (pp. 94–113). Shrinking Cities Project. Federal Cultural Foundation, Germany; Gallery for Contemporary Art Leipzig; Bauhaus Fundation Dessau and the Journal Archplus.

Brown, B. B., Werner, C. M., Amburgey, J. W., & Szalay, C. (2007). Walkable route perceptions and physical features converging evidence for en route walking experiences. *Environment and Behavior, 39*(1), break 34–61.

Brundtland Report. (1987). *Our common future. Report of the World Commission on Environmentand Development*. United Nations. http://www.un-documents.net/our-common-future.pdf.

Bunch, D. S., Bradley, M., Golob, T. F., Kitamura, R., & Occhiuzzo, G. P. (1993). Demand for clean-fuel vehicles in California: A discrete-choice stated preference pilot project. *Transportation Research Part A: Policy and Practice, 27*(3), 237–253.

Burgess, M., King, N., Harris, M., & Lewis, E. (2013). Electric vehicle drivers' reported interactions with the public: Driving stereotype change? *Transportation Research Part F: Traffic Psychology and Behaviour, 17*, 33–44.

Buttimer, A. (1972). Social space and planning of residential areas. *Environment and Behavior, 4*(3), 279–318.

Campbell, A., & Converse, P. E. (1972). *The human meaning of social change*. New-York: Russel Sage Foundation.

Campbell, A., Converse, P. E., & Rodgers, W. L. (1976). *The quality of American life: Perceptions, evaluations, and satisfactions*. New York: Russell Sage.

Canter, D. (1969). Should we treat buildings as subject or object? In D. Canter (Ed.), *Architectural psychology. Proceedings of the Conference Held in Dalandui, UK* (pp. 11–18). London: Royal Institut of British Architecture.

Canter, D., & Rees, K. (1982). A multivariate model of housing satisfaction. *Applied Psychology, 31*(2), 185–207.

Cappelo, R., Nijkamp, P., & Pepping, G. (1999). *Sustainable cities and energy policies*. Berlin: Springer.

Carro, D., Valera, S., & Vidal, T. (2010). Perceived insecurity in the public space: Personal, social and environmental variables. *Quality & Quantity, 44*(2), 303–314.

Casas, F. (1996). *Bienestar social: una introducción psicosociológica* [Welfare: A psychosociological introduction]. Barcelona: PPU.

Cassidy, T. (1997). *Environmental psychology: Behaviour and experience in context*. London: Taylor & Francis.

CCCB (1998). *La ciutat sostenible* [The sustainable city] (Barcelona's contemporary culture centre). Barcelona: CCCB.

Cerdà, I. (1859). *Teoría de la Construcción de las Ciudades* [Theory of city construction]. Reedition 1991 by Ministerio para las Administraciones Públicas, and Barcelona City Hall. English summary at https://www.google.es/search?q=Cerda+Theory+of+City+Construction%22,+1859&ie=utf-8&oe=utf-8&gws_rd=cr&ei=KJiaVfrlDILWU9mzgcAN. Accessed 23 Apr 2015.

Cervero, R., & Kockelman, K. (1997). Travel demand and the 3Ds: Density, diversity, and design. *Transportation Research Part D: Transport and Environment, 2*(3), 199–219.

Cervero, R., Sarmiento, O. L., Jacoby, E., Gomez, L. F., & Neiman, A. (2009). Influences of built environments on walking and cycling: Lessons from Bogotá. *International Journal of Sustainable Transportation, 3*(4), 203–226.

Christens, B. D. (2012). Targeting empowerment in community development: A community psychology approach to enhancing local power and well-being. *Community Development Journal, 47*(4), 538–554.

Christens, B. D. (2013). In search of powerful empowerment. *Health Education Research, 28*(3), 371–374.

Clayton, S. D. (2012). *The Oxford handbook of environmental and conservation psychology*. New York: Oxford University Press.

Clayton, S., & Myers, G. (2009). *Conservation psychology: Understanding and promoting human care for nature*. New York: Wiley.

Cohen, L. (2007). Buying into downtown revival: The centrality of retail to postwar urban renewal in American cities. *The Annals of the American Academy of Political and Social Science, 611*, 82–95.

Corral-Verdugo, V., Tapia-Fonllem, C., Ortiz-Valdez, A., & Fraijo-Sing, B. (2013). Las virtudes de la humanidad, justicia y moderación y su relación con la conducta sustentable. *Revista Latinoamericana de Psicología, 45*(3), 361–372.

Corral-Verdugo, V., Frías, M., Gaxiola, J., Tapia, C., Fraijo, B., & Corral, N. (2015). *Ambientes positivos. Ideando entornos sostenibles para el bienestar humano y la calidad ambiental*. Mexico City: Pearson.

Cummins, R. A. (2000). Personal income and subjective well-being: A review. *Journal of Happiness Studies, 1*(2), 133–158.

Cutter, S. (1982). Residential satisfaction and the suburban homeowners. *Urban Geography, 3*(4), 315–327.

Davis, M., & Levine, S. (1967). Toward a sociology of public transit. *Social Problems, 15*(1), 84–91.

De Neve, J.E., Diener, L.T., Cody, X. (2014). The objective benefits of subjective well-being. In John Helliwell, Richard Layard, Jeffrey Sachs (Eds.) *World Happiness Report 2013* (pp. 54–79). Chap 4, United Nations. Available at: http://unsdsn.org/wp-content/uploads/2014/02/WorldHappinessReport2013_online.pdf. (May 28, 2015).

Di Masso, A., Dixon, J., & Pol, E. (2011). On the contested nature of place:'Figuera's Well', 'The Hole of Shame'and the ideological struggle over public space in Barcelona. *Journal of Environmental Psychology, 31*(3), 231–244.

Diener, E. (1984). Subjective well-being. *Psychological Bulletin, 9*(3), 542–575.

Diener, E., & Seligman, M. E. (2002). Very happy people. *Psychological Science, 13*(1), 81–84.

Dodman, D. (2009). Blaming cities for climate change? An analysis of urban greenhouse gas emissions inventories. *Environment and Urbanization, 21*(1), 185–201.

Downton, P. F. (2009). *Ecopolis: Architecture and cities for a changing climate*. Co-published by Springer Science+Business Media B.V., Dordrecht and CSIRO Publishing, Collingwood.

Dunn, H. L. (1959). What high-level wellness means. *Canadian Journal of Publich Health, 50*(11), 447–457.

Easterlin, R. A. (1974). Does economic growth improve the human lot? Some empirical evidence. In P. A. David & M. W. Reder (Eds.), *Nations and households in economic growth: Essays in honor of Moses Abramovitz* (pp. 89–125). New York: Academic.

Easterlin, R. A. (2005). Feeding the illusion of growth and happiness: A reply to Hagerty and Veenhoven. *Social Indicators Research, 74*(3), 429–443.

Easterlin, R. A., McVey, L. A., Switek, M., Sawangfa, O., & Zweig, J. S. (2010). The happiness-income paradox revisited. *Proceedings of the National Academy of Sciences, 107*(52), 22463–22468.

Egbue, O., & Long, S. (2012). Barriers to widespread adoption of electric vehicles: Analysis of consumer attitudes and perceptions. *Energy Policy, 48*, 717–729.

Ehrlich, P. (1968). *The population bomb*. New York: Ballantine.

Engels, F. (1845). *The condition of the working class in England*. London: Penguin (first translation from German to English in 1886. Consulted edition, 2009).

European Comission (EC). (2006). *Measuring progress towards a more sustainable Europe. 2007 monitoring report of the EU sustainable development strategy*. Eurostat Books. http://ec.europa.eu/eurostat/publications/collections/statistical-books.

Evans, G. W., & Lepore, S. J. (1993). Non-auditory effects of noise on children. *Children's Environments, 10*(1), 42–72.

Everett, P. B., & Watson, B. G. (1987). Psychological contributions to transportation. In D. Stokols & I. Altman (Eds.), *Handbook of environmental psychology* (pp. 987–1008). New York: Wiley.

Ewing, R., & Cervero, R. (2010). Travel and the built environment: A meta-analysis. *Journal of the American Planning Association, 76*(3), 265–294.

FAD. (2009). *La U urbana. El libro blanco de las calles de Barcelona. Fomento de las Artes y el Diseño (FAD) y Ayuntamiento de Barcelona*. [The Urban U. The White Book of the streets of Barcelona. Fostering Arts and Design (FAD) and Barcelona City Council]. http://issuu.com/ecourbano/docs/la-u-urbana. Accessed 9 Mar 2015.

Ferriss, A. L. (2004). The quality of life concept in sociology. *The American Sociologist, 35*(3), 37–51.

Festinger, L. (1957). *A theory of cognitive dissonance*. New York: Harper.

Frank, L. D., Sallis, J. F., Conway, T., Chapman, J., Saelens, B., & Bachman, W. (2006). Multiple pathways from land use to health: Walkability associations with active transportation, body mass index, and air quality. *Journal of the American Planning Association, 72*(1), 75–87.

Galster, G. C., & Hesser, G. W. (1981). Residential satisfaction compositional and contextual correlates. *Environment and Behavior, 13*(6), 735–758.

García, I.; Giuliani, F. y Wiesenfeld, E. (1994/2000). El lugar de la teoría en psicología comunitaria: comunidad y sentido de comunidad. En M. Montero (Comp.). *Psicología social comunitaria* (pp. 75–101). Jalisco: Universidad de Guadalajara.

García-Mira, R., & Vega, P. (Eds.). (2009). *Sostenibilidad, valores y cultura ambiental* [Sustainability, values and environmental culture]. Madrid: Ediciones Pirámide.

Gardner, P. C., & Stern, G. T. (1996). *Environmental problems and human behavior*. Boston: Alin and Bacon (2nd edition 2002, Boston: Pearson).

Gehl, J. (1971). Social dimention of architecture. In B. Honikman (Ed.), *Proceedings of the architectural psychology conference at Kingston polytechnic 1970* (pp. 62–64). Kingston: Kingston Polytechnic and RIBA.

Glatzer, W., & Mohr, H. M. (1987). Quality-of-life-concepts and measurement. *Social Indicators Research, 19*(1), 15–24.

Goldsmith, E., Allen, R., Allaby, M., Davoll, J., Lawrence, S. (1972). A blueprint for survival. *The Ecologist*,

2(1) Entire issue. Republished in book form by Penguin Books, London, 1972. Available at http://www.theecologist.info/page34.html. (May 27, 2015).

Golinska, P., & Hajduk, N. (Eds.). (2012). *European union policy for sustainable transport system: Challenges and limitations*. New York: Springer.

Golob, T. F., Canty, E. T., Gustafson, R. L., & Vitt, J. E. (1972). An analysis of consumer preferences for a public transportation system. *Transportation Research, 6*(1), 81–102.

Hagerty, M. R., & Veenhoven, R. (2003). Wealth and happiness revisited: Growing national income does go with greater happiness. *Social Indicators Research, 64*(1), 1–27.

Hartig, T. (2004). Restorative environments. In C. Spielberger (Ed.), *Encyclopedia of applied psychology* (pp. 273–279). New York: Academic/Elsevier.

Harvey, D. (2012). *Rebel cities: From the right to the city to the urban revolution*. London: Verso Books.

Haussmann, G. E. (1852). *The Bourgeoisification of Paris. Haussmann's (1809–1892) renovation of the city.* https://www.mtholyoke.edu/courses/rschwart/hist255-s01/mapping-paris/Haussmann.html.

Helliwell, J. F., & Wang, S. (2014). Happiness: Trends, explanations and distribution. In J. Helliwell, R. Layard, & J. Sachs (Eds.), *World happiness report 2013 (WHR)* (pp. 8–32). New York: UNSDSN-The Earth Institute, Columbia University. http://unsdsn.org/wp-content/uploads/2014/02/WorldHappinessReport2013_online.pdf.

Hellpach, W. (1911). *Geopsyche*. Leipzig: W. Engelmann.

Hellpach, W. (1924). Psychologie der Umwelt. In E. Aberhalden (Ed.), *Handbuch der biologischen Arbeitsmethoden* (pp. 109–218). Berlin: Urban und Schwarzenberg.

Hessel, S. (2011a). *Time for Outrage!* [Indignez-vous!]. New York: Hachette Book.

Hessel, S. (2011b). *Engagez-vous ! Entretiens avec Gilles Vanderpooten* [Get Involved! Interviews with Gilles Vanderpooten].LaTour-d\T1\textquoterightAigues: Éditions de l'Aube.

Hogg, M. A., & Abrams, D. (1988). *Social identifications: A social psychology of intergroup relations and group processes*. New York: Taylor & Frances/Routledge.

Hollander, J. (2010). Moving toward a shrinking cities metric: Analyzing land use changes associated with depopulation in Flint, Michigan. *Cityscape, 12*(1), 133–152.

Howard, E. (1898). *To-morrow: A peaceful path to real reform*. Reprinted in 1902 as *Garden Cities of To-Morrow*. London: S. Sonnenschein & Co., Ltd.

Hur, M. H. (2006). Empowerment in terms of theoretical perspectives: Exploring a typology of the process and components across disciplines. *Journal of Community Psychology, 34*(5), 523–540.

Inglehart, R., Foa, R., & Peterson, C. (2008). Development, freedom, and rising happiness. *Perspectives on Psychological Science, 3*, 264–285.

IUCN, UNEP, WWF. (1991). *Caring for the earth: A strategy for sustainable living*. IUCN (World Conservation Union), UNEP (United Nations Environment Programme) and WWF (World Wide Fund for Nature). London: Earthscan.

Jacobs, J. (1961). *The death and life of great American cities*. New York: Random House.

Javaloy, F. (2007). *Bienestar y Felicidad de la Juventud Española* [Welfare and happiness of the Spanish Youth]. Madrid: Edición Instituto de la Juventud (Injuve). http://www.injuve.es/ca/observatorio/economia-consumo-y-estilos-de-vida/bienestar-y-felicidad-de-la-juventud-espanola. Accessed 1 Feb 2015.

Klein, S., Wener, R., & Lehman, S. (Eds.). (1985). *Environmental change/social change: Proceedings of the sixteenth annual conference of the Environmental Design Research Association – EDRA*. Washington, DC: Environmental Design Research Association.

Krier, L. (1984). The city within the city. *Architectural Design, 54*, 70–105.

Krier, L. (1998). *Architecture: Choice or fate*. Windsor: Andreas Papadakis Publisher.

Land, K. C., Michalos, A. C., & Sirgy, M. J. (Eds.). (2011). *Handbook of social indicators and quality of life research*. New York: Springer.

Lane, B. (2011). *Market delivery of ultra-low carbon vehicles in the UK: An evidence review for the RAC Foundation*. Ecolane Transport Consultancy. http://design.open.ac.uk/documents/Market_delivery_of_ULCVs_in_the_UK-Ecolane.pdf.

Lee, T. (1969). Do we need a theory? In D. Canter (Ed.), *Architectural psychology. Proceedings of the conference held in Dalandui, UK* (pp. 18–25). London: Royal Institut of British Architecture.

Levi, L., & Anderson, L. (1974). *Psychosocial stress: Population, environment, and quality of life*. New York: Spectrum (Used Spanish version: *La tensión psicosocial. Población, ambiente y calidad de vida*. México: Manual Moderno, 1980).

Living Planet Report. (2012). *Biodiversity, biocapacity and better choices*. Oakland: WWF International, Gland, Switzerland; Global Footprint Network. http://wwf.panda.org/about_our_earth/all_publications/living_planet_report/2012_lpr/.

Lofland, L. H. (2007). *The public realm. Explorin the city's quintessential social territory*. New Brunswick: Aldine Transaction.

Lyubomirsky, S., Sheldon, K. M., & Schkade, D. (2005). Pursuing happiness: The architecture of sustainable change. *Review of General Psychology, 9*(2), 111–131.

Maragall, P., Benach, N., & Bohigas, O. (2004). *Transforming Barcelona*. New York: Routledge.

Marans, R. W., & Rodgers, W. (1975). Toward an understanding of community satisfaction. In A. Hawley & V. Rock (Eds.), *Metropolitan America in contemporary perspective* (pp. 299–352). New York: Halstead Press.

Martelle, S. (2012). *Detroit: A biography*. Chicago: Chicago Review Press.

Maslow, A. H. (1954). *Motivation and personality*. New York: Harper & Row.

Matthies, E., Kuhn, S., & Klöckner, C. A. (2002). Travel mode choice of women. The result of limitation, ecological norm, or weak habit? *Environment and Behavior, 34*(2), 163–177.

Max-Neef, M. (1995). Economic growth and quality of life: A threshold hypothesis. *Ecological Economics, 15*, 115–118.

McCormick, E. J. (1976). *Human factors in engineering design*. New York: McGraw-Hill.

McGreal, Ch. (2010, December 17). Detroit mayor plans to shrink city by cutting services to some areas. *The Guardian.*. http://www.theguardian.com/world/2010/dec/17/detroit-shrinking. Accessed 3 Mar 2015.

McGregor, D. (1960). *The human side of enterprise*. New York: McGraw-Hill.

McMillan, D. W., & Chavis, D. M. (1986). Sense of community: A definition and theory. *Journal of Community Psychology, 14*(1), 6–23.

Meadows, D. H., Meadows, D. L., Randers, J., & Behrens, W. W. (1972). *The limits to growth*. New York: Potomac Associates.

Mey, G., & Günther, H. (Eds.). (2015). *The life space of the urban child. Perspectives on Martha Muchow's classic study*. New Brunswick: Transaction Publishers.

Michalos, A. C. (1985). Multiple discrepancies theory (MDT). *Social Indicators Research, 16*(4), 347–413.

Michelson, W. (1983). *The impact of changing women's roles on transportation needs and usage. Executive summary*. Final Report. Washington, DC: Urban Mass Transportation Administration.

Mitchell, W. J. (Ed.). (1972). *Environmental design: Research and practice*. Proceedings of the EDRA 3/AR 8 conference. Los Angeles: University of California.

Montaner, J. M. (2006). Vulnerabilidades urbanas: separar, olvidar, deshabitar [Urban vulnerabilities: separate, forget vacate]. In J. Nogué & J. Romero (Eds.), *Las otras geografías* (pp. 353–368). Valencia: Tirant lo Blanch.

Montaner, J. M., & Muxí, Z. (2015). *Arquitectura y política* [Architecture and politics]. Barcelona: Gustavo Gili.

Montero, M. (1994). *Construcción y crítica de la psicología social* [Construction and criticism of social psychology]. Barcelona: Anthropos.

Moore, C. B. (1982). *Travel patterns and transit needs of women*. Final Report, Vol. 1. Washington, DC: Urban Mass Transportation Administration.

Muchow, M., & Muchow, H. H. (1935/1998). *Der Lebensraum des Großstadtkindes*. [The living space of urban child]. Weinheim: Juventa. (First publication in 1935, by Riegel, Hamburg).

Myers, D. G. (2000). The funds, friends, and faith of happy people. *American Psychologist, 55*(1), 56–57.

Nefs, M., Alves, S., Zasada, I., & Haase, D. (2013). Shrinking cities as retirement cities? Opportunities for shrinking cities as green living environments for older individuals. *Environment and Planning A, 45*(6), 1455–1473.

Nelson, G., & Prilleltensky, I. (Eds.). (2010). *Community psychology: In pursuit of liberation and well-being*. New York: Palgrave Macmillan.

Nicolaidis, G. C. (1975). Quantification of the comport variable. *Transportation Research, 9*(1), 55–66.

Ogburn, W. F. (Ed.). (1943). *American society in wartime*. Chicago: University of Chicago Press.

Ouellette, J. A., & Wood, W. (1998). Habit and intention in everyday life: The multiple processes by which past behavior predicts future behavior. *Psychological Bulletin, 124*(1), 54.

Pallagst, K. (2009). *The future of shrinking cities: Problems, patterns and strategies of urban transformation in a global context*. Berkeley: Institute of Urban & Regional Development (IURD) Monograph Series. http://escholarship.org/uc/item/7zz6s7bm. Accessed 15 Mar 2015.

Parsons, T. (1958). Definition of health and illness in light of American values and social structure. In J. E. Gartly (Ed.), *Pattiens, physitians, and illness: A sourcebook in behavioral science and health* (pp. 165–187). New York: Free Press.

Patterson, A. H., & Ralston, P. A. (1983). *Fear of crime and fear of public transportation among the elderly*. Final Report. Washington, DC: Urban Mass Transportation Administration.

Pigou, A. C. (1932). *The economics of welfare*. London: McMillan&Co.

PlaNYC. (2007). *Inventory of New York city greenhouse gas emissions*. New York: Mayor's Office of Operations, New York City (Since 2007 to present, available at: http://www.nyc.gov/html/planyc/downloads/pdf/NYC_GHG_Inventory_2014.pdf).

Pol, E. (2006). Blueprints for a history of environmental psychology (I): From first birth to American transition. *Medio Ambiente y Comportamiento Humano 2006, 7*(2), 95–113.

Pol, E. (2007). Blueprints for a history of environmental psychology (II): From architectural psychology to the challenge of sustainability. *Medio Ambiente y Comportamiento Humano, 8*(1y2), 1–28.

Pol, E. (2009). Sostenibilidad, ciudad y medio ambiente. Dinámicas urbanas y construcción de valores ambientales. In R. García-Mira & P. Vega (Eds.), *Sostenibilidad, Valores y Cultura Ambiental*. Cap. 9 (pp. 143–163). Madrid: Pirámide.

Proshansky, H. M., Ittelson, W. H., & Rivlin, L. G. (Eds.). (1970). *Environmental psychology: People and their physical settings*. New York: Holt, Rinehart & Winston. (Used Spanish version: *Psicología Ambiental*. México: Trillas, 1878)

Putnam, R. D. (2000). *Bowling alone: The collapse and revival of American community*. New York: Simon & Schuster.

Rapoport, A. (1969). An approach to the study of environmental quality. In H. Sanoff & S. Cohn (Eds.), *Environmental design research association: 1st annual conference proceedings*, Chapel Hill, June 1969 (pp. 1–13). Stroudsburg: Dowden, Hutchinson and Ross.

Rappaport, J. (2003). U.S. urban decline and growth, 1950 to 2000. *Economic Review: Federal Reserve Bank of Kansas City, 3*, 15–44.

Redmond, L. S., & Mokhtarian, P. L. (2001). The positive utility of the commute: Modeling ideal commute time and relative desired commute amount. *Transportation, 28*(2), 179–205.

Register, R. (1987). *Ecocity Berkeley: Building cities for a healthy future*. Berkeley: North Atlantic Books.

Register, R. (1990). *First international ecological city conference (Ecocity 1). Enthusiasm born of fresh innovation*. Ecocity Builders Org. http://www.ecocitybuilders.org/. Accessed 23 Mar 2015.

Richardson, H. W., & Nam, C. W. (Eds.). (2014). *Shrinking cities: A global perspective*. New York: Routledge.

Rioux, L., Le Roy, J., Rubens, L., & Le Conte, J. (2013). *Le confort au travail – Que nous apprend la psychologie environnementale ?* [Comfort at work – What can we learn from environmental psychology?]. Québec: Presses Universitaires de Laval.

Rosenbloom, S. (2001). Sustainability and automobility among the elderly: An international assessment. *Transportation, 28*(4), 375–408.

Rotter, J. B. (1954). *Social learning and clinical psychology*. Englewood Cliffs: Prentice Hall.

Rueda, S. (1995). *Ecologia urbana* [Urban ecology]. Barcelona: Beta Editorial, 1995.

Saelens, B. E., & Handy, S. L. (2008). Built environment correlates of walking: A review. *Medicine and Science in Sports and Exercise, 40*(7), S550–S566.

Saelens, B. E., Sallis, J. F., & Frank, L. D. (2003). Environmental correlates of walking and cycling: Findings from the transportation, urban design, and planning literatures. *Annals of Behavioral Medicine, 25*, 80–91.

Salingaros, N. A. (2001). *The future of cities: The absurdity of modernism*. Nikos Salingaros interview Leon krier. Planetizen, Monday, 5 Nov 2001. http://www.planetizen.com/node/32. Accessed 13 May 2015.

Salingaros, N. A. (2005). *Principles of urban structure*. Amsterdam: Techne Press.

Sánchez, A. (1991). *Psicología Comunitaria. Bases conceptuales y operatives. Métodos de intervención*. Barcelona: PPU.

Sarason, S. B. (1974). *The psychological sense of community: Prospects for a community psychology*. San Francisco: Jossey-Bass.

Sartori, G. (2001). *La sociedad multiétnica* [Multiethnic society]. Madrid: Taurus.

Sassen, S. (1988). *The mobility of labour and capital*. Cambridge: Cambridge University Press.

Sassen, S. (1996). *Losing control? Sovereignty in an age of globalization*. New York: Columbia University Press.

Sassen, S. (2014). *Expulsions: Brutality and complexity in the global economy*. Cambridge: Belknap.

Schett, S. (2011). *An analysis of shrinking cities*. Innsbruck: Institut für Städtebau und Raumplanung Universität.

Schilling, J., & Logan, J. (2008). Greening the rust belt: A green infrastructure model for right sizing America's shrinking cities. *Journal of the American Planning Association, 74*(4), 451–466.

Schmuck, P., & Schultz, P. W. (Eds.). (2002). *Psychology of sustainable development*. Norwell: Kluwer.

Seligman, M. E. (1975). *Helplessness: On depression, development, and death*. New York: WH Freeman/Times Books/Henry Holt & Co.

Seligman, M. E., & Csikszentmihalyi, M. (2000). Positive psychology: An introduction. *American Psychologist, 55*, 5–14.

Sennett, R. (1970). *The uses of disorder: Personal identity and city life*. New York: Knopf.

Sennett, R. (1999). *The corrosion of character*. New York: Norton.

Seoane, G., Rodríguez, M., & Arce, C. (2000). Comparación de modelos de evaluación del servicio de autobús urbano para usuarios reales y potenciales. *Psicothema, 12*(2), 522–525.

Simmel, G. (1903). *Die Grosstädte und das Geistesleben*. Dresden: Petermann [*The Metropolis and Mental Life*. The Sociology of Georg Simmel' New York:Free Press, 1976].

Sirgy, M. J. (2012). *The psychology of quality of life: Hedonic well-being, life satisfaction, and eudaimonia* (Social indicators research series 50). Berlin: Springer Science & Business Media.

Sorkin, M. (Ed.). (1992). *Variations on a theme park*. New York: Farrar, Straus and Giroux, LLC.

Steg, L., & Gifford, R. (2005). Sustainable transportation and quality of life. *Journal of Transport Geography, 13*, 59–69.

Stern, P. C. (1992). Pychological dimensions of global environment change. *Annual Review of Psychology, 43*, 269–302.

Stohr, K. (2004, February 5). Shrinking city syndrome. *New York Times*. http://www.nytimes.com/2004/02/05/garden/shrinking-city-syndrome.html. Accessed 23 Mar 2015.

Stokes, B. (2007, July 24). Happiness is increasing in many countries: But why? *National Journal*. http://www.pewglobal.org/2007/07/24/happiness-is-increasing-in-many-countries-but-why/. Accessed 16 Feb 2015.

Stokols, D. (1976). The experience of crowding in primary and secondary environments. *Environment and Behavior, 8*, 49–86.

Stokols, D. (1987). Conceptual strategies of environmental psychology. In D. Stokols & I. Altman (Eds.), *Handbook of environmental psychology* (Vol. 1, pp. 41–70). New York: Wiley.

Stokols, D. (1996). Translating social ecological theory into guidelines for community health promotion. *American Journal of Health Promotion, 10*, 282–298.

Stokols, D., & Novaco, R. W. (1981). Transportation and well-being. In I. Altman, J. F. Wohlwill, & P. B. Everett (Eds.), *Transportation and behavior* (pp. 85–130). New York: Plenum Press.

Stouffer, S. A. (1949). *The American soldier*. Princeton: Princeton University Press.

Stringer, P. (1969). Architecture, psychology, the game's the same. In D. Canter (Ed.), *Architectural psychology. Proceedings of the conference held in Dalandui, UK* (pp. 7–11). London: Royal Institut of British Architecture.

Sugrue, T. J. (1996). *The origins of the urban crisis: Race and inequality in postwar Detroit*. Princeton: Princeton University Press.

Tajfel, H. (1978). *Differentiation between social groups: Studies in the social psychology of intergroup relations*. Oxford: Academic.

Tajfel, H. (1981). *Social identity conflicte and stereotypes: Studies in intergroup behavior*. Cambridge: Cambridge University Press.

Talarn, A. (2007). *Globalización y salud mental* [Globalization and mental health]. Barcelona: Herder.

Tönnies, F. (1887). *Gemeinschaft und Gesellschaft* [Community and society]. Leipzig: Fues's Verlag.

Turner, J., & Grieco, M. (2000). Gender and time poverty: The neglected social policy implications of gendered time, transport and travel. *Time and Society, 9*(1), 129–136.

UN-HABITAT. (2007). *Enhancing urban safety and security: Global report on human settlements, 2007*. http://unhabitat.org/books/global-report-on-human-settlements-2007-enhancing-urban-safety-and-security/.

UN-HABITAT. (2014). *World urbanization prospects. The 2014 revision*. Published by the United Nations. Department of Economic and Social Affairs of the United Nations. http://www.un.org/en/development/desa/population/publications/.

Uzzell, D., & Moser, G. (2006). Environment and quality of life. *Revue Européenne de Psychologie Appliquée, 56*, 1–4.

Valera, S., & Guàrdia, J. (2014). Perceived insecurity and fear of crime in a city with low-crime rates. *Journal of Environmental Psychology, 38*, 195–205.

Veenhoven, R. (1999). Quality-of-life in individualistic society. *Social Indicators Research, 48*(2), 159–188.

Veenhoven, R. (2000a). Freedom and happiness: A comparative study in 44 nations in the early 1990's. In E. Diener & E. M. Suh (Eds.), *Culture and subjective wellbeing* (pp. 257–288). Cambridge, MA: MIT Press. ISBN 0 26204182 0.

Veenhoven, R. (2000b). Well-being in the welfare state: Level not higher, distribution not more equitable. *Journal of Comparative Policy Analysis: Research and Practice, 2*(1), 91–125.

Veenhoven, R. (2005). Inequality of happiness in nations. *Journal of Happiness Studies, 6*(4), 351–355.

Veenhoven, R. (2011). Greater happiness for a greater number: Is that possible? If so, how? In K. M. Sheldon, T. B. Kashdan, & M. F. Steger (Eds.), *Designing positive psychology: Taking stock and moving forward* (pp. 396–409). New York: Oxford.

Veenhoven, R., & Hagerty, M. (2005). Rising happiness in nations 1946–2004: A reply to Easterlin. *Social Indicators Research, 79*, 421–436.

Veenhoven, R., & Vergunst, F. (2013). *The Easterlin illusion: Economic growth does go with greater happiness*. EHERO working papers: Erasmus Happiness Economics Research Organization, vol. 2013, No 2013/1 (23 Jan 2013). http://mpra.ub.uni-muenchen.de/43983/.

Verne, P. (2009). Happiness, freedom and control. *Journal of Economic Behavior & Organization, 71*, 146–161.

Verplanken, B., & Holland, R. W. (2002). Motivated decision making: Effects of activation and self-centrality of values on choices and behavior. *Journal of Personality and Social Psychology, 82*(3), 434.

Verplanken, B., Walker, I., Davis, A., & Jurasek, M. (2008). Context change and travel mode choice: Combining the habit discontinuity and self-activation hypotheses. *Journal of Environmental Psychology, 28*(2), 121–127.

Vlek, C. (2002). Common dilemmas, cultural development and quality of life. What can we do, what do we want, what shall we achieve? In R. Garciá-Mira, J. M. Sabucedo, & J. Romay, (Eds.), *Culture, quality of life – Problems and challenges for the new millennium (IAPS 17 conference proceedings)*, 23–27 July 2002.

Weiss, K., & Girandola, X. (2010). *Psychologie et développement durable* [Psychology and Sustainability]. Paris: In Press.

WHO (1948). *Preamble to the constitution of the World Health Organization*. Adopted by the International Health Conference, New York, 19–22 June 1946 and entered into force on 7 Apr 1948. http://www.who.int/about/definition/.

WHP. (2012). *World happiness report 2012*. New York: United Nations-The Earth Institute, Columbia University. http://worldhappiness.report/ed/2012/.

WHP. (2013). *World happiness report 2013*. New York: United Nations-The Earth Institute, Columbia University. http://worldhappiness.report/ed/2013/.

WHP. (2015). *World happiness report 2015*. New York: United Nations-The Earth Institute, Columbia University. http://worldhappiness.report/ed/2015/.

Wolf, C. P. (1974). Social impact assessment: The state of the art. In D.H. Carson (Ed.), *Proceedings EDRA-5 Vol 2 social impact assessment*. http://www.edra.org/sites/default/files/publications/EDRA05-v2-Wolf-1-44.pdf. Accessed 25 Mar 2015.

Zimmerman, M. A. (1995). Psychological empowerment: Issues and illustrations. *American Journal of Community Psychology, 23*(5), 581–599.

Zimmerman, M. A. (2000). Empowerment theory: Psychological, organizational, and community levels of analysis. In J. R. E. Seidmann (Ed.), *Handbook of community psychology* (pp. 43–63). New York: Kluwer Academic/Plenum.

Some Cues for a Positive Environmental Psychology Agenda

Sergi Valera and Tomeu Vidal

3.1 Introduction

Manfred Gnädiger was born on January 27th 1936 in Germany. At 25, he arrived on the coast of Galicia, in Finisterre, and settled near the sea, in Camelle. There he reorganized his life; he built himself a shack and, with his hands, he designed his own sculptural and pictorial universe by taking advantage of the natural environment, living like a hermit, wearing a loincloth as his only outfit, calling himself just Man. The locals became accustomed to his presence and his peculiar museum became a part of this unusual, harsh Galician landscape.

On November 13th 2002, the oil tanker "Prestige" was carrying about 77 thousand tons of low-quality fuel. At about 28 miles off Cape Finisterre, it sprung an enormous leak in the midst of a storm, which left it adrift in 6-meter waves and winds of force 8. A large fuel spill, of about 10 thousand tons, devastated the area. The environmental consequences of this disaster were terrible, as was its impact in terms of the perception of quality of life (for a detailed social and environmental analysis, see García-Mira et al. 2006; García-Mira 2013). At Man's house, the tar destroyed the whole world he had created over the years, and on December 28th 2002, Manfred died. The opinion of the neighbors of Camelle about Manfred summarizes his way of living and dying:

> He lived as he had chosen to live. The tar had ruined the work of his life; Manfred died of melancholy, Camelle is going to preserve his house-museum because it is one of the town's landmarks. Manfred was a free man.

What is fascinating for those of us who study the relationship between the physical environment and psychology is realizing how, in a way, a natural space had turned into a built space, transformed physically and psychologically to become a home, a significant place and how, in a way, that space ended up becoming a town landmark, a place full of meaning beyond its physical or structural characteristics. *Topogenesis* and *sociogenesis* combine to generate a socio-physical space with a strong psychological impression, as rightly spotted by Muntañola from an architectural viewpoint (1974, 1979a, b, 1980) or, in our day, by Zárate (Zárate and Muntañola 2000; Zárate 2001, 2004, 2010). So strong is that impression that its alteration, affectation, or destruction has a strong psychological impact. In Man's case, this impact is expressed in a truly dramatic way. In any case, though, it is a recurring phenomenon in cases of post-occupational assessment or POE (Zimring and Reizenstein 1980), refugees analysis (Knudsen and Hanafi 2011), disaster situations (Gould 2009; White and Frew 2013), or any other type of forced displacement (Fullilove 2014; Lees et al. 2015). When a psychologically significant environment is altered, so are the relationships between the people, groups, or communities and the physical environments they inhabit. This is so because it alters all the scenes where people

S. Valera (✉) • T. Vidal
Department of Social Psychology, University of Barcelona, Barcelona, Spain
e-mail: svalera@ub.edu; tvidal@ub.edu

lead their everyday lives, where they satisfy their needs, where they know other people, where they also recognize themselves. And when that happens, people suffer from it. It is what Fullilove (2014) proposes to call "the frayed knot hypothesis" and the subsequent "root shock" that derives from the loss of the person's own emotional ecosystem.

Thinking about the human being, we realize that people make great psychological efforts to maintain compatibility between the environment and goals and expectations for development. Although sometimes environmental factors are not compatible with this, we persist in seeking opportunities for personal growth, creativity and the acquirement of well-being and happiness (Stokols et al. 2009).

In this way, the socio-physical space can be considered either a generator of well-being and positive experiences or the context in which people can experience positive personal or social situations (Stokols 2003). In this chapter, different theoretical developments of environmental psychology are presented and proposed for inclusion in a Positive Environmental Psychology agenda. On the one hand – space as a generator of well-being – we analyze the aesthetic quality of the landscape (Berlyne 1974; Galindo and Corraliza 2000; Galindo and Hidalgo 2005; Kaplan 1995); the restorative capacity of environments (Korpela et al. 2001; Korpela and Harting 1996; Korpela and Ylen 2007); and place identity and place attachment developments (Altman and Low 1992; Di Maso et al. 2008; Hidalgo and Hernández 2001; Lewicka 2011; Manzo and Devine-Wright 2014; Proshansky et al. 1983; Sarbin 1983; Scannell and Gifford 2010; Twigger-Ross and Uzell 1996). On the other hand – space as a context where people experience positive situations – we suggest some reflections on the processes of urban placemaking, such as the tradition of Placemaking (PPS), Community Participation and Planning (Manzo and Perkins 2006), Community Design (Hester 2006), or Socially Restorative Urbanism (Thwaites et al. 2013), which aims to restore social well-being and the sense of belonging to urban environments.

When people participate and are involved in urban and community design processes, they are changing not only the physical features of the place, but also the symbolic aspects of space for the generation of personal or social identities (Stokols 2003; Valera and Pol 1994; Valera and Guàrdia 2002) and place attachment and place identity are changed. These "socio-physical" changes are relevant to generating well-being and positive experiences. Thus, place is a context where people experience positive situations and is a generator of well-being.

3.2 Towards a Positive Environmental Psychology

When we reflect on human nature, we realize that we make a great psychological effort to try to feel good about ourselves and our lives, whatever life situations we must face. The search for positive distinctiveness of self in processes of social categorization (Ellemers and Barreto 2006; Turner et al. 1987), the hedonic level theory (hedonic treadmill model) (Brickman and Campbell 1971; Diener et al. 2006), or the processes of reduction of cognitive dissonance (Festinger 1957; Harmon-Jones and Mills 1999) are examples of how psychology has dealt with the subject of the conquest of well-being and its paradoxes.

Similarly, we try to maintain an adequate compatibility between ourselves and our environments – despite the fact that these do not always present favorable characteristics for our development – by looking for opportunities for personal growth, creativity, or well-being and happiness (Stokols et al. 2009). Thanks to this, human beings have learned to adapt themselves and their environment to inhabit practically every possible type of scene, surpassing, in their transforming capability, any other living being.

However, psychology has taken a long time to delve into the subjects of well-being, happiness, or satisfaction with life, and still today these topics appear incipiently. Attention to the aspects that influence negatively and cause discomfort has been predominant in what we can call reaction-oriented psychology, as opposed to promotion-oriented psychology, which emphasizes and raises the positive aspects of the psychological experience. The change is occurring at a time when several psychologists have stopped talking about "preserving" to talk about "promoting" well-being. This is the change in what has come to be known as Positive Psychology (Diener and Seligman 2004; Fredrickson 2001; Linley and Joseph 2004; Linley et al. 2006; Seligman 1999; Seligman and Csikzentmihalyi 2000).

Actually, this is not an original idea. Already in the late 1960s, Bradburn developed his affective or emotional balance theory, analyzing how people experienced states of subjective well-being or discomfort and how it influenced their level of satisfaction with life (Bradburn 1969). In his conclusions, later corroborated and clarified (Kim and Mueller 2001; McDowell and Praught 1982; Watson et al. 1988), he refuted – to a certain extent – the intuitive hypothesis of an inverse correlation between experiences with positive emotional valence and those with negative emotional valence. In other words, the decrease in unhappiness does not necessarily entail the increase in happiness. Consequently, it is as important, or more so, to decrease unhappiness as it is to promote happiness. These are two processes that may happen in parallel, but which do not necessarily occur in the opposite sense. This is, in general terms, the basis on which Positive Psychology stands, as opposed to Traditional Psychology, which was presented formally in Martin Seligman's (1999) opening speech as the president of APA and in the 2000 monographic issue of American Psychologist.

We will not linger on this point, but Corey Keyes has analyzed 13 dimensions reflecting mental health as flourishing (Keyes 2007). Grouped into emotional well-being, psychological well-being, and social well-being, these dimensions could summarize the main anchorage points of Positive Psychology (see Table 3.1). They will be useful in the last part of this chapter.

For its part, Environmental Psychology has made great efforts to understand how we react and adapt ourselves to our physical environs, especially when they present unfavorable conditions. The environmental stress theory (Selye 1956, 1974), the environmental control theory (Averill 1973), the study of overcrowding (Baum and Epstein 1978), the perception of risk (Slovic 1987), or, in general terms, research focusing on the paradigm of adaptation (Saegert and Winkel 1990) would find this context "reactive". However, as will be seen in this chapter, it is possible to recognize in many topics or research lines of a consolidated psycho-environmental tradition, the goal of generating knowledge for a better adaptation of person-environment relationships, as well as the analysis of the dimensions that can generate environments where people and social groups can develop their potential, satisfy their needs, and obtain social and psychological well-being. Thus, for instance, Cattell et al. (2008) explored the relationship between the public space and the well-being of people. To that end, they turned to environmental psychology and confirmed the positive benefits of the environment through concepts such as place, identity, sense of attachment or residential satisfaction, even though they mainly focused on stress-related mechanisms. In turn, the forms in which people report and locate their life experiences may reveal the contexts in which they experience their well-being.

For all these reasons, what we are defending here is the fact that Environmental Psychology must undergo a transformation similar to that of this century's psychology. Such a transformation supposes an opening towards a paradigm change when understanding the psychological and social experiences of people and the phenomena they entail (Sheldon and King 2001). We understand that it is possible to adopt a psychologically positive point of view in the ecological analysis of human behavior. Stokols' approach (2003) follows this line when he proposes the ecology of human strengths. For this author, it is necessary to overcome some obstacles still present in environmental psychology, which can be summarized in three points: (a) the incomplete conceptualization of environmental contexts as they influence well-being, (b) the disproportionate emphasis on positive emotional states and under emphasis on

Table 3.1 Factors and 13 dimensions reflecting mental health as flourishing (Source: Keyes 2007, p. 98)

Dimension	Definition
Positive emotions (i.e., emotional well-being)	
Positive affect	Regularly cheerful, interested in life, in good spirits, happy, calm and peaceful, full of life
Avowed quality of life	Mostly or highly satisfied with life overall or in domains of life
Positive psychological functioning (i.e., psychological well-being)	
Self-acceptance	Holds positive attitudes toward self, acknowledges, likes most parts of self, personality
Personal growth	Seeks challenge, has insight into own potential, feels a sense of continued development
Purpose in life	Finds own life has a direction and meaning
Environmental mastery	Exercises ability to select, manage, and mold personal environs to suit needs
Autonomy	Is guided by own, socially accepted, internal standards and values
Positive relations with others	Has, or can form, warm, trusting personal relationships
Positive social functioning (i.e., social well-being)	
Social acceptance	Holds positive attitudes toward, acknowledges, and is accepting of human differences
Social actualization	Believes people, groups, and society have potential and can evolve or grow positively
Social contribution	Sees own daily activities as useful to and valued by society and others
Social coherence	Interested in society and social life and finds them meaningful and somewhat intelligible
Social integration	A sense of belonging to, and comfort and support from, a community

the temporal links between individuals' positive and negative experiences, and (c) the incomplete assessment of the threshold levels at which exposure to environmental constraints either enhances or undermines the development of psychological strengths. (op.cit., p. 331).

From this perspective, we can call positive environments those whose socio-physical characteristics generate environmental configurations which, generally speaking, condition us in favor of the development and growth of people and their potential, while also favoring the psychological experience of physical, mental, and social well-being, of satisfaction with life, and of positive emotional states. In turn, these environments can improve or reduce the negative consequences of our exposure to unpleasant or stressful socio-physical environments, or even pathological environments, while they also help physical, psychological, or physical restoration. The home, the neighborhood, the town, the institutional environment, or the natural environment should all be analyzed, or else intervened, in the light of these positive parameters. This way of understanding what we have called positive environments is part of a socio-physical view of space, which we will deal with below.

3.3 The Socio-physical Conception of Space

The physical space must be understood – from Environmental Psychology – as a socio-physical space, a crossroads of physical features, of psychological experiences, and of social and cultural meanings that define socio-spatial configurations with which human beings establish dialogues and interactions. This derives from a simple, fundamental idea: human beings are permanently located, permanently related to their environs and, for this reason, we should analyze them as though we were fish studying the water (Sommer 1990). In fact, it only takes a little reflection to see that we need to relate actively to our environs and to discover their potential, their functionalities, to attach ourselves to places, to appropriate places, to feel good in them. These bonds are articulated through the experiences and meanings that turn the space into a place. Whatever the complex nature of these bonds, there is an affective, an emotional, and an essential component. This affective bond with the environment is an important factor for the development of psychological well-being. That is why, when human beings suffer material loss due to disasters – e.g. tsunamis, flooding, etc. – and they lose their homes, apart from the material loss, it is very hard to recover from the feeling of emptiness. The relocation and regeneration of one's own spatial, psychological, and social life is extremely difficult (Brown and Perkins 1992; Fullilove 2014).

Any space, understood as a unit of meaning for a person or group of people, takes its validity criterion from the set of meanings – past, present, or potential – that it is capable of supporting, as well as from the set of significant psychological configurations that it is capable of elucidating among the people who had, have, or will have contact with them. In this way, the space acquires meaning inasmuch as the person signifies it and, in turn, the person acquires meaning inasmuch as they place themselves in a space significant for them. However, whereas the products of such transactions are always individual – i.e. subjective – the origins of these meanings are eminently social, and both levels are always susceptible to being placed on the plane of intersubjectivity. The reciprocity of the transaction is also placed on the plane of intersubjectivity. Accordingly, we can recognize ourselves as well as our spaces in a social context, which is a melting pot of the set of socially elaborate, negotiated, shared meanings, or "perceived social field" in Stokols' words (Stokols and Shumaker 1981).

From all of the above, we can infer the central idea that will drive this chapter: placing ourselves adequately in our psycho-social-environmental world is essential for establishing positive bonds with our environment and thus obtaining elements that contribute to our well-being. Identity, as we will see below, appears then as a key element for our subjective and social well-being.

Once the subject has been outlined in these terms, the socio-physical space may be considered either a generator of well-being and positive experiences or the context where people can experience positive personal or social situations (Stokols 2003). Closer to the former case (the space as a generator of well-being), we find developments such as the aesthetic quality of the landscape, as discussed in the following section.

3.4 The Physical Environment as a Source of Aesthetic Appreciation

In the paragraphs above, we pointed out the need for human beings to relate to physical spaces and how this has important consequences on the attainment of our well-being.

The evolutionist perspective in the study of environmental perception has established firm bases to understand the scope of this need, and has enabled us to connect the functional aspects to the aesthetic aspects in the processes of environmental appreciation, while providing interesting elements for analyzing the links between the relationship with the environment, the satisfaction of needs and the promotion of human well-being.

In Küller's words (1992): "To understand the underlying significance of pleasantness, we must consider the fundamental biological urge to survive, grow, and multiply. (...). Seen in this perspective, pleasantness may be considered as a projection onto the environment of an assessment process based on three crude values — good, harmless, and bad— refined into what is commonly known as the hedonic or evaluative emotional dimension" (pp. 118–119).

This phylogenetic basis explains, for example, why we find environments that stimulate us perceptively to be pleasanter, i.e. they allow us to maintain a positive, exploratory, and inquisitive attitude. To sum up, we find those environments pleasant that enable us to become actively involved in them, since that attitude has allowed us to survive as a species in extraordinarily diverse environments. In this sense, the studies by Daniel Berlyne are central.

Following Gustav Fechner, Berlyne assumes the motivational principles of behavior, according to which the search for pleasure has a determining role in aesthetic appreciation processes. Berlyne updated and expanded Fechner's ideas through psychology's own experimental and correlational methodology into what he called *New Aesthetics in Psychology*. From Berlyne's early works (1949, 1960), he tried to discover the motivational aspects that explain our interest in the environment. Thus, he defined a special class of drive with three categories: (a) variation due to satiation, which involves the need to maintain an interesting activity to avoid boredom; (b) "curiosity" as an "active drive" to find new sensations, experiences and knowledge, which is translated into a natural drive to explore the environment around us; and (c) the "aesthetic interest" as a quality of the environment regardless of its representational content.

Despite the criticism caused by Berlyne's studies – see, for example, Cupchik (1986) – two themes derive from his work that are essential to our purposes: the relationship between environmental perception and psychological arousal processes, and the relationship between psychological arousal and the aesthetic appreciation of the environment.

According to Berlyne, there is abundant evidence to link hedonic value to fluctuations in arousal. He claims that aesthetic patterns produce their hedonic effects by influencing arousal – be it through arousal-boost mechanisms, or arousal-reduction mechanisms upon unpleasant states. In any case, the magnitude of the arousal when it works as a reinforcement mechanism depends on multiple factors, such as the intensity of the environmental stimulus, the link to other biologically significant events, as well as on the *collative* properties of the environment (Berlyne 1974).

On the other hand, environmental stimuli also have the ability to modify arousal, the person's level of activation, and to generate experiences that cause different affective responses, as well as different voluntary exploration activities. As previously mentioned, one of the features pointed out by these authors is the person's permanent need to become involved and find coherence in the environment, to be able to "read an environment" significantly. In this sense, this activation, this exploratory capability of becoming involved in the environment, is favored according to variables such as the environment's novelty or familiarity, complexity and simplicity, surprise or predictability, ambiguity or clarity, congruence or incongruence, etc.

Therefore, certain properties of the environment imply a level of activation and exploration that can become a hedonic or well-being value (Fig. 3.1).

Once again, following Küller's words: "The attention or orientation reaction is accompanied by a temporary increase in arousal, that is, a phasic arousal reaction, which is likely to be sustained and eventually canalized if orientation leads to exploration, conflict, approach, or withdrawal. Frequently the phasic arousal reaction will be accompanied by mental feelings of curiosity, interest, and the like. On the other hand, when habituation occurs, there is no increase in arousal and no feeling of interest" (Berlyne 1971, 1974; Küller 1992, p. 116).

It is Berlyne himself who links the levels of arousal generated by environmental stimuli and the aesthetic appreciation of the environment. This aesthetic appreciation, to Berlyne, is not as idiosyncratic, subjective, and individual as people tend to believe: "the aesthetic reactions of differing individuals turn out to have an appreciable degree of consistency underlying the undeniable differences" (1974, p. 22). Once again, the evolutionist perspective appears to offer a plausible explanation: the studies on aesthetic appreciation prove that human beings appreciate aesthetically those environments where our species could phylogenetically develop and solve their most basic or elementary needs more efficiently. Therefore, environments which generically provide shelter, protection, food, etc. are those whose symbolically referenced direct elements lead us to that phylogenetic past and, accordingly, to that positive aesthetic appreciation.

This explains the fact that one of the main findings in this type of study is that we tend to appreciate natural environments better than built ones. We are drawn to and stimulated by nature. This preference for natural environments, sustained in the phylogenetic argument, is usually referred to as a biophilia hypothesis (Kellert and Wilson 1995). This is a term proposed by Wilson (1984), which assumes that this inclination towards nature is an essential feature of the human race, while at the same time it binds us to other living beings. To support this idea, we should mention Ulrich's classic study (1984) conducted in a hospital environment. He found that surgery patients in rooms with views of natural environments recovered more quickly and with less medication than the control group. Such evidence is related to the Stress Reduction Theory by Ulrich himself (1984), the Attention Restoration Theory (Kaplan and Kaplan 1989; Kaplan 1995), and, in general, the restorative capability of natural environments (Hartig et al. 1991; Korpela and Staats 2014), as will be seen in the following section.

Therefore, according to Pluta (2012), "immersion in nature promotes intrinsic motivations, which in turn leads to ecological behavior, personal well-being, and cooperative and prosocial behavior" (...). Further research shows how immersion in nature contributes to mental, physical, and community health and well-being." (op. cit, p. 11). In fact, recent studies have shown the relationship between subjective well-being (SWB) and ecologically responsible behavior (ERB) (Brown and Kasser 2005; Corral-Verdugo et al. 2011), as well as pro-environment values and attitudes (Corral-Verdugo 2012; Schultz et al. 2005). Likewise, Weinstein et al. (2009) proved how people immersed in natural environments tended to develop intrinsic aspirations (linked to pro-sociality and behaviors focused on others), as well as decisions based on generosity. Meanwhile, people immersed in non-natural environments tended to develop extrinsic aspirations (or self-centered behaviors).

The relationship between the level of arousal and experiencing pleasant sensations has been developed by several authors in the sphere of the study of the affective relations of people with their physical environments. One of the most classic proposals in this respect is that by Russell and Pratt (Russell 1980; Russell et al. 1981, 1989; Russell and Pratt 1980) and their bidimensional model based on activation/non-activation and pleasant/unpleasant dyads, on which a set of emotional states rest (Fig. 3.2).

This type of classic proposal in environmental psychology is by no means far from other more recent ones in positive psychology. For example, Csikszentmihalyi's Flow theory (1990, 1998) is also based on a dual model where the axes are defined by high/low challenge perception and high/low skill perception dyads (Fig. 3.3). Interesting parallelisms can be observed on both charts. In fact, the different emotional states, which for Russell and Pratt are caused by our affective relations with the environment, are located in quadrants practically identical to the mental states that, according to Csikszentmihalyi, our activities provoke in terms of challenges and required skills. Thus, it seems as if those environments requiring little activation and involving unpleasant emotions bear little relevance to challenges in terms of psychological well-being and few requirements in terms of skills: they result in apathy and boredom. On the other hand, unpleasant activation would be related to high challenges and little coping capability, and would result in nervousness and stress. In contrast, weak

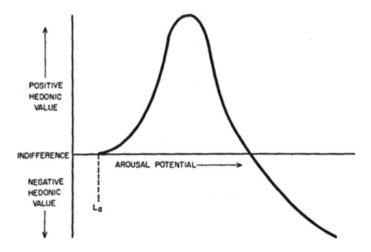

Fig. 3.1 Wundt's curve reinterpreted to analyze the relationship between hedonic value and arousal (Berlyne 1974, p. 10)

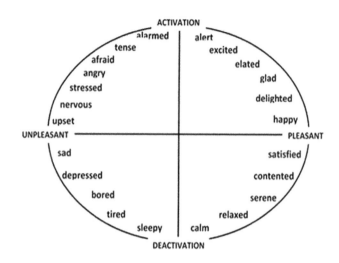

Fig. 3.2 A graphical representation of the Circumplex Model of Affect with the horizontal axis representing the valence dimension and the vertical axis representing the arousal or activation dimension (Based on Russell (1980) and Russell et al. (1989))

Fig. 3.3 Mental state in terms of challenge level and skill level, according to Csikszentmihalyi's Flow Model (Csikszentmihalyi 1998)

but pleasant activation would be related to mental states related to poor challenges and high skills, which would result in peacefulness, relaxation, and satisfaction states. Finally, the most positive mental states and the most intense emotional states occur in environments whose high activation generates well-being, i.e. those where mental states of control prevail and, especially, mental states of flow, where we become most psychologically involved while putting at risk our best psychological qualities and skills.

To sum up, there is an interesting link between environments capable of providing pleasant emotional experiences and activities that entail positive mental states and, therefore, contribute to states of well-being and happiness. It is no wonder, then, that the environments that can most easily generate flow, control, or relaxation states are the most psychologically beneficial and, consequently, the ones with the most restorative effects for the person. The next section is devoted to this. There is a more detailed approach in Chap. 7 of this handbook.

3.5 The Physical Environment as a Psychological Restorer

As seen in the section above, numerous studies have shown the beneficial role of natural environments over built ones. We tend to prefer the former over the latter; exposure to natural landscapes contributes to our well-being and prevents physical and mental disorders. Therefore, it is no wonder that one of the main goals of contemporary Environmental Psychology is the analysis of the relationship between human well-being and landscape, with an emphasis on the way our life environments can foster or hinder the well-being of people.

The biophilia hypothesis and its effects on well-being are complemented by the ability of natural environments to restore our physical and psychological faculties in the face of everyday hardship. At the beginning of this chapter, we discussed how people make great efforts to relate adequately to their physical and social environments, often in difficult situations or those presenting obstacles. We must, therefore, find adequate strategies to sustain this effort and face the additional tasks our interactions involve. One possible strategy lies in the possibility of recovering these psychological resources through contact with the nature we find in urban parks or natural areas outside the cities (Berto 2005; Ulrich 1984).

The Stress Recovery Theory (SRT) of Ulrich (1983, 1984; Ulrich et al. 1991), the Attention Restoration Theory (ART), of the Kaplans (Kaplan and Kaplan 1989; Kaplan 1995), along with Hartig's contribution (Hartig et al. 1991, 2003) are the most commonly cited referents in the development of restorative environments. Their studies have shown the benefits of contact with nature, such as vegetation and water, on the physical and mental well-being of people. According to the Stress Recovery Theory (SRT), the perception of certain qualities and contents of a landscape may help us recover from stress, from both psychological and physiological viewpoints. Moderate depth, moderate complexity, the presence of a focal point, gross structural qualities, and natural content such as vegetation and water can evoke positive emotions, sustain non-vigilant attention, restrict negative thoughts, and thus aid autonomic arousal to return to more moderate levels. In line with the previously mentioned phylogenetic argument (Wilson 1984), Ulrich views humans as biologically prepared to respond positively to environmental features that signal possibilities for survival, and so assumes an evolutionary basis for aesthetic and restorative responses to some natural scenes (Hartig et al. 2003).

Taking the studies on information overload (Milgram 1970) as a precedent, the Attention Restoration Theory (ART) claims that certain places can reduce the fatigue caused by people's everyday activities and thus facilitate the restoration of certain cognitive skills.

> According to ART, restoration from directed attention fatigue occurs with psychological distance from routine mental contents (*being away*) in conjunction with effortless, interest-driven attention (*fascination*), sustained in coherently ordered environments of substantial scope (*extent*) when the person's inclinations match the demands imposed by the environment as well as the environmental supports for intended activities (*compatibility*). Kaplan and Kaplan (1989) argue that these four factors commonly hold at high levels in natural environments (Hartig et al. 2003).

There is strong evidence to support ART, which suggests that natural environments are usually better restorers than built, urban environments (Herzog et al. 2003). In spite of this, other contributions have researched the repairing capability of different typologies of cityscapes (Abkar et al. 2011; Tenngart Ivarsson and Hagerhall 2008), considering a more complex pattern of features and choices. As Joye and van den Berg (2012) point out, both approaches must not be seen as excluding, but as complementary explanations, since they focus on different aspects of psychological restoration processes. In this sense, Hartig et al. (2003) compared psycho-physiological stress recovery and directed attention restoration in natural and urban field settings. Their results enable us to suggest, as practical implications, that easy pedestrian and visual access to natural settings can produce preventive benefits. Accordingly, they propose that public health strategies with a natural environment component may have particular value in this time of growing urban populations, exploding health care expenditure, and deteriorating environmental quality.

The phenomenon of emotional self-regulation applies environmental, mental, physical, and social strategies to start a process by which people keep a balance between pleasant and unpleasant emotions. The strategies include the use of places as well as convictions and affections related to those places. However, apart from what has been discussed so far, it seems pertinent to add three further considerations to the study of the environment as a psychological restorer.

Firstly, if, on the one hand, the environment as a source of psychological restoration is seen from a reactive perspective – i.e. the environment can work as an antidote in negative psychological states – it is true that the study of the environment's restorative qualities yields enough clues to think that we can design environments whose features can favor the strengthening of people's capabilities and thus favor their positive states. Although introducing natural elements into urban environments can favor the being-away factor (Fornara 2011) or specific places of cult that cause soft fascination (Herzog et al. 2010), the design of a good public space can clearly have an impact on extent and compatibility, especially when the public space can be explored on a human scale and with all its possible sense (Gehl 1987, 2010). In addition, including art in the public space can be an essential element both for improving aesthetic appreciation of the landscape and having a restorative effect on the person (Blackman 2014; Gonçalves Siebra 2012).

Secondly, apart from the restoration-nature relationship, ART is especially relevant in the urban medium too. Because most life experiences take place in cities, people are more prone to develop identity bonds with urban landscapes and environments. The studies including urban environments are ideal for researching place identity and its positive influence on restoration. In two studies focused on the restorative capabilities of façades of different types of architecture (Nenci et al. 2005, 2006), historic buildings were compared to non-historic ones. In the first study, buildings considered part of the national architecture heritage were chosen, whereas the second study included buildings from foreign countries. The first study demonstrated that historic buildings had a greater restorative potential than ordinary buildings, while no differences were found in the second study. Although the influence of place identity was not directly included in these studies, the results led to a series of new studies focusing on the place of the experience. These applied the taxonomy devised by Galindo and Hidalgo (2005), who identified three types of attractive urban places that can be distinguished in terms of aesthetic preference: historic-cultural, recreational (leisure and/or walking), and those with scenic views.

Lastly, although natural environments are what humans are most fond of – biophilia – in addition to playing an important role in reducing stress (Ulrich 1984), and in recovery from the attention fatigue that we often suffer from (Kaplan 1995), other contributions have proved their role in strengthening self-esteem and self-regulation (Korpela 1989, 1992). This issue suggests studying the capability of environments as a source of identity as the study of restoration processes, which is related to emotional problems, and includes properties in the person-environment interaction that may be involved in developing place identity.

> Theory and research dealing with place identity and restorative environments have for the most part proceeded independently. Assuming that emotional- and self-regulation are processes underlying the development of place identity, and that a person's favorite place is an exemplar of environments used in such regulation processes, the present study goes beyond preliminary observations about restorative aspects of favorite places to consider how individuals evaluate their favorite places using terms set out in the restorative environments theory. (Korpela and Hartig 1996)

The next section is devoted to the relationship between place identity and psychological well-being.

3.6 The Physical Space as a Source of Identity and Attachment

The act of giving meaning to the space is, perhaps, the first principle that determines our socio-spatial relationships universally. This universalism, however, is not to be taken as a formalized law of human spatial behavior. Nevertheless, we do claim it can be considered an axiological principle that regulates the development of the phenomena that relate people to their environments and that are the object of study of Environmental Psychology. With no intention of establishing universal patterns or laws of human behavior, we can certainly turn to our own experience and to common sense to comment on some evidence.

1. The environment in which we develop as people is more than a set of physical variables and objects arranged in a specific order and structure. Beyond that, the physical environ-

ment is determined by a set of meanings that we people attach to them. These meanings are based on our experiences with the place and its psychological impact, especially when they are socially elaborate and attributed meanings that configure the socio-physical universe.

2. Consequently, human beings tend to establish identity bonds with their significant environments, especially with the most relevant ones for their history, their daily life, and for their development as people. In this sense, it would not be absurd to think in terms of social need, on the same level as the need to establish significant social contacts with those around us.
3. Due to their own human condition, these bonds are articulated on the basis of the meanings we construct and with which we tinge the physical spaces that, as a result of that operation, go from being *spaces* to being *places*.
4. Fourthly, when these spaces, these places, are violated, assaulted, or destroyed, people suffer, which reveals this bond we were discussing as largely an affective bond.
5. Lastly, this affective bond with the environment is an important factor in the development of the psychological and psycho-social well-being of people and in generating attachment to place in different ways.

This evidence related to the meanings, identities, and attachment to spaces is correlated in several concepts specially developed from environmental psychology. Without delving too deeply, we mean mainly place attachment (Altman and Low 1992; Hernández et al. 2014; Lewicka 2011; Scannell and Gifford 2010) and place identity (Proshansky 1978; Proshansky et al. 1983). Recent reviews on the concepts of place identity (Casakin and Bernardo 2012; Devine-Wright and Clayton 2010; Droseltis and Vignoles 2010; Vidal et al. 2012) and place attachment (Lewicka 2011; Manzo and Devine-Wright 2014; Scannell and Gifford 2010), reveal how relevant both are, attachment especially, to understanding the bonds between people and environment. The relationship between place identity and place attachment has led to numerous studies (Hernández et al. 2007; Williams 2014), although the link between both concepts offers no clear consensus. One proposal for approaching this debate is that offered by Di Masso and Dixon (see Chap. 5) in this manual, to which they add the relationship with the concepts of residential satisfaction (Amérigo and Aragonés 1997; Christensen and Carp 1987; Fleury-Bahi et al. 2008) and place satisfaction (Ramkissoon et al. 2013; Stedman 2002). In addition to specifying some of the debates around the boundaries and overlapping of these concepts and a few attempts to integrate them, such as the tripartite model of Scannell and Gifford (2010), these authors propose reorienting the epistemological perspective in the study of person-environment bonds, while they remark on the discursive construction of environmental categories and related psychological reactions and the political dimension of people-environment relationships.

Environmental Psychology introduces the subject of identity from an essentially individualistic perspective, through the concept of place identity (Proshansky 1978; Proshansky et al. 1983). This is considered a substructure of self-identity (i.e. the image we generate of ourselves) and it comprises a set of cognitions referring to places or spaces where the person leads their daily life and according to which they can establish emotional bonds and bonds of belonging to specific environments. These bonds are at least as important as those established with the different social groups with which the person has a relationship. At the base of this structure lies the person's "environmental past", as well as socially elaborate meanings referring to these spaces that they have integrated into their spatial relationships. This "cognitive tank" that configures place identity – which, according to Proshansky, the individual is not aware of except when they feel their identity is being threatened – allows the person to recognize properties in the new environment that are related to their "environmental past". It also favors a sense of familiarity and the perception of stability in the environment, it indicates to them how to behave, it determines the degree of appropriation or the capability of modifying the environment, and, lastly, it favors a sense of environmental control and security.

However, the individual levels of identity are not enough to account for the phenomenon completely. It is necessary to turn to the social dimension. Tajfel has already put forward that social identity, based on the fact that an individual belongs to certain groups or categories, implies the perceptive emphasis of the similarities with the group itself and the differences from it with respect to other groups. This comparative perspective is what relates social categorization to social identity (Tajfel 1983). Turner picks up this notion to conduct a reconceptualization of the social group. He considers it a collection of individuals who perceive themselves to be members of a specific social category and who are capable, therefore, of distinguishing themselves from other collections of individuals on the basis of the dimensions associated with this categorization (Turner 1987). To sum up, the configuration of the group's social identity is defined both by the perception of similarities within the endogroup and by the perception of endogroup-exogroup differences. These processes are based on certain categorical dimensions that are relevant to the group and that usually generate a positive image of its members. In this way, social identity is configured through the strengthening of self-esteem.

This approach involves considering that urban environments can be understood as categorizations of self in a specific level of group abstraction. Therefore, we can talk about urban social identity (Valera and Pol 1994). The sense of belonging to certain social categories also comprises the sense of belonging to certain significant urban environments for the group. Behind this idea lies seeing the urban environment as something more than the physical stage where the lives of individuals take place. It is a social product resulting from the symbolic interaction between the people who share a specific urban environment. Thus, the urban environment goes beyond the physical dimension to acquire a symbolic, social dimension. However, urban identity also plays another essential function: it enables us to internalize the special features of the place based on a collection of attributions that configure its specific image – in a very similar way to Stokols' and Shumaker's "social imaginability" (1981). This image determines the attribution of a group of features to the individuals, grants them a certain type of personality: "feeling resident of a town confers a number of quasi-psychological qualities to the people related to it" (Lalli 1988, p. 305). In turn, urban identity provides the person with positive evaluations of self and with a subjective feeling of time continuity that enables an identity-generation link related to the urban environment.

Later, Twigger-Ross and Uzzell (1996) developed these classic contributions based on the principles of distinctiveness, self-esteem, self-efficacy, and time continuity – reference and congruence – which, according to Breakwell (1992), guide the processes of assimilation/accommodation and evaluation that make up identity. Among these four principles, time continuity has raised more interest in the literature judging by the large number of contributions regarding identity and place (Devine-Wright and Lyons 1997; Feldman 1990). However, apart from its individual dimension, just like the other principles, time continuity is reflected in memory, through the narration of places, as a constant social reconstruction of shared meanings. In sum, this implies a social practice (Vázquez 2001).

Lewicka (2014) remarks that the interest in memory (of places and people) is now usually presented as contradictory to one of today's most characteristic global features: mobility. However, for this author they are not opposite terms. On the contrary, the interest in the past strengthens the sense of time continuity of the person and the place, thus facilitating the attachment to new places among individuals with mobility patterns. Her argument is sustained by the need not to conceive place attachment as a uniform concept. Lewicka refers to Hummon's contributions (1992) with respect to everyday and ideological rootedness and the different types of nonattachment (alienation, place relativity, and placelessness). Thus, according to the type of attachment (active or traditional), the different forms of memory (procedural, episodic, declarative, and declarative semantic) can contribute to people being linked to places based on their habits and routines acquired through living in a place, nostalgia, and family and place history respectively.

The latest literature (Manzo and Devine-Wright 2014) considers place attachment as the collection of bonds people establish with places, thus reviving the multidimensional character of earlier definitions (Altman and Low 1992). Such is the case of the tridimensional model proposed by Scannell and Gifford (2010), which comprises person (individual, groups), process (affect, cognition, behavior) and place (physical, social) and shows its applied relevance. From the most purely psychological tradition, place attachment is understood as "an affective bond that people establish with a specific place where they tend to stay, feel comfortable and safe" (Hidalgo and Hernández 2001, p. 274). This bond can develop toward places of different scales, although most studies have focused on the neighborhood level (Lewicka 2011), highlighting a few variables related to attachment, such as the time of residence and the expectations of remaining in the current place (Riger and Lavrakas 1981), or the number of previous homes, and local participation (Cuba and Hummon 1993). It is measured through questionnaires whose respondents generally focus on a feeling of emotional attachment to the place. An example is the proposal by Hidalgo and his collaborators (Hernández et al. 2007; Hidalgo and Hernández 2001). Their view highlights the affection for the place, feeling of belonging to the place, feeling happy to go back to the place, feeling proud of living in the place, and the intention to continue living in the place.

Meanwhile, other less restrictive notions of place attachment (Manzo and Devine-Wright 2014) incorporate aspects such as memory (Lewicka 2014), mobility (Gustafson 2014), multiple and simultaneous place attachments (Scannell and Gifford 2010, 2014), and the ways in which a series of emotions and experiences contribute to the attachment to communities and their role in developing trust in civil society and cooperation between institutions and citizens, as Perkins et al. (2002) define the social capital (Mihaylov and Perkins 2014).

Similarly, we consider that the concept of place identity requires its development and conceptualization to be completed by approaching certain elements of tension, which are essential in light of current social and environmental changes (Stokols et al. 2009). These elements can be summarized by two premises: on the one hand, the need to go beyond the individual dimension to adopt a social and communal view, just like the individualistic concept of well-being has required new, more social developments (Keyes 1998); and, on the other hand, the need to contemplate a place identity that is increasingly multiplied, displaced, and dislocated as a sign of new, liquid times (Di Maso et al. 2008; Dixon and Durrheim 2000; Vidal et al. 2010).

In spite of this, we can claim that the concept of place identity – and also place attachment – has been one of the topics that has given rise to more literature in the sphere of environmental psychology during recent decades. Especially with respect to place attachment, recent contributions (Manzo and Devine-Wright 2014) show the relevance of the bonds of identity and attachment to place, in order to understand spaces not only as generators of well-being, but also as contexts where people can experience positive situations in the light of some applications where place attachment – and, in our understanding, place identity – appears in the most recent literature, like, for example, pro-environmental engagement, social housing, and community design (Manzo and Devine-Wright 2014). In this sense, we have already mentioned the proposal by Korpela and Hartig (1996) regarding ART and its possibilities for researching the relationships between restoration and place identity, given that they are based on the interests, objectives, goals, purposes, and environmental limitations that interact with life and the activities of individuals. This interaction, as mentioned at the beginning of this section, we consider to stem from the principle that seems to rule our socio-spatial relations: the act of giving meaning to space.

This act requires making it "ours", turning physical space into a place with meaning (Páramo

2011). This is a daily act immanent to human activity. It derives from an epistemic exercise, i.e. obtaining knowledge from the environment that is valid and significant as well as locational – that is, from a location in a socially built environment. When giving meaning to a space, we become attached to places emotionally, we feel safe and we obtain psychological well-being, we transform the space to our functional and symbolic interest, we delimit it, we manage and defend it, we relate to it, it binds us socially or as a group, and we incorporate it as another element in our social interaction. Turning the physical space into "our" place with meanings is an "appropriation of the space", as seen in the following section.

3.7 The Appropriation of Space as a Generator of Well-Being and Meaning

The concept of appropriation of space (Graumann 2002; Korosec-Serfaty 1976; Pol 1996; Vidal et al. 2004; Vidal and Pol 2005; Serfaty-Garzon 2003) is comprised in a phenomenological approach to people-environment studies, as Graumann pointed out (2002) in the Handbook of Environmental Psychology, edited by Bechtel and Churchman; or more recently, Seamon (2014) regarding place attachment. For Benages-Albert et al. (2015), this concept allows us to re-integrate, within the same temporal process, place-related psychological experiences, such as place identity, place attachment or place preference, rather than treating such constructs as neatly separated entities somehow interrelated in a part-whole relationship. The concept of appropriation enables the redefinition of such constructs as different experiential moments within the same ongoing process of territorial and symbolic relationship with places.

Pol (1996, 2002) turns to a dual model of appropriation based on two complementary paths: action-transformation and symbolic identification. The former, compartmentally-based, supposes an action on the environment developed by the person and the group, changing the space, leaving a "footprint". Action-transformation refers to the use, defense, and signaling characteristic of territoriality. The latter path involves the incorporation of action into the space through cognitive, affective, and interactive processes that turn the space into a place, and where the person or social group becomes identified with the environment. Symbolic identification refers to social categorization processes such as processes that allow the continuity of self and group cohesion (Pol 2002; Vidal and Pol 2005; Vidal et al. 2004). Action-transformation and symbolic identification define a continuous cyclical process, which is present in the entire life cycle of people. However, depending on the stage, one path may be more relevant than the other. In turn, through appropriation, the space becomes a place, a space with meaning and identity to self and to the group, in addition to communicating to the others such place identity (Pol 2002; Vidal and Pol 2005).

For Serfaty-Garzon (2003) – previously Korošec-Serfaty –, the editor of the minutes of the 1976 international conference on this concept, the appropriation of space is a complex process. Through appropriation, people create themselves through their own actions, hence the process of socialization in a socio-cultural, historical context. It is also the domain of the significances of the appropriated object or space, regardless of its legal property. It is not adapting, but rather mastering, an aptitude, the capability of appropriation. It is a time-related phenomenon, which means considering the person's change over time. It is, all in all, a dynamic process of interaction of the person with the medium.

Returning to Pol's dual model of appropriation (2002), the predominance of one path or the other also depends on the degree of control, more or less shared, and the social rules, more or less formal, which define space management. Therefore, a private space, such as one's own home, is more likely to be transformed, like when someone paints the walls and changes the decoration after moving in to make the home "their own". Other spaces such as a street or a square, ornamented by neighbors on local holidays, allow ephemeral or temporary transformations. However, the transformation of urban or public spaces can also include a certain prominence of neighbors and residents. Such is the case of urban planning processes with community participation (Horelli 2002; Sanoff 2000), united under different forms and names.

In the words of Sanoff himself (2006), community design is a movement for discovering how to make it possible for people to be involved in shaping and managing their environment. Community architecture is the activist term used in England (social architecture is used in the United States), which encompasses community planning, community design, community development and other forms of technical community aid. Community participation, on the other hand, covers all the scales and techniques but refers to the processes involving professionals, families, community groups and government officials in shaping the environment. Facilitation is another approach that has emerged, which uses participatory methods for both problem definition and design solution generation through design assistance techniques. In contrast to the political activist role assumed by community design, facilitation is a means of bringing people together to determine what they wish to do and to help them find ways to work together in deciding how to do it.

The history of urban planning – see an approach to it in the chapter of this manual by Romice et al. – shows several contributions (Jacobs, Davidoff, advocacy planning, Turner, new urbanism, etc.) defending the involvement of people in the design of urban environments under several labels (community or social architecture, community planning, community participation, facilitation, and placemaking). The early stages of some of these critical views, faced with the separation between the physical and social dimensions of urban environments, are contemporary with contributions such as that of Lefebvre, with his studies on the city and the social space (Lefebvre 1968, 1972). His approaches are close to the view proposed by the concept of space appropriation. In this context, it is not strange that environmental psychology was called "psychology of architecture", according to Pol (2006) in the 1960s, a decade characterized by greater social orientation, especially in Europe.

Each and every one of the above labels includes the participation and involvement of people in modifying an urban environment, that is, their intention of acquiring transformation actions and control of the spaces oriented to functional improvement or to an adaptation to their users' needs, bearing in mind technical criteria. In other words, although the main goal is usually to improve the physical environment, the participative process generated to that end usually has an effect on the well-being of people through the creation of a sense of place. In order to understand the relationship between well-being and sense of place (and place attachment and place identity), the concepts of space appropriation (Pol 2002; Vidal and Pol 2005) and people empowerment (Rappaport 1986) are relevant. This classic community psychology concept, close to appropriation, refers to the process by which people (organizations and communities) take control of and acquire authority over topics of their own interest.

Zimmerman (1995) summarized three levels of interdependent analyses with which empowerment is presented: psychological, organizational, and communal. The interdependence of the three levels enables us to understand the maintenance and improvement of a community's quality of life. Participation in urban design supposes an opportunity to improve quality of life. Involvement in urban design has an impact on a greater perception of control and critical understanding of the socio-political environment (from a psychological point of view), it provides necessary mutual support (from the organizational level), and, overall, it helps improve a community's quality of life (communal level).

People's participation and involvement in processes of urban improvement of their close environment (community planning, placemaking, etc.) offer opportunities of space appropriation and people empowerment (from individual and social points of view) through which senses of belonging and community are generated towards the socio-physical environment (Remesar 2008; Remesar et al. 2012). These bonds between people and places have been explained by different concepts (sense of community, sense of place, place identity, place attachment). However, the most relevant aspects are, on the one hand, the psycho-social dimension of these types of processes, which are based on the social interaction generated in the improvement of the urban environment, and, on the other hand, the personal and collective well-being that is generated through place attachment (Lewicka 2011, 2014; Scanell and Gifford 2014) and the generation of trust in others and in the institutions or social capital (Mihaylov and Perkins 2014; Perkins et al. 2002) thanks to the participative urban design process. We mean, in sum, that it is basically a psycho-social view of well-being, in line with Keyes' contributions (2007) and the five dimensions of social well-being, as seen in Table 3.1: integration, contribution, coherence, actualization and acceptance.

Oriented towards psychology and the individual, another contribution to understanding the implications of action on the environment is Habraken's (1998), defended by Thwaites et al. (2013) among others. His concept of form, place and understanding provides a framework that relates physical structure and spatial organization directly to patterns of human behavior. [...] Place and understanding are essentially concepts that can be related to human well-being, via issues of territorial expression, development of self-esteem and the experience of belonging (Thwaites et al. 2013, p. 37).

The relationship between participation and appropriation seems evident to us. Participation is understood as the development, in the closest environment, of the person's action-transformation scope, which affects their perception of control and their involvement in their own environment; in other words, in their appropriation of it. Through "participation" in designing the environment, it is transformed leaving a footprint and is actively incorporated into cognitive and affective processes. Conversely, through symbolic identification, the appropriated space becomes a factor of continuity and stability of self, as well as a one of identity stability and group cohesion. Additionally, this also generates a bond with the place, thus facilitating responsible conduct and involvement and participation in one's own environment.

To sum up, one of the main conclusions in the so-called Community Participation and Planning (Manzo and Perkins 2006), Community Development by Design (Hester 2006), Community Design (Toker 2007), or Socially Restorative Urbanism (Thwaites et al. 2013) is the relationship between participation in urban design to restore social well-being and the sense of belonging in urban environments. From our understanding, the key is to enable re-appropriation of the urban space. This can be considered a positive intention from the psychological, social, and environmental points of view.

3.8 Conclusions

In this chapter, we have presented some of the developments of Environmental Psychology most directly related to what we could call Positive Environmental Psychology. We have tried to find out to what extent the physical environment can be an essential element for establishing psychological bonds, having an impact on the well-being of people and social groups, and for the development of their positive capabilities. Overall, we have endeavored to discover to what extent the relationship with the physical environment contributes to psychological well-being. This should enable us to come closer to a proposal we could call Positive Environmental Psychology.

To sum up, the physical environment (sociophysical) can favor well-being and, in general, positive states of people (positive environments) inasmuch as it provides aesthetic experiences related to positive emotions. This is also the case when the psycho-environmental relationship with it activates and stimulates us to make use of our best, utmost psychological capabilities in a pleasant and satisfactory way. Likewise, the environment can be positive when our relationship (especially with natural environments) involves us and allows us to recover from our daily efforts. In addition, it can also be positive when it allows us to establish psychologically significant bonds with it in terms of identity, appropriation, or attachment. Lastly, positive environments can be this way, not only for people, but also for groups and communities, especially when they enable social participation in their design or use, thereby generating processes of community empowerment or urban social identity and facilitating social integration, the perception of contribution, or psycho-socio-environmental coherence. All in all, looking at Keyes' dimensions in Table 3.1, we can see how the physical environment can contribute to the flourishing of emotional, psychological, and social well-being.

In conclusion, firstly, it is interesting to verify that many of these psycho-environmental developments present mutually theoretical connections. Korpela takes the place identity concept as a starting point (Korpela 1989) to reach the study of people's favorite environments (Korpela 1992; Korpela and Ylen 2007) and, from there, restorative environments (Korpela and Harting 1996; Korpela and Staats 2014). For his part, Kaplan reaches his attentional restoration theory (Kaplan 1995), from, among others, the works of Berlyne (1960) and evolutionist psychology (Berlyne 1971; Kaplan and Kaplan 1989), by way of the studies on landscape preference (Kaplan 1979). Furthermore, the pioneering studies on space appropriation (Korošec-Serfaty 1976; Pol 1996) find clear links with the concept of place identity (Proshansky 1976, 1978; Proshansky et al. 1983), the study of identity processes in the environment (Lalli 1988, 1992; Valera and Pol 1994; Twigger-Ross and Uzzell 1996) and, finally, the most recent development of the notion of attachment (Altman and Low 1992; Gustafson 2014; Hidalgo and Hernández 2001), with links, in turn, to restorative environments (Devine-Wright and Howes 2010). Despite these obvious links, the general impression is often that these topics have followed, to a certain extent, independent paths. Recovering and verifying these links thus facilitates the visualization of a true Positive Environmental Psychology from the early stages of the discipline. The recent link between positive psychology and sustainability-oriented behavior (Corral-Verdugo 2012; Corral-Verdugo et al. 2011) is a clear example of this trend.

Secondly, certain psycho-environmental topics, despite being framed within a reactive environmental psychology, can very well be approached from a proactive perspective. Such is the case of the concept of the restorative capabilities of environments. People go to their favorite places to relax, calm down and clear their heads after facing emotionally negative events. In addition, these places provide experiences of beauty, control, freedom of speech and escape from social pressures, thus having a restorative "function". In fact, whereas the concept of a restorative environment implies a previous psychologically negative state that can be environmentally counteracted to restore well-being (Herzog and Hector 2009), similar to therapeutic environments (Gesler 2003; Williams 1999), the aesthetic experiences that favor positive emotional states or the safety and sense of bonding to the psychological, social, environmental world that provides place identity become positive environmental elements in themselves. In this line of thought, it is interesting to explore these developments regarding such sensitive concepts for well-being as the perception of urban insecurity or fear of crime (Valera and Guàrdia 2014).

On the other hand, it seems interesting to study the role of historic landscapes in promoting the restorative experience of their inhabitants. Fornara and Troffa (2009) carried out a study to verify whether restorative properties were related to the "historic" or the "panoramic" dimension, or rather a combination of the two. The results showed that historic places can trigger restorative experiences and feelings of pleasantness and relaxation. These outcomes suggest a possible influence of symbolic and identitarian meanings and elements of the landscape on influencing psychophysical restoration and well-being. It has been suggested that some typologies of the urban built environment can help people to cope with environmental stress and cognitive fatigue, and to recover their physical and mental resources. People's relationships and transactions with historic landscapes, in particular, are characterized by a prominent role of symbolic aspects, which can represent key components of urban identity.

Thirdly, certain developments of environmental psychology and person-environment studies – initially proactive and aiming to improve urban design functionally – such as people's involvement in designing the environment where they live or work, entail other "results" like, for example, the restoration of well-being and the generation of the sense of place, place attachment, and place identity. It is this relationship between sense of place (and related concepts such as place attachment and place identity) and well-being and health, as pointed out by De Miglio and Williams (2008) among others, which we consider particularly relevant in this chapter. Well-being and place are explained, precisely, from the person-space interaction, be it through experiences stemming from daily interaction in the place, or those derived from the improvement or transformation of their urban design. The latter experiences, grouped into Community Participation and Planning (Manzo and Perkins 2006), Community Development by Design (Hester 2006), Community Design (Toker 2007), or Socially Restorative Urbanism (Thwaites et al. 2013), provide opportunities of re-appropriation of urban space and civic empowerment. In addition to a proactive view of the user/resident in managing their closest environment, this also implies a positive view of the disciplines involved, other than environmental psychology.

In any case, a new way of approaching the psycho-environmental agenda prevails. Despite all its controversies and limitations, positive psychology has traced a fundamental path for our discipline, environmental psychology, which should lead us – in a global context where stress, fear, and risk prevention are commonplace – towards a new conceptualization of positive environments where these negative states have to be faced with trust, hope, well-being, and the promotion of health in its full sense.

References

Abkar, M., Kamal, M. S., Maulan, S., Mariapan, M., & Davoodi, S. R. (2011). Relationship between the preference and perceived restorative potential of urban landscapes. *Hort Technology, 21*(5), 514–519.

Altman, I., & Low, S. (1992). *Place attachment* (Human behavior and environment, Vol. 12). New York: Plenum.

Amérigo, M., & Aragonés, J. I. (1997). A theoretical and methodological approach to the study of residential satisfaction. *Journal of Environmental Psychology, 17*(1), 47–57.

Averill, J. R. (1973). Personal control over aversive stimuli and its relationship to stress. *Psychological Bulletin, 80*(4), 286–303.

Baum, A., & Epstein, Y. M. (1978). *Human response to crowding*. Hillsdale: Lawrence Erlbaum Associates.

Benages-Albert, M., Di Masso, A., Porcel, S., Pol, E., & Vall-Casas, P. (2015). Revisiting the appropriation of space in metropolitan river corridors. *Journal of Environmental Psychology, 42*, 1–15.

Berlyne, D. E. (1949). 'Interest' as a psychological concept. *British Journal of Psychology, 39*(4), 184–195.

Berlyne, D. E. (1960). *Conflict, arousal and curiosity*. New York: McGraw-Hill.

Berlyne, D. E. (1971). *Aesthetics and psychobiology*. New York: Appleton Century Crofts.

Berlyne, D. E. (1974). *Studies in the new experimental aesthetics. Steps toward an objective psychology of aesthetic appreciation*. New York: Halstead.

Berto, R. (2005). Exposure to restorative environments helps restore attentional capacity. *Journal of Environmental Psychology, 25*(3), 249–259.

Blackman, M. (2014). *The angel of the north: Public art and wellbeing*. PhD. Thesis. School of Applied Social Sciences, Durham University. http://etheses.dur.ac.uk/10927/1/Maeve_Blackman_PhD_Thesis_-_Angel_of_the_North_and_Wellbeing.pdf.

Bradburn, N. M. (1969). *The structure of psychological well-being*. Chicago: Aldine Publishing.

Breakwell, G. M. (1992). Social representations and social identity. *Papers on Social Representations [Textes sur les Représentations Sociales], 2*(3), 1–217.

Brickman, P., & Campbell, D. (1971). Hedonic relativism and planning the good society. In M. H. Apley (Ed.), *Adaptation-level theory: A symposium* (pp. 287–302). New York: Academic.

Brown, K. W., & Kasser, T. (2005). Are psychological and ecological well-being compatible? The role of values, mindfulness, and lifestyle. *Social Indicators Research, 74*, 349–368.

Brown, B. B., & Perkins, D. D. (1992). Disruptions in place attachment. In I. Altman & S. Low (Eds.), *Place attachment* (Human behavior and environment, Vol. 12, pp. 279–304). New York: Plenum.

Casakin, H., & Bernardo, F. (Eds.). (2012). *The role of identity in the perception, understanding, and design of the built environment*. Bussum: Bentham Science Publishers.

Cattell, V., Dines, N., Gesler, W., & Curtis, S. (2008). Mingling, observing, and lingering: Everyday public spaces and their implications for well-being and social relations. *Health & Place, 14*, 544–561.

Christensen, D. L., & Carp, F. M. (1987). PEQI-based environmental predictors of the residential satisfaction of older women. *Journal of Environmental Psychology, 7,* 45–64.

Corral-Verdugo, V. (2012). The positive psychology of sustainability. *Environment, Development and Sustainability, 14,* 651–666.

Corral-Verdugo, V., Montiel, M., Sotomayor, M., Frıas, M., Tapia, C., & Fraijo, B. (2011). Psychological wellbeing as correlate of sustainable behaviors. *International Journal of Hispanic Psychology, 4,* 31–44.

Csikszentmihalyi, M. (1990). *Flow. The psychology of optimal experience.* New York: Harper & Row.

Csikszentmihalyi, M. (1998). *Finding flow: The psychology of engagement with everyday life* (Masterminds series). New York: Basic Books.

Cuba, L., & Hummon, D. M. (1993). A place to call home: Identification with dwelling, community, and region. *The Sociological Quarterly, 34*(1), 111–131.

Cupchik, G. C. (1986). A decade after Berlyne. New directions in experimental aesthetics. *Poetics, 15,* 345–369.

DeMiglio, L., & Williams, A. (2008). Sense of place, a sense of well-being. In J. Eyles & A. Williams (Eds.), *Sense of place, health and quality of life* (pp. 15–30). Surrey: Ashgate.

Devine-Wright, P., & Clayton, S. (2010). Introduction to the special issue: Place, identity and environmental behavior. *Journal of Environmental Psychology, 30*(3), 267–270.

Devine-Wright, P., & Howes, Y. (2010). Disruption to place attachment and the protection of restorative environments: A wind energy case study. *Journal of Environmental Psychology, 30,* 271–280.

Devine-Wright, P., & Lyons, E. (1997). Remembering pasts and representing places: The construction of national identities in Ireland. *Journal of Environmental Psychology, 17,* 33–45.

Di Maso, A., Vidal, T., & Pol, E. (2008). La construcción desplazada de los vínculos persona-lugar. Una revisión teórica. *Anuario de Psicología, 39*(3), 371–385.

Diener, E., & Seligman, M. (2004). Beyond money: Toward an economy of well-being. *Psychological Science in the Public Interest, 5*(1), 1–31.

Diener, E., Lucas, R. E., & Scollon, C. N. (2006). Beyond the hedonic treadmill. Revising the adaptation theory of well-being. *American Psychologist, 61*(4), 305–314.

Dixon, J., & Durrheim, K. (2000). Displacing place-identity: A discursive approach to locating self and other. *British Journal of Social Psychology, 39,* 27–44.

Droseltis, O., & Vignoles, V. L. (2010). Towards an integrative model of place identification: Dimensionality and predictors of intrapersonal-level place preferences. *Journal of Environmental Psychology, 30*(1), 23–34.

Ellemers, N., & Barreto, M. (2006). Categorization in everyday life: The effects of positive and negative categorizations on emotions and self-views. *European Journal of Social Psychology, 36,* 931–942.

Feldman, R. M. (1990). Settlement-identity psychological bonds with home places in a mobile society. *Environment and Behavior, 22*(2), 183–229.

Festinger, L. (1957). *A theory of cognitive dissonance.* Stanford: Stanford University Press.

Fleury-Bahi, G., Félonneau, M.-L., & Marchand, D. (2008). Processes of place identification and residential satisfaction. *Environment and Behavior, 40*(5), 669–682.

Fornara, F. (2011). Are "attractive" built places as restorative and emotionally positive as natural places in the urban environment? In M. Bonaiuto, M. Bonnes, A. M. Nenci, & G. Carrus (Eds.), *Urban diversities: Environmental and social issues* (Advances in people-environment studies, Vol. 2, pp. 159–170). Gottingen: Hogrefe & Huber.

Fornara, F., & Troffa, R. (2009). Restorative experiences and perceived affective qualities in different built and natural urban places. In H. Turgut Yildiz & Y. Ince Guney (Eds.), *Revitalising built environments: Requalifying old places for new uses* (pp. 1–10). Istanbul: Istanbul Technical University. Proceedings of the IAPS-CSBE & Housing Networks International Symposium.

Fredrickson, B. L. (2001). The role of positive emotions in positive psychology: The broaden-and-build theory of positive emotions. *American Psychologist, 56*(3), 218–226.

Fullilove, M. T. (2014). "The Frayed Knot". What happens to place attachment in the context of serial forced displacement? In L. C. Manzo & P. Dewine-Wright (Eds.), *Place attachment: Advances in theory, methods and applications* (pp. 141–153). Oxon: Routledge.

Galindo, M. P., & Corraliza, J. A. (2000). Environmental aesthetics and psychological well-being. *Psychology in Spain, 4*(1), 13–27.

Galindo, M. P., & Hidalgo, M. C. (2005). Aesthetic preferences and attribution of meaning. Environmental categorization processes in the evaluation of urban scenes. *International Journal of Psychology, 40*(1), 19–27.

García Mira, R. (Ed.). (2013). *Lecturas sobre el desastre del Prestige Contribuciones desde las Ciencias Sociales.* A Coruña: Instituto de Estudios e Investigación Psicosocial "Xoán Vicente Viqueira".

García Mira, R., Real, J. E., Uzzell, D., San Juan, C., & Pol, E. (2006). Coping with a threat to quality of life: The case of the "Prestige" disaster. *European Review of Applied Psychology [Revue Européenne de Psychologie Appliquée], 56,* 53–60.

Gehl, J. (1987). *Life between buildings.* Washington, DC: Island Press.

Gehl, J. (2010). *Cities for people.* Washington, DC: Island Press.

Gesler, W. M. (2003). *Healing places.* Lanham: Rowman & Littlefield.

Gonçalves Siebra, L. M. (2012). *Percepçao da arte pública da cidade de Barcelona e seus significados simbólicos.* PhD Thesis. Department of Sculpture. University of Barcelona.

Gould, C. W. (2009). The right to housing recovery after natural disasters. *Harvard Human Rights Journal, 22*(2), 36.

Graumann, C. F. (2002). The phenomenological approach to people-environment studies. In R. Bechtel & A. Churchman (Eds.), *Handbook of environmental psychology* (pp. 95–113). New York: Wiley.

Gustafson, P. (2014). Place attachment in an age of mobility. In L. C. Manzo & P. Dewine-Wright (Eds.), *Place attachment: Advances in theory, methods and applications* (pp. 37–48). Oxon: Routledge.

Habraken, N. J. (1998). *The structure of the ordinary: Form and control in the built environment*. Cambridge: MIT Press.

Harmon-Jones, E., & Mills, J. (1999). An introduction to cognitive dissonance theory and an overview of current perspectives on the theory. In E. Harmon-Jones & J. Mills (Eds.), *Cognitive dissonance: Progress on a pivotal theory in social psychology* (Science conference series, pp. 3–21). Washington, DC: American Psychological Association.

Hartig, T., Mang, M., & Evans, G. W. (1991). Restorative effects of natural environment experiences. *Environment and Behavior, 23*(1), 3–26.

Hartig, T., Evans, G. W., Jamner, L. D., Davis, D. S., & Gärling, T. (2003). Tracking restoration in natural and urban field settings. *Journal of Environmental Psychology, 23*(2), 109–123.

Hernández, B., Hidalgo, M. C., Salazar-Laplace, M. E., & Hess, S. (2007). Place attachment and place identity in natives and non-natives. *Journal of Environmental Psychology, 27*(4), 310–319.

Hernández, B., Hidalgo, M. C., & Ruíz, C. (2014). Theoretical and methodological aspects of research on place attachment. In L. C. Manzo & P. Dewine-Wright (Eds.), *Place attachment: Advances in theory, methods and applications* (pp. 136–155). Oxon: Routledge.

Herzog, T. R., & Hector, A. E. (2009). Perceived danger and judged likelihood of restoration. *Environment and Behavior, 41*(3), 387–401.

Herzog, T. R., Maguire, C. P., & Nebel, M. B. (2003). Assessing the restorative components of environments. *Journal of Environmental Psychology, 23*, 159–170.

Herzog, T. R., Ouellette, P., Rolens, J. R., & Koenigs, A. M. (2010). Houses of worship as restorative environments. *Environment and Behavior, 42*(4), 395–419.

Hester, R. (2006). *Design for ecological democracy*. Cambridge: MIT Press.

Hidalgo, M. C., & Hernández, B. (2001). Place attachment: Conceptual and empirical questions. *Journal of Environmental Psychology, 21*(3), 273–281.

Horelli, L. (2002). A methodology of participatory planning. In R. Bechtel & A. Churchman (Eds.), *Handbook of environmental psychology* (pp. 607–628). New York: Wiley.

Hummon, D. M. (1992). Community attachment: Local sentiment and sense of place. In I. Altman & S. M. Low (Eds.), *Place attachment* (pp. 253–278). New York: Plenum Press.

Joye, Y., & van den Berg, A. E. (2012). Restorative environments. In L. Steg, A. E. van den Berg, & J. I. M. de Groot (Eds.), *Environmental psychology: An introduction* (pp. 57–66). Chichester: Wiley-Blackwell.

Kaplan, S. (1979). Perception and landscape: Conceptions and misconceptions. In *Proceedings of our National Landscape Conference* (pp. 241–248). USDA Forest Service General Technical Report PSW 35.

Kaplan, S. (1995). The restorative benefits of nature: Toward an integrative framework. *Journal of Environmental Psychology, 15*(3), 169–182.

Kaplan, R., & Kaplan, S. (1989). *The experience of nature: A psychological perspective*. Cambridge (NY): Cambridge University Press.

Kellert, S. R., & Wilson, E. O. (Eds.). (1995). *The biophilia hypothesis*. Washington, DC: Island Press.

Keyes, C. L. M. (1998). Social well-being. *Social Psychology Quarterly, 61*, 121–140.

Keyes, C. L. M. (2007). Promoting and protecting mental health as flourishing. *American Psychologist, 62*(2), 95–108.

Kim, K. A., & Mueller, D. J. (2001). To balance or not to balance: Confirmatory factor analysis of the affect-balance scale. *Journal of Happiness Studies, 2*, 289–306.

Knudsen, A., & Hanafi, S. (Eds.). (2011). *Palestinian refugees: Identity, space and place in the levant*. Milton Park/London: Routledge.

Korosec-Serfaty, P. (1976). *Appropriation of space. Proceedings of the Strasbourg conference. IAPC-3*. Strasbourg-Louvain La Neuve: CIACO.

Korpela, K. M. (1989). Place-identity as a product of environment self-regulation. *Journal of Environmental Psychology, 9*, 241–256.

Korpela, K. M. (1992). Adolescents' favorite places and environmental self-regulation. *Journal of Environmental Psychology, 12*, 249–258.

Korpela, K., & Harting, T. (1996). Restorative qualities of favorite places. *Journal of Environmental Psychology, 16*, 221–233.

Korpela, K., & Staats, H. (2014). The restorative qualities of being alone with nature. In R. J. Coplan & J. C. Bowker (Eds.), *The handbook of solitude: Psychological perspectives on social isolation, social withdrawal, and being alone* (pp. 351–367). Hoboken: Wiley.

Korpela, K., & Ylen, M. (2007). Perceived health is associated with visiting natural favorite places in the vicinity. *Health & Place, 13*(1), 138–151.

Korpela, K., Harting, T., Kaiser, F. G., & Fuhrer, U. (2001). Restorative experience and self-regulation in favorite places. *Environment & Behavior, 33*(4), 572–589.

Küller, R. (1992). Environmental assessment from a neuropsychological perspective. In T. Gärling & G. Evans (Eds.), *Environment, cognition, and action: An integrated approach* (pp. 111–147). New York: Oxford University Press.

Lalli, M. (1988). Urban identity. In D. Canter (Ed.), *Environmental social psychology* (NATO ASI series). Dordrecht: Behavioral and Social Sciences.

Lalli, M. (1992). Urban-related identity: Theory, measurement, and empirical findings. *Journal of Environmental Psychology, 12*, 285–303.

Lees, L., Shin, H. B., & López-Morales, E. (2015). *Global gentrifications: Uneven development and displacement*. Bristol: Policy Press.

Lefebvre, H. (1968). *Le droit à la ville*. Paris: Anthropos.

Lefebvre, H. (1972). *Espace et politique: le droit à la ville II*. Paris: Anthropos.

Lewicka, M. (2011). Place attachment: How far have we come in the last 40 years? *Journal of Environmental Psychology, 31*(3), 207–230.

Lewicka, M. (2014). Place inherited or place discovered? Agency and communion in people-place bonding. *Estudios de Psicología, 34*(3), 261–274.

Linley, P. A., & Joseph, S. (Eds.). (2004). *Positive psychology in practice*. New York: Wiley.

Linley, P. A., Joseph, S., & Word, A. M. (2006). Positive psychology: Past, present, and (possible) future. *The Journal of Positive Psychology, 1*, 3–16.

Manzo, C., & Devine-Wright, P. (Eds.). (2014). *Place attachment. Advances in theory, methods and applications*. Oxon: Routledge.

Manzo, C., & Perkins, D. (2006). Finding common ground: The importance of place attachment to community participation and planning. *Journal of Planning Literature, 20*(4), 335–350.

McDowell, I., & Praught, E. (1982). On the measurement of happiness. An examination of the Bradburn scale in the Canada Health Survey. *American Journal of Epidemiology, 116*(6), 949–958.

Mihaylov, N., & Perkins, D. D. (2014). Community place attachment and its role in social capital development in response to environmental disruption. In L. Manzo & P. Devine-Wright (Eds.), *Place attachment: Advances in theory, methods and research* (pp. 61–74). Oxon: Routledge.

Milgram, S. (1970). The experience of living in cities. *Science New Series, 167*(3924), 1461–1468.

Muntañola, J. (1974). *La arquitectura como lugar*. Barcelona: Gustavo Gili.

Muntañola, J. (1979a). *Topogénesis* (Vol. 1). Vilassar de Mar: Oikos Tau.

Muntañola, J. (1979b). *Topogénesis* (Vol. 2). Vilassar de Mar: Oikos Tau.

Muntañola, J. (1980). *Topogénesis* (Vol. 3). Vilassar de Mar: Oikos Tau.

Nenci, A. M., Troffa, R., & Perriccioli, M. (2005). Efectos restauradores de Tipologías urbanas contemporaneas. In R. García Mira, Á. Fernandez González, M. D. Losada Otero, & M. Goluboff Scheps (Eds.), *Psicología Ambiental, Comunitaria y de la educación*. Madrid: Biblioteca Nueva.

Nenci, A. M., Troffa, R., & Carrus, G. (2006). The restorative properties of modern architectural styles. In M. Tolba, S. Soliman, & A. Abdel-Hadi, *Abstracts of the 19th IAPS Conference on Environment, Health and Sustainable Development* (p. 225).

Páramo, P. (2011). *Sociolugares*. Bogotá: Publicaciones Universidad Piloto de Colombia.

Perkins, D. D., Hughey, J., & Speer, P. W. (2002). Community psychology perspectives on social capital theory and community development practice. *Journal of the Community Development Society, 33*(1), 33–52.

Pluta, A. (2012). *Integrated well-being: Positive psychology and the natural world*. University of Pennsylvania Scholarly Commons. http://repository.upenn.edu/mapp_capstone/37. Accessed Feb 2015.

Pol, E. (1996). La apropiación del espacio. In L. Iñiguez & E. Pol (Eds.), *Cognición, representación y apropición del espacio*. Barcelona: Publicacions Universitat de Barcelona, Monografies Psico/Socio/Ambientals n°9.

Pol, E. (2002). El modelo dual de la apropiación del espacio. In R. García-Mira, J. M. Sabucedo, & J. Romay (Eds.), *Psicología y medio ambiente. Aspectos psicosociales, educativos y metodológicos* (pp. 123–132). A Coruña: Asociación Galega de Estudios e Investigacion Psicosocial-Publiedisa.

Pol, E. (2006). Blueprints for a history of environmental psychology (I): From first birth to American transition. *Medio Ambiente y Comportamiento Humano, 7*(2), 95–113.

Proshansky, H. M. (1976). The appropriation and misappropriation of space. In P. Korosec-Serfaty, *Appropriation of space. Proceedings of the Strasbourg conference. IAPC-3*. Strasbourg-Louvain La Neuve: CIACO.

Proshansky, H. M. (1978). The city and self-identity. *Environment and Behavior, 10*(2), 147–169.

Proshansky, H. M., Fabian, A. K., & Kaminoff, R. (1983). Place-identity. Physical world socialization of the self. *Journal of Environmental Psychology, 3*, 57–83.

Ramkissoon, H., Smith, L. D. G., & Weiler, B. (2013). Testing the dimensionality of place attachment and its relationships with place satisfaction and pro-environmental behaviors: A structural equation modeling approach. *Tourism Management, 36*, 552–566.

Rappaport, J. (1986). Collaboration for empowerment: Creating the language of mutual help. In H. Boyte & F. Riessman (Eds.), *The new populism: The politics of empowerment*. Philadelphia: Temple University Press.

Remesar, A. (2008). Civic empowerment: A challenge for public art and urban design. In M. Acciaiouli, J. Cunha Leal, & H. Maia (Eds.), *Arte e Poder* (pp. 421–424). Lisboa: IHA. Estudos de Arte Contemporanea.

Remesar, A., Salas, X., Padilla, S., & Esparza, D. (2012). Inclusion and empowerment in public art and urban design. *On the w@terfront, 24*, 3–32. http://www.raco.cat/index.php/Waterfront/article/view/259235/346497.

Riger, S., & Lavrakas, P. J. (1981). Communities: Patterns of attachment and interaction in urban neighborhoods. *American Journal of Community Psychology, 9*, 55–66.

Russell, J. A. (1980). A circumplex model of affect. *Journal of Personality and Social Psychology, 39*(6), 1161–1178.

Russell, J. A., & Pratt, G. (1980). A description of the affective quality attributed to environments. *Journal of Personality and Social Psychology, 38*(2), 311–322.

Russell, J. A., Ward, L. M., & Pratt, G. (1981). Affective quality attributed to environments: A factor analytic study. *Environment and Behavior, 13*, 159–288.

Russell, J. A., Lewicka, M., & Niit, T. (1989). A cross-cultural study of a circumplex model of affect. *Journal of Personality and Social Psychology, 57*(5), 848–856.

Saegert, S., & Winkel, G. H. (1990). Environmental psychology. *Annual Review of Psychology, 41*, 441–477.

Sanoff, H. (2000). *Community participation methods in design and planning*. New York: Wiley.

Sanoff, H. (2006, winter). Origins of community design. *Progressive planning magazine*. http://www.plannersnetwork.org/2006/01/origins-of-community-design/.

Sarbin, T. R. (1983). Place identity as a component of self: An addendum. *Journal of Environmental Psychology, 3*, 337–342.

Scannell, L., & Gifford, R. (2010). Defining place attachment: A tripartite organizing framework. *Journal of Environmental Psychology, 30*, 1–10.

Scannell, L., & Gifford, R. (2014). Comparing the theories of interpersonal and place attachment. In L. C. Manzo & P. Dewine-Wright (Eds.), *Place attachment: Advances in theory, methods and applications* (pp. 23–36). Oxon: Routledge.

Schultz, P. W., Gouveia, V., Cameron, L., Tankha, G., Schmuck, P., & Franek, M. (2005). Values and their relationship to environmental concern and conservation behavior. *Journal of Cross-Cultural Psychology, 36*, 457–475.

Seamon, D. (2014). Place attachment and phenomenology: The synergistic dynamism of place. In L. C. Manzo & P. Devine-Wright (Eds.), *Place attachment: Advances in theory, methods, and applications* (pp. 11–22). New York: Routledge.

Seligman, M. (1999). The president's address. APA. 1998. Annual report. *American Psychologist, 54*, 559–562.

Seligman, M., & Csikzentmihalyi, M. (2000). Positive psychology. An introduction. *American Psychologist, 55*(1), 5–14.

Selye, H. (1956). *The stress of life*. New York: McGraw-Hill.

Selye, H. (1974). *Stress without distress*. Philadelphia: J. B. Lippincott Co.

Serfaty-Garzon, P. (2003). L'Appropriation. In M. Segaud, J. Brun, & J. C. Driant (Eds.), *Dictionnaire critique de l'habitat et du logement* (pp. 27–30). Paris: Armand Colin. http://www.perlaserfaty.net/texte4.htm.

Sheldon, K. M., & King, L. (2001). Why positive psychology is necessary. *American Psychologist, 56*(3), 216–217.

Slovic, P. (1987). Perception of risk. *Science New Series, 236*(4799), 280–285.

Sommer, R. (1990). A fish who studies water. In I. Altman & K. Christensen (Eds.), *Human behavior studies* (Human behavior and environment, Vol. 11, pp. 31–48). New York: Plenum Press.

Stedman, R. C. (2002). Toward a social psychology of place: Predicting behavior from place-based cognitions, attitude, and identity. *Environment & Behavior, 34*, 405–425.

Stokols, D. (2003). The ecology of human strengths. In L. G. Aspinwall & U. M. Staudinger (Eds.), *A psychology of human strengths. Fundamental questions and future directions for a positive psychology*. Washington, DC: American Psychological Association.

Stokols, D., & Shumaker, S. (1981). People in places: A transactional view of settings. In J. H. Harvey (Ed.), *Cognition, social behavior and the environment* (pp. 441–488). Hillsdale: Erlbaum.

Stokols, D., Misra, S., Runnerstrom, M. G., & Hipp, J. A. (2009). Psychology in an age of ecological crisis. From personal angst to collective action. *American Psychologist, 64*(3), 181–193.

Tajfel, H. (1983). Psicología social y proceso social. In J. R. Torregrosa & B. Sarabia (Eds.), *Perspectivas y contextos de la psicología social*. Barcelona: Hispano Europea.

Tenngart Ivarsson, C., & Hagerhall, C. M. (2008). The perceived restorativeness of gardens. Assessing the restorativeness of a mixed built and natural scene type. *Urban Forestry and Urban Greening, 7*(2), 107–118.

Thwaites, K., Mathers, A. R., & Simkin, I. M. (2013). *Socially restorative urbanism: The theory, process and practice of experiemics*. Oxon: Routledge.

Toker, Z. (2007). Recent trends in community design: The eminence of participation. *Design Studies, 28*(3), 309–323.

Turner, J. C. (1987). *Rediscovering social group*. Oxford: Basil Blackwell.

Turner, J. C., Hogg, M. A., Oakes, P. J., Reicher, S. D., & Wetherell, M. S. (1987). *Rediscovering the social group: A self-categorization theory*. Oxford: Blackwell.

Twigger-Ross, C., & Uzell, D. (1996). Place and identity processes. *Journal of Environmental Psychology, 16*, 205–220.

Ulrich, R. S. (1983). Aesthetic and affective response to natural environment. In I. Altman & J. F. Wohlwill (Eds.), *Behavior and the natural environment* (pp. 85–125). New York: Plenum Press.

Ulrich, R. S. (1984). View through a window may influence recovery. *Science, 224*(4647), 224–225.

Ulrich, R. S., Simons, R. F., Losito, B. D., Fiorito, E., Miles, M. A., & Zelson, M. (1991). Stress recovery during exposure to natural and urban environments. *Journal of Environmental Psychology, 11*(3), 201–230.

Valera, S., & Guàrdia, J. (2002). Urban social identity and sustainability. Barcelona's Olympic Village. *Environment and Behavior, 34*(1), 54–66.

Valera, S., & Guàrdia, J. (2014). Perceived insecurity and fear of crime in a city with low crime rates. *Journal of Environtmental Psychology, 38*, 195–205.

Valera, S., & Pol, E. (1994). El concepto de identidad social urbana: Una aproximación entre la psicología social y la psicología ambiental [The concept of urban social identity: An approach between social psychology and environmental psychology]. *Anuario de Psicología, 62*(3), 5–24.

Vázquez, F. (2001). *La memòria como acción social. Relaciones, significados e imaginario*. Barcelona: Paidós.

Vidal, T., & Pol, E. (2005). La apropiación del espacio: una propuesta teórica para comprender la vinculación entre las personas y los lugares. [The appropriation of space: A theoretical proposal for understanding the link between people and places]. *Anuario de Psicología, 36*, 281–297.

Vidal, T., Pol, E., Guàrdia, J., & Peró, M. (2004). Un modelo de apropiación del espacio mediante ecuaciones estructurales [An appropriation of space model using structural equations]. *Medio Ambiente y Comportamiento Humano, 5*(1–2), 27–52.

Vidal, T., Valera, S., & Peró, M. (2010). Place attachment, place identity and residential mobility in undergraduate students. *Psyecology, 1*(3), 291–307.

Vidal, T., Troffa, R., Valera, S., & Fornara, F. (2012). Place identity as a useful psychological construct for approaching modern social challenges and new people-environment relations: Residential mobility, restorative environments, and landscape. In H. Casakin & F. Bernardo (Eds.), *The role of identity in the perception, understanding, and design of the built environment*. Bussum: Bentham Science Publishers.

Watson, D., Clark, L. A., & Tellegen, A. (1988). Development and validation of brief measures of positive and negative affect: The PANAS scales. *Journal of Personality and Social Psychology, 54*(6), 1063–1070.

Weinstein, N., Przybylski, A. K., & Ryan, R. M. (2009). Can nature make us more caring? Effects of immersion in nature on intrinsic aspirations and generosity. *Personality and Social Psychology Bulletin, 35*, 1315–1329.

White, L., & Frew, E. (Eds.). (2013). *Dark tourism and place identity: Managing and interpreting dark places*. Oxon: Routledge.

Williams, A. (1999). *Therapeutic landscapes: The dynamic between health and place*. Lanham: University Press of America.

Williams, D. R. (2014). "Beyond the Commodity Metaphor", revisited. Some methodological reflections on place-attachment research. In L. C. Manzo & P. Dewine-Wright (Eds.), *Place attachment: Advances in theory, methods and applications* (pp. 90–99). Oxon: Routledge.

Wilson, E. O. (1984). *Biophilia*. Cambridge: Harvard University Press.

Zárate, M. (2001). *Perspectivas cognoscitivas y proyectuales posibles para un urbanismo ambiental alternativo. Indagación en el problema metodológico de un conocimiento holista y una aproximación especialista desde un enfoque sociofísico al desarrollo sustentable*. Tesis doctoral presentada en la Escuela Técnica Superior de Arquitectura de Barcelona, Universidad Politécnica de Cataluña. Barcelona: Universidad Politécnica de Cataluña.

Zárate, M. (Ed.). (2004). *Urbanismo ambiental alternativo. Selección de textos teóricos y propuesta*. Barcelona: Edicions UPC.

Zárate, M. (2010). El lugar urbano como estrategia de conocimiento proyectual en urbanismo. *Arquitectonics, 19/20*.

Zárate, M., & Muntañola, J. (2000). *El lugar, la arquitectura y el urbanismo, elementos teóricos para la proyectación del ambiente sociofísico*. Colección Polis Cientifica, 2, Santa Fe: Facultad de Arq. Diseño y Urbanismo de la Universidad Nacional del Litoral.

Zimmerman, M. A. (1995). Psychological empowerment: Issues and illustrations. *American Journal of Community Psychology, 23*(5), 581–599.

Zimring, C. M., & Reizenstein, J. E. (1980). Post-occupancy evaluation. An overview. *Environment & Behavior, 2*(4), 429–450.

Linking People-Environment Research and Design. What Is Missing?

Carole Després and Denise Piché

4.1 Introduction

The design of the built environment is a favorable means to enhance the quality of people's lives. However, its success depends on the attention paid to serving the needs of users of cities and buildings throughout the process. It is thus fundamental to teach future designers approaches and methods for understanding people and the environment as part of the same system. As environmental psychologist Gifford puts it: "Wherever you go, there you are [...]. We are always embedded in a place" (2014: 541). Professors of interior design, architecture, urban design, landscape architecture and planning in many universities around the world are responsible for teaching the human aspects of design in theory classes as well as for training future designers to use concepts and empirical evidence to support their decisions in design studios.

Following the two decades of the construction boom after the Second World War, a fair number of these concepts were put forward by environmental psychologists,[1] as part of a movement toward creating a new interdisciplinary field of research: people-environment relations. These concepts were the outcome of their close collaboration with architects and planners, in a common search for more livable environments, and of research methods solidly anchored in the materiality of the environment, at a time when urban and architectural morphology as an analytical approach to the built environment was only emerging in France, Italy, the US and the UK.[2] International and interdisciplinary associations, such as EDRA and IAPS, were born out of this movement, as well as several graduate programs. With more inclusive labels than environmental psychology, programs specializing in People-Environment (P-E) or Environment-Behavior (E-B) studies welcomed designers into this area of research. In return, graduates from these programs brought P-E theories, concepts and knowledge into the curriculum of designers.[3]

---

[1] For historical accounts of people-environment studies, see Després et al. (2012a), Gifford (2014), Giuliani and Scopelitti (2009), Gunther (2009), Noschis (2015), and Pol (2007).

[2] Conzen (1960) in the UK, Rossi (1966) in Italy, Alexander et al. (1977) in the US and Panerai et al. (1977) in France were pioneers in this field of research that gained momentum in the 1980s.

[3] The PhD in EBS at the University of Wisconsin-Milwaukee in the US is a good example of a program located in a Faculty of Architecture and Planning that was successful in attracting students trained in environmental

---

C. Després (✉) • D. Piché
École d'architecture, Université Laval, Québec, Canada
e-mail: carole.despres@arc.ulaval.ca; denise.piche@arc.ulaval.ca

Although P-E research and education developed on the margins of dominant debates in design, this body of knowledge continues to make various types of contributions to the quality of buildings and cities.

Yet, environmental psychology in the last 25 years has proved to be a much less fertile ground for harvesting P-E concepts and knowledge useful or readily accessible to designers. Pol refers to "a certain 'disciplinary' confinement of Environmental Psychology" and to the "disillusionment – and the abandonment of interdisciplinary work with psychologists" by architects who did not receive the answers they had hoped for (2007: 12). Giuliani and Scopelliti's comparative longitudinal content analysis of the Journal of Experimental Psychology and of Environment and Behavior suggests that taking into account the physical settings in which behaviors take place has regressed: "with references to methods, a progressive decline of the innovative emphases on both the 'social' and the 'physical' side of the socio-physical environment, also by means of observation and field research, was observed. New methods hardly emerged, overwhelmed by approaches and methodologies borrowed from other sub-fields of psychology, such as social psychology (. . .) [T]he environment often appears to be confined to the background, missing any specific definition of its nature or functional character, except for a quite generic ecological function" (2009: 385). The result is a worrying widening gap between researchers and designers at a time when working together has never been more important.

Indeed, the twenty-first century announced its colors from the outset, with immeasurable challenges associated with sustainability in its social, environmental and economic dimensions. New conditions call for important changes in the built environment, but these are just too complex to be dealt with by designers alone. Urban sprawl and public health issues, green gas emissions and climate change, massive migrations, poverty and social inequities are a few examples of the severe problems that countries on every continent are facing. Designers are invited to contribute to the mission of building or rebuilding cities, neighborhoods and buildings to support more sustainable behaviors while assuring the wellbeing and harmonious cohabitation of all citizens. In parallel, in several countries, most certainly in the northern hemisphere, public buildings and infrastructures inherited from the post-war boom have worn out and need to be either replaced or retrofitted. These include schools, hospitals, hospices, correctional centers, churches, and also public amenities such as streets, parks, pools, community centers, etc. At the same time, in the global South, particularly in Africa, and in our own Fourth World, the provision of housing, public amenities and infrastructures cannot keep up with the rapidly growing population. Although this context represents a rare opportunity to make cities and buildings more sustainable and more livable, the risks of rebuilding and building are enormous without relevant people-environment knowledge and the input of multidisciplinary taskforces. Beyond interdisciplinary research collaborations aimed at *understanding* these complex problems, it is urgent to initiate multisectoral collaborations to identify avenues for *solving* them.

This chapter advocates maintaining collaborations between environmental psychologists and designers and, above all, renewing and enriching them. This requires that researchers and professionals in P-E relations, including environmental psychologists, be prepared and ready to get their hands dirty in tackling multidimensional problems with other social actors, and become involved in designing solutions contextually. The argument developed is based on our combined teaching and research experience over the last 40 years at the School of Architecture, University of Laval (Quebec, Canada). The first section traces back the contribution of P-E knowledge and methods to environmental design and to designers' academic training, as illustrated with local examples. In the second section, we focus on the design process itself to show how and where researchers in P-E relations can best articulate

---

psychology as well as in design (see Ahrentzen et al. 2012). For a relevant discussion in the context of the UK, see also Mikellides (2007).

Characteristics of Environmental Psychology
1. It is interactionist in perspective, interested in the dynamic interchange between people and their environments.
2. It refers mainly to human beings in their everyday, intact settings.
3. It is multidisciplinary in character.
4. It is concerned with social problems.
5. It studies basic psychological processes such as perception, cognition, growth and development in this real world context.
6. It acknowledges the crucial role of norms, values and attitudes that people bring to their environment.
7. It is concerned with the builder, that is to say "with the problems of conceiving and designing an environment that is functional, in the practical sense, yet humanly satisfying".

Assumptions for Understanding People-Environment Relations
1. The environment is experienced as a unitary field.
2. The person has environmental properties as well as individual ones.
3. There is no physical environment that is not embedded in and inextricably related to a social system.
4. The degree of influence of the physical environment on behavior varies with the behavior in question.
5. The environment frequently operates below the level of awareness.
6. The 'observed' environment is not necessarily the 'real' environment.
7. The environment is cognized as a set of mental images.
8. The environment has a symbolic value […], this is the quality of the environment that provides man with the sense of 'place identity' […].

Fig. 4.1 Characteristics of environmental psychology and assumptions for understanding people-environment relations (Adapted from Ittelson et al. 1974: 5–6, 12)

their contribution to the future of the built environment. Finally, we attempt to interpret why environmental psychologists and designers moved away from each other and to open up avenues to strengthen the link between P-E researchers and designers. Although this chapter considers the field of P-E research as a whole, where possible it highlights the particular contribution of environmental psychology.

4.2 Looking Back to P-E Relations Research and Its Contribution to Design

From the outset, environmental psychology has made a fruitful contribution to our understanding of the transactions between people and their environment. Many of its original concepts have also infiltrated architectural training and practice. In fact, human aspects of architecture are sometimes taught with concepts whose origins remain unknown to students, and sometimes to professors alike. For instance, comfort and the importance of the fit between people and their environment are now part of courses on environmental systems, while wayfinding and the continuum between privacy and social encounters are now considered in how space organization is addressed and assessed in projects. Teaching people-environment relations is now part of the curricula of designers in many universities around the world.[4] In North America, it is even part of the learning objectives imposed by national accreditation boards for architectural programs.

Every principle, assumption, topic and method laid down in the first major textbook in environmental psychology by Ittelson et al. (1974)[5] is still relevant today (see Fig. 4.1), as are all the topics covered in the second edition of Gifford's Environmental Psychology:

---

[4]The short biographies of 35 alumni from the doctorate program in Environment-Behavior Studies at the University of Wisconsin-Milwaukee, more specifically their teaching responsibilities, illustrate our point (Ahrentzen et al. 2012: 284–354).

[5]Several other textbooks, more limited in breadth, were produced during the same period, namely by Mehrabian and Russell (1974), Thornberg (1974), and Canter (1974).

Principles and Practices (1996). A large number of concepts from or developed within environmental psychology are applicable across all scales of the environment, whether in interior design, architecture, urban design or landscape architecture. Their common characteristic consists of considering human and physical environments simultaneously. Imageability (Lynch 1960), place attachment (Fried 1963), proxemics (Hall 1966), socio-spatial schema (Lee 1968), behavior settings (Barker 1968), personal space (Sommer 1969), crowding (Stokols 1972), privacy and territoriality (Altman 1975), place identity (Proshansky 1978), affordances (Gibson 1977), appropriation (Korosec-Serfaty 1976), and environmental stress (Evans 1984) are a few examples of these long-standing concepts.[6]

Moreover, other theories, concepts and research areas from psychology, almost forgotten today,[7] are still extremely useful in explaining contemporary transactions between people and their environment, some of them finding their way back into environmental design after having been repackaged and reshaped for and by environmental considerations. For instance, the Gestalt, that was so influential in the design experiments of the Bauhaus, was comprehensively applied to the language of architecture by Hesselgren (1969)[8] and, more recently, served as the main foundation for an introductory textbook on architecture by von Meiss (1986). Today, it is rarely discussed in environmental psychology, although it explains perceptual P-E transactions operating below the level of awareness. As a counterpoint, the information theory of perception as applied to environmental aesthetics by Berlyne (1960) and Kaplan and Kaplan (1982) is significant for students in architecture, all the more so through the pattern language for landscape by Kaplan et al. (1998). In other areas, we could mention Barker and Gump's manning theory (1964), an excellent guide for the architecture and renovation of large-scale institutions such as secondary schools and hospitals.

For these concepts to be usable as decision aids in the design process of a given building or place, they must be reorganized into overarching concepts and methods. It is not surprising that this type of work is generally done by professors of design themselves.

At the architectural scale, *home* is a striking example of a meta-concept that has made its way into design, teaching and research.[9] Brought to the attention of designers by the landscape architect Cooper (1974), it integrates many powerful P-E concepts such as privacy, territoriality, control, personalization, appropriation and identity, allowing designers to think of housing beyond bricks and mortar or strictly in terms of functional and aesthetic considerations. Serfaty has also devoted her career to the multiple facets of home as experience (see 2003, for an example). Housing As If People Mattered (1988), later published by Cooper and Sarkissian, was conceived as a set of design guidelines. It made these concepts even more accessible to designers, presenting them in a logical sequence to accompany the design process. In Housing, Dwellings and Homes: Design Theory, Research and Practice (1991), the architect Lawrence moved the field of P-E relations research a huge step forward, with a holistic approach in which housing (economics and physical commodities), dwelling processes (design, construction and then the ongoing personalization of residential environments) and

---

[6] Our listing of these concepts and their authorship is inevitably biased by our North American context of research and teaching. We wish to apologize to subsequent generations of researchers who have developed variations of these concepts and whose names are not listed. Finally, several P-E concepts originating from environmental anthropology, sociology, geography, and urban history, also very useful in teaching design, have been left aside to focus on the contribution of environmental psychology.

[7] Pol (2007) did a great service to our memory, by situating the origins of environmental psychology in the first decades of the twentieth century.

[8] Hesselgren developed his approach in a PhD thesis much earlier, but it was only made available in English in 1969, and in a shorter version in 1977.

[9] As a testimony to the interest in the concept, it is worth mentioning the new interdisciplinary journal, *Home Cultures: The Journal of Architecture, Design and Domestic Space*, created in 2003.

homes (meaning and usage) are discussed, along with a historical perspective. This author thereby fully integrates the production of the environment in specific societal and historical contexts in our conceptualization of P-E transactions. This type of conceptual integration, social contextualization and exploration of environmental types is making an important contribution not only to our understanding but also to more responsive design.

Of the many examples available, we can mention the evolution and enrichment of the production of institutional environments from environmental psychologists Rivlin and Wolfe's pioneer work on institutional environments and their impact on children's lives (1985) to Robinson's major work on the differences between institutions and homes as dwelling places (2008, 2015), as well as the seminal work by White on the social life of small urban places (1980) and by Gehl on the spaces between buildings (1971), integrating many fundamental concepts already mentioned and led by the objective of translating empirical findings into directions for more human and sociable public spaces.

This current section illustrates how rich and fruitful past associations between environmental psychologists and designers have been. The following gray insert provides insights into our personal experiences of teaching some of these concepts to future architects and urban designers at Laval University (Quebec, Canada).

> **Teaching Design with People in Mind at Laval University (Quebec, Canada): 1. Understanding People-Environment Relations**
>
> The first step in teaching human aspects of design is to make the students aware of the indivisible transactions between people and everyday settings, as well as of the basic concepts explaining the modalities of these interrelations, along with a capacity to operationalize these in built environment analyses.
>
> At the undergraduate level, the mandatory course *Human Aspects of Architecture*, offered to students in their second year, plays this role. It outlines a diversity of approaches used to understand people-environment relations with a focus on both quality of life and quality of the built environment. It covers a variety of concepts associated with environmental perception, cognition and meaning, spatial behaviors and social life with regards to the quality of housing, work, institutional, commercial, leisure and natural environments. Practical assignments enable the students to experience the needs and feelings of different user groups. Students are required to observe and document people's behavior in given settings, for instance: (1) living in a wheelchair for a full weekday; (2) prolonged observation of users' behaviors at a bus stop, during Quebec City's harsh and snowy winter; (3) defining space use patterns in public libraries. Until 2001, the course *Housing: Forms, Uses and Regulations* was offered to students in their third and fourth year of training, and was mandatory for those taking the advanced housing option. The course material covered knowledge about the morphological properties of housing, the uses and meanings of these spaces, as well as the design methods at the architectural and urban scales. Practical assignments consisted of evaluating the quality of various housing complexes in the city of Quebec, analyzing floor plans, regulations and other related documents, visiting the building itself, and interviewing residents.
>
> When the degree of Master of Architecture was required to have access to the profession across Canada, this course was replaced by a graduate seminar *Urban Form and Cultural Practices*, mandatory for all students who wanted to include ur-

(continued)

ban design in their curricula.[10] The content covers key concepts in people-environment relations, including several inherited from environmental psychology but also urban sociology, anthropology and geography, as well as architecture and planning: imageability, wayfinding, affordances, collective memory, settlement-identity, perceived density and crowding, and housing behaviors (preferences, aspirations, satisfaction, meaning and experience). The pedagogical approach favors a weekly discussion of key writings and recent studies covering each of these concepts, one per week, along with two practical assignments to be carried out during the semester. The course is structured so that students can hierarchize among a list of P-E concepts those most relevant to a given urban design problem, interpret cutting-edge research or locate experts, as well as rapidly access complementary local knowledge, all of which inform the design decision.

Each assignment—designed for teams doing the equivalent of a 2-week mandate for a professional in a private firm—has proved successful in teaching students to be efficient in helping orient a design decision with people in mind, where to locate an urban function, how to minimize the perceived density of a housing development, how to make its new housing forms socially acceptable, how to induce social cohesion, etc. They develop their abilities to document important human aspects of an urban design project using different means of data collection, namely videos, structured observations and behavioral mapping, interviews, focus groups, and on-line surveys. An example of a half-term assignment is one aimed at identifying avenues to strengthen the identity of Lac-Beauport, a municipality and ski resort located about 25 km from downtown Quebec City. Students interviewed residents as well as visitors, asking them to draw cognitive maps. Interestingly, the lake turned out to be missing from several drawings. An objective urban analysis further showed that all accesses to the lake, except for one belonging to the parish, were private and reserved for lakeside homeowners. One of the recommendations was the acquisition by the city of a right of way to assure public access to the lake. Another assignment consisted of learning about people's mobility with regard to food supply behaviors. Data were collected through an on-line survey about the various shops frequented, the frequency with which they were visited, with whom and using which transport means, as well as the social behaviors and meanings attached to the different places. As a last example, an assignment asked students to compare the degree of fitness between people's residential location in the Quebec metro area, their childhood home location, and their housing aspirations, with regard to their social-spatial representations of the city, suburb and country. In all of these exercises, the students had to formulate recommendations to inform urban design.

4.3 How Designers Work and Use People-Environment Knowledge

This section discusses the nature of design, the approaches used by designers to make their work more responsive to people, and how we teach and train future designers for this purpose. If

---

[10] Knowledge about the morphological properties of built environments became part of an advanced seminar on *Morphology and Syntax of the Built Environments*, and about design methods, part of another seminar on *Urban Design: Concepts and Methods*.

Fig. 4.2 Modes of reasoning involved in solving design problems (Inspired by March 1976)

we want to see how environmental psychologists could be trained to work with designers, to conduct research with design purposes in mind, and to report on research in ways that can be used to inform design, it is clearly important to cover this topic. Environmental psychologists and designers must understand each other, and be made aware of their respective contributions and working processes.

4.3.1 On the Nature of Design

Simon (1969), Gutman (1972), and Rittel and Webber (1973) were the first generation of academics to point out the discrepancy between research problems and design problems (the design of both artifacts and public policies). In The Sciences of the Artificial (1969), Simon defined design as an endeavor aimed at producing an artifact, a goal-oriented and context-specific functional object. This "amazing and tangled process of human mind activities, aiming at developing drawings and projects to finally produce a contingent object" (Le Moigne 1991a:125, authors' translation) has been the subject of many studies and theories in architecture, but it is the modes of reasoning that will be particularly enlightening for the purpose of the present discussion.

Scientists learn to see the world through the hypothetico-deductive method and are thus concerned with *what the reality must be* (deduction) and by *what the reality is* (induction). However, as March (1976) showed, designers operate with a third mode of reasoning, *abduction*, (Fig. 4.2), aimed at *what the reality may be*.[11] This mode of reasoning is especially useful to address the complexity and uncertainty of reality as an open system on which Le Moigne insists (1991b).[12] This is the case for design where even the smallest projects require making connections between a variety of factors, which calls for different types of knowledge and actors, and where there is no good definitive answer.[13] To inform the decision process in the making of an artifact, designers have access to various types of knowledge. They must be trained to work with different types of experts, to retrieve useful information from various sources, including about people-environment relations, and to make sense of it. Design is thus a multidisciplinary and action-oriented process, capable of a constructive dia-

---

[11] For discussions on the nature of design, see *The Reflective Practitioner* by Schön (1982).

[12] Closer to P-E studies, see the discussion of this mode of reasoning by the Swedish architect Johansson (2010).

[13] See *How designers think* (2005, 4th edition) and *What designers know* (2004) by Lawson.

logue with other domains of knowledge, including the natural and social/human sciences (Després et al. 2011). It may very well be useful to environmental psychology also in its applied mission to contribute to solving socio-environmental problems.

4.3.2 Methods and Processes to Design with People in Mind

Following what could be named the golden years of People-Environment research, a few researchers put together an approach to address behavioral issues in design in more systematic ways. Some intended to inform the design process with existing or original P-E knowledge (post-occupancy evaluation and evidence-based design), some made sure the users concerned were involved in the process (participatory design and collaborative planning) while others aimed at accompanying architectural or urban programming and design. Although not exhaustive and mutually exclusive, these 'methods and processes' will be briefly discussed to illustrate the 'designerly way of knowing'[14] toward more humane environments.

4.3.2.1 Building Evaluation

It is interesting to begin with the architect Zeisel (1984), who was probably the first to approach design as a mode of inquiry in environment-behavior research and to propose a toolbox entirely related to the design process.[15] He presents each stage of the complete design cycle – that is to say programming, designing, building, use adaptation and evaluation – as being occasions for collaboration between design and research. Moreover, he calls for studies that would embrace the whole process from the intentions of a project to its post-occupancy evaluation, with the results being fed back into another cycle for a new project. By definition, research as well as design questions then stem from behavior settings, building types or urban components in their environmental characteristics as well as their behavioral ones. They are thus methodologically challenging, requiring interdisciplinary collaboration as well as tools and concepts facilitating a common understanding. To this end, Zeisel's toolbox proposes using several sources of data: annotated plans, second-hand data, observation of physical traces and behaviors, focused interviews and questionnaires.

Examples of studies using these methods can be found in the analyses of public spaces mentioned earlier, to which could be added the research tradition of Post-Occupancy Evaluation (POE) (e.g. Preiser et al. 1988, 1991; Vischer 1995). An interesting instance accumulating knowledge from this type of study is the Center for Health Design, which aims to develop a body of evidence linking the physical environment with safety and quality outcomes for patients and staff in order to design better hospitals (Zimring and Bosch 2008). The emergence of environmental gerontology to study how to adapt the environment to old age is another growing field of research. Nevertheless, much remains to be done, considering the variety of building types and their diverse expressions according to cultural and social contexts. There is certainly food for thought in environmental psychology as to how to best prepare future professionals for this type of work.

4.3.2.2 Collaborative Design

Scientific examination of the transaction between people and their environment by environmental types at different stages of the design process is one potent source of information for designing the human environment, but this is insufficient. Stakeholders have views and knowledge about how the environment should be shaped and a right to influence the direction it will take in the future.[16] Designers and researchers must vali-

---

[14] This expression is borrowed from Cross (2006).

[15] We believe his book would make a fine contribution to the training of environmental psychologists.

[16] The contribution of local stakeholders to collaborative research has been prompted by pressure from user groups, research on urban and environmental activism, peace and conflict research, international cooperation, and women's studies (see Elzinga 2008).

date their hypotheses against such knowledge and views. Participatory Design and Collaborative Planning are two design approaches that call for the inclusion of stakeholders in all phases of the design process. Unfortunately, scientific publications on participation are largely dominated by descriptive case study reports. Although a fair number of handbooks have been published in the last 10 years, their goal is mainly to provide toolkits. Evaluations of participation methods are often limited to ad hoc suggestions and criticisms about the advantages and disadvantages of various techniques. Participation theory is underdeveloped and in need of a clear framework to integrate its resulting forms of knowledge.

This being said, the work of several planners has brought participation to new analytical levels. Considering the public dimension of large-scale environments, it is not surprising to observe a deliberative or collaborative trend in planning studies. In their respective book, The Deliberative Practitioner (1999) and Collaborative Planning (2005), Forester and Healy made significant contributions to help schools of planning and urban design revise their curricula. The collaborative approaches they propose actively involve stakeholders as legitimate decision-makers in the planning process, the ultimate goal being to reach a consensus or at least an acceptable compromise.[17]

At the building scale, although participation is less theorized, many architects have been working closely with inhabitants, users and stakeholders since the early 1960s. Sanoff has experimented with and published on participation over the past 40 years (see, for instance, 2000). Peña and Parshall (2001) developed many participatory methods for the programming phase of design. Blundell Jones et al. (2005) also edited a reader on a variety of case studies. Nevertheless, apart from these rare exceptions, the literature on P-E relations only pays lip service to participation: it is often mentioned, but very little studied, and very few people are trained to develop and apply this approach.

4.3.2.3 Evidence-Based Design

More recently, evidence-based design approaches have developed to help interior designers, architects and facility managers make better decisions in the planning, design and construction of specialized types of building (Brandt et al. 2010; Hamilton and Walkins 2008; Kopec et al. 2012). Relatively new, these are inspired by evidence-based medicine, using credible proof to support design. So far, they have mostly been used in healthcare design to improve patient and staff wellbeing, patient healing, stress reduction and safety, with evidence collected from building evaluations and systematic literature reviews, but their scope is growing.[18] To operate in such a way, designers must be able to access credible evidence from the scientific literature, experts and previous analyses. This evidence must somehow be translated into either qualitative objectives to be reached, quantifiable performance criteria to be respected, or design concepts to be explored. This requires a great deal of knowledge translation, which some designers must be trained to do, as well as certain ways of reporting research to make it accessible/understandable.

4.3.2.4 Architectural and Urban Programming and Design Methods

Beyond understanding concepts of People-Environment Relations and acquiring an operational capacity to document them, the next step is for designers to integrate this knowledge systematically in various forms in the design process. Some architects have developed systemic approaches to programming and design as the outcome of a fine understanding of the various phases involved in the design process and their iterative sequence.

With regard to urban design, Responsive Environments: A Manual for Designers (1985)

---

[17] See also Innes and Booher (2010).

[18] The Center for Health Design focuses on EBD practices, their uses and application to each step of the healthcare design process. More than 600 studies with environmental-design relevance have been identified. See also Verderber (2005).

Fig. 4.3 The seven keys to urban design quality (Bentley et al. 1985:9)

by Bentley and his colleagues from Oxford Brookes University remains to this day, on the basis of our experience, the best textbook ever written to teach urban design with people in mind. The method, which Bentley calls a "Practical tool for creative use," is organized in a concise user-friendly sequence of seven concepts, each of which relates to people-environment relations and is supported by a strong body of knowledge: permeability, variety, legibility, visual appropriateness, robustness, richness and personalization (Fig. 4.3). Their sequential use is organized in order to approach the design problem from the larger to the smaller environmental scale.

At the architectural scale, Duerk has managed the same feat with her textbook Architectural Programming: Information Management for Design (1993). Figure 4.4 illustrates the systemic approach she developed to integrate the wide range of data necessary to inform the design process and, most importantly, to make sense of them as the decision-making process advances. To collect original behavioral information, she proposes data collection methods borrowed from post-occupancy evaluation (POE), namely prolonged sessions of observations with behavioral mapping, along with interviews with staff and users. The following gray insert illustrates how Duerk's approach to architectural programming is taught to architects at Laval University in Quebec, Canada.

Teaching Design with People in Mind at Laval University (Quebec, Canada): 2. Using People-Environment Relations in Design

At Laval University, a specialized training course of 360 h is offered to graduate students in "Urban Design" and "Programming and Design". Each consists of a design studio for two thirds and a concomitant theory seminar for the other third of the time. These two one-semester specialties bring the application of P-E theories, concepts and methods to a higher level. Here, students are taught systemic design methods addressing behavioral issues, which include Post-occupancy Evaluation, Evidence-Based Design, and Collaborative Design.

(continued)

Programming and Design has been taught at Laval University since the early 1980s (see Després and Piché 2011). In the 15 consecutive editions of our graduate Programming and Design studio so far, we have used Duerk's textbook to guide our teaching. Ten editions have been dedicated to hospital environments—pediatrics, nephrology, obstetrics, psychiatrics, acute care, operating room, ambulatory clinics—, three to school environments, and two others to elderly housing.

In the context of a real negotiated and financed collaboration with an institutional milieu, students are taught collaborative methods to work with local experts, namely facility managers and technical resources, on-site professionals, and also private and public architects involved in the evaluation, design or planning of such facilities. They are asked to analyze architectural precedents from the point of view of P-E transactions, to search databases for scientific evidence about relevant P-E concepts and to document user behaviors through prolonged observation sessions, interviews, focus groups and/or on-line surveys, according to time and legal possibilities and restrictions. Practical exercises of translation into performance criteria and design concepts follow, in direct relationship with the building concerned. Figure 4.5 illustrates the three components of the Programming and Design elective specialty.

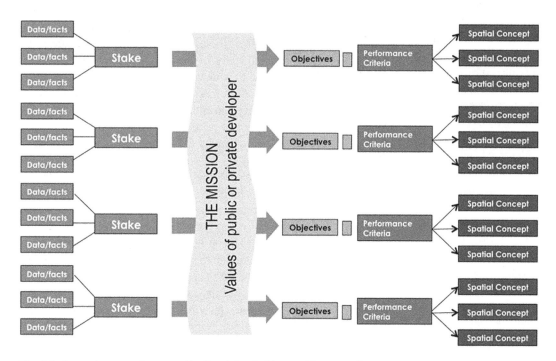

Fig. 4.4 Systemic approach proposed by Duerk for Architectural Programming

Fig. 4.5 Components of the programming and design studio (Laval University, Quebec, Canada)

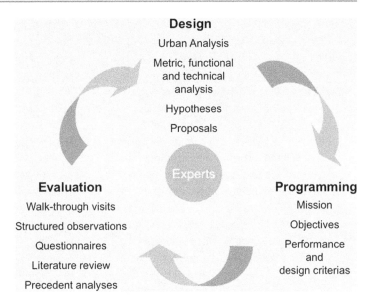

4.4 When and Why Did We Lose Track of Our Common Mission?

Although the complexity of societal problems is far better understood today than it was 40 years ago, and a considerable amount of P-E research is available, and operational collaborative design methods have been experimented with, the gap between environmental psychologists and designers seems to be widening. Why is this happening?

One hypothesis is that the growing and widespread interest in sustainability has led environmental psychologists to focus more on self-reported behaviors and attitudes about vast territories, with a loss of interest in specific everyday settings (Pol 2007). In his discussion of environmental psychology, Günther (2009) makes the same observation: "…EP increasingly deals with 'global' issues such as pro-environmental behavior, climate conservation and sustainability" (2009: 363). Giuliani and Scopelliti's (2009) examination of studies published in Environment & Behavior (E&B) and the Journal of Environmental Psychology (JEP) from their foundation to 2005 confirms a shift toward green psychology.[19]

In addition, scientific evidence emanating from this research is often reported in ways that makes it difficult to use to inform design. As the environmental psychologist Noschis (2015) puts it: "for the architect, designer and urban planner, these contributions do not address the complexity of their projects including the user who they deal with during the design process" (p. 21). A large number of recent doctoral theses in environmental psychology consist of extensive paper and pencil surveys, with little concern for the materiality of the built environment (Després 2005).

Part of this general trend may also be attributed to the interest in predicting behavioral research in the natural and health sciences, where researchers have recently embarked on people-environment research with regard to sustainability and public health. Although this recent interest should be applauded, we must also be wary of the hard science paradigms they come from and against which the scientificity of research is evaluated. For example, the systemic or meta-reviews conducted are looking for generalizable scientific

---

[19] For an alternative critical position on the definition of *environment* in environmental psychology, see Depeau and Ramadier (2014).

evidence from research designs assessed against quality criteria borrowed from medical or natural sciences research. Most often, only studies published in high-impact factor journals and based on random samples, with controlled groups and randomized trials, are retained. The chances are that findings reported in these reviews also refer to studies based on quantitative analyses derived from large surveys, census track or epidemiological studies, which are difficult to use to inform context-specific design problems. Attention must be paid to this hazardous reduction of the complexity of P-E relations to a limited number of quantifiable variables, of which the built environment is a small part, if any.

Our long-lasting experience teaching P-E relations has shown us that generalizable evidence cannot always be translated into performance or design criteria, and is sometimes hardly usable to shed light on specific design problems. In contrast, converging evidence from qualitative research and case studies conducted in comparable political, economic and cultural contexts is often useful to help design decisions. Yet, a considerable part of these research results will mostly be published in book chapters or *gray literature* such as research reports, and will often be left out of systemic literature reviews.

It should be remembered that many long-lasting and useful P-E concepts emerged from qualitative research with limited sample sizes. The credibility of these concepts developed over the years not because of the generalizable character of the original findings but because the results were replicated in different cultural contexts and with different research designs, refining them and bringing them to a higher level. Triangulation has made these concepts more resilient over the years. On this issue, Lincoln and Guba (1985), and more recently Cohen and Crabtree (2008), have prepared the ground for more productive and respectful collaborations between researchers engaged in *hard* or *soft* sciences, proposing that different sets of scientific quality criteria be adopted to evaluate their respective research (Fig. 4.6). Both subgroups of researchers have to learn to understand each other, and be able to identify good scientific research from whatever paradigms it originates.

It is extremely important to train the next generation of P-E researchers to be able to locate and evaluate quality research, whatever the paradigm, and to be able to work in a productive manner with experts from all research fields. Research handbooks published in the last decade or so advocate the recognition of both well-designed quantitative and quantitative research designs, the valuing of mixed methodologies and triangulation, and rigorous case studies (Creswell 2003; Groat and Wang 2002; Yin 2003).

It is also time for P-E academics to teach new generations of professionals to conduct research and implement knowledge transfer in ways that are useful to inform design (Kent and Thompson 2014). Despite worldwide academic discourse on the importance of interdisciplinary and intersectoral research collaborations, there is still a strong tendency for university professors to

Fig. 4.6 Scientificity criteria for *Hard* and *Soft* sciences (Proposed by Lincoln and Guba 1985)

Scientificity Criteria	Experimental & Quasi-Experimental Research	Correlative & Qualitative Research
1. Truth	Internal Validity	Credibility
2. Applicability	External Validity	Transferability
3. Consistency	Reliability	Dependability
4. Neutrality	Objectivity	Confirmability

work in disciplinary silos, and candidates with multidisciplinary profiles still have a hard time getting hired or published.

A second hypothesis is that a significant proportion of future designers around the world are currently trained within the studio culture, where students are grouped in small 'families' and work under the supervision of one professor. This apprenticeship system is a partial legacy of the French Beaux-Arts school, where students organized in *ateliers* (in French) followed the rules imposed by a renowned practicing architect, (Jann 2010; Stevens 2010).[20] Even though interdisciplinary knowledge provided through theory classes now accounts for more than half the time dedicated to teaching (at Laval University, it represents 60 % of the credits), the studio remains the most valued form of learning in architecture schools.

Moreover, despite the fact that valuable and successful teaching alternatives have developed worldwide outside of this mold, namely various attempts to link design to research and/or to community services, recent directions given to design-related programs suggest that the importance of teaching the human aspects of architectural and urban forms is losing strength, with the focus being put on green buildings, construction-related topics and computer-generated design.

This conjuncture might have contributed to widening the gap between people-environment research and design at a time when the multiple challenges of adapting the built environment to induce more sustainable behaviors and healthier lifestyles must be met. Professors teaching P-E relations and design need to combine their efforts to engage their students in a more productive dialogue and, for this to happen, they have to train them differently.[21] Indeed, for P-E studies to contribute to solving societal problems, academic reforms must take place regarding how research is taught, conducted and reported, while problem-seeking and problem-solving must be taught to designers.

4.5 New Avenues for Linking Design and Research

To link research and design better, we propose three modus operandi that should guide our teaching and research strategies.

The first is to endorse the transdisciplinary paradigm (for more, see the special issue of *Futures* on transdisciplinarity, Lawrence and Després 2004). Beyond *multidisciplinarity*, which looks at a problem from different disciplinary angles, or *interdisciplinarity*, which focuses on disciplinary intersections in terms of theory, methods and concepts, *transdisciplinarity* looks for common ground but also respects what is unique to each discipline, even if this means considering contradictory knowledge, as long as it is helpful to understand a problem and identify solutions. It also gives priority to solving contextual problems over developing theory.

This lack of connection between research and design encouraged one of us to create, with other colleagues in 2000, the Interdisciplinary Research Group on Suburbs (*Groupe interdisciplinaire de recherche sur les banlieues*, in French, or GIRBa). Based at the School of Architecture at Laval University in Quebec City, the group is made up of ten professors in architecture, urban design and planning, sociology and anthropology from three different universities, with additional collaborators in Quebec, France, Switzerland and Italy. The group welcomes annually about 30 graduate students conducting relevant research projects in the context of their Master of Sciences or their PhD, as well as about 50 students from the Masters of Architecture and of Urban Design programs involved in design research. GIRBa's program of research, design and action aims to identify sustainable solutions with regard to Quebec City's sprawling metropolitan territory and low-density occupation (for discussions of

---

[20]In France, schools of architecture are not necessarily part of universities but attached to the Ministry of Culture. This disciplinary orientation makes it difficult for these professors to build bridges between design and other disciplines.

[21]It is especially challenging in North American schools of architecture where professors are almost systematically trained in design. This is the result of professional associations or accreditation boards that dictate the content of architecture education being centered on studios.

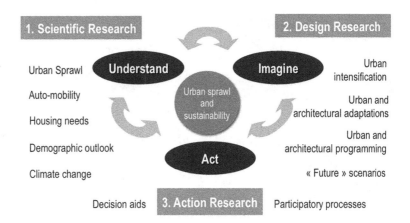

Fig. 4.7 A transdisciplinary program of research and action (GIRBa, Laval University, Quebec Canada)

our research and teaching methods, see Després et al. 2011, 2012b) (Fig. 4.7). By having students in architecture, planning and social sciences conducting research on P-E relations interact with designers around complex and contextual problems, we want to ensure that these future professionals learn to exchange different forms of disciplinary knowledge and concerns in useful ways, to develop a shared understanding, and to identify creative solutions.

A second modus operandi consists of improving the abilities of P-E researchers and designers to operate research translation so that they can build bridges between research and design. There is an enormous quantity of useful research knowledge available and powerful search engines to access it. At Laval University, we share the responsibility of training students from the Masters programs in Architecture and in Urban Design, and also from research-oriented Masters and PhD programs, to become knowledge brokers. It is not only a matter of developing their abilities to identify empirical evidence but of translating it into accessible forms to help make design decisions. Figure 4.8 illustrates the process used for this purpose in a graduate studio in Programming and Design in the context of designing elderly housing facilities.

The last modus operandi consists of teaching P-E researchers and design students to work in collaborative ways with stakeholders, and to organize such participatory processes. The enduring design studio, to which our students dedicate 18 h of their work time weekly, represents a unique opportunity to initiate these pedagogical changes. For example, the studio can be linked to specific agendas of research groups, such as we have been doing for several years at GIRBa, or to a special expertise request from a local community or organization.

Among the actors with whom students must learn to work, there will be experts and scientists but also spokespeople for other types of knowledge. Our introspection on design teaching and research within GIRBa led us to develop our own model of design collaboration to tackle complex urban problems related to urban sprawl and sustainability. Inspired by Habermas's Theory of Communicative Action (1985), it proposes that bearers of four types of rationality and knowledge must be brought together: (1) *scientific rationality and knowledge* or what is generally held as "what is true" (usually the result of empirical research); (2) *instrumental rationality and knowledge*, which refers to practicality or "what is possible", the knowledge of how to go about things; (3) *ethical rationality and knowledge* or "what is good", which is linked to customs, beliefs, values and past experiences that help people to determine what is wrong and what is right about a specific issue; (4) *aesthetic rationality and knowledge*, or "what is beautiful", which comprises images and refers to aesthetic judgment and experience, as well as tastes, preferences and feelings about the built environment (Fig. 4.9 illustrates this model). By bringing together stakeholders of these four types of rationality and knowledge in a face-to-face interaction, a fifth type gradually

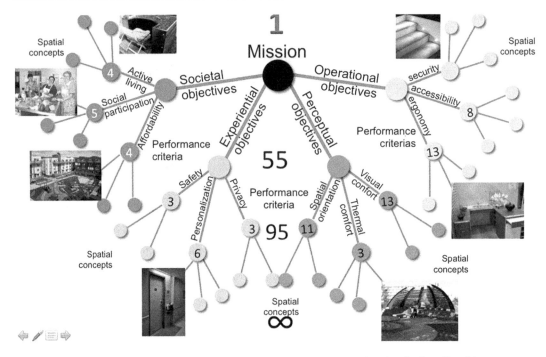

Fig. 4.8 Systemic approach to programming housing for the elderly (Laval University, Quebec, Canada)

Fig. 4.9 Model of collaborative design research (GIRBa, Laval University, Quebec, Canada)

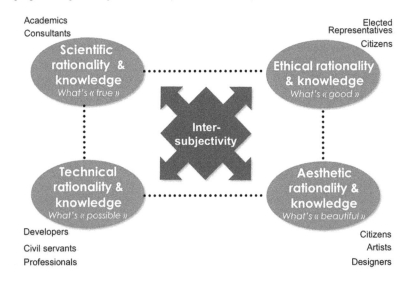

emerges that is more than the sum of the four others since inconsistencies in thought and arguments are revealed and collectively overcome. By developing a unique type of "prospective" knowledge through collaborative design, they can also identify issues that might otherwise have been overlooked.

4.6 Conclusion/Challenges

To link research and design, we have the collective mission to inflect the way professionals are trained within universities or design schools. For this, we must first find ways to get involved in teaching P-E relations in professional curricula of all horizons (health sciences, natural sciences, social and human sciences, arts and humanities). Indeed, if the fact that human behaviors do not exist outside built or natural settlements is recognized in research, the contributions of their material properties is not always well understood. Second, at the graduate level, we should train students to not only conduct innovative research on people-environment relations, but also, to become knowledge translators, capable of interpreting empirical evidences in accessible and useful ways for designers. Integrating qualitative research will be a challenge, especially with health and natural scientists involved in research on current societal problems. By doing so, the next generations P-E researchers could play a more active role in helping society meet urgent and complex environmental, social and economic challenges. Third, designers should be welcome as legitimate contributors within multidisciplinary research teams and programs, as they can shed light on what solutions are possible. In schools of architecture, this might imply resisting the tendency to focus on the materiality of buildings and their fabrication and neglect the uses and users, with the development of digital or green technologies, or else, on architecture as an art that does not need to be fed by sciences. Finally, the scarcity of funds for the financing of research and of universities on all continents calls for imaginative solutions to put transdisciplinarity into practice. Indeed, the operationalization of such an approach to research, within large intersectoral teams, takes time and energy, and thus money. In time of economic austerity, it might also be tempting to fall back on more conventional research.

Our reflection to identify what is missing for linking P-E research and design, as well as our proposals to bridge the gap between the two, may look like déjà vu for several of our P-E colleagues who are operating with the same goals in mind. Our perspective might also sound ethnocentric; we certainly do not have an understanding of the way design is taught or how P-E researchers are trained in all parts of the world. Our aim is to open up the debate and welcome positions challenging our critical look at P-E and design cultures, and our institutional positions and individual experiences on avenues to bridge the gap between research and design. As professors teaching in a school of architecture, we would also like to hear from P-E researchers in others disciplines about the academic contexts in which they operate. Maybe it is time to share good and bad practices, our successes and failures at keeping the relations alive between People-Environment researchers and designers.

References

Ahrentzen, S., Després, C., & Schermer, B. (Eds.). (2012). *Building bridges, blurring boundaries. The Milwaukee school in environment-behavior studies*. Montreal: SARUP, University of Wisconsin-Milwaukee/VRM. http://dc.uwm.edu/sarup_facbooks/1. Accessed 5 May 2015.

Alexander, C., Ishikawa, S., & Silverstein, M. (1977). *A pattern language: Towns, buildings, construction*. New York: Oxford University Press.

Altman, I. (1975). *The environment and social behavior: Privacy, personal space, territory, crowding*. Monterey: Brooks/Cole.

Barker, R. (1968). *Ecological psychology. Concepts and methods for studying the environment of human behavior*. Stanford: Stanford University Press.

Barker, R., & Gump, P. (1964). *Big school, small school: High school size and student behavior*. Stanford: Stanford University Press.

Bentley, I., Alcock, A., Murrain, P., McGlynn, S., & Smith, G. (1985). *Responsive environments. A manual for designers*. Oxford: Architectural Press.

Berlyne, D. (1960). *Conflict, arousal and curiosity*. New York: McGraw-Hill.

Blundell Jones, P., Petrescu, D., & Till, J. (2005). *Architecture and participation*. London: Routledge.

Brandt, R., Chong, G. H., & Martin, W. M. (2010). *Design informed: Driving innovation with evidence-based design*. New York: Wiley.

Canter, D. (1974). *Psychology for architects*. New York: Wiley.

Cohen, D. J., & Crabtree, B. F. (2008). Evaluative criteria for qualitative research in health care: Controversies and recommendations. *Annals of Family Medicine*, 6(4), 331–339. doi:10.1370/afm.818.

Conzen, M. R. G. (1960). *Alnwick, Northumberland: A study in town-plan analysis* (Transactions and papers, no.27). London: Institute of British Geographers. doi:10.2307/621094.

Cooper, C. (1974). The house as a symbol of the self. In J. Lang, C. Burnette, W. Moleski, & D. Vachon (Eds.), *Designing for human behavior* (pp. 130–146). Stroudsburg: Dowden, Hutchinson & Ross.

Cooper, C., & Sarkissian, W. (1988). *Housing as if people mattered. Site design factors in the design for medium-density family housing*. Berkeley: University of California Press.

Creswell, J. W. (2003). *Research design. Qualitative, quantitative and mixed methods approaches* (2nd ed.). Thousand Oaks: Sage.

Cross, N. (2006). *Designerly ways of knowing*. Berlin: Springer.

Depeau, S., & Ramadier, T. (2014). Une psychologie environnementale critique est-elle possible? In D. Marchand, S. Depeau, & K. Weiss (Eds.), *L'individu au risque de l'environnement. Regards croisés sur la psychologie environnementale* (pp. 113–143). Paris: In press.

Després, C. (2005, Fall). Understanding complexity in people-environment research: Theoretical considerations. Losing sight of complexity in people-environment research? In R. Mira (Ed.), *Bulletin of people-environment studies*. http://www.udc.es/dep/ps/grupo/bulletin/bulletin27.pdf. Accessed 20 June 2015.

Després, C., & Piché, D. (2011). Programmer pour mieux innover. *ARQ: Architecture Québec, 155*, 14–17.

Després, C., Vachon, G., & Fortin, A. (2011). Implementing transdisciplinarity: Architecture and urban planning at work. In I. Doucet & N. Janssens (Eds.), *Transdisciplinary knowledge production in architecture and urbanism. Towards hybrid modes of inquiry* (pp. 33–49). Dordrecht: Springer.

Després, C., Ahrentzen, S., & Schermer, B. (2012a). Genealogy of the Milwaukee school. In S. Ahrentzen, C. Després, & B. Schermer (Eds.), *Building bridges, blurring boundaries. The Milwaukee school in environment-behavior studies* (pp. 261–275). Montreal: SARUP, University of Wisconsin-Milwaukee/VRM. http://dc.uwm.edu/sarup_facbooks/1.

Després, C., Vachon, G., & Fortin, A. (2012b). Requalifying aging suburbs to counter urban sprawl: The contribution of GIRBa to cultural sustainability. In R. J. Lawrence, H. Turgut Yildiz, & P. Kellett (Eds.), *Requalifying the built environment. Challenges and responses* (pp. 135–160). Göttingen: Hogrefe and Huber.

Duerk, D. P. (1993). *Architectural programming. Information management for design*. New York: Wiley.

Elzinga, A. (2008). Participation. In G. Hirsch Hadorn et al. (Eds.), *Handbook of transdisciplinarity* (pp. 345–359). Dordrecht: Springer.

Evans, G. W. (Ed.). (1984). *Environmental stress*. Cambridge: Cambridge University Press.

Forester, J. (1999). *The deliberative practitioner*. Cambridge: MIT Press.

Fried, M. (1963). Grieving for a lost home. Psychological cost of relocation. In J. Q. Wilson (Ed.), *Urban renewal. The record and controversy* (pp. 359–379). Cambridge: MIT Press.

Gehl, J. (1971). *Life between buildings. Using public space* (J. Koch, Trans.). New York: Van Nostrand Reinhold, 1987.

Gibson, J. J. (1977). The theory of affordances. In R. Shaw & J. Bransford (Eds.), *Perceiving, acting, and knowing. Toward an ecological psychology* (pp. 127–143). Hoboken: Wiley.

Gifford, R. (1996). *Environmental psychology. Principles and practices* (2nd ed.). Boston: Allyn & Bacon.

Gifford, R. (2014). Environmental psychology matters. *Annual Review of Psychology, 65*, 541–579. doi:10.1146/annurev-psych-010213-115048.

Giuliani, M. V., & Scopelliti, M. (2009). Empirical research in environmental psychology. Past, present, and future. *Journal of Environmental Psychology, 29*(3), 375–386. doi:10.1016/j.jenvp.2008.11.008.

Groat, L., & Wang, D. (2002). *Architectural research methods*. New York: Wiley.

Gunther, H. (2009). The environmental psychology of research. *Journal of Environmental Psychology, 29*(3), 358–365. doi:10.1016/j.jenvp.2009.02.004.

Gutman, R. (1972). *People and buildings*. New York: Basic Books.

Habermas, J. (1985). *The theory of communicative action*, vol. 1–2 (T. A. McCarthy, Trans.). Boston: Beacon Press.

Hall, E. T. (1966). *The hidden dimension*. Garden City: Doubleday.

Hamilton, D. K., & Watkins, D. H. (2008). *Evidence-based design for multiple building types*. New York: Wiley.

Healy, P. (2005). *Collaborative planning. Shaping places in fragmented societies* (2nd ed.). New York: Palgrave Macmillan.

Hesselgren, S. (1969). *The language of architecture*. Lund: Studentlitteratur.

Hesselgren, S. (1977). *Man's perception of man-made environments. An architectural theory*. New York: Wiley.

Innes, J. E., & Booher, D. E. (2010). *Planning with complexity. An introduction to collaborative rationality for public policy*. New York: Routledge.

Ittelson, W. H., Proshansky, H. M., Rivlin, L. G., & Winkel, G. H. (1974). *An Introduction to environmental psychology*. New York: Holt, Rinehart and Winston.

Jann, M. (2010). Revamping architectural education. Ethics, social service, and innovation. *International Journal of Arts and Sciences, 3*(8), 45–89.

Johansson, R. (2010). How to handle complexity in the design process. Paper presented at the Nordic Association of Architectural Research Annual Seminar, Tampere University of Technology, Tampere, Finland.

Kaplan, R., & Kaplan, S. (1982). *Humanscape. Environments for people*. Ann Arbor: Ulrich's Books.

Kaplan, R., Kaplan, S., & Ryan, R. L. (1998). *With people in mind. Design and management of everyday nature*. Covelo: Island Press.

Kent, J. L., & Thompson, S. (2014). The three domains of urban planning for health and well-being. *Journal of Planning Literature, 29*(3), 239–256. doi:10.1177/0885412214520712.

Kopec, D., Sinclair, E., & Matthes, B. (2012). *Evidenced-based design. A process for research and writing*. New York: Prentice-Hall.

Korosec-Serfaty, P. (Ed.). (1976). *Appropriation de l'espace*, Actes de la Conférence internationale de psychologie de l'espace construit (3rd IAPS), Strasbourg, France.

Lawrence, R. J. (1991). *Housing, dwellings and homes: Design theory, research and practice*. New York: Wiley.

Lawrence, R., & Després, C. (2004). Futures of transdisciplinarity. In R. J. Lawrence & C. Després (Eds.), *Transdisciplinarity in theory and practice. Futures, 36*(4), 397–405.

Lawson, B. (2004). *What designers know*. Oxford: Architectural Press.

Lawson, B. (2005). *How designers think. The design process demystified* (4th ed.). Oxford: Architectural Press.

Le Moigne, J. L. (1991a). Voir la nature avec les yeux de l'art. In F. Tinland (Ed.), *Systèmes naturels et systèmes artificiels*. Seyssel: Champ-Vallon.

Le Moigne, J. L. (1991b). *La modélisation des systèmes complexes* (2nd ed., 1995). Paris: Dunod.

Lee, T. (1968). Urban neighbourhood as a socio-spatial schema. *Human Relations, 21*(3), 241–267. doi:10.1177/001872676802100303.

Lincoln, Y. S., & Guba, E. G. (1985). Establishing trustworthiness. In *Naturalistic inquiry* (pp. 289–331). New York: Sage.

Lynch, K. (1960). *The image of the city*. Cambridge: MIT Press.

March, L. (1976). The logic of design and the question of value. In *The architecture of form* (pp. 1–40). Cambridge: Cambridge University Press.

Mehrabian, A., & Russell, J. A. (1974). *Approach to environmental psychology*. Cambridge: MIT Press.

Mikellides, B. (2007). Environmental psychology: 1969–2007. *Brookes eJournal of Learning and Teaching, 2*(2). http://bejlt.brookes.ac.uk/paper/architectural_psychology_19692007-2/. Accessed 5 May 2015.

Noschis, K. (2015). Preface. In R. J. Lawrence & G. Barbey (Eds.), *Rethinking habitats. Making sense of housing* (pp. 21–24). Gollion: In Folio.

Panerai, P., Castex, J., & Depaule, J. C. (1977). *Formes urbaines. De l'îlot à la barre*. Paris: Dunod.

Peña, W., & Parshall, S. A. (2001). *Problem seeking. An architectural programming primer*. New York: Wiley.

Pol, E. (2007). Blueprints for a history of environmental psychology. From architectural psychology to the challenge of sustainability. *Medio Ambiente y Compotamiento Humano, 8*(1y2), 1–28.

Preiser, W. F. E., Rabinowitz, H. Z., & White, E. T. (Eds.). (1988). *Post-occupancy evaluation*. New York: Van Nostrand Reinhold.

Preiser, W. F. E., Vischer, J., & White, E. T. (Eds.). (1991). *Design intervention. Toward a more humane architecture*. New York: Van Nostrand Reinhold.

Proshansky, H. M. (1978). The city and self-identity. *Environment and Behavior, 10*(2), 147–169.

Rittel, H. W. J., & Webber, M. W. (1973). Dilemmas in a general theory of planning. *Policy Sciences, 4*(2), 155–169.

Rivlin, L. G., & Wolfe, M. (1985). *Institutional settings in children's lives*. New York: Wiley.

Robinson, J. (2008). *Institution and home. Architecture as a cultural medium*. Amsterdam: Techne Press.

Robinson, J. (2015). The challenge of institution and home. In R. J. Lawrence & G. Barbey (Eds.), *Rethinking habitats. Making sense of housing* (pp. 135–133). Gollion: In Folio.

Rossi, A. (1966). *L'architecture de la ville* (F. Brun, Trans.). Paris: L'Équerre.

Sanoff, H. (2000). *Community participation methods in design and planning*. New York: Wiley.

Schön, D. (1982). *The reflective practitioner. How professionals think in action*. New York: Basic Books.

Serfaty, P. (2003). *Chez-soi. Les territoires de l'intimité*. Paris: Armand Colin.

Simon, H. A. (1969). *The sciences of the artificial*. Cambridge: MIT Press.

Sommer, R. (1969). *Personal space. The behavioral basis of design*. Englewood Cliffs: Prentice-Hall.

Stevens, G. (2010). *A history of architectural education in the west, architectural blatherations*.http://www.archsoc.com/kcas/Historyed.html. Accessed 7 Mar 2015.

Stokols, D. (1972). On the distinction between density and crowding. Some implications for future research. *Psychological Review, 79*(3), 275–277.

Thornberg, J. M. (1974). Children's conception of places to live in. In D. H. Carson (Ed.), *Man-environment interactions* (pp. 178–190). Stroudsburg: Dowden, Hutchinson & Ross.

Verderber, S. (2005). Compassion in architecture. Evidence-based design for health. Louisiana, Compassioninarchitecture.EvidencebaseddesignforhealthinLouisiana. Lafayette: Center for Louisiana Studies (distributed by UL Press, Lafayette).

Vischer, J. C. (1995). *Workspace strategies: Environment as a tool for work*. New York: Chapman and Hall.

Von Meiss, P. (1986). *De la forme au lieu*. Lausanne: Presses polytechniques et universitaires romandes.

White, W. H. (1980). *The social life of small urban spaces*. Washington, DC: Conservation Foundation.

Yin, R. K. (2003). *Case study research. Design and methods*. Thousand Oaks: Sage.

Zeisel, J. (1984). *Inquiry by design. Tools for environment-behavior research*. New York: Cambridge University Press.

Zimring, C., & Bosch, S. (2008). Building the evidence base for evidence-based design. *Environment & Behavior, 40*(2), 147–150. doi:10.1177/0013916507311545.

Place Attachment, Sense of Belonging and the Micro-Politics of Place Satisfaction

Andrés Di Masso, John Dixon, and Bernardo Hernández

5.1 Introduction

This chapter tackles the political dimension of place satisfaction, both as a psychological experience and as part of a conceptual approach in environmental psychological research. In the first section, we locate place satisfaction within a broader set of concepts traditionally used to account for how people feel, think and act towards places, stressing in particular the complex relationships between place satisfaction, place attachment, place identity and appropriation of space. In the following sections, we argue for a shift from cognition-centered approaches towards a pragmatic-constructionist perspective. In order to develop this argument, we briefly discuss early studies on the discursive framing of environmental evaluations as well as research on the political dimension of people-environment bonds. We then summarize the main theoretical tenets of the discursive approach in psychology as an epistemological framework that offers a new perspective of human-environment relations. The next section develops and illustrates some of the discursive-psychological principles by discussing emerging work in environmental psychology that has addressed the "shadow side" (Manzo 2013) of people's bonds to places, especially work that has emphasized their political dimensions by locating satisfaction-related experiences of place in broader contexts of inter-group conflict, political inequality and social exclusion. Finally, in the last section of the chapter, we briefly summarize and discuss the main implications and limitations of this politically-sensitive approach to place satisfaction, attachment and belonging. In so doing, we seek to expand current approaches in a way that highlights the socially constructed nature of place experiences, reconnecting them to the structural processes that shape them and that are relevant to the reproduction of a given social order.

5.2 Quality of Life and People-Environment Studies

The importance attributed to the environment in the development of a personal sense of well-being has featured in most of the standard approaches to quality of life since the United Nations initial system of indicators of life conditions in 1954 (Moreno and Pol 1996). Among

A. Di Masso (✉)
Universitat de Barcelona, Barcelona, Spain
e-mail: adimasso@ub.edu

J. Dixon
Open University, Milton Keynes, United Kingdom
e-mail: john.dixon@open.ac.uk

B. Hernández
Universidad de La Laguna, San Cristóbal de La Laguna, Spain
e-mail: bhdezr@ull.edu.es

the classic evaluative dimensions (e.g., health, education, work, etc.), the UN's approach included objective measurements of dwelling and environmental conditions. The increase in concern for the qualitative dimensions of personal well-being justified the OECD's 1971 proposal of eight major quality of life areas, including subjective appraisals of the individual's physical environment. Nowadays, the OECD's "Better Life Index[1]" is calculated at an international scale comparing countries across 11 major variables, including "life satisfaction" (an overall reflective evaluation of one's life conditions and circumstances), partly related to "housing" (based on "housing expenditure", "dwellings with basic facilities" and "rooms per person") and the "environment" ("water quality" and "air pollution"). Similarly, EUROSTAT's quality of life indicators[2] include objective measurements and subjective assessments of the "natural and living environment" (i.e., air, water, and noise pollution), on the grounds that personal well-being also depends on the protection of the environment. At a national scale, quality of life evaluations that are sensitive to environmental conditions usually consider specific factors, such as waste management and recycling, energy use, perceptions of the city, traffic and transport (e.g., New Zealand's Quality of Life Project[3]), or perception of one's neighborhood and urban green areas close to one's dwelling (e.g., Argüeso et al. 2013, for the Spanish National Institute of Statistics). On the whole, it seems reasonable to state that personal well-being is derived, at least in part, from the material properties and subjective appraisals of one's physical surroundings.

Nevertheless, residential satisfaction and environmental assessments have arguably been underplayed in mainstream psychological research on quality of life, life satisfaction and subjective well-being (e.g., Csikszentmihalyi 1990; Diener 2000; Kahneman et al. 1999; Sirgy 2012). For obvious reasons, this has not been the case within the sub-discipline of environmental psychology, which has devoted significant efforts to clarify the role played by our everyday environments in the development of the individual's experience of "feeling good". The majority of research conducted in this area can be broadly grouped into two main strands: the classic approach to "residential satisfaction" (e.g., Carp and Carp 1982) and more recent perspectives on "place satisfaction" (e.g., Stedman 2002; Ramkissoon et al. 2013).

5.2.1 Modeling Residential Satisfaction

Traditional research on residential satisfaction has been summarized by Amérigo and Aragonés (1997), who consider that it is a component of general life satisfaction defined by "a positive affective state which the individual experiences towards his/her residential environment and which will cause him/her to behave in certain ways intended to maintain or increase congruence with that environment" (p. 48). The authors develop their argument based on their own empirical evidence, as well as on Canter and Rees's (1982) original model of residential quality, which focuses on the evaluation of three main elements (neighborhood, house, and neighbors). Amérigo and Aragonés compared existing conceptualizations of residential satisfaction and then proposed a theoretical framework organized across two main dimensions: physical-social environment and objective-subjective properties. According to this framework, residential satisfaction depends, for instance, on aspects such as the maintenance of the neighborhood and apartment evaluations (subjective-physical dimensions), the relationship with neighbors and the attachment to the residential area (subjective-social), the noise level (objective-physical) and the length of residence in the neighborhood (objective-social). In this framework, the role of emotional attachment to the residential environment is highlighted as a significant predictor of satisfaction, as well as the importance of reference groups, normative standards and social and cultural status in shaping subjective appraisals of the residential area.

---

[1] http://www.oecdbetterlifeindex.org/
[2] http://epp.eurostat.ec.europa.eu/statistics
[3] http://www.qualityoflifeproject.govt.nz/built.htm

Similar attempts to systematize and empirically test multidimensional approaches to residential satisfaction are found in environmental psychology (e.g., Amérigo and Aragonés 1990; Christensen and Carp 1987; Fleury-Bahi et al. 2008; Wiesenfeld 1995). To cite but a few, Bonaiuto et al. (1999) explored the influence of residential satisfaction upon neighborhood attachment, defining the former as a multifaceted perception of residential quality covering four areas: architectural and town-planning features (e.g., building aesthetics, accessibility), social relations (e.g., presence of social relationships, threatening people), network services (e.g., social and health services, cultural activities and meeting places) and context features (e.g., lifestyle, pollution or maintenance). Aiming to find specific predictors of neighborhood attachment, Bonaiuto et al. concluded that all areas were relevant but hierarchically organized, such that contextual features had the highest impact upon attachment, followed by social relations and architectural features, with services having the smallest influence. Along similar lines, but with a specific interest in residential satisfaction among the elderly, Rioux and Werner (2011) argued that evaluations of the home are embedded in assessments of the broader residential environment. They identified a multidimensional structure of residential satisfaction encompassing appraisals of the local area, accessibility of services, and relationships with neighbors and the home. Following the same logic of modeling, but from the perspective of "neighborhood satisfaction", Hur et al. (2010) proposed a three-component framework based on physical measurements of environmental attributes (e.g., vegetation level, building density), perceived environmental attributes (e.g., perceived naturalness and openness) and evaluation of those attributes (e.g., satisfaction with presence of trees, amount of open spaces and density of housing).

All in all, most of the existing theoretical models of residential satisfaction have been guided by the underlying assumption that it is possible to identify the main determinants of subjective well-being, which are derived specifically from the individual's relationship with his/her environment, enriching an overall sense of life satisfaction. However, these models are marked by their individual focus of analysis (to the detriment of in-group and inter-group social processes), their lack of exploration of the social practices through which places are evaluated, and a general neglect of the political value that may underlie people's relationships with places. These issues are particularly relevant from a critical perspective, as we discuss later in the chapter.

5.2.2 Place Satisfaction and People-Place Bonds

In addition to their willingness to find generalizable predictors of residential satisfaction as the result of an individual, cumulative process of environmental perception and evaluation, a common feature in most of the studies discussed so far is the deliberate aim to relate environmental satisfaction to specific psychological bonds to the place. Amérigo and Aragonés (1997) considered attachment to the area to be a relevant predictor of environmental satisfaction, while Bonaiuto et al. (1999) treated residential satisfaction as an antecedent of neighborhood attachment. In fact, the theoretical relationship between environmental evaluations and people-place bonds has been straightforwardly addressed in other studies, thereby connecting research on residential satisfaction to classic conceptual debates in environmental psychology on the psychological organization of inter-related constructs like place attachment, place identity and place dependence.

Mesch and Manor (1998) explicitly differentiated place attachment from residential satisfaction, defining the latter as "the evaluation of features of the physical and social environment (…) based on the perception of place as a relatively nice place to live in terms of people, housing and noise", and the former as "a sentiment (…), a more affectual level of attraction to place or remorse to move out" (p. 509). In line with Feldman (1990) and Bonaiuto et al. (1999), these authors concluded that residential satisfaction predisposes individuals to place attachment, increasing their likelihood of maintaining residence and environmental commitment in the future. Likewise, Stedman (2002) refers to place satisfaction as "a multidimensional summary judgment of the perceived quality of a setting" (p. 564), therefore treating this construct as an attitude and something different from place attachment, in turn operationalized as "identity salience". Extending these theorizations of place satisfaction, Ramkissoon et al. (2013) tested a model that treats place satisfaction as being influenced by place identity and place dependence (the two main dimensions of place attachment), place affect and place social bonding. Thus, here, in line with Amérigo and Aragonés (1997) but unlike Stedman's (2002) and Mesch and Manor's (1998) models, people-place psychological bonds are viewed as influencing place satisfaction and not the other way around. This same treatment of residential/place satisfaction as a dependent variable influenced by psychological people-place bonds features in Fleury-Bahi et al.'s (2008) conclusion that length of residence increases place identification and that this, in turn, strengthens residential satisfaction, especially its social components – social relationships and social image (see also Marcouyeux and Fleury-Bahi 2011). However, other studies have treated place identity as an ongoing process of environmental bonding, which is not reduced to an antecedent of place satisfaction but is rather entwined with it (e.g., Jorgensen et al. 2007). This position is closer to the dialectic perspective on place identification and environmental involvement as described in earlier studies of appropriation of space (Pol 1996, 2002; Vidal and Pol 2005).

This exploration of the relationships between environment-related experiences of satisfaction and psychological bonds to places has had fruitful but ambivalent consequences at a theoretical level. On the one hand, as just mentioned, it has effectively located accounts of residential/place satisfaction within core debates on the psychology of place attachment, identity and related concepts. In this respect, future research might be developed around the idea that place satisfaction is psychologically related to place-based accounts of who we are, how we feel towards everyday environments, and how we behave in relation to such environments. On the other hand, this conceptual connection has dragged place satisfaction towards a theoretical territory defined by confusion, lack of agreement and frequent contradiction. As in other fields of knowledge, as Hernández et al. (2013) have argued, the vast proliferation of conceptualizations and measurements of place attachment and related constructs, often incompatible, has been of limited help in understanding and explaining the psychology of people's bonds with their life spaces. The problem here is not limited to the difficulty of ascertaining whether place satisfaction is an outcome or a determinant of place attachment, place identity or place dependence. It also involves a debate about the very nature of attachment, identity and dependence (i.e., as cognitive entities, affective patterns, attitudinal dispositions, etc.), and their particular forms of "internal" organization and relationships (i.e., hierarchical, with place attachment encompassing identity and dependence or the other way around; or at the same psychological level, being separate components of an overall sense of place, etc.). As long as this is the case, research integrating place satisfaction and place attachment, place identity and categories alike will suffer from the same conceptual complexity, confusion and contradiction.

This ambivalence opens up at least two different paths to overcome conceptual "messiness". One option is to refine and accommodate existing theoretical models that try to integrate experiences of the environment that are assumed to be substantially different in their psychological nature and form organization, hoping that novel conceptualizations will be more accurate and insightful than older ones. This is the case, for instance, of Scannell and Gifford's (2010) "tripartite model" of place attachment, one of the most promising frameworks to date (see also Hernández et al. 2013 and Lewicka 2011, for details). The other option is to change the epistemological perspective on people-place relations in a way that makes "messiness" and confusion a secondary issue, while re-signifying contradiction and disagreement as an inescapable circumstance bound to the structural variability of people's accounts of their experiences of places. This second option implies problematizing (i.e., questioning, rethinking) *the psychological nature of people-place bonds*, as well as exploring how people's subjective assessments of places can be informed and regulated by *wider social processes* beyond the individual. We can find at least two significant steps in this direction in early experimental studies on the discursive construction of environmental categories and related psychological reactions, on the one hand, and in loosely interrelated explorations on the political dimension of people-environment relations, on the other hand.

5.3 The Discursive Framing of Environmental Evaluations

A first critical approach to traditional conceptions of environmental appraisals is represented by a series of experimental studies aimed at demonstrating how the discursive framing or definition of "nature" and environmental categories triggers different preferences and evaluative responses in individuals confronted with an environmental topic (Aiello and Bonaiuto 2003). As a theoretical starting point, as Bonaiuto and Bonnes (2000) put it, "If the concern becomes the production of environmental representations or versions, and both their cognitive effects and social implications, the focus of the analysis can be turned towards the discursive strategies through which these representational practices are concretely realized" (p. 75). Among these "environmental representations or versions" are the kinds of evaluations, assessments and expressions of environmental preference that construct "place satisfaction".

A classic study in this area is Macnaghten et al.'s (1992) experiment on the effects of rhetoric upon people's evaluation of, and preferences for, "natural" settings. Their preliminary study showed that people exhibit a significantly more positive evaluative response towards things (cannabis, in their study) when they are framed as "natural". This conclusion extended previous evidence on how labels affect evaluative responses depending on the general positive or negative social value associated with the label (Eiser 1990). It also showed how this effect was specifically produced by categories with an implicit meaning of "naturalness" (Hodgson and Thayer 1980). Moreover, in a second study, they concluded that not only did the "nature" category affect evaluative responses, but that the acceptance of the same environmental transformation depended on the way "nature" itself was defined (i.e. discursively framed). In this second study, questionnaire responses confirmed that when nature was defined as a "virgin territory", non-human changes were more accepted than human-fitting and human-unfitting changes, whereas when nature was defined as "visual harmony", both human-fitting and human-unfitting changes were rated as more acceptable.

What Macnaghten et al.'s twofold study shows is that people's environmental evaluations, perceptions and preferences, as well as their positive acceptance of environmental changes, are contingent upon variable definitions of the meaning of the environment. Therefore, the discursive strategies that shape different versions of the environment are highlighted as having a critical analytic relevance. In the context of this chapter, insofar as place satisfaction is defined as a multidimensional evaluation of one's environment (e.g., Amérigo and Aragonés 1997; Mesch and Manor 1998; Stedman 2002), and evaluations depend on discursive constructions of the environment, we argue that place satisfaction depends on the discursive dynamics that shape environmental meanings. In other words, what this study suggests is that place satisfaction results from dynamic discursive processes that shape its very psychological meaning – processes that are located not inside the individual, but in the social practices of linguistic meaning-making.

A similar study by Reicher et al. (1993) underlined Macnaghten et al.'s conclusions by stressing the rhetorical management of naturalness through the analysis of "real" arguments. In this case, a first sub-study showed that evaluations of radiation from nuclear power production were more positive when radiation was categorized as "natural" rather than "man-made" or undefined.

More interestingly, the authors explored in the second sub-study the extent to which the naturalness of radiation was itself a relevant topic in the real-life debate around nuclear power. Pro-nuclear materials categorized nuclear radiation as natural, in order to make it more positively evaluated and benevolent (in line with the first sub-study and with Macnaghten et al.'s results) and reduce the perceived impact of nuclear emissions. Conversely, anti-nuclear arguments countered this attempt to put both types of radiation in the same category (i.e., "natural") by highlighting its "man-made" and noxious aspects. Reicher et al. concluded that "choices are a matter of values and priorities" (p. 107), confirming the ideological nature of signifying radiation as both "natural" and specifically "artificial and noxious, not natural". Similar effects of linguistic framing on environmental evaluations were explored by Eiser et al. (1993) to clarify the role of additional factors in individuals' responses. These authors concluded that defining the same scene (a beach) using a language of "pollution" (condition 1), "dirtiness" (condition 2) or "danger" (condition 3) had an influence upon judgments of environmental quality in combination with the set size and severity of specific pollutants rated in a questionnaire. As seen in the previous section, "pollution", "dirtiness" and "danger" are concrete instances of typical dimensions of residential satisfaction (environmental quality, physical maintenance and threat perception), therefore rhetorical constructions of these aspects can be expected to affect environmental satisfaction in a significant way.

Taken together, these studies demonstrate the critical relevance of the discursive and argumentative activities involved in the construction of the meaning of environmental issues, activities that lead to variable judgments and evaluations of environmental changes. Applied to the analysis of environmental experiences of satisfaction, these studies enable the individual's subjective assessments of his/her environment (e.g., "I like the place where I live") to be re-conceptualized not as a stable and neutral psychological judgment, but rather as a "subject position" (Hollway 1984) within broader discursive dynamics of meaning-making, which are socially contested and can have significant political implications (see Reicher et al. 1993, on ideological debates about nuclear energy). Consequently, by focusing on "the discursive strategies used to specifically realize different representations of an environmental issue, which in turn implicitly or explicitly favor different interpretations and different meanings attributed to the environmental issue" (Aiello and Bonaiuto 2003, p. 255), the discursive approach in environmental psychology introduces the everyday politics of environmental/place construction as a relevant research focus.

5.4 The Political Dimension of People-Environment Relations

A second group of studies that has problematized ordinary definitions of environmental meanings and evaluations has diverted attention from the individual, to focus instead on the socio-political processes shaping individuals' experiences of place. The political significance of human-environment relations has received little attention in environmental psychology. Among the exceptions to this general trend, Hubbard (1996) used the theory of social representations to explore competing interpretations of architectural pieces by professionals and lay people. Highlighting that environmental preferences and tastes are socially constructed and "embedded in the structures of power and dominance which exist in capitalist society" (p. 77), he concluded that the psychological study of place meanings should deal with conflict as a central issue. Interested in the topic of inter-group conflict, Mazumdar and Mazumdar (1997) showed how power and domination over the religious minority of the Zoroastrians in Iran are reflected through architectural arrangements (e.g., heights of buildings, orientations, spatial segregation, lack of ornaments, etc.). In this case, conflict is expressed by the abuse of Muslim vandals invading Zoroastrian residences due to their obligation to erect only low walls as a symbol of inferiority, by their own violation of the architectural regulations and by defensive devices in their homes aimed at achieving a relative spatial protection. From a different perspective, Devine-Wright and Lyons (1997) studied the symbolic properties of places with a historical value and their relationships with national identities in Ireland. They obtained questionnaire responses from individuals who were asked to rate their feelings towards, values attributed to, and perceived importance of four historical Irish sites. Results indicated that "traditional" and "non-traditional" groups "construct Irish history in different ways" (p. 43), based on different place-embedded social memories that reflect conflicting national identity projects. For instance, for the "traditional" group, the General Post Office meant the 1916 uprising to free Ireland from the British, whereas for the "non-traditional" group it seemed to recall the Irish Republican Army (IRA) violence over the previous two decades. More recently, Possick (2004) referred to "ideological place attachment" as a kind of cultural place attachment, expressing the bond established by a group of Jewish settlers in the West Bank who were evicted by the Israeli government in 1979. The episode of eviction was summarized by one of the interviewees as an "ideological trauma for the whole idea of settlement" (p. 61), conceiving settlement as a collective pathway to redemption via land occupation – an emotional feeling that is clearly embedded, one could argue, in a broader structural conflict involving the geopolitics of division between Israel and Palestine (see Mazumdar 2005, for a commentary on Possick 2004).

Beyond this handful of studies, the general disconnection between the psychological experience of the environment and its political underpinnings has not been explicitly noticed until recently. However, as Lynne Manzo (2003 p. 54) puts it, "it is not possible to adequately consider people's emotional relationships with places without recognizing the significant political implications of such a phenomenon – that who we are can have a real impact on where we find ourselves and where we feel we belong.

A proper understanding of people's emotional relationships to places, then, must include a contextualized – and politicized – view of these relationships". Accordingly, in Manzo's work, emotional bonds towards places are found to be not only positive, but also ambivalent or even profoundly negative. She questions the traditional focus on the residential environment as the main site for the study of place attachment and challenges ordinary views of the "home as haven". Manzo argues that much research on the home has been conducted reproducing a dominant ideology of the private sphere that imposes a normative experience of the home as the true site for privacy, emotionality, authenticity and sensibility, as opposed to the "coldness" of the public sphere and the marketplace, a division rooted in the original capitalist split between domestic life and work life and clearly related to the sexual division of labor and the territorialization of gender relations. Manzo expands this rationale in a qualitative study of people's emotional relationships to places in New York, showing how experiences of place preference, belonging and feeling safe (all of them included in most of the classic models of place satisfaction and environmental quality assessment) are governed by broader politics of identity, which convey social relations of dominance and exclusion. As Manzo (2005 p. 79) states, "participants' race, gender and sexuality influenced their experiences of places and created different potentials and restrictions on their use and enjoyment of space, thereby influencing their ability to be themselves. In this way, identity, and the socio-political underpinnings of it, makes a critical difference in how we use and view place".

It should be noted that, while these studies highlight the politics of place and the ideologically rooted and contested character of people-environment relations, they are still located within an epistemological paradigm that assumes the existence of an inner psychological experience separated from the practices that produce it. The sense of this experience is certainly seen to derive from, or to be shaped within, the broader milieu of political power that frames it, but the experience itself is defined as something *substantially* psychological. In this respect, the researchers' position regarding the psychological nature of people-place bonds does not differ from that assumed by classic studies on residential and place satisfaction. Therefore, conceptual confusion between people-place psychological constructs is still potentially an unsolved issue. In the majority of these studies, the particular ongoing dynamics of environmental meaning-making, which channel, reproduce and reformulate the political value of people-place bonds, as well as the variable construction of the psychological meaning of place (in the sense of Macnaghten et al.'s and Reicher et al.'s studies), are still out of analytical focus. In order to approach the everyday politics of people-environment construction, and to relate them to discursive accounts of place satisfaction, recent research has been conducted under the constructionist paradigm, which entails an epistemological shift with respect to traditional approaches.

5.5 Common Epistemological Assumptions in the Psychology of Place

Environmental psychological research on people-place bonds and residential/place satisfaction has been mainly conducted under a common set of ontological and epistemological assumptions. By "ontological assumptions", we mean a set of normalized beliefs about the *psychological nature* of people-environment relations, whereas "epistemological assumptions" refer to equally naturalized modes of building knowledge on the basis of such beliefs. At the ontological level, common assumptions revolve around the general idea that there is a "psychological realm" inside the individual, an objective domain of the material environment "out there", and "real" patterns of relationships between the former and the latter (in the form of interactions, influences, transactions, aspects of holistic units, part-whole relationships, etc.; see Altman and Rogoff 1987). While acknowledging the differences, and even incommensurability, between a cognitivist and a phenomenological approach (e.g., Seamon 1983), both assume that there is an "inner world" of experiences within the individual, a psychological "substance" that is forged through peoples' relationships with their surroundings, in the form of either cognitions, emotions, attitudes, identities, memories, preferences, etc., or genuine, singular and ineffable feelings and profound existential motions. Similarly, at the epistemological level, while there is a broader diversity in terms of philosophical traditions, research programs and scientific paradigms (Patterson and Williams 2005), there is a common belief that it is possible to access the individual's interior psychology of place, by either asking or watching and then measuring or interpreting the outcomes as pathways to, and reflections of, the individual's "inner world". In both cases, it is unquestioned that the individual's place-related measurements (e.g., surveys), accounts (e.g., interviews) and spatial-environmental behavior (e.g., observation) somehow reflect the individual's *inner* and *real* psychology of place.

Applied to the research discussed in the earlier section, these ontological and epistemological assumptions require that place satisfaction, place identity and place attachment, whatever their specific relationships might be, are treated as substantial components of the internal psychological landscape of the individual; a psychological landscape that is *in there* and that can be effectively *mapped* (no matter how biased) onto place-related measurements, accounts and observations. Hence, statements like "I belong in this city" would be an external manifestation of an internal place identity, as much as "I would never leave my town" and "I like my neighbors very much" would be interpreted as a reflection of the individual's place attachment and residential satisfaction, respectively.

In the specific case of research on residential/place satisfaction, as in most studies on people-place bonds, this "psychologistic" epistemology of people-environment relations fits into a typically positivistic conception of knowledge. As shown in the studies discussed in the previous section, this means that the exploration of the internal psychology of place is guided by the principles of verification, reliability, congruency, explanatory linearity, generalization and predictability. Surveys and interviews examining people's residential satisfaction operate as reality-check devices aimed at "discovering" objective psychological entities that are "truly" located inside the individual's mind (i.e., verification). Given that the inner experience is *what it is*, it is expected that different observers will "find" the same results (i.e., reliability), and that the subject's answers to questions dealing with the same constructs will be stable and congruent (i.e., congruency). At a psychological level, place-related cognitions and perceptions are expected to cause, or significantly correlate with, specific emotions and evaluations, and these in turn influence place behaviors, and vice versa. When contradictions appear, other intermediate variables explaining contradictions must be postulated (i.e., explanatory linearity). Finally, research aspirations aim to establish patterns of truth regarding specific relationships between psychological entities and behavioral correlates (e.g., "People more satisfied with their residential environments will be more attached to their neighborhood"), that can be generalized to the wider population, to other samples and/or to similar contexts (i.e., generalization), making it possible to anticipate and control what will happen if certain psychological/behavioral/environmental conditions are present at a given time (i.e., predictability).

This logic of knowledge-construction works as a "regime of truth" (Foucault 1975), which is still dominant not only in environmental psychology, but also broadly in the social sciences. The question posed at this point is less related to its legitimacy than to the possibility of thinking and building knowledge about people-environment relations from a different research-logic and epistemological position. The constructionist-discursive approach in psychology is suited to this purpose.

5.6 Constructionism, Discursive Approach and Environmental Psychology

An emerging strand of research in environmental psychology is interested in the social construction of environmental meanings and people-place relations (e.g., Aiello and Bonaiuto 2003; Di Masso et al. 2011; Dixon and Durrheim 2000; Stokoe and Wallwork 2003). Building on the "discursive turn" in social psychology (Harré and Gillet 1994; Potter and Wetherell 1987), this approach proposes a shift of "the *analytic* and *explanatory* focus from cognitive processes and entities to discursive practices and the resources they draw on" (Potter 1998 pp. 235–236, italics in the original). This shift of focus entails assuming the linguistic nature of both psychological reality and the world "out there" (Edwards and Potter 1992), meaning that verbal or textual productions about psychological states and other realities outside the individual are treated less as faithful representations of an "internal" or "external" world than as context-bound versions of reality, produced with the discursive resources provided by our shared culture and functional in a given social situation. From this perspective, psychological entities such as people-place bonds do not pre-exist our culturally shared ways of talking and arguing about them: they are actively *created* through socially organized linguistic practices and repertoires that we employ in our daily interactions. By means of these discursive practices, accounts of our relationships with the environment re-shape our "private" experience of places as much as they produce particular social, moral and even political effects. In this way, the discursive approach replaces the study of the "inner", individual and psychological nature of our mental states with that of their "interactional", socially embedded and discursive nature.

The discursive-psychological epistemology is therefore constructionist and considers itself "agnostic" (Edwards and Potter 1992) regarding the existence of a psychological inner world, an external objective reality and "true" relationships between them. Instead, reality is defined as an emerging and variable outcome of discursive practices organized around that which interacting individuals orient toward and treat as "real" in their situated interactions. The main theoretical tenets of the discursive-psychological approach can be summarized as follows (see also Di Masso et al. 2013):

1. Discursive researchers are sensitive to how people flexibly construct different versions of reality (descriptions, explanations, evaluations, etc.) drawing on culturally shared discursive resources (e.g., "It's normal that people don't want to leave this place, because they like it very much");
2. They pay attention to the discursive strategies through which such versions of reality are depicted to be perceived as factual and real-seeming, i.e., as objective realities independent of the speaker's stakes or motivations (e.g., "*I talked to* the local inhabitants and *all of them* told me that *everybody* liked the place very much");
3. They are careful to pinpoint the kinds of social actions and interaction work done through specific ways of constructing reality (e.g., "Everybody likes the place very much and no one wants to leave", acting as a description that *blames* the mayor of a city who promotes an urban plan that entails the removal of local inhabitants and their relocation to another neighborhood);
4. They are interested in the rhetorical organization of the accounts (Billig 1987), meaning that versions of reality are often strategically designed to undermine, or resist being discounted by, alternative versions (e.g., "Despite the fact that local inhabitants are really satisfied with their place, the local administration wants to transform the area to make money with the new dwellings", against "The local administration wants to improve the area and give even better residential conditions to the neighbors");
5. They identify how the same people produce variable versions of reality at different times, as this variability depends on the shifting social functions accomplished by different accounts depending on the context and time of interaction (e.g., "People like their neighborhood very much" working to justify resistance to relocation, but then "people don't like the neighborhood and would easily leave" to claim more infrastructure and local facilities if staying);
6. They are sensitive to the speakers' accountability and orientation towards "what is going on" in the discursive interaction (e.g., "We understand that local inhabitants like their neighborhood and are upset with our proposal, but things have to be done soon", said by a promoter of the urban relocation program who is being careful about the possibility of being accused of not being responsive to the local inhabitants' claims).
7. Related to the rhetorical organization of the accounts, discursive psychologists pay close attention to the moral and ideological implications of certain ways of depicting reality, meaning that specific ideas and beliefs are functional to legitimize or discount specific power relations and normative arrangements of the social order (e.g., "Local inhabitants will rapidly be satisfied with their new environment and will like it even more", as a statement that supports the interests of private capital in the regeneration of a part of the city).

The examination of people-environment bonds as discursive practices means stating that there is no psychological substance underlying our relationships with places, but rather an ongoing, interaction-driven social practice that articulates psychology-like accounts in order to make sense of people's experience (of place) and to perform interaction work, thereby provoking situated social effects. In this framework, conceptual boundaries allegedly mirroring psychologically differentiated "internal" feelings, cognitions and assessments about places (i.e., place attachment, place identity, residential satisfaction) are less insightful than the meaning-making practices through which people actively create variable *versions* and *evaluations* of self-in-place, which often carry moral, relational, political and material implications and consequences.

The sole focus on people's discursive involvement in the construction, negotiation and contestation of the meaning of reality requires a concept of knowledge-construction that differs significantly from the positivistic paradigm summarized earlier. Since there is no objective reality, but rather discursive accounts claiming factuality about real-seeming "things", verification is replaced by interpretation, argumentation and credibility, so research outcomes themselves become persuasive accounts that are expected to be validated by other researchers who confirm or contest their plausibility (and not their reliability) in the light of theoretical principles, the empirical focus and the research question (Willig 2008). Given that variability is a defining tenet of the discursive approach, incongruence and contradictions in people's accounts of reality are expected and desirable as they enable their local, interactional functions to be analyzed in a given social context. The idea that psychological language is not a neutral mirror of psychological reality but a social practice that creates it leads to a lack of analytic interest in cognitions, emotions, preferences, etc. as "internal" causes or effects of "external" behavior following a two-way linear path. Instead, they are understood as publically available discursive resources that people flexibly deploy in their daily lives to account for their experiences and to *do things* in their social relations. Finally, the context-bound nature of discursive practices makes generalizability and predictability irrelevant in this framework, given that what is at stake is the ongoing, dynamic construction of meanings driven by the emerging properties of local and situated social interactions.

5.7 Environmental Discourse, People-Place Bonds and the Politics of Place Evaluation

To date, the application of the discursive approach to environmental issues has been empirically developed in the frame of the "spatial turn" in social psychology (e.g., Dixon et al. 1994; Dixon and Durrheim 2000; Benwell and Stokoe 2006; Taylor 2010), and has been addressed theoretically in a few textbooks (see Aiello and Bonaiuto 2003; Bonaiuto and Bonnes 2000; Di Masso et al. 2013). Within the disciplinary boundaries of environmental psychology, the attention given to discourse appears less as an epistemological "turn" and more as an emerging area of research sensitive to the performative, relational and ideological nature of environmental discourse.

In the brief summary of discursive psychology's main theoretical principles, we underlined the interest in the moral, relational and ideological implications of specific ways of depicting reality. This analytic concern emphasizes not just the varying *formulations* and *negotiations* of the meaning of reality (e.g., 'what is nuclear energy?'), but also the *contested nature* of meaning-making practices (e.g., opposing definitions of nuclear energy exist), their *normative value* (e.g., each partisan definition of nuclear energy aspires to be the adequate, appropriate and correct one vis-à-vis the others), and their potential to warrant and legitimize, or to challenge and undermine, specific *arrangements of social relations*, which are *power-driven* and are relatively central to the *social order* (e.g., normative constructions of nuclear energy are functional to support or discount dominant environmental policies and their opposing stakeholders). The contested, normative and ideological nature of human-environment discursive constructions demands attention be paid to their political potential and implications. This implies the need to treat environmental meanings, human-environment bonds and place-satisfaction experiences as discursive practices potentially embedded in wider social, historical and political processes of conflict, inequality and exclusion. In doing so, early experimental studies on the constructive properties of environmental discourse and research on the political dimension of people-environment relations are brought together from a discursive-psychological perspective.

The first systematic attempt to bring together discursive social psychology and environmental psychology in a politically sensitive way can be found in John Dixon and Kevin Durrheim's research program on desegregation and inter-group relations in post-apartheid South Africa. From a discursive perspective, the authors analyze the ways in which racist ideologies and beliefs are mapped onto a language of spatiality and people-place relations, thus reproducing racist assumptions and practices (Dixon et al. 1994, 1997; Dixon and Durrheim 2000, 2004; Durrheim and Dixon 2001, 2005). In this series of studies, a common theme is that people construct varying versions and evaluations of their residential or leisure environments, and of themselves in those environments (i.e., place identity), in ways that subtly perform a stubborn resistance to racial desegregation. For the purpose of this chapter, these and later studies are useful to discuss and empirically illustrate a political re-interpretation of the main psychological components of traditional approaches to residential/place satisfaction described in the first section of the chapter. The main aspects highlighted for each textual extract included in the next section in no case represent a discursive analysis – this would demand significantly more space to develop the interpretation properly. However, we believe that their inclusion in this handbook is valuable to show the main contribution of the discursive-psychological approach.

5.7.1 Evaluations of Environmental "Naturalness"

In an initial study, Dixon et al. (1994) analyzed a body of textual data obtained from letters submitted to local newspapers, from 1991 to 1993, by white residents in a coastal village called Hout Bay in Cape Province (South Africa). In these letters, the white residents complained about a squatted area growing in Hout Bay's surroundings. This area was inhabited by black settlers after the Cape Provincial Administration decided to extend an original settlement from 8 to 18 ha. The analysis identified interpretative repertoires (Potter and Wetherell 1987) or "systems of images, figures of speech, descriptions and narrative forms that can be employed to serve ends within discourse" (Dixon et al. 1994 p. 280). Throughout the analysis, Dixon et al. found a common and persistent "ecological repertoire" that served to construct the inappropriateness of the squatter camp. This eco-repertoire combined three features: (1) a positive view of nature (confirming Macnaghten et al. 1992 and Reicher et al. 1993); (2) a rhetoric of degradation (e.g., pollution, waste, overpopulation) and (3) human-nature interdependence as a unique eco-system. Extract 1 below exemplifies this repertoire:

Extract 1
When will the ongoing destruction of the environment stop? The Hout Bay Valley was so named due to the forests that used to exist here and now the Cape Provincial Administration wish to take yet more trees away to make room for services to an enlarged squatter community. As residents of Penzance estate, we have been affected by the squatter settlement, about 500 metres away, more than other persons living in Hout Bay. We have learnt to live with the problem of theft, insults, insecurity and property devaluation. Now we understand that the squatter camp is to be enlarged and that thousands of trees are to be uprooted. We chose to live in the hillside in Hout Bay because the trees are here and they give one a sense of stillness and solitude. Seeing the squirrels jumping through the trees, the owls at dusk and our noisy Egyptian geese with all their quacking makes the area a delight. The squatter community in its enlarged form will now be only 140 m from our home, a home we used to enjoy before our lives were shattered by some 'official' in the Cape Provincial Administration who decided that this was the one and only place these unwanted people could go. I would take a bet that this official lives nowhere near to Hout Bay. Our family, like many Hout Bay residents, would like to register total unacceptance of this plan of destruction and rape of our environment. (*Cape Times*, 25 September 1992:6)

From the perspective of residential satisfaction, it seems obvious that the white residents complaining in Hout Bay are not satisfied with the place they live in at the moment they are speaking. Using a traditional approach to residential satisfaction, this discomfort would be the outcome of a series of environmental evaluations based on the decrease in perceived naturalness, problems with the maintenance of the area, poor relationships with neighbors and dubious safety due to the presence of threatening people (Amérigo and Aragonés 1997; Bonaiuto et al. 1999; Hur et al. 2010; Rioux and Werner 2011). Based on recent theorizations of place satisfaction ("the evaluation of features of the physical and social environment", Mesch and Manor 1998; "a multidimensional summary judgment of the perceived quality of a setting", Stedman 2002), it seems clear that this letter is a clear-cut example of lack of place satisfaction.

However, Dixon et al.'s discursive analysis argues that this residential evaluation can be interpreted not as a faithful expression of psychological distress caused by an allegedly objective environmental destruction, but as a race-related evaluative practice that deploys the "ecological repertoire" to reject the proximity of black neighbors. Dixon et al. discuss three rhetorical effects provoked by the use of an ideology of ecological concern, in which white residents are shielded in order to make subtle racist claims while remaining protected from accusations of racism. First, assuming that ecological arguments are culturally linked to an image of progressiveness, a priori incompatible with racism, people thereby protected themselves from accusations of racism. Second, the "non-anthropocentric" qualities of ecological arguments rhetorically removed claims from personal interest. Third, the ecological repertoire became "a site for the regeneration of the imagery of the Other" (p. 289) through the lexical use of "disease", "degradation" and "overpopulation". These acted as metaphors extending the ideology of black people as an underclass. In sum, the authors showed "the re-articulation of racist imagery – submerged, transfigured, cast in the semiotics of ecologism" (p. 291), which surreptitiously warranted resistance to racial desegregation. Ultimately, accounts of residential/place satisfaction appear strategically connected here to the reproduction of a racist ideology.

5.7.2 Place Identity and Environmental Satisfaction

Two later studies extended this idea. Dixon and Durrheim (2000, 2004) addressed the discursive and contested nature of place identity and place attachment, again in the frame of environmental evaluations by white residents. As discussed earlier, both place identity and place attachment have been consistently related to place satisfaction either as antecedents (Ramkissoon et al. 2013), consequences (Bonaiuto et al. 1999) or separate experiences (Mesch and Manor 1998; Stedman 2002). Dixon and Durrheim assert that environmental psychology has largely focused on place identity as an individual cognitive property in people's minds and as a socially non-problematic experience. In contrast, they approach place identity as: (1) a collective construction jointly achieved by people through talk; (2) a publicly available discursive resource orientated toward the realization of multiple actions (e.g., justifying, blaming, excluding, etc.); and (3) a particular instance of broader ideological traditions. Consequently, experiences of place identity disruption, threat or challenge are re-defined as rhetorical devices that are also available to perform ideological work through the language of place.

Dixon and Durrheim (2000) illustrated these ideas again by analyzing letters sent to newspaper editorials and interviews related to the Hout Bay controversy in Cape Province. In their empirical data, white residents expressed a psychological sense of loss of the place's properties, drawing on arguments around the "despoliation of the natural environment", "nostalgia for the place in the past", "presence of 'alien' groups not conformed to local values" and "a space for family activity degraded by pollution or lack of proper manners". All of these arguments can again be related to specific components of many of the multidimensional models of residential/place satisfaction discussed so far (see extract 2):

Extract 2
On occasions during the past years I have taken my little children to the beachfront paddling pools and nearby beach and have always come away with a feeling of warmth and contentment. It was my misfortune to expect the same when I ventured there on January 2. We were revolted at the filth and stench around the paddling-pool. Garbage was on the sidewalks and the water in the pool was brown. Some of the bathers using the pools were half-naked, while others were fully clothed... Durban beachfront and its amenities are forever lost to whites. Never again will I take my family near the place. (Dixon and Durrheim 2000 p. 35–6)

As in their previous study, Dixon and Durrheim suggested that these expressions might not (only) represent a way of externalizing a subjective sense of disruption of place belonging, connected to a negative evaluation of the area and leading to an overall sense of place dissatisfaction. Instead, they could be interpreted as rhetorical strategies reproducing the racial ideology of spatial segregation. In their analyses, accounts of place identity disruptions were framed as racially driven responses to geopolitical changes in South Africa, warranting resistance of white residents to "transgressive presences" (i.e., black people) in "their" territory. Discourses of loss of place identity in this case worked re-establishing white territorial entitlements that had been challenged by desegregation: the latter defied long-established racialized patterns of place belonging, undermining traditional place identities tied to the spatial exclusion of black people. As shown in extract 2, negative evaluations of the place in terms of environmental maintenance ("filth", "garbage", "the pool was brown") and of controversial spatial behavior of some (black) groups of people ("half naked", "fully clothed") were rhetorically useful to account for a sense of loss of an environmental configuration that, prior to desegregation policies, positively supported white racial identity ("Durban beachfront and its amenities are forever lost to whites"). Hence, environmental dissatisfaction works here as a discursive device re-connecting place, racial identity and white people's privileges against the sudden physical presence of black people constructed as being "out-of-place". As in their previous study, Dixon and Durrheim relate the psychology of place identity, attachment and satisfaction to an intergroup political conflict to dominate and undermine everyday spaces of racial contact and exclusion.

5.7.3 Place Attachment and Environmental Appraisals

The idea that environmental evaluations and related accounts of place satisfaction can draw on place attachment repertoires carrying political implications has been illustrated in other contexts. Di Masso et al. (2011) reported an analysis of discourses around the appropriation of an urban open space in Barcelona. The "Hole of Shame" (*Forat de la Vergonya*), a strip of public land located in the Casc Antic area of Barcelona's old town, was from 1985 and for over 20 years the site of numerous official and unofficial attempts at redevelopment, involving struggles between a range of stakeholders including local residents, neighborhood associations, squatters, urban development agencies, and the local administration. In 2000, a group of local inhabitants denounced that the local administration was willing to gentrify the area, attracting private capital and transforming the place into a "trendy" neighborhood with expensive dwellings and tourist amenities against the needs and demands of the old, working class, impoverished population. For this reason, they encroached on the area and created a green space for the community, triggering a long, bitter, sometimes violent struggle for control over the space. The struggle involved not only territorial actions to "appropriate" (Pol 1996) and control the physical design and layout of the space, but also competing definitions of its environmental, social and psychological meaning, as seen in extracts 3 and 4 below:

Extract 3
J: It's [the park] a symbolic element upon which quite a lot is projected. It represents something that has been made and achieved and as if it were something of one's own, that no one wants it to be replaced. There is an instinct of protection that sometimes is not very rational, is it? And this creates distrust also towards any kind of, of attempt at transformation.

Extract 4
S: I think that the people aren't attached to that space because of the way it is. I think that once the problem is solved, I believe that yes, because, well, it's an important part of the neighborhood, and it's also a place that's permanently walked through, and very, very... (...) there is mud.
(...)
G: When it rains it's destroyed of course. You aren't likely to cross it, then, surely anyone individually would take another route. What happens is obvious. People will get attached when they see something more dignified.

Extract 3 is taken from an interview with a representative of a group that "squatted" the Hole of Shame and attempted to create a self-managed "park" by planting an orchard and installing facilities for children. Extract 4 is from an interview with two representatives of a group of neighbors who opposed this occupation and wanted the park "redeveloped". Reading these extracts, we could come to a straightforward conclusion: some neighbors were strongly attached to the Hole of Shame because they had invested their efforts in creating it ("it represents something that has been made and achieved"; extract 3), whilst others were not attached to the place due to its negative environmental qualities ("there is mud", "when it rains it's destroyed"; extract 4). Place-attached neighbors would be satisfied with the "park" as the outcome of a larger process of "symbolic appropriation" involving territorial occupation, transformation and place identification (Pol 1996), whereas non-attached neighbors would not be "place-satisfied" given the physical state of the area. These psychological reactions could in turn be used to explain interviewees' contrasting orientations toward the prospect of environmental change (i.e., urban regeneration). After all, people who establish deep psychological bonds with a place are typically anxious when change arrives, whereas people who make negative appraisals are more likely to support such change.

A discursive re-reading of these extracts would proceed from a quite different angle. In extract 3, for example, the interviewee describes the territorial feelings that arise when individuals actively create a new environment. They come to perceive such an environment "as if it were one's own". By implication, they develop an "instinct of protection" towards it that cannot be reduced to rational calculations ("it is not very rational"). In this context, discursive researchers would argue that expressions of connection to place are doing more than merely articulating a psychological state. They are also contributing to a rhetoric of resistance to environmental change. By appealing to the culturally shared notion of a "protective" instinct, for example, the interviewee in extract 3 is able to define the urban regeneration plan as a "threat" – as opposed to, say, an "opportunity" – and to portray resistance to such change as a natural and accountable reaction.

In sharp contrast, extract 4 is rhetorically designed to warrant the *necessity* of the official regeneration plan. Discursive researchers might ask, for instance, what is accomplished by "S's" use of the generic pronoun "the people" in the opening line of the extract. They might also ask why "G" provides a personal narrative of his negative experiences whilst crossing the park and why he describes himself as "having" to do so. They might further ask why he later uses a so-called "extreme case formulation" (Pomerantz 1986) when describing how "*anyone* would take another route" rather than cross the park after a period of rainfall. Perhaps most importantly, they might explore how arguments promoting the creation of a more "dignified" public space are warranted here not only by a listing of negative environmental features, but also by an explanation of how such features have prevented local users of the park from forming the kinds of emotional attachments and satisfaction experiences that would normally occur within such a fundamental city space. Place attachment and symbolic appropriation work in extract 3 as discursive constructions at the service of a political project of resistance to an urban regeneration program that some people felt was profoundly exclusionary, whereas place detachment and environmental dissatisfaction in extract 4 normalize and legitimize this program and the actual spatial removal of its occupants.

5.7.4 Socio-Spatial Trouble, Threatening Social Groups and Place Satisfaction

In the extracts discussed so far, part of the rhetoric of environmental evaluation and place (dis)satisfaction included not only place attachment and place identity formulations, but also environmental qualities ("filth", "garbage", "there is mud") and social presences ("half-naked", people "protecting the space") in an equally significant manner. As envisaged in traditional models of residential satisfaction, environmental physical properties and social relationships in the residential area are key evaluative components shaping the subject's experience of "feeling good" in their place of residence. The rhetorical and political value of physical-spatial conditions and social presences justifying negative environmental appraisals can be further illustrated in another example, taken from research on perceived insecurity and fear of crime in Barcelona (see Di Masso et al. 2014):

Extract 5
Yes, there's a lot of insecurity and I won't say that the police don't do anything. They sell drugs right here. There are robberies all the time, all the time trouble, always brawls (...) this is always filthy, the people, the people who have brought us here, this is a ghetto, people who have come, it might be their culture, or whatever, they are used to throwing away what they don't need. They don't throw the rubbish inside [the bin], they throw it outside, but this is nonsense, but bits of nonsense make a huge nonsense, and the neighborhood is shit.

The man (45 years old) in extract 5 is answering the interviewer's question about his opinion of an alleged state of urban insecurity in the neighborhood (Raval, in Barcelona's city centre). The man produces an account that makes an unambiguously negative evaluation of the neighborhood he lives in ("the neighborhood is shit"), based on real-seeming evidence of criminal activities ("selling drugs", "robberies"), physical properties ("filthy", "rubbish") and poor civic manners ("throwing the rubbish outside [the bin]") by disruptive groups of people of different "cultures" ("the people who have brought us here", "the ghetto"). These elements would serve as evidence to support any existing hypothesis on the main factors influencing residential satisfaction, especially those dealing with perceived safety and the presence of threatening people (Bonaiuto et al. 1999). In contrast, from a discursive perspective, one may ask why the question about insecurity triggers an appraisal about physical properties and spatial behaviors other than criminal activities, what social functions may be served by relating this appraisal to "other cultures", the "ghetto" and "people who have brought us here", and what sort of interaction work these metonymic depictions of the immigrant population accomplish. In this analytic frame, "throwing rubbish outside" is essentialized as a cultural characteristic of an out-group, which is ultimately blamed for a state of urban messiness that rhetorically equates urban incivilities with criminal activities, both explaining perceived insecurity and justifying a negative overall appraisal of the neighborhood. Immigration is never brought explicitly to the fore because it would expose the speaker to being accused of racism, but it rhetorically works by implication as the threatening Other submerged in a dominant narrative of insecurity that criminalizes immigrants defined as trouble-makers. According to this interpretation, environmental evaluations channel here xeno-racist assumptions that actualize a broader ideology of native supremacy, based on the pre-ordained right of "autochthonous" people (those who "belong *here*") to judge from a vantage point who is different and, above all, to warrant prejudiced claims against immigrants (see Di Masso et al. 2014, for details). Again, the language of environmental dissatisfaction performs ideological work far beyond its cognitive and emotional, largely non-problematic, correlates.

5.8 Conclusion

As Manzo (2003) has argued, "although our experiences in places are felt on a deeply personal level, they are products of a larger political, social and economic reality" (p. 54). Developing this idea, in this chapter we have advocated a constructionist approach to people-environment relations as being especially suited to underscore the political and ideological implications of particular ways of depicting the environment and people's psychological bonds with places. In this epistemological framework, individuals construct descriptive and evaluative versions of their relationships with their environments in ways that can be socially strategic to warrant or contest broader relational patterns of inter-group conflict, social exclusion and political struggle (e.g., racial contact and segregation, privatization of public spaces, native-immigrant relations, etc.). We have illustrated this approach by discussing the kinds of evaluations that are typically involved in people's subjective assessments of their residential environments, traditionally labeled in environmental psychology as overall experiences of "residential satisfaction" or "place satisfaction". As a result, we have discussed research on how the discursive-psychological interpretation re-specifies the cognitive and emotional elements that typically explain place-satisfaction experiences (e.g., place identity, place attachment), redefining them as situated discursive resources that can perform ideological work.

This re-conceptualization of subjective environmental evaluations as ideologically framed discursive practices enables an exploration of the "micro-politics" of place satisfaction. This means that everyday environmental appraisals can be quite effective to channel normative views of place and of people-place relations that can be controversial from a political perspective. Hence, ordinary evaluations such as "I like this place" or "I feel satisfied with my neighborhood" may act socially as discursive small gestures echoing wider ideological tensions about who belongs and who does not belong to the place, what the space is for and who has the right or privilege to decide upon it.

To the extent that environmental appraisals and subjective residential satisfaction are being increasingly considered to be relevant dimensions of overall judgments of quality of life, this politically sensitive approach to place satisfaction can be useful to problematize standard definitions of quality of life and life satisfaction measurements. While acknowledging their powerful descriptive and explanatory potential, existing theoretical models of residential and place satisfaction as multidimensional constructs, on the one hand, and quality of life surveys including operationalizations of these dimensions, on the other hand, may eventually support a largely individualized, de-contextualized and non-problematic concept of life satisfaction and quality. As argued throughout this chapter, insofar as residential evaluations can rhetorically serve ideological purposes, for instance warranting patterns of racial segregation or dominant trends of privatization of the public space, related quality of life assessments may channel, instantiate and even mask wider ideologies of exclusion and social inequality. A critical environmental psychological approach to quality of life would probably contribute to existing theorizations by exploring how class-, race- or gender-based assumptions, among others, may underlie people's subjective evaluations of their life conditions, including their life spaces.

Finally, the constructionist-discursive approach is clearly not free from controversies and limitations. As discussed in Di Masso et al. (2013), few environmental psychologists would accept a sole focus on language to explore people's psychological experiences of places. Ultimately, what people feel and think is something fundamentally unquestionable. Regarding this issue, the discursive approach would argue that it does not deny the existence of a private psychological experience inside the individual, but it rather sidesteps it as the focus of analysis to examine how this experience is discursively constructed and what social actions and ideological purposes such an appeal to an inner psychological experience may serve. At a different level, existing constructionist debates, critical of the excessive attention given to language, have been sensitive to other kinds of meaning-making practices that also shape reality (e.g., Cromby and Nightingale 1999), and people-environment relations in particular, such as geographical arrangements, territorial behavior, embodied practices and affective patterns beyond language (see Di Masso and Dixon 2015; Dixon and Durrheim 2003; Durrheim and Dixon 2005; Durrheim et al. 2013).

References

Aiello, A., & Bonaiuto, M. (2003). Rhetorical approach and discursive psychology: The study of environmental discourse. In M. Bonnes, T. Lee, & M. Bonaiuto (Eds.), *Psychological theories for environmental issues* (pp. 235–270). Aldershot: Ashgate.

Altman, I., & Rogoff, B. (1987). World views in psychology: Trait, interactional, organismic and transactional perspectives. In D. Stokols & I. Altman (Eds.), *Handbook of environmental psychology* (Vol. I, pp. 7–40). New York: Wiley.

Amérigo, M., & Aragonés, J. I. (1990). Residential satisfaction in council housing. *Journal of Environmental Psychology, 10*, 313–325.

Amérigo, M., & Aragonés, J. I. (1997). A theoretical and methodological approach to the study of residential satisfaction. *Journal of Environmental Psychology, 17*, 47–57.

Argüeso, A., Escudero, T., Méndez, J., & Izquierdo, M. J. (2013). *Alternativas en la construcción de un indicador multidimensional de calidad de vida.* Documento de trabajo 1/2013. Instituto Nacional de Estadística. http://www.ine.es. Accessed 17 Oct 2014.

Benwell, B., & Stokoe, E. (2006). *Discourse and identity.* Edinburgh: Edinburgh University Press.

Billig, M. (1987). *Arguing and thinking. A rhetorical approach to social psychology.* Cambridge: Cambridge University Press.

Bonaiuto, M., & Bonnes, M. (2000). Social-psychological approaches in environment-behaviour studies. Identity theory and the discursive approach. In S. Wapner, J. Demick, T. Yamamoto, & H. Minami (Eds.), *Theoretical perspectives in environment-behaviour research* (pp. 67–78). New York: Kluwer Academic/Plenum Publishers.

Bonaiuto, M., Aiello, A., Perugini, M., Bonnes, M., & Ercolani, A. (1999). Multidimensional perception of residential environment quality and neighbourhood attachment in the urban environment. *Journal of Environmental Psychology, 19*, 331–352.

Canter, D., & Rees, K. (1982). A multivariate model of housing satisfaction. *International Review of Applied Psychology, 31*, 185–208.

Carp, F. M., & Carp, A. (1982). Perceived environmental quality of neighborhoods: Development of assessment scales and their relation to age and gender. *Journal of Environmental Psychology, 2*, 295–312.

Christensen, D. L., & Carp, F. M. (1987). PEQI-based environmental predictors of the residential satisfaction of older women. *Journal of Environmental Psychology, 7*, 45–64.

Cromby, J., & Nightingale, D. (1999). *Social constructionist psychology: A critical analysis of theory and practice.* Buckingham: Open University Press.

Csikszentmihalyi, M. (1990). *Flow: The psychology of optimal experience.* New York: Harper & Row.

Devine-Wright, P., & Lyons, E. (1997). Remembering pasts and representing places: The construction of national identities in Ireland. *Journal of Environmental Psychology, 17*, 33–45.

Di Masso, A., & Dixon, J. (2015). More than words: Place, discourse and the struggle over public space in Barcelona. *Qualitative Research in Psychology, 12*, 45–60.

Di Masso, A., Dixon, J., & Pol, E. (2011). On the contested nature of place: 'Figuera's Well', 'The Hole of Shame', and the ideological struggle over public space in Barcelona. *Journal of Environmental Psychology, 31*, 231–244.

Di Masso, A., Dixon, J., & Durrheim, K. (2013). Place attachment as discursive practice. In L. Manzo & P. Devine-Wright (Eds.), *Place attachment: Advances in theory, methods and applications* (pp. 75–86). New York: Routledge.

Di Masso, A., Castrechini, A., & Valera, S. (2014). Displacing xeno-racism: The discursive legitimation of native supremacy through everyday accounts of 'urban insecurity'. *Discourse & Society, 25*(3), 341–361.

Diener, E. (2000). Subjective well-being. The science of happiness and a proposal for a national index. *American Psychologist, 55*(1), 34–43.

Dixon, J., & Durrheim, K. (2000). Displacing place-identity: A discursive approach to locating self and other. *British Journal of Social Psychology, 39*, 27–44.

Dixon, J., & Durrheim, K. (2003). Contact and the ecology of racial division: Some varieties of informal segregation. *British Journal of Social Psychology, 42*, 1–23.

Dixon, J., & Durrheim, K. (2004). Dislocating identity: Desegregation and the transformation of place. *Journal of Environmental Psychology, 24*, 455–473.

Dixon, J., Foster, D., Durrheim, K., & Wilbraham, L. (1994). Discourse and the politics of space in South Africa: The 'squatter crisis'. *Discourse & Society, 5*, 277–296.

Dixon, J., Reicher, S., & Foster, D. (1997). Ideology, geography and racial exclusion: The squatter camp as 'blot on the landscape'. *Text, 17*, 317–348.

Durrheim, K., & Dixon, J. (2001). The role of place and metaphor in racial exclusion: South Africa's beaches as sites of shifting racialization. *Ethnic and Racial Studies, 24*, 433–450.

Durrheim, K., & Dixon, J. (2005). Studying talk and embodied practices: Toward a psychology of materiality of 'race relations'. *Journal of Community & Applied Social Psychology, 15*, 446–460.

Durrheim, K., Rautenbach, C., Nicholson, T., & Dixon, J. (2013). Displacing place-identity: Introducing an analytics of participation. In B. Gardener & F. Winddance-Twine (Eds.), *Geographies of privilege* (pp. 43–70). Cambridge: Cambridge University Press.

Edwards, D., & Potter, J. (1992). *Discursive psychology.* London: Sage.

Eiser, J. R. (1990). *Social judgement.* Buckingham: Open University Press.

Eiser, J. R., Reicher, S., & Podpadec, T. (1993). What's the beach like? Context effects in judgements of environmental quality. *Journal of Environmental Psychology, 13*, 343–352.

Feldman, R. (1990). Settlement-identity: Psychological bonds with home places in a mobile society. *Environment and Behavior, 22*, 183–229.

Fleury-Bahi, G., Félonneau, M. L., & Marchand, D. (2008). Processes of place identification and residential satisfaction. *Environment and Behavior, 40*(5), 669–682.

Foucault, M. (1975). *Discipline and punish: The birth of the prison*. New York: Vintage Books.

Harré, R., & Gillett, G. (1994). *The discursive mind*. Thousand Oaks: Sage.

Hernández, B., Hidalgo, M., & Ruiz, C. (2013). Theoretical and methodological aspects of research on place attachment. In L. Manzo & P. Devine-Wright (Eds.), *Place attachment: Advances in theory, methods and applications* (pp. 125–137). New York: Routledge.

Hodgson, R., & Thayer, R. (1980). Implied human influence reduces landscape beauty. *Landscape Planning, 7*, 171–179.

Hollway, W. (1984). Gender difference and the production of subjectivity. In J. Henriques, W. Hollway, C. Urwin, C. Venn, & V. Walkerdine (Eds.), *Changing the subject: Psychology, social regulation and subjectivity* (pp. 227–263). London: Methuen.

Hubbard, P. (1996). Conflicting interpretations of architecture: An empirical investigation. *Journal of Environmental Psychology, 16*, 75–92.

Hur, M., Nasar, J., & Chun, B. (2010). Neighborhood satisfaction, physical and perceived naturalness and openness. *Journal of Environmental Psychology, 30*, 52–59.

Jorgensen, A., Hitchmough, J., & Dunnett, N. (2007). Woodland as a setting for housing-appreciation and fear and the contribution to residential satisfaction and place identity in Warrington New Town, UK. *Landscape and Urban Planning, 79*, 273–287.

Kahneman, D., Diener, E., & Schwarz, N. (Eds.). (1999). *Well-being: The foundations of hedonic psychology*. New York: Russell Sage.

Lewicka, M. (2011). Place attachment: How far have we come in the last 40 years? *Journal of Environmental Psychology, 31*, 207–230.

Macnaghten, P., Brown, R., & Reicher, S. (1992). On the nature of nature: Experimental studies in the power of rhetoric. *Journal of Community & Applied Social Psychology, 2*, 43–61.

Manzo, L. (2003). Beyond house and haven: Toward a revisioning of place attachment. *Journal of Environmental Psychology, 23*, 47–61.

Manzo, L. (2005). For better or worse: Exploring multiple dimensions of place meaning. *Journal of Environmental Psychology, 25*, 67–86.

Manzo, L. (2013). Exploring the shadow side. Place attachment in the context of stigma, displacement and social housing. In L. Manzo & P. Devine-Wright (Eds.), *Place attachment: Advances in theory, methods and applications* (pp. 178–190). New York: Routledge.

Marcouyeux, A., & Fleury-Bahi, G. (2011). Place-identity in a school setting: Effects of the place image. *Environment and Behavior, 43*(3), 344–362.

Mazumdar, S. (2005). Religious place attachment, squatting, and "qualitative" research: A commentary. *Journal of Environmental Psychology, 25*, 87–95.

Mazumdar, S., & Mazumdar, S. (1997). Intergroup social relations and architecture: Vernacular architecture and issues of status, power, and conflict. *Environment and Behavior, 29*, 374–421.

Mesch, G., & Manor, O. (1998). Social ties, environmental perception, and local attachment. *Environment and Behavior, 30*, 504–520.

Moreno, E., & Pol, E. (1996). *Nociones psicosociales para la intervención y la gestión ambiental* (Monografies socioambientals, 14). Barcelona: PUB.

Patterson, M., & Williams, D. (2005). Maintaining research traditions on place: Diversity of thought and scientific progress. *Journal of Environmental Psychology, 25*, 361–380.

Pol, E. (1996). *La apropiación del espacio* (Monografies Socioambientals, 9). Barcelona: PUB.

Pol, E. (2002). The theoretical background of the City-Identity-Sustainability (CIS) network. *Environment and Behavior, 34*, 8–25.

Pomerantz, A. (1986). Extreme case formulations: A new way of legitimating claims. *Human Studies, 9*, 219–230.

Possick, C. (2004). Locating and relocating oneself as a Jewish settler on the west bank: Ideological squatting and eviction. *Journal of Environmental Psychology, 24*, 53–69.

Potter, J. (1998). Discursive social psychology: From attitudes to evaluative practices. *European Review of Social Psychology, 9*, 233–266.

Potter, J., & Wetherell, M. (1987). *Discourse and social psychology. Beyond attitudes and behaviour*. London: Sage.

Ramkissoon, H., Weiler, B., & Smith, L. (2013). Place attachment, place satisfaction and pro-environmental behaviour: A comparative assessment of multiple regression and structural equation modelling. *Journal of Policy Research in Tourism Leisure and Events, 5*(3), 215–232.

Reicher, S., Podpacec, T., Macnaghten, P., Brown, R., & Eiser, J. R. (1993). Taking the dread out of radiation? Consequences of and arguments over the inclusion of radiation from nuclear power production in the category of the natural. *Journal of Environmental Psychology, 13*, 93–109.

Rioux, L., & Werner, C. (2011). Residential satisfaction among ageing people living in place. *Journal of Environmental Psychology, 31*, 158–169.

Scannell, L., & Gifford, R. (2010). Defining place attachment: A tripartite organizing framework. *Journal of Environmental Psychology, 30*, 1–10.

Seamon, D. (1983). The phenomenological contribution to environmental psychology. *Journal of Environmental Psychology, 3*, 109–111.

Sirgy, M. (2012). *The psychology of quality of life: Hedonic well-being, life satisfaction, and eudaimonia*. Dordrecht: Springer Publishers.

Stedman, R. (2002). Toward a social psychology of place: Predicting behavior from place-based cognitions, attitude, and identity. *Environment and Behavior, 34*, 561–581.

Stokoe, E., & Wallwork, J. (2003). Space invaders: The moral-spatial order in neighbour dispute discourse. *British Journal of Social Psychology, 42*, 551–569.

Taylor, S. (2010). *Narratives of identity and place*. New York: Routledge.

Vidal, T., & Pol, E. (2005). La apropiación del espacio: una propuesta teórica para comprender la vinculación entre las personas y los lugares. *Anuario de Psicología, 36*, 281–297.

Wiesenfeld, E. (1995). *La vivienda: Su evaluación desde la psicología ambiental*. Caracas: Universidad Central de Venezuela.

Willig, C. (2008). *Introducing qualitative research in psychology*. Maidenhead: McGraw-Hill.

Part II

People-Environment Relations and QoL – *Restorative Environments*

Self, Nature and Well-Being: Sense of Connectedness and Environmental Identity for Quality of Life

6

Pablo Olivos and Susan Clayton

6.1 Role of the Environment in the Construction of Positive and Lasting Self-Identities in Modernity

From the study of the dimensions of self that underlie environmental concerns, a relatively new research issue in environmental psychology has been raised. This approach studies the bond or connection between humans and the natural environment, specifically when people incorporate nature into their psychological image of themselves. As developed later, many different concepts and measures have been proposed to address the study of this phenomenon. They include emotional affinity toward nature (Kals et al. 1999), inclusion of nature in self (Schultz 2001), environmental identity (Clayton 2003), connectedness with nature (Mayer and Franz 2004), nature relatedness (Nisbet et al. 2009), and love and care for nature (Perkins 2010). Different adaptations of these have also been developed, such as connectivity with nature (Dutcher et al. 2007), the scale of environmental preferences (Sánchez et al. 2009), the environmental connectedness scale (Beery 2012), and inclusion of environment in self (Olivos and Aragonés 2014). In general, all these procedures have tried to describe how people believe that the image they have of themselves (self or identity) is linked to or separate from nature (Brügger et al. 2011; Tam 2013), and have given directions to more recent research about the relationship of nature with psychological well-being.

However, in some ways, this field of study, as well as other classic theoretical contributions in the environmental field of social sciences, has been built on vague, ambiguous and unspecific ideas about the natural environment and the self (Leary and Tangney 2003; Lee 1976; Seagert 1987). As pointed out by Canter and Craig (1981), this happened in environmental psychology to avoid constricting the development of the specialty with an overly narrow definition. However, adopting one definition or another involves significant epistemological and methodological implications to place the scope of studies in environmental and social psychology. Thus, in this section, we briefly develop the argument that the social science perspective has been based on a particular conceptualization of these two concepts.

The modern definition of nature typically adopted in social sciences is the result of

P. Olivos (✉)
Psychology Department, University of Castilla-La Mancha, Ciudad Real, Spain

School of Labor Relations & Human Resources, Plaza de la Universidad, 02071 Albacete, Spain
e-mail: pablo.olivos@uclm.es

S. Clayton
Psychology Department, College of Wooster, 930 College Mall, Wooster, Ohio 44691, USA
e-mail: SClayton@wooster.edu

conceptual distinctions that reflect the dualism "nature/culture" or "human/non-human world" (Dove and Carpenter 2008). According to analysis by historians (Kahler 1943; Williams 1983), this distinction had its heyday in the secular relationship of humans with nature, which characterized the emergence of the bourgeois spirit and tendency to live in a community during the Reformation, and in the lifestyles of medieval German towns in the sixteenth century.

Anthropological studies have shown that this kind of categorization has been culturally transmitted from the past, with the help of complex cognitive elaborations (Descola and Palsson 1996; Ingold 2000; Ellen 2001). These distinctions adopt forms of identification modes that provide distinctions between nature as a kind of separate thing or non-human space and as an interior aspect of humanness and a common essence of life. For example, totemism endows human groups with properties of the natural world (signs), from empirically observable specificities; animism attributes human social features to the natural world; and, according to naturalism, nature exists independently of fate and human will. Nothing happens without a reason or cause, linked to a higher order or a metapersonal self (Arnocky et al. 2007; Leary et al. 2008; Olivos and Aragonés 2014). These anthropological worldviews have also been reproduced as modes of relationship with nature, as rapacity or protection. For example, in terms of social representations of nature (Callaghan et al. 2012; Navarro 2013), it is easy to understand how scientists now define a change of era called the Anthropocene (Latour 2013; Steffen et al. 2007; Zalasiewicz et al. 2010), according to patterns of relationships characterized by intense human predation of the environment.

As Descola and Palsson (1996) pointed out, understanding nature involves reconstructing models or patterns of social practice, represented as mental guidelines that direct people's interactions with the natural environment. In this regard, one of the most common modes of identification is naturalization, which Franklin (2002) defined as the need for individuals and groups to reconnect with aspects of the natural world, something like Western culture embeddedness in nature.

This is not meant to suggest a proliferation of environmentalism, or widespread environmental awareness, but rather a process of identification with and through the natural world, which appeals to the family kinship among its members to convey different ideas. According to Franklin, the most relevant subject of study for social sciences in the relationship between nature and society is not the connection but the disconnection between nature and the contemporary Western world, due to the development of an urban lifestyle, and how the society develops desperate attempts to reunite with nature. The ancient and desired connection with nature has not gone; it has been relocated and redefined in a way that involves new social objects, new practices and new cultural needs. The way in which advertisers try to sell products by marketing their supposed connection with nature, or having a house in the countryside, are emblems of a kind of nature connection while still representing a social status of inclusion in the network of a consumer society.

There are studies about the relationships between humans and animals that could be interpreted in one cultural mode of anthropomorphic attribution (e.g. Opotow 1993; Sevillano et al. 2007), but there is a scarcity of studies in environmental psychology that explicitly apply anthropological categories to research relationships between self and nature, or develop a sociological analysis of the extent of their results. An example of this type of approach is the research of Tam et al. (2013), who carried out experiments with students from Singapore and Hong Kong, showing that anthropomorphism of nature fosters conservation behavior and increases feelings of connection with nature. Moreover, these authors suggested that a sense of connectedness could mediate the association between the anthropomorphism of nature and conservation behavior.

Something similar to the historical path of the concept of nature occurred in the development of the concept of self and identity. The historical origins of modern identity are located in different processes of objectification (Kahler 1943; Siedentop 2014; Taylor 1989). The first began with a physical separation from the universe, from the concept of the forces of nature as different categories of living things, transforming relationships with nature and other members of the tribe into practices that became rituals over time. The second was the expansion of the idea of the universe as an all-encompassing spiritual being, and humanity as a community of equal human beings under a universal God and a common destiny. This second objectification was made possible by: (i) the Greek concept of a spiritual life accessible through speculative reflection and democracy, with its concept of a community of free and equal citizens in the city-state; (ii) the Judeo-Christian concept of a human being made in the image and likeness of a universal and spiritual God; and (iii) the expansion of both ideas (universal religion and city-states) under Roman imperialism, which helped others to turn citizens into subjects.

The third and last objectification that gave rise to modern identity corresponded to a change in the relationship between human beings and nature. It was derived from the total split between the old covenant of earthly (emperors) and spiritual (pope) forces, and a gradual process of secularization. According to Taylor (1989), the features of modern identity were developed by ideas transmitted into foundational revolutions consolidated from the eighteenth century, such as the ideals of equality, universal rights, the work ethic, the recognition of sexual love and family, and so on. The liberal secularism that originated in the Christian West, according to Siedentop (2014), initiated the concept of individual liberalism that prevails today in the West.

In social psychology, despite the early advances in the study and application of the contemporary use of the concept of self and identity, thanks to authors such as William James, Charles Cooley and George Mead, the use of the word *self* and different theoretical approaches to understand the identity processes have provoked much discussion and controversy (e.g. Baumeister 1987; Hogg et al. 1995). Some of the controversy is derived from the theoretical-epistemological antagonism between different traditions of psychosocial thinking, and some is only the result of a lax use of the term. The latter is probably the major source of discord.

After an extensive review of the scientific use of the concept, in which they identified at least five different uses, Leary and Tangney (2003) suggested that researchers should define more precisely what they mean when they study phenomena under the broad and ambiguous concept of self. They further recommended that researchers should reserve the concept for the study of the cognitive mechanisms that give rise to reflective thinking about ourselves. However, none of the most current and comprehensive reviews on the study of the self (Leary and Tangney 2012; Schwartz et al. 2011) devotes a comprehensive review to the role of environment in the construction of identity, much less of the natural environment.

Thus, the historical background of the cultural construction of the concepts of nature and self provides a foundation for analyzing how the human-environment relationship has been a key in the formation of modern identity, centered on the concept of well-being anchored in the anthropocentric relationship of humans with the environment. The frameworks of the description of this relationship, in environmental psychology, have been studied in terms of self-determination theory, attachment theory, and restoration theory. We develop some research findings according to these perspectives throughout the chapter.

6.2 Connectedness with Nature

A growing body of research attests to the relevance of the natural environment in promoting a positive identity that contributes to well-being. The burgeoning recognition of the importance of environment to identity has led to the development of a number of ways of assessing the connection. We review some of the principal concepts and main measures. Because the measures have been found to be highly intercorrelated, we refer to specific measures when describing specific studies, but may describe the general construct as a sense of connection or relationship with nature or as an environmental identity.

Despite an overall similarity among the measures, there are important distinctions that differentiate them. One is format. As described below, most of the measures simply ask respondents to indicate their agreement with a number of statements on a Likert-type scale; however, Schultz and colleagues have developed one measure based on reaction time (the Implicit Associations Test or IAT-Nature) and another that uses a visual analogue of the idea of interdependence (the Inclusion of Nature in the Self or INS scale). Another distinction relates to the type of construct: some involve emotion and some cognition; some explicitly concern self-concept and others are more attitudinal.

Although emotional response does not by itself indicate that nature is relevant to identity, it is consistent with this idea. In general, issues that are self-relevant evoke a greater response, and empathic responses are typically facilitated by a perception of similarity or shared identity. Another concept is more cognitive; the idea of a self-concept in which the construct of nature and the construct of self are not independent but are joined together and possibly overlap. Because the self-concept is constructed from experiences, a cognitive connection between self and nature would be facilitated by memories of self in nature (Olivos et al. 2013; Schroeder 2007; Thomashow 1995). Most measures assess this conceptual connection between self and nature in some way. Similarly, a third concept relevant to the connection between self and nature is behavior. Experiences and activities in the natural world, or which evince care for the natural world, are likely to demonstrate the perception of an environmental identity. Many measures also include a self-report behavioral component, which has demonstrated strong links to connection with nature measures. In a study of 306 undergraduate students, who completed a questionnaire composed of the Connectedness to Nature Scale (CNS), the Environmental Identity scale (EID), the Pro-environmental behaviors scale (PEB), and a series of questions designed to measure their commitment to eco-friendly activities of information and conservation (Pro-environmental Commitment Index, PCI), Olivos et al. (2014) noticed that those participants with a high level of PCI were more connected with nature and obtained higher scores in environmental identity, and that PEB was predicted by the environmentalism dimension of EID and CNS.

6.2.1 Relevant Measures

6.2.1.1 Emotional Affinity Toward Nature

Kals and colleagues (Kals et al. 1999) tapped into the psychological significance of the natural world with their Emotional Affinity toward Nature scale, which includes questions about "feelings of oneness with nature" as well as about love for nature and feelings of freedom and security associated with nature. This recognition that emotional responses to nature are connected to a sense of relatedness was an early acknowledgement of the ways in which self-concept might play a role in environmental attitudes. As this scale indicates, the emotional response to nature is one of the key concepts that has been included in research on connection. Kals and Ittner (2003) describe emotional affinity as an "indicator" of environmental identity, and suggest that it is based on the *biophilia* hypothesis, a concept proposed by Wilson (1984) to address people's sense of emotional connection with the natural world, implying that an innate tendency to focus on life and life processes is part of our genetic inheritance.

6.2.1.2 Inclusion of Nature in the Self

The Inclusion of Nature in the Self (INS) scale, developed by Schultz (2001, 2002), was perhaps the first measure to emphasize connections between the natural environment and the self-concept. Based on Aron et al.'s (1992) Inclusion of Other in Self scale, the INS measures the extent to which people perceive themselves as distinct from, related to, or commensurate with the natural environment by asking them to select from a series of seven pictures. Each picture depicts two circles, one labeled "self" and the other labeled "nature." They start side by side and then overlap to an increasing extent until they are entirely the same. This measure visually represents the metaphor by which people define their relationship with nature, and appears to be easy for people to understand. Notably, it relates directly to one of the core motives from Self-Determination Theory, namely, the desire for connection or belongingness. Just as Aron et al.'s measure was designed to examine the extent to which one has successfully expanded one's sense of self to include others, the INS assesses the extent to which one has expanded one's sense of self to include the natural world.

6.2.1.3 Environmental Identity

With the explicit goal of examining the way in which a connection with the environment can be part of an individual's identity, and inspired by work on group identities such as racial and gender identity, Clayton (2003) developed a 24-item environmental identity (EID) scale. More recently, a shorter form of only 11 items has been proposed (see Clayton 2012). Items on the scale reflect the extent to which a person tends to interact with elements of the natural world, whether a person rates nature as important, whether the person thinks of him/herself as part of nature, and whether the natural environment evokes positive emotions. The EID scale also includes items related to competence, autonomy, belongingness, and engaging in outdoor activities. It should be distinguished from some other measures that are designed to assess whether someone self-identifies as an *environmentalist* (e.g., Van der Werff et al. 2013a, b). An environmentalist identity refers to one's self-perception as someone who is actively involved in protecting nature, but such an identity does not necessarily imply a sense of connection with the natural world.

Reliabilities for the EID scale are high in samples from various countries, and it has shown the expected pattern of correlations with attitudes, values, and worldview (Clayton 2003). Importantly, it also seems to reflect the internal cognitive structure of information related to environmental issues: in one study, people with a high score in EID found it slightly easier to make decisions about environmental dilemmas and were more confident in their decisions. In research in a zoo setting, EID was related to a reported sense of connection with the animals and a perception of the animals as similar to humans. Zoo members also scored higher than non-members on a measure of environmental identity (Clayton et al. 2011).

Although in a factor analysis of the scale among college students Clayton (2003) found only a single dominant factor, Olivos and Aragones (2011) found evidence of five factors in a Spanish sample. The first factor, accounting for by far the largest proportion of variance (32.8 %, with the second factor accounting for only 7.2 %), was described as "environmental identity"; additional factors (after eliminating one that was represented by a single item) were identified as assessing time in nature, appreciation of nature, and environmentalism. Each subscale was correlated with pro-environmental behavior, but only the "environmentalism" and "environmental identity" subscales contributed unique variance in a regression analysis.

6.2.1.4 IAT-Nature

If biophilia is inherited, as proposed by Wilson (1984), a sense of connection with nature may be rooted in an instinctual response rather than an explicit one. Consistent with this idea, Schultz suggested that connections with nature might be implicit, operating outside conscious awareness. Schultz and colleagues (Schultz et al. 2004; Bruni and Schultz 2010) developed a form of the Implicit Associations Test that uses a reaction-time paradigm to assess the extent to which people have strong cognitive connections between themselves and nature. This test requires people to decide rapidly and accurately if two paired terms belong together or not. For example, does an item like "flower" belong with "me or nature" or "other or built"? After rounds of this sort, the game requires choices of whether "flower" (and other terms) belong with "me or built" or "other or nature"? A person with a mental network connecting the self to nature will answer the former items more rapidly because there is a cognitive connection between "me or nature", whereas it will take more deliberation to discern that "flower" belongs with "other or nature" because "other" and "nature" are not cognitively associated. The time delay to answer questions correctly provides an estimate of the degree of how closely the person implicitly identifies with nature or with the built environment. The advantage of such an approach is that it does not rely on self-report; people may not have conscious access to their own sense of interconnectedness with nature, or their report may be biased by social desirability. Schultz found that a high degree of implicit connectedness with nature was associated with more biospheric, and fewer egoistic, concerns. This measure, however, does not always correlate with explicit measures related to identity or environmental concern (Bruni and Schultz 2010; Brügger et al. 2011; Olivos and Aragonés 2013). Further research is needed to assess the difference between implicit and explicit measures.

6.2.1.5 Connectedness to Nature

Mayer and Frantz (2004) developed a Connectedness to Nature Scale (CNS), also theoretically anchored in the concept of biophilia, and defined as a measure of emotional connectedness with the natural world. Similar to the EID in some ways, it is shorter and focuses primarily on an affective response to nature. It correlates with environmentalist identity and worldviews among college and community samples. Like attitude measures, CNS is likely to be more responsive to situational manipulations than identity measures would be, and is thus useful as a measure of contextual variability in perceived nature-relatedness (Frantz et al. 2005). In fact, more recent research has differentiated between "trait" and "state" CNS (Mayer et al. 2009). For the purposes of the present chapter, we are more interested in connection as it relates to identity, and thus as a relatively stable trait. There is some ambiguity about whether the CNS is primarily a measure of affective or cognitive response to nature (Perrin and Benassi 2009). Cervinka et al. (2012) found stronger relationships with a single-item measure of connection than with the CNS scale, and suggested that CNS only measures some of the aspects of connection with nature.

6.2.1.6 Nature Relatedness

In their Nature Relatedness (NR) scale, Nisbet et al. (2009) identified three components of a relationship with nature: affective, cognitive, and experiential. Although they do not refer to it as an identity measure, they do describe it as an internalized identification with nature and as indicating the strength of an individual's connection with nature (Nisbet et al. 2011). Nisbet et al. argue that individuals have a need to connect with nature, so that the extent to which this need is met should predict well-being. Indeed, the measure seems to capture the personal benefits of associations with nature (Nisbet et al. 2011; Zelenski and Nisbet 2014). Relevant to Self-Determination Theory, high NR scores were associated with greater autonomy and environmental mastery as well as with positive relationships with others.

Tam (2013) compared the above explicit measures (not the IAT-Nature) and other similar measures (Connectivity with Nature, Dutcher et al. 2007, which refers to a perception of similarity between self and nature; and Commitment to Nature, Davis et al. 2009, which conceptualizes the link between people and nature as similar to an interpersonal relationship). Using samples from both Hong Kong and the U.S., Tam examined the relationship between the diverse variables as well as to behavioral and dispositional criteria. He found strong convergence around one common factor, a latent construct of connection with nature. All measures were highly correlated, and all were correlated with the criterion variables such as environmental attitudes and behavior, but EID and NR had the highest external validity. Olivos and Aragones (2013) also found strong intercorrelations between the IAT-Nature, INS, EID, and CNS among Spanish college students.

In addition to having significant correlations with happiness and satisfaction with life, Tam (2013) found that EID and NR had significant relationships with the Big 5 personality factors and with Schwartz's value dimensions, suggesting that it makes sense to think of EID as tied to fundamental dispositional characteristics. People with high EID or NR scores were more extravert, agreeable, conscientious, and open to experience, and less neurotic.

6.2.2 The Social Context for Environmental Identity

Not surprisingly, environmental identity is facilitated by experiences in nature. For example, Tam (2013) found that it was highly correlated to both past and present contact with nature. Chawla (e.g., Chawla and Derr 2012) has argued that early experiences of nature are important for a grounded self as well as for the development of pro-environmental concern, but she notes that the social context within which the experiences occur is also important. Environmental identity develops through interactions not only with the natural environment but also, crucially, with the social environment. Identity formation is intrinsically related to social interactions (Wortham 2006). Although some nature experiences are solitary, people also experience nature in more social settings, as members of social groups and in their communities. Experiences of nature in the company of others affect our understanding of what nature signifies as well as the way we conceptualize our own relationship with nature.

A relationship with the natural world may affect how a person affiliates or conflicts with others, or how other people define and perceive a person. As a specific example, Abell (2013) described the way in which volunteers working for an environmental cause formed a collective identity based on shared values and a sense of connection. Similarly, Fraser et al. (2009) explored the way in which interactions among zoo volunteers served to strengthen their commitment to environmental causes and their self-definitions as environmentalists.

In the reverse direction, the psychological relationship with nature is also related to positive social relations. The Nature-Relatedness Scale shows strong relations with agreeableness and humanitarianism (Nisbet et al. 2009). In a study by Kaiser and Byrka (2011), 90% of environmentalists had prosocial personalities, compared with 65% of those with low environmental scores; they were also more likely to act cooperatively. Zhang et al. (2014a) found that those more prone to perceive natural beauty scored higher on agreeableness, perspective-taking and empathy. In experimental work, Zhang et al. exposed participants to more beautiful examples of nature (images, plants) or less beautiful examples and found that the former evoked more generous sharing, more trust, and more helping behavior. In a field experiment, Guegen and Stefan (2014) found more helping behavior among people who had just been immersed in an urban green park. In a "virtuous cycle," the sense of connection with nature may help people to feel more connected to other people as well, improving well-being by strengthening social bonds.

6.3 Well-Being, Environment and Connectedness

Dating from around the 1970s, well-being has been studied in different areas of the social sciences, including economics (e.g. Easterlin 1974), sociology (e.g. Levy and Guttman 1975), psychology (e.g., Frankenhaeuser 1977) and others. Because of the introduction of a positive psychology framework, it has also been one of the most popular variables analyzed in environmental studies, as we describe below. However, there is little consensus about how it may be identified, measured and achieved because it lacks a clear conceptual base (Baumeister et al. 2013; Cronin et al. 2005; Mathews and Izquierdo 2010).

One of the most important challenges in measuring and describing well-being is validity. The international well-being group (http://www.deakin.edu.au/research/acqol/iwbg/) has developed an International Well-being Index (IWI). According to Cummins (2010, 2011), the homeostatically protected mood provides people with an affective and positive view of the self, generated genetically, which could vary individually by a sense of contentment, happiness and arousal. Gullone and Cummins (2002) argue for the universality of subjective well-being indicators, according to a homeostasis theory of a normal distribution of level of subjective well-being, measuring eight life domains (community life; health; life achievement; living standard; personal relationships; safety; security; and spirituality) in 50 countries, under normal living circumstances. However, other authors argue that the study of well-being must consider that people's lives do not only take place under normal circumstances, and the study of well-being has acquired its value due to the differences in the very meaning of well-being, happiness and quality of life in different cultures (Mathews and Izquierdo 2010; Selin and Davey 2012; Thin 2010). For example, studies conducted in China, in contrast to Western countries, show significantly higher levels of national well-being in the domain of national security than environment, probably because of the salience of a culture regulation of the social order (Chen and Davey 2009).

6.3.1 Well-Being and Environmental Psychology Studies

Despite the existence of well-known theories used to study well-being in environmental psychology, such as the "self-determination theory" (Ryan and Deci 2000) and "attention restoration theory" (Kaplan and Kaplan 1989), different concepts have been used as equivalent in these studies, such as happiness, welfare, health, satisfaction, and attention. For example, although some authors have proposed that satisfaction and well-being are equivalent and interchangeable concepts (e.g. Di Tella et al. 2001; Vennhoven 1997, 2004; Welsch 2006; Brereton et al. 2008) discussed the ways in which they are distinct, because of the influence of sociodemographic variables on the latter.

Consequently, various scales have been developed to measure each one of the proposed concepts. Some of the most popular in the study of well-being are the Positive and Negative Affective Schedule (PANAS, Watson et al. 1988), the Satisfaction With Life Scale (SWLS, Diener et al. 1985), the Psychological Well-being Scale (PWS, Ryff 1989), and the General Health Questionnaire (GHQ, Goldberg and Blackwell 1970). There are arguments against the use of self-reports to measure well-being, due to the biases involved. However, this is more problematic when well-being is an independent variable. Because of this, several physiological indicators of psychological restoration, such as attention, heart rate, eye tracking, etc., have also been used. In any case, providing empirical measures of individual well-being is more valuable than avoiding its measurement (Blanchflower and Oswald 2004; Brereton et al. 2008).

Studies carried out in environmental psychology about bonds between well-being and environment reflect this diversity of approaches, as is developed in different chapters of this book. When investigating the effects of contact with nature and other environmental stimuli, some have focused on health, others on the meaning of life and others on positive emotions.

In general, studies show positive effects of contact with green environments, built or natural, on health and well-being (Astell-Burt et al. 2014; Cerina and Fornara 2011; Gidlöf-Gunnarsson and Öhrström 2007; Pijanowski et al. 2011). Contact with nature has also been observed to increase the quality of life, happiness and mental restoration (e.g., Burns 2005; Dasgupta 2001; Gatersleben 2008; Hartig et al. 1991; Kaplan 1973, 1983, 1995, 2001; Kaplan and Kaplan 1989; Kaplan and Talbot 1983; Knopf 1983; Sagar 2007; Staats et al. 1997; Wallenius 1999). Similar results have been observed even for watching images or videos of nature, and some of this research has focused on the positive effects of the physical conditions of environments studied, like a high level of naturalness in an urban environment (Carrus et al. 2013; Nordh et al. 2010), spring landscapes (Falsten 2014), mountains, trees, forests, and valleys (Hinds and Sparks 2011), reduced wind speed, slight temperature rise, proximity to coastal areas (Brereton et al. 2008), and air quality (Ferrer-i-Carbonell and Gowdy 2007; Welsch 2006).

Restoration theory has been used to explain some of the most important cognitive benefits, but focusing on only the cognitive effects of nature misses some of the important mechanisms through which it has a positive impact. When environmental studies focus their research on subjective and psychological well-being, most of them support the positive relationship between contact with nature and positive emotions such as happiness (e.g. Corral-Verdugo et al. 2011; Detweiler et al. 2008; Diette et al. 2003; Horsburgh 1995; Ulrich 1984). In the same way, some studies have observed a positive relationship between this kind of well-being and personality traits such as extraversion, friendliness and openness (Deneve and Coper 1998; Falsten 2014; Merrero and Carballeira 2010). Others show positive relationships between perfectionist and control lifestyles and well-being (Ragip et al. 2012).

Another important conceptual distinction in the study of well-being is the difference between hedonic and eudaimonic well-being. The first concept has been assigned an affective dimension that emphasizes the pursuit of pleasure (Diener 1984). Eudaimonic well-being, in contrast, focuses on developing human potential, giving meaning and direction to life through the promotion of personal strengths and virtues (e.g., Ryff 1989; Waterman et al. 2008). Although emotions play an important role in the construction of self (e.g. Church et al. 2014), this is not the sense in which it has been addressed in most investigations of restoration theory. Furthermore, the eudaimonic dimension of well-being is most closely linked to the development of positive and complex identities (Ryff and Singer 2013). A study of participants undergoing addiction treatment (Pryor et al. 2006), who participated in an Outdoor Experience program called "bush adventure therapy", suggested that social environment, physical challenges and contact with nature were the factors that participants identified as key in increasing their physical and mental health and well-being.

Hinds and Sparks (2011) observed that people who have spent their childhood in a rural environment, and who have great experience of outdoor activities, experience more positive (eudaimonia) and less negative (apprehension) emotions when asked to imagine their emotions in a natural environment. In another study, which recorded eudaimonic and hedonic behaviors in 100 field notebooks throughout 4 days of observation, Henderson et al. (2013) observed that hedonic well-being, when controlling eudaimonic, predicted positive affect, negative affect, life satisfaction, carelessness, vitality, depression and stress. Eudaimonic well-being, when controlling hedonic, only predicted the meaning of life and elevation. Both types of well-being predicted flourishing.

6.3.2 Connection to Nature and Well-Being

As described above, connectedness with nature has become popular in environmental psychology studies, but its relationship with well-being has been studied for a relatively short time. However, most of the researchers who have explored this area assessed connectedness with CNS or NR scales, and well-being with emotional (e.g. PANAS) and eudaimonic (e.g. Psychological Well-being Inventory) scales. In this section, we summarize the results of some papers, published in the last 5 years, which have studied this issue specifically.

Correlational studies, carried out with CNS, concur in finding positive correlations between connectedness and different scales of well-being (Cervinka et al. 2012; Howell et al. 2011; Wolsko and Lindberg 2013) and the psychological health dimension (Cervinka et al. 2012; Kamitsis and Francis 2013), in different populations, an effect sometimes mediated by other variables such as higher engagement with natural beauty (Zhang et al. 2014b). A recent meta-analysis (Capaldi et al. 2014) incorporating research that operationalized nature connectedness in a variety of ways showed a small but consistent relationship with happiness, operationalized as life satisfaction, positive affect, and vitality. The relationship was not mediated by age, gender, or publication year, but it was stronger for the INS measure than for the CNS or NR scales.

It is particularly interesting to examine studies that try to determine causal relationships by employing some kind of experimental manipulation. Mayer et al. (2009) conducted one of the first of these. Due to some criticism of the connectedness scale (e.g. Perrin and Benassi 2009), they distinguished between a trait and state version of connectedness. They observed a close relationship between them, because trait CNS scores predicted state scores; however, trait scores tended to remain at their original levels despite the positive effect of contact with nature on state CNS scores. During three studies, they tested the mediational effect of connectedness on the relationship between contact with different settings (nature, urban and images) and well-being, measured with PANAS and attentional capacity. Statistical analyses showed that the positive impacts were due to the sense of connection with nature, rather than to stress reduction or attention.

Other studies show the same trend of positive associations between connectedness and well-being, measuring similar concepts, grouped under the label of happiness. For example, Nisbet and Zelenski (2011) carried out two experiments to study the relationship between connectedness (INS) and well-being (PANAS), as well as relaxation, arousal states, fascination, curiosity, and interest. Comparing differences between an indoor and an outdoor walk in a group of Canadian students, in two different studies with similar conditions, they observed that the latter obtained higher scores in positive affect, relaxation, and fascination, and less negative affect, than those who walked indoors. Outdoor walks produced greater state nature relatedness and this effect was mediated by positive affect, taking into account that the participants who walked outdoors underestimated the hedonic benefits of the walk, while those who walked indoors overestimated their moods. In another study, Nisbet et al. (2011) studied the relationship between natural relatedness and well-being, through three studies that provided supporting evidence for positive relationships with the psychological well-being inventory (PWB), PANAS, satisfaction with life scale, and vitality, in different populations. They found that levels of NR were maintained over time in a group who participated in activities of environmental education, although they suggested that the effect might have been due to the weather in which they conducted the experiment (autumn-winter).

Recently, these authors have obtained the same results with a shorter version of the NR scale (Nisbet and Zelenski 2013), comprising self (sense of identification) and experience (contact with nature) dimensions. Through different studies, in multiple samples (students, community members, business people) and longitudinally, they found associations of NR (long and short versions) with sustainable behaviors, environmental concern, nature contact and happiness, the latter measured with PANAS, the Psychological Well-being Inventory (stronger with autonomy and personal growth dimensions) and the Satisfaction with Life scale. Positive correlations between NR and emotional, psychological and social well-being have been observed by other researchers, as well as positive correlations with awareness and acceptance dimensions of mindfulness (Howell et al. 2011).

The perception that one is connected to nature affects more than just hedonic well-being. Scores on the Connectedness to Nature Scale have been found to correlate with psychological resilience among a self-selected but diverse sample from the northwestern U.S. (Ingulli and Lindbloom 2013). In the studies by Nisbet et al. (2011), those higher in nature relatedness also showed greater purpose in life, self-acceptance, autonomy, personal growth and positive relationships. Connectedness to nature is also related to a sense of meaningfulness in life (Cervinka et al. 2012). Finally, environmental identity has been found to correlate with a sense of self-efficacy about the possibility of effecting environmental change (Clayton et al. 2014).

Based on the research, there seem to be two types of phenomena associated with an environmental identity. One can be described as a sort of self-expansion, related to the idea that environmental identity might be motivated by the desire for a larger sense of self. Wolsko and Lindberg (2013) talk about the way in which a sense of relatedness to nature can be part of a process of "cultivating a larger identity" (p. 81) that transcends egoism. This would include experiences described as "transcendent", which are usually considered to be moments when one feels elevated or lifted out of oneself to experience a sense of connection to something larger, or sometimes a sense of how small one's individual self is. Associated emotions are awe and humility.

Hoot and Friedman (2011) regard identification with nature as a type of self-expansion; they found that CNS scores were significantly correlated with a measure of self-expansiveness. Leary et al. (2008) describe a sense of personal connection to others, either with other people or with the natural world, as enabling people to overcome egocentrism; they find that people with a high score in this type of expanded identity are more concerned about other people as well as about ecological issues, and inclined toward spiritual experiences. In this way, Davis and Gatersleben (2013) compared the influence of the trait connectedness to nature, in emotional appraisal and transcendent experiences, between visitors of wild (cliff) and manicured (garden) settings. This study was carried out in summer time, in Ireland. They found differences in the transcendent experiences of participants between settings; negative deep flow transcendent for gardens, and positive diminutive transcendent for cliffs. They also noted that high levels of connectedness to nature at the wild cliffs predicted transcendent and awe-inspiring experiences, even though gender (female) and previous experience with this setting had significant effects on these variables and calming experiences. Similarly, Kamitsis and Francis (2013) found that CNS scores were associated with a measure of spirituality, which included items such as "I have had an experience in which something greater than myself seemed to absorb me". Notably, the relationship between CNS and well-being was insignificant when spirituality was included. Although this work is in its early stages, it suggests that the positive impacts of connection to nature may be due to its ability to link the self-concept to something larger than the self.

A second correlate of environmental identity may reflect the ability of natural environments to enhance self-knowledge. As noted above, studies have found correlations between a sense of connection and a sense of meaningfulness (Cervinka et al. 2012). Passmore and Howell (2014) observed that trait connectedness to nature predicts levels of well-being – in terms of positive affect and elevation (PANAS) – but independently of nature involvement (experimental condition), but they found no moderator effect of CNS on the effect of contact with nature on well-being. In research on "environmental epiphanies," Vining and Merrick's (2012) participants described experiences that led to a new understanding of their relationship with nature. One person described it this way: "I understood my place in the universe" (p. 485). In the study by Howell et al. (2011), a confirmatory factor analysis showed that different well-being scales loaded significantly on a well-being latent factor, and significant associations were found between this variable and connectedness with nature, and between well-being and mindfulness, but not between connectedness and mindfulness. However, all the variables fitted when loaded into one single latent variable composed by Allo-Inclusive Identity items – concerning connectedness between oneself and nature (Leary et al. 2008), CNS and NR. Wolsko and Lindberg, collapsing the psychological well-being latent variable from observed flourishing, subjective vitality and positive and negative emotions, found that CNS correlated with it, and with engaging in more appreciative outdoor recreational activities. These results suggest the convenience of using a set of connectedness measures or well-being measures as a whole to study the relationship between both variables, as did Wolsko and Lindberg (2013) and Leary et al. (2008), who found a relationship between connectedness and internal state awareness.

This may be related to a general association between nature and cognitive activity. Research showing that time in natural settings restores attentional capacity (e.g., Kaplan 1995) and promotes cognitive functioning suggests that it may provide an appropriate context for reflecting on one's own identity, goals, and values, and thus contribute to self-understanding. Indeed, people report that natural settings are desirable places for self-reflection (Korpela et al. 2001). Experiencing a connection between oneself and nature may provide additional cognitive benefits: Leong et al. (2014) found that people with high CNS or NR scores demonstrated more innovative and holistic thinking.

6.3.3 Pro-environmental Behavior and Well-Being

Connections to the natural environment may also have an indirect effect on well-being through pro-environmental behavior. A number of studies have found that behaviors with a low environmental impact are associated with happiness or satisfaction. Frugality has been described as intrinsically satisfying (De Young 1996) and related to greater personal well-being (Brown and Kasser 2005), while materialism predicts lower levels of happiness (Kasser 2002). Corral-Verdugo (Corral-Verdugo et al. 2011) defined the construct of "sustainable behavior" reflecting frugality, altruism, equity, and pro-environmental behavior. In research with over 600 Mexican undergraduates, he found that sustainable behavior, encompassing each of these four components, was a significant predictor of happiness. The reasons for this are not fully clear, but Corral-Verdugo et al. suggest that the emphasis on intrinsic rather than extrinsic rewards is part of the explanation. A focus on doing good for the planet may provide a sense of purpose and meaning that contributes to eudaimonic well-being (Venhoeven et al. 2013). In studies reported by Taufik et al. (2014), people who were told that they were acting pro-environmentally literally perceived a warmer temperature, compared to those who received no message or who were told they were behaving in a relatively harmful way, as a consequence of feeling good about themselves.

6.4 Conclusions

To conclude, we would like to develop three main ideas arising from the literature review. First, we consider the methodological challenges for studying the relationship between connection with nature and well-being; second, we describe the theoretical and conceptual challenges inherent in its study, and finally we discuss some issues for the future.

As described above, correlational studies concur in finding positive relationships between a sense of connection with nature and different scales of well-being. Sometimes, a sense of connectedness is described as a mediator variable between contact with nature and well-being, while at other times it is considered a moderator of the relationship between nature and well-being effects. However, using connection with nature as a mediator or a moderator variable is not the same. A moderator variable affects the direction and/or strength of the relationship between an independent or predictor variable and a dependent or criterion variable, while a mediator variable explains how external physical events take on internal psychological significance (Baron and Kenny 1986). In other words, "whereas moderator variables specify when certain effects will hold, mediators speak of how or why such effects occur" (p. 1176).

This methodological distinction is important according to the epistemological concept of the bond between self and nature. Connectedness in the biophilic sense of being in nature could be considered a mediator variable because of its possible phylogenetic origin and a potential universal explanation of the sense of belonging to nature. That is, if this connection is something everyone develops, it could be the mechanism through which experiences with nature have positive effects. However, according to a more sociocultural sense of belonging to nature, connection with nature could be better conceptualized as a moderator variable, because of the variability of its salience according to different cultural settings. Because people vary in the extent to which they experience connection with nature, based on their culture and experiences, it could moderate the extent to which experiences in nature have a large or small effect on well-being. Researchers should clarify whether the connection with nature is intended as a mediator or a moderator, as this reflects a fundamental difference in the way in which the construct is conceptualized.

Another observation resulting from research is the use of experimental procedures to test causal effects in different settings. These kinds of studies are necessary, because they provide evidence that mere exposure to natural environments is not enough to obtain well-being, as these effects are amplified with high levels of connectedness. Despite the growing number of experimental studies in this field, there remains a lack of longitudinal studies examining the stability of change or resistance to change after exposure to natural settings. Research into the conditions that encourage lasting change would enable the development of evidence-based interventions to promote a feeling of connection with nature.

We also observe two types of cultural studies. On the one hand, research conducted to achieve an international validation of a universal idea of well-being, also tends to make assumptions about universal definitions of nature. On the other hand, some research aims to study the specific cultural sense of well-being and nature in different contexts. It is easy to find an artificial antagonism between these two perspectives, derived from basic research that seeks psychobiological explanations of human behavior through a reductive and analytical approach to the object of study, against constructivist social research, which abounds in inductive descriptions of symbolic components of culture observed in individual belief systems of members of certain social groups, with no intention to generalize results and explanations. We can share an anecdote that is a very good example of this. In a recent conversation with a well-known international environmental anthropologist, he expressed his disappointment in working with environmental psychologists because they always ask him about his "data", and he always asks them about the "human sense" of their research. He asked, "Why can't a sample of one case be a good sample?".

The study of well-being in environmental psychology must consider an integrated approach from both perspectives, as a complementary framework to study the relationship between environment, self and well-being. In other words, the study of psychological configurations (cognitive processes for example) to provide consistency to the human being whom our society and ecosystem claim we have become. An alternative could be to study the synergy between general psychological cognitive (or sociological) structures and the contents and processes displayed to build ideas of nature, self, and happiness from cultural distinctions naturalized as normal ways to perceive human interactions about them, as some authors in our field have pointed out (e.g. Droseltis and Vignoles 2010; Falk and Balling 2010; Palma-Oliveira 2011; Tam et al. 2013). We need to study not only ways in which to measure hedonic levels as part of explaining how to feel good in nature, but also the sense of the relationship between nature and self in order to pursue a happiness that comes from being in harmony with the environment.

Before trying to develop creative methodological approaches, it is necessary to discuss the concepts and processes involved. First, we address the idealization of the concepts of nature and the emotional tone of well-being in our culture.

The ecological foundation of our existence and our inseparable relationship with the elements of nature (solar radiation, water, air, etc.) are evident. However, in recent years, the environment has attracted the attention of scientific research because of the intense anthropogenic effects that human beings have on it, seriously threatening the ecological balance (Melnick et al. 2005). Thus, the difference is not between the study of phylogenetic bonds with nature in contrast with a sociocultural construction of the sense of nature, self and the self-nature relationship from the perspective of symbolic mediation of human actions on the environment. In this way, it is important to understand the idealization of nature and the emotional ties between nature and well-being as part of the sociocultural construction of our sense of nature and well-being. For example, even as part of a phylogenetic bond, negative emotions are deeply adaptive and functional, so that connectedness with nature must not only be biophilic but biophobic, as displayed by the emotional attributions of fear that small children express when they see wild animals (Felipe and Olivos 2014), or when people are exposed to natural environments low in prospect and high in refuge, such as dense wooded areas (Gatersleben and Andrews 2013).

As we have seen from the literature review, one of the shortcomings of environmental psychology studies about nature and well-being is the fact that concepts of well-being that are essentially different tend to be grouped together in a single category (Baumeister et al. 2013). The studies understand well-being in at least three different ways. The best known and most popular approach in the field of environmental psychology has been the study of well-being as a physiological positive feeling, according to the framework of restoration theory. This contribution has undeniably helped to locate the relevance of the role of nature in the psychological health of the human being. At the same time, it has provided a new anthropocentric attention to the relationship of people with nature focused on the personal benefits it provides.

From this approach, the study of well-being emerged as a collection of positive emotions, often described as happiness, outweighing negative ones. Despite the great contribution of positive psychology to the understanding of the psychological experiences of contact with nature, reducing happiness to a positive emotion is like confusing laughter with a smile, that is confusing an explosive instant of emotion with a relatively permanent state of feelings. We need research that explores well-being over a longer period of time. In addition, almost all the environmental psychology research consulted has focused on a Western population, so the next generation of research needs to raise the question of what happiness is for Western culture, and whether this happiness is in tune with the ecosystem dynamics, lifestyles, and even health and psychological well-being itself (Diener and Suh 2000).

The third concept of well-being defines it as a sense of personal growth, and has been labeled as eudaimonic well-being or satisfaction with life. In this way, studies show links between connection with nature and mystic variables, such as mindfulness and spirituality, and the metapersonal self. This is more complex than the emotional concept of well-being, because it must be linked to the symbolic production of positive and permanent identity. Thus, nature should not be studied as only a source of pleasure and immediate feelings.

Finally, we suggest some issues for future research in our discipline, especially in light of ongoing environmental degradation. As the world is increasingly affected by a changing climate, we need to consider the possible impacts on people's sense of connection with nature and on their well-being. How will identities be affected by environmental change? The extinction of experience means we may have less ability to form environmental identities. Not only do people spend less time outdoors than they did in previous generations, even children's books, magazines, and Disney films seem to show fewer and less vivid depictions of nature (e.g., Prévot-Julliard et al. 2014). We know that experiences in nature are important for the development of environmental identity. In the absence of such experiences, will people find alternative sources of meaning and connection? Or will there be a general decline in positive identity? The cultural emphasis on materialism (as an alternative source of identity) can encourage identities that are tied to continued consumption and disposable goods rather than to stable natural features (Kasser 2002).

To the extent that place continues to be important for identity, a contaminated environment can be a source of stigma. Those who reside in environments that are known to be toxic experience a range of effects, including threats to their social identity, as they recognize not only the negative perceptions others have of their communities but also the limits of their own sense of self-control and personal security (Edelstein 2002). Clayton et al. found that people with a high environmental identity score reported greater negative affect in response to negative environmental phenomena, such as climate change or the explosion of the Deepwater Horizon rig (Clayton et al. 2013). Some authors have argued that a general tendency toward decreased psychological well-being stems from our attempts to construct a sense of self within degraded and industrialized environments (Jordan 2009; Kidner 2007).

The more we understand the importance of a connection with the natural world for human well-being, the more we need to promote policies that preserve opportunities for people to develop that connection through experiences in healthy natural environments.

References

Abell, J. (2013). Volunteering to help conserve endangered species: An identity approach to human–animal relationships. *Journal of Community and Applied Social Psychology, 23*, 157–170.

Arnocky, S., Stroink, M., & DeCicco, T. (2007). Self-construal predicts environmental concern, cooperation, and conservation. *Journal of Environmental Psychology, 27*, 255–264.

Aron, A., Aron, E., & Smollan, D. (1992). Inclusion of the other in the self scale and the structure of interpersonal closeness. *Journal of Personality and Social Psychology, 63*, 596–612.

Astell-Burt, T., Mitchell, R., & Hartig, T. (2014). The association between green space and mental health varies across the lifecourse. A longitudinal study. *Journal of Epidemiology Community Health, 0*, 1–6. doi:10.1136/jech-2013-203767.

Baron, R. M., & Kenny, D. A. (1986). The moderator-mediator variable distinction in social psychological research: Conceptual, strategic, and statistical considerations. *Journal of Personality and Social Psychology, 51*, 1173–1182.

Baumeister, R. F. (1987). How the self became a problem: A psychological review of historical research. *Journal of Personality and Social Psychology, 52*, 163–176.

Baumeister, R. F., Vohs, K. D., Aaker, J. L., & Garbinsky, E. N. (2013). Some key differences between a happy life and a meaningful life. *The Journal of Positive Psychology: Dedicated to Furthering Research and Promoting Good Practice, 8*(6), 505–516. doi:10.1080/17439760.2013.830764.

Beery, T. H. (2012). Establishing reliability and construct validity for an instrument to measure environmental connectedness. *Environmental Education Research, 19*(1), 1–13. doi:10.1080/13504622.2012.687045.

Blanchflower, D. G., & Oswald, A. J. (2004). Well-being over time in Britain and USA. *Journal of Public Economics, 88*, 1359–1386.

Brereton, F., Clinch, J. P., & Ferreira, S. (2008). Happiness, geography and the environment. *Ecological Economics, 65*, 386–396. doi:10.1016/j.ecolecon.2007.07.008.

Brown, K. W., & Kasser, T. (2005). Are psychological and ecological well-being compatible? The role of values, mindfulness, and lifestyle. *Social Indicators Research, 74*, 349–368.

Brügger, A., Kaiser, F. G., & Roczen, N. (2011). One for all? Connectedness to nature, inclusion of nature, environmental identity, and implicit association with nature. *European Psychologist, 16*(4), 324–333. doi:10.1027/1016-9040/a000032.

Bruni, C. M., & Schultz, P. W. (2010). Implicit beliefs about self and nature: Evidence from an IAT game. *Journal of Environmental Psychology, 30*, 95–102. doi:10.1016/j.jenvp.2009.10.004.

Burns, G. W. (2005). Naturally happy, naturally healthy: The role of the natural environment in well-being. In F. A. Huppert, N. Baylis, & B. Keverne (Eds.), *The science of well-being* (pp. 405–431). New York: Oxford University Press.

Callaghan, P., Maloney, G., & Blair, D. (2012). Contagion in the representational field of water recycling: Informing new environment practice through social representation theory. *Psyecology, 22*, 20–37. doi:10.1002/casp.1101.

Canter, D., & Craig, K. H. (1981). Environmental psychology. *Journal of Environmental Psychology, 1*, 1–11. doi:10.1016/S0272-4944(81)80013-8.

Capaldi, C. A., Dopko, R. L., & Zelenski, J. M. (2014). The relationship between nature connectedness and happiness: A meta-analysis. *Frontiers in Psychology, 5*, 976. doi:10.3389/fpsyg.2014.00976.

Carrus, G., Lafortezza, R., Colangelo, G., Dentamaro, I., Scopellitic, M., & Sanesi, G. (2013). Relations between naturalness and perceived restorativeness of different urban green spaces. *Psyecology, 4*(3), 227–244. doi:10.1174/217119713807749869.

Cerina, V., & Fornara, F. (2011). The psychological determinants of attitudes toward relocation in the elderly: A survey study in urban and rural environments. *Psyecology, 2*(3), 335–348. doi:10.1174/217119711797877744.

Cervinka, R., Röderer, K., & Hefler, E. (2012). Are nature lovers happy? On various indicators of well-being and connectedness with nature. *Journal of Health Psychology, 17*(3), 379–388. doi:10.1177/1359105311416873.

Chawla, L., & Derr, V. (2012). The development of conservation behaviors in childhood and youth. In S. Clayton (Ed.), *Oxford handbook of environmental and conservation psychology* (pp. 527–555). New York: Oxford University Press.

Chen, Z., & Davey, G. (2009). Subjective quality of life in Zhuhai city, south China: A public survey using the international wellbeing index. *Social Indicators Research, 91*, 243–258. doi:10.1007/s11205-008-9280-1.

Church, A. T., et al. (2014). Relating self-concept consistency to Hedonic and Eudaimonic well-being in eight cultures. *Journal of Cross-Cultural Psychology*. Published online before print March 19, 2014. doi:10.1177/0022022114527347.

Clayton, S. (2003). Environmental identity: A conceptual and an operational definition. In S. Clayton & S. Opotow (Eds.), *Identity and the natural environment. The psychological significance of nature* (pp. 45–65). Cambridge: The MIT Press.

Clayton, S. (Ed.). (2012). *Handbook of environmental and conservation psychology*. New York: Oxford University Press.

Clayton, S., Fraser, J., & Burgess, C. (2011). The role of zoos in fostering environmental identity. *Ecopsychology, 3*, 87–96.

Clayton, S., Koehn, A., & Grover, E. (2013). Making sense of the senseless: Justice, identity, and the framing of environmental crises. *Social Justice Research, 26*, 301–319.

Clayton, S., Prévot, A. C., Germain, L. (2014). *Promoting positive attitudes toward protection of biodiversity: The role of personal experience*. Paper presented at the meeting of the international conference of Applied Psychology, Paris.

Corral-Verdugo, V., Mirles-Acosta, J., Tapia-Fonllem, C., & Fraijo-Sing, B. (2011). Happiness as correlate of sustainable behavior: A study of pro-ecological, frugal, equitable and altruistic actions that promote subjective wellbeing. *Human Ecology Review, 18*(2), 95–104.

Cronin, A., Backett-Milburn, K., Parry, O., & Platt, S. (2005). Understanding and researching wellbeing: Its usage in different disciplines and potential for health research and health promotion. *Health Education Journal, 64*(1), 70–87.

Cummins, R. A. (2010). Subjective well-being, homeostatically protected mood and depression: A synthesis. *Journal of Happiness Studies, 11*, 1–17.

Cummins, R. A. (2011). Comparison theory in economic psychology regarding the easterlin paradox and decreasing marginal utility: A critique. *Applied Research Quality Life, 6*, 241–252. doi:10.1007/s11482-011-9151-9.

Dasgupta, P. (2001). *Human well-being and the natural environment*. New York: Oxford University Press.

Davis, N., & Gatersleben, B. (2013). Transcendent experiences in wild and manicured settings: The influence of the trait connectedness to nature. *Ecopsychology, 5*(2), 92–102. doi:10.1089/eco.2013.0016.

Davis, J. L., Green, J. D., & Reed, A. (2009). Interdependence with the environment: Commitment, interconnectedness, and environmental behavior. *Journal of Environmental Psychology, 29*, 173–180. doi:10.1016/j.jenvp.2008.11.001.

De Young, R. (1996). Some psychological aspects of reduced consumption behavior: The role of intrinsic satisfaction and competence motivation. *Environment and Behavior, 28*, 358–409.

Deneve, K. M., & Cooper, H. (1998). The happy personality: A meta-analysis of personality traits and subjective well-being. *Psychological Bulletin, 124*(2), 197–229.

Descola, P., & Pálsson, G. (Eds.). (1996). *Nature and society. Anthropological perspectives*. London: Routledge.

Detweiler, M. B., Murphy, P. F., Myers, L. C., & Kim, K. Y. (2008). Does a wander garden influence inappropriate behaviors in dementia residents? *American Journal of Alzheimer's Disease and Other Dementias, 23*, 31–45.

Di Tella, R., MacCulloch, R. J., & Oswald, A. J. (2001). Preferences over inflation and unemployment: Evidence from surveys of happiness. *American Economic Review, 91*, 335–341.

Diener, E. (1984). Subjective well-being. *Psychological Bulletin, 95*(3), 542–575.

Diener, E., & Suh, E. M. (2000). *Culture and subjective well-being*. Cambridge: The MIT Press.

Diener, E., Emmons, R. A., Larsen, R. J., & Griffin, S. (1985). The satisfaction with life scale. *Journal of Personality Assessment, 49*(1), 71–75. doi:10.1207/s15327752jpa4901_13.

Diette, G. B., Lechtzin, N., Haponik, E., Devrotes, A., & Rubin, H. R. (2003). Distraction therapy with nature settings and sounds reduces pain during flexible bronchoscopy. *Chest, 123*, 941–948.

Dove, M. R., & Carpenter, C. (2008). Introduction: Major historical currents in environmental anthropology. In M. R. Dove & C. Carpenter (Eds.), *Environmental anthropology. A historical reader* (pp. 1–85). Malden: Blackwell.

Droseltis, O., & Vignoles, V. L. (2010). Towards an integrative model of place identification: Dimensionality and predictors of intrapersonal-level place preferences. *Journal of Environmental Psychology, 30*(1), 23–34. doi:10.1016/j.jenvp.2009.05.006.

Dutcher, D., Finley, J. C., Luloff, A. E., & Johnson, J. B. (2007). Connectivity with nature as a measure of environmental values. *Environment and Behavior, 30*, 474–493.

Easterlin, R. (1974). Does economic growth improve the human lot? Some empirical evidence. In P. A. David & M. W. Reder (Eds.), *Nations and households in economic growth: Essays in honour of Moses Abramovitz* (pp. 89–125). New York: Academic.

Edelstein, M. (2002). Contamination: The invisible built environment. In R. Bechtel & A. Churchman (Eds.), *Handbook of environmental psychology* (pp. 559–588). New York: Wiley.

Ellen, R. (2001). La geometría cognitiva de la naturaleza. Un enfoque contextual. In P. Descola & G. Pálsson (Eds.), *Naturaleza y sociedad, perspectivas antropológicas* (pp. 124–146). Mexico: Siglo XXI.

Falk, J. H., & Balling, J. D. (2010). Evolutionary influence on human landscape preference. *Environment & Behavior, 42*(4), 479–493. doi:10.1177/0013916509341244.

Falsten, G. (2014). Personality predicts perceived potential for attention restoration of natural and urban scenes. *Psyecology, 5*(1), 37–57. doi:10.1080/21711976.2014.881663.

Felipe, B., & Olivos, P. (2014). *Madre Naturaleza, nos enseña: Conectividad, emoción, y significado de la naturaleza en menores de 6 años*. Communication XII National Conference of Social Psychology, 20–22 November, Seville, Spain.

Ferrer-i-Carbonell, A., & Gowdy, J. M. (2007). Environmental degradation and happiness. *Ecological Economics, 60*, 509–516. doi:10.1016/j.ecolecon.2005.12.005.

Fiske, S. T., Gilbert, D. T., & Lindzey, G. (2010). *Handbook of social psychology*. New York: Wiley.

Frankenhaeuser, M. (1977). Job demands, health and wellbeing. *Journal of Psychosomatic Research, 21*(4), 313–321.

Franklin, A. (2002). *Nature and social theory*. Great Britain: Sage.

Frantz, C., Mayer, F. S., Norton, C., & Rock, M. (2005). There is no "I" in nature: The influence of self-awareness on connectedness to nature. *Journal of Environmental Psychology, 25*, 427–436. doi:10.1016/j.jenvp.2005.10.002.

Fraser, J., Clayton, S., Sickler, J., & Taylor, A. (2009). Belonging at the zoo: Retired volunteers, conservation activism, and collective identity. *Ageing and Society, 29,* 351–368.

Gatersleben, B. (2008). Humans and nature: Ten useful findings from environmental psychology research. *Counselling Psychology Review, 23,* 24–34.

Gatersleben, B., & Andrews, M. (2013). When walking in nature is not restorative: The role of prospect and refuge. *Health & Place, 20,* 91–101. doi:10.1016/j.healthplace.2013.01.001.

Gergen, K. (1991). *The saturated self. Dilemmas of identity in contemporary life.* New York: Basic Books, Harper Collins Publishers.

Giddens, A. (1991). *Modernity and self-identity. Self and society in the late modern age.* Cambridge: Polity Press, & Basil Blackwell.

Gidlöf-Gunnarsson, A., & Öhrström, E. (2007). Noise and well-being in urban residential environments: The potential role of perceived availability to nearby green areas. *Landscape and Urban Planning, 83,* 115–126. doi:10.1016/j.landurbplan.2007.03.003.

Goldberg, D. P., & Blackwell, B. (1970). Psychiatric illness in general practice: A detailed study using a new method of case identification. *British Medical Journal, 1,* 439–443.

Guegen, N., & Stefan, J. (2014). Green altruism: Short immersion in natural green environments and helping behavior. *Environment and Behavior.* doi:10.1177/0013916514536576.

Gullone, E., & Cummins, R. A. (2002). *The universality of subjective wellbeing indicators. A multi-disciplinary and multi-national perspective.* Dordrecht: Kluwert. doi:10.1007/978-94-010-0271-4.

Hartig, T., Mang, M., & Evans, G. W. (1991). Restorative effects of natural environment experiences. *Environment and Behavior, 23*(1), 3–26. doi:10.1177/0013916591231001.

Henderson, L. W., Knight, T., & Richardson, B. (2013). An exploration of the well-being benefits of hedonic and eudaimonic behaviour. *The Journal of Positive Psychology: Dedicated to Furthering Research and Promoting Good Practice, 8*(4), 322–336. doi:10.1080/17439760.2013.803596.

Hinds, J., & Sparks, P. (2011). The affective quality of human-natural environment relationships. *Evolutionary Psychology, 9*(3), 451–469.

Hogg, M. A., Terry, D. J., & White, K. M. (1995). A tale of two theories: A critical comparison of identity theory with social identity theory. *Social Psychology Quarterly, 58,* 255–269.

Hoot, R. E., & Friedman, H. (2011). Connectedness and environmental behavior: Sense of interconnectedness and Pro-environmental behavior. *International Journal of Transpersonal Studies, 30*(1–2), 89–100.

Horsburgh, C. R. (1995). Healing by design. *The New England Journal of Medicine, 333,* 735–740.

Howell, A. J., Dopko, R. L., Passmore, H.-A., & Buro, K. (2011). Nature connectedness: Associations with well-being and mindfulness. *Personality and Individual Differences, 51,* 166–171. doi:10.1016/j.paid.2011.03.037.

Ingold, T. (2000). *The perception of the environment. Essays in livelihood, dwelling and skill.* Great Britain: Routledge.

Ingulli, K., & Lindbloom, G. (2013). *Ecopsychology, 5*(1), 52–55. doi:10.1089/eco.2012.0042.

Jordan, M. (2009). Nature and self: An ambivalent attachment? *Ecopsychology, 1,* 26–31. doi:10.1089/eco.2008.0003.

Kahler, E. (1943). *Man the measure.* New York: Pantheon Books.

Kaiser, F. G., & Byrka, K. (2011). Environmentalism as a trait: Gauging people's prosocial personality in terms of environmental engagement. *International Journal of Psychology, 46,* 71–79.

Kals, E., & Ittner, H. (2003). Children's environmental identity: Indicators and behavioral impacts. In S. Clayton & S. Opotow (Eds.), *Identity and the natural environment* (pp. 135–157). Cambridge: MIT Press.

Kals, E., Schumacher, D., & Montada, L. (1999). Emotional affinity toward nature as a motivational basis to protect nature. *Environment and Behavior, 31,* 178–202.

Kamitsis, I., & Francis, A. (2013). Spirituality mediates the relationship between engagement with nature and psychological wellbeing. *Journal of Environmental Psychology, 36,* 136–143. doi:10.1016/j.jenvp.2013.07.013.

Kaplan, R. (1973). Some psychological benefits of gardening. *Environment and Behavior, 5,* 145–152. doi:10.1177/001391657300500202.

Kaplan, S. (1983). A model of person-environment compatibility. *Environment and Behavior, 5,* 311–332. doi:10.1177/0013916583153003.

Kaplan, S. (1995). The restorative benefits of nature: Toward an integrative framework. *Journal of Environmental Psychology, 15*(3), 169–182. doi:10.1016/0272-4944(95)90001-2.

Kaplan, R. (2001). The nature of the view from home psychological benefits. *Environment and Behavior, 33,* 507–542. doi:10.1177/00139160121973115.

Kaplan, R., & Kaplan, S. (1989). *The experience of nature: A psychological perspective.* Cambridge: Cambridge University Press.

Kaplan, S., & Talbot, J. F. (1983). Psychological benefits of a wilderness experience. In I. Altman & J. F. Wohlwill (Eds.), *Behavior and the natural environment* (pp. 163–203). New York: Plenum.

Kasser, T. (2002). *The high price of materialism.* Cambridge: MIT Press.

Kidner, D. (2007). Depression and the natural world: Toward a critical ecology of psychological distress. *Critical Psychology, 19,* 123–146.

Knopf, R. C. (1983). Recreational needs and behavior in natural settings. In I. Altman & J. F. Wohlwill (Eds.), *Behavior and the natural environment* (pp. 205–240). New York: Plenum Press.

Korpela, K. M., Hartig, T., Kaiser, F. G., & Fuhrer, U. (2001). Restorative experience and self-regulation in favorite places. *Environment and Behavior, 33*, 572–589.

Latour, B. (2013). *Políticas de la Naturaleza*. Barcelona: RBA Libros.

Leary, M. R., & Tangney, J. P. (2003). The self as an organizing construct in the behavioural and social sciences. In M. R. Leary & J. P. Tangney (Eds.), *Handbook of self and identity* (pp. 3–14). New York: The Guilford Press.

Leary, M. R., & Tangney, J. P. (Eds.). (2012). *Handbook of self and identity* (2nd ed.). New York: The Guildford Press.

Leary, M. R., Tipsord, J. M., & Tate, E. B. (2008). Allo-inclusive identity: Incorporating the social and natural worlds into one's sense of self. In H. A. Wayment & J. J. Bauer (Eds.), *Transcending self-interest: Psychological explorations of the quiet ego. Decade of behavior* (pp. 137–147). Washington, DC: American Psychological Association.

Lee, T. R. (1976). *Psychology and the environment*. London: Methuen Co.

Leong, C., Fischer, R., & McClure, J. (2014). Are nature lovers more innovative? The relationship between connectedness with nature and cognitive styles. *Journal of Environmental Psychology, 40*, 57–63. doi:10.1016/j.jenvp.2014.03.007.

Levy, S., & Guttman, L. (1975). On the multivariate structure of wellbeing. *Social Indicators Research, 2*(3), 361–388.

Mathews, G., & Izquierdo, C. (2010). Anthropology, happiness, and well-being. In G. Mathews & C. Izquierdo (Eds.), *Pursuits of happiness: Well-being in anthropological perspective* (pp. 1–19). New York: Berghahn Books.

Mayer, F. S., & Frantz, C. M. (2004). The connectedness to nature scale: A measure of individuals' feeling in community with nature. *Journal of Environmental Psychology, 24*, 503–515. doi:10.1016/j.jenvp.2004.10.001.

Mayer, F. S., Frantz, C. M., Bruehlman-Senecal, E., & Dolliver, K. (2009). Why is nature beneficial? The role of connectedness to nature. *Environment and Behavior, 41*, 607–643. doi:10.1177/0013916508319745.

Melnick, D., McNeely, J., Kakabadse, Y., Schmidt-Traub, G., & Sears, R. R. (2005). *Environment and human wellbeing: A practical strategy*. New York: UN Millennium Project Task Force on Environmental Sustainability.

Merrero, R. J., & Carballeira, M. (2010). Contact with nature and personal well-being. *Psychology, 1*(3), 371–381. doi:10.1174/217119710792774825.

Navarro, O. (2013). Representación social del medio ambiente y de la contaminación del aire: Efecto de imbricación de dos objetos. *Revista CES Psicología, 6*(1), 104–121.

Nisbet, E. K. L., & Zelenski, J. M. (2011). Underestimating nearby nature: Affective forecasting errors obscure the happy path to sustainability. *Psychological Science, 22*(9), 1101–1106. doi:10.1177/0956797611418527.

Nisbet, E. K. L., & Zelenski, J. M. (2013). The NR-6: A new brief measure of nature relatedness. *Frontiers in Psychology, 4*, 1–11. doi:10.3389/fpsyg.2013.00813.

Nisbet, E. K. L., Zelenski, J. M., & Murphy, S. A. (2009). The nature relatedness scale: Linking individuals' connection with nature to environmental concern and behavior. *Environment and Behavior, 41*, 715–740. doi:10.1177/0013916508318748.

Nisbet, E. K. L., Zelenski, J. M., & Murphy, S. A. (2011). Happiness is in our nature: Exploring nature relatedness as a contributor to subjective well-being. *Journal of Happiness Studies, 12*, 303–322. doi:10.1007/s10902-010-9197-7.

Nordh, H., Hagerhall, C. M., & Holmqvist, K. (2010). Exploring view pattern and analysing pupil size as a measure of restorative qualities in park photos. In G. Prosdocimi Gianquinto, & F. Orsini (Eds.), *Proceedings of the 2nd international conference on landscape and urban horticulture*, 9–13 June 2010, Bologna, Italy.

Olivos, P., & Aragones, J. I. (2011). Psychometric properties of the Environmental Identity Scale (EID). *Psyecology, 2*(1), 65–74. doi:10.1174/217119711794394653.

Olivos, P., & Aragonés, J. I. (2013). Test de asociaciones implícitas con la naturaleza: aplicación en España del "IAT-Nature". *Revista de Psicología Social, 28*(2), 237–245. doi:10.1174/021347413806196672.

Olivos, P., & Aragonés, J. I. (2014). Medio ambiente, self y conectividad con la naturaleza. *Revista Mexicana de Psicología, 31*(1), 71–77.

Olivos, P., Aragonés, J. I., & Navarro, O. (2013). Educación ambiental: Itinerario en la naturaleza y su relación con conectividad, preocupaciones ambientales y conducta. *Revista Latinoamericana de Psicología, 45*(3), 501–511. doi:10.14349/rlp.v45i3.1490.

Olivos, P., Talayero, F., Aragonés, J. I., & Moyano, E. (2014). Dimensiones del comportamiento proambiental y su relación con la conectividad e identidad ambientales. *Psico, 45*(3), 369–376.

Opotow, S. (1993). Animals and the scope of justice. *Journal of Social Issues, 49*, 71–85. doi:10.1111/j.1540-4560.1993.tb00909.x.

Palma-Oliveira, J. (2011). Homo sapiens versus homo artiflex: ou a inevitabilidade (?) da destruição ambiental. In B. Fernandez-Ramirez, C. Hidalgo, C. Salvador, & M. Martos (Eds.), *Psicologia ambiental: Entre los estudios urbanos y el análisis de la sustentabilidad* (pp. 23–40). Almeria: Universidade de Almeria, PSICAMB.

Passmore, H.-A., & Howell, A. J. (2014). Eco-existential positive psychology: Experiences in nature, existential anxieties, and wellbeing. *The Humanistic Psychologist, 42*(4), 370–388. doi:10.1080/08873267.2014.920335.

Perkins, H. E. (2010). Measuring love and care for nature. *Journal of Environmental Psychology, 30*, 455–463. doi:10.1016/j.jenvp.2010.05.004.

Perrin, J. L., & Benassi, V. A. (2009). The connectedness to nature scale: A measure of emotional connection to nature? *Journal of Environmental Psychology, 29*, 434–440. doi:10.1016/j.jenvp.2009.03.003.

Pijanowski, B. C., et al. (2011). Soundscape ecology: The science of sound in the landscape. *BioScience, 61*, 203–216. doi:10.1525/bio.2011.61.3.6.

Prévot-Julliard, A.-C., Julliard, R., & Clayton, S. (2014). Historical evidence for nature disconnection in a 70-year time series of Disney animated films. *Public Understanding of Science, 10*, 1–9. doi:10.1177/0963662513519042.

Pryor, A., Townsend, M., Maller, C., & Field, K. (2006). Health and well-being naturally: 'contact with nature' in health promotion for targeted individuals, communities and populations. *Health Promotion Journal of Australia, 17*(2), 114–123.

Ragip, A., Yuzel, I., & Sezer, F. (2012). Investigating psychological well-being of university students according to lifestyles. *Procedia – Social and Behavioral Sciences, 47*, 256–262. doi:10.1016/j.sbspro.2012.06.648.

Ryan, R. M., & Deci, E. L. (2000). Self-determination theory and the facilitation of intrinsic motivation, social development and well-being. *American Psychologist, 55*(1), 68–78. doi:10.1037/0003-066X.55.1.68.

Ryan, R. M., & Deci, E. L. (2003). On assimilating identities to the self: A self-determination theory perspective on internalization and integrity within cultures. In M. Leary & J. Tangney (Eds.), *Handbook of self and identity* (pp. 253–272). New York: Guilford.

Ryff, C. D. (1989). Happiness is everything, or is It? Explorations on the meaning of psychological well-being. *Journal of Personality and Social Psychology, 57*(6), 1069–1081. doi:10.1037/0022-3514.57.6.1069.

Ryff, C. D., & Singer, B. H. (2013). Know thyself and become what you are: A eudaimonic approach to psychological well-being. In A. D. Fave (Ed.), *The exploration of happiness. Present and future perspectives* (pp. 97–116). Netherlands: Springer. doi:10.1007/978-94-007-5702-8.

Sagar, A. (2007). Review of ecology and human well-being: Nature and society in Himachal Pradesh. *Journal of the Indian Academy of Applied Psychology, 33*, 301–302.

Sánchez, M. P., de la Garza, A., & López, E. O. (2009). La identidad y actitud hacia el medio ambiente en estudiantes de biología y psicología. *Revista PsicologiaCientifica.com, 11*(7). http://www.psicologiacientifica.com/identidad-actitud-hacia-el-medio-ambiente-estudiantes.

Schroeder, H. W. (2007). Place experience, gestalt, and the human-nature relationship. *Journal of Environmental Psychology, 27*, 293–309. doi:10.1016/j.jenvp.2007.07.001.

Schultz, P. W. (2001). The structure of environmental concern: Concern for self, other people, and the biosphere. *Journal of Environmental Psychology, 21*, 327–339. doi:10.1006/jevp.2001.0227.

Schultz, P. (2002). Inclusion with nature: Understanding the psychology of human-nature relations. In P. Schmuck & P. Schultz (Eds.), *The psychology of sustainable development* (pp. 61–78). New York: Kluwer.

Schultz, P., Shriver, C., Tabanico, J., & Khazian, A. (2004). Implicit connections with nature. *Journal of Environmental Psychology, 24*, 31–42. doi:10.1016/S0272-4944(03)00022-7.

Schwartz, S. J., Luyckx, K., & Vignoles, V. L. (Eds.). (2011). *Handbook of identity theory and research*. New York: Springer.

Seagert, S. (1987). Environmental psychology and social change. In D. Stokols & I. Altman (Eds.), *Handbook of environmental psychology*. New York: Wiley.

Selin, H., & Davey, G. (2012). *Happiness across cultures. Views of happiness and quality of life in non-western cultures*. Netherlands: Springer. doi:10.1007/978-94-007-2700-7.

Sevillano, V., Aragonés, J. I., & Schultz, P. W. (2007). Perspective taking, environmental concern, and the moderating role of dispositional empathy. *Environment and Behavior, 39*, 685–705. doi:10.1177/0013916506292334.

Siedentop, L. (2014). *Inventing the individual. The origins of western liberalism*. London: Penguin Group.

Staats, H., Gatersleben, B., & Hartig, T. (1997). Change in mood as a function of environmental design: Arousal and pleasure on a simulated forest hike. *Journal of Environmental Psychology, 17*, 283–300. doi:10.1006/jevp.1997.0069.

Steffen, W., Crutzen, P. J., & McNeill, J. R. (2007). The anthropocene: Are humans Now overwhelming the great forces of nature. *AMBIO: A Journal of the Human Environment, 36*(8), 614–621. doi:10.1579/0044-7447.

Tam, K.-P. (2013). Concepts and measures related to connection to nature: Similarities and differences. *Journal of Environmental Psychology, 34*, 64–78. doi:10.1016/j.jenvp.2013.01.004.

Tam, K.-P., Lee, S.-L., & Chao, M. M. (2013). Saving Mr. Nature: Anthropomorphism enhances connectedness to and protectiveness toward nature. *Journal of Experimental Social Psychology, 49*, 514–521. doi:10.1016/j.jesp.2013.02.001.

Taufik, D., Bolderdijk, J. W., & Steg, L. (2014). Acting green elicits a literal 'warm glow'. *Nature Climate Change, 5*, 37–40. doi:10.1038/nclimate2449.

Taylor, C. (1989). *Sources of the self: The making of the modern identity*. Cambridge: Harvard University Press.

Thin, N. (2010). Why anthropology can ill afford to ignore well-being. In G. Mathews & C. Izquierdo (Eds.), *Pursuits of happiness: Well-being in anthropological perspective* (pp. 23–44). New York: Berghahn Books.

Thomashow, M. (1995). *Ecological Identity, becoming a reflective environmentalist*. London: MIT Press.

Ulrich, R. (1984). View through a window may influence recovery from surgery. *Science, 224*, 420–421.

Van der Werff, E., Steg, L., & Keizer, K. (2013a). The value of environmental self-identity: The relationship between biospheric values, environmental self-identity and environmental preferences, intentions and behavior. *Journal of Environmental Psychology, 34*, 55–63. doi:10.1016/j.jenvp.2012.12.006.

Van der Werff, E., Steg, L., & Keizer, K. (2013b). It is a moral issue: The relationship between environmental self-identity, obligation-based intrinsic motivation and pro-environmental behavior. *Global Environmental Change, 23*, 1258–1265. doi:10.1016/j.gloenvcha.2013.07.018.

Veenhoven, R. (2004). *World database of happiness*. www.eur.nl/fws/research/happiness, 16 Sept 2004.

Venhoeven, L., Bolderdijk, J.-W., & Steg, L. (2013). Explaining the paradox: How pro-environmental behavior can both thwart and foster well-being. *Sustainability, 5*, 1372–1386. doi:10.3390/su5041372.

Vennhoven, R. (1997). Advances in the understanding of happiness. *Revue Québécoise de Psychologie, 18*(2), 29–74.

Vining, J., & Merrick, M. (2012). Environmental epiphanies. In S. Clayton (Ed.), *Oxford handbook of environmental and conservation psychology* (pp. 485–508). New York: Oxford University Press.

Wallenius, M. (1999). Personal projects in everyday places: Perceived supportiveness of the environment and psychological well-being. *Journal of Environmental Psychology, 19*, 131–143. doi:10.1006/jevp.1998.0118.

Waterman, A., Schwartz, S., & Conti, R. (2008). The implications of two conceptions of happiness (hedonic enjoyment and eudaimonia) for the understanding of intrinsic motivation. *Journal of Happiness Studies, 9*, 41–79. doi:10.1007/s10902-006-9020-7.

Watson, D., Clark, L. A., & Tellegen, A. (1988). Development and validation of brief measures of positive and negative affect: The PANAS scales. *Journal of Personality and Social Psychology, 54*(6), 1063–1070. doi:10.1037/0022-3514.54.6.1063.

Welsch, H. (2006). Environment and happiness: Valuation of air pollution using life satisfaction data. *Ecological Economics, 58*, 801–813. doi:10.1016/j.ecolecon.2005.09.006.

Williams, R. (1983). *Keywords*. London: Fontana Paperbacks.

Wilson, E. O. (1984). *Biophilia*. Cambridge: Harvard University Press.

Wolsko, C., & Lindberg, K. (2013). Experiencing connection with nature: The matrix of psychological well-being, mindfulness, and outdoor recreation. *Ecopsychology, 5*(2), 80–91. doi:10.1089/eco.2013.0008.

Wortham, S. (2006). *Learning identity: The joint emergence of social identification and academic learning*. New York: Cambridge University Press.

Zalasiewicz, J., Williams, M., Steffen, W., & Crutzen, P. (2010). The new world of the anthropocene. *Environmental Science & Technology, 44*(7), 2228–2231. doi:10.1021/es903118j.

Zelenski, J. M., & Nisbet, E. K. (2014). Happiness and feeling connected: The distinct role of nature relatedness. *Environment and Behavior, 46*(1), 3–23. doi:10.1177/0013916512451901.

Zhang, J. W., Piff, P. K., Iyer, R., Koleva, S., & Keltner, D. (2014a). An occasion for unselfing: Beautiful nature leads to prosociality. *Journal of Environmental Psychology, 37*, 61–72. doi:10.1016/j.jenvp.2013.11.008.

Zhang, J. W., Howell, R. T., & Iyer, R. (2014b). Engagement with natural beauty moderates the positive relation between connectedness with nature and psychological well-being. *Journal of Environmental Psychology, 38*, 55–63. doi:10.1016/j.jenvp.2013.12.013.

Restorative Environments and Health

Silvia Collado, Henk Staats, José Antonio Corraliza, and Terry Hartig

7.1 Introduction

The populations of many countries today are concentrated in urban environments, which offer residents more diverse cultural amenities compared to rural areas. These include a greater variety of entertainment options, easier access to educational and medical centers and more opportunities for mixing with different kinds of people in lively public places. Yet, city inhabitants also suffer from unfavorable conditions like air pollution, visual blight, crowding and noise. Growing up in cities has been linked to an increased risk of chronic stress (Steinheuser et al. 2014), which can lead to illnesses such as anxiety disorders, depression and coronary heart disease (Chauvet-Gélinier et al. 2013; Lederbogen et al. 2011).

For many years, people have had the intuition that contact with nature has psychologically restorative benefits for people living in urbanized societies (Kaplan and Kaplan 1989). This intuition has been corroborated by empirical studies conducted during the past several decades (e.g. see Hartig et al. 2014). Researchers have demonstrated experimentally that spending time outdoors and/or in visual contact with nature can offer greater restorative benefits than other everyday urban environments, and this can promote physical and mental health in the long run. As cities continue to expand, however, urban residents may have fewer possibilities for contact with nature, just as the potential for realizing restorative benefits of contact with nature increases, given the increasingly stressful urban circumstances. Residents' physical, psychological and social resources are thus in danger of becoming overly taxed as they try to meet everyday demands in the urban context. If they cannot readily renew these resources, their health and well-being will eventually suffer.

Researchers and policy-makers have highlighted the need to design citizen-friendly urban spaces that promote well-being, health and quality of life in cities. One way of doing this is by ensuring access to nature at different scales. This has been a key point in urban planning, be

S. Collado (✉)
Department of Psychology and Sociology, University of Zaragoza, Facultad de Ciencias Sociales y Humanas, Universidad de Zaragoza, Campus de Teruel, 44003 Teruel, Aragón, Spain
e-mail: scollado@unizar.es

H. Staats
Centre for Energy and Environmental Research, Social and Organizational Psychology, Leiden University, Leiden, The Netherlands
e-mail: staats@fsw.leidenuniv.nl

J.A. Corraliza
Department of Social Psychology and Methodology, Universidad Autónoma de Madrid, Madrid, Spain
e-mail: josea.corraliza@uam.es

T. Hartig
Institute for Housing and Urban Research, Uppsala University, Uppsala, Sweden
e-mail: Terry.Hartig@ibf.uu.se

it in the form of green belts around cities (Amati 2008), large recreation areas such as Central Park (Olmsted 1865/1952) or smaller spaces like pocket parks (Kaplan 1983; Nordh et al. 2009). Time spent in a park, a garden, a room with plants or looking at street trees from a window can help a person gain psychological distance from the demands of daily life and engage instead with something pleasant and interesting. Research on restoration and restorative environments focuses on the physical and psychological benefits of such respites from everyday cares and worries. The research is, however, not limited to contact with nature as a support for restoration; the urban context offers many other restorative opportunities as well.

In this chapter, we review the literature on restorative environments. More than half of the human population now lives in cities, and this proportion is expected to increase markedly in the coming decades (UN-Habitat 2013). For these reasons, our focus in this chapter is mainly on people in urbanized societies, their restoration needs, and the settings in which they can achieve restoration. Before moving on to describe restorative environments and how they promote health, we set out some basic concepts; we provide definitions of health and restoration, summarize the main theories regarding environmental supports for restorative processes, and consider the distinction between restorative outcomes achieved on single occasions and those achieved cumulatively over time. We then review the empirical literature on the physical and psychological benefits that can be gained in different environments, indicating the characteristics that make these environments restorative under given conditions. The literature is quite extensive, and our review should be considered illustrative rather than exhaustive. Finally, we outline some issues of implementation and suggest possible lines for future research.

7.1.1 Main Concepts

Before describing restorative environments and the implications of restorative experiences for health, it is necessary to provide some background on what the terms "health", "restoration" and "restorative environment" refer to and on the theories that guide the study of environmental supports for restorative processes.

The World Health Organization defines *health* in the following way: "Health is a state of complete physical, social and mental well-being, and not merely the absence of disease or infirmity. Health is a resource for everyday life, not the object of living. It is a positive concept emphasizing social and personal resources as well as physical capabilities" (World Health Organization 1996).

This definition may understandably strike some as overly idealistic and vague, yet it is widely appreciated because it acknowledges the importance of the subjective aspects of health, its multi-dimensionality, the framing of health in positive terms, and, concomitantly, the possibility that people may, in some respects, have poor physical health yet nonetheless feel that they enjoy good health in other, more meaningful ways (cf. Hartig et al. 2014).

Health is shaped by many processes, and a general one that works on an individual level is restoration. By *restoration*, we mean the renewal or recovery of adaptive resources or capabilities that have become depleted in meeting the demands of everyday life (cf. Hartig 2004). As such, the term serves as a rubric; restoration seen as a general process can involve multiple, specific processes through which particular depleted resources become restored. The resources of interest may be physical, for instance the capacity for physiological activation, drawn on when hurrying from one activity to another, as when rushing from work to collect the children from school. They can also be psychological, such as the ability to concentrate on tasks even when there are many distractions around. Resource availability can also be regarded in social terms. If one person's resources are depleted, then support from another may be necessary. The ability of the other to lend a helping hand or to recognize that help is needed may be undermined when physical energy and the ability to direct attention are depleted. Even if the other has the ability to provide support, the exchange is typically predicated on the existence of a relationship between the giver and the recipient. This relational resource exists between people, not in one or the other person alone, and it also can become depleted, for example through a lack of reciprocity in the giving of support (Hartig et al. 2013). As people use physical, psychological and social resources on an everyday basis, and so pay the cost of adapting to demanding circumstances, the need for restoration arises regularly. Therefore, restoration is an important process, as new demands will come along and resources will have to be renewed to face them. If proper restoration is not achieved and adaptive costs accumulate, a person may develop mental and physical health problems. For instance, if a person continuously fails to restore adequately after stressful days at work, and so remains unable to cope effectively with subsequent stressful daily events, then the ensuing chronic stress may, over time, translate into serious illnesses such as depression and coronary heart disease.

Restoration always occurs in the context of an activity that involves some form and degree of engagement with the sociophysical environment. In experimental and observational research, walking has received substantial attention as a fundamental form of human movement in the environment (Gatersleben and Andrews 2013; Hartig et al. 1991, 2003; Faber Taylor and Kuo 2009; Johansson et al. 2011; Staats and Hartig 2004). Attention has also been given to other activities, including running (Bodin and Hartig 2003), other forms of exercise (Pretty et al. 2005), meditation (Ouellette et al. 2005), gardening (Van den Berg and Custers 2011), driving (Parsons et al. 1998), and simply sitting and looking out of a window (Hartig et al. 2003; Faber Taylor et al. 2002) or looking at indoor plants (Shibata and Suzuki 2001). Some environments support the restorative potential of specific activities better, and in this chapter we focus on the settings and environmental characteristics that promote restoration within the given activity, rather than the characteristics of the activity per se.

Since its origins in the 1960s, research on restorative environments has increasingly become organized around psychoevolutionary theory, which concerns stress reduction (Ulrich 1983; Ulrich et al. 1991), and attention restoration theory (Kaplan and Kaplan 1989; Kaplan 1995), which concerns the renewal of a capacity for directed attention. Both these theories make assumptions about the lasting significance of human evolution in environments now described as natural, considering that people will restore better in environments that have characteristics that were relevant for survival during early evolution. In addition, both of them necessarily specify an antecedent condition that involves resource depletion. They also describe a process through which resources become restored and the characteristics of environments that promote the given process. Like the definition of health given above, the definition of such environments is in positive terms; restorative environments do not only allow restoration, but also promote it (Hartig 2004). That is, a restorative environment is defined as such not only because it imposes relatively few demands on depleted resources, but also because it has positive features that enable a faster, more complete renewal of depleted resources. These two theories differ in their specification of the antecedent condition from which a person becomes restored and so point out the potential for different restorative benefits, but the possibility of their integration has long been recognized (Ulrich et al. 1991; Hartig and Evans 1993; Kaplan 1995).

We first focus on the psychoevolutionary theory, which for pedagogic purposes we refer to hereinafter as stress reduction theory (SRT). The author of SRT, Roger Ulrich (1983), assumes that an individual's first level of response toward the environment is mainly an affective, automatic one, evoked by visual properties of a scene. For Ulrich, the restorative effects of certain environments are reflected in physiological processes as well as in emotional changes when recovering from stress. Stress, understood as a response to a situation perceived as threatening to one's well-being, is thus seen as the antecedent condition. According to Ulrich, certain natural environments help individuals restore more quickly and completely from acute stress because they possess, to a relatively suitable degree, those characteristics that, perceived in an automatic, almost unconscious way, rapidly provoke positive emotions, capture non-vigilant attention, block negative thoughts, and allow physiological activation to sink to more moderate levels. Because some natural environments have visual stimulus characteristics and contents that people have been biologically prepared to appreciate through evolution, spending time in a natural environment or viewing a natural scene can help people reduce stress more adequately than spending time in urban settings. In addition to the absence of threat and the presence of survival-relevant contents, such as water and particular configurations of vegetation, these characteristics include the presence of a focal point, a moderate degree of depth in the scene, and a moderate level of visual richness, among others. According to SRT, in the course of the stress recovery process, people will come to show more positive emotions, less negative emotions (e.g., anger) and a decline in physiological parameters such as blood pressure (Ulrich et al. 1991).

The second theory that has guided much research on restorative environments is attention restoration theory (ART; Kaplan and Kaplan 1989). This focuses on the capacity to direct attention as the resource to be restored and directed attention fatigue as the antecedent condition. Unlike SRT, it emphasizes a cognitive resource and mechanism rather than an affective-physiological one. Based on the work of James (1892) and Kaplan and Kaplan (1989) proposed that attention has two modes, and that these involve different degrees of willful effort: directed attention is effortful, and involuntary attention is effortless. People employ directed attention when attending to tasks or situations that do not catch and hold their attention by themselves but to which it is important to pay attention nonetheless, as when writing a paper, studying for an exam or driving when there is a lot of traffic. Directed attention, according to Kaplan (1995), plays a central role in achieving focus, with the control of distraction through the use of inhibition.

Because of its effortful character, the use of directed attention intensively and/or for an extended period of time will diminish a person's ability to inhibit distraction and remain focused. This does not imply that the person dislikes the task at hand; "even a thoroughly enjoyable project is likely to lead to this same outcome (*directed attention fatigue*)" (Kaplan 1995 p. 170). This aside, the importance of directed attention for everyday functioning is reflected in the consequences of its fatigue: errors in performance, being inefficient at problem-solving, not being able to inhibit impulses, becoming easily distracted, experiencing difficulties in planning and executing a plan, and behaving in inappropriate or unhelpful ways. Negative feelings such as irritability can also appear as a consequence of directed attention fatigue. According to ART, directed attention can be recovered when a person can engage in activities that draw primarily on involuntary attention or, as the Kaplan and Kaplan (1989) call it, fascination. Fascination can be engaged by processes, such as exploration of an environment or following unfolding events, as with a sunrise. It can also be engaged by environmental contents, such as wild animals, trees, and unusual architectural features of buildings.

Kaplan and Kaplan (1989) proposed three other qualities of a restorative experience in addition to fascination. Like fascination, they are not part of the environment or part of the person, but rather refer to the transaction between them. One, compatibility, refers to the congruence or match between an individual's purposes, the environmental supports for the pursuit of that purpose, and the demands imposed by the environment. The second, extent, refers to the experience of the environment as coherently structured and with substantial scope for involvement. Thirdly, being away refers to gaining physical and/or psychological distance from routine mental contents and everyday worries and demands. Kaplan and Kaplan (1989) argue that restoration is enhanced to the degree that these four components characterize the person's encounter with the environment. Of these four, fascination is believed to play a more important role in the restoration process (Staats 2012); it is the cognitive mechanism of restoration, whose operation is supported by the other components of the restorative experience.

SRT and ART both predict that natural environments in general will be more restorative than non-natural ones. They also complement each other in their predictions of specific effects. Because of this, most of the studies described in the next sections concern the restorative effects of contact with nature, as represented by both cognitive and affective measures (e.g., Faber Taylor and Kuo 2009; Johansson et al. 2011) and in some cases physiological measures as well (Hartig et al. 2003; Kelz et al. 2015; Park and Mattson 2009a). In addition to measuring the changes that actually occur in people given an opportunity for restoration, or the perception of these changes (i.e., perceived restoration), researchers have paid attention to the likelihood of restoration that people estimate to exist in a certain environment and their perceptions of environments in terms of theoretically specified restorative qualities, such as being away and fascination (i.e., perceived restorativeness). These research approaches assume that a person's estimate of restoration likelihood or restorative quality in a setting is based on his/her past experiences of restoration in the given environment, ones like it, and other, different environments (Collado and Corraliza 2015; Hartig et al. 2001).

Before moving on to describe how different environments can support restoration and health, we need to clarify a final point; the distinction between restorative outcomes achieved on a single occasion and those achieved cumulatively with repeated occasions over time. There is a relative wealth of empirical research on the outcomes achieved on a single occasion, or what Hartig (2007) has referred to as a discrete restorative experience. Such an experience is considered in isolation; it is the experience of a person in need of restoration, during time available for restoration, in an environment available for restoration. In a common experimental approach, the discrete restorative experience follows efforts to meet demands that have potentiated restoration. For example, Gatersleben and Andrews (2013) had their experimental participants perform the Stroop task to induce attentional fatigue, and then had them walk along two different nature trails on different occasions or watch video simulations of walks along the same trails, also on different occasions. Among their findings, performance on an attentional task and self-reported affect improved following a walk along a trail with more open views.

As this experiment and numerous others (Johansson et al. 2011; Karmanov and Hamel 2008; Ulrich et al. 1991) reveal, restorative benefits can appear quickly. These experiments commonly involve a uniform administration of demands to induce an antecedent condition (e.g., directed attention fatigue), as well as a uniform activity (e.g., a walk in one or another comparison environment), a uniform duration for this activity (e.g., 10 min), and uniform social circumstances (e.g., alone). The experimental research done on discrete restorative experiences has helped in outlining the time course of restorative processes in more detail, showing not only that some restorative effects of environments can appear quickly but also that some can dissipate rather quickly, while others may persist beyond the time in the environment (Hartig et al. 2003). The experimental research has also contributed to an understanding of the environmental conditions that support different restorative processes, documenting differences in outcomes not only with comparisons between natural and urban environments (e.g., Hartig et al. 1991) but also with comparisons involving different kinds of natural environments (e.g., Gatersleben and Andrews 2013; Martens et al. 2011).

These experimental insights are helpful in a variety of ways; however, one restorative experience isolated in time will ordinarily not yield long-lasting health benefits. Accordingly, attention has also been paid to cumulative effects (Hartig 2007) or, in other words, effects achieved with repeated experiences by people with depleted resources in environments thought to promote restoration. In contrast to studies of discrete restorative experiences, the researcher has less control in studies of cumulative effects; the time spent in a given environment, the specific setting characteristics, the activities conducted, the frequency of "exposure" and the specific conditions under which the person accesses the environment cannot be manipulated. In spite of this, such studies are of special importance as they reflect the medium- and long-term benefits of having access to environments thought to promote restoration.

Now that we have given an overview of the concepts and theories that guide research on restorative environments, we can move on to describe the existing evidence relating restorative environments to health.

7.2 Restorative Environments and Health

The next sections give an illustrative overview of the literature on environments that tend to enhance restorative experiences and so promote health. Natural environments are not the only ones that enhance restoration, and not all natural environments are necessarily restorative; however, the majority of studies in this research area to date have considered whether contact with nature in various forms (e.g., wild areas, indoor plants, urban green space) has a more positive impact on people's functioning, health and wellbeing than other environments. The emphasis on natural environments reflects the early origins of this area of inquiry in studies of the psychological benefits of wilderness experiences and in efforts to manage forests and other large-scale natural areas better for recreation as well as for timber and other natural resources. Such research has affirmed the restorative value of large-scale, wild natural environments (e.g., Cole and Hall 2010; Talbot and Kaplan 1986; Williams and Harvey 2001; for an early but still useful review, see Knopf 1987); however, the restorative effects of visits to them are likely to vary with familiarity with wild nature, individual time constraints, and the possibility of disengaging from the occupational role while in the setting (Staats 2012; Von Lindern et al. 2013). Important here is that getting to such places usually requires substantial travel time. Noting that wild areas are not readily available for the mundane restoration needs of many urban residents, we focus here on the work done regarding common encounters with nature in people's daily lives.

As we describe the empirical evidence on restorative environments and health, we will point out whether we refer to short-term effects of discrete experiences, medium/long-term effects achieved cumulatively with multiple experiences, or the potential for restoration perceived in a setting. When discussing medium/long-term effects, we include studies ranging from hospital stays, which typically are relatively short, to residing in certain places for long periods.

The work is organized into four categories: residential context, work and school settings, care settings, and other settings. Several forms of nature presentation (e.g., pictures, plants, gardens) and types of contact with nature (e.g., window view, walks, more intensive green exercise) are discussed in each category of settings; moreover, research not concerned with nature is also described in each category.

7.2.1 Residential Context

The home and its surroundings are places where people spend many of their waking hours and most of their sleeping ones. Therefore, the restorative qualities of these environments have received considerable attention from the scientific community (e.g. see Hartig and Lawrence 2003; Hartig 2012). Here, we are particularly concerned with the research on the residential context as a constellation of settings, which supports restoration during waking hours.

Numerous studies conducted in the residential context speak of the potential for restoration-mediated health benefits. With regard to long-term or cumulative physical benefits, different studies have found that children's and adults' physical activity is positively related to the presence of nature within the neighborhood (De Vries et al. 2011). For example, Saelens et al. (2003) studied neighborhood environmental characteristics that could be linked to physical activity, including what they called "attractive natural sights" (p. 1554). Their results suggested that characteristics such as aesthetics, walking facilities or land use access made neighborhoods more walkable and this, in turn, was associated with less overweight among residents. Similarly, Liu et al. (2007) found that the amount of vegetation surrounding children's urban residences was associated with childhood overweight. More than 7,000 children between 3 and 18 years of age participated in the study. After controlling for individual sociodemographic characteristics and neighborhood socioeconomic status, more neighborhood vegetation was associated with lower childhood weight. A positive link between living in relatively green areas and physical activity has also been highlighted for old people (Broekhuizen et al. 2013).

One could think that the link between green spaces and active lifestyles has nothing to do with the restoration process. However, some researchers have argued that the attraction felt by people towards being physically active in green environments is linked to their expectations of restorative benefits (Bodin and Hartig 2003; Staats et al. 2003; Hartig 2007; Hartig et al. 2014), and studies of discrete restorative experiences point in that direction. In further developing a line of research on what they labeled "green exercise", Pretty et al. (2005) carried out a laboratory study in which people had to exercise for 20 min on a treadmill looking at either a rural (pleasant or unpleasant) or an urban (pleasant or unpleasant) collection of pictures or at a blank power point (control group). Their results showed that physical activity per se (control group) reduced blood pressure and increases self-esteem. When the same activity (running on a treadmill) was conducted looking at pleasant urban or rural pictures, both including natural elements, the benefits were significantly greater. The authors call this a synergetic relationship between the benefits produced by physical activity and those produced by looking at nature. The same conclusion was drawn by Grahn and Stigsdotter (2003), who found that people who suffered more stress expressed wanting to visit urban outdoor green areas more often than those less stressed. Staats and Hartig (2004) asked their participants to indicate whether they preferred to walk for an hour in a forest or an urban center once they imagined being mentally fatigued, as compared to their attentional capabilities being fully charged. They found that the prospect of walking in a forest was preferred over walking in an urban center to a greater degree when participants where attentionally fatigued. In fact, the natural vs. urban preference differential was approximately twice as large with the fatigued vs. rested condition. The same results were found with students with actual instead of imagined mental fatigue (Hartig and Staats 2006). All of this suggests that people do indeed prefer to conduct physical activity in near-home areas perceived as more restorative, anticipating that psychological restoration may occur in the course of their physical activity.

Moving on to the cumulative psychological health benefits of restorative environments within the residential context, the work conducted by Kuo and colleagues (Kuo 2001; Kuo and Sullivan 2001) is of special interest. These researchers had the opportunity to assess the restorative effects of green surroundings in a low-income housing estate in Chicago, Illinois, USA. People living on this housing estate were believed to suffer chronic attentional fatigue due to the demands imposed by poverty. As suggested by Kaplan (1995), this may lead, in turn, to aggression and violence problems. Kuo and Sullivan (2001) had their team interview 145 residents who had been assigned by local authorities through a quasi-random process to apartments in architecturally identical buildings. This offered an opportunity to assess the effects of nature in real-life situations while ruling out self-selection into apartments and other possible confounding factors. Homes were classified by the researchers as *barren* or *green*, according to the amount of nearby nature (e.g., trees, grass) outside the apartment buildings. Participants' capacity to direct attention was registered with a behavioral measure, together with their self-reported levels of aggression, violent behavior, mood and stress level, among other variables. The researchers' results suggested that the presence of nearby nature helped these people to maintain a capacity to direct attention and that this in turn positively influenced social conduct, reflected in less aggressive behavior.

The contents of residential window views apparently matter to people's quality of life. For example, Kaplan (2001) found greater well-being among people who could look at nature from a window at home. She asserted that brief periods spent looking out provide micro-restorative experiences, the positive effects of which can accumulate over time. Similar results were obtained in a university dormitory, where students live temporarily and need to adapt to a new environment in a short period of time. Tennessen and Cimprich (1995) found that students who could look out at nature through their dorm room window performed better in attention-demanding tasks than those with a built view. The authors suggested that the natural view from their window improved students' capacity to concentrate on their academic work and, therefore, helped to boost their academic performance.

Access to greenery in the residential context also appears to have positive psychological health benefits for children. For example, the cognitive capabilities of 7–12 year-olds from low-income families improved after they were relocated to residential areas considered more restorative because of better opportunities for contact with nature (Wells 2000). Participants were included in a self-help housing program, which consisted of relocating the families from poor housing to better housing. This design enabled Wells (2000) to study the characteristics of the houses, including the amount of nature present in the residence and its surroundings before and after the families moved, and to relate this to the children's cognitive functioning. Families were visited in their initial house and in the new one, with approximately 1 year between the two measurements. After controlling for the children's pre-move cognitive functioning score, it was found that the change in naturalness (an increase from pre-move to post-move) was a significant predictor of the post-move attentional capacity. The change in the quality of the house itself did not predict post-move attention performance. Consistent with these findings, self-discipline in a sample of urban children was found to be better when they had more nature available in their residential area (Faber Taylor et al. 2002). Parents of 7–12 year-olds rated the views from their house windows in terms of naturalness. At the same time, children's self-discipline (e.g., capability to inhibit impulses) was registered. Faber Taylor et al. found that, for girls, living in houses with natural views appeared to help them maintain greater self-discipline, which is thought to depend on directed attention.

Access to nature in the residential context may also have a protective effect against the negative health consequences of stressful events (Van den Berg et al. 2010; Wells and Evans 2003). Wells and Evans (2003) collected observer ratings of registered vegetation in and around houses in rural areas of the state of New York. Children approximately 7–11 years-old living in the houses reported stress due to adverse situations. The frequency of exposure to stressful events was a negative predictor of the children's self-reported self-worth and a positive predictor of their psychological distress as rated by their mothers. In contrast, the amount of nature in the residential area was a positive predictor of the children's self-worth and a negative predictor of their psychological distress. Interestingly, the availability of nearby nature in the residential area buffered the psychological distress related to stressful events. The psychologically negative effects produced by events such as moving to another house, being punished or listening to parents' arguments was lower in children who had more nature in the residential context. The most vulnerable children, those who suffered stressful events more often, were the ones who obtained more benefits from having more access to nature.

Research on cumulative restorative experiences with children diagnosed with Attention Deficit Disorder (ADD) or Attention Deficit Hyperactivity Disorder (ADHD) is also relevant here. A weak ability to direct attention is the regular state of children who suffer from ADD or ADHD. Faber Taylor et al. (2001) conducted a study on the effects of spending time in nature with children from 7 to 12 years-old. The parents of these children reported on two activities that improved their child's functioning and two that made it worse. The activities were then coded into those likely to take place in a natural setting (green activities) and those unlikely to take place in a natural environment (not green). At the same time, they rated a series of after-school and weekend activities in terms of the after-effects that these activities had on their child's attention deficit symptoms. Parents also reported on the greenness of their everyday surroundings as well as how severe their children's symptoms were when not on medication. Contact with nature was found to be related to fewer ADD/ADHD symptoms. For instance, those activities that parents saw as better for their child's ADD/ADHD symptoms generally took place in green areas and, in contrast, those that appeared to aggravate their children's ADD/ADHD symptoms were conducted in non-green environments. In addition, the greener the place where the children usually played, the less severe their ADD/ADHD symptom were when not medicated.

Similar findings were obtained by Kuo and Faber Taylor (2004) in a study conducted with children diagnosed with ADHD. As in the previous research, parents were interviewed. The effects of after-school and weekend activities on children's symptoms were assessed. Parents rated common after-school and weekend activities proposed by the researchers, including activities carried out in green outdoor settings (e.g., parks, farms or green backyards) and those conducted in built outdoor (e.g., parking lots, downtown areas or neighborhood space) or indoor settings (e.g. bedroom, sitting room). For example, the activity *reading* was evaluated in different settings (indoors, green outdoor setting, built outdoor setting). Parents were asked to rate the after-effects of the weekend activities on four ADHD symptoms: (1) difficulty in remaining focused on unappealing tasks, (2) difficulty in completing tasks, (3) difficulty in listening to and following directions, and (4) difficulty in resisting distractions. According to the authors, parents indicated that green outdoor activities reduced the ADHD symptoms more than activities conducted in other settings.

More recently, Faber Taylor and Kuo (2011) evaluated the effects that routine (every day or nearly every day) exposure to green spaces has on children and teenagers with ADD/ADHD. Parents or legal guardians of children from 5 to 18 years-old diagnosed with ADD/ADHD participated in an Internet-based survey. Their results showed that children and teenagers with attention deficits usually played outdoors and that playing in green settings, especially open grassy areas, was related to milder ADHD symptoms, as reported by the parents.

A final study worth mentioning here concerned the value for health of a second home in natural surroundings for residents of urban areas. With longitudinal data for more than 42,000 Swedish adults, Hartig and Fransson (2009) estimated the likelihood of taking early retirement for health reasons as a function of leisure home ownership. They posited that a leisure home would serve as a psychologically restorative resource, helping people over time to cope better with the demands of working life in the city. All of the people included in the study had paid employment and had maintained the same primary residence during an 8-year period of follow-up. Hartig and Fransson anticipated that those who owned a leisure home would be less likely to take early retirement for health reasons after the follow-up period than those who did not own a leisure home. The particular form of early retirement in focus was a social benefit provided after a medical examination that led to a determination of a permanently diminished capacity for work. Leisure home ownership was associated with lower odds of this form of early retirement, but only for men. Among women, and particularly among highly educated and highly paid women, leisure home ownership was instead accompanied by higher odds of early retirement. This may be because the leisure home, although in natural surroundings, was nonetheless a domestic setting, and these women still had relatively more responsibility for indoor tasks such as cooking and cleaning. The study illustrates the complexities that can challenge the study of restorative experiences and health in the residential context, and the ways in which cumulative restorative benefits may be contingent on sociodemographic characteristics and social roles.

7.2.2 Work and School Settings

Adults spend a large amount of time in their workplace, as do students in college and children in day care, primary schools, and secondary schools. These environments and the activities organized within them can impose heavy demands on physical, psychological and social resources. Accordingly, as with research on restoration in the residential context, a substantial amount of research has considered restorative experiences in work and school settings. Some, but not all, of this research has considered contact with nature a support for restoration.

7.2.2.1 Work Settings

Most of the studies regarding the health benefits of contact with nature in work settings have been experiments conducted in laboratories or simulated environments and have therefore documented short-term effects (e.g., Felsten 2009; Kweon et al. 2008). Overall, the evidence suggests that passive contact with nature (e.g., views from a window, ornamental plants, pictures of nature, posters) can yield psychological benefits such as increased productivity, relaxation, less anger or less stress (e.g., Chang and Chen 2005; Kweon et al. 2008). The evidence is, however, mixed with regard to some forms of contact with nature, such as indoor plants (see Bringslimark et al. 2009).

In an early contribution to this research area, Kaplan (1993) completed a survey study on the cumulative benefits of "micro-restorative experiences" in the workplace, premised on the idea that the cognitive resources needed for intellectual work may be restored more effectively during repeated brief respites if a worker sitting at a desk can look out from a window onto natural features such as trees and vegetation. The results suggested that a view onto natural features was indeed associated with higher levels of satisfaction with the window view from one's work site, which in turn was associated with enthusiasm for the job and with general health. Kaplan noted that breaks taken away from the desk or workstation were important, but that micro-restorative experiences had while sitting at one's desk might be particularly important for reducing attentional fatigue, because the worker had to face that immediate work setting more continuously.

Bringslimark et al. (2007) also considering cumulative restorative experiences in the workplace, reported cross-sectional survey evidence that indoor plants were negatively associated with office workers' frequency of sick leaves and positively associated with their productivity. The results held with adjustment for psychosocial workplace factors (e.g., job demands), physical characteristics of the environment (e.g., temperature) and personal factors (e.g., gender). The authors suggested that the associations were due to the support provided by plants for restoration over time. Although weak, the reported associations are nonetheless potentially practically relevant, given the large amounts of time spent by large numbers of people in office settings.

Bringslimark et al. (2011) used data from the same survey to test the hypothesis that workers bring plants and nature pictures into their office space to compensate for the lack of a window view. They found that the employees in windowless offices were roughly five times more likely to bring plants into their work place than those with a window view, and were three times more likely to decorate their office space with a picture of nature. These results held up with control for factors such as space personalization, gender, and work load. According to the authors, bringing representations of nature into the office can be viewed as a compensatory strategy that offsets the lack of access to the outdoors provided by windows, which would otherwise afford psychological benefits (e.g., restoration for those in need of it).

7.2.2.2 School Settings

A considerable part of a child's waking life is spent at school. Thus, the restorative qualities of school buildings and their surroundings can be important for the effective functioning, health and well-being of school-age children. Here, too, possibilities for contact with nature can be particularly important (e.g., Bell and Dyment 2008; Lindholm 1995).

Most of the research about health and schools has involved real-life settings and considered cumulative or long-term effects. With regard to physical health, Grahn et al. (1997) carried out a study in Sweden in two different day care settings, with 3–7 year-olds. One of the centers was considered typically urban, surrounded by tall buildings, including low plants and a brick path for cycling. At the other center, the playground was surrounded by nature and children played outside a considerable amount of time each day, regardless of the weather. The results showed that children attending the day care center with more natural features had better motor coordination, were less often sick and had better attentional concentration capabilities than the children attending the urban day care center. Unfortunately, perhaps because of the small size of the sample, the authors did not control for potentially potent confounders, including characteristics of the residence and household.

Following up on these indicative results, Fjørtoft (2001) conducted a pre-test/post-test quasi-experimental study on the motor ability of 5–7 year-old children who spent between 1 and 2 h per day playing in a small forest next to their kindergarten. Children who attended similar kindergartens, and who spent the same amount of time playing in their traditional outdoor playground, participated in a control group. After controlling for possible confounding factors, such as family living conditions, their results showed that these children improved their motor coordination and balance abilities after a 9-month period, compared to children in the control group. As with studies of green space and physical activity in adults, the anticipation of a restorative experience may have helped to motivate the outdoor activities, which could explain the physical benefits, including quiet play activities done alone.

Similarly, Ozdemir and Yilmaz (2008) randomly selected five public schools located in different districts of Ankara, Turkey. The researchers found that the physical characteristics of the school play yard, including the presence of trees and water, were associated with children's Body Mass Index (BMI) as well as their perception and satisfaction with the playground. Children who played in larger school playgrounds had lower BMI than those who played in smaller ones. Moreover, children preferred larger yards with vegetation and described their ideal school playground as one with green elements such as trees and lawn areas. Unfortunately, the results were not controlled for pupils' socioeconomic status, which could have influenced the results. Again, considering the studies conducted with adults, anticipation of restoration could have motivated the children to be more physically active in the greener environments (cf. Staats et al. 2003), and micro-restorative experiences may have enhanced their satisfaction with the play yard (Kaplan 1993).

More recently, Kelz et al. (2015) had the opportunity to assess the physical and psychological health benefits of improvements made to school grounds. The inclusion of more greenery around the school was one of the improvements provided for in the renovation plan. The researchers conducted a quasi-experimental study in which physiological (e.g., blood pressure), psychological (e.g., self-reported well-being), executive functioning and restorativeness measures were obtained from adolescents before and after the renovation. In addition, two control schools were considered. After 6–7 weeks, children in the intervention school showed lower blood pressure and reported higher psychological well-being than those in the control schools. They also perceived their schoolyard as being more compatible with their needs and goals, and more fascinating than before the renovation occurred.

Using an ecological study design to assess the psychological benefits of time spent in school settings with presumably more restorative characteristics, Matsuoka (2010) found a positive relationship between students' non-problematic (e.g., non- aggressive) behaviors, academic performance and the amount of nature within the school area. He evaluated the physical characteristics of more than a hundred public high schools. Students' exposure to nature was registered in three ways: (1) views of nature from the school cafeteria and classrooms,

(2) vegetation present in the school grounds (e.g., greenery density), and (3) students' potential access to vegetated areas (e.g., whether or not students were allowed to eat lunch outdoors). Data were also collected for students' aggregate performance (e.g., graduation rate, percentage intending to study at college) and aggregate problem behavior (e.g., social tension, bullying, criminal activity) for each school over a 1-year period. With adjustment for school socioeconomic status, ethnicity, number of students enrolled and building age, the availability of contact with nature during the school day was positively related to student academic performance. Participants in schools with more nature obtained higher grades, more often planned to study for a 4-year degree, obtained more merit awards and showed fewer criminal activities than those in less natural ones. A relevant question with regard to these results concerns the extent to which the differences in the school's physical characteristics reflected differences between the homes of the students, and so the extent to which the associations found depend on the schools alone versus the combination of residential context and school.

Natural elements have also been described as a protective factor against stress in primary schools (Corraliza and Collado 2011). These researchers considered the potential effect that contact with nature during school time would have on children's ability to cope with stressful events. Data were collected in four schools varying in the amount of nature in the playground and its surroundings, from non-natural to very natural. Children's frequency of exposure to stressful events and perceived stress level were registered. The authors concluded that children attending the more natural school were better able to cope with stressful events in daily life, such as having their parents arguing in front of them, than children in less natural schools. In line with Wells and Evans' (2003) suggestions of possible explanatory mechanisms, it may be that having restorative respites at school helps children maintain their attention capabilities, which in turn promotes thinking more clearly, putting problems into perspective, and dealing with stressful events in a more competent way (cf. Kaplan 1995).

7.2.3 Care Settings

The category of care settings includes but is not limited to hospitals, rehabilitation centers, and nurseries. Over their life course, many people spend a substantial amount of time in such settings, be it as a patient or a visitor. Many people also work in care settings, and for some a care setting is also their residential context, as with elderly people in assisted living facilities. Whether the provision of care aims to help healing or coping with unavoidable decline, the psychological and physical well-being of people receiving care in hospitals and other institutions is important. Accordingly, researchers have worked hard to identify factors that will enhance their quality of life, as well as that of visitors and staff.

As in the residential context and in work and school settings, contact with nature appears to be beneficial in health care settings. Research has demonstrated that natural window views (Ulrich 1984), landscape images together with nature sounds (Diette et al. 2003), indoor plants (Park and Mattson 2008), and accessible gardens (Whitehouse et al. 2001) can enhance people's health and well-being in care settings. In some circumstances, natural elements may yield such benefits by supporting anxiety and stress reduction in connection with procedures involving discomfort or pain. For instance, watching a nature program on television in a hospital room entailed a lower heart rate in people waiting to donate blood, compared to watching a videotape of an urban setting, watching some ordinary daytime television program, or just waiting without watching television (Ulrich et al. 2006). More recently, Diette et al. (2003) demonstrated that viewing a natural image and listening to natural sounds while going through a bronchoscopy procedure helped patients manage pain better than patients in a control group.

Contact with nature may also help to reduce pain over longer periods of care. The two preceding examples were inspired by a pioneering study of Ulrich (1984), who assessed the influence of the window view from a patient's room on recovery from surgery. The sample was formed by 23 patients assigned to a room with a view of trees and 23 matching patients assigned to a room that had a view of a brick wall. In addition to the kind of operation performed, patients were matched for factors such as age and smoking. Ulrich's results showed that the patients with the tree views had, on average, shorter postoperative hospital stays, needed to take fewer doses of strong analgesics, and had more positive comments from their nurses than those with a brick wall view.

In other circumstances, natural elements in care settings may prove beneficial by supporting restorative respites, enabling the renewal of resources needed to cope with the stressful situations faced in the setting (Hartig and Cooper Marcus 2006). For example, cumulative positive effects of having a natural window view have also been reported for participants in a demanding rehabilitation program at a residential facility (Raanaas et al. 2012). The researchers found that an unobstructed natural bedroom view supported patients' physical and mental health better during their demanding multi-week rehabilitation program than having a view either partially or totally blocked by buildings. The benefits that the residents obtained from a natural window view differed according to gender and the particular condition for which the patients were going through rehabilitation. Like the findings regarding leisure home ownership and early retirement mentioned previously, this finding illustrates an important point, namely, that the restorative value of contact with nature experiences may be contingent on personal and contextual factors.

Aside from what can be seen outside from the window, contact with nature within the patient room may be important. Park and Mattson (2009a) assessed the possible influence of ornamental indoor plants on people recovering from surgery. Ninety patients were randomly assigned to one of ten rooms on the same hospital floor, the only difference being whether they contained indoor plants. The researchers reported that ornamental indoor plants had positive effects on patients recovering from surgery (e.g., lower systolic blood pressure, lower anxiety and more positive feelings) compared to the control group.

The possible cumulative effects of indoor plants have also been evaluated in other health settings. In an intervention study, Raanaas et al. (2010) considered whether placing plants in the common areas (e.g., TV rooms, lounges) of a rehabilitation center would be beneficial for its residents. Patients' self-reports about physical and psychological health, subjective well-being and emotional states were collected at the time of arrival at the center, 2 weeks after arrival and at the end of the program (3–4 weeks). This was done for 11 months before and 11 months after the introduction of a large number of indoor plants of varying size. The subjective well-being of patients who suffered from asthmatic diseases or chronic obstructive pulmonary disease increased more during the rehabilitation programs run after the plant intervention than before it, and the residents reported greater satisfaction with the center after the intervention. In contrast to the results of previous studies in hospital rooms (e.g., Park and Mattson 2009a), however, no increment in physical health improvement could be attributed to the intervention. According to the authors, this could be due to the fact that residents in the rehabilitation center were more mobile (and thus less exposed to the plants placed in the common areas) than those in hospitals, who mainly stay in a single room. It may also be that any effects of the plant intervention were effaced by other restoration opportunities within and around the center.

Acknowledging the benefits of contact with nature for patients, visitors and staff, many health care facilities have a natural area placed close to the buildings where care is provided. These areas are often called "healing gardens", although the terms "restorative" or "therapeutic gardens" have also been used. These terms refer to the intertwining of a place (i.e., a garden in a health care facility) and a process (i.e., the recovery or improvement of physical, mental and/or social health). Healing gardens can be designed for a specific group (e.g., people suffering from Alzheimer's disease or in need of physical rehabilitation) or to provide general benefits to the broader population of visitors, staff, and patient groups through opportunities for relaxation, reflection, solace and joy (Hartig and Cooper Marcus 2006). For example, Whitehouse et al. (2001) found that the healing garden installed in a children's hospital was used by parents, children and staff to obtain distance from stressful situations, to relax, to think more clearly and to take a rest. Satisfaction with the hospital increased due to the presence of the garden. When asked about physical changes they would recommend in the garden, fifty percent of the parents would like to see more trees, vegetation and greenery.

7.2.4 Other Settings

Up to this point, we have presented evidence of the restorative potential of three common categories of settings (residential areas; workplaces, including schools; care settings). In general, including natural elements in these settings enhances their restorative quality which, in turn, may contribute to their users' health and wellbeing. Nevertheless, it is important to keep in mind that many settings without prominent natural features, such as monasteries (Ouellette et al. 2005), museums (Kaplan et al. 1993) and plazas (Abdulkarim and Nasar 2014), can also be restorative. Although the amount of research done in this area is relatively small, the findings of several studies suggest that certain characteristics of predominantly built settings make them restorative and, considering that most people live in cities, their assessment is important in order to promote health in urban populations. We refer here to both discrete and cumulative restorative experiences, as well as to the perceived restorativeness or potential for restoration in certain places.

There is some empirical evidence that some non-natural settings within the public space of cities offer significant restorative benefits. For instance, one study found that when comparing urban and natural scenes, citizens rated some urban scenes – especially those with some openness and no traffic – as more conducive to restoration than some of the natural ones, in which vegetation obstructed longer views (Herzog et al. 2003). Karmanov and Hamel (2008) evaluated the possible positive effects of discrete restorative experiences in, according to the authors, a well-designed urban neighborhood, compared to a nature-reserve and partially agricultural landscape in The Netherlands. Both environments were described as being outstandingly beautiful. The urban landscape did not have a great deal of vegetation in it but other natural elements, such as water, were abundant. After taking an exam and therefore considered to be attentionally fatigued, students' affective states and feelings (e.g., anger, tension and depression symptoms) were registered. Then, they watched a 10-min video of either the natural or the built environment and rated the setting in terms of attractiveness, interestingness and naturalness. Their affective states and feelings were then measured a second time. The authors found that the built neighborhood was rated as being highly attractive and interesting, while the natural environment was perceived as more attractive and the urban one as more interesting. Both environments were equally restorative in terms of pre-test/post-test change in affect. Participants' scores on anger and tension were lower after watching one of the two videos, without significant differences between them. However, viewing a natural scene reduced the participants' depression scores more than watching the urban one. These results reflect the ways in which different urban settings can serve restoration. Nevertheless, even though Karmanov and Hamel (2008) described the neighborhood as a built environment, it still had some natural features (mainly water).

With regard to another widely used urban environment, Abdulkarim and Nasar (2014) checked whether the elements that increase visitability in a plaza (availability of seating, presence of a sculpture and access to food) also increase their restorativeness. The idea behind the authors' hypotheses was that seats, sculptures and access to food would be related to restorative components described in ART. For instance, they reasoned that adding seats, a sculpture or access to food in a plaza would increase participants' sense of being away, fascination and compatibility, as compared to plazas without those elements. Slide simulations were made in which plazas without any of these elements, with one of them or with a combination of several were shown to the participants. Their results showed that placing a sculpture in plazas increased the participants' restorativeness ratings, as did including seats. Access to food did not seem to have an effect on perceived restorative potential. Moreover, the larger the plaza, the more restorative it was perceived. Openness, the availability of facilities for leisure, and places for meeting people were also found to be relatively strong correlates of restorative quality in the urban settings identified by the participants surveyed by Galindo and Hidalgo (2005) (see also Lindal and Hartig 2013).

Religious settings have also been considered places where cumulative restorative benefits can be obtained. Ouellette et al. (2005) studied the retreat experience of middle-aged men in a monastery, anticipating that certain characteristics of this experience would match the qualities of a restorative experience. The authors argued that participants in the retreat situation experienced a sense of being away while in the monastery, that they found being there compatible with their purposes, and that they found the experience interesting and suitable for reflection. As such, it helped them to recover attentional capabilities, reflect on personal matters, and put their problems in perspective. Following up on this study, Herzog et al. (2010) assessed the restorative potential of houses of worship (e.g., churches). These settings are more commonly used than monasteries and, therefore, their restorative properties may benefit a wider range of people. The researchers found that the motives to visit a house of worship regularly were similar to those expressed by the monastery retreat participants. They included being away from daily responsibilities and fascination enhanced by the setting, among others. Visitors to these places said they went there to conduct activities related to religion (e.g., listen to spiritual talks) but they also reported non-religious activities (e.g., letting one's mind wander). As for the outcomes, participants in both studies claimed to obtain peace and more effective functioning and to renew their ability to focus as results of their experiences, all of which may contribute to better health in the long run.

7.3 Issues of Implementation and Future Research

All human activity takes place in some sociophysical environment. The various characteristics of these environments have diverse effects on these activities and the people performing them. Some of these effects can be understood through reference to restoration and evaluated in light of the fundamental importance of restoration for human health and well-being. Thus, restorative quality can sometimes be considered an important basis for the evaluation of environmental design.

How then can residential areas, workplaces, schools, care settings and plazas be best designed to help people meet the restoration needs they may face in them or bring to them? Every specific real environment is unique due to its location, dimensions, physical characteristics, population of visitors/users and other factors. Moreover, an environment becomes known as restorative in relation to other environments in which restoration needs are potentiated. The uniqueness of environments and the relational character of the restorative environment concept make it difficult to state specific design guidelines that can serve as a recipe for restorative environments. Some researchers have made design recommendations concerned with boosting restorative quality (Kaplan et al. 1998), while others have studied how variations in specific physical variables amenable to design manipulations can affect perceived restorative quality (Nordh et al. 2009; Lindal and Hartig 2013). Here, we outline some implementation ideas informed by findings presented in this chapter. In doing so, we describe some practical issues in designing commonly used places, as well as other potential benefits of design measures, such as the promotion of pro-social behavior through restorative experiences. In addition, we refer to topics that, in our opinion, require further scholarly attention.

7.3.1 Implementation Ideas

The provision of certain physical elements, notably trees and other greenery, appears to be a way of promoting restoration within the urban environment. Few researchers have yet gone into detail in evaluating how different elements and their combinations may serve restoration; however, calls for such research have been made (e.g., Velarde et al. 2007) and researchers have begun to address this knowledge gap. For example, Faber Taylor and Kuo (2011) concluded that children's ADHD symptoms decrease when playing in natural environments characterized by open grassy areas, but not in those where big trees are predominant. Nordh et al. (2009) assessed the restorative potential of small parks and open spaces in Scandinavian cities. Using a set of photographs, participants rated 74 parks in terms of likelihood of restoration. Park size and the amounts of grass, trees and bushes were the strongest predictors for restorativeness. Other elements, such as the presence of water or flowering plants, seemed to be less important in terms of restorative opportunities. More recently, instead of using photographs, Nordh et al. (2011) had participants choose from successive pairs of texts that described park alternatives with systematically varied combinations of attributes. For each pair of alternatives, the participants had to choose the best setting when looking for restoration. Their findings showed that grass, trees and other people were among the most influential components when looking for a restorative respite. In general, the more grass, trees or bushes, the more often a park was chosen for taking a rest. With regard to people, the presence of a few people was the most preferred option.

Although often important, greenery is not the only element that promotes restoration. The study of Karmanov and Hamel (2008) illustrates how water, in this case in the form of canals, can contribute to perceiving a neighborhood as restorative. According to their results, a carefully designed neighborhood, with water bodies and houses facing them, can be as restorative as a spectacular wild natural area (see also White et al. 2010). However, when water is present but not prominent in a setting, as with small fountains or ponds, its restorative value may be lower than that of vegetation (Nordh et al. 2011).

Another issue to consider when aiming to promote restoration through contact with nature in cities and, therefore, when designing urban spaces, is that the densification of urban areas limits the possibilities for retaining extensive green areas in central areas, hampering access for many residents. As opportunities for entering larger natural areas diminish, the inclusion of natural elements in smaller, frequently used urban areas becomes an important means of ensuring that residents have some access to restorative nature. For example, Nordh et al. (2009) considered how small (or pocket) parks might be designed to support restorative experiences in cities (see also Kaplan 1983). Well-designed small green spaces distributed throughout urban areas can be accessed and serve the restoration needs of a large number of people. Keeping in mind that the use of urban green areas is related to the distance between a person's home and the green space as well as to the time needed to get there (Grahn and Stigsdotter 2003), innovative design ideas that bring green spaces closer to where people live – pocket parks, green lines, roof gardens, and more – can become important for ensuring opportunities for restoration. Moreover, simply requiring more green areas may only spur sprawling development, which in turn may undermine public health (Richardson et al. 2012). Whatever provisions are made for restorative contact with nature in the urban context should be coordinated with provisions for housing and services such as transport.

When spending time outdoors is not easy, including natural elements such as indoor plants, posters or other nature representations indoors may support restorative respites. When in hospital, contact with nature through indoor vegetation allocated to patients' rooms and common areas as well as looking at a natural view may speed the healing process. At the workplace, where stress and tension often accumulate, indoor plants as well as natural views and images of nature may help workers cope with daily demands. Similarly, indoor plants and other kinds of nearby nature in residential and school settings may enhance people's ability to cope with stressful events (Corraliza and Collado 2011; Van den Berg et al. 2010; Wells and Evans 2003).

In addition, it is necessary to keep in mind that natural environments should not be equated with restorative environments, and vice versa (Hartig and Staats 2003). Many settings can be valued for their restorative quality despite an absence of natural features. For example, Abdulkarim and Nasar (2014) demonstrated that the availability of seats and the presence of a sculpture in a plaza, as well as the combination of seats and sculptures, increased the possibilities of obtaining restoration.

The results described above provide support for design ideas meant to enhance restorative quality. However, we doubt that it is possible to create a set of environments that always offer support for all types of restoration. The achievement of restoration will depend on the sociophysical attributes of the setting in relation to the antecedent condition from which a person needs to recover. One way to improve knowledge about which contents in which combinations are conducive to particular kinds of restoration is to create environmental simulations and assess the attributes and combinations that boost restorative experiences. These simulations can give an impression of what the restorative experience would be like in a certain setting with controlled sociophysical characteristics. This could be further explored in real-life situations (see e.g., Lindal 2013).

Overall, more research is needed in order to improve our knowledge about suitable contents and arrangements of natural and other physical elements when designing restorative environments. This could prove useful to policy makers, planners, architects and others involved in the design of cities, homes, workplaces, schools, hospitals, and other settings. These professionals must deal with a range of tradeoffs when making design decisions. For example, including large trees and greenery in an already built area will probably require more effort and expense over time than including non-natural elements such as seats and sculptures. Better knowledge about what works can help the actors involved to make such tradeoffs.

7.3.2 Other Topics for Further Research

Aside from the topics identified above, we see many opportunities for further research. Not least, we think that the social aspects of restoration need more consideration than they have received to date in terms of research and application. For example, more attention should be paid to how the immediate social context facilitates or hinders restorative experiences. The current results suggest that, when safety is not an issue, people in need of restoration prefer to be alone; however, if a setting is perceived as unsafe, the company of another appears to be conducive to restoration (Staats and Hartig 2004). Staats et al. (2010) found that people prefer to be alone in a park when attentionally fatigued, but they prefer the company of familiar others when spending time in an urban environment. This may be due to safety issues, to the desire to comply with a social norm of being with someone in a public place, or reasons outside the scope of their study. More recently, Johansson et al. (2011) concluded that feelings of revitalization increased more during a 40-min walk in a park if the participant was alone, while a comparable walk in an urban center boosted feelings of revitalization more if it was done in the company of a friend. These findings support the theoretical assumption that people in need of restoration do not like to be in the company of others, perhaps wanting to avoid social feedback and demands for maintaining socially appropriate behavior. Therefore, ensuring users' perceptions of safety should be kept in mind when designing city spaces aimed at boosting restoration. We believe that more insights are needed into the role played by the social context in the restorative process.

Thus, the physical characteristics of a setting should be studied in combination with the social ones.

The social aspects of restoration extend beyond the immediate social context and its role in the renewal of one individual's depleted resources. More recent research has called attention to the relational resources that many people hold in common, the ways in which the shared resources constituted by relationships can become depleted and restored, and how the sociophysical environment can play a role in these resource dynamics. In an initial test of a theory of collective restoration, Hartig et al. (2013) found evidence of a curvilinear relationship between vacation-taking and the dispensation of antidepressants in Sweden. Monitoring a period of more than 12 years, the more people were on vacation during a given month, the larger the decline in dispensation to the population. As further evidence of the spread of benefits, they found that the number of workers taking a vacation was also related to the dispensation of antidepressants to people beyond working age. They discuss the importance of vacation time for the convergence of people in preferred social constellations in preferred locations.

Other issues that deserve further scholarly attention concern the factors and processes that lead children to consider nature a restorative resource, and the circumstances under which such learning occurs. Contact with nature does not necessarily lead to restoration. Children's personal factors, such as different kinds of daily interaction with nature, may influence the potential restorative benefits they could obtain while in a natural setting (Collado et al. 2015). The study of the developmental aspects of children's restorative experiences is truly in its infancy. A deeper understanding of the role played by familiarity with various nearby natural settings (e.g., agricultural, coastal, forests) as well as different ways of interacting with nature (e.g., play vs. compulsory time spent outdoors) is needed. Moreover, the possible influence of the relationships of role models with nature (e.g., professional vs. recreational) on children's perceptions regarding restorative quality awaits research. In addition, it would be interesting to assess whether these early restorative experiences have a lasting effect on adults, shaping the way they perceive and experience restoration in different environments.

Apart from the social and developmental aspects of restorative experiences, the relationship between restorative experiences and pro-social behavior deserves scientific attention. The limited research conducted on this issue shows that individual restorative experiences can enable or enhance healthy social outcomes, such as less violent behavior directed at members of one's family (Kuo and Sullivan 2001), improved social networks (Kuo et al. 1998) and reduced occurrence of criminal behaviors in school (Matsuoka 2010). Kuo and Sullivan (2001) found that the relationship between exposure to greenery and aggressive behavior was completely mediated by people's attention capacity, perhaps reflecting more complete restoration from attention fatigue in those individuals for whom nature is more readily available. It would be interesting to see whether and how restorative experiences encourage pro-social behavior toward other people in different settings, such as communities, schools or prisons.

Aside from helping people to manage the ordinary demands of everyday life better, spending time in restorative environments may serve as an inexpensive, complementary treatment for different illnesses and as an aid to rehabilitation (Faber Taylor and Kuo 2011; Liu et al. 2007; Gonzalez et al. 2010; Hartig and Cooper Marcus 2006; Park and Mattson 2009a, b; Raanaas et al. 2012). The various cost reductions that restorative experiences may provide for health care systems, employers and families (e.g., due to less consumption of medication, less sick leave, less psychological treatment) are important and under-researched topics.

Finally, designing settings with a greater restorative potential is not enough. Researchers and practitioners should also look at ways of encouraging the use of such spaces (e.g., Korpela and Ylén). Encouraging doctors and other health care facilitators to recommend regular contact with nature is one often-discussed option, but such recommendations may have little effect on people who are caught in circumstances that allow little time for restoration. To date, the literature on restorative environments has hardly considered the implications of time limitations (though see Hartig et al. 2013).

7.4 Conclusion

The stress-reducing effects, the enhancement of salutary behaviors and the increase in positive mood provided by restorative environments in urban areas are important means to compensate, reduce or even replace some of the negative aspects of city life. Workplaces, schools, homes, hospitals, and plazas are some of the key settings in which demanding physical, psychological and social tasks are faced every day. These settings can be designed with a restorative potential in mind, to help people meet their recurrent needs for restoration.

The literature discussed in this chapter shows potentially important implications for policy, design, and research. Given the densification of urban spaces, which makes access to nature more difficult, and the parallel increase in lifestyle-related illnesses such as stress, anxiety, depression and overweight, research on the relationship between restoration and health is of pressing importance. We acknowledge that the health benefits associated with restorative environments may not seem impressive when compared to other factors such as socioeconomic status. Yet, the importance of urban restorative environments for health becomes significant when considering the cumulative benefits these can offer to large numbers of people over long periods of time. Moreover, like other environmental amenities, access to restorative environments is a scarce resource more readily available to people of economic means; it is one more pathway through which socioeconomic status can operate on the distribution of health and illness in populations. In light of the findings reviewed above, it is apparent that opportunities for restoration must be available as part of everyday life, not only in terms of time but also in terms of environmental opportunities. Such opportunities can involve visual as well as physical access to environments likely to support restoration. Interdisciplinary work uniting policy makers, design professionals, city planners, health professionals and researchers should be encouraged in an effort to improve our knowledge about how to create restorative opportunities within urban spaces. After all, the more restorative we make our cities, the healthier they will be.

Acknowledgement Partial support came from the Spanish Ministry of Economy and Competitiveness (PSI-2013-44939).

References

Abdulkarim, D., & Nasar, J. (2014). Are livable elements also restorative? *Journal of Environmental Psychology, 38*, 29–38.

Amati, M. (Ed.). (2008). *Urban green belt in the 21st century*. London: Ashgate.

Bell, A. C., & Dyment, J. E. (2008). Grounds for health: The intersection of green school grounds and health-promoting schools. *Environmental Education Research, 14*, 77–90.

Bodin, M., & Hartig, T. (2003). Does the outdoor environment matter for psychological restoration gained through running? *Psychology of Sport and Exercise, 4*, 141–153.

Bringslimark, T., Hartig, T., & Patil, G. G. (2007). Psychological benefits of indoor plants in workplaces: Putting experimental results into context. *HortScience, 42*, 581–587.

Bringslimark, T., Hartig, T., & Patil, G. G. (2009). The psychological benefits of indoor plants: A critical review of experimental literature. *Journal of Environmental Psychology, 29*, 422–433.

Bringslimark, T., Hartig, T., & Patil, G. G. (2011). Adaptations to windowlessness: Do office workers compensate for a lack of visual access to the outdoors? *Environment and Behavior, 29*, 422–433.

Broekhuizen, K., de Vries, S. I., & Pierik, F. H. (2013). *Healthy aging in a green living environment: A systematic review of the literature* (TNO rep. R10154). Leiden: TNO.

Chang, C. Y., & Chen, P. K. (2005). Human response to window views and indoor plants in the workplace. *HortScience, 40*, 1354–1359.

Chauvet-Gélinier, J., Trojak, B., Vergès-Patois, B., Cottin, Y., & Bonin, B. (2013). Review on depression and coronary heart disease. *Archives of Cardiovascular Diseases, 106*, 103–110.

Cole, D. N., & Hall, T. E. (2010). Experiencing the restorative components of wilderness environments: Does congestion interfere and does length of exposure matter? *Environment and Behavior, 42*, 806–823.

Collado, S., & Corraliza, J. A. (2015). Children's restorative experiences and self-reported pro-environmental behaviors. *Environment and Behavior, 47*, 38–56.

Collado, S., Corraliza, J. A., Staats, H., & Ruiz, M. (2015). Effect of frequency and mode of contact with nature on children's self-reported ecological behaviors. *Journal of Environmental Psychology, 41*, 65–73.

Corraliza, J. A., & Collado, S. (2011). La naturaleza cercana como moderadora del estrés infantil [Nearby nature as a moderator of children's stress]. *Psicothema, 23*, 221–226.

De Vries, S., Classen, T., Hug, S. M., Korpela, K., Maas, J., Mitchell, R., et al. (2011). Contributions of natural environments to physical activity: Theory and evidence base. In K. Nilsson, M. Sangster, C. Gallis, T. Hartig, S. De Vries, K. Seeland, et al. (Eds.), *Forests, trees, and human health* (pp. 205–243). Dordrecht: Springer.

Diette, G., Haponik, E., & Rubin, H. (2003). Distraction therapy with nature sights and sounds reduces pain during flexible bronchoscopy. *Chest, 12*, 941–948.

Faber Taylor, A., & Kuo, F. (2009). Children with attention deficits concentrate better after walk in the park. *Journal of Attention Disorders, 12*, 402–409.

Faber Taylor, A., & Kuo, F. (2011). Could exposure to everyday green spaces help treat ADHD? Evidence from children's play settings. *Applied Psychology: Health and Well-being, 3*, 281–303.

Faber Taylor, A., Kuo, F., & Sullivan, W. (2001). Coping with ADD: The surprising connection to green play settings. *Environment and Behavior, 33*, 54–77.

Faber Taylor, A., Kuo, F., & Sullivan, W. (2002). Views of nature and self-discipline: Evidence from inner city children. *Journal of Environmental Psychology, 22*, 49–63.

Felsten, G. (2009). Where to take a study break on the college campus: An attention restoration theory perspective. *Journal of Environmental Psychology, 29*, 160–167.

Fjørtoft, I. (2001). The natural environment as a playground for children: The impact of outdoor play activities in pre-primary school children. *Early Childhood Education Journal, 29*, 111–117.

Galindo, M. P., & Hidalgo, M. C. (2005). Aesthetic preferences and the attribution of meaning: Environmental categorization processes in the evaluation of urban scenes. *International Journal of Psychology, 40*, 19–26.

Gatersleben, B., & Andrews, M. (2013). When walking in nature is not restorative. The role of prospect and refuge. *Health & Place, 20*, 91–101.

Gonzalez, M. T., Hartig, T., Patil, G. G., Martinsen, E. W., & Kirkevold, M. (2010). Therapeutic horticulture in clinical depression: A prospective study of active components. *Journal of Advanced Nursing, 66*, 2002–2013.

Grahn, P., & Stigsdotter, U. A. (2003). Landscape planning and stress. *Urban Forestry and Urban Greening, 2*, 1–19.

Grahn, P., Martensson, F., Lindblad, B., Nilsson, P., & Ekman, A. (1997). *Ute på dagis [Outdoors at the daycare center] [Stad and Land Nr. 145]*. Alnarp: Movium, Swedish University of Agricultural Sciences.

Hartig, T. (2004). Restorative environments. In C. Spielberger (Ed.), *Encyclopedia of applied psychology* (pp. 273–279). San Diego: Academic.

Hartig, T. (2007). Three steps to understanding restorative environments as health resources. In C. Ward Thompson & P. Travlou (Eds.), *Open space: People space* (pp. 163–179). London: Taylor and Francis.

Hartig, T. (2012). Restorative housing environments. In S. J. Smith (Ed.), *The international encyclopedia of housing and home (environments section)* (pp. 144–147). Oxford: Elsevier.

Hartig, T., & Cooper Marcus, C. (2006). Essay: Healing gardens – places for nature in health care. *Lancet, 368*, S36–S37.

Hartig, T., & Evans, G. W. (1993). Psychological foundations of nature experience. In T. Gärling & R. G. Golledge (Eds.), *Behavior and environment: Psychological and geographical approaches* (pp. 427–457). Amsterdam: North-Holland.

Hartig, T., & Fransson, U. (2009). Leisure home ownership, access to nature, and health: A longitudinal study of urban residents in Sweden. *Environment & Planning A, 41*, 82–96.

Hartig, T., & Lawrence, R. J. (2003). Introduction: The residential context of health. *Journal of Social Issues, 59*, 455–473.

Hartig, T., & Staats, H. (2003). Guest editors' introduction: Restorative environments. *Journal of Environmental Psychology, 2*, 103–107.

Hartig, T., & Staats, H. (2006). The need for psychological restoration: A determinant of environmental preference. *Journal of Environmental Psychology, 26*, 215–226.

Hartig, T., Mang, M., & Evans, G. W. (1991). Restorative effects of natural environment experience. *Environment and Behavior, 23*, 3–26.

Hartig, T., Kaiser, F., & Bowler, P. (2001). Psychological restoration in nature as a positive motivation for ecological behavior. *Environment and Behavior, 33*, 590–607.

Hartig, T., Evans, G. W., Jamner, L. D., Davis, D. S., & Garling, T. (2003). Tracking restoration in natural and urban field settings. *Journal of Environmental Psychology, 23*, 109–123.

Hartig, T., Catalano, R., Ong, M., & Syme, S. L. (2013). Vacation, collective restoration, and mental health in a population. *Society and Mental Health, 3*, 221–236.

Hartig, T., Mitchell, R., De Vries, S., & Frumkin, H. (2014). Nature and health. *Annual Review of Public Health, 35*, 207–228.

Herzog, T. R., Maguire, C. P., & Nebel, M. B. (2003). Assessing the restorative components of environments. *Journal of Environmental Psychology, 23*, 159–170.

Herzog, T. R., Ouellette, P., Rolens, J. R., & Koening, A. M. (2010). Houses of worship as restorative environments. *Environment and Behavior, 42*, 395–419.

James, W. (1892). *Psychology: The briefer course*. New York: Holt.

Johansson, M., Hartig, T., & Staats, H. (2011). Psychological benefits of walking: Moderation by company and outdoor environment. *Applied Psychology: Health and Well-being, 3*, 261–280.

Kaplan, R. (1983). The role of nature in the urban context. In I. Altman & J. F. Wohlwill (Eds.), *Human behavior and environment: Advances in theory and research* (Behavior and the Natural Environment, Vol. 6, pp. 127–161). New York: Plenum Press.

Kaplan, R. (1993). The role of nature in the context of the workplace. *Landscape Urban Planning, 26*, 193–201.

Kaplan, S. (1995). The restorative benefits of nature: Toward an integrative framework. *Journal of Environmental Psychology, 15*, 169–182.

Kaplan, R. (2001). The nature of the view from home: Psychological benefits. *Environment and Behavior, 33*, 507–542.

Kaplan, R., & Kaplan, S. (Eds.). (1989). *The experience of nature: A psychological perspective*. Cambridge: Cambridge University Press.

Kaplan, S., Bardwell, L. V., & Slakter, D. B. (1993). The museum as a restorative environment. *Environment and Behavior, 25*, 725–742.

Kaplan, R., Kaplan, S., & Ryan, R. L. (1998). *With people in mind: Design and management of everyday nature*. Washington, DC: Island Press.

Karmanov, D. Y., & Hamel, R. (2008). Assessing the restorative potential of contemporary urban environment (s): Beyond the nature versus urban dichotomy. *Landscape and Urban Planning, 88*, 15–25.

Kelz, C., Evans, G. W., & Röderer, K. (2015). The restorative effects of redesigning the schoolyard: A multi-methodological, quasi-experimental study in rural Austrian middle schools. *Environment and Behavior, 47*, 119–139.

Knopf, R. (1987). Human behavior, cognition, and affect in the natural environment. In D. Stockols & J. Altman (Eds.), *Handbook of environmental psychology* (pp. 786–826). New York: Wiley.

Kuo, F. E. (2001). Coping with poverty: Impacts of environment and attention in the inner city. *Environment and Behavior, 33*, 5–34.

Kuo, F. E., & Faber Taylor, A. (2004). A potential natural treatment for attention-deficit/hyperactivity disorder: Evidence from a national study. *American Journal of Public Health, 94*, 1580–1586.

Kuo, F. E., & Sullivan, W. C. (2001). Environment and crime in the inner city: Does vegetation reduce crime? *Environment and Behavior, 33*, 343–367.

Kuo, F. E., Sullivan, W. C., Coley, R. L., & Brunson, L. (1998). Fertile ground for community: Inner-city neighborhood common spaces. *American Journal of Community Psychology, 26*, 823–851.

Kweon, B. S., Ulrich, R. S., Walker, V. D., & Tassinary, L. G. (2008). Anger and stress: The role of landscape posters in an office setting. *Environment and Behavior, 40*, 355–381.

Lederbogen, F., Kirsch, P., Haddad, L., Streit, F., Tost, H., Schuch, P., et al. (2011). City living and urban upbringing affect neural social stress processing in humans. *Nature, 474*, 498–501.

Lindal, P. J. (2013). *Restorative environmental design for densifying cities*. Doctoral dissertation, Faculty of Architecture, Design & Planning, University of Sydney: Sydney, Australia.

Lindal, P. J., & Hartig, T. (2013). Architectural variation, building height, and the restorative quality of urban residential streetscapes. *Journal of Environmental Psychology, 33*, 26–36.

Lindholm, G. (1995). Schoolyards: The significance of place properties to outdoor activities in schools. *Environment and Behavior, 27*, 259–293.

Liu, G., Wilson, J., Qi, R., & Ying, J. (2007). Green neighborhoods, food retail and childhood overweight: Differences by population density. *The Science of Health Promotion, 21*, 317–325.

Martens, D., Gutscher, H., & Bauer, N. (2011). Walking in "wild" and "tended" urban forests: The impact on psychological well-being. *Journal of Environmental Psychology, 31*, 36–44.

Matsuoka, R. (2010). Student performance and high school landscapes: Examining the links. *Landscape and Urban Planning, 97*, 273–282.

Nordh, H., Hartig, T., Hagerhall, C. M., & Fry, G. (2009). Components of small urban parks that predict the possibility for restoration. *Urban Forestry & Urban Greening, 8*, 225–235.

Nordh, H., Alalouch, C., & Hartig, T. (2011). Assessing restorative components of small parks using conjoint methodology. *Urban Forestry & Urban Greening, 10*, 95–103.

Olmsted, F. L. (1865/1952). The Yosemite valley and the Mariposa big trees: A preliminary report: With an introductory note by Laura Wood Roper. *Landscape Architecture, 43*, 12–25.

Ouellette, P., Kaplan, R., & Kaplan, S. (2005). The monastery as a restorative environment. *Journal of Environmental Psychology, 25*, 175–188.

Ozdemir, A., & Yilmaz, O. (2008). Assessment of outdoor school environments and physical activity in Ankara's primary schools. *Journal of Environmental Psychology, 28*, 287–300.

Park, S. H., & Mattson, R. H. (2008). Effects of flowering and foliage plants in hospital rooms on patients recovering from abdominal surgery. *Horttechnology, 18*, 563–568.

Park, S. H., & Mattson, R. H. (2009a). Ornamental indoor plants in hospital rooms enhanced health outcomes of patients recovering from surgery. *The Journal of Alternative and Complementary Medicine, 15*, 975–980.

Park, S. H., & Mattson, R. H. (2009b). Therapeutic influences of plants in hospital rooms on surgical recovery. *HortScience, 44*, 102–105.

Parsons, R., Tassinary, L. G., Ulrich, R. S., Hebl, M. R., & Grossman-Alexander, M. (1998). The view from the road: Implications for stress recovery and immuniza-

tion. *Journal of Environmental Psychology, 18*, 113–140.
Pretty, J., Peacock, J., Sellens, M., & Griffin, M. (2005). The mental and physical health outcomes of green exercise. *International Journal of Environmental Health Research, 15*, 319–337. doi:10.1080/09603120500155963.
Raanaas, R. K., Patil, G. G., & Hartig, T. (2010). Effects of an indoor foliage plant intervention on patient well-being during a residential rehabilitation program. *HortScience, 45*, 387–392.
Raanaas, R. K., Patil, G. G., & Hartig, T. (2012). Health benefits of a view of nature through the window: A quasi-experimental study of patients in a residential rehabilitation center. *Clinical Rehabilitation, 26*, 21–32.
Richardson, E. A., Mitchell, R., Hartig, T., de Vries, S., Astell-Burt, T. E., & Frumkin, H. (2012). Green cities and health: A question of scale? *Journal of Epidemiology and Community Health, 66*, 160–165.
Saelens, B. E., Sallis, J. F., Black, J. B., & Chen, D. (2003). Neighborhood-based differences in physical activity: An environmental scale evaluation. *American Journal of Public Health, 93*, 1552–1558.
Shibata, S., & Suzuki, N. (2001). Effects of indoor foliage plants on subject's recovery from mental fatigue. *North American Journal of Psychology, 3*, 385–396.
Staats, H. (2012). Restorative environments. In S. Clayton (Ed.), *The Oxford handbook of environmental and conservation psychology* (pp. 445–458). New York: Oxford University Press.
Staats, H., & Hartig, T. (2004). Alone or with a friend: A social context for psychological restoration and environmental preference. *Journal of Environmental Psychology, 24*, 199–211.
Staats, H., Kieviet, A., & Hartig, T. (2003). Where to recover from attentional fatigue. An expectancy-value analysis of environmental preference. *Journal of Environmental Psychology, 23*, 147–157.
Staats, H., Van Gemerden, E., & Hartig, T. (2010). Preference for restorative situation: Interactive effects of attentional state, activity-in-environment, and social context. *Leisure Sciences, 32*, 401–417.
Steinheuser, V., Ackermann, K., Schönfeld, P., & Schawabe, L. (2014). Stress and the city: Impact of urban upbringing on the (re)activity of the hypothalamus-pituitary-adrenal axis. *Psychosomatic Medicine, 76*, 678–685.
Talbot, A., & Kaplan, S. (1986). Perspectives on wilderness: Re-examining the values of extended wilderness experience. *Journal of Environmental Psychology, 6*, 177–188.
Tennessen, C., & Cimprich, B. (1995). Views to nature: Effects on attention. *Journal of Environmental Psychology, 15*, 77–85.
Ulrich, R. S. (1983). Aesthetic and affective response to natural environment. In I. Altman & J. F. Wohlwill (Eds.), *Behavior and the natural environment* (pp. 85–125). New York: Plenum Press.
Ulrich, R. S. (1984). View through a window may influence recovery from surgery. *Science, 224*, 420–421.
Ulrich, R. S., Simon, R., Losito, B., Fiorito, E., Miles, M., & Zelson, M. (1991). Stress recovery during exposure to natural and urban environments. *Journal of Environmental Psychology, 11*, 201–230.
Ulrich, R. S., Simons, R. F., & Miles, M. A. (2006). Effects of environmental simulations and television on blood donor stress. *Journal of Architectural and Planning Research, 20*, 38–47.
UN-Habitat. (2013). *State of the world's cities 2012–2013. Prosperity of cities*. New York: Routledge.
Van den Berg, A. E., & Custer, M. H. G. (2011). Gardening promotes neuroendocrine and affective restoration from stress. *Journal of Health Psychology, 16*, 3–11.
Van den Berg, A. E., Maas, J., Verheij, R. A., & Groenewegen, P. P. (2010). Green space as a buffer between stressful life events and health. *Social Science and Medicine, 70*, 1203–1210.
Velarde, M. D., Fry, G., & Tveit, M. (2007). Health effects of viewing landscapes: Landscape types in environmental psychology. *Urban Forestry and Urban Greening, 6*, 199–213.
Von Lindern, E., Bauer, N., Frick, J., Hunziker, M., & Hartig, T. (2013). Occupational engagement as a constraint on restoration during leisure time in forest settings. *Landscape and Urban Planning, 118*, 90–97.
Wells, N. (2000). At home with nature: Effects of "Greenness" on children's cognitive functioning. *Environment and Behavior, 32*, 775–795.
Wells, N., & Evans, G. (2003). Nearby nature. A buffer of life stress among rural children. *Environment and Behavior, 35*, 311–330.
White, M. P., Smith, A., Humphries, K., Pahl, S., Snelling, D., & Depledge, M. (2010). Blue space: The importance of water for preference, affect and restorativeness ratings of natural and built scenes. *Journal of Environmental Psychology, 30*, 482–493.
Whitehouse, S., Varni, J. W., Seid, M., Cooper-Marcus, C., Ensberg, M. J., Jacobs, J., & Mehlenbeck, R. S. (2001). Evaluating a children's hospital garden environment: Utilization and consumer satisfaction. *Journal of Environmental Psychology, 21*, 301–314.
Williams, K., & Harvey, D. (2001). Transcendent experience in forest environments. *Journal of Environmental Psychology, 21*, 249–260.
World Health Organization. (1996). Constitution and by-laws. In *WHO Basic Documents* (41st edn). Geneva: WHO, Office of Publications.

8 Green Exercise, Health and Well-Being

Ana Loureiro and Susana Veloso

8.1 Physical Activity, Health and Well-Being

The evolutionary perspective of physical activity, fitness and health states that human anatomy and physiology have remained relatively unchanged over the past 40,000 years (Astrand 1994). In this sense, the relationship between energy intake, energy expended and physical activity required has essentially persisted the same since the Stone Age (Spence and Lee 2003). For prehistoric man, who depended on hunting, fishing and exploitation of wild resources to survive, physical activity played a major role in his daily life. In fact, we are now living our lives in totally different ways from what we have done as humans for more than 99 % of our existence (Biddle and Mutrie 2008). Since the industrial revolution, people have reduced their physical activity, reaching this huge contradiction: a human body biologically prepared for high levels of energy expenditure left at the mercy of modernization with an increasingly sedentary lifestyle (Spence and Lee 2003). Motorized transport, all kinds of work done by machines (that was once manual work), modern forms of entertainment such as television, movies, videos, and computers, have all brought humans to the point of living every day in an almost fully sedentary way. In fact, most people preferably perform mental and non-physical work (Sallis and Owen 1999).

Epidemiologic studies, like the Eurobarometer survey, report that 41 % of EU citizens exercise or play sport at least once a week, while a significant proportion of them (59 %) never or seldom do so (European Commission 2014). At least once a week, 48 % do some form of other physical activity (such as cycling, dancing or gardening), while 30 % never do this kind of activity at all. Adults spend 50–60 % of their day in sedentary pursuits. Gender differences are favorable for men, who engage in more physical activity than women. However, this is more evident in the younger group (15–24 years old) where boys tend to exercise on a regular basis (74 %) more than girls (55 %). Physical activity tends to decrease with age, reaching about 70 % in people over 55 years old. In general, citizens in the Northern part of Europe (e.g. Sweden, Denmark, and Finland) are more active than in the Southern member states (e.g. Bulgaria, Malta, Portugal, and Italy) (European Commission 2014). These decreasing trends in physical activity are reflected in changing bodies, contributing to the growing

A. Loureiro (✉)
School of Psychology and Life Sciences, Lusófona University, Lisbon, Portugal

COPELABS, Lusófona University, Lisbon, Portugal
e-mail: ana.loureiro@ulusofona.pt

S. Veloso
Faculty of Physical Education and Sport, Lusófona University, Lisbon, Portugal
e-mail: veloso.susana@gmail.com

epidemic of obesity in the world. Although the USA is the leader in obesity levels, its worldwide prevalence nearly doubled between 1980 and 2008. According to World Health Organization estimates for 2008, over 50 % of both men and women in the European region were overweight, and roughly 23 % of women and 20 % of men were obese (http://www.euro.who.int/en/health-topics/ noncommunicable-diseases/obesity/data-and-statistics).

This sedentary lifestyle results in a large cost to health, while reducing the quality and quantity of life. Country-specific estimates of economic costs attributable to physical inactivity range from 1.2 % to 2.5 % of annual health care expenditure. The longest sedentary time compared with the shortest was associated with a 49 % increase in the risk of all-cause mortality (Katzmarzyk 2011). In fact, the project Designed to Move is based on the current evidence that "today's children are the first generation to have a shorter life expectancy than their parents" (designedtomove.org). This is an action-project that gives urgent priority to increasing the world's commitment to physical activity. Solutions must be put into practice, and the change-makers must know "what" needs to be done and "how" to do it.

In 1985, Caspersen and colleagues defined physical activity as any bodily movement produced by the contraction of skeletal muscles that results in a substantial increase in caloric requirements over resting energy expenditure (American College of Sport Medicine – ACSM 2013). Aiming to clarify the concept of physical activity further, it is useful to distinguish other constructs such as physical exercise and sports, which are not synonymous. Exercise is a subgroup of physical activity, defined as planned, structured, and repetitive bodily movements done to improve and/or maintain one or more components of physical fitness. This leads to the concept of physical fitness, which is defined as a set of attributes or characteristics that individuals have or achieve that relates to their ability to perform physical activity. These characteristics are usually separated into health-related (e.g. cardiorespiratory endurance, muscular strength, flexibility, body composition) and skill-related (e.g. agility, coordination, balance, speed) components of physical fitness (ACSM 2013). Sport is an even more specific structured form of physical activity; competitive, and characterized by achievement, luck and strategy (Kaplan et al. 1993).

In addition to defining physical activity and exercise, it is important to define clearly the wide range of intensities that help distinguish between active and sedentary individuals, as each can cause different health outcomes. However, measuring the physical activity required for a healthy quality of life is a difficult and complex task. Physical activity can take a huge variety of forms: it can be accomplished in formal and informal contexts, including the most routine tasks of everyday life (walking, housekeeping activities, gardening); it may be practiced in intense, moderate or light forms; for very short periods (a few seconds or minutes) or extended periods (hours); with a high or low frequency, regular or irregular; and alone, in a group or accompanied by someone (Kaplan et al. 1993).

The relationship between health and physical activity has been the subject of research for more than 25 years, and many national health services (e.g. American College of Sport Medicine and Center for Disease Control and Prevention in the US, National Health Service in the UK, World Health Organization) have established guidelines to clarify for people and professionals (of public health, health/fitness, clinical exercise, and health care) the amount and intensity of physical activity needed to improve health, lower susceptibility to disease (morbidity), and decrease premature mortality.

The global recommendations of physical activity for health resumed by the World Health Organization (World Health Organization – WHO 2010) for adults aged 18–64 are: (1) at least 150 min of moderate-intensity aerobic physical activity throughout the week or at least 75 min of vigorous-intensity aerobic physical activity throughout the week or an equivalent combination of moderate-vigorous intensity activity; (2) aerobic activity should be performed in bouts of at least 10 min duration; (3) for additional health benefits, adults should increase their moderate-intensity aerobic physical activity to 300 min per week, or engage in 150 min of vigorous-intensity physical activity per week, or an equivalent combination of moderate-and-vigorous intensity activity; (4) muscle-strengthening activities should be done involving major muscle groups on 2 or more days a week. Unless specific medical conditions indicate the contrary, these guidelines are relevant to all healthy adults and could be applied in leisure time or transportation (e.g. walking or cycling), in occupational time (i.e. work, gardening), in household chores, play, games, sports or planned exercise, in the context of daily, family and community activities (WHO 2010). In addition, the last US physical activity guidelines in 2008 made age-specific recommendations targeted at older adults (>65 years), children and adolescents (6–17 years), and younger children (<6 years) (U.S. Department of Health and Human Services 2008).

These physical activity guidelines have recently been complemented by a new paradigm of sedentary behavior. Physical and sedentary activities are not viewed as opposite behaviors, but as different constructs with independent effects on the health and disease process (Yates et al. 2011b). Epidemiological studies show this independent effect, since a strong association was found between TV viewing time and the risk of type two diabetes and independently of physical activity (Hu et al. 2003). Sedentary behavior, defined as an MET of 1.5 or less (metabolic equivalent units of energy cost of resting quietly), corresponds to activities undertaken while lying or sitting, such as watching TV and other forms of screen time. Thus, any standing activity (unless absolutely still) is classified as non-sedentary (Yates et al. 2011a). This may be an opportunity for new recommendations based on simply sitting less and standing more, which are expected to revolutionize health promotion (Yates et al. 2011b). In the recent ACMS guidelines (ACSM 2013), the complementary advice "long periods of sitting should be avoided" is already included.

Physical inactivity or a sedentary lifestyle is the greatest risk factor for the most common causes of death (e.g., being inactive doubles the risk of cardiovascular disease), meriting the same level of concern as tobacco consumption, cholesterol and obesity. In turn, participation in regular physical activity increases life expectancy, prevents diseases, and has multiple beneficial effects on many body systems (Sallis and Owen 1999; ACSM 2013).

There is a large body of research about the benefits of physical activity and exercise. The immune and nervous systems and many parts of the body (heart, skeletal muscles, bones, blood) can reduce risk factors for non-communicable diseases (NCDs – often referred to as chronic diseases) (C3 Collaborating for Health 2011). This is important because these major NCDs account for 68 % of the 56 million deaths annually, a number that is expected to increase from 38 million in 2012 to 52 million by 2030 (WHO 2014). Some of the risk factors are blood pressure, cholesterol level, and body mass index (BMI), which influence chronic diseases such as type two diabetes, heart disease and many cancers. When regular physical activity is performed in youth, the benefits are, on one hand, reduced levels of adiposity, blood pressure and lipids, cardiovascular risk factors, injury, and mental health concerns like depression and, on the other hand, increased strength, fitness and bone health (Janssen and LeBlanc 2010).

The mental benefits of physical activity are less well documented than the physical effects. However, many studies and clinical trials have shown specific benefits, including improving mood, reducing symptoms of stress, anger and depression, alleviating anxiety and slowing cognitive decline (Babyak et al. 2000). A review of the research literature on the role of physical activity in a wide range of parameters of well-being, such as anxiety, depression, mood and affect, health-related quality of life, cognitive function, and self-esteem, concluded that there is a remarkable consistency in the evidence for a positive association between exercise and well-being; however, the quality of the evidence, for the most part, is not optimal (Ekkekasis and Backhouse 2009). Specific studies support exercise as a first-line treatment for mild to moderate depression, compared to antidepressant medication, and also to improve depressive symptoms when used as an adjunct to medication (Carek et al. 2011). However, for major depression disorders, of mild-to-moderate severity, only aerobic exercise at a dose consistent with public health recommendations is an effective treatment; a lower dose is comparable to a placebo effect (Dunn et al. 2005). Although not as extensively studied as depression, exercise has been shown to be an effective and cost-efficient alternative treatment for a variety of anxiety disorders (Carek et al. 2011).

Research on exercise and well-being frequently discusses the paradox – "If exercise makes most people feel better, why are most people physically inactive or inadequately active?" Backhouse and colleagues suggest that this might be an artifact because research over the past three decades has established that exercise can make people "feel better" (e.g., during walking, during more vigorous exercise among certain participants, and during recovery from vigorous exercise among nearly all participants), but has tended to discount, or not measure, the negative effects of exercise. These authors highlighted the importance of examining the complex exercise–affect relationship and considering whether diverse affective responses could account for part of the variability in physical activity behavior and adherence (Backhouse et al. 2007).

The Self-Determination Theory (SDT), founded by Deci and Ryan in 1985, has proved useful in explaining the antecedents and processes that underpin exercise behavior and adherence (Deci and Ryan 1985; Ryan and Deci 2000; Hagger and Chatzisarantis 2008; Ryan et al. 2009). The aim of Exercise Psychology is to explain why people adopt physically active versus inactive lifestyles. The psychological SDT proposes that all humans possess three basic psychological needs: autonomy, which reflects a desire to engage in activities of one's choosing; competence, which implies a desire to interact effectively with the environment; and relatedness, which involves feeling connected to others or feeling that one belongs in a given social environment (Edmunds et al. 2009). When these psychological needs are satisfied, more autonomous forms of regulation guide behavior (e.g., intrinsic motivation and motivations guided by values) and adaptive exercise outcomes are expected (e.g., exercise adherence and enjoyment). In contrast, thwarted needs and more controlling forms of regulation (e.g., external and introjected) are expected to result in non-optimal outcomes (e.g., dropout and dissatisfaction). To resume, SDT suggests that the psychological needs and the type of motivation guiding behavior determine what kind of exercise-related outcome will occur (Ryan and Deci 2000; Edmunds et al. 2009; Veloso et al. 2012).

Although SDT has recently provided greater understanding of physical activity adherence and how to motivate people to adopt an active lifestyle, the research about physical activity correlates has increased scientific knowledge for decades. The factors associated with children's and adolescents' physical activity, reviewed by Biddle and colleagues, could be demographic/biological, psychological, behavioral, social or environmental (Biddle et al. 2011). Age and gender are the demographic factors (boys and younger children/adolescents have greater levels of physical activity). The positive psychological correlates of physical activity are positive body image, good intentions, feelings of competence and confidence, and a motivational style centered on effort and self-improvement, while a negative factor is the presence of barriers to physical activity. Previous practice and sport participation are the positive behavioral correlates of physical activity, with smoking and sedentary behavior the negative ones. The social/cultural correlates of physical activity are parental influence and social support. Finally, supportive environments are associated with greater physical activity, such as access to facilities, a minor distance from home to school, more time spent outside, and less local crime (Biddle et al. 2011).

For adults, the social correlates with physical activity associated with more practice are high levels of education and socioeconomic status. Overweight and obesity are inversely correlated with physical activity, but a healthy diet is directly associated. The positive psychological correlates are enjoyment, expected benefits, intention, perceived health, self-motivation, self-efficacy, a high stage of behavior change, and self-schemata for exercise, while the negative correlates are mood disturbance and barriers (Biddle and Gorely 2012). There is strong evidence for self-efficacy and enjoyment. It is important to relate the evidence of enjoyment to intrinsic motivation as SDT has demonstrated (Ryan and Deci 2000; Edmunds et al. 2009). The environmental correlates with physical activity are access to facilities, an environment with enjoyable scenery and neighborhood safety (Biddle and Gorely 2012).

Research about correlates of physical activity in older adults is understudied (Biddle and Gorely 2012). A review of studies, including mostly healthy volunteers, who probably do not express the physical activity correlates of those living with chronic illness or disabilities, showed evidence for initiation and maintenance of physical activity (Van Stralen et al. 2009). Physical health status, exercise habits and physical activity at the baseline level are behavioral positive correlates; self-efficacy, intention, action planning, motivational readiness to change, outcome expectations, and perceived benefits are the psychological ones. The physical environmental correlates are perceived access, safety from crime, and program format (home). The social correlates of social support from significant others and social norms have the least evidence (Van Stralen et al. 2009).

Given the great impact that physical inactivity has on people's health and national economies, the problem could be seen as social rather than just individual. The WHO, in its global strategy for diet and physical activity promotion, recognized this fact (WHO 2010). This is consistent with the ecological approach, which demands population-based, multi-sectorial, multi-disciplinary, and culturally relevant strategies (Biddle and Mutrie 2008). In fact, living in cities with more cars, greater urbanization, and lack of play spaces contributes to decreasing physical activity. On the other hand, more structured activity facilities, like new paths for walking or cycling, more pedestrian zones in urban areas and parks for playing or walking the dog, could all contribute to increased physical activity. In this context, the concept of green exercise becomes relevant, due to the growing interest in the physical environment and its influence on involvement in physical activity.

The environmental context, including access to active opportunities, the weather, perceived safety and aesthetics of place, has the potential to influence activity levels and this could interact with psychosocial variables in determining physical activity adherence and promotion (Biddle and Mutrie 2008). Research on environmental and exercise psychology could be integrated to provide evidence for policy-making and the design of relevant environmental changes.

8.2 Contact with Nature, Health and Well-Being

Throughout human history, nature has always been of great importance to the lives of individuals. Becoming innate, this bond, connection and tendency to affiliate with and focus on the natural environment are the main claims of the biophilia hypothesis proposed by Wilson (1984). Even today, when people live further away from other living species, there is a wide recognition of the need to be close to nature and of its benefits, namely those related to physical and mental health (Gullone 2000).

One of the most common reasons why individuals search for and, in many circumstances, prefer natural environments is the resulting improvements in health state and well-being. Wanting to escape from routine and the pressure of daily stress, or to experience calm and stimulation are some of the psychosocial benefits that motivate people to seek natural places (Home et al. 2012; Loureiro 1999). Feelings of being away, relaxation, or reduced negative mood are also mentioned as contributing to the choice of green and natural settings as people's favorite places, and their preference for natural rather than urban places (Hartig and Staats 2006; Korpela 2003; Korpela et al. 2001).

It is expected that by 2050, the great majority of the world's population or even almost the whole population (if developed regions are considered) will live in urbanized areas (UN-Habitat 2011). Our current age is characterized by a growing urbanizing world, where humans live mainly in urban and closed environments and have fewer opportunities to access natural settings. Given the significance of natural experiences for people's lives, an increase in occasions to experience nature or natural elements could be crucial for health status and quality of life (Frumkin 2001; Hartig et al. 2010; Maas et al. 2006; Van den Berg et al. 2007). According to the observed pervasiveness of the psychological and physical benefits of contact with nature, researchers promote the enhancement of human health that can be reached by increasing access to natural settings (Morris 2003).

Environmental psychologists, adopting different theoretical and empirical approaches, have focused on the outcomes for people of their experiences of contact with nature (Hartig et al. 1991; Kaplan 1995; Kaplan and Kaplan 1989; Ryan et al. 2010; Ulrich 1984). These studies have demonstrated several health and well-being outcomes from different types of experiences, such as walking in an urban park, trekking or camping in a national park, looking through a window, or contemplating a coastal landscape. Natural experiences in different contexts and at different levels foster positive emotions, better attention focus, vitality, and reduced signals of physiological arousal (for reviews, see Hartig et al. 2010, 2014).

On one hand, the effects of nature experiences on health and well-being promotion are described as due to the restorative characteristics of these environments. These theories emphasize the restoration of some affected capabilities, such as cognitive ability to focus attention or a stress mood recovery (Kaplan and Kaplan 1989; Ulrich et al. 1991).

On the other hand, other approaches present different physical and psychological results of nature-based experiences, which do not necessarily follow a previous state of some compromised capacities and without stressing this aspect. Instead, they focus on the enhancement of positive states (Marrero and Carballeira 2010; Ryan et al. 2010).

Focusing on the restorative components of natural environments, Kaplan and Kaplan developed the attention restoration theory, suggesting that natural environments allow human beings to refresh and restore their cognitive function from fatigue derived from the need to direct attention to environmental stimuli (Kaplan and Kaplan 1989; Kaplan 1995). According to this theory, the psychological costs of information management or mental fatigue stem from a limited ability to direct and focus attention, which can be recovered in certain environments, such as those that provide an opportunity to be away from the place that causes fatigue, fascination, and compatibility between environmental characteristics and motivations of individuals. Each individual has his/her restorative environments, which may be a playground, a trip to the countryside or waterside, an urban square or a cultural place (Adevi and Grahn 2011; Ashbullby et al. 2013; Collado et al. 2013; Grahn and Stigsdotter 2010; Korpela et al. 2010; Packer and Bond 2010).

Settings that provide contact with nature correspond to very good opportunities to restore psychological functioning, namely in its cognitive aspects, due to their particular features. Taking a few minutes to walk in a garden, listen to the motion of the leaves, look at the clouds, and stroll slowly along a pathway might be an important action to recover cognitive functioning and psychological well-being (Kaplan 1995). This power of nature also explains why people generally prefer natural environments to urban ones. It occurs when the balance between the setting's characteristics is perceived to provide individuals with the ability to process information and in which this process is effective (Kaplan and Kaplan 1989).

Individuals rate their preference for natural settings according to the setting's ability to give them the opportunity to experience more positive emotions, less stress, and emotional regulation (Korpela et al. 2001). The psycho-physiological evolutionary stress recovery theory argues that health benefits derived from contact with nature occur because experiencing natural scenes initiates the physiological and psychological responses that support recovery from stress (Ulrich

1984; Ulrich et al. 1991). Negative emotions and physiological arousal may be decreased after viewing, being exposed to or moving in natural contexts because these environments promote physiological recovery and relaxation from situations that threaten well-being. Within a natural environment, an individual's negative affect is replaced by a positive affect, negative thoughts are inhibited and autonomic arousal decreases.

These theoretical approaches have been the background for several studies that aim to demonstrate the links between exposure to natural settings and the positive outcomes related to recovery from mental fatigue or from stressful events (Berto 2014; Hartig et al. 2014). The evidence comes from studying different types of virtual or real environments, and the use of several measures such as self-reported measures of mood and stress, attention tests and physiological indicators of stress.

Laboratory experiments using exposure to virtual images and environments have provided evidence for the benefits of natural virtual environments on the reported mood or performance, attentional tasks, or physiological measures such as salivary cortisol, skin conductance, pupillometry, eye-tracking and heart rate (e.g. Alvarsson et al. 2010; Brown et al. 2013; Depledge et al. 2011; Haluza et al. 2014; Hartig et al. 2003; Kort et al. 2006). Data on the subject also comes from studies investigating the benefits of exposure to real natural environments, which found a decline in blood pressure and salivary cortisol, better performance in attentional tasks, positive mood and emotion reports, lower self-reported stress, a sense of well-being, or school course ratings as significant outcomes of experiencing window views of natural settings in residential or clinical locations, nature near to public housing and residential places, or walks in natural areas (e.g. Beil and Hanes 2013; Benfield et al. 2015; Kaplan 2001; Kuo and Sullivan 2001; Raanaas et al. 2011; Roe et al. 2013; Taylor et al. 2001).

Following the research on restorative environments, some studies have sought to identify the features of natural environments, as well the quantity of natural elements, which could elicit the positive outcomes related to exposure to nature. The number of trees, percentage of grass covering the ground surface area, the possibility of seeing bushes, the setting size, the presence of flowers and plants, and water features predicted the likelihood of restoration identified by individuals (Nordh et al. 2009; Nordh and Ostby 2013). Individuals rate places that have more natural features as more restorative (Carrus et al. 2013) while viewing spreading trees is associated with positive emotions and happiness (Lohr and Pearson-Mims 2006). In a recent study, researchers found a reverse U curve for stress reduction related to exposure to medium-density tree canopy (Jiang et al. 2014). However, more research is needed to continue to identify the specific settings and their key characteristics that explain the benefits in relation to restoration and well-being (Joye and Van den Berg 2011; Velarde et al. 2007).

Previously, an association was found between several experiences with nature and their physical and physiological signs of short-term benefits for individuals' well-being. These benefits included a better recovery after surgery, lower blood pressure, lower heart rate, lower electrodermal activity, or changes in electromyographic (EMG) activity (e.g. Hartig et al. 2003; Laumann et al. 2003; Lohr and Pearson-Mims 2006; Parsons et al. 1998; Ulrich 1984; Ulrich et al. 1991).

More recently, some studies have suggested that natural spaces are vital to health and well-being, whether it is a personal garden, the presence of trees on the street, a state forest or an urban park. This is something that people recognize as they perceive themselves as being healthier when they are more exposed to environments that have more natural and green elements. In fact, there is a correlation between the number of natural features in an individual's living environment and the level of general health perceived by these individuals (Maas et al. 2006).

A direct relationship between the existence of natural elements, such as trees, in the environment where people live and the level of human health is receiving increased attention and support from research evidence and epidemiological data (Donovan et al. 2013; Hartig et al. 2014; Takano et al. 2002). Despite the need for more evidence on the effects of natural spaces on health and well-being, namely urban green spaces (Lee and Maheswaran 2011; Richardson et al. 2012), the idea of instorative besides restorative effects of the natural environment is receiving growing attention (Joye and Van den Berg 2012). If the deviation from nature has negative health effects, then the change in current patterns of relationships with nature may contribute to greater human vitality and health (Stilgoe 2001).

Being in a natural setting has vitalizing effects, promoting an energized and positively toned state (Plante et al. 2006; Ryan et al. 2010). Outdoor green environments are more revitalizing, more stimulating, and decrease tiredness, particularly when people are performing some kind of activity. On the other hand, viewing virtual natural environments may contribute to relaxation and less tension, even when people are exercising (Plante et al. 2006). This positive impact of natural experiences on subjective vitality was observed in a group of studies using different methodologies, and supports the idea that contact with nature is a way of promoting well-being and physical health, namely by increasing levels of subjective vitality (Ryan and Deci 2008; Ryan et al. 2010). This is evidence for the link between natural experiences and subjective well-being. General satisfaction with life and specific satisfaction with sentimental life and leisure are associated with opportunities to be in contact with nature activities (Marrero and Carballeira 2010). Developing personal projects in natural settings induces positive affect and also a sense of the project's efficacy, support and meaning, which together contribute to personal well-being (Roe and Aspinall 2012). Living in a greener neighborhood is associated with more residential satisfaction and reported happiness (Van Herzele and De Vries 2012).

Different experiences with nature foster psychological and physical well-being and these benefits may come from an experiential sense of unity and harmony with the natural environment that individuals may develop while being in nature during their lives (Bell et al. 2014; Olivos et al. 2011). Feeling that one belongs and is embedded in nature may partly explain the positive benefits of experiences in the natural world (Mayer et al. 2009). In fact, individuals who are more related and connected to nature report a greater perception of a restorativeness capacity from forest settings (Tang et al. 2014). People more related to nature also tend to look for more experiences with nature, and benefit from the well-being outcomes from those experiences, such as feelings of positive mood, happiness or vitality (Nisbet et al. 2011; Zelenski and Nisbet 2014).

Contact with nature may even have an impact beyond well-being and health outcomes. For example, experiencing nature also results in people having feelings of autonomy and intrinsic aspirations. Immersion in natural settings promotes higher intrinsic aspirations, related to prosocial value orientations, and lower extrinsic aspirations, which can lead to more prosocial actions such as generous decision-making (Weinstein et al. 2009). This effect of immersion in a natural context was also found for helping behavior (Guéguen and Stefan 2014). Thus, being in contact with nature could be associated with not only personal well-being but also social well-being.

People in different phases of personal development may benefit from frequent exposure to natural environments. The evidence presented by research in the domain of health and well-being outcomes from the experiences of contact with nature has been an important motive and argument for taking the opportunity to provide people with these experiences in different settings such as schools, playgrounds, work offices, residential and urban spaces, homes for the elderly or healthcare environments, and within the context of different activities like education, work, treatment of physical or psychological diseases, rest, leisure, or physical activity (Bird 2007; Bloom et al. 2014; Corazon 2012; Gladwell et al. 2013; Godbey 2009).

8.3 Green Exercise and Outdoor Physical Activity

The term "green exercise" was proposed by Pretty and colleagues (Pretty et al. 2003) from Essex University, and the first peer-review paper was published in 2005 (Pretty et al. 2005). These authors sought to describe the synergistic benefit to health that occurs when exercising whilst being directly exposed to nature (Gladwell et al. 2013; Pretty et al. 2003). Green exercise is defined as "a physical activity in green places that may bring both physical and mental health benefits" (Pretty et al. 2003, p. 7), or as the exercise or physical activity that occurs in the presence of nature, such as cycling in the countryside or walking in an urban park (Barton and Pretty 2010).

As previously noted, physical activity has positive effects on physical and psychological health, and exposure to nature is also good for mental health and well-being. Thus, the health benefits of green exercise come simultaneously from physical activity and contact with nature. The relationship between the natural environment and health has received wide interest for decades, fostering initiatives in civil and scientific communities both to promote public health and to conserve biodiversity (Bowler et al. 2010). In fact, although most citizens currently live in urban environments, disconnected daily from nature, and there is an increase in sedentary lifestyles in the majority of populations, people tend to appreciate the benefits of protecting the environment (Pretty et al. 2003). Some of these initiatives are: membership of environmental and wildlife organizations; visits to the countryside and the growth in national and international ecotourism; membership of gymnasiums and of sports and outdoor organizations (Pretty et al. 2003). The Conservation Volunteers Green Gym, developed by a British charitable organization, is a program that aims to provide people with a way to enhance their fitness and health while taking action to improve the outdoor environment. The invitation on the program's website, "*Want to improve your health and well-being but not too keen on running machines or lycra?*" summarizes their assumptions (http://www.tcv.org.uk/greengym/).

Several associations between the natural environment and health and well-being have been identified (Hartig et al. 2014). The natural environment (e.g., urban parks, species diversity, and the number of trees near home) is directly associated with air quality and stress, which in turn benefit health and well-being (e.g. performance, subjective well-being, physiological changes, mobility, mortality and longevity). However, the natural environment is also related to contact with nature (e.g. frequency, duration, activity such as viewing or walking), which in turn is associated with air quality and stress, but also with physical activity and social contacts, variables also related to health and well-being. In other words, individuals or groups who consciously engage with nature, simply for viewing or for practicing a physical activity, could amplify the impact of the natural environment on their health and well-being, through promoting psychical activity levels (walking for recreation and outdoor play) and/or social contact (e.g. interacting with neighbors and a sense of community). Of course, all these relationships are subject to modification by the characteristics of the people or the context, and there is also a reciprocal relationship between these variables (air quality, physical activity, social contacts, and stress). This model can support the role of green exercise in health promotion, showing its impact at the personal, social, community and public level. In fact, people engage in physical activity firstly because it helps them to feel good in the short term and then because it will benefit their health in the long term. Thus, people regularly seeking natural spaces for restoration could engage in some form of physical activity to amplify the benefits (Hartig et al. 2014).

Empirical studies have aimed to show the benefits of exercise in nature, arguing that being active in green spaces may yield health benefits over and above the positive effects of physical activity in other environments, such as indoors and without nature elements (Hug et al. 2009; Pretty et al. 2005; Thompson Coon et al. 2011; Mitchell 2013).

Natural settings, such as a park or riverside, providing an added outdoor setting in an urban context, may promote public health as they offer an additional environmental context for physical activity besides indoor spaces. Given the benefits presented previously related to the improvements in health and well-being derived from contact with nature, we may argue for the additional and increased positive outcomes for public health provided by the performance of physical activity in natural environments, as it contributes to both physical and mental health.

The research about the specific benefits of green exercise is growing. A recent systematic review summarizes a wide range of health and well-being outcomes, such as higher positive and lower negative emotions, after exercising in a natural rather than a more synthetic environment (e.g. non-green outdoor built environments and indoor environments) (Bowler et al. 2010). Physiological outcomes, such as healthy levels of blood pressure and cortisol, are less supportive of consistent positive evidence. There is also some support, but again not very strong, for greater attention and concentration after practicing in a natural environment (Bowler et al. 2010). Another review summarized how the great outdoors can promote physical activity and health in the general population, exploring the impact of green exercise on psychological and physiological health markers, and also the mechanisms by which green exercise has an impact on health (Gladwell et al. 2013). Outdoor natural environments, beyond the benefits of simple exposure, may facilitate adherence to physical activity, through lower levels of perceived effort, stress and mental fatigue, leading to improved mood (e.g. reducing tension, anger and depression), self-esteem and perceived health state. Green exercise also promotes physiological functioning, including health markers, such as heart rate, blood pressure and autonomic control, and endocrine markers, such as noradrenaline, adrenaline and cortisol.

Moreover, green exercise can facilitate adherence to physical activity through promoting attention to an external pleasant and green environment, which consequently distracts from and reduces awareness of physiological sensations and negative emotions, thus minimizing the perception of effort (Gladwell et al. 2013). Studies comparing indoor versus outdoor physical activity in natural environments show greater feelings of revitalization and engagement in outdoor settings (Thompson Coon et al. 2011). The difference is not in the quality of the practice of indoor and outdoor exercise, but in the wider benefits that accrue from exposure to an outdoor environment. For example, health clubs and similar establishments have a cost, a closing time, and are more likely to hassle, and this discourages many individuals from adhering to practice (Toftager et al. 2011; Parachin 2011). A person's access to green spaces could thus be one of the important resources of the living environment to enhance physical activity contributions, to reduce obesity and improve health (Lachowycz and Jones 2011). A study of 11,649 exercise participants (54 % outdoors, 18 % indoors and 28 % practicing in both environments) found that outdoor practitioners dealt better with stress and depression, and had a better knowledge of health maintenance (Puett et al. 2014). In Denmark, a study of a random sample of 21,832 adults showed a relationship between a shorter distance between residences and green spaces and a higher level of physical activity and related lower rates of obesity (Toftager et al. 2011). This association will probably not be equal elsewhere. For example, 56.6 % of 514 residents in Philadelphia (USA) were considered active, and of these 64 % were indoor practitioners, 22.6 % were outdoor practitioners and 13.4 % practiced in both environments (Hillier et al. 2014). In Portugal, a study with 282 practitioners of outdoor and indoor physical exercise analyzed the relationship between outdoor physical exercise and well-being and observed that participants with outdoor activity or who combined outdoor with indoor physical exercise (56.4 %) reported more positive emotions and well-being associated with exercise, and that their connectedness to nature was a significant predictor of well-being, also negatively predicting psychological distress. The same association was not found for the group who only performed physical exercise in indoor environments (43.6 %) (Loureiro and Veloso 2014).

Positive outcomes of green exercise for individuals' mental health improvements are observed even for short periods of practice. These effects on self-esteem and mood are independent of location, duration, intensity, gender, age, and health status (Barton and Pretty 2010). Taking this into account, specific recommendations for greater efficacy of green exercise are proposed for duration, intensity and type of green space. Only 5 min of green exercise results in self-esteem and mood improvements, less than 60 min produces a smaller effect, and an active whole day results in great improvements in mood and self-esteem (duration); self-esteem only increases with a light green exercise activity; however, mood increases with both light and vigorous activity (intensity); both health markers improve in green environments, but the presence of water generates greater improvements for near waterside practice (e.g. beach or river) or participation in water-based activities (a type of green/natural space). Green exercise brings improvements in self-esteem for both genders; however, men show a better mood. Younger people report more improved self-esteem after green exercise and the middle-aged group report a better mood. Mentally ill people should be encouraged to undertake green exercise because they experience the greatest changes in self-esteem (Barton and Pretty 2010).

The impact on different subgroups of the population is a subtle point that future studies should consider. Green exercise potentially increases the level of physical activity across the whole population; however, larger individual benefits seem to occur in specific populations (Thompson Coon et al. 2011). For example, a study found that mortality rates of cardiovascular and respiratory diseases decreased with increasing access to natural environmental places, but this only occurred in males (Richardson and Mitchell 2010).

Although fewer people are regularly present in natural settings, many seek out nature for outdoor recreational activities and some look for challenging outdoor activities. Paradoxically, there is a large population with insufficient physical activity levels for the recommendations that ensure health (Gladwell et al. 2013). How might the environment help to motivate and facilitate physical activity? A green environment may foster increased physical activity through decreasing perceptions of effort and improving motivation (Gladwell et al. 2013). Adherence to physical activity could be promoted by extrinsic motivation through relationships between green exercise and health, driven by external factors such as pressure from significant others; however, this is not likely to affect everyone, much less over the long term. The engagement in physical activity by intrinsic motivation, driven by enjoyment or excitement about the challenge, is more likely to occur and be maintained over a long term (Ryan and Deci 2000). Some people engage for health benefits, whereas others adhere for social reasons. However, the social and enjoyment benefits of physical activity appear to be more successful than the health benefits at persuading individuals to participate in physical activity (Gladwell et al. 2013). In this sense, green exercise can help to promote physical activity through the fun and escape from the routine of daily life that it offers, satisfying both social and pleasure reasons for practice adherence.

Another advantage of green exercise is some evidence suggesting that exercise in a natural environment may be perceived as easier to perform. An experimental study comparing brief indoor and outdoor walks found that participants reported a greater intention to engage in future outdoor walks, and this was accompanied by a higher level of enjoyment and positive affect after outdoor walks (Focht 2009). Therefore, the combination of exercise and exposure to nature could be a useful tool to improve physical activity motivation and human physical and psychological health. The epidemiological problem of sedentary people, who fail to achieve the recommended daily amounts of physical activity, could benefit from green exercise, as a vehicle for driving physical activity promotion. In fact, green exercise could be a pleasant activity leading to the fulfillment of the *Healthy People 2010 Guidelines* (U.S. Department of Health and Human Services 2008) which encourage people to select an appropriate dose of activity that is enjoyable.

Although natural environments tend to facilitate physical activity adherence and health benefits, some disengagement with nature has been observed, especially in children and adolescents, due to a reduced relationship and connectedness with nature. For example, in England, only 10 % of today's youth has regular contact with nature, compared to 40 % of adults who did so when they were young (Natural England 2009). The parental fears of traffic, strangers and criminal activity restrict young people from accessing nature (Ward Thompson et al. 2008). Knowing that the amount of time spent outdoors is associated with physical activity in both children and adolescents, access to nature could be a powerful instrument to combat sedentary lifestyles and promote healthy ones (Cleland et al. 2008; Frost n.d.).

The suitability and attractiveness of spaces for certain types of physical activity may influence levels of physical activity and rhythm of practice (Hartig et al. 2014). Access to facilities, an enjoyable scenery and neighborhood safety are important environmental correlates of physical activity (Biddle and Gorely 2012). The quality of urban spaces can also influence the level of physical activity. A Brazilian study of 2,046 participants (over 16 years old) practicing for at least 150 min per week showed a relationship between the level of physical activity and the accessibility of footpaths or spaces for physical activity (Hallal et al. 2012). Studies in different countries, such as Japan, Scandinavia and the Netherlands, showed that access to green space was associated with longevity and a decreased risk of mental illness (Gladwell et al. 2013). Sometimes, running or walking in certain urban streets involves exposure to unpleasant, inadequate and noisy environments, which probably reduces the benefits of the physical activity itself. For example, outdoor exercise in a busy urban environment may have less effect on mental well-being and adherence than an aesthetically appealing and supportive indoor environment (Gladwell et al. 2013). The quality of green space perception may be associated with physical and psychological health benefits (Thompson Coon et al. 2011). The quality of the natural environment could be a moderator in the associations between access to green space and physical activity (Jones et al. 2009). Coastal areas seem to provide more physical activity initiatives, encouraging and facilitating outdoor activity (Thompson Coon et al. 2011). A European study showed a relationship between living in a greener environment and the level of physical activity (three times more likely) and the chance of being overweight or obese (40 % lower chance) (Ellaway et al. 2005). However, more evidence is needed about the association between access and quality of urban green space, physical activity and health (Hillsdon et al. 2006; Maas et al. 2008).

Data about the cumulative effects of experiences in nature strongly suggest that the continued practice of green exercise can enhance the restorative effects of natural environments, and thus result in very significant gains in the health and well-being of the population (Marselle et al. 2013). Nevertheless, more research and evidence is crucial to support the relationship between contact with nature, physical activity and human health and well-being (Hartig et al. 2014).

8.4 Implications of Green Exercise for QOL and Health Promotion

There is a wide recognition of the relevance of physical activity in the promotion of health and quality of life. Physical inactivity levels are rising in many countries, particularly in the more developed regions, and are presently identified as the fourth leading risk factor for global mortality. They contribute significantly to the prevalence of non-communicable diseases (NCDs) and their major implications for the general health of the population worldwide (WHO 2010). This is why WHO recommendations stress the need to increase support actions to raise physical activity levels across all age groups (WHO 2010, 2014).

Physical activity is crucial for the prevention of NCDs and the improvement of general level of public health, helping to address public health challenges faced by humankind. As described in the previous sections of this chapter, nature based physical activity may potentiate these benefits. Green exercise and other forms of outdoor recreational activities foster physical and psychological health and well-being in several ways (Bowler et al. 2010; Pretty et al. 2005; Thompson Coon et al. 2011). Accordingly, the combination of physical activity and exposure to nature in green exercise may be useful for the prevention of NCDs and the promotion of health levels worldwide (Gladwell et al. 2013; Haluza et al. 2014; Pretty et al. 2011).

Declining physical activity levels, especially in the developed world, are significantly associated with a decrease in natural experiences and relatedness with nature. This nature disengagement often begins in childhood and usually leads to an unhealthy life pathway (Pretty et al. 2009). The current younger generation, mostly in developed countries, is extremely deprived of contact with nature, as they have less access to outdoor environments or have become less willing to visit and experience nature. Therefore, this generation's detachment from the natural environment and consequent less real and active contact with nature may be associated with the increase in NCDs in the adult population (Gladwell et al. 2013).

Improving and increasing the availability of settings and supporting access to green exercise in particular, and contact with nature activities in general, would have substantial positive outcomes on the health of the whole population, as these contexts are important supportive environments helping people to be more physically active and encouraging the adoption of healthier lifestyles (Bedimo-Rung et al. 2005; Barton 2009; Pretty et al. 2003). People whose living space has a more natural environment available usually have higher levels of physical activity in different forms besides sport, such as walking, playing or gardening (Calogiuri and Chroni 2014; WHO 2014).

Providing access to a natural environment was the main objective of the design and construction of the first urban parks, driven by the urban park movement in England and North America. Still today, the health benefits of nature and associated healthy lifestyles are a central question in health and quality of life promotion. As access to nature is essential to improve mental and physical health, it should be a main concern in land use policy (American Public Health Association – APHA 2013; Ward Thompson 2011). The implications for public and urban policy and design are widely emphasized, and can be achieved by different types of measures of urban planning and public space design, transport policy, education environments, and campaigns stressing contact with nature as a motive for green exercise practice (APHA 2013; Calogiuri and Chroni 2014; Gladwell et al. 2013). However, it is important to differentiate health outcomes from nature-based physical activity experiences from those related to other activities and interventions like diet and physical activity in itself (Lee and Maheswaran 2011).

WHO recommendations of physical activity for health cover the whole life span and are specific for different phases (WHO 2010). In accordance with this, types of activity and related environments that contribute to promoting health and well-being, namely those that are natural and outdoors, can be identified for people of all ages. Moreover, APHA policy statements reinforce that efforts should be made to incorporate nature in urban and land policies, due to evidence of gains in health and well-being for children, young people, adults and the elderly who have more contact with nature (APHA 2013). Nature in the form of urban parks, gardens, greenways, naturalized schoolyards and playgrounds, and natural landscaping around homes and workplaces give people of all ages the opportunity to experience nature in different ways, such as contemplation or engaging in outdoor physical activity.

When considering the implications for health and well-being across a life span, all features must be integrated into public space design. Environmental factors are potential physical activity promoters and affordances but can also be barriers. Regarding green exercise promotion, the integration of natural features such as trees, plants, and greenways must be considered together with other elements such as street and path type, access points, permeability, views, sound, light, maintenance and surveillance (Pikora et al. 2003). Although it might be advocated that it would be difficult and unrealistic to provide people with access to large park systems, especially in an urban context, contact and experience with nature is affordable by different means, such as planting trees, greening alleys, cultivating gardens in schools, communities, and hospitals, or creating greenways for pedestrians and cyclists (APHA 2013). Moreover, these interventions are effective in giving people more proximity to engage in nature-based physical activity.

In childhood, play and transportation are the main activities that may give a child the chance to be physically active. Public or private gardens, such as school playgrounds, provide children with very good opportunities to engage in activities that are physically demanding, and when the settings are rich in natural features and elements children can gain both psychological and physiological benefits from these activities (Collado et al. 2013; Hodges et al. 2013; Pretty et al. 2009). Outdoor green environments such as neighborhood parks, promoting gardening, play and recreation also have the potential to engage less active children in physical activity (Godbey 2009; Moore and Cosco 2014; Reed et al. 2013). Urban design that encourages access to nature when walking or cycling to school on a greenway or crossing a park is essential to enhance children's physical activity and nature experiences (Moore and Cooper Marcus 2008). For teenagers and young people, the search for natural areas is associated with engaging in play and adventure combined with social play and interaction (Staempfli 2009).

Adults can gain great health and well-being advantages from living in a natural environment, or in an urban context with natural elements, as these can encourage active lifestyles and higher levels of physical activity (Hartig et al. 2014). The potential to combine nature health benefits with physical activity outcomes may be achieved in different types of activities and settings. Green exercise may be practiced on a regular daily basis, as when individuals walk or cycle to work, or jogging in a park at weekends, or on a non-regular basis as when they spend their holidays trekking in a national park. Wilderness recreation and tourism is increasing with more people planning their annual holiday in national parks and wilderness areas looking for adventure and nature-based experiences (Buchell and Eagles 2007). People often look to combine green exercise with other aims such as socializing or enjoying landscape (Miller et al. 2014). Furthermore, setting features are important factors in facilitating or inhibiting levels of participation in green exercise or other recreational activities in natural environments. These features include safety perception, proximity, leisure time and design (Godbey 2009).

Engaging in physical activity in green spaces such as woods and forests lowers the risk of poor mental health more than exercising in a gym or in the streets (Miller et al. 2014). Greenways or urban streets with trees and plants are also especially motivating for pedestrians and green exercise practitioners (Calogiuri and Chroni 2014). Besides offering direct food safety and supply for an urban population, as well as better environmental quality, urban agriculture is an opportunity for people to be more physically active. Involving urban citizens in gardening and horticulture projects increases physical activity levels and fitness and thus contributes to weight management in particular and public health in general. City farmers participating in food growing and gardening community projects experience social connections and reduced stress (Schmutz et al. 2014).

For adults, the workplace is an important setting for health promotion and disease prevention. The feedback provided by pedometer interventions at work, combined with other components such as a diary, a website for records, sharing behaviors or communication between participants in a work setting program, the dissemination of health promotion information, counseling sessions, or group activities motivate individuals to increase and maintain their physical activity over time (Freak-Poli et al. 2013). Promoting green exercise experiences among employees, combining the benefits of being physically active with those of exposure to nature, is a promising way to cultivate a healthier company workforce.

More attention is being given to the implementation of outdoor running and walking group programs as extended measures of public health promotion as they can reach large groups of the population at the same time. The evaluations of these programs find that people taking part show greater positive affect and mental well-being and a decrease in depression, perceived stress and negative affect (Marselle et al. 2013).

Despite the importance of physical activity for disease prevention and maintenance of quality of life in the elderly, there is a lack of knowledge about levels of physical activity that are needed in this population (Sun et al. 2013). Park-based leisure time is associated with health indicators and reduced perceived stress. When older people perceive they have a good physical health state and are accompanied during outdoor experiences, they tend to spend more time in these settings, such as parks, and these walkable green spaces may be responsible for greater longevity among this group of the population (Orsega-Smith et al. 2004). Environmental design and features can be an important source of encouragement for walking and other physical activities among elderly people and thus contribute to their healthy lives.

More directly related to health and quality of life promotion in general, and prevention of diseases such as NCDs in particular, programs aiming to intervene in these areas have a great potential to reduce social and economic costs associated with illness and loss of quality of life. The economic investment in programs to promote physical activity among children, young people, adults and the elderly is less than that needed to treat and heal health problems such as those related to obesity or cardiovascular diseases (WHO 2010). Engaging people in programs of exercise in outdoor and natural environments, such as integrating outdoor running groups in gyms, or trekking activities during ecotourism holidays or leisure time, provide people with natural experiences that can contribute to better psychological states and the relief of stress and, through this, improve their attitudes toward physical activity. This process can be a route to increasing and reinforcing people's intentions to engage in physical activities (Calogiuri and Chroni 2014).

Combining natural experiences with physical activity has provided good opportunities to obtain positive outcomes in mental health treatment (Barton and Pretty 2010; Maller et al. 2005). Exercise and other types of physical activities in natural settings can be therapeutic in contexts such as child attention deficit and hyperactivity, or severe and enduring adult mental illness. Landscape therapy, horticulture therapy, wilderness therapy, nature or animal therapy, therapeutic gardening or healing gardens are different types of treatment with a nature-based approach in common that are receiving more attention from mental health professionals and social services (Maller et al. 2008).

Green exercise programs, combining physical activity, nature and social components, are effective in enhancing well-being, self-esteem and positive mood levels in individuals with mental illnesses (Barton et al. 2012). Nature-based mental health interventions, where people are placed in safe outdoor natural settings, separate them from daily negative influences and give them access to self-characteristics usually more difficult to perceive (Hine et al. 2011). Improvements in self-esteem and mood may induce decreases in depression and anxiety and therefore result in better mental health for individuals participating in therapeutic green exercise. These direct outcomes for mental health conditions occur simultaneously with increased feelings of connection with nature and progress in individuals' physical state such as a better Body Mass Index (Hine et al. 2011).

The positive influence of green exercise goes beyond the direct outcomes for individuals' mental and physical health. The connection with nature resulting from the increased contact with the natural environment can be a way to develop more environmental values and attitudes, and thus have an effect on behaviors and decisions with an environmental impact for individuals and societies (Collado et al. 2015; Hartig et al. 2007). Contact with nature, through the practice of physical activity such as green exercise, may thus be also considered a path to more long-term changes in attitudes and relationships with nature and the environment (Pretty et al. 2003). Associating the individual's health and well-being benefits as a result of environmental actions, framing environmental behaviors as health behaviors, or using health and well-being motivations to promote sustainable values and actions, is a promising approach to the sustainability challenges faced by humankind (Nisbet and Gick 2008). The benefits for present and future societies may come from different paths toward changing values and actions that support a social, economic and environmentally sustainable development.

References

Adevi, A. A., & Grahn, P. (2011). Attachment to certain natural environments: A basis for choice of recreational settings, activities and restoration from stress? *Environment and Natural Resources Research, 1*, 36–52. doi:10.5539/enrr.v1n1p36.

Alvarsson, J. J., Stefan Wiens, S., & Nilsson, M. E. (2010). Stress recovery during exposure to nature sound and environmental noise. *International Journal of Environmental Research and Public Health, 7*, 1036–1046. doi:10.3390/ijerph7031036.

American College of Sports Medicine, Pescatello, L. S., Arena, R., Riebe, D., & Thompson, P. D. (2013). *ACSM's guidelines for exercise testing and prescription* (9th ed.). Philadelphia: Wolters Kluwer/Lippincott Williams & Wilkins Health.

American Public Health Association. (2013). *Policy statement 20137 – improving health and wellness through access to nature*.http://www.apha.org/policies-and-advocacy/public-health-policy-statements/policy-database?q=20137&y=2013.

Ashbullby, K. J., Pahl, S., Webley, P., & White, M. P. (2013). The beach as a setting for families' health promotion: A qualitative study with parents and children living in coastal regions in Southwest England. *Health and Place, 23*, 138–147. doi:10.1016/j.healthplace.2013.06.005.

Astrand, P. O. (1994). Physical activity and fitness: Evolutionary perspective and trends for the future. In C. Bouchard, R. J. Shephard, & T. Stephens (Eds.), *Physical activity, fitness, and health: International proceedings and consensus statement* (pp. 98–105). Champaign: Human Kinetics.

Babyak, M., Blumenthal, J. A., Herman, S., Khatri, P., Doraiswamy M., Moore, K., ... Krishnan, K. R. (2000). Exercise treatment for major depression: Maintenance of therapeutic benefit at 10 months. *Psychosomatic Medicine, 62*, 633–638.

Backhouse, S. H., Ekkekakis, P., Biddle, S. J., Foskett, A., & Williams, C. (2007). Exercise makes people feel better but people are inactive: Paradox or artefact? *Journal of Sport & Exercise Psychology, 29*, 498–517.

Barton, H. (2009). Land use planning and health and well-being. *Land Use Policy, 26S*, S115–S123. doi:10.1016/j.landusepol.2009.09.008.

Barton, J., & Pretty, J. (2010). What is the best dose of nature and green exercise for improving mental health? A multi-study analysis. *Environmental Science & Technology, 44*, 3947–3955. doi:10.1021/es903183r.

Barton, J., Griffin, M., & Pretty, J. (2012). Exercise, nature and socially interactive based initiatives improve mood and self-esteem in the clinical population. *Perspectives in Public Health, 132*, 89–96. doi:10.1177/1757913910393862.

Bedimo-Rung, A. L., Mowen, A. J., & Cohen, D. A. (2005). The significance of parks to physical activity and public health: A conceptual model. *American Journal of Preventive Medicine, 28*, 159–168. doi:10.1016/j.ampre.2004.10.024.

Beil, K., & Hanes, D. (2013). The influence of urban natural and built environments on physiological and psychological measures of stress: A pilot study. *International Journal of Environmental Research and Public Health, 10*, 1250–1267. doi:10.3390/ijerph10041250.

Bell, S. L., Phoenix, C., Lovell, R., & Wheeler, B. (2014). Green space, health and wellbeing: Making space for individual agency. *Health & Place, 30*, 287–292. doi:10.1016/j.healthplace.2014.10.005.

Benfield, J. A., Rainbolt, G. H., Bell, P. A., & Donovan, G. H. (2015). Classrooms with nature views: Evidence of differing student perceptions and behaviors. *Environment and Behavior, 47*, 140–157. doi:10.1177/0013916513499583.

Berto, R. (2014). The role of nature in coping with psycho-physiological stress: A literature review on restorativeness. *Behavioral Sciences, 4*, 394–409. doi:10.3390/bs4040394.

Biddle, S. J., & Gorely, T. (2012). Physical activity interventions. In S. Murphy (Ed.), *The Oxford handbook of sport and performance psychology* (pp. 660–675). New York: Oxford University Press.

Biddle, S. J., & Mutrie, N. (2008). *Psychology of physical activity: Determinants, well-being & interventions* (2nd ed.). London: Routledge.

Biddle, S. J., Atkin, A., Cavill, N., & Foster, C. (2011). Correlates of physical activity in youth: A review of quantitative systematic reviews. *International Review of Sport and Exercise Psychology, 4*, 25–49. doi:10.1080/1750984X.2010.548528.

Bird, W. (2007). *Natural thinking: A report for the royal society for the protection of birds*.http://www.rspb.org.uk/Images/naturalthinking_tcm9-161856.pdf.

Bloom, J., Kinnunen, U., & Korpela, K. (2014). Exposure to nature versus relaxation during lunch breaks and recovery from work: Development and design of an intervention study to improve workers' health, well-being, work performance and creativity. *BMC Public Health, 14*, 488. doi:10.1186/1471-2458-14-488.

Bowler, D. E., Buyung-ali, L. M., Knight, T. M., & Pullin, A. S. (2010). A systematic review of evidence for the added benefits to health of exposure to natural environments. *BMC Public Health, 10*, 456. doi:10.1186/1471-2458-10-456.

Brown, D. K., Barton, J. L., & Gladwell, F. V. (2013). Viewing nature scenes positively affects recovery of autonomic function following acute-mental stress. *En-*

vironmental Science & Technology, 47, 5562–5556. doi:10.1021/es305019p.
Buchell, R., & Eagles, P. F. J. (Eds.). (2007). *Tourism and protected areas: Benefits beyond boundaries*. Wallingford: CABI Pub.
C3 Collaborating for Health. (2011). Review: The benefits of physical activity for health and well-being. www.c3health.org/wp-content/uploads/2009/09/C3-review-of-physical-activity-and-health-v-1-20110603.pdf.
Calogiuri, G., & Chroni, S. (2014). The impact of the natural environment on the promotion of active living: An integrative systematic review. *BMC Public Health, 14*, 873. doi:10.1186/1471-2458-14-873.
Carek, P. J., Laibstain, S. E., & Carek, S. M. (2011). Exercise for the treatment of depression and anxiety. *The International Journal of Psychiatry in Medicine, 41*, 15–28. doi:10.2190/PM.41.1.c.
Carrus, G., Lafortezza, R., Colangelo, G., Dentamaro, I., Scopelitti, M., & Sanesi, G. (2013). Relations between naturalness and perceived restorativeness of different urban green spaces. *Psyecology, 4*, 225–336. doi:10.1174/217119713807749869.
Cleland, V., Crawford, D., Baur, L. A., Hume, C., Timperio, A., & Salmon, J. (2008). A prospective examination of children's time spent outdoors, objectively measured physical activity and overweight. *International Journal of Obesity, 32*, 1685–1693. doi:10.1038/ijo.2008.171.
Collado, S., Staats, H., & Corraliza, J. A. (2013). Experiencing nature in children's summer camps: Affective, cognitive and behavioural consequences. *Journal of Environmental Psychology, 33*, 37–44. doi:10.1016/j.jenvp.2012.08.002.
Collado, S., Corraliza, J. A., Staats, H., & Ruíz, M. (2015). Effect of frequency and mode of contact with nature on children's self-reported ecological behaviors. *Journal of Environmental Psychology, 41*, 65–73. doi:10.1016/j.jenvp.2014.11.001.
Corazon, S. S. (2012). *Stress, nature & therapy* (Forest & Landscape Research No. 49-2012). Frederiksberg: Forest & Landscape Denmark. http://forskning.ku.dk/find-en-forsker/?pure=files%2F38099918%2Fstress_nature_therapy_web_pag_2.pdf.
Deci, E. L., & Ryan, R. M. (1985). *Intrinsic motivation and self-determination in human behavior*. New York: Plenum Press.
Depledge, M. H., Stone, R. J., & Bird, W. J. (2011). Can natural and virtual environments be used to promote improved human health and wellbeing? *Environmental Science & Technology, 45*, 4660–4665. doi.org/10.1021/es103907m.
Donovan, G. H., Butry, D. T., Michael, Y. L., Prestemon, J. P., Liebhold, A. M., Gatziolis, D., & Mao, M. Y. (2013). The relationship between trees and human health; evidence from the spread of the emerald ash borer. *American Journal of Preventive Medicine, 44*, 139–145. doi:10.1016/j.amepre.2012.09.066.
Dunn, A. L., Trivedi, M. H., Kampert, J. B., Clark, C. G., & Chambliss, H. O. (2005). Exercise treatment for depression: Efficacy and dose response. *American Journal of Preventive Medicine, 28*, 1–8. doi:10.1016/j.amepre.2004.09.003.
Edmunds, J., Ntoumanis, N., & Duda, J. L. (2009). Helping your clients and patients take ownership over their exercise: Fostering exercise adoption, adherence, and associated well-being. *Health & Fitness Journal, 13*, 20–25.
Ekkekasis, P., & Backhouse, S. H. (2009). Exercise and psychological well-being. In R. Maughan (Ed.), *The Olympic textbook of science in sport* (pp. 251–271). Hoboken: Wiley-Blackwell.
Ellaway, A., Macintyre, S., & Bonnefoy, X. (2005). Graffiti, greenery and obesity in adults: Secondary analysis of European cross-sectional survey. *BMJ*. doi:10.1136/bmj.38575.664549.F7.
European Commission. (2014). *Special Eurobarometer 412: Sport and physical activity*. doi:10.2766/73002.
Focht, B. C. (2009). Brief walks in outdoor and laboratory environments. *Research Quarterly for Exercise and Sport, 80*, 611–620. doi:10.1080/02701367.2009.10599600.
Freak-Poli, R. L. A., Cumpston, M., Peeters, A., & Clemes, S. A. (2013). Workplace pedometer interventions for increasing physical activity (review). *Cochrane Database of Systematic Reviews, 4*, CD009209. doi:10.1002/14651858.CD009209.pub2.
Frost. (n.d.). *Back to nature and the emerging child saving movement: Restoring children's outdoor play*. http://www.childrenandnature.org/downloads/LWS_Vol1_03.pdf.
Frumkin, H. (2001). Beyond toxicity: Human health and the natural environment. *American Journal of Preventive Medicine, 20*, 234–240. doi:S0749-3797(00)00317-2.
Gladwell, V. F., Brown, D. K., Wood, C., Sandercock, G. R., & Barton, J. L. (2013). The great outdoors: How a green exercise environment can benefit all. *Extreme Physiology & Medicine, 2*, 3. doi:10.1186/2046-7648-2-3.
Godbey, G. (2009). *Outdoor recreation, health, and wellness: Understanding and enhancing the relationship* (Resources for the future DP 09–21). http://www.rff.org/documents/RFF-DP-09-21.pdf.
Grahn, P., & Stigsdotter, U. K. (2010). The relation between perceived sensory dimensions of urban green space and stress restoration. *Landscape & Urban Planning, 94*, 264–275. doi:10.1016/j.landurbplan.2009.10.012.
Guéguen, N., & Stefan, J. (2014). "Green altruism": Short immersion in natural green environments and helping behavior. *Environment and Behavior*. doi:10.1177/0013916514536576.
Gullone, E. (2000). The biophilia hypothesis and life in the 21st century: Increasing mental health or increasing pathology? *Journal of Happiness Studies, 1*, 293–321. doi:10.1023/A:1010043827986.
Hagger, M., & Chatzisarantis, N. (2008). Self-determination theory and the psychology of exercise. *International Review of Sport and Exercise Psychology, 1*, 79–103. doi:10.1080/17509840701827437.

Hallal, P. C., Andersen, L. R., Bull, F. C., Guthold, R., Haskell, W., & Ekelund, U. (2012). Global physical activity levels: Surveillance progress, pitfalls, and prospects. *The Lancet, 380*, 247–257. doi:10.1016/S0140-6736(12)60646-1.

Haluza, D., Schönbauer, R., & Cervinka, R. (2014). Green perspectives for public health: A narrative review on the physiological effects of experiencing outdoor nature. *International Journal of Environmental Research and Public Health, 11*, 5445–5461. doi:10.3390/ijerph110505445.

Hartig, T., & Staats, H. (2006). The need for psychological restoration as a determinant of environmental preferences. *Journal of Environmental Psychology, 26*, 215–226. doi:10.1016/j.jenvp.2006.07.007.

Hartig, T., Mang, M., & Evans, G. W. (1991). Restorative effects of natural environment experience. *Environment and Behavior, 23*, 3–26. doi:10.1177/0013916591231001.

Hartig, T., Evans, G. W., Jamner, L. D., Davis, D. S., & Garling, T. (2003). Tracking restoration in natural and urban field settings. *Journal of Environmental Psychology, 23*, 109–123. doi:10.1016/S0272-4944(02)00109-3.

Hartig, T., van den Berg, A. E., Hagerhall, C. M., Tomalak, M., Bauer, N., Hansmann, R., & Waaseth, G. (2010). Health benefits of nature experience: Psychological, social and cultural processes. In K. Nilsson, M. Sangster, C. Gallis, T. Hartig, S. De Vries, K. Seeland, & J. Schipperijn (Eds.), *Forest, trees and human health* (pp. 127–167). Dordrecht: Springer Science Business and Media.

Hartig, T., Kaiser, F. G., & Strumse, E. (2007). Psychological restoration in nature as a source of motivation for ecological behaviour. *Environmental Conservation, 34*, 292–299. doi:10.1017/S0376892907004250.

Hartig, T., Mitchell, R., de Vries, S., & Frumkin, H. (2014). Nature and health. *Annual Review of Public Health, 35*, 207–228. doi:10.1146/annurev-publhealth-032013-182443.

Hillier, A., Tappe, K., Cannuscio, C., Karpyn, A., & Glanz, K. (2014). In an urban neighborhood, who is physically active and where? *Women and Health, 54*, 194–211. doi:10.1080/03630242.2014.883659.

Hillsdon, M., Panter, J., Foster, C., & Jones, A. (2006). The relationship between access and quality of urban green space with population physical activity. *Public Health, 120*, 1127–1132. doi:10.1016/j.puhe.2006.10.007.

Hine, R., Wood, C., Barton. J., & Pretty, J. (2011). *The mental health and wellbeing effects of a walking and outdoor activity based therapy project: Report for discover quest*. http://www.greenexercise.org/Previous%20community%20projects%20page%202011.htm.

Hodges, E. A., Smith, C., Tidwell, S., & Berry, D. (2013). Promoting physical activity in preschoolers to prevent obesity: A review of the literature. *Journal of Pediatric Nursing, 28*, 3–19. doi:10.1016/j.pedn.2012.01.002.

Home, R., Hunziker, M., & Bauer, N. (2012). Psychosocial outcomes as motivations for visiting nearby urban green spaces. *Leisure Sciences: An Interdisciplinary Journal, 34*, 350–365. doi:10.1080/01490400.2012.687644.

Hu, F. B., Li, T. Y., Colditz, G. A., Willett, W. C., & Manson, J. E. (2003). Television watching and other sedentary behaviors in relation to risk of obesity and type 2 diabetes mellitus in women. *JAMA, 289*, 1785–1791.

Hug, S., Hartig, T., Hansmann, R., Seeland, K., & Hornung, R. (2009). Restorative qualities of indoor and outdoor exercise settings as predictors of exercise frequency. *Health & Place, 15*, 971–980. doi:10.1016/j.healthplace.2009.03.002.

Janssen, I., & LeBlanc, A. G. (2010). Systematic review of the health benefits of physical activity and fitness in school-aged children and youth. *International Journal of Behavioral Nutrition and Physical Activity, 7*, 40. doi:10.1186/1479-5868-7-40.

Jiang, B., Chang, C., & Sullivan, W. C. (2014). A dose of nature: Tree cover, stress reduction, and gender differences. *Landscape and Urban Planning, 132*, 26–36. doi:10.1016/j.landurbplan.2014.08.005.

Jones, A., Hillsdon, M., & Coombes, E. (2009). Greenspace access, use, and physical activity: Understanding the effects of area deprivation. *Preventive Medicine, 49*, 500–505. doi:10.1016/j.ypmed.2009.10.012.

Joye, Y., & Van den Berg, A. (2011). Is love for green in our genes? A critical analysis of evolutionary assumptions in restorative environments research. *Urban Forestry & Urban Greening, 10*, 261–268. doi:10.1016/j.ufug.2011.07.004.

Joye, Y., & Van den Berg, A. (2012). Restorative environments. In L. Steg, A. Van den Berg, & J. I. M. De Groot (Eds.), *Environmental psychology: An introduction* (pp. 58–66). Oxford: Wiley.

Kaplan, S. (1995). The restorative benefits of nature: Towards an integrative framework. *Journal of Environmental Psychology, 15*, 169–182. doi:10.1016/0272-4944(95)90001-2.

Kaplan, R. (2001). The nature of the view from home: Psychological benefits. *Environment and Behavior, 33*, 480–506. doi:10.1177/00139160121973115.

Kaplan, R., & Kaplan, S. (1989). *The experience of nature: A psychological perspective*. New York: Cambridge University Press.

Kaplan, R. M., Sallis, J. F., & Patterson, T. L. (1993). *Health and human behavior*. New York: McGraw-Hill.

Katzmarzyk, P. T. (2011). Cost-effectiveness of exercise is medicine. *Current Sports Medicine Reports, 10*, 217–223.

Korpela, K. M. (2003). Negative mood and adult place preference. *Environment & Behavior, 35*, 331–346. doi:10.1177/0013916503035003002.

Korpela, K. M., Hartig, T., Kaiser, F., & Fuhrer, U. (2001). Restorative experience and self- regulation in favorite places. *Environment & Behavior, 33*, 572–589. doi:10.1177/00139160121973133.

Korpela, K. M., Ylén, M., Tyrvainen, L., & Silvennoinen, H. (2010). Favorite green, waterside and urban environments, restorative experiences and perceived health in Finland. *Health Promotion International, 25*, 200–209. doi:10.1093/heapro/daq007.

Kort, Y. A., Meijnders, A. L., Sponselee, A. A., & IJsselsteijn, W. A. (2006). What's wrong with virtual trees? Restoring from stress in a mediated environment. *Journal of Environmental Psychology, 26*, 309. doi:10.1016/j.jenvp.2006.09.001.

Kuo, F. E., & Sullivan, W. C. (2001). Aggression and violence in the inner city: Effects of environment via mental fatigue. *Environment & Behavior, 33*, 543–571. doi:10.1177/00139160121973124.

Lachowycz, K., & Jones, A. P. (2011). Greenspace and obesity: A systematic review of the evidence. *Obesity Reviews, 12*, 183–189. doi:10.1111/j.1467-789X.2010.00827.x.

Laumann, K., Garling, T., & Stormark, K. M. (2003). Selective attention and heart rate responses to natural and urban environments. *Journal of Environmental Psychology, 23*, 125–134. doi:10.1016/S0272-4944(02)00110-X.

Lee, A. C., & Maheswaran, R. (2011). The health benefits of urban green space: A review of the evidence. *Journal of Public Health, 33*, 212–222. doi:10.1093/pubmed/fdq068.

Lohr, V. I., & Pearson-Mims, C. H. (2006). Responses to scenes with spreading, rounded, and conical tree forms. *Environment and Behavior, 38*, 667–688. doi:10.1177/0013916506287355.

Loureiro, A. (1999). *Espaço público e identidade: Visitantes e residentes do Parque Natural de Montesinho [Public space and identity: Visitors and residents of the Montesinho Natural Park]* (Master dissertation). Lisbon: ISPA.

Loureiro, A., & Veloso, S. (2014). Outdoor exercise, well-being and connectedness to nature. *PSICO, 45*, 299–304. doi:10.15448/1980-8623.2014.3.19180.

Maas, J., Verheij, R. A., Groenewegen, P. P., de Vries, S., & Spreeuwenberg, P. (2006). Green space, urbanity, and health: How strong is the relation? *Journal of Epidemiology & Community Health, 60*, 587–592. doi:10.1136/jech.2005.043125.

Maas, J., Verheij, R. A., Spreeuwenberg, P., & Groenewegen, P. P. (2008). Physical activity as a possible mechanism behind the relationship between green space and health: A multilevel analysis. *BMC Public Health, 8*, 206. doi:10.1186/1471-2458-8-206.

Maller, C., Townsend, M., Pryor, A., Brown, P., & Leger, L. S. (2005). Healthy nature healthy people: 'Contact with nature' as an upstream health promotion intervention for populations. *Health Promotion International, 21*, 45–54. doi:10.1093/heapro/dai032.

Maller, C., Townsend, M., Leger, L. S., Henderson-Wilson, C., Pryor, A., Prosser, L., & Moore, M. (2008). *Healthy parks, healthy people. The health benefits of contact with nature in a park context: A review of relevant literature*. Melbourne: Deakin University and Parks Victoria. http://parkweb.vic.gov.au/__data/assets/pdf_file/0018/313821/HPHP-deakin-literature-review.pdf.

Marrero, R. J., & Carballeira, M. (2010). Contact with nature and personal well-being. *Psyecology, 1*, 371–381. doi:10.1174/217119710792774807.

Marselle, M. R., Irvine, K. N., & Warber, S. L. (2013). Walking for well-being: Are group walks in certain types of natural environments better for well-being than group walks in urban environments? *International Journal of Environmental Research Public Health, 10*, 5603–5628. doi:10.3390/ijerph10115603.

Mayer, F. S., Frantz, C. M., Bruehlman-Senecal, E., & Dolliver, K. (2009). Why is nature beneficial? The role of connectedness to nature. *Environment and Behavior, 41*, 607–643. doi:10.1177/0013916508319745.

Miller, D., Morrice, J., Aspinall, P., Brewer, M., Brown, K., Cummins, R., ... Wang, C. (2014). *Green health final report*.http://www.hutton.ac.uk/research/projects/green-health.

Mitchell, R. (2013). Is physical activity in natural environments better for mental health than physical activity in other environments? *Social Science & Medicine, 91*, 130–134. doi:10.1016/j.socscimed.2012.04.012.

Moore, R. C., & Cooper Marcus, C. (2008). Healthy planet, healthy children: Designing nature into the daily spaces of childhood. In S. R. Kellert, J. Heerwagen, & M. Mador (Eds.), *Biophilic design: The theory, science, and practice of bringing buildings to life*. Hoboken: Wiley.

Moore, R., & Cosco, N. (2014). Growing up green: Naturalization as a health promotion strategy in early childhood outdoor learning environments. *Children Youth and Environments, 24*, 168–191. doi:10.7721/chilyoutenvi.24.2.0168.

Morris, N. (2003). *Health, well-being and open space: Literature review*. http://www.openspace.eca.ed.ac.uk/pdf/healthwellbeing.pdf.

Natural England. (2009). *Childhood and nature: A survey on changing relationships with nature across generations*. Cambridgeshire: Natural England.

Nisbet, E. K., & Gick, M. L. (2008). Can health psychology help the planet? Applying theory and models of health behaviour to environmental actions. *Canadian Psychology, 49*, 296–303. doi:10.1037/a0013277.

Nisbet, E. K., Zelenski, J. M., & Murphy, S. A. (2011). Happiness is in our nature: Exploring nature relatedness as a contributor to subjective well-being. *Journal of Happiness Studies, 12*, 303–322. doi:10.1007/s10902-010-9197-7.

Nordh, H., & Ostby, K. (2013). Pocket parks for people: A study of park design and use. *Urban Forestry & Urban Greening, 12*, 12–17. doi:10.1016/j.ufug.2012.11.003.

Nordh, H., Hartig, T., Hagerhall, C., & Fry, G. (2009). Components of small urban parks that predict the possibility for restoration. *Urban Forestry & Urban Greening, 8*, 225–235. doi:10.1016/j.ufug.2009.06.003.

Olivos, P., Aragonés, J. I., & Amérigo, M. (2011). The connectedness to nature scale and its relationship with environmental beliefs and identity. *International Journal of Hispanic Psychology, 4*, 5–19.

Orsega-Smith, E., Mowen, A., Payne, L., & Godbey, G. (2004). The interaction of stress and park use in psycho-physiological health in older adults. *Journal of Leisure Research, 36*, 232–256.

Packer, J., & Bond, N. (2010). Museums as restorative environments. *Curator: The Museum Journal, 53*, 421–436. doi:10.1111/j.2151-6952.2010.00044.x.

Parachin, V. (2011). Green exercise: Get out! get fit! *American Fitness, 29*, 44–45.

Parsons, R., Tassinary, L. G., Ulrich, R. S., Hebl, M. R., & Grossman-Alexander, M. (1998). The view from the road: Implications for stress recovery and immunization. *Journal of Environmental Psychology, 18*, 113–139. doi:10.1006/jevp.1998.0086.

Pikora, T., Giles-Corti, B., Bull, F., Jamrozik, K., & Donovan, R. (2003). Developing a framework for assessment of the environmental determinants of walking and cycling. *Social Science & Medicine, 56*, 1693–1703. doi:10.1016/S0277-9536(02)00163-6.

Plante, T. G., Cage, C., Clements, S., & Stover, A. (2006). Psychological benefits of exercise paired with virtual reality: Outdoor exercise energizes whereas indoor virtual exercise relaxes. *International Journal of Stress Management, 13*, 108–117. doi:10.1037/1072-5245.13.1.108.

Pretty, J., Griffin, M., Sellens, M., & Pretty, C. (2003). *Green exercise: Complementary roles of nature, exercise and diet in physical and emotional well-being and implications for public health policy* (CES Occasional Paper 2003–1). Colchester: University of Essex.

Pretty, J., Peacock, J., Sellens, M., & Murray, G. (2005). The mental and physical health outcomes of green exercise. *International Journal of Environmental Health Research, 15*, 319–337. doi:10.1080/09603120500155963.

Pretty, J., Angus, C., Bain, M., Barton, J., Gladwell, V., Hine, R., ... Sellens, M. (2009). Nature, childhood, health and life pathways. Interdisciplinary Centre for Environment and Society Occasional Paper 2009–02. University of Essex, UK. www.greenexercise.org/pdf/nature%20childhood%20and%20lifepathways.pdf.

Pretty, J., Barton, J., Colbeck, I., Hine, R., Mourato, S., Mackerron, G., & Wood, C. (2011). *Health values from ecosystems*. The UK national ecosystem assessment technical report. UK National Ecosystem Assessment, UNEP-WCMC, Cambridge. http://uknea.unep-wcmc.org/.

Puett, R., Teas, J., España-Romero, V., Artero, E. G., Lee, D. C., Baruth, M., ... Blair, S. N. (2014). Physical activity: Does environment make a difference for tension, stress, emotional outlook, and perceptions of health status? *Journal of Physical Activity & Health, 11*, 1503–1511. doi:10.1123/jpah.2012-0375.

Raanaas, R. K., Patil, G. G., & Hartig, T. (2011). Health benefits of a view of nature through the window: A quasi-experimental study of patients in a residential rehabilitation center. *Clinical Rehabilitation, 26*, 21–32. doi:10.1177/0269215511412800.

Reed, K., Wood, C., Barton, J., Pretty, J. N., Cohen, D., & Sandercock, G. R. H. (2013). A repeated measures experiment of green exercise to improve self-esteem in UK school children. *PLoS ONE, 8*, e69176. doi:10.1371/journal.pone.0069176.

Richardson, E. A., & Mitchell, R. (2010). Gender differences in relationships between urban green space and health in the UK. *Social Science & Medicine, 71*, 568–575. doi:10.1016/j.socscimed.2010.04.015.

Richardson, E. A., Mitchell, R., Hartig, T., De Vries, S., Astell-Burt, T., & Frumkin, H. (2012). Green cities and health: A question of scale? *Journal of Epidemiology & Community Health, 66*, 160–165. doi:10.1136/jech.2011.137240.

Roe, J. J., & Aspinall, P. A. (2012). Adolescents' daily activities and the restorative niches that support them. *International Journal of Environmental Research and Public Health, 9*, 3227–3244. doi:10.3390/ijerph9093227.

Roe, J. J., Ward Thompson, C., Aspinall, P. A., Brewer, M. J., Duff, E. I., Miller, D., ... Clow, A. (2013). Green space and stress: Evidence from cortisol measures in deprived urban communities. *International Journal of Environmental Research and Public Health, 10*, 4086–4103. doi:10.3390/ijerph10094086.

Ryan, R. M., & Deci, E. L. (2000). Self-determination theory and the facilitation of intrinsic motivation, social development, and well-being. *American Psychologist, 55*, 68–78. doi:10.1037/110003-066X.55.1.68.

Ryan, R. M., & Deci, E. L. (2008). From ego depletion to vitality: Theory and findings concerning the facilitation of energy available to the self. *Social and Personality Psychology Compass, 2*, 702–717. doi:10.1111/j.1751-9004.2008.00098.x.

Ryan, R. M., Williams, G. C., Patrick, H., & Deci, E. L. (2009). Self-determination theory and physical activity: The dynamics of motivation in development and wellness. *Helenic Journal of Psychology, 6*, 107–124.

Ryan, R. M., Weinstein, N., Bernstein, J., Brown, K. W., Mistretta, L., & Gagné, M. (2010). Vitalizing effects of being outdoors and in nature. *Journal of Environmental Psychology, 30*, 159–168. doi:10.1016/j.jenvp.2009.10.009.

Sallis, J. F., & Owen, N. (1999). *Physical activity & behavioural medicine*. Thousand Oaks: Sage Publication.

Schmutz, U., Lennartsson, M., Williams, S., Devereaux, M., & Davies, G. (2014). *The benefits of gardening and food growing for health and wellbeing*. http://www.sustainweb.org/growinghealth/.

Spence, J. C., & Lee, R. E. (2003). Toward a comprehensive model of physical activity. *Psychology of Sport and Exercise, 4*, 7–24. doi:10.1016/S1469-0292(02)00014-6.

Staempfli, M. B. (2009). Reintroducing adventure into children's outdoor play environments. *Environment and Behavior, 41*, 268–280. doi:10.1177/0013916508315000.

Stilgoe, J. R. (2001). Gone barefoot lately? *American Journal of Preventive Medicine, 20*, 243–244. doi:10.1016/S0749-3797(00)00319-6.

Sun, F., Norman, I. J., & While, A. E. (2013). Physical activity in older people: A systematic review. *BMC Public Health, 13*, 449. doi:10.1186/1471-2458-13-449.

Takano, T., Nakamura, K., & Watanabe, M. (2002). Urban residential environments and senior citizens' longevity in megacity areas: The importance of walkable green spaces. *Journal of Epidemiology and Community Health, 56*, 913–918. doi:10.1136/jech.56.12.913.

Tang, I. C., Sullivan, W., & Chang, C. Y. (2014). Perceptual evaluation of natural landscapes: The role of the individual connection to nature. *Environment and Behavior*. doi:10.1177/0013916513520604.

Taylor, A. F., Kuo, F. E., & Sullivan, W. C. (2001). Coping with ADD: The surprising connection to green play settings. *Environment and Behavior, 33*, 54–77. doi:10.1177/00139160121972864.

Thompson Coon, J., Boddy, K., Stein, K., Whear, R., Barton, J., & Depledge, M. H. (2011). Does participating in physical activity in outdoor natural environments have a greater effect on physical and mental wellbeing than physical activity indoors? A systematic review. *Environmental Science & Technology, 45*, 1761–1772. doi:10.1021/es102947t.

Toftager, M., Ekholm, O., Schipperijn, J., Stigsdotter, U., Bentsen, P., Gronbaek, M., . . . Kamper-Jorgensen, F. (2011). Distance to green space and physical activity: A Danish national representative survey. *Journal of Physical Activity & Health, 8*, 741–749.

U. S. Department of Health and Human Services. (2008). *2008 physical activity guidelines for Americans.* www.health.gov/paguidelines/pdf/paguide.pdf.

Ulrich, R. S. (1984). View through a window may influence recovery from surgery. *Science, 224*, 420–421.

Ulrich, R. S., Simons, R. F., Losito, B. D., Fiorito, E., Miles, M. A., & Zelson, M. (1991). Stress recovery during exposure to natural and urban environments. *Journal of Environmental Psychology, 11*, 201–230.

UN-Habitat. (2011). *Cities and climate change: Global report on human settlements 2011.* London: Earthscan.

Van den Berg, A. E., Hartig, T., & Staats, H. (2007). Preference for nature in urbanized societies: Stress, restoration, and the pursuit of sustainability. *Journal of Social Issues, 63*, 79–96. doi:10.1111/j.1540-4560.2007.00497.x.

Van Herzele, A., & De Vries, S. (2012). Linking green space to health: A comparative study of two urban neighbourhoods in Ghent, Belgium. *Population & Environment, 34*, 171–193. doi:10.1007/s11111-011-0153-1.

Van Stralen, M. M., De Vries, H., Mudde, A. N., Bolman, C., & Lechner, L. (2009). Determinants of initiation and maintenance of physical activity among older adults: A literature review. *Health Psychology Review, 3*, 147–207.

Velarde, M. D., Fry, G., & Tveit, M. (2007). Health effects of viewing landscapes: Landscape types in environmental psychology. *Urban Forestry & Urban Greening, 6*, 199–212. doi:10.1016/j.ufug.2007.07.001.

Veloso, S. M., Matos, M. G., Carvalho, M., & Diniz, J. A. (2012). Psychosocial factors of different health behaviour patterns in adolescents: Association with overweight and weight control behaviours. *Journal of Obesity*. ID 852672. doi:10.1155/2012/852672.

Ward Thompson, C. (2011). Linking landscape and health: The recurring theme. *Landscape and Urban Planning, 99*, 187–195. doi:10.1016/j.landurbplan.2010.10.006.

Ward Thompson, C., Aspinall, P., & Montarzino, A. (2008). The childhood factor: Adult visits to green places and the significance of childhood experience. *Environment & Behavior, 40*, 111–143. doi:10.1177/0013916507300119.

Weinstein, N., Przybylski, A. K., & Ryan, R. M. (2009). Can nature make us more caring? Effects of immersion in nature on intrinsic aspirations and generosity. *Personality and Social Psychology Bulletin, 35*, 1315–1329. doi:10.1177/0146167209341649.

Wilson, E. O. (1984). *Biophilia: The human bond with other species.* Cambridge: Harvard University Press.

World Health Organization. (2010). *Global recommendations on physical activity for health.* www.who.int/dietphysicalactivity/factsheet_recommendations/en/.

World Health Organization. (2014). *Global status report on noncommunicable diseases 2014.* www.who.int/nmh/publications/ncd-status-report-2014/en/.

Yates, T., Wilmot, E. G., Davies, M. J., Gorely, T., Edwardson, C., Biddle, S., & Khunti, K. (2011a). Sedentary behavior what's in a definition? *American Journal of Preventive Medicine, 40*(6), 33–34. doi:10.1016/j.amepre.2011.02.017.

Yates, T., Wilmot, E. G., Khunti, K., Biddle, S., Gorely, T., & Davies, M. J. (2011b). Stand up for your health: Is it time to rethink the physical activity paradigm? *Diabetes Research and Clinical Practice, 93*, 292–294. doi:10.1016/j.diabres.2011.03.023.

Zelenski, J. M., & Nisbet, E. (2014). Happiness and feeling connected: The distinct role of nature relatedness. *Environment and Behavior, 46*, 3–23. doi:10.1177/0013916512451901.

Part III

People-Environment Relations and QoL – *Ecological Behavior*

Sustainable Behavior and Quality of Life

Cesar Tapia-Fonllem, Victor Corral-Verdugo, and Blanca Fraijo-Sing

9.1 Introduction: Sustainable Development and Quality of Life

In its widespread expression, the concept of sustainable development represents an evolutionary coordination of several concerns, such as the social, cultural, economic and environmental ones. These concerns had been expressed conventionally, especially in academia, as separate items in the predominant forms of analysis. People flattering the concept of system sustainability have defined it from several perspectives: those that emphasize the natural limits of Earth (Pearce 1988), those that indicate the conditions of social systems and structural factors (Barbier 1987; Simon 1989) and others, like Redclift (1987), that focus on the meaning of the prevailing structures of the international economic system. Some attempts to describe the elements of sustainability have pointed the importance that contextual, spatial and temporal variables have, leading to the need of reassessing imported models that are applied directly to various physical and temporal realities (Brown et al. 1987; Dovers 1990).

C. Tapia-Fonllem (✉) • V. Corral-Verdugo •
B. Fraijo-Sing
University of Sonora (Mexico), Hermosillo, Mexico
e-mail: cesartapia@sociales.uson.mx;
victorcorral@sociales.uson.mx; blancafraijo@gmail.com

The idea that economic development should be sustainable implies the recognition that natural resources are susceptible to exhaustion, so that they impose a limit on socioeconomic activities. Therefore, the concept extends ideologically to the cultural and social relations involved in the sustainable development processes, including those that affect human wellbeing and quality of life. In all cases, the dialectic has been between economy and ecology. The economic and conservationist visions have tried to highlight the difficulty of moving toward the future with a permanently growing population and a reduced or limited availability of natural resources and their ability to reproduce (Sandbach 1978). These visions have influenced the use of resources and the consumption patterns of the population. By necessity those visions have also affected the way we conceive of human quality of life.

9.1.1 Contradictions and Limitations of the "sustainable development" Concept

As it is known, the term "sustainable development" became popular since the "Our Common Future" report in 1987 (Keiner 2004; Fergus and Rowney 2005), which defined it as "a lifestyle that meets current needs without compromising the ability of future generations to satisfying their own needs" (WECD 1987). The idea that human development can be sustainable has been described by some as an apparent contradiction (Redclift 1987; Pearce 1988; Simon 1989; Shearman 1990). Developmental models of the industrial age conceived development as a logical result of economic growth. How can there be development if society has not an endless supply of resources and their use is maximized to accelerate economic growth? How can people achieve quality of life without economic growth? Since quality of life is equated with consumption and an extensive use of natural resources, the conclusion is that human wellbeing depends on the exploitation of nature. This idea, product of the dominant paradigms over the last two centuries of industrial development, appears among the main elements of analysis in environmental education: the economic-ecological conflict that is integrated into the equation; the need to value the services that nature offers, putting them in balance against the benefits of economic growth.

9.1.2 Semantics of "sustainable development"

Schmuck and Shultz (2002) reported that, at the beginning of the twenty-first century, there were over 300 definitions of the term sustainable development. According to Dobson (2000), the origin of the concept goes back to the German term "Nachhaltigkeit" (Schmuck and Shultz 2002: 5), which initially was translated into English as "sustainable yield" and later as "sustainability" (Held 2000). Lélé (1991: 609) asked, "What do these specific connotations of sustainability imply?" The sustainability concept originated having in mind renewable resources such as forests or fishing industries, and has been widely adopted as a motto for the environmental movement. Most of the authors addressing sustainability defined it as "the existence of ecological conditions necessary to support human life at a level that ensures the well-being through future generations," which is interpreted as ecological sustainability.

Lélé (1991) and Mitchan (1995) establish that the Sustainable Development term has become a "cliché", applied to almost any functional and environmental process. This fact justifies the need to understand the differences between economic growth and sustainability without economic growth. The establishment of the semantic roots that include meaning and structure may help to explain the contradictory notions implicit in the sustainable development concept. Within the conceptual framework of Lélé, sustainability has a literal meaning and an ecological and social sense. The literal meaning refers to the possibility of continuity. The ecological sense implies the maintenance of the ecological basis of human activities within a time frame, indicating concern for both the future and the present. In describing the social meaning of sustainable development, Lélé (1991: 610) uses Barbier's (1987) notions of social meaning, which focus on keeping "desired social values, institutions, and cultures". These notions imply a fundamental philosophical difference to the developmental objectives of the structure of society. The second part of the semantic deconstruction of sustainable development examines the word "development," either when it refers to a process that means growth and change or when it refers to a goal that includes the satisfaction of basic needs (Lélé 1991). These fundamental meanings according to Lélé result in two different interpretations of sustainable development: (1) sustained growth, which he describes as "contradictory and trivial", and (2) achieving the basic objectives of human needs' satisfaction.

The Sustainable Development definition has been widely criticized because its vagueness and imprecision, by not determining the meaning of "needs" and because it does not specify the mechanisms for achieving a sustainable society (Norgaard 1994; Solow 1993). The concept of sustainability stresses the idea of lifestyles that allow present and future humans to meet their needs without exceeding the capacity of nature to restore the extracted resources (Fergus and Rowney 2005; Glasby 2002; Keiner 2004; Lumley and Armstrong 2004; WCED 1987). Those lifestyles involve psychological tendencies and behaviors revealing a concern for the conditions of the physical environment and for the integrity of the social milieu.

9.1.3 Sustainability and Environmental Psychology

Psychology, with its specialized area of environmental psychology, is the science responsible for explaining the determinants of sustainable behaviors and lifestyles. This science is also committed to study the mental, behavioral and environmental factors involved in the achievement of quality of life. Concerning sustainable behaviors, until recently, the interest of environmental psychology focused only on the aspects of the physical environment that are impacted by pro-environmental behaviors (Corral-Verdugo and Pinheiro 2004). The notion of sustainability has modified this conceptual approach, which nowadays involves social aspects as effects of sustainable behavior and not only as determinants of this behavior (Schmuck and Schultz 2002). In fact, a growing tendency is detected for using the "sustainable behavior" term (which implies concern for the socio-physical environment) instead of "pro-environmental behavior" (focusing rather on the conservation of the physical environment).

Therefore, the challenges of sustainability reached psychology demanding a commitment to address environmental and quality of life issues in combination. Environmental psychology may contribute to this endeavor with methods and models assessing how a sustainable lifestyle might influence human wellbeing without degrading the environment. In this regard, sustainable behaviors would be conceived as actions that contribute to the quality of life of present and future generations without compromising the resources of the biosphere.

In a first approach, environmental psychology addressed environment-behavior issues from the perspective of promoting pro-ecological behavior (PEB). This behavior was conceived as a set of human activities intended at the protection of natural resources or, at least, the reduction of environmental degradation (Corral-Verdugo 2001; Grob 1995; Hess et al. 1997). According to Corral-Verdugo and Pinheiro (2004), the social environment and, therefore, the human needs were not considered in a specific way as potential impacts of PEB. In any case, if these impacts showed up, they were indirect. For example, the effects of altruism on pro-environmental behavior (Ebreo et al. 1999; Schultz 2001) were studied. In fact, Schultz (2001) considered altruism as a motivation to conserve the environment. Yet, although altruism has other persons as depositories, in the classical models PEB only included actions that had an impact on the integrity of the physical environment. The study of PEB determinants considered aspects of prosocial behavior; however, the interest of PEB models ultimately focused on the effect that these prosocial factors had on the nonsocial aspects of the environment. As a consequence, quality of life was not addressed as a special target of PEB research.

Things changed at the beginning of the twenty-first century. The notion of pro-environmental behavior began to be replaced by the concept of sustainable behavior (Schmuck and Schultz 2002), reflecting the particular interest of environmental psychology in the concept of sustainability. The effects of this behavior should not be evaluated only in terms of the bio-physical dimension of the environment, but also considering the economic, social and political benefits of sustainable behavior; that is, sustainable behavior started to be assessed considering its impact on human wellbeing and quality of life. According to the theorists of sustainable behavior, social levels of sustainability are inextricably intertwined with the purely physical aspects: it is not possible to solve problems of the latter without taking care of the former and vice versa (Gouveia 2002; Winter 2002; Schmuck and Schultz 2002). Corral-Verdugo and Pinheiro (2004: 10) suggested a definition of sustainable behavior as "the set of effective, deliberate and anticipated actions that result in the preservation of natural resources, including the integrity of animal and plant species, as well as wellbeing -both individual and social- for current and future human generations." This definition clearly implies that sustainable behavior positively influences the physical environment as much as it affects human quality of life.

While it is true that the different facets of pro-environmental behavior (i.e., special kinds of behavior such as recycling, energy and water conservation, etc.) and their diverse levels of complexity (i.e., degrees of difficulty) made it appear as heterogeneous in forms and functionalities (Corral-Verdugo et al. 2004; Kaiser 1998), overall, this behavior was conceived uniform in its impacts. That is, SB was thought as only (or mainly) influencing the physical environment without especially considering its impact on human quality of life. By including social, economic and institutional/political effects of sustainable behavior, marked differences resulted between such behavior and PEB, not only in their morphologies and functional complexities, but also in the level of that impact on quality of life (Gouveia 2002). For example, recycling has a high economic and social impact, but its positive effect on the physical environment goes from moderate to low, since recycling involves industrial restructuring and a degree of pollution (Corral-Verdugo 1996; De Young 1991). Alternatively, reduced-consumption behaviors (also known as frugal behaviors) have a higher positive impact on the integrity of nature (De Young 1991) but their effects on economic well-being (employment, generation of economic wealth, etc.) can be negative to some extent. This means that the concept of sustainable behavior is more complex; its measurement is more demanding and requires greater multi and interdisciplinary collaboration.

This also implies that one of the first tasks that the psychology of sustainability must undertake is the classification of sustainable actions and its impact on human wellbeing, a classification that has not been developed. Such classification should determine what actions may be considered sustainable, to what extent they are sustainable, and how they affect quality of life. One additional aspect to address is the assessment of the effect of all sustainable actions at different levels -environmental, physical, social, political/institutional, and economic.

Another aspect considers the establishment of the psychological dimensions of sustainability. To this end, the wealth of information that exists concerning the psychological predictors of pro-environmental behavior may be used. Since it is very difficult to be sustainable without being environmentally friendly, the determinants of pro-environmentalism should predict the way of being sustainable. This information has been collected for over 30 years of psycho-environmental research. Yet, researchers also must consider that sustainable behavior influences the social environment, in addition to its affecting the physical environment; this consideration requires paying attention to factors not previously studied. For example, the apparent contradictions between economic development and environmental conservation must be resolved guaranteeing that sustainable behaviors, human wellbeing and environmental quality go together.

Under the approach of sustainable behavior, this construct is conceived as encompassing pro-ecological, frugal, altruistic, and equitable behavior (Corral-Verdugo 2012). Pro-ecological behavior is a set of actions aimed at protecting natural resources; frugal behavior implies moderation in the consumption of those resources and avoidance of waste. Altruistic behavior considers actions intended at the care of other people without expecting reciprocity; and equitable behavior implies a fair distribution of resources and treating other individuals without biases, regardless of their biological, ethnic, or socio-economic characteristics (Tapia et al. 2013a). The combination of these four sets of behavior allegedly guarantees the conservation of both types of (social, physical) environments. In recent times, a fifth behavioral dimension has been proposed: self-protective behavior (Pato and Corral-Verdugo 2013), since no one is able to protect other people or the natural environment if he or she is unable to first meet his/her own (physical, psychological, spiritual) needs.

Among the psychological factors leading to sustainable behavior that have been under thorough examination, competence, deliberation, future orientation, emotions towards the environment, and affinity towards diversity, are preeminent (Corral-Verdugo et al. 2009; Schultz 2002). They all refer to propensities or individual capacities (i.e., dispositional factors that predict behavior). These dimensions have been investigated in their relationship with pro-environmental behavior, and the study of their impact on the social aspects of sustainability is still incipient but solid. Sustainable behavior is promoted by environmental education in formal and non-formal programs, by companies and governments around the world (Dobson 2007; Roseland 2012). Sustainable behavior has also been positively associated with human virtues and psychological strengths (Corral-Verdugo et al. 2013, in press). Such association leads the authors to assure that a virtuous nature exists in sustainable behavior. Some of the psychological determinants of sustainable behavior constitute also indicators of quality of life, as in the case of competence, character strengths, environmental emotions and affinity towards diversity, which constitute components of mental health, human capacity and subjective quality of life (Diener and Suh 1997). Virtues generate individual satisfaction and other psychological benefits as intrinsic motivation, happiness and personal well-being (Corral-Verdugo 2012).

A number of studies (Corral-Verdugo et al. 2010; Fraijo et al. 2007; Tapia et al. 2013a) indicate that significant covariances exist among pro-ecological, frugal, altruistic and equitable actions, indicating the presence of a higher-order factor of "sustainable behavior." This behavioral factor correlates with the dispositional factors suggesting the presence of a second-order factor, which Corral-Verdugo et al. (2009) call "pro-sustainability orientation. This higher-order factor indicates that congruence exists between pro-sustainable propensities, pro-environmental capacities and sustainable behaviors. Evidence also exists showing that pro-sustainability orientation is strongly associated to quality of life, as the following sections demonstrate.

9.2 Quality of Life and Sustainable Behavior

Quality of life includes a series of indicators of a desirable society and "good life." These indicators encompasses environmental, social, economic and subjective factors (Diener and Suh 1997). Quality of life is achieved by societies that enjoy a conserved natural and built environment, good governance, physical and economic health, and subjective wellbeing. The psychological indicators are as important as the objective ones in determining the level of quality of life. The phrase "meeting present and future needs" of the sustainable development concept opens space to the dimension of individual and collective well-being. The interaction between human beings and their physical and social environment should generate high levels of satisfaction of those needs, and also wellbeing and happiness, if such interaction is pro-sustainable. A pro-sustainable relationship with the sociophysical environment results in the satisfaction of human needs and the conservation of that environment (Moser 2009a). Leff (1999) believes that quality of life and sustainable behavior are simultaneously possible because there has been not only a conceptual but also an attitudinal development leading to a sustainable world. He also stands out that current generations may achieve a balance between fulfilling personal and social goals. More recently, Corral-Verdugo et al. (2015) have extensively developed the notion of "positive environment," defining it as a context that simultaneously meets human needs and instigates environmentally protective behaviors (in both, social and natural scenarios). These authors assure that a positive environment is, by definition, a sustainable environment. They also equate positive environments with quality of life and environmental quality.

Uzzell and Moser (2006) set an attribute of congruence between individuals and the environment in determining indicators of quality of life. They notice that a congruent people-environment relationship is bi-directional in providing satisfaction (for people) and benefits and mutual care for humans and the environment. This means that quality of life is achieved if individuals interact with their environment in a respectful manner, if the environment does not threaten or obstacle what the individual considers quality of life, and if it allows the individual to satisfy their needs. This interactive notion is also addressed in the New Human Interdependence Paradigm, which conceive people in a relationship of mutual dependence with their environment (Garling et al. 2003; Corral-Verdugo et al. 2008).

9.2.1 Positive Consequences of Sustainable Behavior

It is a fact that, by conserving the physical and social environment, sustainable behavior contributes to human quality of life. This is not common sense, but a demonstrated reality. Proecological, frugal, altruistic, and equitable actions make possible a responsible use of resources, and lead to a fair allocation of those resources among all people (Lélé and Jayaraman 2011). Interestingly, sustainable behaviors can also contribute to human quality of life by producing positive psychological consequences in those individuals who engage in them. Those psychological consequences are a component of what Diener and Suh (1997) call the subjective factors of quality of life. Contrary to what is expected, the practice of sustainable behaviors may result in positive psychological states (positive emotions, psychological wellbeing, happiness) and intrinsic motivations (satisfaction, sense of self-efficacy) that add up to the list of environmental benefits that are produced by environmentally-protective behaviors. In this section of the chapter, we will present some of the studied positive consequences of sustainable behaviors that might contribute to quality of life.

9.2.2 Pro-ecological Behavior

Corral-Verdugo (2001) defines pro-ecological behavior as the set of deliberate and effective actions that result in the conservation of natural resources. Recycling, water conservation, energy saving, ecosystem conservation, composting, and use of public transportation are instances of this kind of behavior. Pro-ecological behaviors contribute to quality of life by making possible the preservation of nature's resources that are required for sustaining human life. In addition, these behaviors often are associated with positive psychological states. For instance, people who engage in pro-ecological practices report higher levels of subjective well-being or happiness. Brown and Kasser (2005) and Bechtel and Corral-Verdugo (2010) found that levels of happiness are significantly higher in individuals that engage in environmentally protective behaviors. De Young (2000) argues that many individuals consider engaging in pro-ecological activities due to satisfaction and pleasure resulting from these actions. He also report the presence of motivation competence (i.e., a feeling of self-efficacy that instigates behavior) among people who practice pro-ecological behaviors. Hernandez et al. (2009), in turn, found that the satisfaction due to engaging in recycling highly and significantly predict such behavior.

In addition, psychological wellbeing, which encompasses high levels of self-acceptance, purpose in life, environmental mastery, personal growth, autonomy, and positive relations, seems to be higher in individuals who frequently engage in pro-ecological actions (Corral-Verdugo et al. 2011). Psychological restoration is one more process associated with proecological behavior and quality of life. Restorative experiences are defined as those involving the renewal of depleted psychological resources: attention, positive moods, and mental health (Hartig et al. 2001). Those resources are normally lost because some deficit background conditions (environmental stress). People seek to recover their cognitive resources and capacity to psychophysiologically responding to daily demands (Van den Berg et al. 2007). Corral-Verdugo et al. (2012) found a positive relationship between sustainable behaviors and psychological restoration. In turn, Hartig et al (2001) report that one of the motivations for engaging in pro-ecological behaviors is to conserve the environment in order to enjoy it and obtain restorative experiences. More recently, Collado and Corraliza (2014) found that fascination -a component of restorative experiences- predict pro-ecological behavior.

9.2.3 Frugality

Frugality involves deliberately avoiding unnecessary personal consumption of resources. It is defined as the cautious use of resources and the interest to avoid waste. According to De Young (1991), frugality can become a lifestyle, a type of cautious and conservative behavior that is characteristic of successful organisms living in an uncertain world, for example, in a world where a continued and full access to resources cannot be taken for granted. Frugality also implies a low consumption style that is not based on materialism and avoids waste and resource pillage. Baldi and García (2006), in discussing the sustainable dimension of frugal behaviors, assure that frugality beliefs emphasize the need to limit the consumption of resources. Iwata (2002) uses the term "style of life of voluntary simplicity," which he defines as a lifestyle of low consumption, and finds a positive and significant correlation between environmentally responsible consumption and voluntary simplicity. Simplicity living predicts sustainable behavior in both physical and social dimensions of the environment (Corral-Verdugo et al. 2010).

Brown and Kasser (2005) argue that it is possible for individuals to experience high levels of subjective well-being without excessive consumption. Evidence also has been found that once basic needs are satisfied, a substantial increase in income and consumption does not translate in a substantial increase of happiness (Stutz 2006). Apparently, the excessive consumption of modern societies is not the ideal way of happiness, nor the path towards sustainability (O'Brien 2008). Frugality has also been linked with the experience of positive intrinsic consequences. De Young (1996), for example, found that people report states of intrinsic satisfaction derived from the practice of frugal behaviors.

9.2.4 Altruism

Altruism is considered a part of pro-social behavior (Eisenbeg and Miller 1987) but also as a component of sustainable behavior. In the definition of sustainable behavior the need of protecting the social environment is recognized, through actions that are intended to meet the needs of other people (Tapia et al. 2013a). Altruistic behavior is one of those actions. Hopper and Nielsen (1991) mention that people with pro-environmental concerns are not necessarily looking for economic interest but for the satisfaction of knowing that they are doing something good for others. Altruism refers to the operation by which people selflessly act in favor of their fellows; this is, without the expectation of a reciprocal action of gratification (García et al. 2007). Psychological altruism makes that acts of assistance be accompanied by motivation to do something good for others: these acts will be altruistic only if the actor thinks about the well-being of others as remote objects (Sober and Wilson 1998). The altruistic, according to this meaning, does not only help, but has also intends to do so and anticipates that this assistance will generate long-term benefits.

Altruism generates conditions for quality of life. By taking care of others' needs, altruistic people enhances trust and quality of social relations. Moreover, altruistic individuals obtain intrinsic gratifications when engage in their uninterested actions. Moll et al. (2006) found what seems to be the key to explaining why people behave altruistically: when participants in their study made an altruistic decision, the mesolimbic area of their brain become activated. This is the same area that is activated when individuals engage in sexual activity or when they receive money. In conclusion, altruistic actions produce pleasure.

9.2.5 Equity

Equity can be understood as the justice that corresponds with human rights or the laws of nature, more specifically with getting rid of bias or favoritism (Corral-Verdugo et al. 2010). It involves distributing natural and social resources fairly and treating others without biases derived from considering their demographic or physical characteristics. Equity does not only refer to the distribution of natural resources. This concept involves social equity, equitable access to health, education, opportunities and quality of life. One of the most harmful manifestations of the lack of sustainability is inequity, this is, the unfair distribution of resources and benefits so that some have a lot, and others have a little and that risk and environmental damage fall more in some than in others (Corral-Verdugo et al. 2010). As we have seen, equity is a fundamental part of sustainability and brings benefits to the environment. Schmuck and Shultz (2002) argue that equity and redistribution are the true paths towards sustainability.

There are three types of equity. The first type is intergenerational equity, suggested in the definition of sustainable development within the Brundtland Report. This implies considering developmental costs of the present *vis-à-vis* the demand of future generations (Dalziel et al. 2009). The second type is intra-generational equity, which involves including disadvantaged groups (e.g., poor women, children, and the disabled) in the decision-making that affects ecology, society and economy (Anand and Sen 2000). The third type is equity between countries, which implies the need of changing the abuse of power practiced by the more developed countries in detriment to those that are less developed (Artaraz 2002).

Equitable practices provide multiple psychological benefits; happiness is one of them. Amato et al. (2007) demonstrate that egalitarian marriages report higher levels of subjective wellbeing than those that are not egalitarian. According to Veenhoven (2006) the most equitable societies and individuals tend to be happier. Equitable people tend to behave more pro-environmentally. Corral-Verdugo et al. (2010) found that people who exhibit equitable behaviors tend to engage in pro-environmental actions, care for a fair distribution of resources, avoid these resources' overexploitation and ensure its availability for the present and the future.

This review shows that, on one hand, all sustainable behaviors are required in the effort to conserve natural and social resources for present and future generations. On the other hand, the count indicates that these sustainable actions have a positive effect on the individual who practices it in the form of psychological benefits, such as satisfaction, happiness and intrinsic positive consequences.

9.2.6 Social Capital

Environmental psychology traditionally has studied individual behaviors that result in both quality of life for people and environmental quality. However, it is clear that focusing on the individual is not enough. Researchers must transcend the person-environment approach and incorporate social issues, especially those that address collective commitments, social and community integration and socio-contextual factors generating the adoption of sustainable behaviors (Moser 2009b).

Putman (1993), introducing his notion of social capital stands out how organizational characteristics in society, networks, norms and trust, provide cooperation for mutual benefit. In this sociological view, the study of citizenship and society, of relationships between individuals and voluntary associations and networks to which they belong, allow understanding how a better social working can be achieved. Coleman (1990) assures that social capital is a resource of basic action for people who can affect the ability of acting on the environment and the perception of quality of life. Social capital can be also considered an indicator of quality of life because communities with empowered, cooperative and trustful networks are more conducive to the satisfaction of human needs.

Social capital considers that interactions between social groups lead to connections that foster cooperation and trust among their members, seeking mutual benefit. The networks of civic agreement promote cooperation because they increase the costs of not cooperating; they make communication easier and improve flow of information; reinforce norms of reciprocity, represent a history of collaboration and provide a plan for future cooperation as an effect of delayed reciprocity (Knight 2001).

The study on the benefits of social interaction and its impact on certain components of sustainable behavior has been addressed indirectly by Coleman (1988), Helliwell (2006), and Bartolini et al. (2009). Tapia et al. (2013b), in a study aimed at exploring this relationship (social-behavior, sustainable capital), report evidence of a positive correlation between social variables such as participation in democracy, volunteerism and civic participation with the four components of sustainable behavior: equity, altruism, frugality and pro-environmental behavior. A possible interpretation of this finding is that sustainable behaviors promote the conservation of social capital.

9.3 Concluding Remarks

Sustainable behaviors contribute to quality of life in more instances than expected. The logical thinking is that those behaviors lead to human wellbeing by protecting the natural resources that are necessary to meet people's needs. A conserved environment also provides conditions for psychological restoration, health and enjoyment; these are elements of quality of life. Yet, sustainable behaviors are something more than actions aimed at conserving the natural environment. They also result in the protection of the social environment since individuals oriented towards sustainability practice altruistic and equitable behaviors, in addition to those (pro-ecological, frugal) behaviors intended at conserving the natural milieu. Altruism and equity are important contributors in the creation and maintenance of social capital, an indicator of quality of life associated to cooperation, trust, and social network operation. Most of the correlates of sustainable behavior (emotions towards the environment, affinity towards diversity, character strengths and virtues, pro-environmental competence and abilities, environmental knowledge, etc.) are indicators of quality of life. This situation seemingly points to the fact that sustainability and quality of life are highly and significantly interrelated. Moreover, contrary to the expected, sustainable behaviors are followed by a series of psychological positive consequences that also constitute indicators of quality of life. Those consequences include satisfaction, feelings of self-efficacy, intrinsic motivation, psychological wellbeing and restoration, happiness and pleasure. Since the practice of sustainable behaviors conduce in so many ways to quality of life, the promotion of those behaviors should be prescribed in educational, public policy and social programs. In addition, the continued study of sustainable behavior and its impact on positive social practices that conduce to a better life is a present and future challenge for environmental psychology. Such a study should be supplemented by theoretical contributions offered by diverse social sciences and other areas of scientific inquiry.

References

Amato, P., Booth, A., Johnson, D., & Rogers, S. (2007). *Alone together: How marriage in America is changing*. Cambridge: Harvard University Press.

Anand, S., & Sen, A. (2000). Human development and economic sustainability. *World Development, 28*, 2029–2050.

Artaraz, M. (2002). Teoría de las tres dimensiones de desarrollo sostenible [Theory of the three dimensions of sustainable development]. *Ecosistemas, 11*, 2.

Baldi-López, G., & García-Quiroga, E. (2006). Una aproximación a la psicología ambiental. *Fundamentos en Humanidades, 7*(1), 157–168.

Barbier, E. B. (1987). The concept of sustainable economic development. *Environmental Conservation, 14*, 101–110.

Bartolini, S., Bilancini, E., & Sarracino, F. (2009). *Sociability predicts happiness: Worldwide evidence from time series*. Siena: Department of Economics University of Siena.

Bechtel, R. B., & Corral, V. (2010). Happiness and sustainable behavior. In V. Corral, C. García, & M. Frías (Eds.), *Psychological approaches to sustainability*. New York: Nova.

Brown, K., & Kasser, T. (2005). Are psychological and ecological well-being compatible? The role of values, mindulness, and lifestyle. *Social Indicators Research, 74*, 349–368.

Brown, B., Hanson, M., Liverman, D., & Merideth, R. (1987). Global sustainability: Toward definition. *Environmental Management, 11*(6), 713–719.

Coleman, J. S. (1988). Social capital in the creation of human capital. *American Journal of Sociology, 94*, 95–120.

Coleman, J. S. (1990). *Foundations of social theory*. Cambridge: Harvard University Press.

Collado, S., & Corraliza, J. A. (2014). Children's restorative experiences and self-reported environmental behaviors. *Environment & Behavior, 47*, 38–56.

Corral-Verdugo, V. (1996). A structural model of reuse and recycling in Mexico. *Environment & Behavior, 28*, 665–696.

Corral-Verdugo, V. (2001). *Comportamiento Proambiental: una introducción al estudio de las conductas protectoras del ambiente [Proenvironmental behavior: An introduction to the study of environmentally protective behaviors]*. Santa Cruz de Tenerife: Ed. Resma.

Corral-Verdugo, V. (2012). *Sustentabilidad y Psicología Positiva. Una visión optimista de las conductas protectoras del ambiente. [Sustainability and positive psychology: An optimistic vision of environmentally protective behaviors]*. Mexico: El Manual Moderno.

Corral-Verdugo, V., & Pinheiro, J. Q. (2004). Aproximaciones al estudio de la conducta sustentable [Approaches to the study pf sustainable behavior]. *Medio Ambiente y Comportamiento Humano, 5*, 1–26.

Corral-Verdugo, V., Varela, C., & González, D. (2004). O papel da Psicologia Ambiental na promoção de competência pró-ambiental [The role of environmental psychology in the promotion of pro-environmental competence]. In E. Tassara, E. Rabinovich, & M. C. Guedes (Eds.), *Psicologia e Ambiente*. São Paulo: EDUC.

Corral-Verdugo, V., Carrus, G., Bonnes, M., Moser, G., & Sinha, J. (2008). Environmental beliefs and endorsement of sustainable development principles in water conservation: Towards a "New Human Interdependence Paradigm" scale. *Environment & Behavior, 40*, 703–725.

Corral-Verdugo, V., Bonnes, M., Tapia, C., Fraijo, B., Frías, M., & Carrus, G. (2009). Correlates of pro-sustainability orientation: The affinity towards diversity. *Journal of Environmental Psychology, 29*, 34–43.

Corral-Verdugo, V., García, C., Castro, L., Viramontes, I., & Limones, R. (2010). Equity and sustainable lifestyles. In V. Corral, C. García, & M. Frías (Eds.), *Psychological approaches to sustainability*. New York: Nova.

Corral-Verdugo, V., Montiel, M., Sotomayor, M., Frías, M., Tapia, C., & Fraijo, B. (2011). Psychological wellbeing as correlate of sustainable behaviors. *International Journal of Hispanic psychology, 4*, 31–44.

Corral-Verdugo, V., Tapia, C., García, F., Varela, C., Cuen, A., & Barrón, M. (2012). Validation of a scale assessing psychological restoration associated with sustainable behaviours. *Psyecology, 3*, 87–100.

Corral-Verdugo, V., Tapia, C., Ortiz, A., & Fraijo, B. (2013). Las virtudes de la Humanidad, Justicia y Moderación y su relación con la conducta sustentable [Virtues of humanity, justice and temperance and their relationship with sustainable behavior]. *Revista Latinoamericana de Psicología, 45*, 361–372.

Corral-Verdugo, V., Frías, M., Gaxiola, J., Tapia, C., Fraijo, B., & Corral, N. (2015). *Ambientes positivos [Positive environments]*. Mexico: Pearson.

Corral-Verdugo, V., Tapia, C., & Ortiz, A. (in press). On the relationship between character strengths and sustainable behavior. *Environment & Behavior*.

Dalziel, P., & Saunders, C., with Fyfe, R., & Newton, B. (2009). *Sustainable development and equity* (Official statistics research series, 6).http://www.statisphere.govt.nz/officialstatisticsresearch/series/default.htm.

De Young, R. (1991). Some psychological aspects of living lightly: Desired lifestyle patterns and conservation behavior. *Journal of Environmental Systems, 20*, 215–227.

De Young, R. (1996). Some psychological aspects of a reduced consumption lifestyle: The role of intrinsic satisfaction and competence motivation. *Environment & Behavior, 28*, 358–409.

De Young, R. (2000). Expanding and evaluating motives for environmentally responsible behavior. In L. Zelezny, & P. W. Schultz (Eds.), Promoting environmentalism. *Journal of Social Issues, 56*, 509–526.

Diener, E., & Suh, E. (1997). Measuring quality of life: Economic, social, and subjective indicators. *Social Indicators Research, 40*, 189–216.

Dobson, A. (2000). Drei Konzepte ökologischer Nachhaltigkeit [Three concepts of ecological sustainability]. *Natur und Kultur — Transdisziplinäre Zeitschrift für ökologische Nachhaltigkeit, 7*, 62–85.

Dobson, A. (2007). Environmental citizenship: Towards sustainable development. *Sustainable Development, 15*, 276–285.

Dovers, S. (1990). Sustainability in context: An Australian perspective. *Environmental Management, 14*(3), 297–305.

Ebreo, A., Hershey, J., & Vining, J. (1999). Reducing solid waste: Linking recycling to environmentally responsible consumerism. *Environment & Behavior, 31*, 107–135.

Eisenberg, N., & Miller, P. A. (1987). The relation of empathy to prosocial and related behaviors. *Psychological Bulletin, 94*, 100–131.

Fergus, A. H. T., & Rowney, J. I. A. (2005). Sustainable development: Lost meaning and opportunity? *Journal of Bussines Ethics, 60*, 17–25.

Fraijo, B., Tapia, C., Corral-Verdugo, V., & Mireles, J. (2007). Orientación hacia la sustentabilidad en estudiantes universitarios: un estudio diagnóstico [Orientation towards sustainability in undergraduate students: A diagnostic study]. In D. González & M. Maytorena (Eds.), *Estudios Empíricos en Educación Superior*. Mexico: Universidad de Sonora-Conacyt.

García, M., Estévez, I., & Letamendía, P. (2007). El CUIDA como instrumento para la valoración de la personalidad en la evaluación de adoptantes, cuidadores, tutores y mediadores [CUIDA as instrument to assess personality in the evaluation of adopting parents, tutors and mediators]. *Intervención Psicosocial, 16*, 393–407.

Gärling, T., Biel, A., & Gustafsson, M. (2003). The new environmental psychology: The human interdependence paradigm. In R. Bechtel & A. Churchman

(Eds.), *Handbook of environmental psychology*. New York: Wiley.

Glasby, G. (2002). Sustainable development: The need for a new paradigm. *Environment, Development and Sustainability, 4*, 333–345.

Gouveia, V. (2002). Self, culture and sustainable development. In P. Schmuck & P. W. Schultz (Eds.), *Psychology of sustainable development*. Norwell: Kluwer.

Grob, A. (1995). Meinungen im umweltbereich und umweltgerechtes Verhalten. Ein psychologisches ursachenntzmodell. Unpublished doctoral dissertation. Bern: University of Bern.

Hartig, T., Kaiser, F. G., & Bowler, P. A. (2001). Psychological restoration in natureas a positive motivation for ecological behavior. *Environment and Behavior, 33*, 590–607.

Held, M. (2000). Geschichte der Nachhaltigkeit. *Natur und Kultur-Transdisziplinäre Zeitschrift für ökogishe Nachhaltigkeit, 1*, 17–31.

Helliwell, J. F. (2006). Well-being and social capital: Does suicide pose a puzzle? *Economic Journal, 116*, C34–C45.

Hernández, B., Tabernero, C., & Suarez, E. (2009). Psychosocial motivations and self-regulation processes that actívate environmentally responsible behavior. In J. Valentin & L. Gamez (Eds.), *Environmental psychology: New developments*. New York: Nova.

Hess, S., Suárez, E., & Martínez-Torvisco, J. (1997). Estructura de la conducta ecológica responsable mediante el análisis de la teoría de facetas [Structure of responsibly ecological behavior through the analysis of facet theory]. *Revista de Psicología Social Aplicada, 7*, 97–112.

Hopper, J. R., & Nielsen, J. M. (1991). Recycling as altruistic behavior: Normative and behavioral strategies to expand participation in a community recycling program. *Environment & Behavior, 23*, 195–220.

Iwata, O. (2002). Some psychological determinants of environmentally responsible behavior. *The Human Science Research Bulletin of Osaka Shoin Women's University, 1*, 31–41.

Kaiser, F. (1998). A general measure of ecological behavior. *Journal of Applied Social Psychology, 28*, 395–442.

Keiner, M. (2004). Re-emphasizing sustainable development-the concept of 'evolutionability'. *Environment, Development and Sustainability, 6*, 379–392.

Knight, J. (2001). Social norms and the rule of law: Fostering trust in a socially diverse society. In K. S. Cook (Ed.), *Trust in society*. New York: Russell Sage.

Leff, E. (1999). La racionalidad ambiental y el fin del naturalismo dialéctico [Environmental rationality and the end of dialectic naturalism]. Persona y sociedad. Special issue. Santiago de Chile.

Lélé, S. M. (1991). Sustainable development: A critical review. *World Development, 19*, 607–621.

Lélé, S., & Jayaraman, T. (2011). *Equity in the context of sustainable development, Note for UN-GSP*. New Delhi: Ministry of Environment & Forests, Government of India.

Lumley, S., & Armstrong, P. (2004). Some of the nineteenth century origins of the sustainability concept. *Environment, Development and Sustainability, 6*, 367–378.

Mitcham, C. (1995). The concept of sustainable development: Its origins and ambivalence. *Technology in Society, 17*, 311–326.

Moll, J., Krueger, F., Zahn, R., Pardini, M., Oliveira, R., & Grafman, J. (2006). Human fronto-mesolimbic networks guide decisions about charitable donations. *Proceedings of the National Academy of Sciences of the United States of America, 103*, 15623–15628.

Moser, G. (2009a). Quality of life and sustainability: Toward person-environment congruity. *Journal of Environmental Psychology, 29*, 351–357.

Moser, G. (2009b). *Psychologie environnementale*. Bruxelles: de Boeck.

Norgaard, R. B. (1994). *Development betrayed. The end of progress and a coevolutionary revisioning of the future*. London: Routledge.

O'Brien, C. (2008). Sustainable happiness: How happiness studies can contribute to a more sustainable future. *Canadian Psychology, 49*(4), 289–295.

Pato, C., & Corral-Verdugo, V. (2013). *Self-care behaviors and sustainable behaviors*. Paper presented at the XXXIV international congress of psychology, Brasilia.

Pearce, D. (1988). Economics, equity and sustainable development. *Futures, 20*(6), 595–602.

Putnam, R. (1993). *Making democracy work: Civic traditions in modern Italy*. Princeton: Princeton University Press.

Redclift, M. (1987). *Sustainable development: Exploring the contradictions*. London: Methuen.

Roseland, M. (2012). *Toward sustainable communities: Solutions for citizens and their governments*. Gabriola Island: New Society.

Sandbach, F. (1978). Ecology and the 'Limits of Growth' debate. *Antipeze, 10*(2), 22–32, University of Kent.

Schmuck, P., & Schultz, P. W. (2002). *Psychology of sustainable development*. Dordrecht: Kluwer.

Schultz, P. W. (2001). The structure of environmental concern. Concern for self, other people, and the biosphere. *Journal of Environmental Psychology, 21*, 327–339.

Schultz, P. W. (2002). Inclusion with nature: The psychology of human-nature relations. In P. Schmuck & P. W. Schultz (Eds.), *Psychology of sustainable development*. Dordrecht: Kluwer.

Shearman, R. (1990). The meaning and ethics of sustainability. *Environmental Management, 14*, 1–8.

Simon, D. (1989). Sustainable development: Theoretical construct or attainable goal? *Environmental Conservation, 16*(1), 41–48.

Sober, E., & Wilson, D. S. (1998). *Unto others: The evolution and psychology of unselfish behavior*. Cambridge: Harvard University Press.

Solow, R. (1993). Sustainability: An economist's perspective. In R. Dorfman & N. S. Dorfman (Eds.), *Economics of the environment*. New York: Norton.

Stutz, J. (2006). *The role of well-being in a great transition* (GTI paper series N°10). Tellus Institute. www.gtinitiative.org/documents/PDFFINALS/10WellBeing.pdf.

Tapia, C., Corral-Verdugo, V., Fraijo, B., & Durón, F. (2013a). Assessing sustainable behavior and its correlates: A measure of pro-Ecological, frugal, altruistic and equitable actions. *Sustainability, 5*, 711–723.

Tapia, C., Corral-Verdugo, V., Fraijo B., & Durón, M. (2013b). *Construcción y prueba de una escala del capital social: Correlatos con la conducta sustentable*. Revista Mexicana de Psicología, Número Especial. Memorias del XXI Congreso Mexicano de Psicología. Sociedad Mexicana de Psicología.

Uzzell, D. L., & Moser, G. (2006). Environment and quality of life. *European Review of Applied Psychology, 56*, 1–4.

Van den Berg, A., Hartig, T., & Staats, H. (2007). Preference for nature in urbanized societies: stress, restoration, and the pursuit of sustainability. *Journal of Social Issues, 63*, 79–96.

Veenhoven, R. (2006). Is life getting better? How long and happy people live in modern society. *European Psychologist, 10*, 330–343.

WCED, World Comission on Environment and Development. (1987). *Report of the world commission on environment and development: Our common future*. New York: Oxford University Press.

Winter, D. (2002). Gendering sustainable development. In P. Schmuck & P. W. Schultz (Eds.), *Psychology of sustainable development*. Dordrecht: Kluwer.

10

Self-Determined, Enduring, Ecologically Sustainable Ways of Life: Attitude as a Measure of Individuals' Intrinsic Motivation

Florian G. Kaiser, Alexandra Kibbe, and Oliver Arnold

When people lead more self-determined, intrinsically motivated, ecologically sustainable lives, they tend to be healthier, more personally fulfilled, and happier (e.g., Corral Verdugo 2012; De Young 1985–1986). Even without such impressive benefits, self-determined behavior is expected to bring some truly remarkable advantages. For example, it does not require supervision or continual governance as people implement self-determined behavior for themselves (e.g., Deci and Ryan 1985; Ryan and Deci 2000). Whatever the person is doing seems to be inherently worth doing from the person's own perspective. Hence, self-determined behavior is maintained even in the absence of apparent reinforcement (i.e., without incentives) and can thus be expected to be rewarding and gratifying in and of itself. Predictably, self-determined (e.g., environmentally protective) behavior is durable and can withstand the test of time (e.g., Pelletier and Sharp 2008). Not surprisingly, self-determined sacrifices of personal commodities or conveniences to protect the environment (strictly speaking, self-determined ecologically sustainable performance patterns and ways of life) are seen as critical for an ultimate change for the better (e.g., Otto et al. 2014).

Although self-determination and attitude theory are commonly recognized as two separate theories within environmental psychology (e.g., Vining and Ebreo 2002), in this chapter, we question the disparity of the two concepts, "motivation toward the environment" (e.g., Pelletier et al. 1998) and "attitude toward environmental protection" (e.g., Kaiser et al. 2013). By contrast, we argue that identical formal attributes are applied when psychologists gauge the extent of a person's intrinsic motivation and the intensity with which a person embraces a certain attitude. The disparity of the two concepts is, as we demonstrate, more technical than psychological.

Furthermore, our analysis reveals a lack of evidence for how to promote effectively individuals' self-determined motivation to protect the environment. Without such strategies, however, the fundamental self-determined lasting changes in people's ways of life that are needed cannot be achieved. We show that the behavior-change strategies currently applied to advance the environmentally protective performance of

---

Author Note This chapter was written as part of the Helmholtz Alliance ENERGY-TRANS with a grant from the Helmholtz Society and the German State of Saxony-Anhalt. We wish to thank Jane Zagorski for her language support and Christina Werker, Franziska Körner, Madeleine Breitkreutz, Franz X. Bogner, and Christoph Fricke for their comments on earlier versions of this chapter. The section on "Promoting Intrinsic Motivation to Protect the Environment" is adapted with permission from: Otto et al. (2014). doi: 10.1027/1016-9040/a000182. © 2014 Hogrefe Publishing (www.hogrefe.com). All rights reserved.

F.G. Kaiser (✉) • A. Kibbe • O. Arnold
Institute of Psychology, Otto-von-Guericke-University, 4120, 39016 Magdeburg, Germany
e-mail: florian.kaiser@ovgu.de

individuals generally focus on enticements extrinsic to environmental protection or on structural, and thus again extrinsic, restraints (e.g., Steg and Vlek 2009; Thøgersen 2014). Before examining these strategies, we begin by discussing the attributes that are employed when the extent of intrinsic motivation behind a person's environmentally protective performance is assessed.

10.1 Motivation to Protect the Environment

Although self-determination (e.g., Deci and Ryan 1985) and attitude theory differ in much of their psychological wrapping, they employ—at least when attitude theory is implemented within the Campbell paradigm as outlined below (e.g., Kaiser et al. 2010)—the same logic for assessing a person's level of intrinsic motivation toward protecting the environment and the extent to which a person embraces an attitude toward environmental protection. First, the behavioral means must correspond to the actor's underlying goal. Second, the behavior must arise from within the actor and must therefore be intentional.

10.1.1 Behavioral Means-Goal Correspondence

According to self-determination theory, a behavior is intrinsically motivated when it is an end in and of itself (e.g., Pelletier et al. 1998). By contrast, a behavior is extrinsically motivated when it is implemented for instrumental reasons (e.g., Green-Demers et al. 1997; Villacorta et al. 2003).

A typical example is when the reuse of towels to save energy is depicted as either the commended behavior (i.e., the injunctive norm) or the convention (i.e., the descriptive norm; see e.g., Goldstein et al. 2008). By reusing towels (i.e., an environmentally protective behavior), one can gain social approval (i.e., attain a social rather than an environmental protection goal). Another example is car use when it results in a fine (e.g., Jakobsson et al. 2002). By refraining from car use (i.e., an environmentally protective behavior), one can avoid the monetary penalty (i.e., achieve a financial goal). In other words, the reasons behind the environmentally protective behaviors in these examples are something other than environmental protection. Such extrinsically motivated behavior is thereby instrumental to a, strictly speaking, mismatched goal (see Fig. 10.1).

Intrinsically motivated behavior, by contrast, is aligned with its goal (e.g., Green-Demers et al. 1997; Pelletier and Sharp 2008); for example, when a person avoids excessive energy consumption by reusing a towel to improve the environment. In other words, if an apparently environmentally protective behavior is implemented to attain an environmental protection goal, the behavior corresponds to its goal. Accordingly, distinguishing intrinsically from extrinsically motivated behavior is straightforward as the two motivation types are the endpoints of a continuum representing the extent of correspondence between a behavior and the goal at which it is aimed (see Fig. 10.1). The second attribute that is applied when assessing the extent to which a person is intrinsically motivated is how intentional the behavior is.

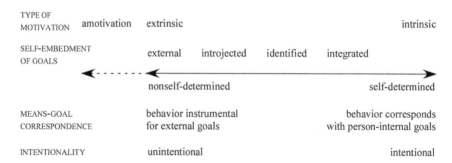

Fig. 10.1 The continuum from amotivation to extrinsic and intrinsic motivation and its distinguishing attributes: self-embedment of goals, behavioral means-goal correspondence, and intentionality (cf. Ryan and Deci 2000)

10.1.2 Degrees of Intentionality

According to self-determination theory, a behavior is self-determined when it is performed for internal (i.e., person-inherent) reasons; strictly speaking, when it is *intentional*—that is, when behavior is grounded in a personal reason or goal that has arisen from within the actor—and when gratification is derived from its practice. At the other extreme of the continuum are non-self-determined behaviors. These occur when a behavior is *unintentional*—that is, when there is no personal reason for the engagement (e.g., due to a lack of options) or when the personal reasons are externally controlled (e.g., by means of financial enticements or legal restraints). If a behavior lacks a personal reason, it is said to be without motivation (i.e., *amotivated*).

When there is a personal reason for a behavior, the degree to which the behavior is externally controlled decreases as the self-embedment of the reason increases (see Fig. 10.1). Depending on the extent to which a personal reason is integrated into the actor's self (i.e., the way he/she sees him/herself), a behavior can be regarded as increasingly more intentional and, concurrently, less extrinsically motivated. Hence, for motivation to be intrinsic, the behavioral reason must represent an integral part of the actor's self-concept (Ryan and Deci 2000). Alternatively, reasons can be *external*, *introjected*, *identified*, or *integrated* into an actor's self.

People's motivation, and thus the extent of their intention to protect the environment, is typically determined by studying their self-reflections on the six types of personal reasons for "doing things for the environment" (Pelletier et al. 1998). For example: (1) When people express that they do not know why they engage in environmentally protective behavior, they are amotivated. (2) When they express reasons such as wanting to avoid upsetting other people, they are externally motivated. (3) When they engage in environmentally protective behavior due to feelings of guilt, they are introjectedly motivated. (4) When they see environmentally protective behavior as the reasonable thing to do, they are identifiedly motivated. (5) When they see environmentally protective behavior as an integral part of their lives, they are integratedly motivated. (6) Finally, gaining pleasure from protecting the environment renders people intrinsically motivated.

Whether or not a behavior is intrinsically motivated depends on the extent to which the behavior is intentional that is, the extent to which the reason or goal for action is profound and has arisen from within the actor. In turn, the degree of intentionality is maximal when a behavior is aimed toward a goal that is an integral part of an actor's self. Such intrinsically motivated behavior has some remarkable consequences, which have been observed for environmentally protective behavior as is discussed next.

10.1.3 Consequences of Self-Determined Environmentally Protective Behavior

Behavior that arises from within the actor and is thus inherently rewarding and gratifying has predictably been found to be durable (i.e., to last over time: e.g., Pelletier and Sharp 2008) and can be expected to be more probable. Strong relationships—as pronounced as $r = 0.40$ and $r = 0.60$—have repeatedly been reported between intrinsic motivation and self-reports of mundane environmentally protective behavior, such as recycling, reusing, purchasing environmentally friendly products, and learning about issues relevant to environmental protection (e.g., De Young 1985–1986; Green-Demers et al. 1997). Furthermore, the relationship between self-determined motivation and self-reported environmentally protective behavior seems to become stronger as the behavior appears subjectively more difficult (Green-Demers et al. 1997).

In the following, an overview is given of an alternative approach that we believe can also be applied to appraise people's intrinsic motivation to protect the environment. Within what Kaiser et al. (2010) call the Campbell paradigm, the extent of self-determination or intentionality can be derived from self-reports of environmentally protective behavior without exploring the self-relevance of the reasons behind the behavior. As shown below, the evidence from the two research traditions, self-determination and attitude theory, converges greatly.

10.2 Attitude Toward Environmental Protection

The degree to which a person disregards the difficulty of a behavior is another way of assessing his/her degree of *self-determination* to protect the environment. Such a measure is grounded in statistical tests of the assumed *behavioral means-goal* correspondence and of the derived estimates of *people' extent of intentionality*. In other words, instead of a measurement model based on assumptions about the self-relevance of personal reasons, which can, at best, be supported by circumstantial evidence, the proposed measurement model directly tests the purported formal attributes of self-determination: means-goal correspondence and degrees of intentionality (see Fig. 10.1).

10.2.1 Disregarding Behavioral Difficulties

According to self-determination theory, actors can be more or less detached from their reasons for protecting the environment. Only if environmental protection represents an actor's own personal goal (is personally intended by the actor) is the environmentally protective behavior intrinsically motivated. Instead of linking goals with actors' views of themselves (i.e., with their self-concept), it can just as well be presumed that all possible behavioral reasons are essentially intrinsic, but to varying degrees across individuals. In other words, every person is to some extent personally determined, or intends, to protect the environment, save money, or gain social approval as a few examples; the only question is, how much?

This alternative approach for exploring the extent of a person's self-determination—although not directly presented in this way—was proposed by Donald Campbell (1963). According to Campbell, a person's determination to attain a particular goal (e.g., protect the environment) becomes obvious in the face of increasing difficulties (e.g., painful sacrifices that come with an action). Thus, if people are determined to protect the environment, then they can be expected to engage in a set of activities that reflect this very goal. Moreover, the more obstacles a person overcomes and the more effort he/she expends on implementing the environmental protection goal, the more determined the person is to protect the environment; or, in the terminology of the Campbell paradigm, the more pronounced is the person's attitude toward environmental protection (Kaiser et al. 2010). Why would people recycle glass, commute by bike in inclement weather, endure lower temperatures at home, and give money to environmental organizations if they did not intend to protect the environment? Conversely, when the slightest inconvenience is enough to stop people from taking the appropriate behavioral steps, their personal determination, and thus their intrinsic motivation to protect the environment (i.e., environmental attitude), is probably rather weak (see Fig. 10.2).

10.2.2 Recognizing Behavioral Means-Goal Correspondence

Originally, Campbell (1963) proposed that the Guttman model (Guttman 1944) be used to implement his measurement idea. Kaiser et al. (2010) viewed this proposal as unrealistically stringent and, as an alternative, they proposed the Campbell paradigm, which uses the Rasch model (Rasch 1960/1980: see Formula 10.1) to represent Campbell's original idea more realistically.

$$\ln\left(\frac{p_{ki}}{1-p_{ki}}\right) = \theta_k - \delta_i \quad (10.1)$$

According to the Rasch model, the natural logarithm of the ratio of the probability (p_{ki}) that person k engages in a specific behavior i

Fig. 10.2 Offsetting increasing behavioral demand with progressively more pronounced intrinsic motivation to protect the environment. The *horizontal bars* display a prototypical distribution of people's motivation to protect the environment from low to high. Correspondingly, environmentally protective behaviors (behavioral examples mentioned in the text are displayed in caps) are ordered according to their demand-based difficulty on a scale ranging from easy to difficult

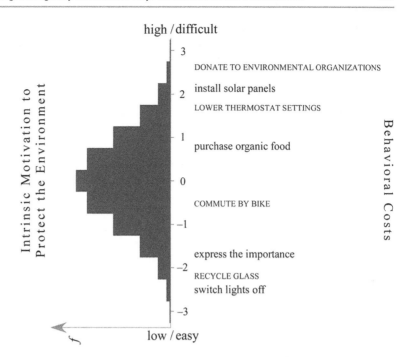

relative to the probability that person k does not engage in behavior i ($1-p_{ki}$) is equated with the arithmetic difference between person k's intrinsic motivation or self-determination or intentionality (conceptually correct attitude) to act in a particular way (θ_k) and the composite of the behavioral demands or costs epitomized by the estimate of the difficulty of behavior i (δ_i). Correspondingly, people are distinguishable with respect to the extent of their intrinsic motivation to act in a particular way, and behaviors can be distinguished by costs, that is, how demanding the behaviors are to implement.

In this model, a personal goal and its behavioral indicators are fused into a means-end relationship. The more personally important a goal (e.g., environmental protection), the more behaviors a person will engage in to attain the goal reflected by the behaviors (see Fig. 10.2).

10.2.3 Degrees of Intentionality Inferred from Behavior

People's intrinsic motivation to protect the environment—the extent of intentionality or self-determination—is directly derived from what persons do to protect the environment (see Fig. 10.2). The behavior from which a person's motivation is inferred, however, does not need to be comprised of only observed overt acts. Behavior can also consist of verbal acts, such as self-reports of past performance or expressions of appreciation for environmental protection. Thus, in order to measure the extent of a person's self-determined motivation, it is essential to establish composite measures that reflect a transitive structure of the behavioral indicators sorted by their difficulty (see Fig. 10.2). This transitive structure—corroborated by a successful Rasch model test—in turn allows for differences in people's motivation to be recognized (e.g., Kaiser and Byrka 2015).

Kaiser and colleagues have successfully performed such Rasch model tests with various sets of self-reports of environmentally protective behavior (e.g., Kaiser and Wilson 2004). In addition, Byrka (2009) confirmed that evaluative statements in the form of verbal expressions of appreciation for various environmentally protective behaviors, along with straightforward self-reports of past engagement in these behaviors, represent a single transitively ordered class of indicators. A similar conclusion based on a related measurement model can be found in Andrich and Styles (1998).

10.2.4 Evidence of Self-Directedness

Behavior that is goal-directed by design (e.g., Kaiser and Wilson 2004) can necessarily be expected to arise from within the actor, and is thus predictably intentional. Not surprisingly, a Campbellian measure of a person's motivation to protect the environment has been found to overlap by as much as 95 % with conventional measures of a person's intention to engage in activities to protect the environment (e.g., Kaiser et al. 2005, 2007). With such inherently intentional behavior, a lack of spontaneous fluctuation in people's motivation would not be surprising either. Correspondingly, we found that a Campbellian measure of people's motivation to protect the environment almost perfectly predicted ($\beta = .99$) their motivation to protect the environment more than 2 years later (Kaiser et al. 2014). This finding impressively corroborates the durability that accompanies individuals' self-directed performances (e.g., Pelletier and Sharp 2008).

Because a person's motivation to protect the environment within the Campbell paradigm is a property that is typically inferred from answers to behavioral self-reports (e.g., "Do you reuse shopping bags?"), the connection between an individual's motivation and these behavioral self-reports is a formal one (see Formula 10.1). Accordingly, even a strong empirical correlation between behavioral self-reports from which the respective motivation was originally inferred and the very self-reports would be tautological and, thus, empirically trivial (e.g., Campbell 1963).

However, it would certainly not be trivial if a person's intrinsic motivation to protect the environment as derived from self-reports could be shown to affect real environmentally protective behavior (e.g., using a smart meter to restrict a household's use of electricity). It would also not be trivial if a person's intrinsic motivation to protect the environment as derived from self-reports was linked to the annual consumption of electricity.

In a retrospective study, we explored 237 households that had previously been equipped with smart meters, which are electronic devices that provide real-time and historical electricity consumption feedback in an online platform. It turned out that households that had opted to use their smart meters (by registering them online: the smart-meter users) saved significantly more electricity than households that opted not to use them (the nonusers) but only if the users were comparatively high in their intrinsic motivation to protect the environment (see Fig. 10.3).

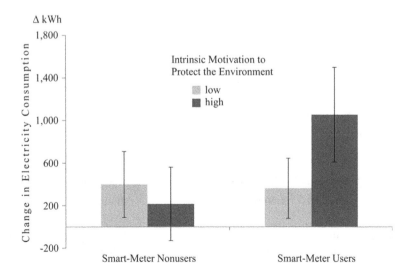

Fig. 10.3 Electricity savings (in Δ kWh) as a function of people's intrinsic motivation to protect the environment and their active use of smart meters

Fig. 10.4 Annual electricity consumption (in kWh) and people's intrinsic motivation to protect the environment (in logits) for regular and *"green"* customers. Logits stand for the natural logarithm of the engagement/non-engagement probability ratio across the entire response vector of a person. The smaller a logit value, the lower a person's intrinsic motivation

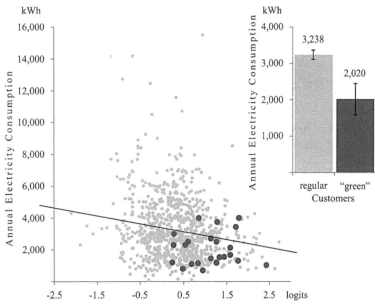

Obviously, the availability of a smart meter alone was insufficient for smart-meter-based feedback to be effective; only in combination with a person's motivation to protect the environment was feedback successful in helping households reduce their electricity consumption.

After pooling the sample of smart-meter users and nonusers with two other independently collected samples ($N = 893$), we found that a Campbellian measure of intrinsic motivation to protect the environment was significantly inversely related to electricity consumption (see Fig. 10.4 or Arnold et al. 2015). Remarkably, this negative correlation held even when we controlled for income. Although the association between motivation and consumption appeared rather small in magnitude (i.e., explaining approximately 4 % of the variance), motivational differences nevertheless resulted in rather pronounced electricity savings. By comparing customers enlisted in a certified but more expensive green electricity program (i.e., "green" customers) with a convenience sample of regular customers, we found that the "green" customers consumed one third less electricity per year than the regular customers (see Fig. 10.4).

We have already found a fair amount of evidence in contemporary environmental psychology that intrinsic motivation to protect the environment—measured as a person's attitude toward environmental protection within the Campbell paradigm—is relevant to both behavior and impact. From a practical point of view, knowledge about how to promote people's intrinsic motivation to protect the environment is therefore essential. Unfortunately, the literature shows that research has so far focused primarily on extrinsic—social or financial—enticements rather than on how to promote people's intrinsic motivation, as is shown next.

10.3 Promoting Intrinsic Motivation to Protect the Environment

The preferred approach to behavior change involves targeting one behavior or behavioral type at a time (e.g., eco-driving) (e.g., Schultz 2014). Here, an overview is given of the motivation-based strategies that are applied to advance individuals' engagement in environmental protection (e.g., Steg and Vlek 2009). We argue that the common practice of employing—social or financial—enticements will not be able to increase individuals' intrinsically motivated commitment to protect the environment (e.g., Pelletier and Sharp 2008).

10.3.1 Extrinsically Motivated Changes for the Better

People commonly opt for behaviors with comparatively greater benefits; thus, behavior change can be instigated by removing structural restraints (e.g., eliminating the unavailability of curbside recycling) or creating advantages (e.g., designating carpool lanes; e.g., Geller 2002). A prominent motivation-based strategy is to provide monetary enticements routinely used in commerce (i.e., free access to products or services) and administration (i.e., subsidizing solar panels, refunding beverage cans). Despite their undeniable effectiveness in advancing behavior, monetary enticements have several shortcomings. For example, they are expensive, must be protected against exploitation, and seem to carry the potential to undermine, due to overjustification, an already existing intrinsic motivation to protect the environment (e.g., De Young 2000; Lehman and Geller 2004). Not surprisingly, alternative enticements to advance specific behaviors have become more popular in environmental psychology.

Environmentally protective behavior can also be promoted by offering benefits that are grounded in people's social motives (e.g., Joule et al. 2007). Such social enticements involve the prospect of public praise, increased reputation or status, and avoidance of repercussions. By complying with a commitment or social norm, people can either avoid social sanctions (e.g., reproaches, public embarrassment) or hope for additional personal benefits (e.g., praise, prestige). By providing information about what others do (descriptive norms) and what is considered proper behavior (injunctive norms), Schultz and colleagues successfully increased the household recycling rate and the reuse of towels by hotel guests (Schultz 1999; Schultz et al. 2008). People can also be bound to certain courses of action (e.g., saving electricity, reducing mileage) by making promises in public (e.g., Joule et al. 2007). Apparently, a publicly-signed contract binds people to honor the contract because they wish to avoid the poor social reputation that comes with hypocrisy, or the practice of committing oneself to a course of action without complying (e.g., Dickerson et al. 1992).

Nevertheless, managing individual behavior with enticements generally comes with one unavoidable shortcoming: people engage in the act for instrumental and thus extrinsic reasons (i.e., because they wish to obtain the financial or social incentive) rather than for attaining the environmental protection goal that is reflected by the behavior. Strictly speaking, the particular environmentally protective behavior is not intrinsically motivated (see Fig. 10.1). Rather, enticements essentially counterbalance some of the costs of the promoted behavior (see Fig. 10.2) and in turn increase the chance that people will engage in the behavior but without enhancing their intrinsic motivation to protect the environment (see Formula 10.1).

10.3.2 Spillover Behavior Effects

Motivating behavior extrinsically by complementing a specific environmentally protective behavior with some extra benefit generally works only as long as the enticement remains in place (e.g., Bolderdijk et al. 2011). Even worse, extrinsic reasons for environmental protection can trigger increased use. For example, after drivers purchased a new, heavily subsidized "green" car, Klöckner et al. (2013) found that driving increased at the expense of more environmentally protective modes of transportation (e.g., walking, bicycling). Thus, financial incentives, disincentives, social sanctions, and gains in social reputation are not real options to create the enduring motivation in individuals to aim ultimately to protect the environment of their own accord.

Predictably, changing people's consumptive ways of life will only be possible when interventions in environmental psychology succeed in motivating people to protect the environment self-determinedly (Otto et al. 2014). In other words, interventions are needed that affect a multitude of specific behaviors simultaneously, a phenomenon commonly referred to as spillover behavior effects (e.g., Thøgersen 1999). Unfortunately, there is limited research in contemporary environmental psychology on measures that are capable of promoting people's intrinsic motivation to protect the environment (De Young 2000). Pelletier and Sharp (2008) emphasize persuasive communication, whereas others stress that information relevant to environmental protection (Otto and Kaiser 2014) and appreciation of nature (e.g., Roczen et al. 2014) have the potential to promote effectively more intrinsic such motivation in individuals. Still, more systematic empirical exploration is required to shed light on this issue.

10.4 Conclusion

In this chapter, we have identified the behavioral means-goal correspondence and degree of intentionality as the two decisive formal features of intrinsic motivation (e.g., Ryan and Deci 2000). Hence, behavior is thought to be extrinsically motivated when it is unintentional and implemented for reasons that fail to match the apparent goal of the behavior. By contrast, a behavior is intrinsically motivated when it reflects the personal goal at which it is apparently aimed and when it is inherently intended (see Fig. 10.1). We have argued that a Rasch-model-based test (a) of the behavioral means-goal correspondence and (b) of actors' degree of intentionality is technically preferable to assessing the self-embedment of people's reasons for protecting the environment. This is because such a Rasch-model test does not demand that people self-reflect on the reasons behind their actions, and does not rely on a synthetically derived, empirically untestable, classification system of reasons (e.g., Pelletier et al. 1998).

In the suggested alternative approach for exploring the extent of a person's self-determination, we presume that any reason can essentially be intrinsic, be it protecting the environment, saving money, or gaining social approval. If so, the core question becomes a quantitative one: how much? On the basis of what Kaiser et al. (2010) call the Campbell Paradigm, people's levels of intrinsic motivation become obvious in the face of the obstacles they ignore. Thus, if people are determined to protect the environment, they will engage in a set of increasingly demanding activities aimed at the environmentally protective goal (see Fig. 10.2). In other words, a person's motivation to protect the environment can be appraised by means of composite measures, which allow the very motivation to be derived from self-reports of environmentally protective behavior (e.g., Kaiser and Byrka 2015). Such composite measures simultaneously reflect the extent of people's inherent self-determination or intention to protect the environment and their more or less consumptive ways of life.

As expected of a measure of motivation that is assumed to be intentional, that is, based on behavior instrumental to goals that have arisen from within the actor, behavior-based motivation measures have been found to overlap markedly with conventional intention measures (Kaiser et al. 2007). Moreover, and not surprisingly, people's motivation to protect the environment has turned out to be extremely durable. We discovered quite a lack of inter- or intrapersonal variability over a period of more than 2 years (Kaiser et al. 2014). Moreover, as one would hope, people's motivation to protect the environment shows in various overt behaviors, such as the effective use of smart meters (see Fig. 10.3). As expected, Arnold et al. (2015) also corroborated that people's motivation to protect the environment was related to their overall electricity consumption (see Fig. 10.4).

Unfortunately, the strategies that are currently applied to advance individuals' environmentally protective performance tend to focus on enticements and structural restraints (e.g., Steg and Vlek 2009; Thøgersen 2014), both of which are extrinsic to environmental protection. Extrinsically motivated people are, however, not a real solution to create lasting change in individuals' behavior (Otto et al. 2014). Novel strategies are needed that can help advance people's intrinsic motivation to protect the environment and that will eventually encourage more ecologically sustainable performance patterns and, hence, ways of life.

References

Andrich, D., & Styles, I. M. (1998). The structural relationship between attitude and behavior statements from the unfolding perspective. *Psychological Methods, 3*, 454–469.

Arnold, O., Kibbe, A., Hartig, T., Kaiser, F. G. (2015). *Capturing the environmental impact of individual lifestyles: Evidence of the criterion validity of the general ecological behavior scale.* Manuscript submitted for publication.

Bolderdijk, J. W., Knockaert, J., Steg, E. M., & Verhoef, E. T. (2011). Effects of pay-as-you-drive vehicle insurance on young drivers' speed choice: Results of a Dutch field experiment. *Accident Analysis & Prevention, 43*, 1181–1186.

Byrka, K. (2009). *Attitude-behavior consistency: Campbell's paradigm in environmental and health domains.* Unpublished doctoral dissertation. Eindhoven University of Technology, Eindhoven, the Netherlands.

Campbell, D. T. (1963). Social attitudes and other acquired behavioral dispositions. In S. Koch (Ed.), *Psychology: A study of a science* (Vol. 6, pp. 94–172). New York: McGraw-Hill.

Corral Verdugo, V. (2012). The positive psychology of sustainability. *Environment, Development and Sustainability, 14*, 651–666.

De Young, R. (1985–1986). Encouraging environmentally appropriate behavior: The role of intrinsic motivation. *Journal of Environmental Systems, 15*(4), 281–292.

De Young, R. (2000). Expanding and evaluating motives for environmentally responsible behavior. *Journal of Social Issues, 56*, 509–526.

Deci, E. D., & Ryan, R. M. (1985). *Intrinsic motivation and self-determination in human behavior.* New York: Plenum.

Dickerson, C. A., Thibodeau, R., Aronson, E., & Miller, D. (1992). Using cognitive dissonance to encourage water conservation. *Journal of Applied Social Psychology, 22*, 841–854.

Geller, E. S. (2002). The challenge of increasing proenvironmental behavior. In R. B. Bechtel & A. Churchman (Eds.), *Handbook of environmental psychology* (pp. 525–540). New York: Wiley.

Goldstein, N. J., Cialdini, R. B., & Griskevicius, V. (2008). A room with a viewpoint: Using social norms to motivate environmental conservation in hotels. *Journal of Consumer Research, 35*, 472–482.

Green-Demers, I., Pelletier, L. G., & Ménard, S. (1997). The impact of behavioural difficulty on the saliency of the association between self-determined motivation and environmental behaviours. *Canadian Journal of Behavioural Science, 29*(3), 157–166.

Guttman, L. (1944). A basis for scaling qualitative data. *American Sociological Review, 9*, 139–150.

Jakobsson, C., Fujii, S., & Gärling, T. (2002). Effects of economic disincentives on private car use. *Transportation, 29*, 349–370.

Joule, R.-V., Girandola, F., & Bernard, F. (2007). How can people be induced to willingly change their behavior? The path from persuasive communication to binding communication. *Social and Personality Psychology Compass, 1*, 493–505.

Kaiser, F. G., & Byrka, K. (2015). The Campbell paradigm as a conceptual alternative to the expectation of hypocrisy in contemporary attitude research. *The Journal of Social Psychology, 155*, 12–29.

Kaiser, F. G., & Wilson, M. (2004). Goal-directed conservation behavior: The specific composition of a general performance. *Personality and Individual Differences, 36*, 1531–1544.

Kaiser, F. G., Hübner, G., & Bogner, F. X. (2005). Contrasting the theory of planned behavior with the value-belief-norm model in explaining conservation behavior. *Journal of Applied Social Psychology, 35*, 2150–2170.

Kaiser, F. G., Schultz, P. W., & Scheuthle, H. (2007). The theory of planned behavior without compatibility? Beyond method bias and past trivial associations. *Journal of Applied Social Psychology, 37*, 1522–1544.

Kaiser, F. G., Byrka, K., & Hartig, T. (2010). Reviving Campbell's paradigm for attitude research. *Personality and Social Psychology Review, 14*, 351–367.

Kaiser, F. G., Hartig, T., Brügger, A., & Duvier, C. (2013). Environmental protection and nature as distinct attitudinal objects: An application of the Campbell paradigm. *Environment and Behavior, 45*, 369–398.

Kaiser, F. G., Brügger, A., Hartig, T., Bogner, F. X., & Gutscher, H. (2014). Appreciation of nature and appreciation of environmental protection: How stable are these attitudes and which comes first? *European Review of Applied Psychology [Revue Européenne de Psychologie Appliquée], 64*, 269–277.

Klöckner, C. A., Nayum, A., & Mehmetoglu, M. (2013). Positive and negative spillover effects from electric car purchase to car use. *Transportation Research Part D: Transport and Environment, 21*, 32–38.

Lehman, P. K., & Geller, E. S. (2004). Behavioural analysis and environmental protection: Accomplishments and potential for more. *Behavioural and Social Issues, 13*, 13–24.

Otto, S., & Kaiser, F. G. (2014). Ecological behavior across the lifespan: Why environmentalism increases as people grow older. *Journal of Environmental Psychology, 40*, 331–338.

Otto, S., Kaiser, F. G., & Arnold, O. (2014). The critical challenge of climate change for psychology: Preventing rebound and promoting more individual irrationality. *European Psychologist, 19*, 96–106.

Pelletier, L. G., & Sharp, E. (2008). Persuasive communication and proenvironmental behaviours: How message tailoring and message framing can improve the integration of behaviours through self-determination. *Canadian Psychology, 49*, 210–217.

Pelletier, L. G., Tuson, K. M., Green-Demers, I., Noels, K., & Beaton, A. M. (1998). Why are you doing things for the environment? The Motivation Toward the Environment Scale (MTES). *Journal of Applied Social Psychology, 28*, 437–468.

Rasch, G. (1960/1980). *Probabilistic models for some intelligence and attainment tests*. Chicago: University of Chicago Press (Original work published 1960).

Roczen, N., Kaiser, F. G., Bogner, F. X., & Wilson, M. (2014). A competence model for environmental education. *Environment and Behavior, 46*, 972–992.

Ryan, R. M., & Deci, E. D. (2000). Self-determination theory and the facilitation of intrinsic motivation, social development, and well-being. *American Psychologist, 55*, 68–78.

Schultz, P. W. (1999). Changing behavior with normative feedback interventions: A field experiment on curbside recycling. *Basic and Applied Social Psychology, 21*, 25–36.

Schultz, P. W. (2014). Strategies for promoting proenvironmental behavior: Lots of tools but few instructions. *European Psychologist, 19*, 107–117.

Schultz, P. W., Khazian, A. M., & Zaleski, A. C. (2008). Using normative social influence to promote conservation among hotel guests. *Social Influence, 3*, 4–23.

Steg, L., & Vlek, C. (2009). Encouraging pro-environmental behaviour: An integrative review and research agenda. *Journal of Environmental Psychology, 29*, 309–317.

Thøgersen, J. (1999). Spillover processes in the development of a sustainable consumption pattern. *Journal of Economic Psychology, 20*, 53–81.

Thøgersen, J. (2014). Unsustainable consumption: Basic causes and implications for policy. *European Psychologist, 19*, 84–95.

Villacorta, M., Koestner, R., & Lekes, N. (2003). Further validation of the Motivation Toward the Environment Scale. *Environment and Behavior, 35*, 486–505.

Vining, J., & Ebreo, A. (2002). Emerging theoretical and methodological perspectives on conservation behavior. In R. B. Bechtel & A. Churchman (Eds.), *Handbook of environmental psychology* (pp. 541–558). New York: Wiley.

Commitment and Pro-Environmental Behaviors: Favoring Positive Human-Environment Interactions to Improve Quality of Life

Christophe Demarque and Fabien Girandola

11.1 Introduction

Environmental psychology focuses on people-environment relations. These relations are often considered in adaptative terms: how can one avoid degradations, limit the overexploitation of natural resources, or cope with some urban annoyances (noise, pollution, etc.)? From a different perspective, Weiss and Girandola (2009, 2010a, b) suggested developing a positive psychology of sustainable development, dealing with social and dispositional factors that enable satisfaction, quality of life and, even more, individual well-being. Thus, the environment should no longer be considered a constraint requiring efforts to adapt to it but more a potential source of positive emotions, as shown in works on restorative environments or place attachment (e.g., Lewicka 2011; Staats 2012). Because of the perceptions, attitudes and representations it arouses, the environment is a major source of influence on individuals' well-being and quality of life. But under what conditions? According to Uzzell and Moser (2006) "… *a sustainable quality of life is only achieved when people interact with the environment in a respectful way, on one hand, and when that environment in turn is not impeding or threatening what the individual considers as their 'quality of life', on the other, this renders possible the capacity for the individual to satisfy their needs.*" (p. 3). Thus, individuals' environmental quality of life will depend on the evaluation of "people-environment congruity" through the subjective assessment of a combination of objective material factors of the daily environment. Fleury-Bahi et al. (2013) distinguished four dimensions that should be taken into account: physical and spatial environment, social context, environmental annoyances and local facilities/services.

In line with this concept, we consider behavioral commitment a way to contribute to quality of life, by improving these objective indicators. In this sense, this chapter shows that commitment enables the introduction of new positive relations with the environment in a more or less direct way. Commitment produces cognitive (in terms of attitudes, emotions, representations) and behavioral effects likely to improve the perception of people-environment congruity. To illustrate this idea, we provide a review of the commitment theory (e.g., Lokhorst et al. 2013) and of the effects of binding communication (Girandola and Joule 2012; Joule and Beauvois 2014; Joule et al. 2007b) on the adoption of pro-environmental behaviors such as waste sorting, recycling, non-activist behaviors in the public sphere and energy saving, likely

C. Demarque (✉)
University of Toulouse 2, Le Mirail, France
e-mail: demarque@univ-tlse2.fr

F. Girandola
Aix-Marseille University, Aix-en-Provence, France
e-mail: fabien.girandola@univ-amu.fr

to be positively perceived by the individual and contribute to quality of life.

First, the basis of commitment theory and its effects on the adoption of pro-environmental behaviors are described and the hypocrisy and self-fulfilling prophecies are presented. Then, we focus on the binding communication paradigm (i.e., theoretical and practical interest) and its objectives: to optimize awareness and information campaigns in order to favor behavioral adhesion. The effects of binding communication on behaviors and attitudes toward the environment are highlighted, including how these effects strengthen the positive characteristics of interactions between humans and the environment and, thereby, quality of life. Finally, action research based on binding communication in the field of promoting pro-environmental behaviors is reviewed. Uzzell and Moser's proposals (2006) are used to show that this paradigm has been applied at different levels of human-environment interactions, as defined by Moser (2003): (1) the private space level (household), (2) the proximal environment level, and (3) the public space level.

Finally, the current limitations of works on commitment, and especially the lack of precise measures of quality of life following research actions, are described. Most of the research presented in this chapter originated from institutional requests. Following logics that could be labeled as "top-down", the expected behavioral changes are rarely discussed with the populations concerned. We show the interest of doing such work beforehand. In particular, the role of social representations in binding communication and associated techniques (e.g., foot-in-the-door) is discussed, using studies showing that commitment strategies are more effective when they take into account some elements of social representations.

11.2 Commitment Theory and Environmental Issues

11.2.1 The Commitment Theory

Behavioral intention is the closest cognitive element for the effective fulfillment of a behavior (e.g., Ajzen and Fishbein 1980; Azjen 1991). Even so, the "intention to do something" does not usually predict the associated effective behavior (e.g., Webb and Sheeran 2006; Schultz and Kaiser 2012). For instance, Bickman (1972) showed that 95 % of the participants in an experiment declared that they were willing to pick up litter from the ground. However, when they were observed in this situation, less than 2 % actually did it. Likewise, information or awareness campaigns rarely lead to effective behaviors. They are necessary but not sufficient to provoke changes (e.g., Kaiser 2014; Schultz 2014; Schultz and Kaiser 2012; Steg and De Groot 2012; Steg et al. 2013).

Since Lewin (1947), it has been known that a behavioral change requires an act and/or a decision beforehand. In line with this idea, research on commitment reveals the influence techniques likely to generate a behavior change and their fallout at the cognitive and behavioral levels (e.g., Cialdini 2006; Kiesler 1971). The literature on behavioral changes in the field of sustainable development considers commitment an effective way to lead to the expected changes (e.g., McKenzie 2011; McKenzie-Mohr et al. 2012). In a meta-analysis focused on 19 studies carried out between 1976 and 2010, Lokhorst et al. (2013) confirmed this effectiveness.

Joule and Beauvois (1998) suggested bringing together the research on commitment and its behavioral and cognitive effects in the paradigm of free will compliance (Joule et al. 2007b). This concerns the study of techniques likely to lead someone to modify his/her behaviors voluntarily. Free will compliance highlights the essential role of preparatory acts in obtaining later commitments (Girandola 2003; Joule and Beauvois 2014; Girandola and Joule 2013). These generally consist of inexpensive acts (i.e., easy to perform) such as filling out a pro-environmental questionnaire (Freedman and Fraser 1966), giving time to a bystander in the street (Harris 1972), wearing a badge (Pliner et al. 1974), etc. These preparatory acts have a double function: on the one hand, they make individuals more sensitive to arguments or information diffused in the later persuasive message (Kiesler 1971); on the other hand, they increase the probability that these individuals will agree to perform consistent but more costly acts (e.g., foot-in-the-door, Burger 1999; Freedman and Fraser 1966), such as signing a petition or recycling waste (Dufourq-Brana et al. 2006). How can commitment be obtained? According to Joule and Beauvois (1998), it depends on:

1. Action visibility and importance. This category includes five factors: (a) the public nature of the action: an action carried out in public is more binding than one in private; (b) the irrevocability of the action: an irrevocable action is more binding than one that is not; (c) the repetition of the action: an action that is repeated is more binding than an action carried out once; (d) the consequences of the action: an action is more binding if it has important consequences; (e) the cost of the action: an action is more binding if it is costly in money, time, energy, etc.
2. The reasons for the action and the context of freedom. This category includes two factors: (a) the reasons imputed to the action may be external (i.e., situation or circumstances) or internal (i.e., personal values). External reasons reduce commitment: the greater the reward or the punishment, the more the action is justified. Thus, external reasons weaken the link between the individual and his/her action. On the contrary, internal reasons (e.g., *"you really are a generous person"*) strengthen the bond between the individual and his/her actions; (b) the context in which the individual operates must give him/her the status of a "free" individual (e.g., *"you are free to accept or refuse"*). Commitment theorists consider freedom of choice the main commitment factor (e.g., Guéguen et al. 2013).

11.2.2 Commitment and Pro-environmental Behaviors

As mentioned by Dwyer et al. (1993), the 1960s in the United States marked the evolution of the 100-year-old conservation movement – focused on the protection of wildlife and natural settings – toward an environmental movement, stressing the idea that environmental quality itself was threatened, partially because of human behavior. In fact, a notable series of environmental incidents occurred in the 1960s and 1970s (the Torrey Canyon and Amoco-Cadiz oil spills in 1967 and 1978, the Seveso dioxin cloud in 1976, the Three Mile Island nuclear accident in the USA), contributing to an ecological awareness. Several laws were also approved in the 1970s (the Clean Air Act, Clean Water Act, etc.), and a number of institutions were created; for instance, the Environmental Protection Agency (EPA) in the United States in 1970, the French Ministry of Environment in 1971, and the United Nations Conference on Human Environment held in Stockholm in 1972. In their review of 54 behavioral interventions to preserve the environment, Dwyer et al. (1993) observed the development in this period of research in the field of psychology, adapting the techniques of applied behavior analysis to environmentally relevant behaviors. They highlighted the relation between this socio-political context and the number of research studies in the environmental field, with a peak at the end of the 1970s and a steady decline through the 1980s. Their main conclusion was that antecedent conditions (i.e., activators or prompting strategies) using commitment, demonstrations and goal-setting strategies were effective in favoring pro-environmental behavior. They also observed that consequence conditions (i.e., feedback, rewards and penalties) were effective during the experiments. However, they underlined the methodological limitations of some of the studies (no comparison condition, few follow-up measures or little maintenance of the behavioral changes).

More recently, Lokhorst et al. (2013) conducted a meta-analysis on commitment in the environmental field, which confirmed its effectiveness overall. Research was mainly centered on recycling and energy conservation behaviors, through the use of two main techniques: the foot-in-the-door (Freedman and Fraser 1966) and the public pledge.

The foot-in-the-door technique consists of getting a person to agree to a modest request, in order to increase the likelihood that he/she will accept a larger request, that is, the target behavior. In the framework of commitment theory, we consider the first behavior a committing (or binding) preparatory act (cf. Joule and Beauvois 1998, 2014). The technique was first used in the environmental field by Arbuthnot et al. (1976/1977) whose research focused on recycling metal cans. The experimental design included three independent preparatory acts: a measure of the participants' knowledge about recycling, saving and sorting cans for 1 week, and writing a letter explaining the interest of the action 1 week after the cans were collected. The results notably showed a recycling rate of more than 80 % in the condition combining the three preparatory acts, and also in the one combining saving cans and the letter. These results highlight the possible additive effects of the combined use of different preparatory acts, all the more so when they are consistent with the final request. This is in line with Kiesler's theoretical proposals (1971) about the continuous character of commitment. It should also be noted that the observed effects lasted 18 months after the experiment. The foot-in-the-door method has also been used to lead individuals to reduce their energy consumption (Katzev and Johnson 1983). Participants were asked to reduce their electricity consumption by 10 %. In the foot-in-the-door condition, this request was preceded by a low-cost preparatory act: answering a short questionnaire about energy conservation. During the 12 weeks of post-experimental measures, the highest number of participants who reduced their consumption was in the foot-in-the-door condition, by comparison with the 4-week experimental period. In another experiment on energy conservation, the authors showed that the foot-in-the-door was more effective over the long term than an incentivizing strategy, such as giving money to participants (Katzev and Johnson 1984).

Another commitment procedure, often found in the literature, is the public pledge, which was mainly used in studies to promote recycling behaviors. Pardini and Katzev (1983/1984) showed the effectiveness of this technique compared to a verbal commitment or a classic information procedure. Wang and Katzev (1990) observed that signing a commitment form, bearing the name of the participant's reference group, increased the average quantity of paper for recycling in a retirement home by almost 50 %, a new dynamic maintained during the 4-week post-experimental phase. For their part, Werner et al. (1995) showed that adding one's signature to a list of people interested in a recycling program was more effective in increasing commitment to this program than information (brochure) or persuasion (face-to-face interaction) procedures. These results are consistent with a series of previous observations (Katzev and Pardini 1987/1988; Wang and Katzev 1990; Katzev and Wang 1994). The results also revealed that the act of signing strengthened the participants' attitudes toward recycling at the same time. Pallak et al. (1980) observed the same kind of behavioral effects in the field of energy conservation.

11.2.3 When Commitment Meets Dissonance: Hypocrisy and Self-Fulfilling Prophecies

Experiments carried out on the hypocrisy paradigm suggest an interpretation in terms of cognitive dissonance (Festinger 1957; Fointiat et al. 2013a). It is the situation of hypocrisy itself that triggers the dissonance arousal, because it leads to a situation where the individuals becoming aware of the gap between their personal standards or their morally good self (i.e., "*I see myself as sincere and honest*") and their behavior (i.e., "*I'm preaching a behavior I don't practice myself*"). Dissonance, experienced as a state of mental stress or discomfort, motivates attempts to reduce it by changing behavior (e.g., Stone and Cooper 2001). From an operational point of view, hypocrisy, and thus dissonance, is induced through the combination of two successive steps. During step 1 ("commitment"), individuals make a normative speech (e.g., writing arguments) about a behavior (e.g., "*Don't waste water*"). Generally, this is a behavior that is not always performed by these individuals. During step 2 ("mindfulness"), previous transgressions are made prominent (e.g., earlier water waste), by asking these same individuals when, in the recent past, they did not act in accordance with the behavior they have just defended. This two-step strategy (i.e., commitment + mindfulness) arouses dissonance when the individual becomes aware that his/her earlier behaviors contradict his/her values and norms. A way of reducing dissonance is to perform pro-normative behaviors (i.e., saving water). Dickerson et al. (1992), for instance, carried out an experiment about water conservation in a hot dry Californian zone (Santa Cruz) near a pool. During step 1 (commitment), half the participants signed a flyer about water conservation, just before entering the locker room. The other half did not sign anything. During step 2 (mindfulness), half the participants filled in a questionnaire, supposedly about setting up a program against water waste. Actually, the questionnaire was used to make participants aware of their numerous wastes of water (i.e., recall of previous transgressions). The other half did not fill in this questionnaire (i.e., no recall of previous transgressions). Shortly afterwards, another experimenter measured the time spent under the shower by each participant. The results confirmed the hypotheses. Participants in the hypocrisy condition (steps 1 + 2), took shorter showers (221 s on average), as a way of reducing their dissonance, than participants without recall of their transgressions (302 s on average). They also closed the tap more often during their shower (14 times on average) than in the control condition (7 times on average). Overall, hypocrisy can be considered an efficient technique for promoting pro-environmental and other pro-social behaviors (e.g., Fointiat et al. 2013b; Lopez et al. 2011; Stone and Fernandez 2008; Stone and Focella 2011).

Another paradigm deals with self-fulfilling prophecies. Asking someone if he/she will perform a behavior or not (i.e., "*Ask yourself*: *Will I recycle?*") increases the probability of it happening. Predicting a realization emphasizes the discrepancy between the normative beliefs linked to the behavior (e.g., *I should recycle*) and the behavioral transgressions (e.g., *I don't recycle*). By predicting his/her behavior, an individual becomes aware, on the one hand, of what he/she does or does not do (previous behavior) and, on the other hand, of what he/she thinks he/she should do, that is, what is socially approved and morally good to do. Thus, the discrepancy between these two elements (i.e., past behavior and normative beliefs) is a source of dissonance (Spangenberg et al. 2003, 2012). The motivation to reduce the dissonance leads the individual to perform a behavior consistent with his/her prediction. For instance, Sprott et al. (1999) asked students to make a prediction about their future recycling behavior. As expected, those who made the prediction recycled more than those in a control group. Studies in this domain (Spangenberg et al. 2003) showed that a prediction request presented as a message for the general public, and consequently with no direct contact with anybody, increased the recycling rate by 12 %. In the same perspective, Spangenberg et al. (2003) reported several studies showing that self-fulfilling prophecies produced an effect on recycling behavior even in a situation of mass communication. One of these involved students of Washington State University and was divided into three phases: (1) a pre-experimental phase: before the campaign, (2) a campaign phase, during which an electronic board presenting the prediction request was installed, and (3) a post-experimental phase: after the campaign. The recycling behavior for each of these three phases was measured using containers. As expected, the recycling rate was higher during (27.6 %) and after (28.2 %) the prediction campaign than before (15.8 %). Sprott et al. (2003) manipulated the realization of a prediction (control vs. prediction) and the strength of normative beliefs (strong vs. weak). Their results confirmed that the effect of prophecies was greater when normative beliefs associated with the predicted behavior were strong. In another experiment, Rodrigues and Girandola (2014a, b) asked participants to predict their recycling behavior and then their normative beliefs and recycling habits were measured. Results showed that participants felt dissonance if, and only if, their recycling habit was weak and their normative beliefs were strong. After 30 years of research, studies show that self-fulfilling prophecies produce behavioral effects (Spangenberg and Greenwald 1999, 2001) with a moderate size effect (i.e., Cohen's $d = 0.39$ or $r = 0.19$).

11.2.4 Toward a Binding Communication

These results are very consistent. If, under certain conditions, persuasion influences attitudes, free will compliance procedures make it possible to obtain lasting behavioral effects in a context of freedom and favor commitment to a series of actions. However, despite their effectiveness for behavioral change, it should be noted that the great majority of these studies focused on the private sphere and at an individual level. Moreover, the dominant approaches of policy makers in this field are based on the idea that environmental issues can be solved by marginal changes in our way of life, an idea which is also widely held by the public. However, we agree with the following assessment: "*The cumulative impact of large numbers of individuals making marginal improvements in their environmental impact will be a marginal collective improvement in environmental impact*" (Thogersen and Crampton 2009). To go further and obtain deeper transformations of the physical and spatial environment, and thus favor quality of life more effectively, we consider the binding communication paradigm a promising approach. Lokhorst et al. (2013) showed that commitment was more effective when it was combined with other interventions. In this line, the binding communication paradigm suggests bringing together the fields of persuasion and commitment. It is presented in the following section along with empirical illustrations at different levels of spatiality.

The binding communication paradigm offers the possibility of working at the intersection of research conducted in the field of persuasive communication and those of commitment and free will compliance. The central assumption is that communication will be more effective when the exposure to a persuasive message is preceded by a preparatory act in a context of freedom (Girandola and Joule 2012). Thus, in the binding communication paradigm, as in the framework of classic persuasion (cf. Girandola 2003; Girandola and Joule 2012), pertinent questions remain: "*What type of information should be conveyed?*", "*What are the best arguments to offer?*", "*What are the most appropriate channels of information?*", to which we could add another important one: "*Which preparatory actions must I obtain from those I want to rally?*". A binding communication approach is distinguished from a more "traditional" one by taking into account this last question, and by conferring the status of "actor" on the target rather than that of a mere passive receiver (Girandola and Joule 2012, 2013; Joule et al. 2007; Michelik et al. 2012; Demarque et al. 2013a).

The interest of binding communication in the environmental field can be illustrated by a study that aimed to measure the effects of an electronic commitment on purchasing low-energy light bulbs by customers of a DIY store (Bernard et al. 2010). The commitment concerned installing a low-energy light bulb, instead of a classic one, at home. Overall, the results showed that customers who carried out the preparatory acts (i.e., clicks to replace a light bulb) bought significantly more low-energy light bulbs and expressed a more favorable attitude toward these bulbs 15 days later. Significantly more customers who signed an electronic commitment replaced a classic light bulb at home than those who did not sign the commitment. This experiment highlights the interest of leading individuals to preparatory actions in a binding communication situation. These make it easier to accept behaviors favoring energy conservation and, in addition, participate in improving household comfort.

This example shows the potential effectiveness of the binding communication paradigm. However, it is still situated at a strictly individual level and at the household level. In order to show how commitment and binding communication could improve quality of life in a significant way, the next section deals with the main research carried out with the binding communication paradigm at different levels of human-environment interactions.

11.3 Favoring Quality of Life Through Commitment and Binding Communication: Examples of Research at Three Different Space Levels

People-environment relations involve problems of different natures according to the type and extent of the space concerned. These relations are located at several scales of spatial reference. Moser (2003) distinguished four levels, involving distinctive physical and social aspects (cf. Table 11.1). Here, another research study situated at level I is presented, which is noteworthy for its originality and innovative approach to quality of life. Furthermore, action research at levels II and III is described to show that commitment and binding communication should not be conceived only as techniques aiming at the adoption of individual behavior that would be an end in itself, but also as means to encourage collective dynamics. The importance of this dimension in leading to significant changes in individuals' life environment, a synonym of quality of life improvement, is discussed.

11.3.1 At the Household Scale (Level I)

The limitations of a focus on the household level were mentioned above. Therefore, it could seem paradoxical to begin this third section with the description of a research project centered on this level. However, we highlight its novelty and ambition to create a user community.

The Sensomi project was conducted from April 2011 to June 2013 in the Provence-Alpes-Côte d'Azur (PACA) region in France. An ambitious part of the project was the development of a new type of device for obtaining commitment: an online cooperative video game, called Kwaan, specially created for the study (Demarque et al. 2013b). The first aim was to make individuals aware of the effects of their way of life on energy consumption, and then to foster effective energy-saving behaviors. Based on our understanding of quality of life, we began with the premise that it is necessary to recreate a positive bond between individuals and the physical reality of the terrestrial environment on which they depend, and for which energy is probably a more abstract issue. Thus, the project emphasized the idea of "hybridization" of reality and virtuality. In the framework of the commitment theory, an online game appeared a relevant way to favor commitment in a series of actions. In Kwaan, the players were collectively in charge of the life of an imaginary tree in an online community experience, based on an autonomous and collective learning process. Since their actual electricity consumption had an influence on the virtual tree, energy was no longer seen as an abstract aspect of reality. Moreover, the community dimension enabled

Table 11.1 Socio-spatial levels of analysis

	Physical aspect of the environment	Social aspect of the environment	Type of space and control
Level I	Micro-environment	Individual	Private spaces
	Private space/housing	Family	Extensive control
	Work space		
Level II	Proximal environments, neighborhood	Interindividual	Semi-public spaces
		Community	Mediated control
	Open private spaces	Users, customers	
Level III	Public environments	Inhabitants	Public spaces
	Cities, villages	Groups of individuals	Mediated control
Level IV	Global environment	Society	Country, nation, planet
		Population	Hypothetical control

From Moser 2003, p. 17

one player to perceive what the other players were doing over time, which should reduce the feeling of being alone to act and commit the player in relation to the other players. One hundred and forty-eight psychology students at Aix-Marseille University participated in the study. In short, the results partially confirmed our hypotheses. They showed, on one hand, a marginally significant effect of the game on meter-reading behaviors in the free choice situation and, on the other hand, a significant positive effect of the game on the evolution of considering future consequences for the players who carried out at least 1 m-reading behavior. However, these results should be considered with caution, considering the sample size and the weakness of the effects.

In their study, Fleury-Bahi et al. (2013) went beyond the private sphere and identified physical space as one of the dimensions of environmental quality of life. In the next section, we show the effectiveness of commitment and binding communication at higher socio-spatial levels on individual behaviors, which positively transform collective spaces and improve quality of life.

11.3.2 At the Proximal Environment Scale (Level II)

11.3.2.1 In Schools

In this section, a binding communication procedure conducted in 11 primary schools in the south of France, in the framework of the European project ALTENER (Joule 2004), is described. The aim of the project was to encourage 9- and 10-year-old children, as well as their families, to adopt pro-environmental behaviors. The action involved 700 families and 28 teachers. During the school year, the teachers had to inform and convince their pupils of the importance of environmental protection and energy conservation (e.g., lessons, activities in the classroom), and lead them to perform preparatory actions (e.g., taking notes about their family's habits that could be changed without a loss of comfort, filling in a questionnaire with their parents about energy conservation at home, etc.). Each child was thus encouraged by his/her teacher to make an oral and written commitment and change his/her habits. Parents were also invited to make a commitment (e.g., leaving the car at home for short distances, replacing ordinary light bulbs by low-energy ones, etc.).

The results showed that the majority of the children and parents (up to 100 % in some classes) committed themselves to perform actions such as decreasing their energy consumption. Numerous studies have shown that a written commitment (or public pledge) usually results in the acceptance and performance of costly behaviors (Girandola and Roussiau 2003; Katzev and Wang 1994; Pallack et al. 1980; Wang and Katzev 1990). In our case, this campaign also led to improvements in some schools, such as replacing ordinary bulbs by low-energy bulbs or installing recycling bins. Some pupils wrote to their mayor, asking for the installation of timers for the lights in the school halls.

11.3.2.2 Promoting Litter Recycling and Sorting at a Highway Rest Area

In this study, the authors used binding communication procedures at a highway rest area in the south of France in order to encourage users to sort their trash (Blanchard and Joule 2006). In agreement with the highway operating company, two decisions were made: (1) to eliminate all isolated classic bins, and (2) to reduce the overall number of places where trash could be thrown away. This last decision led users to perform a preparatory action: to move with their trash. Once they had made this effort, they found three containers for sorting and a classic bin. They thus had a choice to make: sorting, using the special containers, or not sorting (putting everything in the classic bin). Results showed that the amount of sorted trash was multiplied by 3.5, compared to a control situation, without any deterioration in the cleanliness of the site. However, sorting was of lower quality. In order to improve this, a message was put on the bins: "*I sort. For the planet, for my children and for my children's children*". This message enabled users to identify their sorting at a high level of identification (Vallacher and Wegner 1985). It produced an improvement in the quality of sorting, which was controlled during visual checks of the bins. One of the novelties of the research was the absence of any direct contact with the users.

11.3.3 At the Public Space Scale (Level III)

11.3.3.1 Protecting Beaches

Joule et al. (2007a) tested the effectiveness of two communication procedures to lead beachgoers to keep beaches clean, notably by encouraging them not to throw their cigarette butts around or bury them in the sand. The first procedure consisted of exposing beachgoers to classic communication media (posters and brochures). The second procedure used binding communication: beachgoers were exposed to the same communication media, and were free to choose the commitment they wanted to make from a list of a dozen possible commitments, (e.g., not throw their cigarette butts in the sand). The research was conducted on a beach in Marseilles, which was divided into three zones comparable in surface area and in visitors. In the first zone, beachgoers were exposed to a classic communication procedure (i.e., persuasion). In the second zone, they were exposed to a binding communication procedure, and in the third zone, they were not exposed to any communication procedure (i.e., control zone). As expected, the observers detected, on a daily basis, 90 cigarette butts on average in the sand in the binding communication zone, 176 butts in the classic communication zone and 162 butts in the control zone. The beachgoers in the binding communication zone were observed without their knowledge, before and after their commitment: 77 % put their cigarette butts in the sand before the intervention; 40 % did so after the procedure.

11.3.3.2 Mediterranean Coastline Conservation

This action research was conducted on the Mediterranean coast along the Côte d'Azur (Joule et al. 2006). It concerned boaters (amateur sailors and professionals) and was carried out with an environmentalist association and about 50 "sea ambassadors" who met with boaters during the summer. The aim was to encourage amateur sailors to improve their behaviors and knowledge of sea conservation. The boaters were invited to perform freely different preparatory actions (accept a short interview about sea conservation, give their opinion about the most relevant advice to give to sailors, take a free booklet containing information and advice about sea conservation). Next, the ambassadors gave the sailors a commitment form that included a list of behaviors. They could freely commit to adopt one or several of these behaviors (e.g., do not anchor in the Posidonia meadow, use natural soaps). Finally, they could fly the association campaign flag on their boat, to serve as an example of environmental conservation: this action strengthened commitment because of its public character. More than 3000 boaters were solicited and almost all agreed to the interview, and then committed themselves to modify one or more behaviors by signing the commitment form. Based on questionnaires and direct observations, the results showed that the committed sailors had better knowledge and adopted more sea-friendly behaviors, in comparison with the non-committed ones.

11.3.3.3 At a Town Scale

Following a request from the environmental service of the PACA Region, a campaign was implemented to include two entire towns with all their inhabitants. It aimed to lead inhabitants to modify their behaviors regarding energy consumption (Joule 2004). Two towns of 8,000 inhabitants sharing common characteristics (i.e., size, social composition, climate conditions and environment) were selected in the south of France. Each of these two towns (A and B) used a different communication campaign.

In town A, a classic communication campaign was used based on the usual media diffusion (posters, quiz, brochure with advice about energy conservation and a media plan of press articles in regional newspapers and municipal bulletins).

Binding communication was used in town B. This campaign used the same tools as in town A, as well as some independent actors (i.e., researchers trained in the commitment theory) and people living in the city who served as go-betweens (e.g., councilors, teachers, organizers, storekeepers, etc.) who encouraged actions favoring energy conservation. The actions were made public during an event day (exhibitions, plays, movies, debates, etc.) in order to get the inhabitants to make concrete commitments. They received a commitment form with a list of actions (e.g., walking rather than using the car for short trips, buying low-energy light bulbs). The inhabitants freely chose one or several actions to perform, and then signed the form. Each commitment was symbolized by a paper sun, cut out by school children and attached to a large net set up in the Town Hall square. Consequently, each inhabitant was able to follow the progression of the number of commitments. The results showed the effectiveness of binding communication: it produced a greater impact than the classic communication campaign. Average annual consumption per household increased less in comparison with the previous year in the town exposed to binding communication than in the town exposed to a classic communication campaign (6 % versus 14 %). Overall, binding communication enabled the development of actions by the population and the public authorities. One of the interests of this project was the support of mediators in this procedure. The presence of the mediators probably maintained the innovative dynamics and the collective ambition developed during the project, which was aimed at improving the quality of life in an entire town.

11.4 Perspectives and Conclusion

The research described in this chapter outlines the different facets of the applications of the commitment theory in pro-environmental behaviors. At a time when pro-environmental communication campaigns are still searching for the right words, the commitment theory enables not only attitudinal changes but also behavioral ones. We have shown that this behavioral change can be on three different levels of space (private spaces, proximal environments and public spaces) and thus can potentially have a significant impact on the quality of life of the communities concerned. Although most of the studies focus on individual changes, we believe that this capacity to create a collective dynamic constitutes the future of research on commitment and binding communication.

In this perspective, the works about the link between social representations and commitment constitute a promising pathway. Eyssartier et al. (2009) suggested optimizing the effects of commitment by taking into account the social representations of a given object. A social representation is "*a form of knowledge, socially produced and with a practical function, namely to contribute to the construction of a reality shared by a social group or entity*" (Jodelet 1989, p. 36). From a structural point of view, social representations are regulated by central and peripheral systems (Abric 2001). The central system is fundamental to the social representation as it contains the more stable elements over time. It determines the meaning and the organization of the social representation. The peripheral system allows an adjustment to specific contexts and integration of the modulations. In the context of energy conservation, Souchet and Girandola (2013) showed that the study of social representations helps identify the preparatory actions likely to lead to the expected behaviors. These authors asked participants, chosen at random, if they would agree to answer a few questions about energy conservation (first preparatory action) and to write a short text in favor of conservation (second preparatory action). Each preparatory action activated either central elements of the representation (i.e., *making energy savings, it's about conserving the environment*) or peripheral elements (i.e., *making energy savings, it's about using the car less*). They were then asked to keep a diary for 2 weeks, to write down all the actions they carried out in favor of energy conservation, and to send their diary, at their expense, to the psychology department (final request). As expected, significantly more participants whose central elements were activated (e.g., *environmental conservation*) sent their diaries than those exposed to peripheral elements (e.g., *using the car less*). Other studies in the framework of sustainable development showed that the activation of central elements led to stronger effects in a situation of binding communication than the activation of peripheral elements (Zbinden et al. 2011). In addition to its usefulness for constructing persuasive arguments and enabling more effectiveness in terms of behavior (Eyssartier et al. 2009), the work on social representations provides a better understanding of the dynamic exchanges among the target group and, potentially, its expectations in terms of relations to the environment. For projects focusing on level III, a more systematic study of town representations could also prove very useful. Therefore, we suggest more in-depth work at the interface of these two research fields applied to sustainable development.

In conclusion, do the apparent behavioral changes automatically mean a better quality of life? Research on commitment mentioned above does not provide precise information about this aspect, which is one of its main limitations. Future projects need to integrate a more systematic use of environmental quality of life scales after the commitment procedures. Different tools are available depending on the aims and the level of spatiality on which the research focuses. For instance, at level I, administering the Residential Environmental Satisfaction Scale (RESS, Adriaanse 2007) before and after a project could evaluate the effect of a commitment procedure on residential quality of life. Another interesting tool, the Environmental Satisfaction Scale (Pelletier et al. 1996), is composed of two subscales, one measuring satisfaction with government policies, and the other measuring satisfaction with local environmental conditions. The latter subscale could be used for level II studies. Finally, we have already referred to Fleury-Bahi et al.'s scale (2013), which measures the four dimensions mentioned above.

Despite the absence of measures in the studies cited, some clues indicate that commitment and its consequences create satisfaction with the environment and an improvement in quality of life. This is the case of the research conducted in an entire town (Level III) in which the inhabitants were approached to commit themselves to pro-environmental behaviors. This commitment led to a reduction in wasting energy in comparison with a town without commitment procedures. After these first commitments, the project partners and the go-betweens decided to continue the actions. For instance, the town council took the initiative to fund training for heating installers. It also organized an Energy Trophy to reward the best initiatives for energy conservation. Therefore, even in the absence of precise measures, all these actions suggest that the commitment dynamic persists and constitutes a body of evidence in favor of improving environmental satisfaction and quality of life (cf. Uzzell and Moser 2006). Furthermore, we assume that individuals identify their action at a high level (e.g., "*I'm protecting the planet*"), which is meaningful and easy to internalize (e.g., "*It's natural for me to act like this*"). Thus, individuals would finally be led to internalize and accept performing the action as a reflection of a personal value (Beauvois 2001). From this point of view, we can easily imagine a generalization of the initial action to other pro-environmental behaviors of the same kind: positive spillover effects take on great importance here.

This example of a town (level III) shows the value of involving go-betweens in the target population. This dimension seems very important to us and needs to be integrated more in future research. Effectively, most of the projects presented in this chapter originated from institutional demands. They took place in a global context where research is increasingly based on obtaining grants. In our examples, most of the research was funded by the PACA Region. In this framework, policy makers plan the expected behavioral changes beforehand and researchers have to develop experiments in order to achieve this goal. This raises the question of which entity defines what is "socially good" or what is a "socially acceptable" goal. In our view, there is a real necessity to co-develop or co-construct the definition of the intended goals with the populations involved beforehand. However, at present, research on commitment is generally based on approaches that we could qualify as top-down, as we try to answer an institutionally-defined goal (for instance, that of the European Union in the case of ALTENER or the PACA Region for the Sensomi project), usually made without prior consultation. In the future, it seems important to develop more participatory approaches when defining objectives, as is the case in other social science projects.

References

Abric, J.-C. (2001). L'approche structurale des représentations sociales: développements récents. *Psychologie et Société, 4*, 81–103.

Adriaanse, C. C. M. (2007). Measuring residential satisfaction: A residential environmental satisfaction scale (RESS). *Journal of Housing and the Built Environment, 22*, 287–304.

Ajzen, I. (1991). The theory of planned behavior. *Organizational Behavior and Human Decision Processes, 50*, 179–211.

Ajzen, I., & Fishbein, M. (1980). *Understanding attitudes and predicting social behavior*. Englewood Cliffs: Prentice-Hall.

Arbuthnot, J., Tedeschi, R., Wayner, M., Turner, J., Kressel, S., & Rush, R. (1976/1977). The induction of sustained recycling behavior through the foot-in-the-door technique. *Journal of Environmental Systems, 6*, 355–368.

Beauvois, J. L. (2001). Rationalization and internalization: The role of internal explanations in attitude change and the generalization of an obligation. *Swiss Journal of Psychology, 60*, 215–230.

Bernard, F., Halimi-Falkowicz, S., & Courbet, D. (2010). Expérimentation et communication environnementale: la communication engageante et instituante. In D. Courbet (Ed.), *Objectiver l'humain ? Communication et expérimentation* (Vol. 2, pp. 71–113). Paris: Lavoisier.

Bickman, L. (1972). Environmental attitudes and actions. *Journal of Social Psychology, 87*(2), 323–324.

Blanchard, G., & Joule, R. V. (2006). *La communication engageante au service du tri des déchets sur les aires d'autoroutes: une expérience-pilote dans le Sud de la France*. Actes du 2ème Colloque International Ecocitoyenneté : des idées aux actes favorables à l'environnement, Marseille, France, 9–10 November.

Burger, J. M. (1999). The foot-in-the-door compliance procedure: A multiple-process analysis and review. *Personality and Social Psychology Review, 3*(4), 303–325.

Cialdini, R. B. (2006). *Influence: The psychology of persuasion*. New York: HarperBusiness.

Demarque, C., Apostolidis, T., & Joule, R. V. (2013a). Consideration of future consequences and pro-environmental decision making in the context of persuasion and binding commitment. *Journal of Environmental Psychology, 36*, 214–220.

Demarque, C., Charles, M., Bourguignon, D., & Lo Monaco, G. (2013b, October). *Influence of a virtual environment on an energy conservation behaviour: the case of a cooperative online game*. Paper presented at the International Congress of Environmental Psychology, Barcelona, Spain.

Dickerson, C., Thibodeau, R., Aronson, E., & Miller, D. (1992). Using cognitive dissonance to encourage water conservation. *Journal of Applied Social Psychology, 22*, 841–854.

Dufourcq-Brana, M., Pascual, A., & Guéguen, N. (2006). Déclaration de liberté et pied-dans-la-porte. *Revue Internationale de Psychologie Sociale, 19*, 173–187.

Dwyer, W. O., Leeming, F. C., Cobern, M. K., Porter, B. E., & Jackson, J. M. (1993). Critical review of behavioral interventions to preserve the environment: Research since 1980. *Environment and Behavior, 25*, 275–321.

Eyssartier, C., Guimelli, C., & Joule, R. V. (2009). Représentations sociales et engagement. In P. Rateau & P. Moliner (Eds.), *Représentations sociales et processus cognitifs* (pp. 151–163). Rennes: PUR.

Festinger, L. (1957). *A theory of cognitive dissonance*. Stanford: Stanford University Press.

Fleury-Bahi, G., Marcouyeux, A., Préau, M., & Annabi-Attia, T. (2013). Development and validation of an environmental quality of life scale: Study of a French sample. *Social Indicators Research, 113*(3), 903–913.

Fointiat, V., Girandola, F., & Gosling, P. (2013a). *La dissonance cognitive. Quand les actes changent les idées*. Paris: Armand Colin.

Fointiat, V., Priolo, D., Saint-Bauzel, R., & Milhabet, I. (2013b). Justifier nos transgressions pour réduire notre hypocrisie? Hypocrisie induite et identification des transgressions. *Revue Internationale de Psychologie Sociale, 4*, 49–78.

Freedman, J. L., & Fraser, S. C. (1966). Compliance without pressure: The foot-in-the-door technique. *Journal of Personality and Social Psychology, 4*, 195–202.

Girandola, F. (2003). *Psychologie de l'engagement et de la persuasion*. Besançon: PUFC.

Girandola, F., & Joule, R. V. (2012). La communication engageante: aspects théoriques, résultats et perspectives. *L'Année Psychologique, 112*(1), 115–143.

Girandola, F., & Joule, R. V. (2013). Attitude, changement d'attitude et comportement. In L. Bègue & O. Desrichard (Eds.), *Traité de psychologie sociale. La science des interactions humaines* (pp. 221–248). Brussels: De Boeck.

Girandola, F., & Roussiau, N. (2003). L'engagement comme source de modifications à long terme. *Cahiers Internationaux de Psychologie Sociale, 57*, 83–101.

Guéguen, N., Joule, R. V., Halimi-Falkowicz, S., Pascual, A., Fischer-Lokou, J., & Dufourcq-Brana, M. (2013). I'm free but I'll comply with your request: Generalization and multidimensional effects of the "evoking freedom" technique. *Journal of Applied Social Psychology, 43*, 116–137.

Harris, M. B. (1972). The effects of performing one altruistic act on the likelihood of performing another. *Journal of Social Psychology, 88*, 65–73.

Jodelet, D. (1989). *Les représentations sociales*. Paris: Presses Universitaires de France.

Joule, R. V. (2004). *What is the role of energy-education in changing habits?* European congress "Energy education: roles, actors and tools. European experiences", Brussels Belgium.

Joule, R. V., & Beauvois, J.-L. (1998). *La soumission librement consentie*. Paris: Presses Universitaires de France.

Joule, R. V., & Beauvois, J.-L. (2014). *Petit traité de manipulation à l'usage des honnêtes gens* (3rd edition). Grenoble: Presses Universitaires de Grenoble.

Joule, R. V., Masclef, C., & Jarmasson, J. (2006). Communication engageante et préservation de la méditerranée: comment promouvoir de nouveaux comportements de la part des plaisanciers ? Actes du 2ème Colloque International Ecocitoyenneté: des idées aux actes favorables à l'environnement, Marseille, France, 9–10 November.

Joule, R. V., Bernard, F., Laganne, J., & Girandola, F. (2007a). *Promote cleanness of beaches: Communication and committing communication.* 10th European Congress of Psychology, 3–6th July, Prague, Czech Republic.

Joule, R. V., Girandola, F., & Bernard, F. (2007b). How can people be induced to willingly change their behavior? The path from persuasive communication to binding communication. *Social & Personality Psychology Compass, 1*, 493–505.

Kaiser, F. G. (2014). Using cutting-edge psychology to advance environmental conservation. *European Psychologist, 19*(2), 81–83.

Katzev, R., & Johnson, T. R. (1983). A social-psychological analysis of residential electricity consumption: The impact of minimal justification techniques. *Journal of Economic Psychology, 3*, 267–284.

Katzev, R., & Johnson, T. R. (1984). The use of posted feedback to promote recycling. *Psychological Reports, 71*, 259–264.

Katzev, R., & Pardini, A. (1987/1988). The comparative effectiveness of reward and commitment in motivating community recycling. *Journal of Environmental Systems, 17*, 93–113.

Katzev, R., & Wang, T. (1994). Can commitment change behavior? A case study of environmental actions. *Journal of Social Behavior and Personality, 9*, 13–26.

Kiesler, C. A. (1971). *The psychology of commitment*. New York: Academic.

Lewicka, M. (2011). Place attachment: How far have we come in the last 40 years? *Journal of Environmental Psychology, 31*(3), 207–230.

Lewin, K. (1947). Group decision and social change. In E. Swanson, T. M. Newcomb, & E. L. Hartley (Eds.), *Readings in social psychology* (pp. 197–211). New York: Holt.

Lokhorst, A. M., Werner, C., Staats, H., van Dijk, E., & Gale, J. L. (2013). Commitment and behavior change: A meta-analysis and critical review of commitment-making strategies in environmental research. *Environment and Behavior, 45*(1), 3–34.

Lopez, A., Lassarre, D., & Rateau, P. (2011). Dissonance et engagement. Comparaison de deux voies d'intervention visant à réduire les ressources énergétiques au sein d'une collectivité territoriale. *Pratiques Psychologiques, 17*, 263–284.

McKenzie-Mohr, D. (2011). *Fostering sustainable behavior: An introduction to community-based social marketing*. Gabriola Island: New Society Publishers.

McKenzie-Mohr, D., Lee, N. R., Schultz, P. W., & Kotler, P. (2012). *Social marketing to protect the environment: What works*. Washington, DC: Sage.

Michelik, F., Girandola, F., Joule, R. V., Zbinden, A., & Souchet, L. (2012). Effects of the binding communication paradigm on attitudes. *Swiss Journal of Psychology, 71*(4), 227–235.

Moser, G. (2003). Questionner, analyser et améliorer les relations à l'environnement. In G. Moser & K. Weiss (Eds.), *Espaces de vie: Aspects de la relation homme-environnement* (pp. 11–42). Paris: Armand Colin.

Pallak, M. S., Cook, D. A., & Sullivan, J. J. (1980). Commitment and energy conservation. In L. Bickman (Ed.), *Applied social psychology annual* (pp. 235–253). Beverly Hills: Sage.

Pardini, A., & Katzev, R. (1983/1984). The effect of strength of commitment on newspaper recycling. *Journal of Environmental Systems, 13*, 245–254.

Pelletier, L. G., Legault, L. R., & Tuson, K. M. (1996). The environmental satisfaction scale: A measure of satisfaction with local environmental conditions and government environmental policies. *Environment and Behavior, 28*, 5–26.

Pliner, P., Hart, H., Kohl, J., & Saari, D. (1974). Compliance without pressure: Some further data on the foot-in-the-door technique. *Journal of Experimental Social Psychology, 10*, 17–22.

Rodrigues, L., & Girandola, F. (2014a). *Auto-prophéties: une application en faveur du recyclage des déchets*. International Congress of Applied Psychology, Paris, France, 8–13 July.

Rodrigues, L., & Girandola, F. (2014b). *Auto-prophéties : effets du libre-choix et du degré de croyances normatives*. 10ème Colloque International de Psychologie Sociale en Langue Française, Institut de Psychologie, Paris, France, 27–30 August.

Schultz, P. W. (2014). Strategies for promoting proenvironmental behavior: Lots of tools but few instructions. *European Psychologist, 19*(2), 107–117.

Schultz, P. W., & Kaiser, F. G. (2012). Promoting pro-environmental behavior. In S. D. Clayton (Ed.), *The Oxford handbook of environmental and conservation psychology* (pp. 556–580). New York: Oxford University Press.

Souchet, L., & Girandola, F. (2013). Double foot-in-the-door, social representations and environment: Application for energy savings. *Journal of Applied Social Psychology, 43*(2), 306–315.

Spangenberg, E. R., & Greenwald, A. G. (1999). Social influence by requesting self prophecy. *Journal of Consumer Psychology, 8*(1), 61–89. doi:10.1207/s15327663jcp0801_03.

Spangenberg, E. R., & Greenwald, A. G. (2001). Self-prophecy as a behavior modification technique in the United States. In W. Wosinska, R. B. Cialdini, D. W. Barret, & J. Reykowski (Eds.), *The practice of social influence in multiple cultures* (pp. 51–62). Mahawah: Lawrence Erlbaum.

Spangenberg, E. R., Sprott, D. E., Grohmann, B., & Smith, R. J. (2003). Mass-communicated prediction requests: Practical application and cognitive dissonance explanation for self-prophecy. *Journal of Marketing, 67*, 47–62. doi:10.1509/jmkg.67.3.47.18659.

Spangenberg, E. R., Sprott, D. E., Knuff, D. C., Smith, R. J., Obermiller, C., & Greenwald, A. G. (2012). Process evidence for the question-behavior effect: Influencing socially normative behaviors. *Social Influence, 7*(3), 211–228.

Sprott, D. E., Spangenberg, E. R., & Perkins, A. W. (1999). Two more self-prophecy experiments. In L. Scott & E. J. Arnould (Eds.), *Advances in consumer research* (Vol. 25, pp. 621–626). Provo: Association for Consumer Research.

Sprott, D. E., Spangenberg, E. R., & Fisher, R. J. (2003). The importance of normative beliefs to the self-prophecy effect. *Journal of Applied Psychology, 88*, 423–431. doi:10.1037/0021-9010.88.3.423.

Staats, H. (2012). Restorative environments. In S. Clayton (Ed.), *The Oxford handbook of environmental and conservation psychology* (pp. 445–458). New York: Oxford University Press.

Steg, L., & de Groot, J. I. M. (2012). Environmental values. In S. Clayton (Ed.), *The Oxford handbook of environmental and conservation psychology* (pp. 81–92). New York: Oxford University Press.

Steg, L., Van den Berg, A. E., & De Groot, J. I. M. (2013). *Environmental psychology: An introduction*. Chichester: BPS Blackwell.

Stone, J., & Cooper, J. (2001). A self-standards model of cognitive dissonance. *Journal of Experimental Social Psychology, 37*, 228–243.

Stone, J., & Fernandez, N. C. (2008). To practice what we preach: The use of hypocrisy and cognitive dissonance to motivate behavior change. *Social and Personality Compass, 2*, 1024–1051.

Stone, J., & Focella, E. (2011). Hypocrisy, dissonance and the self-regulation processes that improve health. *Self and Identity, 10*, 295–303.

Thøgersen, J., & Crompton, T. (2009). Simple and painless? The limitations of spillover in environmental campaigning. *Journal of Consumer Policy, 32*, 141–163.

Uzzell, D., & Moser, G. (2006). Environment and quality of life. *European Review of Applied Psychology, 56*(1), 1–4.

Vallacher, R. R., & Wegner, D. M. (1985). *A theory of action identification*. Hillsdale: Lawrence Erlbaum Associates.

Wang, T. H., & Katzev, R. D. (1990). Group commitment and resource conservation: Two field experiments on promoting recycling. *Journal of Applied Social Psychology, 20*, 265–275.

Webb, T. L., & Sheeran, P. (2006). Do changing behavioral intentions engender behavior change? A meta-analysis of the experimental evidence. *Psychological Bulletin, 132*, 249–268.

Weiss, K., & Girandola, F. (2009). Qualité de l'environnement et comportements écocitoyens. In J. Lecomte (Ed.), *Introduction à la psychologie positive* (pp. 251–263). Paris: Dunod.

Weiss, K., & Girandola, F. (2010a). Les enjeux de la psychologie face au développement durable. In K. Weiss & F. Girandola (Eds.), *Psychologie et développement durable* (pp. 9–19). Paris: Editions InPress.

Weiss, K., & Girandola, F. (2010b). Pour une psychologie positive du développement durable: vers de nouveaux enjeux et développements. In K. Weiss & F. Girandola (Eds.), *Psychologie et développement durable* (pp. 277–282). Paris: Editions InPress.

Werner, C. M., Turner, J., Shipman, K., Twitchell, F. S., Dickson, B. D., Bruchke, G. V., & von Bismarck, W. B. (1995). Commitment, behavior, and attitude change: An analysis of voluntary recycling. *Journal of Environmental Psychology, 15*, 197–208.

Zbinden, A., Souchet, L., Girandola, F., & Bourg, G. (2011). Communication engageante et représentations sociales: une application en faveur de la protection de l'environnement et du recyclage. *Pratiques Psychologiques, 17*, 285–299.

Pro-environmentalism, Identity Dynamics and Environmental Quality of Life

12

Marie-Line Félonneau and Elsa Causse

The first thing that strikes one when examining the literature on this topic is the impressive number of publications. The second noticeable feature is how recent these studies are, having seen a considerable renewal during the last decade.

Our basic postulate is the existence of a strong link between identity dynamics, perception of quality of life (QOL) and the adoption of pro-environmental behaviors (PEBs). A range of evidence, by no means exhaustive, is presented.

Firstly, the classic socio-psychological determinants of PEBs are reviewed, ultimately showing the importance of normative and ideological contexts. Then the theme of self and place, which defines both "Who am I?" and "Where am I from?", is introduced, revealing the importance of the question of self in general environmental issues and particularly in people-place bonds. In this sense, following the works that will be mentioned further, we believe that there is a kind of psychological interlinking between identity, the perception of QOL and environmental choices.

The extent to which environmental choices can contribute to the definition of self and participate in QOL is then described. Finally, two examples of current studies are provided to illustrate the theoretical approach developed in this chapter.

12.1 What Predisposes an Individual to Be Environmentally Friendly?

12.1.1 Classic Variables But These Are Only Moderately Predictive

The urgency and severity of environmental problems require an understanding of the genesis of pro-environmental behavior (PEB) and environmental concern. What are the factors that trigger environmental concern and PEB or the lack of them? Is there an idiosyncratic and dispositional propensity to protect the environment or do contextual or situational factors determine behavior? What is certain is that the complexity of this issue has been underestimated and assumed predictors often wrongly isolated. For over 25 years, researchers have accumulated analyses and meta-analyses (Bamberg and Möser 2007; Hines et al. 1986). Recently, in 2014, Gifford and Nilsson gave a particularly impressive range of the determinants of PEB in which they included no less than 18 variables.

They began by distinguishing personal and social factors. Personal factors included childhood

M.-L. Félonneau (✉)
Laboratoire de Psychologie (EA 4139), Université de Bordeaux, Bordeaux, France
e-mail: marie-line.felonneau@u-bordeaux.fr

E. Causse
CHROME (EA 7352) Université de Nîmes, Nîmes, France
e-mail: elsa.causse@unimes.fr

experience, knowledge and education, personality and self-construal, sense of control, values, political and world views, goals, feelings of responsibility, cognitive biases, place attachment, age, gender and chosen activities. Social factors included religion, urban-rural differences, norms, social class, proximity to problematic environmental sites, and cultural and ethnic variations.

In this chapter, while not excluding the role of personality variables, more emphasis is placed on positional variables, considering that age and gender cannot be reduced to the status of sociodemographic variables but also express a sense of social belonging. The variables that have been most commonly examined are gender, age, education and income (Saphores et al. 2006).

However, between 1980 and 1990, no studies revealed any clear correlation between PEB and this type of variable (Neiman and Loveridge 1981; Dunlap and Van Liere 1978). Although some results have been published, many contradictions continue to characterize correlational studies focusing on PEB. For Oskamp and colleagues (1991), demographic variables do not really correspond to determinants of PEB. These determinants are complex and cannot be reduced to a linear causality (Gifford and Nilsson 2014).

However, the impact of gender and age on environmental concerns needs to be included. Age is a psychosocial variable. It refers both to the developmental stage of an individual and to his/her group of generational affiliation.

An age effect or a cohort effect? Sociologists have shown that the socio-historical period in which an individual lived his/her youth has a considerable imprint on his/her thought patterns. Individuals in Western societies have gone from the need to ensure a decent life (materialism) to the need for personal achievement (postmaterialism). Elderly individuals, however, can be more ecocentric and therefore more likely to adopt PEB than younger people (Casey and Scott 2006; Jones and Dunlap 1992; Pinto et al. 2011). This result agrees with that of Saphores and colleagues (2006).

A French study (Dupré 2009) showed, for example, that younger people sort much less systematically than their elders, whether it be glass (64 % compared to 89 % of people in their 70s), batteries (53 % compared to 80 %), packaging (54 % compared to 86 %) or newsprint (49 % compared to 84 %). However, it should be remembered that many young people still live with their parents and therefore probably feel that they are less "responsible" for these daily tasks.

Nevertheless, although young people adopt fewer PEBs, they often say that they are more concerned with environmental outcomes than older people (Klineberg et al. 1998), which at least shows that environmentalism is tending to progress but that a deep environmental concern does not necessarily lead to PEB. In contrast, other studies have found no association (Meneses and Palacio 2005; Werner and Makela 1998).

For a long time, empirical evidence of environmental concern among women, supposedly because of their social and domestic roles (care), proved inconsistent. The relationship between gender and environmental concern is uncertain (Davidson and Freudenburg 1996) and, to date, there has been little empirical evidence to suggest that environmental concern is more consistently widespread among women than men. In fact, many studies have shown no correlation between gender and pro-environmental behavior (Gamba and Oskamp 1994; Hines et al. 1986; Werner and Makela 1998). Yet, regardless of environmental concerns, research has shown that women adopt concrete PEB more than men, especially in relation to their homes and in the private sphere (Arcury et al. 1978; Dietz et al. 2002; McStay and Dunlap 1983; Meneses and Palacio 2005).

On the other hand, they are less active than men in their defense of the environment (Mohai 1992), especially in public (McStay and Dunlap 1983). Nevertheless, it should be noted that Saphores and colleagues (2006) found that women are more willing to recycle electronic waste and take it to recycling centers.

Could environmentalism become a female domain as Hau and Swenson (2013) suggest? According to 82 % of Americans, the green movement is feminine (Bennett and Williams 2011). These authors even suggest that we should "*defeminize environmentalism, eliminating the association between women and environmentally friendly behaviours will help make sustainability more appealing to men*" (Hau and Swenson 2013, p. 5).

The link between environmentalism and female gender is also related to the stereotype of the altruistic, others-oriented woman. With regard to the environment, women express more environmental concern while men are more knowledgeable (Gifford and Nilsson 2014). If sustainability is feminine, energy is masculine, so men are seen as energy deciders and women as environmentally responsible (Hau and Swenson 2013).

Thus, the determinants of PEB cannot be reduced to socio-demographic variables but are part of broader societal issues including those illustrated by norms and values.

12.1.2 The Issue of Social Norms and Values

The issue of "environmental concern" which emerged in the 1970s attempted to identify the existence of a more or less shared "ecological vision of the world" (Dunlap et al. 2000). Tension between the individual and the environment is illustrated by the concepts of "social dilemma" (Hardin 1968) and "ecological dilemma" (Kortenkamp and Moore 2001). Can engagement in PEB be explained by adherence to egoistic or altruistic values, for example? Based on the works of Schwartz (1970, 1992) and Rokeach (1973), research showed that core values could explain the beliefs of individuals with regard to the environment and their desire to act in accordance with them (Stern et al. 1995, 1999).

The normative influence on ecological behaviors has also been the subject of numerous studies. Norm-oriented approaches are now well documented, especially since the development of the norm activation theory (Schwartz 1977) and the theory of normative conduct (Cialdini and Reno 1990; De Groot and Steg 2008). Cialdini and Reno (1990) highlight the importance of a normative approach to PEB, in their case, reducing littering in public places. They discriminate between injunctive and descriptive norms. The former are what most others approve or disapprove, the latter describe what is typical or normal i.e. it is what most people do. Many studies show that knowing that others are acting pro-environmentally inclines people to do the same. Thus, descriptive messages (descriptive normative messages) lead people to adopt PEB, for example, to reuse towels in hotels (Shultz et al. 2008). Norm-focus studies (Cialdini and Reno 1990; Reno et al. 1993) demonstrate this effect of norms on ecological behaviors. So, we can agree with Kaiser (1998) that *"especially in the ecological domain, measures are affected by social pressure, moral norms, and, therefore, social desirability"* (Kaiser 1998, p. 400).

Most studies show the normative nature of pro-environmental attitudes and behaviors but the question remains: do people really adhere to the norm or do they just make that claim? In other words, is it a question of normative adherence or normative clear-sightedness? The concept of normative clear-sightedness has been defined as the knowledge (or lack of knowledge) of the socially valorized nature of particular behaviors or assessments (Py and Somat 1991). Generally, normative clear-sightedness does not require normative adherence (Py and Ginet 1999). Félonneau and Becker (2008) sought to demonstrate the existence of normative clear-sightedness regarding the pro-environmental norm. Using the self-presentation paradigm (Jellison and Green 1981), they showed that a majority of the participants demonstrated their full knowledge of the social valorization of ecological attitudes and behaviors. Clearly, individuals showed their ability to manipulate their self-presentation by giving more or less socially valorized answers in order to provide a positive or negative image of themselves. As expected, the two dimensions studied, normative adherence and normative clear-sightedness regarding the pro-environmental norm, appeared independent (behaviors) or weakly related (attitudes).

Beyond the norms, there is also a more abstract influence of values on adopting PEB. It has often been assumed that pro-environmental attitudes and behaviors are related to people's values (Poortinga et al. 2004; Stern 2000). Thus, attitudes and behaviors are, in general, at least partially value-based. Values are viewed as underlying determinants of attitudes, behaviors and beliefs, and as important life goals that serve as guiding principles in life (Rokeach 1973).

Over the past 15 years, a great many studies have explored the link between values and environmental attitudes and behavior, mostly using Schwartz's model of human values (Corraliza and Berenguer 2000; Karp 1996; Schwartz 1992, 1994; Schultz 2001; Schultz et al. 2005). Thus, it seems that self-transcendence values are positively linked to PEB (mostly self-reported), behavioral intentions and attitudes, while self-enhancement values are negatively linked to these variables (Karp 1996; Nordlund and Garvill 2002; Schultz and Zelezny 1999; Stern et al. 1999; Stern et al. 1995). However, the association between values and pro-environmental attitudes and behaviors is somewhat more complex.

Although there is convincing evidence that pro-environmental attitudes are value-based (Poortinga et al. 2004; Schultz and Zelezny 1999; Stern et al. 1995), research seems to demonstrate that these attitudes can be based on different, sometimes contradictory, value orientations (Schultz 2000, 2001; Schultz et al. 2005). For example, "pro-social" and "pr- self" individuals can make similar choices but with different motives (Gifford and Nilsson 2014). Similar, although less clear, observations have been made regarding behavior. It can be concluded that individual values may be related to pro-environmental behavior. One may be inclined to behave pro-environmentally for self-centered reasons (Gatersleben et al. 2012).

To investigate further the apparently rather complex association between values and pro-environmental attitudes and behaviors, Félonneau and Becker (2008) used a socio-normative framework assuming that pro-environmental attitudes and behavior – as dominant norms in certain social contexts – should be referred to mechanisms of social pressure, and their effects upon cognitive processes. Their results suggest the following pattern: when pro-environmental attitudes and behaviors are the expression of normative adherence, they are based on self-transcendence values (mainly universalism), and not on self-enhancement ones. On the other hand, when pro-environmental attitudes and behaviors are the expression of normative clear-sightedness, they are based on self-enhancement values, and not on self-transcendence ones. Normative adherence or normative clear-sightedness may be expressed through the same or equivalent pro-environmental attitudes and behaviors, but are based on different values.

Today, two opposing normative cultures, hyperconsumption versus pro-environmental culture, seem to continue to coexist. Gatersleben et al. (2012) showed with a factor analysis that to be consumerist (fashion-conscious, reckless, self-indulgent, compulsive, and not cautious) is opposed to being "sensible consumers" (health-conscious, green, fitness-conscious, and ethical). However, perhaps environmentally friendly consumption has become the new norm or, at a deeper level, the new value (Becker and Félonneau 2011; Félonneau and Becker 2008). Gifford and Nilsson (2014) present several international studies showing that environmental concern is increasing around the world.

Depending on whether people are in developing countries or wealthier ones, environmental concern can differ. Nevertheless, *"the issue may be how the structure of attitudes differs from society to society rather than differences in level of concern."* (Gifford and Nilsson 2014, p. 10).

12.2 Identity Dynamics Related to Pro-environmentalism: What Are the Most Operative Identities in the Adoption of PEB?

12.2.1 Self and Place: Place Identity and Place Attachment

For many decades, the literature on environmental psychology has considered the place where one lives a fundamental dimension of identity. Who am I? Where am I from?

The concept of identity remains a central theme in the social and human sciences, probably because the identity of the post-modern individual is not clearly defined, or is even threatened (Ehrenberg 1995). Environmental psychology has focused on the spatial component of the construction of self, i.e. how belonging to a territory or region provides an identity. For over 20 years now, environmental psychology has offered a clear questioning of the spatialized self, the general idea being that the definition of self is partially derived from places (Korpela et al. 2009; Stedman 2003, 2006).

Concepts in this area have multiplied and it is difficult to draw the boundaries and overlaps between them: place identity, sense of community, place attachment, and place identification.

The first researchers who emphasized the importance of place in self-construction were Proshansky, Fabian and Kaminoff in 1983. The place where a person lives becomes a symbolic extension of the self. Proshansky and colleagues (1983) chose the term "place identity" to describe the socialization of the self in relation to the physical world. This was conceived as a sub-structure of an individual's self-identity, including cognitions representing *"memories, ideas, feelings, attitudes, values, preferences, meaning and conceptions of behaviour and experiences which relate to the variety and complexity of physical settings that define the day-to-day existence of every human being"* (Proshansky et al. 1983, p. 59). Thus, place identity means those dimensions of self that define the individual's personal identity *"in relation to the physical environment by means of a complex pattern of conscious and unconscious ideas, beliefs, preferences, feelings, values, goals, and behavioural tendencies and skills relevant to this environment"* (Proshansky 1978, p. 155). It should be remembered that place identity is viewed as an aspect of an individual's identity, comparable to other identities such as gender identity, political identity, or ethnic identity (Lalli 1992).

Perhaps because of an overall trend towards globalization, the interest in the link between people and place remains unabated (see the recent book by Manzo and Devine-Wright 2013). Faced with the confusion and lack of agreement due to the numerous concepts, often extremely close but related to several kinds of connections between people and place, some authors have attempted to clarify the theoretical basis of the spatialization of identity. As an example, the complex constructs of place identity and place attachment are often mingled (Félonneau 2004; Droseltis and Vignoles 2010). Hernandez et al. (2007) distinguish them very clearly: *"place-attachment is an affective bond that people establish with specific areas they prefer to remain and where they feel comfortable and safe. Place-identity, however has been defined as a component of personal identity, a process, which, through interaction with places, people describe themselves in terms of belonging to a specific place"* (p. 312). Place identity and place attachment do not necessarily coexist for the same person and the same place (Hernandez et al. 2007).

In the past 5 years, calls for conceptual integration in this area have been renewed. For example, two articles in the same issue of the Journal of Environmental Psychology (2010) claimed the need for an integrated and deeper conceptual approach in this area. Droseltis and Vignoles (2010) identified four theoretical dimensions of place identification in the literature: self-extension (*"the cognitive sense of a place as being part of the extended self-concept"*, p. 24), environmental fit (*"the ecological/environmental sense of the self as fitting into, or belonging to a place"*, p. 24), place-self congruity (the coincidence between self-image and place-image), and emotional attachment (strong emotional bonds or positive affect towards places). However, empirically, the authors found that self-extension could not be separated from emotional attachment, suggesting that the two constructs are indistinguishable in people's minds (Droseltis and Vignoles 2010; see also Félonneau 2004).

The studies of Breakwell (1992), and later those of his successors (Twigger-Ross and Uzzell 1996; Twigger-Ross et al. 2003; Vignoles et al. 2000) postulated four processes related to place identity. Firstly, place-related distinctiveness; secondly, place-referent continuity and place-congruent continuity (an illustration of the relationship between self-identity and the perception of climate can be found in the paper of

Knez (2005)); thirdly, place-related self-esteem; fourthly, place-related self-efficacy.

To incorporate the question of processes further, Scannell and Gifford (2010a, b) aimed to organize the various definitions of person-place bonding by presenting a "person-process-place" model: a tripartite organizing framework postulating that place attachment is a multidimensional concept with person, psychological processes and place dimensions. Unlike Droseltis and Vignoles (2010), they chose the angle of place attachment to account for the link to environment. Place attachment is defined as *"a bond between an individual or group and a place that can vary in terms of spatial level, degree of specificity, and social or physical features of the place, and is manifested through affective, cognitive, and behavioural psychological processes"* (Scannell and Gifford 2010, p. 5).

Even though in the last 40 years, as Lewicka says (2011), the person component has been studied more than the place and process components, researchers who associate paradigms from environmental psychology and social psychology insist that spatial bonds symbolize social bonds and that a place represents a group (Félonneau et al. 2013; Lalli 1992; Scannell and Gifford 2010). Hidalgo and Hernandez (2001) even assert that social attachments are stronger than setting attachments.

12.2.2 Self, Place and Environment

Beyond place attachment and place identity, environmental bases of identity can be observe (Manzo 2005). To what extent is the environment part of the self of an individual? In general, does my relationship to nature define my identity?

Some eco-psychologists use the concept of connectedness with nature to measure the feeling of being emotionally connected to the natural world. Schultz (2002, p. 67) defines it as *"the extent to which an individual includes nature within his/her cognitive representation of self"*. The Inclusion of Nature in the Self scale (INS) (Schultz 2001) is used to operationalize this construct.

Mayer and Frantz (2004), in turn, made this measure more complex by introducing a Connectedness to Nature Scale (CNS), which includes affective, experiential connection to nature (see also Olivos et al. 2011). It is not only a relatively abstract feeling of belonging to nature as a whole but also an identification with some ecological behaviors. For instance, the extent to which you see yourself as an environmentally friendly person is important to understand attitudes and behaviors (Gatersleben et al. 2002). Thus, a true environmental identity can be observed for those for whom environmentalism is a central part of who they are (Korpela 1989). Several notions appear in the literature to explain pro-environmentalism: a "green" identity (Whitmarsh and O'Neill 2010), an energy-saving identity, an environmental self-identity (Van der Werff et al. 2013), a self-identity as a recycler (Mannetti et al. 2004; Terry et al. 1999). Conversely, a consumer identity is negatively correlated with environmental concerns. Studies based on the Materialistic Values Scale (MVS) of Richins (2004) clearly showed that this value is incompatible with pro-environmentalism (Gatersleben et al. 2010).

12.2.3 Self, Place and PEB

The issue of interest here is the potential relationship between identity, at whatever level of definition (from the most general to the most environmental), and the propensity to adopt PEB (sustainability). However, too simplistic an explanation should be avoided as asserted by Devine-Wright and Clayton (2010) in the introduction to the special issue of the Journal of Environmental Psychology, entitled "Place, identity and environmental behavior": *"calling for adaption of a single, shared paradigm in researching behavioural aspects of self-environment relations is probably both unrealistic and unwise"* (2010, p. 269). The relationship between identity and pro-environmental behaviors is not linear. It could even be asserted that *"identity can be examined as both a dependent variable and an independent variable, both an effect and a cause"* (Devine-Wright and Clayton 2010, p. 267).

At the general level of social identity, the relationship between identity and sustainability constituted the central axis of the City-Identity-Sustainability Research Network (Pol 2002; Pol and Castrechini 2002). This CIS model postulates explicitly that sustainability is not possible without social identity, even if this link can vary depending on the local contexts and their characteristics. Several studies conducted on very diverse populations showed that people who have a strong attachment to a place want to protect it (Raymond et al. 2011; Scannell and Gifford 2010). This process is seen again in NIMBY-ism [not in my back yard] (Devine-Wright 2009) which is strongly related to place attachment and place identity. Clearly, the more we identify with a place, the more attached we are to it, and the more motivated we are to defend its integrity. Any change is then seen as a threat to place-related distinctiveness, continuity and self-efficacy. Thus, spatialized identity could explain many place-protective behaviors.

Nevertheless, place attachment or place identity can have a more direct effect on the adoption of concrete PEB. The studies based on the Theory of Planned Behavior, in particular those enriched with the norms focus theory (Terry et al. 1999), support the idea that the adoption of new social norms is determined by the feeling of group-belonging that generates them (Nigbur et al. 2010). The influence of these norms, which characterize self-identity, can be observed as the intention to adopt PEB. It has been demonstrated that there is an association between self-identity as a recycler and intention to participate in recycling (Mannetti et al. 2004; Terry et al. 1999).

The strength with which we identify with pro-environmental norms does not only influence our intention but also our effective PEB. Bissonnette and Contento (2001) showed that *"the identifiers"* transform the pro-environmental norms into personal norms, which gives meaning to the adoption of PEB. According to Nigbur and colleagues (2010), this explains the lay observations of the *"osmotic effects of neighbourhood behaviour"* (p. 281).

For Gatersleben and colleagues (2012), there is significant evidence that identities play a role in explaining and predicting PEB. There are many well-documented studies to show that when an individual considers that environmentalism is a central part of him/herself, he/she increases his/her engagement in pro-environmental acts (Ramkissoon 2012).

12.2.4 A Question of Levels

Several studies have shown that the links between people and place are widely determined by the level at which the individual is placed. When we say self in a place, or self in an environment, who is the self we are referring to and what is the environment we are speaking about? It seems that the question of "where" is as essential as the question of "who" (Fornara et al. 2011).

Each environmental scale, as defined by Moser (2003), could correspond to a self. At the first level of the micro-environment, the most intimate self is found: the "residential self". At the second level of the meso-environment, it is the "neighborhood self". At the third level, it is the "citizen self". At the fourth level, it is the "earthling" self. Progressively, moving from one spatial scale to another, i.e. from one level of categorization to another, from the subordinate to the supra-ordinate level as described by the theory of auto-categorization of Turner and colleagues (1987), the relationship between people and place is transformed. In many cases, individuals have a propensity to want sustainability within the places that they appropriate (CIS) and feel less concerned by the global level. The "residential self" and the "neighborhood self" are more salient than the "citizen self", which is itself less salient than the "earthling self". The theory of auto-categorization shows a functional antagonism between the different levels of categorization. In other words, the activation of a level deactivates the other levels. Thus, a separation can be observed between the perception of the local and the perception of the global (Moser 2003).

The influence of norms on PEB has been widely studied but the level of definition of the norm is essential. Fornara and colleagues

(2011) therefore introduced the construct of "local social norms" to identify the normative influence derived from people sharing the same spatial-physical setting. Here, localized social influence is distinguished from a general social influence: *"The reference group is rooted in a place"* (Fornara et al. 2011, p. 625). Based on the distinction between injunctive and descriptive norms of Cialdini and colleagues (1991), the authors imagined four distinct dimensions (i.e. four kinds of norms) by crossing subjective versus local and injunctive versus descriptive norms. Quoting Goldstein and colleagues (2008), they recall that the effects of social norms refer systematically to the "who" (those who act), forgetting the "where" (in which spatial-physical setting the behavior is performed). It seems that local descriptive norms are a strong predictor of PEB. This is the case, for example, for water consumption (Corral-Verdugo et al. 2002). The behavior of significant others who share the same environment is thus a very strong source of social influence. Witnessing the PEB of others who are spatially close strengthens self-efficacy and perceived control (Fornara et al. 2011).

The importance of spatial levels in the relationship to the environment is vividly illustrated by the research of Gifford and collaborators (2009) entitled Temporal Pessimism and Spatial Optimism in Environmental Assessments. This study undertaken in 18 countries shows that, in the majority of them, temporal pessimism ("things will get worse") generally coexists with a spatial optimism bias ("things are better here than there"). Going back to the concept of comparative optimism, the authors show that there is a comparative environmental optimism. In general, individuals seem to believe that, in environmental terms, they are safer than others. This could probably be called a proxemic model (Moles and Rohmer-Moles 1972), according to which the further we move away, the more unfavorable are our assessments.

The preceding findings might indicate that people have a greater propensity to sustainable behavior only when they act within their own territory. However, several studies have demonstrated the existence of the spillover effects of PEB (Thogersen and Orlander 2003). Thus, the engagement of individuals in PEB in a specific setting increases their propensity to engage in PEB in other domains (Halpenny 2010; Whitmarsh and O'Neill 2010). This transfer from one scale to another may mean that the separation between local and global can be questioned.

12.2.5 Identity Motives and Pro-environmental Behavior

In the last few decades, many studies have attempted to demonstrate the relevance of the link that can be established between identity processes and PEB. Among the various theoretical approaches to identity, the Motivated Identity Construction Theory (MICT, Vignoles 2011; Vignoles et al. 2006) appears a relevant and operational theoretical framework for many reasons. Notably, this model considers identity a motivated process based on several different motives. Thus, it goes beyond the classic idea that the self-esteem motive is the only motive of identity. Vignoles and colleagues (2006) identified six conceptually distinct motives (motivational goals): self-esteem, continuity, distinctiveness, belonging, efficacy and meaning.

These motives guide the construction of identity because people try to satisfy them and avoid their frustration. Thus, in many of their actions in everyday life, they put more emphasis on the elements that provide stronger feelings of self-esteem, continuity, distinctiveness, belonging, efficacy and meaning. Each of these motives has a variety of direct and indirect effects on cognition, affective and behavioral dimensions of identity, which are seen as having reciprocal influences. The self-esteem motive directly influences the definition and enactment of identity: *"Motives for continuity, distinctiveness and meaning contribute directly to identity definition and indirectly – through identity definition – to identity enactment. Motives for belonging and efficacy contribute directly to identity enactment and indirectly – through identity enactment – to identity definition"* (Vignoles et al. 2006, p. 327).

The efficacy and belonging motives are of greater relevance to the behavioral dimension of identity since they have a direct impact on behavior and correspond to a person's relationships with the external world. In fact, efficacy refers to how a person acts on the environment while belonging refers to how a person has a place in the environment. Therefore, *"attempts to increase enactment of particular identities might focus especially on potential benefits for self-esteem, belonging and efficacy"* (Vignoles et al. 2006, p. 329).

Regarding the current barriers to adopting specific behaviors in the area of the environment, the following question should be asked: how far do the messages from different institutions or social actors who encourage PEB present these behaviors as potential threats to particular identity motives? Alternatively, how could we lead people to think that the adoption of PEB can effectively satisfy identity motives such as self-esteem, efficacy and belonging? Finally, there seem to be many situations where PEBs could appear as complex behaviors. This is notably because people do not always understand the meaning of these behaviors (which are sometimes relevant only in the long term and so people have no control over their consequences) or because the implementation of these actions appears difficult for them since they have to follow a certain number of instructions (which often makes people think that they are not competent enough in this domain to carry out these instructions correctly). In both cases, this could represent a threat to the efficacy motive. Similarly, it is not certain that PEB appears as a potential source of satisfaction of the belonging motive. In fact, the dominant norm of hyperconsumerism leads individuals to buy many objects in a short period of time with little consideration for behaviors such as buying second-hand products and making things last as long as possible. Moreover, consumption indicates social belonging and social status. Many objects or products symbolically reflect social belonging, which occurs through the possession and display of these things. Therefore, at first sight, diminishing one's consumption and keeping the same things for a long time might threaten the belonging motive.

However, *"the potential for a given situation or message to threaten identity can be avoided to the extent that it does not undermine satisfaction of these motives and that it provides alternative means of restoring these satisfactions"* (Vignoles et al. 2006, p. 329). The problem is that behavioral changes are often presented in an informative or injunctive form, which removes the possibility of these behaviors becoming a source of satisfaction for the efficacy and belonging motives.

In terms of potential solutions, some changes to the current environmental campaigns could be suggested. Generally, these campaigns indicate a list of gestures that people should (or must) adopt to protect the environment and/or save money. These campaigns should perhaps emphasize that controlling the cost of one's personal energy and having the capability of doing so is one way of acting on the environment, in the widest sense. It might also be useful to highlight the ability of everyone to be efficient in this area. The belonging motive could also be more evident in prevention messages, given that it is known that when people are informed that others in the same situation as them have adopted a specific behavior, a majority of them will adopt the same behavior (Cialdini 2007). Moreover, when your neighbors are environmentally friendly, you tend to change your own behavior and to be more respectful of the environment (Dupré 2009). Finally, although the norm of hyperconsumption is still dominant, a pro-environmental norm is now emerging and gaining more importance in social discourse (Félonneau and Becker 2008). Therefore, campaigns should highlight the idea that adopting PEB is also a way of enhancing social insertion (see Sect. 4.2).

According to Oskamp (2000), future studies need to identify the motivational factors of behavioral change in the area of the environment. Indeed, the research of Gardner and Stern (1996) shows that simply giving information to people, without reinforcing their motivation, has very little effect on behavioral change.

MICT seems to offer some perspectives for reconsidering the ways in which people are asked to change their behaviors. It must be remembered that people prefer to engage in actions that maintain or enhance their feelings of efficacy and belonging and that they would certainly avoid actions that could threaten these feelings. Thus, when asking people to adopt new behaviors, it is particularly important to demonstrate that these can be sources of satisfaction of identity motives, and specifically of efficacy and belonging motives.

12.3 A Mediator Variable: Environmental Quality of Life, Identity and Pro-Environmental Behaviors

12.3.1 Quality of Life: Definitions

The concept of quality of life (QOL) is relatively recent and covers a wide area. It is associated with happiness, well-being, and environmental satisfaction, amongst other things. In 1995, the World Health Organization (WHO) proposed this definition of quality of life: *"an individual's perception of life in the context of the cultural and value system in which they live and in relation to their goals, expectations, standards and concerns"* (World Health Organization Quality of Life assessment, WHOQOL 1995). This addresses perceived QOL and not objective QOL, that is to say, the estimation by individuals of their living environment i.e. the assessment of local environmental quality. Fleury-Bahi and colleagues suggested defining the subjective environmental quality of life as a judgment of the degree of satisfaction with the different components of the daily environment (Fleury-Bahi et al. 2013).

12.3.2 Identity and Environmental Perception

The link between satisfaction and residential identity is quite complex (Fleury-Bahi et al. 2008). In 1988, Lalli discovered a connection between a positive assessment and the degree of identification with the city of residence. Later, Bonaiuto et al. (1996) revealed that a strong local identity leads youths to perceive their city in less negative environmental terms, more specifically, to view it as less polluted. Félonneau (2004) showed that a high level of identification leads to an underestimation of the frequency of the incivilities encountered in everyday life. Conversely, a weak level of identification leads to an overestimation of those same incivilities.

Regarding the identity motives for belonging and continuity, Vignoles and colleagues (2006) suggested that a strong psychological involvement in the daily living place, expressed through a strong feeling of belonging and a perception of the neighborhood as a place of memories and an environment for future plans, could be linked to a favorable assessment of both the environment and, in particular, the people encountered there and with whom satisfying relationships are likely to be cultivated.

The model of cognitive dissonance (Festinger 1957) can be used to hypothesize that an individual would tend toward the highest possible congruity between the image he/she has of the ideal residential environment and the image of his/her own residential situation (Amerigo and Aragones 1997; Premius 1986). Sociological studies have shown that even in highly blighted neighborhoods, some residents are unable to express dissatisfaction explicitly (Avenel 2005; Lepoutre 2000). Stating one's dissatisfaction with one's neighborhood and its inhabitants when one feels identified with them would be the same as assuming a negative image of oneself (Cialdini et al. 1976). In other words, a positive perception of the neighborhood's social image and the relationship potential it offers enhances the awareness of belonging to an emotionally invested localized in-group.

Uzzell et al. (2002) indicated the importance of exploring the relationships between place identification, social cohesion and residential satisfaction.

Independently of the objective characteristics of the environment, individuals can rationalize by positively readjusting their environmental perception and the degree of related satisfaction to maintain a positive self-image. The reduced dissonance then becomes an important element in the relationship between people and place.

12.3.3 Quality of Life and Pro-environmental Behavior

The links between QOL and PEB are complex. A high level of residential satisfaction, often related to a positive place identity, can lead to the adoption of PEB with the more or less explicit aim of preserving a positive social image. Paradoxically, a high level of identification can also modulate the perception of the environment, for example leading to an underestimation of nuisances.

However, a low level of residential satisfaction can also lead to the adoption of PEB in order to change a devalorized social image. Some studies show that a low level of environmental satisfaction predicts engagement in numerous PEBs, notably recycling, conservation and the purchase of environmentally friendly products (Prester et al. 1987; Stedman 2002).

The approach chosen here highlights a strong link between identity dynamics, perceived quality of life and the adoption of PEB. The latter is impacted by the subjective quality of life, which is itself strongly dependent on the definition of self.

12.4 Identity Dynamics, Perceived QOL and the Adoption of PEB: Some Examples of Current Research

12.4.1 Waste

Waste, with its quantitative explosion in recent decades, has become the symbol of the environmental issue. People have a core role here because they can have a strong impact on waste reduction. However, people's behaviors depend on many factors. To provide a better understanding of the difficulty people have in changing their behaviors, the multiple levels of self-representation within space, time and society need to be taken into account.

A major resistance to behavioral change in the environmental domain is that, most of the time, the consequences of one's behavior cannot be directly observed (Denisov and Christoffersen 2001; Geller 2002). Indeed, they are generally the result of an extensive process consisting of different related actions. For instance, a person who sorts his/her waste cannot observe the consequence of his/her action on the preservation of global resources at the moment when he/she acts. Thus, the focus on the close environment, such as the household or the neighborhood, considerably limits the meaning that can be attributed to the behavior of sorting waste in a global perspective. In terms of identity, the activation of self-representation at a particular level excludes the other levels (Turner et al. 1987). The fact that sorting waste is a behavior that first takes place within the household tends to highlight more the self-categorization as a particular individual living in a particular home than that as a human among many others living on Earth. Finally, this raises the question of the instant feedback of the behavior of sorting waste, which is probably more related to the feeling of being normative (Félonneau and Becker 2008) than a real awareness of preserving the planet and its resources in the longer term.

In addition, some resistance to adopting behavior such as sorting waste or reducing it comes from one's own representation of one's social role within society. In Western societies, people tend to think as isolated individuals, which clearly impacts the way they perceive their place in the social system, notably in terms of responsibility. Who is responsible for the situation regarding waste and who should deal with it? When people are asked this question, they first absolve themselves from the responsibility of producing waste (e.g. *"It's not our fault, there is too much packaging on products, we are not responsible for the toxic products used by industries which pollute and create waste"*, Gombert-Courvoisier et al. 2014b). Secondly, they mention a multitude of social stakeholders responsible for this problem, including manufacturers, distributors, farmers, politicians, waste managers and consumers. These two facts make changing practices and promoting PEB complex. On the one hand, the individual does not necessarily perceive himself/herself as being at the root of the current concerns about waste and thus he/she perceives it as unfair to have to change their practices while there are other people involved who are more obviously responsible and who should therefore deal with this problem. On the other hand, the responsibility is seen as shared between different stakeholders in society, which means for the individual that it is not only his/her role to act and improvement of the situation is not entirely based on his/her actions ("*What do the other people do? It's not up to me to change my behaviour, why should I change if other people do nothing?*" Gombert-Courvoisier et al. 2014b).

This may cause some inertia based on the idea that individual behavior does not really change the situation (it is just a drop of water less in the ocean). There is a general feeling of helplessness. People feel powerless regarding not only environmental problems but also, more importantly, the positive impact they can have (Denisov et al. 2005). This idea of the uselessness of their behavior diminishes people's confidence in collective effort and drives them to feel powerless faced with the issue of waste (Denisov et al. 2005). They generally think that their personal efforts will not be followed by enough other people to make a sufficiently significant change (Staats et al. 1996).

This recalls the work on the social dilemma (Dawes 1980) and the social trap (Platt 1973), which showed that individual motivations override collective motivations. The individual is confronted with a social dilemma when he/she has the choice between a behavior that is personally beneficial and another that is less favorable for himself/herself but is of collective interest. It is observed that the individual prefers to act according to his/her personal interests. In the social trap, the individual is faced with the choice between an immediate benefit and a positive behavior for future generations. Here he/she is seen to prefer an immediate benefit while the interest of future generations remains abstract.

This also suggests how difficult it is to set up an action based on collective effort and to have a collective issue in a society that works around individualistic and meritocratic values.

12.4.2 Downward Social Mobility, Pro-environmental Behavior and Perceived Quality of Life

A current study, funded by the French Ministry of the Environment, Sustainable Development and Energy (Gombert-Courvoisier et al. 2014a), suggests the adoption of PEB as a potential solution to overcome the feeling of downward social mobility and restore subjective QOL.

Today, recent developments in society are tracing the outlines of a transitional context at multiple levels (social, economic and environmental), which are leading to the reorganization of people's lifestyles and having an impact on their QOL. A new segment of the population is deprived of some of the goods and services to which it previously had access and risks ending up in a situation of downward social mobility, resulting in problems at multiple levels: psychological, social, economic and health. The current expansion of the socio-economic crisis has directed research towards vulnerable populations who, by obligation and no longer out of conviction, must adopt new patterns of consumption. This raises the question of the relationship between QOL and consumption. There is very little research aimed at understanding the consequences of these forced changes, both in terms of environmental impact and regarding the well-being of vulnerable populations. The literature is also sparse concerning the strategies and adaptive capacities of households trying to maintain a satisfactory level of QOL, relying in particular on the adoption of sustainable consumption practices.

In the current context of transition, which is unstable by definition, the appearance of disruptors in the lives of individuals (such as a lower income or illness) may lead to a switchover that puts them at risk of downward social mobility, i.e. with a lower social status than before. At the root of this phenomenon, there may be a cluster of factors that combine to cause a switchover in the subject's life and that create an absence of or significant reduction in flexibility (e.g. the person can no longer cope with unexpected expenses). This type of downward social mobility is recognized on the basis of objective indicators such as reduced income, indebtedness ratio, and social isolation.

However, it is important to combine this approach with subjective downward social mobility, i.e. the feeling of downward social mobility, which causes a loss of self-esteem and the feeling of being unable to adapt. In addition, this creates a feeling of deterioration in QOL. Indeed, one cannot really talk about downward social mobility and not also feel the situation as one of downward social mobility. Thus, it is both the situation and the perception of it that enable the definition of downward social mobility to be refined. In other words, next to objectively measurable impacts, it is also necessary to identify the psychosocial impacts, particularly in terms of identity. If an individual is able to overcome their downward social mobility by maintaining their positive self-esteem through the adoption of new behaviors, then they will be able to cope.

For several decades, the social norm has been consumerism. The more I consume, the more I possess, the more I am socially valued. Now, this norm prescribes behaviors that run counter to environmental protection, precisely because it leads people to buy newer and more sophisticated objects (or those presented as such) and to throw away many objects that sometimes still function effectively. It also suggests that QOL is synonymous with high consumption of resources and energy. However, at the same time, a pro-environmental norm is emerging that opposes the dominant norm of hyperconsumption. Pro-environmentalism is increasingly seen as a socially desirable norm (Félonneau and Becker 2008). This may be due to the development of prevention campaigns and rising awareness. In general, declaring that one pays attention to the environment and has environmentally friendly behaviors is clearly socially valued today. Therefore, the citizen is facing an ambiguous socio-normative context about environmental issues since these two norms prescribe opposing practices for the individual.

However, if PEBs are the subject of social valorization, they may represent adaptive strategies to overcome downward social mobility. For example, a change in reference group, and therefore the adoption of new values, can lead to perceiving as positive what was previously perceived as a constraint.

In the strict sense, ecological strategies can mean behaviors and consumption patterns that directly foster a reduced environmental impact. In a broader sense, this refers to behaviors and consumption patterns that combine social, economic and cultural dimensions with environmental issues. More specifically, it can include behaviors such as reducing (purchasing, energy consumption), sharing (joint purchasing, community lifestyles, sharing and renting, public housing, roommate, support), substitution (replacement of new products by other cheaper new products, buying used products), self-production ("do it yourself", gardening, repair), or behaviors that might be associated with transgression (undeclared work). The implementation of these strategies depends on the populations and territories, particularly in terms of the resources available in the area to promote adoption of these behaviors (communication, networks, and places for discussion).

These ecological behaviors can satisfy identity motives that directly contribute to identity enactment, such as the motives for efficacy and belonging. In fact, they are related to the acquisition of skills and knowledge. In addition, they offer a sense of control, in the sense that the individual can choose an alternative solution that allows him/her to continue consuming without having economic problems. Some products that are accessed by this type of behavior are now clearly valorized; for example, vintage clothing, antiques, unique items, etc. Regarding the belonging motive, this can be satisfied by these behaviors at multiple levels. Most of them help to develop social ties, which is sometimes what engages people to pursue them. For example, a recent pilot study revealed that the staff of associative structures offering alternative consumer behaviors, such as gardening, repairing, or buying used clothes or objects, found that some people of modest means, who came to these structures primarily for the social relationships they had developed there, gradually became interested in alternative modes of consumption (Gombert-Courvoisier et al. 2014a). This example illustrates the idea that pro-environmental behaviors can be promoted among people in a situation of downward social mobility via identity motives. At a more general level, the adoption of these behaviors helps strengthen the social inclusion of these people in the sense that they fit more directly into a social trend that is spreading (multiplication of prevention messages in the media, environmental issues increasingly present in societal debates, growing number of associative structures offering alternative modes of consumption).

By satisfying identity motives, the adoption of pro-environmental practices can have positive consequences for an individual in a situation of downward social mobility and result in a dynamic process of resilience. It is therefore important to promote PEB by emphasizing the potential for social and personal valorization and the positive impact on the quality of life of vulnerable people (Gombert-Courvoisier et al. 2014a).

12.5 Conclusion

In this chapter, an approach was developed in which identity dynamics, the perception of QOL and environmental choices are strongly linked.

From the inventory of psychological determinants of PEB to their integration at an ideological level, from idiosyncratic characteristics of the self to macro-sociological determinants, the general theme of people/environment relationships needs to be placed in a societal context defined by several spatial scales, from the micro-environment to the global one. The difficulties for the average individual to appropriate questions anchored in a spatiality and temporality often distant from his/her everyday concerns are probably located here.

At a time when geographic mobility is becoming the norm, the issue is to what extent the individual draws on spatial belonging in the construction of his/her identity (Pollini 2005). Many researchers have suggested that information technologies and "time-space compression" contribute to making local identities obsolete (Harvey 1989). In response to the forces of globalization, efforts to strengthen old boundaries and create new ones are observed, often based on "identities of resistance" (Castells 1997).

It is important to remember that this issue must be placed in the perspective of a normative socio-spatial order and in the broader dynamics of inter-group conflict and social and economic regulation.

References

Amerigo, M., & Aragones, J. I. (1997). A theoretical and methodological approach to the study of residential satisfaction. *Journal of Environmental Psychology, 17*, 47–57.

Arcury, T. A., Scollay, S. J., & Johnson, T. P. (1978). Sex differences in environmental concern and knowledge: The case of acid rain. *Sex Roles, 16*, 463–472.

Avenel, C. (2005). *Sociologie des quartiers sensibles* [Sociology of sensitive neighborhoods]. Paris: Armand Colin.

Bamberg, S., & Möser, G. (2007). Twenty years after Hines, Hungerford, and Tomera: A new meta-analysis of psycho-social determinants of pro-environmental behavior. *Journal of Environmental Psychology, 27*, 14–25.

Becker, M., & Félonneau, M. L. (2011). Pourquoi être pro-environnemental? Une approche socio-normative des liens entre valeurs et pro-environnementalisme [Why be pro-environmental? A socio-normative approach to the links between values and pro-environmentalism]. *Pratiques Psychologiques, numéro spécial "Psychologie Sociale appliquée à l'Environnement", 17*(3), 219–236.

Bennett, G., & Williams, F. (2011). *Mainstream green: Moving sustainability from niche to normal*. New York: Ogilvy & Mather.

Bissonnette, M. M., & Contento, I. R. (2001). Adolescents' perspectives and food choice behaviors in terms of the environmental impacts of food production practices: Application of a psychosocial model. *Journal of Nutrition Education, 33*(2), 72–82.

Bonaiuto, M., Breakwell, G. M., & Cano, I. (1996). Identity processes and environmental threat: The effects of nationalism and local identity upon perception of beach pollution. *Journal of Community & Applied Social Psychology, 6*, 157–175.

Breakwell, G. M. (1992). Processes of self-evaluation: Efficacy and estrangement. In G. M. Breakwell (Ed.), *Social psychology of identity and self-concept*. Surrey: Surrey University Press.

Casey, P., & Scott, K. (2006). Environmental concern and behaviour in an Australian sample within an ecocentric-anthropocentric framework. *Australian Journal of Psychology, 58*, 56–67.

Castells, M. (1997). *The power of identity*. Malden: Blackwell.

Cialdini, R. B. (2007). Descriptive social norms as underappreciated sources of social control. *Psychometrika, 72*(2), 263–268.

Cialdini, R. B., & Reno, R. (1990). A focus theory of normative conduct: Recycling the concept of norms to reduce littering in public places. *Journal of Personality and Social Psychology, 58*(6), 1015–1026.

Cialdini, R. B., Borden, R. J., Thorne, A., Walker, M. R., Freeman, S., & Sloan, L. R. (1976). Basking in reflected failure: Three field studies. *Journal of Personality and Social Psychology, 39*, 406–415.

Cialdini, R. B., Kallgren, C. A., & Reno, R. R. (1991). A focus theory on normative conduct: A theoretical refinement and reevaluation of the role of the norms in human behavior. *Advances in Experimental Social Psychology, 24*, 201–234.

Corraliza, J., & Berenguer, J. (2000). Environmental values, beliefs and actions: A situational approach? *Environment and Behavior, 32*, 832–848.

Corral-Verdugo, V., Frías-Armenta, M., Pérez-Urías, F., Orduňa-Cabrera, V., & Espinoza-Gallego, N. (2002). Residential water consumption, motivation for conserving water and the continuing tragedy of the commons. *Environmental Management, 30*, 527–535.

Davidson, D. J., & Freudenburg, W. R. (1996). Gender and environmental risk concerns: A review and analysis of available research. *Environment and Behavior, 28*, 302–339.

Dawes, R. M. (1980). Social dilemmas. *Annual Review of Psychology, 31*, 169–193.

De Groot, J. I. M., & Steg, L. (2008). Value orientations to explain beliefs related to environmental significant behavior. How to measure egoistic, altruistic, and biospheric values orientations. *Environment and Behavior, 40*(3), 330–354.

Denisov, N., & Christoffersen, L. (2001). *Impact of environmental information on decision-making processes and the environment*. UNEP/GRID-Arendal Occasional Paper 01 2001. http://www.grida.no/impact. Accessed Apr 2006.

Denisov, N., Folgen, K., Rucevska, I., & Simonett, O. (2005). *Impact II: Telling good stories* (Grid-Arendal Occasional Paper 01, 2005). Arendal: GRID-Arendal.

Devine-Wright, P. (2009). Rethinking NIMBYism: The role of place attachment and place identity in explaining place-protective action. *Journal of Community & Applied Social Psychology, 19*, 426–441.

Devine-Wright, P., & Clayton, S. (2010). Introduction to the special issue. *Journal of Environmental Psychology, 30*(3), 267–270.

Dietz, T., Kalof, L., & Stern, P. C. (2002). Gender, values and environmentalism. *Social Science Quarterly, 83*(1), 353–364.

Droseltis, O., & Vignoles, V. L. (2010). Towards an integrative model of place identification: Dimensionality and predictors of intrapersonal-level place preferences. *Journal of Environmental Psychology, 30*, 23–34.

Dunlap, R. E., & Van Liere, K. D. (1978). The "New Environmental Paradigm": A proposed measuring instrument and preliminary results. *Journal of Environmental Education, 9*(4), 10–19.

Dunlap, R. E., Van Liere, K. D., Mertig, A. G., & Jones, R. E. (2000). Measuring endorsement of the new ecological paradigm: A revised NEP scale. *Journal of Social Issues, 3*, 425–442.

Dupré, M. (2009). *De l'engagement comportemental à la participation : élaboration de stratégies de communication sur le tri et la prévention des déchets ménagers* [From behavioral commitment to participation: Development of communication strategies on the sorting and prevention of household waste]. Thèse de Doctorat, Université de Rennes 2.

Ehrenberg, A. (1995). *L'individu incertain* [The uncertain individual]. Paris: Calmann-Lévy.

Félonneau, M. L. (2004). Love and loathing of the city: Urbanophilia and urbanophobia, topological identity and perceived incivilities. *Journal of Environmental Psychology, 24*, 43–52.

Félonneau, M. L., & Becker, M. (2008). Pro-environmental attitudes and behavior: Revealing perceived social desirability. *Revue Internationale de Psychologie Sociale, 21*(4), 25–53.

Félonneau, M. L., Lannegrand-Willems, L., & Becker, M. (2008). Les processus de spatialisation identitaire chez les adolescents et les jeunes adultes: le cas du Pays Basque français [The processes of identity spatialization in adolescents and young adults: The case of French Pays Basque]. *Les Cahiers Internationaux de Psychologie Sociale, 80*(4), 59–71.

Félonneau, M. L., Lannegrand-Willems, L., Becker, M., & Parant, A. (2013). The dynamics of sociospatial identity: Comparing adolescents and young adults in two French regions. *Applied Psychology: An International Review, 62*(4), 619–639. Article first published online 12 April 2012.

Festinger, L. (1957). *A theory of cognitive dissonance*. Stanford: Stanford University Press.

Fleury-Bahi, G., Félonneau, M. L., & Marchand, D. (2008). Processes of place identification and residential satisfaction. *Environment and Behavior, 40*, 669–682.

Fleury-Bahi, G., Marcouyeux, A., Préau, M., & Annabi-Attia, T. (2013). Development and validation of an environmental quality of life scale: Study on a French sample. *Social Indicators Research, 113*, 903–913.

Fornara, F., Carrus, G., Passafaro, P., & Bonnes, M. (2011). Distinguishing the sources of normative influence on proenvironmental behaviors: The role of local norms in household waste recycling. *Group Processes and Intergroup Relations, 14*(5), 623–635.

Gamba, R. J., & Oskamp, S. (1994). Factors influencing community residents' participation in commingled curbside recycling programs. *Environment and Behavior, 26*, 587–612.

Gardner, G. T., & Stern, P. C. (1996). *Environmental problems and human behavior*. Boston: Allyn and Bacon.

Gatersleben, B., Steg, L., & Vlek, C. (2002). Measurement and determinants of environmentally significant consumer behavior. *Environment and Behavior, 34*(3), 335–362.

Gatersleben, B., White, E., Jackson, T., & Uzzell, D. (2010). Values and sustainable lifestyles. *Architecture Science Review, 53*, 1–14.

Gatersleben, B., Murtagh, N., & Abrahamse, W. (2012). Values, identity and pro environmental behaviour. *Contemporary Social Science*, (ahead-of-print), 1–19.

Geller, E. S. (2002). The challenge of increasing proenvironmental behaviour. In R. B. Bechtel & A. Churchman (Eds.), *Handbook of environmental psychology*. New York: Wiley.

Gifford, R., & Nilsson, A. (2014). Personal and social factors that influence pro-environmental concern and behaviour: A review. *International Journal of Psychology*. doi:10.1002/ijop.12034.

Gifford, R., Scannel, L., Kormos, C., Smolova, L., et al. (2009). Temporal pessimism and spatial optimism in environmental assessment: An 18-Nation study. *Journal of Environmental Psychology, 29*(1), 1–12.

Goldstein, N. J., Cialdini, R. B., & Griskevicius, V. (2008). A room with a viewpoint: Using social norms to motivate environmental conservation in hotels. *Journal of Consumer Research, 35*, 472–482.

Gombert-Courvoisier, S., Causse, E., Félonneau, M. L., Ribeyre, F., Ballesta, O., Carimentrand, A., Delerue, F., Sennes, V. (2014a). INOVE.COM: *Identité, Norme, Vulnérabilité, pour une approche Ecologique de la Consommation durable des Ménages* [Identity, norm, vulnerability, for an ecological approach to sustainable household consumption]. Programme de recherches financé par le Ministère de l'Environnement, du Développement Durable et de l'Energie.

Gombert-Courvoisier, S., Causse, E., Ribeyre, F., Félonneau, M. L., Carimentrand, A. (2014b). *Rôle de l'écologie familiale dans la réduction des ordures ménagères* [Role of family ecology in the reduction of household waste]. ADEME, Programme Déchets et Sociétés 2010, REFIOM, Rapport Final.

Halpenny, E. A. (2010). Proenvironmental behaviours and park visitors: The effect of place attachment. *Journal of Environmental Psychology, 30*(4), 409–421.

Hardin, G. (1968). The tragedy of the commons. *Science, 162*, 1243–1248.

Harvey, D. (1989). *The condition of post-modernity*. Cambridge: Blackwell Publishers.

Hau, D., & Swenson, A. (2013). Gendered innovations in energy and environmental media: A case study. *Intersect, 6*(2), 1–10.

Hernandez, B., Hidalgo, C., Salazar-Laplace, E., & Hess, S. (2007). Place attachment and place identity in natives and non-natives. *Journal of Environmental Psychology, 27*(4), 310–319.

Hidalgo, M. C., & Hernandez, B. (2001). Place attachment: Conceptual and empirical questions. *Journal of Environmental Psychology, 21*, 273–281.

Hines, J. M., Hungerford, H. R., & Tomera, A. N. (1986). Analysis and synthesis of research on responsible environmental behaviour: A meta-analysis. *Journal of Environmental Education, 18*, 1–8.

Jellison, J. M., & Green, J. (1981). A self-presentation approach to the fundamental attribution error: The norm of internality. *Journal of Personality and Social Psychology, 40*, 643–649.

Jones, R. E., & Dunlap, R. E. (1992). The social bases of environmental concern: Have they changed over time? *Rural Sociology, 57*(1), 28–47. doi:10.1111/j.1549-0831.1992.tb00455.x.

Kaiser, F. G. (1998). A general measure of ecological behavior. *Journal of Applied Social Psychology, 28*, 395–422.

Karp, D. G. (1996). Values and their effect on pro-environmental behavior. *Environment and Behavior, 28*(1), 111–133.

Klineberg, S. L., McKeever, M., & Rothenbach, B. (1998). Demographic predictors of environmental concern: It does make a difference how it's measured. *Social Science Quarterly, 79*, 734–753.

Knez, I. (2005). Attachment and identity as related to a place and its perceived climate. *Journal of Environmental Psychology, 25*, 207–218.

Korpela, K. M. (1989). Place identity as a product of self-regulation. *Journal of Environmental Psychology, 9*, 241–256.

Korpela, K. M., Ylen, M., Tyrväinen, L., & Silvennoinen, H. (2009). Stability of self-reported favorite places and place attachment over a 10-month period. *Journal of Environmental Psychology, 29*, 95–100.

Kortenkamp, K. V., & Moore, C. F. (2001). Ecocentrism and anthropocentrism: Moral reasoning about ecological commons dilemmas. *Journal of Environmental Psychology, 4*, 47–60.

Lalli, M. (1992). Urban related identity: Theory, measurement, and empirical findings. *Journal of Environmental Psychology, 12*, 285–303.

Lepoutre, D. (2000). *Coeur de Banlieue : codes, rites et langages* [Heart of the Suburbs: Codes, rites and languages]. Paris: Odile Jacob.

Lewicka, M. (2011). Place attachment: How far have we come in the last 40 years? *Journal of Environmental Psychology, 31*, 207–230.

Mannetti, L., Pierro, A., & Livi, S. (2004). Recycling/planned and self-expressive behaviour. *Journal of Environmental Psychology, 24*, 227–236.

Manzo, L. C. (2005). For better or worse: Exploring multiple dimensions of place meaning. *Journal of Environmental Psychology, 25*, 67–86.

Manzo, L. C., & Devine-Wright, P. (2013). *Place attachment. Advances in theory, methods and applications*. London: Routledge.

Mayer, F. S., & Frantz, C. M. (2004). The connectedness to nature scale: A measure of individuals' feeling in community with nature. *Journal of Environmental Psychology, 24*, 503–515.

McStay, J., & Dunlap, R. E. (1983). Male-female differences in concern for environmental quality. *International Journal of Women's Studies, 6*(4), 291–301.

Meneses, G. D., & Palacio, A. B. (2005). Recycling behavior. A multidimensional approach. *Environment and Behavior, 37*(6), 837–860.

Mohai, P. (1992). Men, women, and the environment: An examination of the gender gap in environmental concern and activism. *Society & Natural Resources: An International Journal, 5*(1), 1–19.

Moles, A. A., & Rohmer-Moles, E. (1972). *Psychologie de l'espace* [Psychology of space]. Paris: Casterman.

Moser, G. (2003). Questionner, analyser et améliorer les relations à l'environnement ? [Question, analyze and improve relationships with the environment?] In G. Moser & K. Weiss (Eds.), *Espaces de vie. Aspects de la relation Homme-Environnement* [Living spaces. Aspects of the man-environment relationship] (pp. 11–42). Paris: A. Colin.

Neiman, M., & Loveridge, R. O. (1981). Environmentalism and local growth control. A probe into the class bias thesis. *Environment and Behavior, 13*(6), 759–772.

Nigbur, D., Lyons, E., & Uzzel, D. (2010). Attitudes, norms, identity and environmental behaviour: Using an expanded theory of planned behaviour to predict participation in a kerbside recycling program. *British Journal of Social Psychology, 49*(2), 259–284.

Nordlund, A., & Garvill, J. (2002). Values structures behind proenvironmental behavior. *Environment and Behavior, 34*(6), 740–756. doi:10.1177/001391602237244.

Olivos, P., Aragones, J. I., & Amerigo, M. (2011). The connectedness to nature scale and its relationship with environmental beliefs and identity? *IPHJ, 4*(1), 5–19.

Oskamp, S. (2000). Psychological contributions to achieving an ecologically sustainable future for humanity. *Journal of Social Issues, 56*(3), 373–390.

Oskamp, S., Harrington, M., Edwards, T. C., Sherwood, D. L., Okuda, S. M., & Swanson, D. (1991). Factors influencing household recycling behavior. *Environment and Behavior, 23*(4), 494–519.

Pinto, D. C., Nique, W. M., Añaña, E. D. S., & Herter, M. M. (2011). Green consumer values: How do personal values influence environmentally responsible water consumption? *International Journal of Consumer Studies, 35*(2), 122–131.

Platt, J. (1973). Social traps. *American Psychologist, 28*, 642–651.

Pol, E. (2002). The theoretical background of city-identity-sustainability network. *Environment and Behavior, 34*(1), 8–25.

Pol, E., & Castrechini, A. (2002). City-identity-sustainability research network: Final words. *Environment and Behavior, 34*, 150–160.

Pollini, G. (2005). Socio-territorial belonging in a changing society. *International Review of Sociology, 15*(3), 493–496.

Poortinga, W., Steg, L., & Vlek, C. (2004). Values, environmental concern, and environmental behavior. A study into household energy use. *Environment and Behavior, 36*(1), 70–93.

Premius, H. (1986). Housing as a social adaptation process. *Environment and Behavior, 18*(1), 31–52.

Prester, G., Rohrmann, B., & Schellhammer, E. (1987). Environmental evaluation and participation activities: A social psychological field study. *Journal of Applied Social Psychology, 17*, 751–787.

Proshansky, H. M. (1978). The city and self-identity. *Environment and Behavior, 10*(2), 147–169.

Proshansky, H. M., Fabian, A. K., & Kaminoff, R. (1983). Place identity: Physical world socialization of the self. *Journal of Environmental Psychology, 3*, 57–83.

Py, J., & Ginet, A. (1999). Clairvoyance normative et attribution causale dans le cadre d'une situation de formation d'impression [Normative clear-sightedness and causal attribution in the context of an impression-forming situation]. *Revue Internationale de Psychologie Sociale, 12*(1), 7–35.

Py, J., & Somat, A. (1991). Normativité, conformité et clairvoyance: leurs effets sur le jugement évaluatif dans un contexte scolaire [Normativity, conformity and clear-sightedness: their effects on evaluative judgment in a school context]. In J. L. Beauvois, R. V. Joulé & J. M. Monteil (Eds.), *Perspectives cognitives et conduites sociales* [Cognitive and social behavioral perspectives] (Vol. 3, pp. 167–193). Cousset: DelVal.

Ramkissoon, H., Weiler, B., & Smith, L. D. G. (2012). Place attachment and pro-environmental behaviour in national parks: The development of a conceptual framework. *Journal of Sustainable Tourism, 20*(2), 257–276.

Raymond, C. M. M., Brown, G., & Robison, G. M. (2011). The influence of place attachment, and moral and normative concerns on the conservation of native vegetation. *Journal of Environmental Psychology, 31*, 323–335.

Reno, R. R., Cialdini, R. B., & Kallgren, C. A. (1993). The trans-situational influence of social norms. *Journal of Personality and Social Psychology, 64*(1), 104–112.

Richins, M. L. (2004). The material values scale: Measurement properties and development of a short form. *Journal of Consumer Research, 31*, 209–218.

Rokeach, M. (1973). *The nature of human values*. New York: Free Press.

Saphores, J. D., Dixon, H., Ogunseitan, O., & Shapiro, A. A. (2006). Household willingness to recycle electronic waste. An application to California. *Environment and Behavior, 38*(2), 183–208.

Scannell, L., & Gifford, R. (2010a). Defining place attachment: A tripartite organizing framework. *Journal of Environmental Psychology, 30*, 1–10.

Scannell, L., & Gifford, R. (2010b). The relations between natural and civic place attachment and pro-environmental behavior. *Journal of Environmental Psychology, 30*, 289–297.

Schultz, P. W. (2000). Empathizing with nature: The effects of perspective taking on concern for environmental issues. *Journal of Social Issues, 56*, 391–406.

Schultz, P. W. (2001). Assessing the structure of environmental concern: Concern for self, other people, and the biosphere. *Journal of Environmental Psychology, 21*, 327–339.

Schultz, P. W. (2002). Inclusion with nature: Understanding the psychology of human-nature relations. In P. Schmuck & P. Schultz (Eds.), *The psychology of sustainable development* (pp. 61–78). New York: Kluwer.

Schultz, P. W., & Zelezny, L. (1999). Values as predictors of environmental attitudes: Evidence for consistency across 14 countries. *Journal of Environmental Psychology, 19*(3), 255–265.

Schultz, P. W., Gouveia, V. V., Cameron, L. D., Tankha, G., Schmuck, P., & Franek, M. (2005). Values and their relationship to environmental concern and conservation behavior. *Journal of Cross-Cultural Psychology, 36*(4), 457–475.

Schultz, P. W., Khazian, A., & Zelezny, A. (2008). Using normative social influence to promote conservation among hotel guests. *Social Influence, 3*, 4–23.

Schwartz, S. H. (1970). Moral decision making and behavior. In J. Macauley & L. Berkowitz (Eds.), *Altruism and helping behavior* (pp. 127–141). New York: Academic.

Schwartz, S. H. (1977). Normative influence on altruism. In L. Berkowitz (Ed.), *Advances in experimental social psychology* (pp. 221–279). New York: Academic.

Schwartz, S. H. (1992). Universals in the content and structure of values: Theoretical advances and empirical tests of 20 countries. *Advances in Experimental Psychology, 25*, 1–65.

Schwartz, S. H. (1994). Are there universal aspects in the structure and contents of human values? *Journal of Social Issues, 50*(4), 19–45.

Staats, H. J., Wit, A. P., & Midden, C. Y. H. (1996). Communicating the greenhouse effect to the public: Evaluation of a mass media campaign from a social dilemma perspective. *Journal of Environmental Management, 45*, 189–203.

Stedman, R. C. (2002). Toward a social psychology of place: Predicting behavior from place-based cognitions, attitude, and identity. *Environment and Behavior, 34*, 561–581.

Stedman, R. C. (2003). Is it really just a social construction? The contribution of the physical environment to sense of place. *Society and Natural Resources, 16*, 671–685.

Stedman, R. C. (2006). Understanding place attachment among second home owners. *The American Behavioral Scientist, 50*, 187–205.

Stern, P. C. (2000). Toward a coherent theory of environmentally significant behavior. *Journal of Social Issues, 56*(3), 407–424.

Stern, P. C., Dietz, T., Kalof, L., & Guagnano, G. A. (1995). Values, beliefs and proenvironmental action: Attitude formation toward mergent attitude objects. *Journal of Applied Social Psychology, 25*(18), 1611–1636.

Stern, P. C., Dietz, T., Abel, T., Guagnano, G. A., & Kalof, L. (1999). A values-belief-norm theory of support for social movements: The case of environmentalism. *Human Ecology Review, 6*, 81–97.

Terry, D. J., Hogg, M. A., & White, K. M. (1999). The theory of planned behaviour: Self-identity, social identity and group norms. *British Journal of Social Psychology, 38*, 225–244.

Thogensen, J., & Orlander, F. (2003). Spillover of environmentally-friendly consumer behavior. *Journal of Environmental Psychology, 23*, 225–236.

Turner, J. C., Hogg, M., Oakes, A., Reicher, P. J., & Wetherell, M. S. (1987). *Rediscovering the social group: A self-categorization theory*. Oxford: Basil Blackwell.

Twigger-Ross, C. L., & Uzzell, D. L. (1996). Place and identity processes. *Journal of Environmental Psychology, 16*, 205–220.

Twigger-Ross, C. L., Bonaiuto, M., & Breakwell, G. (2003). Identity theories and environmental psychology. In M. Bonnes, T. Lee, & M. Bonaiuto (Eds.), *Psychological theories for environmental issues* (pp. 203–233). Aldershot: Ashgate Publishing Limited.

Uzzell, D., Pol, E., & Badenas, B. (2002). Place identification, social cohesion, and environmental sustainability. *Environment and Behavior, 34*, 26–53.

Van der Werff, E., Steg, L., & Keizer, K. (2013). The value of environmental self-identity: The relationship between biospheric values, environmental self-identity and environmental preferences, intentions and behaviour. *Journal of Environmental Psychology, 34*, 55–63.

Vignoles, V. L. (2011). Introduction: Toward an integrative view of identity. In V. L. Vignoles, S. J. Schwartz, & K. Luyckx (Eds.), *Handbook of identity: Theory and research* (pp. 1–27). New York: Springer.

Vignoles, V. L., Chryssochoou, X., & Breakwell, G. M. (2000). The distinctiveness principle: Identity, meaning, and the bounds of cultural relativity. *Personality and Social Psychology Review, 4*, 337–354.

Vignoles, V. L., Regalia, C., Manzi, C., Golledge, J., & Scabini, E. (2006). Beyond self-esteem: Influence of multiple motives on identity construction. *Journal of Personality and Social Psychology, 90*(2), 308–333.

Werner, C. M., & Makela, E. (1998). Motivations and behaviors that support recycling. *Journal of Environmental Psychology, 18*, 373–386.

Whitmarsh, L., & O'Neill, S. (2010). Green identity, green living? The role of proenvironmental self identity in determining consistency across diverse proenvironmental behaviors. *Journal of Environmental Psychology, 30*(3), 305–314.

World Health Organization. (1995). *Report of the working group on concepts and principles of health promotion*. Copenhagen: WHO.

World Health Organization Quality of Life Assessment (WHOQOL): Position paper from the World Health Organization (1995). *Social Science & Medicine, 41*(10), 1403–1409.

13
Can Engagement in Environmentally-Friendly Behavior Increase Well-Being?

Leonie Venhoeven, Linda Steg, and Jan Willem Bolderdijk

13.1 Introduction

The transition to a sustainable society is an important goal in the coming years. As defined in the Brundtland report, a sustainable society is one in which "the use of goods and services [...] respond to basic needs and bring a better quality of life, while minimizing the use of natural resources, toxic materials and emissions of waste and pollutants over the life cycle, so as not to jeopardize the needs of future generations" (Norwegian Ministry of the Environment 1994). To make a transition to such a society, individual behavior changes are needed (IPCC 2014), as they are an important driver of environmental change (DuNann Winter and Koger 2004; Gardner and Stern 2002; Gifford et al. 2011; Hackmann et al. 2014; Vlek and Steg 2007; Weaver et al. 2014).

As a sustainable society not only focuses on increasing the quality of the environment, but also includes increasing quality of life, one would expect individuals to be very willing to change their behavior accordingly. Although a better quality of life is a crucial part of Brundtland's definition of sustainability, however, many seem to assume that engagement in sustainable or environmentally-friendly behavior will have a negative impact on the individual quality of life or well-being of those who act in this way. Engagement in environmentally-friendly behavior may entail some level of discomfort – for example, taking a short shower or turning down the thermostat – or may involve giving up certain things – for example, not flying or decreasing meat consumption. Because of this, it is often assumed that people see acting in an environmentally-friendly way as "being asked to give up a modern, high-technology existence for an austere, bleak but needed substitute" (De Young 1990–1991, p. 216).

However, there is also a brighter view of sustainable or environmentally-friendly behavior. Several correlational studies show that people who engage in more environmentally-friendly behavior actually experience more well-being (Brown and Kasser 2005; Kasser and Sheldon 2002; Xiao and Li 2011). Furthermore, other research reveals that, overall, people associate environmentally-friendly behavior with positive emotions (Venhoeven et al., under review) and may experience feelings of "warm glow" after acting this way (Taufik et al. 2015). How can we explain that engagement in environmentally-friendly behavior seems to be both positively and negatively related to well-being? Is one of these options simply truer than the other, or is the story more complex? In this chapter, we discuss the relationship between environmentally-friendly

L. Venhoeven (✉) • L. Steg • J.W. Bolderdijk
University of Groningen, Groningen, The Netherlands
e-mail: l.a.venhoeven@rug.nl; e.m.steg@rug.nl; j.w.bolderdijk@rug.nl

behavior and well-being in more depth, and give several explanations for why a positive and negative relationship might exist.

13.2 Why Would Environmentally-Friendly Behavior Affect Well-being?

A possible explanation for why environmentally-friendly behavior seems to be both positively and negatively related to well-being can be found in the definition of well-being itself, and thereby in its related causes (Venhoeven et al. 2013). The well-being literature often makes a distinction between hedonic and eudaimonic well-being. The former concept generally refers to feelings of *pleasure*, while the latter refers to feelings of *meaning* (Deci and Ryan 2008). However, the idea that there are different contents of well-being has been challenged (Kashdan et al. 2008). Notably, it has been argued that rather than different *types* of well-being, there may be different *causes* of well-being. Behavior may contribute to a general feeling of well-being both because it is pleasurable and because it is meaningful to do. Applying this reasoning to environmentally-friendly behavior, both the pleasure associated with the behavior and the meaning of the behavior may lead to its contributing to or, under some circumstances, detracting from well-being.

13.2.1 Environmentally-Friendly Behavior and Pleasure

Although it is not assumed to be the first association people have with engagement in environmentally-friendly behavior, some types of environmental action can very well be perceived as pleasant to do. For example, some people enjoy riding their bicycle on a sunny day or perceive a train ride to be more comfortable than a flight with a low-cost airline. For these people, engagement in these activities thus not only benefits the environment, but also is inherently enjoyable.

However, not all environmentally-friendly behaviors are perceived to be more enjoyable than their environmentally-unfriendly counterparts, and many may even be perceived to be less enjoyable. Turning down the thermostat on a cold winter day, waiting for the bus at a small, windy bus stop, or cycling in the rain may be considered environmentally-friendly but, at the same time, uncomfortable behavior. It is this association that may underlie the assumed negative relationship between engagement in environmentally-friendly behavior and well-being.

Being enjoyable or comfortable is thus not a defining part of environmentally-friendly behavior as such, but rather a byproduct of specific types of behavior. Some strategies therefore target increasing the pleasure aspect of specific environmentally-friendly behavior in order to stimulate engagement (Nordahl 2012; Volkswagen 2011). Examples are making travel by public transport more comfortable by working with taxi companies that bring passengers from the station to their final destination (http://www.ns.nl/reizigers/producten/van-deur-tot-deur/ns-zonetaxi.html), or transforming a bottle bank into an arcade to make separating glass a fun activity (Volkswagen 2011). However, as will be discussed later in the chapter, there may be some drawbacks of using this strategy to increase well-being associated with environmentally-friendly behavior. In the next section, therefore, we focus on meaning as a route to well-being. As we will argue next, this route may prove to be a way to strengthen the link between engagement in environmentally-friendly behavior and well-being *in general*, instead of focusing on specific behaviors like with pleasure, via a route that may still bring pleasure in the end.

13.2.2 Environmentally-Friendly Behavior and Meaning

As described above, pleasure is not inherent to engaging in all types of environmentally-friendly behavior: while some types are more pleasant to engage in than their environmentally-unfriendly counterpart, other types are not, and may even be less pleasant to engage in than their environmentally-unfriendly counterpart. What is inherent to engaging in environmentally-friendly behavior, however, is that it can be positioned as doing the right thing, for example as assuring a better living environment for people now and in the future, all over the world. As a result, engagement in this type of behavior could be seen as meaningful. More often than a pleasant experience, therefore, environmentally-friendly behavior may be perceived as a meaningful experience by those who act in this way.

Many studies show that it feels good to do good. Spending money on others feels better than spending money on yourself (Aknin et al. 2012; Dunn et al. 2008), having the feeling that your work helps or benefits others prevents signs of burnout (Grant and Sonnentag 2010) and prosocial (Andreoni 1989; Andreoni 1990) and environmentally-friendly behavior (Taufik et al. 2015) can make people feel good and elicit feelings of "warm glow". If perceiving your actions as 'right' or meaningful is indeed a route to well-being other than pleasure, the extent to which engagement in environmentally-friendly behavior contributes to well-being probably depends on the extent to which it is perceived as doing the right thing. Below, we first discuss why doing the right thing may contribute to well-being. Following this reasoning, we argue that, in as far as it is considered the right thing to do, environmentally-friendly behavior can enhance well-being. We end with several factors that may influence the extent to which this behavior is perceived as the right thing to do.

13.2.2.1 Why Meaning Provides Well-Being

Self-Concept. One of the reasons why doing the right thing could foster well-being is that it affects how you see yourself: if you are making the effort to engage in good behavior, you must be a good person. Indeed, well-being is greater when one's actions are perceived as doing good instead of doing harm (Grant and Campbell 2007), and how positively people think of themselves is determined by how moral they perceive themselves to be (Dunning 2007; Sachdeva et al. 2009).

Following this reasoning, having a strong environmental self-identity, that is, seeing yourself as an environmentally-friendly person (Van der Werff et al. 2014a), may be beneficial for well-being. Seeing yourself as someone whose actions are environmentally-friendly may have a positive influence on your self-concept: it may mean you see yourself as someone who does good (Bolderdijk et al. 2013). Research shows, for instance, that the higher perceived environmental (e.g., CO_2 emissions) and symbolic (e.g., shows who I am) values of an electric vehicle increase the likelihood that people will be interested in buying this type of car (Noppers et al. 2014). Although this was not explicitly studied, environmental and symbolic attributes may be precisely those attributes that make such a purchase an illustration of how good a person you are, and thereby may make such a purchase feel good. Furthermore, a more direct test of the process of self-concept shows that environmentally-friendly behavior can elicit feelings of "warm glow" *because* this behavior sends a positive self-signal (Taufik et al. 2015). Being able to attribute engagement in environmentally-friendly behavior to yourself, in other words, feeling *you* are someone who does the right thing, may therefore be an important factor in explaining why this engagement could foster well-being.

13.2.2.2 When Environmentally-Friendly Behavior Provides Well-Being

Autonomy. As described in the previous section, engagement in meaningful or right behavior may enhance well-being because it contributes to a positive self-concept: by doing good you show yourself that you are a good person. In order to provide meaning in this sense, it is important that people can credibly attribute their choice to engage in environmentally-friendly behavior to their own volition. One important factor in this respect may be that the choice to engage in this behavior is made autonomously. When people autonomously choose to behave in a certain way, this signals to themselves and to others that they value this autonomously chosen behavior: they are more likely to attribute the choice of engagement to internal instead of external causes (Van der Werff et al. 2014b). Autonomously choosing to engage in environmentally-friendly behavior shows that you are someone who values helping the environment, and this in turn leads you to conclude you are a good person. However, when the choice of this behavior is not autonomous, you may be less likely to attribute its value and goodness to yourself, and thereby gain less well-being from its engagement. In line with this reasoning, the question is whether the meaning that can be derived from pursuing a sustainable lifestyle will be experienced by any person who accidentally happens to act in a pro-environmental way. It is likely that only those who deliberately choose a pro-environmental lifestyle will deem this engagement meaningful (Evans and Jackson 2008), thereby gaining well-being from their engagement.

In addition to attributing the behavior to yourself, engaging in environmentally-friendly behavior should be seen as the right thing to do by those who act in this way, in order to provide meaning. Below, we discuss two factors that may influence the perceived "rightness" of environmentally-friendly behavior: individual values and the perceived environmental impact of the behavior.

Values. The first factor that may influence whether environmentally-friendly behavior is perceived to be the right thing to do is the extent to which this type of behavior matches with individuals' values, that is, with what people find important in their life (De Groot and Steg 2010; Sheldon and Houser-Marko 2001). Values can be defined as "desirable trans-situational goals varying in importance, which serve as a guiding principle in the life of a person or other social entity" (Schwartz 1992, p. 21). The value people place on the ecosystem and biosphere is a particularly important predictor of engagement in environmentally-friendly behavior (De Groot and Steg 2008, 2009, 2010; Steg and De Groot 2012). People who value the ecosystem and biosphere, that is, people with strong biospheric values, place a strong importance on preserving the environment, which they see as a guiding principle in their lives (De Groot and Steg 2008). Therefore, behaviors that contribute to reaching this trans-situational goal may be more meaningful for them than for people with weaker biospheric values. This implies that engagement in environmentally-friendly behavior may contribute more to the well-being of individuals with strong biospheric values.

In general, biospheric values are related to more internalized types of motivation to act in an environmentally-friendly way (De Groot and Steg 2010). People with stronger biospheric values say they enjoy contributing to the environment more (intrinsic motivation), see doing things for the environment as a more integral part of their life (integrated regulation), and think doing things for the environment is a more sensible thing to do (identified regulation). In other words, they see engagement in environmentally-friendly behavior as more meaningful. An interesting side-effect seems to be that, by seeing this behavior as more *meaningful* (integrated and identified regulation), it becomes more *pleasurable* to engage in (intrinsic motivation) as well. When people talk about the 'simple pleasures' associated with environmentally-friendly behaviors, such as an energy-saving ritual, a shift in the way they eat, or a shift in the way they move (Evans and Jackson 2008), it is probably this source of pleasure they are describing.

Yet, people with stronger biospheric values also say that they would feel guiltier if they failed to do things for the environment (introjected regulation) (De Groot and Steg 2010). Similarly, the more strongly people endorse self-transcendent and biospheric values, the more moral obligation they feel to reduce their personal car use (Nordlund and Garvill 2003). This indicates that because they see the behavior as a more moral and thus meaningful cause, they feel more obliged to act accordingly, hence guiltier when not acting accordingly. Building on these findings, one could argue that the effect of biospheric values on the meaning provided by engagement in environmentally-friendly behavior operates like a catalyst of both negative and positive emotions. Firstly, as people with stronger biospheric values place more value on environmentally-friendly behavior, the engagement in this type of behavior may be more meaningful for them and thereby may have a more positive impact on their well-being than for people with weaker biospheric values. For instance, people who act more in line with their values are found to feel their lives are more meaningful (Sheldon and Krieger 2014). Secondly, as people with stronger biospheric values place more value on environmentally-friendly behavior, the *failure* to engage in environmentally-friendly behavior may also be more meaningful for them and thereby may have a more negative impact on their well-being than for people with weaker biospheric values. As mentioned above, people expect to feel guilty if they do not act in line with their values (De Groot and Steg 2010). As stronger biospheric values make engagement in environmentally-friendly behavior more meaningful, this may amplify both the positive emotions elicited by engagement in this type of behavior, and the negative emotions elicited by a failure to engage in it.

Impact. A second factor that may influence whether environmentally-friendly behavior is perceived to be the right thing to do is the impact it has on the environment, for example in terms of the amount of CO_2 emitted. It sounds obvious that the higher the impact of a certain type of behavior on the environment, the more meaningful it would be perceived to engage in that behavior. For instance, people who are saving 500 tons of CO_2 per year via energy reductions could be expected to feel they are making a more meaningful contribution than people who save 500 kg of CO_2 per year. Previous research, however, suggests that the amount and perceived worth of behavior do not always correspond.

One of the reasons often mentioned for this discrepancy is that the value of the commonly used unit to express environmental impact, CO_2 emissions, may be difficult to grasp for most people (Fitzpatrick and Smith 2009; Jain et al. 2013; Vassileva et al. 2012; Zapico et al. 2011). In general, people know that more CO_2 emissions are worse for the environment than fewer CO_2 emissions, but whether a single number of CO_2 emissions is large or small may be difficult for them to judge. When the actual impact of behavior is described in terms of CO_2 emissions, therefore, this may tell people little about whether this is meaningful. Building on this, studies have not found the amount of CO_2 emissions to be related to the perceived meaning or worth of environmentally-friendly behaviors (Dogan et al. 2014). CO_2 emissions may be one indicator people use to evaluate the meaning or worth of behavior, but it is not the only one, and possibly not even the most important.

A more significant reason why the actual impact of behavior and its perceived meaning may not always correspond is that how people feel about certain outcomes is not necessarily related to the size of those outcomes (Hsee and Rottenstreich 2004; Hsee et al. 2005). Therefore, especially when people rely on how they feel about an effect, they can be insensitive to the scope of the effect when judging its perceived value. For instance, previous research illustrates that emphasizing the environmental benefits of several eco-driving behaviors makes engagement in these behaviors more worthwhile than emphasizing their financial benefits, independently of the amount of these benefits (Dogan et al. 2014). Just as people do not necessarily feel better about the idea that they can help 100 people compared to one person (Small et al. 2007), engagement in environmentally-friendly behavior may be considered meaningful regardless of the impact, as any contribution, however small, is a good contribution, and thus diagnostic of a good self-concept.

13.3 Practical Implications

Following the processes discussed in this chapter, there seem to be two routes to increase the positive effect of engagement in environmentally-friendly behavior on well-being: decrease the "discomfort" and make it more "pleasant" to engage in the behavior, or increase the "meaning" of engagement. Although policymakers often use the first route, we now present several reasons why the second route may have additional benefits.

13.3.1 Spillover

One way to make using public transport more comfortable or more fun than using a car is to reshape bus stops into small strawberry houses that shelter passengers from the cold and rain (example derived from Nordahl 2012). Although this may be an effective strategy to increase the comfort or fun of taking the bus, it *only* has this effect. It does not influence how people perceive other environmentally-friendly modes of transport, such as riding a bicycle, or other environmentally-friendly behaviors such as decreasing energy use by taking shorter showers. Thus, in order to contribute to well-being via this route, one would have to increase the comfort or fun of many environmentally-friendly behaviors separately, which would require a huge investment of time and money.

In contrast, the meaning of environmentally-friendly behavior is derived from the mere idea that it *is* environmentally-friendly, i.e. that the behavior belongs to this category. Specific behavior thus gains meaning when it is seen as making a contribution to a better environment, and when making a positive contribution to the environment in general is seen as the right thing to do. By making people aware that it is important to take care of the environment, and by showing them what types of behavior contribute to this goal, whole groups of behavior potentially gain meaning (Evans et al. 2013; Thøgersen and Crompton 2009), and engagement in them can thereby contribute to well-being. This route may therefore be much more efficient as it potentially has a much larger range.

13.3.2 Long-Term Effects

Besides the investments it would take to increase the pleasure or fun of all separate environmentally-friendly behaviors, one could question the endurance of such a strategy. As the effect known as the 'hedonic treadmill' suggests, people undergo a short-lived boost in their well-being when they have a pleasant experience. However, when they get used to the new experience, this boost fades and their well-being returns to its previous levels (but for a revision of the model, see Diener et al. 2006). Increasing the fun of environmentally-friendly behavior, for instance by using a slide instead of stairs to get to your train (http://metro.co.uk/2011/07/22/overvecht-railway-station-installs-childrens-slide-to-help-busy-passengers-88544/), may have only a short-lived effect on the well-being provided by engagement in this behavior, and thereby be only a short-lived motivation for engagement.

Meaning, on the other hand, may prove to be a longer-lasting basis for well-being and engagement. On a general level, an orientation towards "the pleasant life" with a focus on having pleasurable experiences is less strongly related to long-term life satisfaction than either an orientation towards "the engaged life", with a focus on losing oneself in one's activities/experiencing flow, or an orientation towards "the meaningful life", with a focus on having a purpose (Peterson et al. 2005). Applying this to the environmental domain, this suggests that meaning derived from engagement in environmentally-friendly behavior may provide a better basis for long-term well-being than pleasure will. In addition, besides being more strongly linked to long-term well-being, perceived personal meaning can provide a more stable basis for actual engagement in an activity than perceived pleasure can. Having an interest in following politics and finding it inherently enjoyable to do so, for instance, does not necessarily translate into voting behavior, while following politics because it is perceived to be meaningful does lead people to vote (Losier and Koestner 1999). Applying this to the environmental domain, this suggests that following the available knowledge on climate developments because it is inherently enjoyable might not necessarily translate into acting in an environmentally-friendly way, while following climate developments because it is perceived to be meaningful does lead people to act accordingly. Thus, as well as providing a better basis for long-term well-being, meaning may also be a more stable base for engagement in environmentally-friendly behavior in the first place.

13.3.3 Initial Pull Versus Long-Term Effects

As mentioned in the previous sections, increasing the feeling that engagement in environmentally-friendly behavior is meaningful may be a more stable basis for behavior change than making the behavior more pleasant to engage in. However, the latter may be crucial for convincing people to start engaging in this behavior, especially if they do not yet greatly value the environment. The question that arises is whether this means one has to choose between a focus on pleasure, which may motivate people to act in a desirable way in the short-run but may not last, and a focus on meaning, which may have more stable long-term effects but may have a difficult start. Is it possible to obtain both desirable short- and long-term effects?

Studies on environmental self-identity suggest that there are opportunities to combine these two. Framing behavior people have already performed in the past as environmentally-friendly increases their perception of themselves as an environmentally-friendly person, which in turn increases the likelihood of engagement in environmentally-friendly behavior in the future (Cornelissen et al. 2008; Van der Werff et al. 2014a). Making some environmentally-friendly behaviors more pleasant to engage in may increase the initial engagement and well-being derived from these specific behaviors. Then, by making people aware of the positive effects of their engagement on the environment afterwards, the personal meaning they start to attach to the behavior may lead to a longer-lasting well-being. Moreover, their initial engagement in environmentally-friendly behavior may cause spillover to other types of environmentally-friendly behavior, once they link their engagement to personal meaning.

Referring back to the processes discussed in this chapter, making it easy for people to attribute the choice of engagement in this behavior to themselves (for instance, by increasing feelings of autonomous choice), linking environmentally-friendly behavior to the values people hold (for instance, by strengthening biospheric values from a young age onwards), and increasing the emotional rather than the actual impact of engagement (for instance, by projecting a tree on the wall that grows when energy use is low, and withers when energy use is high) may be a good starting point for convincing people of the personal meaning they attach to environmentally-friendly behavior – thereby providing a more stable basis for engagement itself and the well-being derived from it.

References

Aknin, L. B., Dunn, E. W., & Norton, M. I. (2012). Happiness runs in a circular motion: Evidence for a positive feedback loop between prosocial spending and happiness. *Journal of Happiness Studies, 13*(2), 347–355. doi:10.1007/s10902-011-9267-5.

Andreoni, J. (1989). Giving with impure altruism: Applications to charity and ricardian equivalence. *Journal of Political Economy, 97*(6), 1447–1458.

Andreoni, J. (1990). Impure altruism and donations to public goods: A theory of warm-glow giving. *The Economic Journal, 100*(401), 464–477.

Bolderdijk, J. W., Steg, L., Geller, E. S., Lehman, P. K., & Postmes, T. (2013). Comparing the effectiveness of monetary versus moral motives in environmental campaigning. *Nature Climate Change, 3*, 413–416. doi:10.1038/nclimate1767.

Brown, K. W., & Kasser, T. (2005). Are psychological and ecological well-being compatible? The role of values, mindfulness, and lifestyle. *Social Indicators Research, 74*(2), 349–368. doi:10.1007/s11205-004-8207-8.

Cornelissen, G., Pandelaere, M., Warlop, L., & Dewitte, S. (2008). Positive cueing: Promoting sustainable consumer behavior by cueing common environmental behaviors as environmental. *International Journal of Research in Marketing, 25*(1), 46–55. doi:10.1016/j.ijresmar.2007.06.002.

De Groot, J. I. M., & Steg, L. (2008). Value orientations to explain beliefs related to environmental significant behavior: How to measure egoistic, altruistic, and biospheric value orientations. *Environment and Behavior, 40*(3), 330–354. doi:10.1177/0013916506297831.

De Groot, J. I. M., & Steg, L. (2009). Mean or green: Which values can promote stable pro-environmental behavior? *Conservation Letters, 2*(2), 61–66. doi:10.1111/j.1755-263X.2009.00048.x.

De Groot, J. I. M., & Steg, L. (2010). Relationships between value orientations, self-determined motivational types and pro-environmental behavioural intentions. *Journal of Environmental Psychology, 30*(4), 368–378. doi:10.1016/j.jenvp.2010.04.002.

De Young, R. (1990–1991). Some psychological aspects of living lightly: Desired lifestyle patterns and conservation behavior. *Journal of Environmental Systems, 20*(3), 215–227.

Deci, E. L., & Ryan, R. M. (2008). Hedonia, eudaimonia, and well-being: An introduction. *Journal of Happiness Studies, 9*(1), 1–11. doi:10.1007/s10902-006-9018-1.

Diener, E., Lucas, R. E., & Scollon, C. N. (2006). Beyond the hedonic treadmill: Revising the adaptation theory of well-being. *American Psychologist, 61*(4), 305–314. doi:10.1037/0003-066X.61.4.305.

Dogan, E., Bolderdijk, J. W., & Steg, L. (2014). Making small numbers count: Environmental and financial feedback in promoting eco-driving behaviours. *Journal of Consumer Policy, 37*(3), 413–422. doi:10.1007/s10603-014-9259-z.

DuNann Winter, D., & Koger, S. M. (2004). *The psychology of environmental problems*. Mahwah: Lawrence Erlbaum.

Dunn, E. W., Aknin, L. B., & Norton, M. I. (2008). Spending money on others promotes happiness. *Science, 319*(5870), 1687–1688. doi:10.1126/science.1150952.

Dunning, D. (2007). Self-image motives and consumer behavior: How sacrosanct self-beliefs sway preferences in the marketplace. *Journal of Consumer Psychology, 17*(4), 237–249.

Evans, D., & Jackson, T. (2008). *Sustainable consumption: Perspectives from social and cultural theory* (Working Paper No. RESOLVE Working Paper 05–08). Guildford: University of Surrey.

Evans, L., Maio, G. R., Corner, A., Hodgetts, C. J., Ahmed, S., & Hahn, U. (2013). Self-interest and pro-environmental behaviour. *Nature Climate Change, 3*, 122–125. doi:10.1038/nclimate1662.

Fitzpatrick, G., & Smith, G. (2009). Technology-enabled feedback on domestic energy consumption: Articulating a set of design concerns. *Pervasive Computing IEEE, 8*(1), 37–44. doi:10.1109/MPRV.2009.17.

Gardner, G. T., & Stern, P. C. (2002). *Environmental problems and human behavior*. Boston: Pearson Custom Publishing.

Gifford, R., Kormos, C., & McIntyre, A. (2011). Behavioral dimensions of climate change: Drivers, responses, barriers, and interventions. *Wiley Interdisciplinary Reviews: Climate Change, 2*(6), 801–827. doi:10.1002/wcc.143.

Grant, A. M., & Campbell, E. M. (2007). Doing good, doing harm, being well and burning out: The interactions of perceived prosocial and antisocial impact in service work. *Journal of Occupational and Organizational Psychology, 80*(4), 665–691. doi:10.1348/096317906X169553.

Grant, A. M., & Sonnentag, S. (2010). Doing good buffers against feeling bad: Prosocial impact compensates for negative task and self-evaluations. *Organizational Behavior and Human Decision Processes, 111*(1), 13–22. doi:10.1016/j.obhdp.2009.07.003.

Hackmann, H., Moser, S. C., & St. Clair, A. L. (2014). The social heart of global environmental change. *Nature Climate Change, 4*, 653–655. doi:10.1038/nclimate2320.

Hsee, C. K., & Rottenstreich, Y. (2004). Music, pandas, and muggers: On the affective psychology of value. *Journal of Experimental Psychology: General, 133*(1), 23–30. doi:10.1037/0096-3445.133.1.23.

Hsee, C. K., Rottenstreich, Y., & Xiao, Z. (2005). When is more better? On the relationship between magnitude and subjective value. *Current Directions in Psychological Science, 14*(5), 234–237. doi:10.1111/j.0963-7214.2005.00371.x.

IPCC. (2014). *Climate change 2014: Mitigation of climate change. Contribution of working group III to the fifth assessment report of the intergovernmental panel on climate change*. Cambridge: Cambridge University Press.

Jain, R. K., Taylor, J. E., & Culligan, P. J. (2013). Investigating the impact eco-feedback information representation has on building occupant energy consumption behavior and savings. *Energy and Buildings, 64*, 408–414. doi:10.1016/j.enbuild.2013.05.011.

Kashdan, T. B., Biswas-Diener, R., & King, L. A. (2008). Reconsidering happiness: The costs of distinguishing between hedonics and eudaimonia. *The Journal of Positive Psychology, 3*(4), 219–233. doi:10.1080/17439760802303044.

Kasser, T., & Sheldon, K. M. (2002). What makes for a Merry Christmas? *Journal of Happiness Studies, 3*(4), 313–329. doi:10.1023/A:1021516410457.

Losier, G. F., & Koestner, R. (1999). Intrinsic versus identified regulation in distinct political campaigns: The consequences of following politics for pleasure versus personal meaningfulness. *Personality and Social Psychology Bulletin, 25*(3), 287–298. doi:10.1177/0146167299025003002.

Noppers, E. H., Keizer, K., Bolderdijk, J. W., & Steg, L. (2014). The adoption of sustainable innovations: Driven by symbolic and environmental motives. *Global Environmental Change, 25*, 52–62. doi:10.1016/j.gloenvcha.2014.01.012.

Nordahl, D. (2012). *Making transit fun! How to entice motorists from their cars (and onto their feet, a bike, or bus)*. Washington, DC: Island Press.

Nordlund, A. M., & Garvill, J. (2003). Effects of values, problem awareness, and personal norms on willingness to reduce personal car use. *Journal of Environmental Psychology, 23*(4), 339–347. doi:10.1016/S0272-4944(03)00037-9.

Norwegian Ministry of the Environment. (1994). *Symposium: Sustainable consumption. 19–20 January 1994, Oslo, Norway*. Oslo: Norwegian Ministry of the Environment.

Peterson, C., Park, N., & Seligman, M. E. P. (2005). Orientations to happiness and life satisfaction: The full life versus the empty life. *Journal of Happiness Studies, 6*(1), 25–41. doi:10.1007/s10902-004-1278-z.

Sachdeva, S., Iliev, R., & Medin, D. L. (2009). Sinning saints and saintly sinners: The paradox of moral self-regulation. *Psychological Science, 20*(4), 523–528.

Schwartz, S. H. (1992). Universals in the content and structure of values: Theoretical advances and empirical tests in 20 countries. In M. Zanna (Ed.), *Advances in experimental social psychology* (pp. 1–65). Orlando: Academic.

Sheldon, K. M., & Houser-Marko, L. (2001). Self-concordance, goal attainment, and the pursuit of happiness: Can there be an upward spiral? *Journal of Personality and Social Psychology, 80*(1), 152–165. doi:10.1037/0022-3514.80.1.152.

Sheldon, K. M., & Krieger, L. S. (2014). Walking the talk: Value importance, value enactment, and well-being. *Motivation and Emotion, 38*(5), 609–619. doi:10.1007/s11031-014-9424-3.

Small, D. A., Loewenstein, G., & Slovic, P. (2007). Sympathy and callousness: The impact of deliberative thought on donations to identifiable and statistical victims. *Organizational Behavior and Human Decision Processes, 102*(2), 143–153. doi:10.1016/j.obhdp.2006.01.005.

Steg, L., & De Groot, J. I. M. (2012). Environmental values. In S. Clayton (Ed.), *The Oxford handbook of environmental and conservation psychology* (pp. 81–92). New York: Oxford University Press.

Taufik, D., Bolderdijk, J. W., & Steg, L. (2015). Acting green elicits a literal 'warm-glow'. *Nature Climate Change, 5*, 37–40. doi:10.1038/nclimate2449.

Thøgersen, J., & Crompton, T. (2009). Simple and painless? The limitations of spillover in environmental campaigning. *Journal of Consumer Policy, 32*(2), 141–163. doi:10.1007/s10603-009-9101-1.

Van der Werff, E., Steg, L., & Keizer, K. (2014a). I am what I am, by looking past the present. The influence of biospheric values and past behavior on environmental self-identity. *Environment and Behavior, 46*(5), 626–657. doi:10.1177/0013916512475209.

Van der Werff, E., Steg, L., & Keizer, K. (2014b). Follow the signal: When past environmental actions signal who you are. *Journal of Environmental Psychology, 40*, 273–282. doi:10.1016/j.jenvp.2014.07.004.

Vassileva, I., Odlare, M., Wallin, F., & Dahlquist, E. (2012). The impact of consumers' feedback preferences on domestic electricity consumption. *Applied Energy, 93*, 575–582. doi:10.1016/j.apenergy.2011.12.067.

Venhoeven, L. A., Bolderdijk, J. W., & Steg, L. (2013). Explaining the paradox: How pro-environmental behaviour can both thwart and foster well-being. *Sustainability, 5*, 1372–1386. doi:10.3390/su5041372.

Vlek, C., & Steg, L. (2007). Human behavior and environmental sustainability: Problems, driving forces and research topics. *Journal of Social Issues, 63*(1), 1–19. doi:10.1111/j.1540-4560.2007.00493.x.

Volkswagen. (2011). The fun theory. Retrieved from http://www.thefuntheory.com/

Weaver, C. P., Mooney, S., Allen, D., Beller-Simms, N., Fish, T., Grambsch, A. E., et al. (2014). From global change science to action with social sciences. *Nature Climate Change, 4*, 656–659. doi:10.1038/nclimate2319.

Xiao, J. J., & Li, H. (2011). Sustainable consumption and life satisfaction. *Social Indicators Research, 104*(2), 323–329.

Zapico, J. L., Guath, M., & Turpeinen, M. (2011). Kilograms or cups of tea: Comparing footprints for better CO_2 understanding. *PsychNology Journal, 9*(1), 43–54.

Part IV

Well-Being and Daily Environments – *Urban Environments*

Urban Design and Quality of Life

Ombretta Romice, Kevin Thwaites, Sergio Porta, Mark Greaves, Gordon Barbour, and Paola Pasino

Outline From an overview on recent trends in urbanization, we will introduce the notion of control as a key to read the following text and in particular we will:

1. contextualize the concept of control in relation to the fields of both quality of life (QoL) and urban form. In fact, the literature in both domains shows that there is a mutually reciprocal relationship between aspects of quality of life and urban spatial structure;
2. review established and recent research on the relationships between QoL and urban form, structured around metropolitan, neighborhood and pedestrian scales, which illustrates the centrality of control in shaping our cities and allowing quality of life to be fulfilled within them;
3. propose a conceptual framework for socio-spatial urban design, which is sensitive to the relative importance of predictive/structural and loose/flexible urban elements in the production and management of urban space, and their critical role in affording their users a sense of control;
4. suggest the need for a reconceptualization of city form away from an assemblage of material and spatial elements towards a more integrated sense of a city as a mutually defining socio-spatial system.

Implicit in the development of our narrative are two assumptions, which we will aim to highlight throughout our discussion. These are:

- a wealth of literature has accumulated over the past five decades, ranging from the work of Jacobs and Hall, for example, in the 1960s through to Gehl, Dovey and Habraken more recently, which has attempted to connect the form of cities with social processes in various ways. Despite this, effective synthesis of this material has yet to be systematically undertaken and its influence on, and acceptance within, the mainstream of practice remains limited at best;
- there is a continuing corrosive impact of a prevailing disciplinary fragmentation, which perpetuates the separation of the built environment disciplines from those concerned with human social and psychological processes, resulting in communication barriers that obstruct effective cross-disciplinary discourse.

O. Romice (✉) • S. Porta • G. Barbour • P. Pasino
Department of Architecture, University of Strathclyde, Glasgow, UK
e-mail: Ombretta.r.romice@strath.ac.uk

K. Thwaites
Department of Landscape, University of Sheffield, Sheffield, UK
e-mail: k.thwaites@strath.ac.uk

M. Greaves
Mark Greaves Urbanism, Glasgow, UK

14.1 Introduction: An Urbanized Future

Urbanism is a very old term; it has accompanied the development of our cities for centuries, through the skilled and at times grand and intentional, to the ad hoc and piecemeal intervention of development, growth and refinement. Significantly different targeted, widespread, professional and coordinated approaches to urban planning emerged to address a severe public health crisis only when industrialization hit cities in Europe and North America in the second half of the nineteenth and the first half of the twentieth centuries. A new profession was born here that effectively divorced scales of intervention by separating architecture from planning and thus created a gap in the layout and design of our cities. This became evident at a large scale after WWII and has had a significant impact ever since on how we experience them.

Thereafter, the concomitant effects of both World Wars and the aging of the stock built during industrialization called again for large-scale intervention. Healthier cities (physically and morally), more efficient cities, less city-like cities, reformed cities were the ambitions of these early experiments in urban planning, which were broadly translated into a dispersed model and the zoning of functions. Dispersion and zoning combined to shape post-war urbanization, and are still playing a part in our daily environments.

Nevertheless, the potential benefits of density and mix became clear in the early 1970s when the oil crises and greater environmental awareness revealed the un-tenability of a world based on the consumption of finite resources and the need for a different model of development. Between the 1980s and 1990s, advances in technology and globalization brought a very polarized economic growth, changing the form of cities yet again, making it more specialized, with great repercussions on the relationships between regions around the world, resulting in increased social inequalities.

Everyone is affected by the problem of sustainable development: on the one hand, there are areas in the world where population and urban infrastructure are not yet synchronized, that is the scale of urbanization is not yet fully matched by income growth and institutional development, and where the experimented paradigms and planning approaches cited are imported as signs of aspiring modernization. Here, we call these "the becoming cities". On the other hand, there are those countries where planning, policy, technological and scientific advances are available and matched, but the nature of change is profoundly cultural and therefore slow, due to a complex balance of economic, political, social, and environmental interests; we call these "retrofitting cities". These are not fundamentally different problems, but more like two sides of the same coin.

This apparent mismatch between the resources available to deliver sustainability and the scale of the task calls for a different paradigm of creation and delivery of our space, one in which the responsibilities of structuring, equipping, using and managing land are shared between institutions (intended here in the broadest sense) and users, in a way that recognizes that the benefits derived from responsibility can actually become shared benefits – cultural, societal, financial and environmental.

The form of our cities has a role in generating such benefits, in relation to its capacity to afford its users, among other things, control. It embeds cultural values and supports habits but, unlike values and habits, it has, in principle, a longer life span. Life span and adaptability are now the key issue because the cost of remediation for environments that are not fitting and supporting will become increasingly prohibitive. Individual urban forms differ greatly, but the principles that govern and structure them are surprisingly lasting over time and were only significantly challenged after WWII. The capacity of these structures to survive life spans, representing and supporting changing values and habits, may also differ accordingly; we cannot stay in some places, we cannot inhabit them without losing our identity, feeling unsafe, alienated, or threatened, while others have remained with us for centuries, adapting to our transformations, responding to our needs, fulfilling our lives, and allowing a bond to form. Establishing what determines this difference in responsiveness, and what benefits are derived from it, is summarized in the literature review of this chapter, and will lead to more holistic and phenomenological concepts of human-environment relationships as solutions better able to integrate city form and social processes.

14.1.1 The Research/Review Problem

Overviews of cities and their effect on people, presented in handbooks in the area of environment behavior studies, often start by listing the positive and negative traits of cities – mainly in relation to density and opportunities on one hand, and crowdedness, pollution and alienation on the other. Individual studies on single aspects of urban form and their impact on cognition, affection and behavior and attitudes are also very plentiful, with several journals dedicated to this theme, and a fast-growing international portfolio of cases and examples. More recently, encompassing publications have linked the discussion on cities to environmental effects (Speck 2013), and overall quality of life (Montgomery 2013).

Planning, design and social sciences have also benefitted from the more recent interest and activity of data analysts, mathematicians, etc., with great advances in the understanding of how cities function as complex systems, and how socio-economic and environmental aspects of life are linked to form. Even more recently, the study of cities and their character has become popularized, being embraced by entire communities, often through innovations in social media outlets/forums, to observe, record, map and track morphological, behavioral, and usage data (we can now model, use remote sensing and crowd sourcing, and conduct simultaneous morphological comparisons at global, national, metropolitan and local scales). This is significant as it is creating a much broader pool of diverse knowledge than we have ever had, to the point that we can now link advances in quantitative work to the study of trends and patterns at any scale, and make increasingly sophisticated observations of shared, cross-cultural and contextual behavior,

to use both as evidence and as guidance. In theory, with this knowledge at hand, *"Planning and design, when aware of these complex molar systems, can act on city form, to enhance, enable or alleviate immediate and extended relations and behaviours in cities* (Gifford 2007, p. 265)".

The reality is that, with this knowledge and the goal of making life healthier, fairer, more efficient and richer, our cities have, over the past century, been shaped by the dominance of design as a catalyst and instigator of behavior and habits. We have over-professionalized urban place-making, especially at the human scale, with two consequences. First, this has caused the progressive distancing of the design and delivery sectors from the users of their work (Punter 2011) – this was a necessary outcome, due to the sheer scale of development and lately of its success (Thwaites et al. 2007). Secondly, people have been left with the belief that nearly everything about the shape and management of environmental form is a professional problem, whether it belongs to a policy, management, legal, political or planning framework. Thus, today, people are disempowered and discouraged from acting on and taking charge of space for themselves; in an age of increasing interest in localism, this may well no longer be tenable.

The timing is right. Large-scale and sophisticated operations such as global, national and urban observatories are now widely established; they are repositories of data to monitor, compare and guide sustainable growth. Municipalities are extending and sharing their "guarantors of fair development" role to non-profit urban design groups, agreeing to widen roles and responsibilities to the users and the city. On a local level, responsibility for development is taken up by community movements, supported by the locally-oriented and participative agenda of place-keeping research, which explores innovative approaches to designing and managing open space while securing its long-term future by putting the right people, funding, policies and evaluative processes in place (Dempsey and Burton 2012), trying to disentangle change from excessive professionalism and bureaucracy. Knowledge is power, for all these levels. Urban design needs to use this broad pool of knowledge to guide strategic and structural work at metropolitan and neighborhood levels, and accompany all of us in the gradual transformation of small-scale environments.

Urban form is the setting where a more complex sharing of responsibilities needs to occur because, as we will show, shaping, controlling and being able to access the urban realm is significant for our well-being. Morphological structures and control relationships that are capable of better integrating social processes, material form and spatial organization can be found in the literature and require further investigation and development in the context of contemporary urban design and sustainable living challenges (Habraken 1998; Thwaites et al. 2013).

14.1.2 Aim and Rationale of the Review

Cities are many things to everyone; for the purpose of this chapter, we see them as first and foremost sources of behavioral and experiential opportunities, which other environments cannot offer. As such, we look at urban form as shaped by urban design at three main scales: metropolitan, neighborhood and pedestrian (Clifton et al. 2008; Lehrer 2010). We then search for studies that relate domains of QoL to each of these scales, including a focus on objective and subjective indicators. A significant part of our justification for focusing on different scales is that awareness of scale, and the way that this can have a profound influence on human behaviors and experience, lies at the very heart of contemporary place theories, which intimately connect human functioning with the settings of that functioning. Thus, we argue here that human experience of scale in the environment provides a foundation on which to build an understanding of urban settings as integrated socio-spatial systems.

Particularly significant origins for this are found in the work of anthropologist Hall in the 1960s (Hall 1966). Against a background of growing concern about what many perceived as the placeless consequences of modernist planning and design, Hall, and others at the time, began to develop an understanding of space as an elaboration of culture where space becomes place as a consequence of what people do in it. Hall rejected Cartesian concepts of a dualistic human-environment relationship through research that sought to establish that significant aspects of what it is to be human are not confined within a material skin but are manifested as "learned *situational* personalities" (ibid., p. 115) associated with responses to human-environment transactions at intimate, personal, social and public levels of scale.

Hall developed these ideas into the theory of proxemic space, premised on the innate tendencies of humans to band together in mutually supportive, and usually small, social groupings. Space is therefore cultural, rather than geometric, and becomes distinctive through the activities of individuals and groups within this context. This concept was later used by Greenbie (1978) to describe how culturally distinguishable urban villages and city neighborhoods become apparent in large cities, and was extended by introducing the term distemic space, referring to the often large portions of major cities that are shared by a diversity of cultural sub-groups. In broad terms, proxemic space describes the homeground, which necessarily involves high levels of personalization related to cultural needs and preferences. In psychological terms, this represents a place where basic needs such as security and a place of retreat are found. Distemic space, by comparison, is the place of challenge and enrichment offering diversity of experience, but within which opportunities for personalization may be limited. Proxemic and distemic spaces function in a complimentary manner, with both being required to optimize human psychological health.

The relationship of human behavior in response to levels of environmental scale has perpetuated throughout the development of urban design's intellectual core, spearheaded most explicitly in the work of architectural theorist Alexander (Chermayeff and Alexander 1963; Alexander et al. 1977) and more recently in Habraken's (1998) exploration of the structure of ordinary built environments. Similar themes of human-environment integration in urban settings resonate in Dovey's explorations of the phenomenological nature of place (2010a, b, c).

The main areas which we will refer to are Quality of Life (QoL) and Urban Design (UD); these are complex, multifaceted terms, studied in a variety of disciplines.

14.1.3 Urban Design and City Form

Urban design as a distinct academic and professional area emerged in the USA in the late 1940s and 1950s from the cultural cradle of the late Modern Movement in architecture, through the convergence of themes that, though entirely internal to the Modern Movement of the times and initially proposed to expand and reform it (Mumford 2009), contained the seeds of a radical departure from it. By the early 1960s, themes including the "heart of the city", the historical built heritage as an environmental (not just monumental) asset, social engagement and community empowerment, and the sensorial experience of the "townscape", shaped within the area of urban design a radical opposition to the Modern Movement's core principles; for example, around the role of design and the designer in society, the origin of place identity and above all the role of time and history in cities' evolution (Hebbert 2014). The parallel growth of cognate disciplines, such as ethology, psychology, environmental psychology and urban anthropology, legitimized the development of urban design into a more complex area, which found much of its inspiration and ideas in the desire to understand the relationship between people and space. The "giants" of urban design, that is those thinkers who shaped the foundations of the discipline as we know and practice it today (Porta and Romice 2014), were determined to understand critically place and the human experience within it as a prerequisite for design, conscious of the impact that design ideologies were having on quality of life across the globe.

Urban design today has been defined as a "mongrel discipline" (Carmona 2014), which studies and shapes the form of cities as complex, organized systems (Jacobs 1961 after Weaver 1948) of people, spaces and connections (Cowan et al. 2005). It works in the past, present and future; it deals with individuals, groups and society as a whole (Krier 2009). It works for efficiency and satisfaction and is thus centered on "the process of making better places for people than would otherwise be produced" (Carmona et al. 2003, p. 3). This definition contains the notion that places do change in time, within or without the remits of planning, suggesting that urban evolution is a founding principle of our discipline. Urban design deals with structures and values in order to offer rich, coherent experiences (Cowan et al. 2005). It determines our interface with the external world, modulating our interaction with others, our access to choice, and our bonds with space. Moreover, urban design deals with the delivery of urban form, at different scales. In a metastudy of urban form, Clifton et al. (2008) suggested that this is the focus of many different disciplines, which use different scales of investigation, have a different focus of interest and use different methods. We follow on from their classification of scales, and focus our review on the (sub-) metropolitan, neighborhood and pedestrian scales.

14.1.4 Quality of Life

Research on QoL started in the 1970s, in conjunction with the establishment of the journal Social Indicators Review. Its area of investigation spans many disciplines, although its core sector of work is health. Because of the wide-ranging scope of investigation in QoL, there is little agreement on its definitions and approaches (Schalock et al. 2002). Many have identified factors, domains, frameworks, and concepts to clarify and organize its meaning. The World Health Organization (Kuyken et al. 1995) recognizes that the study of QoL is at the same time subjective and weighted on individuals' experience (contentment), objective (financial status, employment) and multidimensional.

Developmental psychologist Ryff sees satisfaction with life not as contentment with the achievement of a status, but rather as "the realization of talent and potential, and the feeling that you are able to make the most of your abilities in life" (Montgomery 2013, p. 35). The spatial organization of our urban habitat must be conducive to supporting and sustaining us through these journeys. Greenbie (1978) offers perhaps one of the earliest attempts to develop an understanding of spatial structure that is integrated with such fundamental human functioning.

Citing World Value Surveys and Gallup World Polls amongst others, which set out to measure QoL from thousands of respondents' overall satisfaction with life on the basis of many components – personal, social, economic, environmental – which they then correlate, Montgomery (2013) suggests how economic status, which for years was deemed the driving element for life satisfaction, is not dominant and that indeed the most prosperous countries and cities in the world do not score higher in these surveys. Rather, education, employment, location and social ties seem to work together in fulfilling one's life. Satisfaction with life does, in turn, positively affect our perception of health, being linked to the feeling of leading a positive and meaningful life (ibid., p. 35). Well-being is therefore multidimensional and context-specific (Rogers et al. 2012).

14.1.4.1 QoL Domains

Several analyses of the literature have identified domains that contribute to the overall perception of QoL. From a review of almost 10,000 abstracts and 2,500 papers (Schalock et al. 2002), identified eight domains, each assessed through three indicators, objective or subjective, for the study of QoL. Subjective views of QoL are linked to cultural and contextual differences, and tend to be related to a smaller scale of investigation (Pacione 1986). Objective indexes are useful at a mesoscale, and a combination of both is used at higher scales, such as national or international surveys. Acknowledging that international comparisons are difficult, these surveys take into account contexts by weighting them, thus revealing important cross-cultural commonly shared values (Schalock 2004).

Pacione (1990) suggested that liveability is a description of this sense of comfort, and represents the interaction between people and place, involving social, economic, environmental and health-related factors (Newman and Kenworthy 1999). The form and character of most places in the city modulate our interaction with others, and with the environment as a whole, triggering emotional, cognitive, effective, and behavioral processes, on a personal and group level. Taking our lead from the work of Schalock and Verdugo, we therefore focus our study of form on those aspects that have resonance with psychological, physical and material well-being and interpersonal relationships. In doing so, we structure our discussion at each level of scale (metropolitan, neighborhood and pedestrian) within the categories of material well-being, emotional and personal development, interpersonal well-being and physical well-being. These domains relate well to cities, as this is where people act more clearly as individuals (Hall 1966) and as social beings (Greenbie 1978), through the modulations afforded by space. Since our focus is on the relationships between cities and well-being, we then concentrate on those aspects of city design and functioning that can play a role in our *realization of potential, and our feeling that we are able to make the most out of our life*. To us, this means looking for aspects of form that contribute to a sense of security, engagement, freedom, choice and control.

A potentially productive way to summarize the essence of these indicators in relation to particular properties of urban form is to consider the relationship between territorial behavior and the achievement of human self-esteem. In their attempt to develop a manifesto for urban design, Jacobs and Appleyard (1987) suggested that *"The urban environment should be an environment that encourages people to express themselves, to become involved, to decide what they want and act on it"* (Jacobs and Appleyard 1987, p. 169). This kind of territorial awareness can be related to human quality of life in terms of the need to achieve self-esteem. Through their mental and physical actions, individuals make their ideas into something permanent and thereby become aware that they have a mind of their own. Furthermore, through having their actions recognized by others, individuals are able to enjoy self-esteem. These ideas are central to the work of Honneth (1995) who identified the importance of recognition as a vital human need. Honneth considers that self-identity depends on developing self-confidence, self-respect and self-esteem. Achieving these requires the recognition of others who share common concerns within a mutually supportive community. The achievement of self-esteem and feeling comfortable in an urban setting, therefore, appear to be intimately related to human interactions with each another and with place. Consequently, urban environments ought to be configured in ways that will encourage and sustain "beneficial" interactions, those capable of sustaining a balance between individual self-expression and conformity with locally-formed norms and values. Urban forms that encourage the formation of communities, neighborhoods and a sense of belonging may therefore be more beneficial to QoL than those that do not. Much of the problem with the prevailing professionalized processes of urban planning and design is that, by excluding the end-users from the process of making decisions (and making in general) regarding their own space, they often excessively reflect the feelings, values and norms of the professional fraternities involved in the development process, leaving little space or incentive for personal expressions or the embedding of cooperative community.

Social functioning, similar to the territorial dynamics studied by Honneth (1995), which can be understood as a generator of the urban order we experience, is central to Habraken's exploration of the structural characteristics of the ordinary built environment (Habraken 1998). What Habraken means by "ordinary" in this context is the wide fabric of the built environment of human habitation, where the routine of daily life occurs, which until relatively recently managed to evolve and be sustained without the sort of professional attention it receives today. *"Ordinary growth processes that had been innate and self-sustaining, shared throughout society, have been recast as problems requiring professional solution"* (ibid., p. 3). For Habraken, these levels of control reflect the need for a balanced approach to the delivery of urban structure, involving a holistic relationship of specialist expertise (form), territorial behaviors (place), and user expression and conformity (understanding). The overlapping relationships between levels of control generate active and continuously shifting patterns of occupation and expression, creating a kind of margin at an indeterminable boundary where the control necessarily exerted by specialists gradually gives way to the social forces of occupants. Although such margins retain a form of stability and coherence over time, they may in fact be in continual change as the patterns of occupation and control ebb and flow with objects placed for short or longer periods, according to local custom, practicality and negotiation between neighbors. What appears visible results from the resolution of tensions between people's biological need to assert their individuality through territorial expression and the wider need for personal assertions to remain within commonly accepted norms: essentially the drivers of Honneth's concept of the recognition necessary for the achievement of self-esteem. Urban development based on large-scale spatial interventions and compressed timescales squeezes such opportunities.

We suggest that control, through form shaping, place understanding and choice management over time, offers the potential to build a more vital link between the physical structure of our cities and our capacity to establish meaningful relationships with others. In particular, the form of cities:

- organizes and links places, people and functions – at metropolitan and sub-metropolitan scales;
- clusters and distributes choice, facilitates movement, orients, and gives character, encouraging challenge, enrichment and relationships – at the medium, neighborhood scale;
- invites, welcomes, protects, engages and satisfies, allows choice and use – at the small, pedestrian scale.

This review covers these three scales, listing research that has shown a link between aspects of urban form and QoL.

14.2 Urban Design and QoL Literature Review

14.2.1 The Metropolitan Scale

Intuitively, quality of life seems easier to relate to the more human scales of urban experience whereas understanding which components of the wider city scale are influential is perhaps more obscure. Nevertheless, we view cities as places with characteristics that enable us to distinguish one from another and form images of them in our mind. This allows us to grow attached to them, organizing them as referents, for directions, for narratives, and to move through them. However, although the effect on QoL, at this scale, is harder to grasp, our experience of them as "wholes" is nevertheless important in the basic lifestyle they allow us to have, not least because it is at this scale that the arrangement of communication networks, land-uses, the distribution of services and access to them can either help or hinder our movement, and generate positive or negative experiences.

14.2.1.1 Material Well-Being

Bettencourt and West (2010) have calculated the increase in urban productivity, urban benefits and negative externalities that accompany city growth, suggesting that these increase faster than population growth, whilst the urban infrastructure required to accommodate such growth is much slower (Bettencourt and West 2010). From an evolutionary perspective, this might suggest that cities can "*reset their carrying capacity over time, and largely avoid (...) social and physical collapse*" (Pagel 2011) through restless innovation, and the continuous production of creative solutions, geared towards *efficiency*. The issue of efficiency, in both environmental and cultural terms, is crucial to conceptualize and develop the fundamental strategic role of urban design.

Because of the predicted pace of urbanization in the next 25 years, we know that there will be a drastic influx in existing cities and the development of more in conditions of informality but, whilst much of this urbanization will be spontaneous, some elements can be controlled. Research at UN-Habitat (Angel et al. 2011) suggests that rigid or inflexible expansion boundaries, for example, will, in the long term, determine poverty for a section of the population, because they will not be able to afford accommodation within them, as prices will be prefixed by these boundaries. On the other hand, the strategic initial conception, not even necessarily followed by immediate development, of carefully spaced infrastructure would allow for natural and fair occupation over time, enabling negotiation to form ordinary environments with manageable degrees of control where needed. Whilst this view, put forward under the term "the making room paradigm", might be one of a few in relation to urban development, it is reported here as very significant, especially when paired with other findings from UN-Habitat, that population increase and land urbanization are non-linear patterns, with the latter being much greater (and faster) than the former by a scale of 2 % in developed and 7 % in developing countries (i.e. Africa and Asia) (Angel et al. 2011). In this sense, it is possible to make predictions about urban growth (population and land) and therefore infrastructure (arterial grid and hierarchy of open spaces) and edge expansion limits.

Availability of land is an issue for developed countries too; the phenomenon of shrinking cities, which is occurring at a different pace in different political geographies, provides both an opportunity, to deal with the scarce resource of brownfield land in cities, and a risk, given that brownfield sites are not a solution (panacea?) per se as they are often linked to issues of social justice, development risk, location difficulties and servicing. An interesting study on the rather sudden and vast availability of brownfield sites in East Germany is supported by an integrated assessment of their character, and could be used to study the feasibility of their reintegration in the retrofitting city (Schetke and Haase 2008), for the negotiable space they add to existing built-up areas.

Recent work in urban morphology has shown that historically, and independently, the structure of cities has been organized around main urban streets, which in turn has generated "sanctuary areas", that is zones of a predominantly residential character bounded by main channels of movement (Mehaffy et al. 2010). Following on, Porta et al. (2014) confirmed this in an extensive geographic and temporal review of cases, subsequently addressing the importance of some structural physical elements at the metropolitan scale for the performance of urban life within social unspoken behavioral rules at the neighborhood scale (Mehaffy et al. 2014). These spontaneous clusters are important for the establishment and maintenance of such rules. The fact that their scale has a rather consistent dimension seems to suggest that, even today, amongst all changes, urban design should acknowledge such consistencies and respect them in new development.

14.2.1.2 Emotional and Personal Development

The morphological work above, which highlights the historic and geographic persistence of coherent urban areas bounded by movement channels up to modern planning, suggests the development of a rather spontaneous but balanced character within each of these areas, proportional to their size (which is remarkably rather regular, in time and space). This was consistent until large-scale professional planning started to predetermine the character of whole areas from the outset, limiting the spontaneous development of the city (Porta et al. 2014). Interestingly, research mentioned above (Bettencourt et al. 2010) has also shown that the organization of the main city elements, and the dynamics within them, are remarkably consistent and predictable, even across sociocultural processes of diversification, migration and overall change. As such, they are robust and lasting. The degree of organization that such elements allow their users changes substantially, according to both the societal context (including policy and planning) and the physical form of places. Habraken's reference to form, place and understanding, and therefore control, is key, making explicit that much of the contemporary mainstream in urban design tends towards the delivery of mostly professionalized urban structure thus limiting, and even obstructing, the more socially-oriented levels of control (place and understanding). These levels of control have a significant part to play in our capability for emotional and personal development because this is where relationships between individuals and groups most actively interact with material and spatial settings.

The degree of organization afforded in space is fundamental to how we inhabit and experience it. An overview of articles from Landscape and Urban Planning over 16 years has identified a number of consistent human needs in urban settings, valid across cultural differences and political contexts: "*Urban residents worldwide express a desire for contact with nature and each other, attractive environments, places in which to recreate and play, privacy, a more active role in the design of their community, and a sense of community identity*" (Matsuoka and Kaplan 2008). Having a degree of control at a metropolitan scale is a societal need expressed through meanings. Castells defines urban meanings as an expression of peoples' values over time (Castells 1983); they are infused in the city's structure. Nevertheless, people change, and with them their values; the city also changes but on different timescales, and yet needs to maintain congruence between meanings and form, to allow coherence and a sense of place to develop. It is enough to think of recent social change in developed societies, how substantial it has been within a relatively short timescale; from the early 1960s, more women started working, marriage occurred later in life, changing family size, and reducing the number of households with children; life expectancy generally grew, and so did disposable income, with a surge in the number of wealthy in retirement. Whilst these changes are primarily societal, economic and cultural, they require physical adaptability to allow our environments to be supportive, conducive, representative and enabling for our emotional and personal development.

Montgomery (2013) gives an interesting example: the typical image that has been depicted in the media for years, that of the American family living in the suburbs, has recently been substituted by more urban lifestyles (i.e. *Friends, Fraser, Sex and the City*). These "mental libraries of stories" contribute to changing our perception of what is desirable, helping us explore life according to different urban rules and pace (ibid., p. 93). The form of cities helps us develop and understand ideals and models, and with them become part of systems of practices. The congruence between form and these systems, some of which are unspoken, is key to our functioning as social beings. It relieves us from stress and gives us confidence to use the city and its parts; Lewicka suggests that the urban scale can participate in place attachment and deserves more attention by future research (2010). Urban form needs to be able to assimilate meanings over time; it is dangerous and costly to expect urban form to help us substitute them every time society demands new ones. The notion of control demands a more negotiative relationship between us and space, a creative, smaller-scale combination of context and subject in which spatial arrangements interpret, absorb and help develop social and cultural rules.

14.2.1.3 Interpersonal Relationships

Cities increase economic activity and productivity, but people flock to them as much for human interaction as for that. This is a double-edged sword. We crave interaction, which we enjoy when it is accompanied by our controlled ability to retreat from it. Moser (2012) calls urban behavior paradoxical, in that individuals must cooperate socially to maintain their anonymity (p. 208). Urbanity must function as a guide to manage social interaction.

Despite their higher efficiency, big cities have been associated with a cultural bias that has long been studied particularly in America. Recent investigations show that big cities tend to score lower than small towns on three scales: poor neighborhood quality, associated with housing conditions; home and neighborhood satisfaction with fair neighborhood characteristics; and the neighborhood quality rating of older long-term residents satisfied with their neighborhood, and young short-term residents not so satisfied with it. In all these instances, small towns scored better than large cities but a variation in the cities studied seems to suggest that those included were also those with a more generally uniform form of neighborhoods, even across varying incomes, whilst other cities where the polycentric nature of form was more evident did not feature (Greenberg and Crossney 2007).

A significant obstacle to beneficial interpersonal relationships in cities is criminality, one of the greatest sources of stress in urbanites. Fear of crime limits our ability to go out (mobility) and interact with others (sociability), two key domains of quality of life. It is also one of the main reasons why people leave the city (sometimes referred to as suburban flight). Research shows that instances of crime and fear of crime are different, the latter in fact not being the consequence of real risk, as summarized by Moser (2012). Concentration of crime is often higher in city centers, which being denser in activities tend to attract greater densities of people; this can be explained on the basis of, amongst others, the principle of de-individuation (Zimbardo 1969), which suggests that when the concentration of strangers is greatest, it is impossible to identify the odd-one-out. The feeling of insecurity that is associated with fear of crime is linked to the feeling of loss of sense of control and territoriality (Taylor 1978).

Incivilities and aggressive behaviors are heightened by physical form (Moser 2012, p. 209), with the sense of civic responsibility, the probability of intervening when witnessing distress, and simple people-people interactions (i.e. looking at a stranger in the eyes whilst walking) being reduced with an increase in density and the number of people around (ibid.). The concept of helpfulness has also been shown to be linked to city size. In general, this is higher in smaller towns than cities, with 300,000 being the threshold above which there is no significant distinction (Sundstrom et al. 1996), and is affected by weather and noise levels (increases in both above certain limits reduce it (Gifford 2007)). Helpfulness can also increase in complex settings – at least for women, not for men – and decrease with the number of people potentially there to help, explained as the overload approach, similar to the deindividuation principle introduced by Rydin et al. (2012) and Zimbardo (1969).

Urban forms that allow for the performance of urban life through the establishment and maintenance of unspoken behavioral rules have crucial implications for the nature of change and adaptability within urban realms: an important concept in the delivery of urban social sustainability. Change and adaptability in this context, and their relationship to resilient sustainable living, can be captured through the conceptual lens of "forgiveness". Here, the action of forgiveness underpins a conciliatory human-environment relationship uniquely able to articulate how environment can "forgive" human interventions and humans can "forgive" constraints that environment may impose. The concept of forgiveness maintains that we will tolerate large amounts of discomfort if we have what is most important to us. This is established within psychology (McCullough et al. 2003) but not in our relationship with environment. The environment is an actor of forgiveness, part of a process of exchange and thus significant as a means to explore connections that enable and constrain forgiveness (Latour 2005). Such connections become visible in human-environment relationships in how people develop perceptions of relationships among themselves, society at large, and the wider natural world. Consistent with this are ideas related to the struggle for recognition, which facilitates forgiveness by connecting past experience with the present through people's socially interactive need to experience themselves as belonging, "recognized" as a focus of concern, a valued contributor, or a responsible agent, as central to achieving self-esteem (Honneth 1995).

From this perspective, the attention of urban design is beginning to shift from purely form towards patterns and the interpersonal relationships that define them, supported in particular by recent debates criticizing the concept of neighborhood as a physical entity associated with that of community (Mehaffy et al. 2014). Whilst these still perceive neighborhoods as important, they interpret them as fluid and variable, changing around individuals, their interests and pursuits. Such fluidity does not negate the contribution of space to shaping social interactions and collective behaviors; on the contrary, the latter seems to self-organize *around* prominent spatial features, for example concentrations of shops and services. The importance of this in the development of environmental competence was discussed earlier, highlighting the significant role played by understanding the environment in terms of proxemic sets (Hall 1966). The concept of proxemic sets is resonant in the work of Spivak (1973) who considered the environment to consist of a finite range of 13 characteristic settings, or archetypal places. Like proxemic sets, which are primarily concerned with context defined in terms of the human-environment experience, archetypal places go beyond physical features and are defined in terms of the human behavior that occurs in them.

Like Hall's proxemic theory, Spivak's archetypal place concept provided a pivotal contribution to the subsequent development of socially responsive approaches to urban design. Its emphasis that social action and social networks are intimately woven together with the spatial and material fabric of the urban environment. This resonates throughout the evolution of urbanist thinking from Lynch (1960) and Cullen (1971) in the 1970s through to the design guidance of Bentley et al.'s Responsive Environments (Bentley et al. 1985), and the ongoing socially-oriented urban research and practice of Gehl (2010). Contrary to common belief, therefore, social networks do not hinder, but rather encourage and support the constitution of physical networks, organizing them in space (Hampton and Wellman 2000). Recent attempts to capture the morphological implications of this, focusing on the development of an anatomy for urban transitional edges as socio-spatial components of urban form, can be found in Thwaites et al. (2013).

Spatial organization, and especially how this influences a city's collective of services for people to access and use, has a significant influence on mobility. The proportional amount, distribution and quality of services are important, as is our capacity to access them, through choices in mobility. The conceptualization of clustering, and access to such services, is therefore an important area contributing to QoL. Several studies have demonstrated that greater density increases trip generation in a given area (Clifton et al. 2008, p. 28) and that greater balance between employment and residential facilities reduces commute time and the use of motorized transport. Urban diversity also stimulates modes of transport, with an increase in walking and cycling (Weeks 2014). The form of the built environment affects the frequency of trips undertaken to and from a certain area but, most of all, the distance covered to access services (Ewing and Cervero 2001). The important issue of time spent accessing services is directly linked to commute time and QoL. Central densities and gradient densities have fallen dramatically over the past 20 years around the world; sprawl, with an increased spending capacity and reduced transport costs, is responsible for this pattern, which is common. Services and retail outlets have consequently adopted different patterns of distribution and access to them has changed.

14.2.1.4 Physical Well-Being

With research on the links between physical activity and chronic health developed since the 1970s (Weeks 2014), we have gained knowledge about the relationship between the sociopsychological characteristics of individuals and exercise, urban density and exercise, and service distribution at the community scale and exercise. More recent integrated approaches to both monitoring and planning are providing important information on how to achieve healthy cities. This is crucial given that, currently, the most widespread cause of preventable death is heart disease (Speck 2013), and this is associated, amongst other things, with weight. Research has shown that weight is linked to inactivity, and inactivity to physical environments; the role of urban design is therefore becoming increasingly important. Speck (2013) reports a bleak trajectory in the increase in obesity in the US, from 10 % of its population in the 1970s to more than 30 % today, with a further third of the population being overweight. He then warns of predictions by the Center for Disease Control that one third of all children born after 2000 will get diabetes, making this the first generation in America predicted to live shorter lives than their parents.

Physical activity has been found to have positive effects on the control and reduction of obesity, and the studies of physical environments in relation to their capacity to encourage such activity are growing in number and sophistication. This issue will be dealt with in more detail in the Neighborhood and Pedestrian Scale sections.

Reducing car dependency to encourage forms of mobility, which can increase exercise and reduce exposure to harmful gases, tends to favor an urban infrastructure that is richer in its provision of urban green space. Urban green spaces have been shown to be positive for ecosystems and human physical and emotional well-being when carefully designed and distributed, but are also associated with an increase in land values around them, which in turn can speed up gentrification processes in the surrounding areas (Wolch et al. 2014). The capacity of green open spaces to deliver restorative benefits to people has been well established by research in environmental psychology. Restorative environment research is a growing field of academic activity as concerns about the health and well-being of urban populations increase. Establishing evidence for the benefits of access to green open spaces has therefore taken on political and economic as well as social significance in recent years. Although varied in detail and approach, restorative environment research is essentially concerned with developing an understanding of environments, in terms of type, scale and quality, which promote the restoration of depleted psychological, physiological and social resources (Kaplan and Kaplan 1989; Ulrich 1979; Hartig et al. 1991; Hartig 2004).

If green open spaces, and particularly those with a naturalistic emphasis, are good for urban inhabitants' QoL, one of the main challenges in urban design is how sufficient amounts of green open space can be accommodated as cities become ever denser and more compact. One consequence might be to reduce the availability of land in urban centers for large tracts of green space, generating instead a need to look to smaller types of public open space for respite and escape. Central to this developing concept is the re-establishment of the street as the urban focus, which provides a web of connections offering people a range of choices and experiences as they move about. Streets, and their capacity to connect a diversity of outdoor rooms, may therefore have potential as components of a reconceptualized urban park in the regenerated and rejuvenated compact city. The idea of a network of small, restorative open spaces in an urban center has been explored before in the context of urban planning, notably in a proposal by the American landscape architect Zion in 1963, who suggested that New York citizens would be better served by thousands of very small parks rather than a few larger ones. Zion's vision was never realized in the form he envisaged, but one of his pocket parks, Paley Park, has since become one of Manhattan's treasures.

Mosaics of small, designed green open spaces may well be part of the solution to the delivery of restorative benefit in cities, but the growing interest in urban agriculture may offer an additional benefit in this respect, particularly given that growing food crops requires proactive involvement and social interaction from participants. Increasing academic interest in this field has explored the implications of augmenting the implementation of various forms of urban food production as a socio-sustainable and ecologically beneficial component of future resilient cities (Ferrai 2014). Through an extensive literature review and an evaluation of European case studies, benefits to urban populations in the form of social cohesion, food security, economy, sustainability and education have been identified. Communication and collaboration between stakeholders and local authorities were found to be significant obstacles that required addressing, along with a change in public perceptions of productive landscapes as part of the city open space aesthetic. In respect of the latter, a study of front garden use in residential settings revealed that certain ethnic groups, particularly the Bangladeshi community in Leeds, seem to be much more open to using front garden spaces for food production, rather than ornamental display. Native UK residents, by comparison, usually see this as something that should be hidden from view in rear gardens or on allotment sites. The outcome of this work was a comprehensive practical manual of guidance to promote the wider use of front gardens for growing food (ibid.); secondary benefits in relation to maintenance, personalization, attachment and ownership, and similarly externalities, could derive from this initiative.

Summing Up

At a metropolitan scale, the quality of life of urban inhabitants is related to the way in which services and facilities are distributed and, by extension, to the infrastructures provided to achieve optimum distribution. From a quality of life perspective, optimum distribution needs to work towards as inclusive a level of accessibility as possible, ensuring that what people need to have contact with in routine daily life can be reached with relative ease and by the most sustainable means possible. As our review shows, this is likely to require urban patterns of distribution and connectivity diametrically opposite to the specialized, functionalist zoning associated with modernist urban planning approaches in favor of mosaics of multiple centers with diverse, mixed uses. Connectivity within and between such multiple centers will require urban public transport infrastructure capable of replacing the present reliance on private car usage. In order to improve and maintain the physical dimensions of quality of life, this will need to work alongside initiatives for greater levels of walkability within urban settings, coupled with radical re-thinking about the provision and distribution of networks of green urban open space. A variety of modeling and predictive methodologies are now available to help planning and design decisions become much better informed by observations of actual patterns of use in urban settings, such as Space Syntax and Multiple Centrality Assessment (Hillier 1996; Porta et al. 2010), for example, making predictions about trends and growth more realistic and therefore reliable.

14.2.2 The Neighborhood Scale

Neighborhoods are social clusters where interactions among members of the cluster are more likely to take place, and in a stronger way, than those involving externals. As such, neighborhoods may occur in space or even develop entirely in the virtual world. The dynamics involving both the "space of flows" and the "space of places" in the network society of our times have been explored by Castells (2000) who maintains the importance of the local form and function of places, where creative economies are increasingly reliant on human face-to-face interaction to generate innovation, attract choice-makers and thrive (Hall 1997). The social and physical (spatial) dimensions of the neighborhood have undergone cyclical waves of attention and neglect in the history of urban planning on one hand and urban sociology or anthropology on the other since the beginning of the twentieth century.

From an urban planning perspective, space has gained momentum in the past generation of scholarship, with urban renaissance and place-making guiding the agenda for a sustainable future in the age of urbanization, starting from the Urban Task Force (1999) and English Partnership (2000) to the wealth of planning and design guidance published internationally. The persistence of Perry's synthesis of the Neighborhood Unit idea (Perry 1929) through the development of the discipline has emphasized the fixed spatial relationship between location of services and gravitation of local social practices, taken as a whole, on the grounds of a notional distance of 400 m (or 5 min walk) from a center. This notion of neighborhood needs review, to take into account the complexity of sociality in the information age, and local communities expanding their role in relation to services, by becoming producers and not only consumers of services in a way that involves entrepreneurship and innovation primarily in the local space (Mehaffy et al. 2014). New forms of inhabiting, from co-housing to LAT (Living Apart and Together), and working, with the expansion of house-working and multiple-working, coupled with the crisis of publicly-subsidized welfare systems, are emphasizing the benefits of adaptability and resilience through local control, as opposed to centralized planning-and-delivery, as an effective response to emergent societal needs.

In this section, we list studies that explore the layout and character of neighborhoods in relation to behavioral patterns, suggesting that the issue of control as an indicator of quality of life can be observed through: (i) investment in the immediate, private and semi-private environment; (ii) instances of crime and antisocial behavior; (iii) social life in streets. As outlined in the introductory sections of this chapter, the experience of a measure of control over the identification, occupation and appropriation of places we favor and use is connected to quality of life by association with our capacity to develop self-esteem through our interactions with others in society. More so than at the metropolitan scale, where distribution and connectivity are principal drivers, the neighborhood scale begins to highlight greater prominence of dimensions of self-expression and how this is mediated through social and spatial interactions.

14.2.2.1 Material Well-Being

Speck (2013) suggests that home investment is about as local an investment as you can get. We use investment (both economic and emotional) as a signifier of interest, commitment and sense of control over our immediate space, as a starting point to discuss form and quality of life at a neighborhood scale. A significant and useful reference is Akbar (1988), who, in describing the modern Muslim city, identified the relationship between ownership, use and control of space as central to the nature and quality of space. For him, every space in a city is definable in terms of the relationships between the parties who own, control and use it, and divided into five types – trusteeship, possessive, permissive, dispersed, unified (ibid., p. 18–19) – each affecting the dynamics in the development, maintenance and transformation of the built environment. When a space is owned, controlled and used by one single party ("unified" form of submission), maintenance is generally good, change is gradual and piecemeal, corresponding to the user's needs, and the overall environment is socially responsive at the most basic level of society. At the other extreme, the space is owned by a party (the state or the local authority), controlled by another (the housing authority) and used by a third (the inhabitants), in a "dispersed" form of submission; here, direct control over the environment is removed from its direct user, and maintenance is more likely downgraded, with limited emotional investment allowed (Porta and Romice 2014). Together, and with all the variations in between, these relationships explain the complexity and variety of urban environments, also linking their form to management, use and maintenance.

Akbar's model shares similarity with Habraken's form, place and understanding control model in that it emphasizes the connection between form and structure in the urban environment, here at the level of the dwelling, and the extent to which occupiers are empowered and incentivized to maintain and adapt where they live. As Honneth (1995) showed, striking the right balance between individual expression and the recognition of that expression within a mutually supportive community is important for the achievement of self-esteem.

In addition to the contribution that patterns of control and ownership make to neighborhood quality of life, there are environmental and economic implications that can be associated with urban form. The form of cities at a neighborhood scale has been the subject of investigation (i) for the environmental and economic benefits that different physical urban models can contribute to energy consumption and electricity generation, for example, suggesting that increases in the latter of up to 50 % can be achieved through careful layout design (Hachem et al. 2011); (ii) for the role of the built environment in the conservation and production of renewables at city level through image processing of digital urban models and remote sensing imagery (Carneiro et al. 2009); and (iii) for the role of form in thermal comfort in both open and enclosed spaces (Mangiarotti et al. 2008).

In general, an earlier generation of modeling tools for optimizing the use of energy resources throughout the production and consumption of houses is now complemented by efforts to analyze the environmental performance of neighborhoods, in a more holistic understanding of zero-carbon futures; these are being used to assess and plan, in contexts both to retrofit and to plan anew.

14.2.2.2 Emotional and Personal Development

Crime, fear of crime and perception of crime have been found to be linked to the perception of loss of territorial control (Bell 1996) and to impact on quality of life, in an indirect way, through the mediated impact of environmental features (Lorenc et al. 2012). Importantly, stress related to perception of crime reduces people's activity (Bell 1996), with effects on personal development and interpersonal relationships.

Perception of crime is linked to the size of the residential area where one lives and its capacity to establish relationships and unspoken social rules/norms of behavior. Neighborhood form and fear of crime have therefore been studied to understand how the former contributes to the latter; in particular, more walkable neighborhoods with access to shops and transit appear to lower the fear of crime thanks to a perceived increase in territorial (informal) guardians, although they may also increase the perceived crime risk due to the increased presence of strangers to the area (Foster et al. 2010). The homogeneity of neighborhoods and their geometry, including the number of main artery roads traversing them and the amount of use of bounding streets, were also found to play a significant part in crime rates, more so than informal territorial control in a study of pairs of low and high crime rates in neighborhoods in Atlanta, Georgia (Greenberg et al. 1982). In particular, residential homogeneity, fewer traversing arteries and fewer travelers on bounding streets were more frequently associated with lower crime rates.

In a study set in Perth, Australia, the degree of neighborhood upkeep was a more important predictor of perception of safety and social capital than features of the built environment (Wood et al. 2008). However, indirectly, the design, and therefore use, control and ownership of space, as illustrated above (Akbar 1988), play a great role in its upkeep. The way space is perceived, in relation to degrees of privacy and publicity, is a key factor determining to a significant extent the awareness of ownership and responsibility, even in situations where no legally defined ownership exists. Orientation and the relationship between the public realm and built fronts establish informal control through the definition of marginal zones where the form of the urban realm often becomes more a matter of social negotiation than design of the physical form. In relation to mixed development environments, which have also been shown to be those more likely to enhance a sense of social capital, this requires the design of the urban environment to support upkeep and maintenance, enabling the marking of clear boundaries of ownership, competence and responsibility, and dealing with territoriality in an inclusive but defined manner. Achieving the optimum balance of material and spatial organization, and the capacity for social processes to play out as they need to, identifies a complex and hard to define relationship between what professional agencies need to deliver and how patterns of user occupation and control need to be empowered.

Instances of crime, fear of crime, and perception of crime risk are different constructs. The form of the built environment affects each in different ways, and since densification and mixed use are solutions that will probably need to be embraced more widely, it is important that urban design tackles physical features to allow a sense of territoriality, even within denser, more mixed, complex and open (to other than residents only) environments. Territoriality in itself is a complex term, including both signs that deter crime by communicating cohesion and care (found to be more frequent in homogeneous neighborhoods with strong social ties), and others that reveal a more defensive attitude towards crime, or social decay (Hunter 1978; Taylor 1978).

Aspects of the social implications of neighborhood upkeep and the modifications and adaptations that people routinely make to their surroundings is highlighted in Martin's work on the potential of the back alley as a community landscape (Martin 1996). Martin discusses the way different configurations of boundary treatment affect social potential in American residential developments. When boundaries are configured to achieve a balance of what Martin describes as "hidden-ness" and "revealing-ness", the back alleys can be transformed from being merely functional conduits into settings rich in social potential, capable of encouraging and sustaining neighborly behavior in residents. Hidden-ness and revealing-ness reflect that people, depending on mood and circumstance, sometimes wish to preserve privacy whilst at other times choose to be more openly available for contact with neighbors. Martin links the development of community spirit in residential settings with the extent to which the built environment allows individuals to control when they wish to hide or reveal themselves as they move about their daily lives. Boundaries of different heights and degrees of transparency, gate orientation, location of outbuildings and bin storage, places for car maintenance, children's play and so on, can become strategically arranged to optimize such control, allowing inhabitants to position themselves according to how sociable or otherwise they feel. There is a question of balance: infrastructures that facilitate too much hidden-ness may obstruct the sort of spontaneous social encounters from which good neighborly relationships often develop, whilst those that are too revealing can lead people to feel themselves oppressively overlooked. The ability to control privacy and sociability is therefore a factor that may contribute to levels of neighborhood satisfaction.

Neighborhood satisfaction has been studied by many, with a focus on perceptive and evaluative aspects; personal, social and psychological factors have been found to play a significant part in satisfaction, with physical attributes – generally considered through ratings rather than measurement – lagging behind in research (Hur et al. 2010). Overall, residential satisfaction is confirmed as a complex matter, with perception and evaluation interrelated with physical characteristics.

A theoretical model for the study of neighborhoods by Churchman and Ginosar (1999) suggests that the complexity of residential neighborhoods ought to be studied through a multidimensional approach. Bonaiuto et al. (1999) established, from an analysis of neighborhood satisfaction in Rome, that contextual factors and the presence of services are the strongest and weakest predictors, respectively, of neighborhood satisfaction, whilst architectural and town planning factors and social relationships fall in the middle range (Bonaiuto et al. 1999). A later work (Bonaiuto et al. 2003) refined the initial study, combining scales of perceived environmental qualities with a scale of neighborhood attachment. These are scales of perceived environmental quality so, whilst contextual and physical factors are taken into account, they are not measured. Combining both is an effective approach which could now, with more robust, spatial, pervasive capacities, be combined to understand the effective impact of types of form on attachment.

Physical, social and cultural factors have been listed as playing a part in neighborhood and residential satisfaction. Amongst the social ones, the fear of crime, the number of traffic accidents occurring, the sense of neighboring felt (Bell 1996), and the access to services (Rioux and Werner 2011) have been studied. On the other hand, research has found that these can be lessened through the use of good design and maintenance; for example, lighting and well-maintained greenery can help lower the fear of crime (Bell 1996).

Personal factors that have been found to affect such satisfaction are, amongst others, the past experiences that we associate with a place; our adaptive behavior to and within such a place, that is our tendency to grow fond of what we have, or the conditions we are given; whether we own or rent our residential environment and our stage in life of occupation of the home in which we live (Brown et al. 2003); this might also be linked to fear of eviction. Lastly, our sense of control – or lack of – and residential mobility also play a part in our degree of satisfaction towards our residential environment.

Hur et al. (2010) found physical features such as the presence of greenery, upkeep/deterioration, the size of the estate, access to facilities and transport, noise, smog, the degree of naturalness and openness, which are also associated with vistas and residential density all contributed to residential satisfaction.

Whilst cultural factors have also been found to affect satisfaction, being subject to cross-cultural differences, they are often shared and universal values overall (Scott Brown 1990), suggesting that the congruence between residents' values and the physical form of the community they occupy is important (Castells 1983).

In short, all the evidence suggests a correlation of psychological and cultural factors with physical and spatial ones, adding great insight to Jacobs's initial observations (1961), which is reflected in levels of satisfaction and is of interest to urban designers and communities at large.

14.2.2.3 Interpersonal Relationships

Public life is recognized as being key in modern life as it is within it that people learn how to deal with complexity, understanding and using unwritten rules and codes of practice (Sennett 1992a, b). Diversity is crucial as it offers accidental and unlimited scenarios for life. Research reviewed in this area relates our likelihood to establish and engage in social relationships, feel a sense of community and use local facilities to well-being (Francis et al. 2012) and focuses on the physical features in which such events take place. Density and spatial configuration in relation to movement, access and distribution of services are two of these features frequently cited.

In a study of residential layouts of different design principles, Hanson (2000) showed how the spatial configuration of modernist layouts does not appear to contribute to larger and more intense human interactions within the neighborhood or indeed between adjacent neighborhoods, decreasing opportunities to mix, and consequently reducing the potential for a vibrant and successful urban life. Through a study of London's morphological change, Hanson (ibid.) concluded that different design theories are connected to specific preconditions for sociability. Housing estates designed on the basis of social theories aimed at creating strong communities expressed through modernist urban layouts have failed in their goal by isolating people from each other, rather than facilitating social relationships (Milun 2007).

The presence of shops and public open spaces in residential environments has a positive effect on reinforcing a sense of community, independently of the frequency of use by respondents (Francis et al. 2012). The proximity of, and access to, such local facilities has a potential impact on the use that elderly residents make of their neighborhood, as it links to their overall emotional, social and physical well-being; reliance on motorized vehicles to access local services reduces their capacity to interact within the neighborhood. Since this is linked to urban form and layout, age-friendly urban design is very important to encourage participation in neighborhood activities (Vine et al. 2012). Other detailed studies show that the use in time of micro-places, transitional zones and "third places" in neighborhoods is very important to encourage the social life of older residents (Gardner 2011).

There has been much work on the study of the relationships between density and social sustainability. Different cultures have different tolerances to density and adopt different coping behaviors, while environments of different structure and density afford different social relationships to form (Moser 2012). The effects of density can be moderated through design by working on the gap between actual and perceived density, with specific physical features contributing to considerably lowering the latter (Bosselmann and Cervero 1994), but contextual knowledge and solutions are required.

Up to certain values, high densities facilitate physical movement and reinforce social capital (Kyttä et al. 2013). Diversity, which generally comes with density, has been found to favor a higher social effectiveness in certain situations (Weiner 1976), although it also appears to correlate with a lower sense of responsibility (ibid., p. 380). Overall, density has been shown to intensify already natural behaviors in people, that is social people will find more opportunities for interactions where density is higher, whilst people who tend to isolate themselves will do so even more where density grows (Freedman 1975, p. 209).

Greater differences in the appearance of others generate more weariness, and generally higher density may lead to "overload" and correspond to more unhelpful behaviors (Bell 1996, p. 380). Density is a complex concept, with many definitions and characterizations and many factors linked to it; hence the suggestion that it should be studied using both "hard" quantitative measures and "soft" qualitative and contextual ones (Boyko and Cooper 2011).

Densification, which now seems widely accepted as a pathway to deal with both urbanization and environmental challenges, is a delicate subject, and one that causes great debate in planning and design. Whilst this might seem the ideal, if not the only, path ahead for policymakers and professionals, there is still significant cultural resistance to it, especially in those areas where the "suburban dream" remains widely embedded in collective images and values. In these instances, density is associated with the fear of losing local life quality, privacy and access to nature with no evident return. On the other hand, other studies, such as those above, have highlighted some positive outcomes on the improvement of services and infrastructure that would follow increases in density, and changes are emerging in the attribution of values to place configurations, which is largely driven by the media, towards the return of a positive notion of urban "buzz", now associated with individual freedom and increased personal opportunities of the techno-professional elite. It is therefore fundamental that discussions on densification involve the immediate users, since they require a cultural shift, especially in the "developed" world, and – at the very least – adaptation and coping strategies in the urbanizing world. "Location-based evidence" becomes essential to offer contextual solutions to ideas of densification, taking into account the experiential, behavioral and evaluative consequences attached to density.

A recent study in Sweden (Kyttä et al. 2013) has opened the door to invaluable, extensive knowledge about these experiential aspects of densification, suggesting that when it needs to occur, this "softer" information is as important as more physical and objective data. In this work, experiential knowledge was paired to a more structural study of the social potential of places, to establish first where densification would be more appropriate; this was done by overlaying use, density, and capacity studies from GIS, in the combination of experiential and quantitative, objective and quantifiable measures.

The urban layout of neighborhoods, including their density, affects children's mobility, particularly in relation to the street layout, its geometry and the quality of experience for walkabilty. A study in Minnesota comparing children walking to school in suburban and new urban, mixed use pilot NEED/ND (a neighborhood with LEED certification) areas found that in the latter, children were more likely to walk unaccompanied, due to the more pleasant, walkable, crime-safe, dense and diverse environment. Moreover, children in suburban cases were confronted with a greater variety of traffic conditions, since cul-de-sacs tended to funnel traffic into arterial roads, so their level of engagement with road traffic had to vary between points of great to little challenge; in contrast, more urban environments tended to expose children to more uniform traffic conditions and accompany them through a more engaging and variable environment where the public realm contributed to the overall experience (Gallimore et al. 2011).

A study in Atlanta showed that black children from the poorest backgrounds were much more likely to die in car accidents than any other child, and this is because the state of public transport in suburban areas is too basic –suburban bus stops are often one mile apart and separated by highways. In the UK, poor children are 28 times more likely to be killed in street accidents than wealthy ones (Montgomery 2013). Similarly, a number of design features of neighborhoods, such as their spatial organization, overall legibility, presence of landmarks, and richness of detail, play an important role in encouraging the elderly to walk within the neighborhood. In particular, the presence of significant buildings is rated more important than signage, and the absence of barriers, such as poor paving, are factors that encourage walking within a neighborhood (Phillips et al. 2013).

Residential preference (the choice of the type of neighborhood in which we live) is also associated with the travel choice we make: residents who live in a neighborhood type of choice (i.e. walkable vs. car-dependent) are more likely to travel by the means afforded by the neighborhood's own form. On the other hand, dissonance between form and preference of neighborhood encourages the use of private means of transport. People who spontaneously choose suburban, car-dependent neighborhoods stay true to their beliefs and use the car (Schwanen and Mokhtarian 2005), while people who choose and live in walkable neighborhoods tend to drive less and walk even more than necessary (Frank et al. 2007). Disadvantaged neighborhoods with good levels of connectivity and access to public transport were found to encourage walking habits for movement, with benefits in terms of offsetting other inequalities and chronic diseases, which has many implications for practice and policy-making (Turrell et al. 2013).

The affordance of an urban environment for walking is an important factor related to self-determination. Much research has now shown that people prefer being and walking where other people are, because they feel safe and in company, therefore attracting further people for the same reasons in a typical "domino effect" (Gehl 1987; Whyte 1980).

This resounds well with research conducted in Barcelona, about the location of primary and secondary services in urban networks; while general common sense would locate main services along main and more central routes, and secondary services in the immediate surroundings, the study demonstrated that primary activities and attractors can sit comfortably on secondary paths and still remain destinations, while secondary services, whose market is mostly created by passers-by, need the highest degree of centrality to survive in an urban competing environment (Porta et al. 2012).

14.2.2.4 Physical Well-Being

Availability of choices to walk is an important part of human self-determination and is significant for physical well-being. The correlation between physical inactivity and chronic health problems has been studied since the 1970s, initially with a psychological and social focus on individuals undertaking recreational activities (Sallis et al. 2004; Weeks 2014, p. 26). Only in the 2000s has the focus started to include an integrated study of environmental correlates to physical activity (Saelens et al. 2003). Physical activity, like diet, operates at the individual scale (Barton et al. 2013). Physical inactivity is associated with a number of undesirable health outcomes, including coronary heart disease, circulatory diseases, diabetes, and hypertension (Bell et al. 2002). Future approaches to city organization and communication infrastructure conducive to human quality of life should not only facilitate travel by walking, but also actively encourage it.

Problems related to a sedentary life can be reduced significantly by a slight increase in moderate physical activity (Frank et al. 2005). The introduction of moderate daily exercise into the lifestyle of people with sedentary lives brings considerably more benefits than for an already active person committing to even more exercise (Katzmarzyk 2010), suggesting that a sedentary life is unnatural for people and that small, non-life-changing adjustments can have great benefits. A study on transport priorities in England, in relation to public health, established that small behavioral changes in relation to exercise in the whole population would be more effective than targeted changes for specific groups (Milne 2012); it is thus very important that these changes in behavior are encouraged by a physical environment that promotes "utilitarian" activity as part of its use.

Many have distinguished between recreational and utilitarian physical activities associated with exercise, the first referring to those undertaken with intention and purpose, and the latter, also producing benefits, being derived from other activities such as going to work, etc. Recreational activities require intention and commitment and are linked to individual personality and behavior, whereas utilitarian activities are an added-on benefit of the completion of different tasks; they are a consequence of other pursuits and depend on environmental conditions (Saelens et al. 2003; Weeks 2014).

Urban form, which combines the pursuit of daily tasks with utilitarian activities, can generate physical benefits through non-purposeful exercise. This is achieved when urban form is walkable, through a density and diversity of uses, the quality and character of streets and street fronts (Gehl 1987) establishing a direct link with public space (Lopez and Van Ness 2007), a permeable and interconnected street network (Jacobs 1961), and policies of traffic calming, especially on main streets to prevent vehicle flows, and particularly speed, from threatening Vulnerable Road Users (IREC 1990; ITE 1993). At a neighborhood scale, research called WalkScore suggests that those living in a more walkable neighborhood are 35 % likely to be overweight compared to 60 % of those living in less walkable neighborhoods. Frank et al. (2005) showed that single-use sprawl is especially inconvenient for families because most activities depend upon chauffeuring children (Weeks 2014, p. 26).

A study extensively observing street life, street quality and street centrality in Tripoli, Libya, suggests that street life is more likely to occur in central streets and that, in these central streets, it is more likely to occur where street fronts have greater levels of different units, functions, transparency, upkeep and richness of details (Remali et al. 2015). These factors contribute to the experience of walking in the city, encouraging or discouraging it. The presence of public open space in neighborhoods is important to stimulate walking in neighborhoods, but their amount and quality are not the only factors involved; the characteristics of the routes to and from them also count (Koohsari et al. 2013). Numerous studies have recently investigated the relationship between dimensions of urban form and walkability, evaluating features such as block size, diversity, density and fear of crime against the likelihood of people walking to access light transit (Werner et al. 2010). A study by Hanlon et al. (2006) of 65 cases across the USA, Canada, the UK, Australia and Japan showed that, all being equal, people walk more in walkable environments (Weeks 2014).

Studies on elderly people's attitudes to walking have shown that elements such as the presence of historic buildings, good upkeep, safety from crime and pleasantness are more likely to encourage them to walk for transportation (Van Cauwenberg et al. 2014). Interestingly, the National Association for Realtors in America revealed that in 2011, six out of ten Americans would rather live in a walkable neighborhood with accessible facilities than in an environment that would force them to drive cars to access the resources they needed in their daily lives. For elderly populations, this inversion of trend is particularly important as the Atlanta Regional Commission suggests that by 2030, one in five residents will be over 60 and therefore the dependence on private transport will only isolate them even more, forcing them indoors and limiting their social interaction (Montgomery 2013). As mentioned earlier, quality of social interaction is a factor in quality of life.

Summing Up

At the neighborhood scale, and referring back to Habraken's control model of form, place and understanding, urban design decision-making is beginning to confront the difficult balance between what requires delivery by professional planning and design agencies and what requires delivery by neighborhood inhabitants, individually and collectively. It seems clear, in relation to the quality of life considerations reviewed in this section, that provision at the neighborhood scale must empower greater levels of Habraken's "place" than might be necessary at the metropolitan scale. This is primarily because quality of life at this scale depends largely on the capacity of people to experience a sense of belonging, security and association with others. It is also important to distinguish a sense of an environment shared and respected as the homeground, for which an individual might experience a sense of collective responsibility in the interests of sustaining investments relevant to material well-being as well as fruitful interpersonal relationships. At this scale, urban design can work towards the provision of services and facilities relevant to establishing and sustaining a sense of neighborhood: the delivery of meaningful public resources, such as shared green open spaces, shops and other community provision. It can also act to ensure that these are designed in ways that are accessible, clearly defined, and amenable to natural surveillance, and can encourage social diversity and interaction where members of other neighborhoods can be welcomed, bringing social and economic vitality, but within constraints that maintain the identity and sense of belonging for those whose neighborhood it is. Territoriality is, therefore, increasingly important at this scale. It needs to work at and be experienced at a range of scales, from that of awareness of the "whole" neighborhood through to the identification and protection of individual and familial territories within it.

14.2.3 The Pedestrian Scale

Human quality of life, at least in relation to what we experience in routine daily life, rests heavily on what happens at the pedestrian scale. This is evident throughout wide-ranging contributions to the literature, from Jacobs in the 1960s through to Gehl and his contemporaries in the present day. In his "Cities for People", Gehl (2010) provides comprehensive accounts of the ways in which city spaces at the pedestrian scale are intrinsically interwoven with human functioning and social processes at the level of the individual and the collective. In addition to spatial organization at this scale, there is the strong message that to access beneficial experiences in urban settings, people must have a measure of control over what they choose to do and where they do it. Perhaps, therefore, more so than at the metropolitan and neighborhood scales, provision of open space that is conducive to quality of life does not rely entirely on the outcomes of professional design interventions. It seems that, at some very difficult point to identify at pedestrian scales, a transition is needed whereby the kind of prescriptive "design", as conventionally understood in the mainstream of current practice, needs to gradually give way to enable patterns of user occupation, control and adaptation to become more prominent in how the urban environment is shaped.

14.2.3.1 Material Well-Being

In terms of urban design decision-making, material dimensions related to quality of life are intimately tied to our capacity to become aware of spaces that we can own, control and experience responsibility for, and those where these apply to others in society. The literature shows that this can be interpreted in terms of spatial attributes that allow us to become aware of the extent of spatial containment, where boundaries between adjacent spaces exist, and the extent to which these can be controlled in order that we can define and protect items important to our material well being.

Such spaces are often associated with a capacity to combine security with surveillance to encourage the personalization of space and facilitate the protection of acts of personalization, and frequently define a zone between two distinguishable realms (Bosselmann 2008). As Nooraddin observes: *"Public and private claims visually and functionally overlap, which creates an identifiable urban space"* (Nooraddin 2002, p. 50). Where the two spaces join there should not be a linear boundary but instead a place in its own right with a certain thickness to it. It should be a realm between realms, in essence, a transitional sub-space between two larger recognizable spaces. Habraken (1998), Bentley et al. (1985), and Biddulph (2007) showed that personalization requires a spatial dimension to flourish. Cooper-Marcus and Sarkissian (1986) and Gehl et al. (2006) also outlined optimum spatial dimensions for the personalization of space. For example, Cooper-Marcus and Sarkissian (1986) highlighted a British study in which the size and shape of the front garden had an influence on its levels of use and personalization. They showed that front yards need to be in spatial balance and *"should be deep enough for privacy but not so large as to inhibit personalization."* (p. 104).

The awareness of enclosure is therefore important for establishing material well-being in space. Frank and Stevens (2007) suggested that spaces with a strong sense of enclosure occur where the private building meets the public space and can be formed by the building façade and other continuous boundaries such as fences, hedges, walls or natural features (Habraken 1998). Many authors have shown a preference for an articulated façade because it creates a series of niches that can be appropriated (Gehl et al. 2006; Macdonald 2005; Alexander et al. 1977; Dee 2011; Buchanan 1988; Cooper-Marcus and Francis 1997; Cooper-Marcus and Sarkissian 1986). Crinkled façades create pockets of semi-enclosed spaces that make the user feel more protected, creating spatially distinct sub-spaces that are easier to identify with. Therefore, this creates a space that has higher levels of social activity, social interaction and aspects of territoriality and personalization. For Cooper-Marcus and Sarkissian (1986), articulated façades have another territorial benefit: *"the more articulated the façade, the more likely are residents to add their own touches to the design"* (p. 68).

Effective personalization and surveillance require space to have a level of transparency, opening up the structure of the urban realm and preventing it from being experienced as a disconnected set of sealed enclosures. Transparency enables people to be aware of places where they are not and therefore opens up future possibilities. Whilst permeability is generally, although not exclusively, associated with issues of physical accessibility, transparency is usually understood as mainly visual. It is probably most readily recognized as a property of the urban environment that enables us to experience the interplay of "here" and "there" by means of features that make us aware of nearby settings other than the one we currently occupy. This is an aspect of place identity central to Cullen's Townscape concept (1971). In *The Concise Townscape*, Cullen highlights a series of ways this sense of "here-ness" and "there-ness" arises in the urban landscape and shows this act of transparency occurring at the edges where adjacent buildings or courtyards meet the street, for example. For Porta and Renne (2005), the visual characteristics of transparency are reinforced through their assignment of it as a measure of the amount of window space fronting the street. In this case, transparency is highlighted as one of seven quantifiable qualities associated with a socially sustainable streetscape. Transparency is, therefore, a desirable characteristic that increases both the social activity (Gehl et al. 2006; Alexander et al. 1977) and the level of perceived and actual security on the street (Jacobs 1993; Biddulph 2007; Carmona 2010; Llewelyn-Davies 2000; Rudlin and Falk 1999; Newman 1976, 1972). Visual access allows the inhabitants of the space to survey their territory from within the building whilst the openings, such as windows, add visual interest, which attracts the "eyes" of the street user and suggests a human presence.

14.2.3.2 Emotional and Personal Development

We mentioned above that experiencing the capability to organize and adapt the places we routinely use, according to personal or collective preferences, tastes and functions, is a vitally important contributor to human quality of life. It is especially significant at pedestrian scales because it is here that people have a more realistic opportunity to make small adaptations and expressions of preference relatively quickly and easily. This optimizes the experience of reward for effort expended in ways that the larger-scale neighborhood and metropolitan scales are less able to offer.

Emotional and personal well-being is intimately connected with territorial impulses. Awareness of the level to which we have control over territories we use is crucial to the extent to which we are empowered to adapt and organize. Territorial awareness at the pedestrian scale in the urban realm is complex and intimately tied to a spectrum between awareness of what is private and what is public, often involving demarcation and personalization as an extended form of boundary regulation. Research indicates that this characteristic is essential for social contact, safety and personal well-being (Hoogland 2000; Buchanan 1988, Habraken 1998; Altman 1975; Cooper-Marcus and Sarkissian 1986; Newman 1972, 1976). Such territorial acts are closely associated with human well-being. Altman (1975) and Honneth (1995), for example, relate territorial activity to the concept of self-identity. This may be because, as Habraken (1998) and Day (2002) have shown, territory is an innate and fundamental part of human nature, suggesting that if we are unable to inhabit and territorialize a geographic space, we are missing out on an important part of what makes us human. Research also indicates that a secondary territorial space is important for fostering social contact (Altman 1975; Hoogland 2000). Acts of personalization make these areas feel more protected and allow conversation and interaction to flourish.

Emotional and personal well-being is also associated with our capacity to interpret our surroundings according to personal preferences and other subjective impulses. Such interpretive capability is linked to a spatial property some have referred to as "looseness": "*People create loose space through their own actions. Many urban spaces possess physical and social possibilities for looseness, but it is people, through their own initiative, who fulfil these possibilities.*" (Frank and Stevens 2007, p. 10). Loose space can best be understood as a realm that is free, ambiguous, accessible and open-ended, according to Dovey and Polakit (2010), involving three distinct components: "*...a conjunction of loose forms (or loose parts), loose practices (behaviours, functions) and loose meanings*" (p. 167). The *loose form* concept can be seen in the work of Dovey and Raharjo (2010b) and Fernando (2007). Their observations show that flexible or semi-fixed items partake in a continuum moving from the least fixed items, in the open space, to the most fixed items, in the private space (Dovey and Polakit 2010). *Loose meanings* are also supported by the work of Madanipour (2003) and Habraken (1998). For them, a finite understanding of urban open space is often difficult to pinpoint because of its indeterminate form created by loose parts and loose functions. For Habraken (1998), this is because such space is where the physical form determined by the designer meets the ambiguous and emergent process of user occupation.

14.2.3.3 Interpersonal Well-Being

Active edges in urban settings are almost ubiquitously acknowledged in the literature for the crucial role they play in encouraging and supporting social vitality and interpersonal relationships in urban areas. Consequently, they are often acknowledged as integrations of social as well as physical and spatial realms (Habraken 1998). If the socio-spatial nature of these marginal zones is to be accepted, then two key challenges become explicit. The first is that delivery of these socio-spatial margins "by design" can only be expected to go so far because professional design disciplines, as they are currently configured, cannot adequately account for the breadth and ambiguity of human behavioral and social functioning in spaces in entirely prescriptive ways (Cuthbert 2007). Second, and related to this, is that these *edge environments*, active or otherwise, currently fall between disciplinary interests. Despite several decades of recognition of their importance to the social well-being of cities, there remains no environmental planning or design discipline with a specific focus on edge design, management and socio-spatial nature.

Since the early 1960s, one of the most notable desirable characteristics associated with diverse social life in cities has been the need for urban spaces to overcome abrupt divisions of private and public spaces, with a smoother public-private continuum that flows from privacy through to the public realm more gradually (Alexander et al. 1977; Altman 1975; Gehl 2010; Carmona and Tiesdell 2007; Madanipour 2003; Frank and Stevens 2007). Here, in this gradient of settings, one can choose the desired level of intimacy by positioning oneself in the appropriate degree of public or private exposure. In this way, the private-public gradient is a spatial quality that transcends the duality between the architecture and the adjacent open space. Madanipour (2003) sees this gradient working across edge environments: "*In practice, public and private spaces are a continuum, where many semi-public or semi-private spaces can be identified, as the two realms meet through shades of privacy and publicity rather than clear cut separation.*" (p. 239). The private-public gradient is not an assemblage of clear spaces but a smooth and complex gradient of subtle changes, in which a wider range of spaces allows greater diversity of intimacy and social interaction.

Short or longer, stationary activities afforded by the kind of spatial arrangements discussed thus far bring people into close proximity and provide the opportunity for encounters, whether fleeting and temporary or more enduring interactions, which may contribute to greater social cohesion and the development of community. One of the main values of social interaction in the public realm is that it can improve and promote a sense of place and feelings of community. Bosselmann (2008) has shown, for example, that certain kinds of spatial configuration can create both a sense of place and a perception of greater intimacy between neighbors. It appears, therefore, as elements of urban form, they have a significant role to play in encouraging and sustaining the social dynamics of the urban realm.

Related to this, as Jacobs (1993) and Martin (1996) demonstrated, people need to be able to exercise a measure of control over when they wish to be private and when to be sociable: "*A good city street neighbourhood achieves a marvel of balance between its people's determination to have essential privacy and their simultaneous wishes for differing degrees of contact, enjoyment, or help from people around.*" (Jacobs 1993, p. 61). Whether explicitly or implicitly stated, a variety of authors concur that, for this to happen, urban spaces need to achieve a fine balance that displays attributes of both privacy and publicity (Jacobs 1993; Hoogland 2000; Sundstrom 1977; Martin 1996; Korosec-Serfaty 1985; Carmona et al. 2003; Gehl et al. 2006). The settings they occupy should therefore be configured in such a way as to enable this choice to be readily made.

14.2.3.4 Physical Well-Being

Cullen is perhaps best known for his concept of "Townscape" (1971) mentioned above. It reflects Cullen's emphasis on the urban experience as an unbroken sequence of spatial experiences, influenced by the way focal points, landmarks, views, openings, etc., work together to draw people through space and to signal the distinction between the experiences of "here" and "there". This stands in stark and deliberate contrast to perceptions of urban environments as assemblages of objects and buildings, and the spaces they define. The experience of spatial sequence also has an explicitly human dimension going beyond what is merely "seen" to something intimately tied to the way people react and develop a sense of place: "...*the whole city becomes a plastic experience, a journey through pressures and volumes, a sequence of exposures and enclosures, of constraints and relief.*" (Cullen 1971, p. 10). For Cullen, urban space is not, therefore, simply volume, but something capable of conveying to us levels of containment felt and, through this, exerting influence on what we experience and how we might react and engage with urban space, encouraging physical interaction through the experience of sequence and continuity, either stationary or mobile.

One of the principal city structures that can support this are the edges where "the city and building meet" (Gehl 2010, p. 79). Gehl observes that there is often seven times more city life in front of an active façade, which encourages a continuous blend of static engagement with specific places and movement between them. This so-called "edge effect" (Gehl 2010), the observation that individuals gravitate to the edges of spaces, has been well documented by authors on the social aspects of urban design (Alexander et al. 1977; Appleton 1996; Bosselmann 2008; Chalfont 2005; De Jonge 1967; Dee 2011; Gehl 1986, 2010; Gehl et al. 1977, 2006; Frank and Stevens 2007; Whyte 1980). Appleton's prospect and refuge theory offers an explanation for this based on human behavioral ancestry, postulating that these edge spaces are aesthetically and spatially favorable to human biological needs of habitation because they provide "*the ability to see without being seen.*" (Appleton 1996, p. 66). This is also noted by Gehl et al. (2006), Gehl (2010), Frank and Stevens (2007) and Dee (2011) and seems to emphasize that people are drawn to the edges of spaces because they are prime spots for sitting or standing to survey the open space whilst also having one's back protected.

Diversifying opportunities for physical interaction with the urban realm relies on its permeability. Permeability is usually understood in terms of physical accessibility but can also include visual (referred to earlier as transparency), olfactory or audible permeability. Research indicates that permeability can have a significant influence on the level of activity in urban spaces. It is therefore desirable to offer as much permeability as the adjacent spaces can permit without compromising its function. Observations and research conducted by Gehl (2006) and Lopez (2003) showed that the level of activity within a street increases with the level of overall permeability between the building space and the street. These observations have been highlighted in other literature, suggesting that these are consequences of the permeable transitional edges (Rudlin and Falk 1999; Whyte 1980, 1988; Biddulph 2007; Frank and Stevens 2007).

Summary

At the pedestrian scale, quality of life seems to be much more intimately connected to our capacity to contribute to and participate in the determination of the identity, character and functionality of places we use. It is important at this scale that we are able to feel most in control of our settings: to participate in their making, use and adaptation, and not merely receive what professional agencies provide. It seems that, at some very difficult point to identify at pedestrian scales, a transition is needed whereby the kind of prescriptive "design", as conventionally understood in the mainstream of current practice, needs to gradually give way to allow patterns of user occupation, control and adaptation to become more prominent in how the urban environment is shaped. At this scale, perhaps more than at other scales, the boundaries between social and spatial dimensions of the urban realm may become more blurred. A priority for urban design at this scale may thus be not so much what to do as what not to do. This is a very significant challenge because, as our review clearly establishes, there are identifiable spatial attributes that need to be present in order for the self-determining empowerment necessary to quality of life at the pedestrian scale to take hold and sustain. Paradoxically, however, too much external control over spatial organization and material provision here can result in obstacles to user self-organization, which in turn can impede expressive activity, which is important to our recognition within social groups and thus to our sense of self-esteem. It may well be, therefore, at this scale in particular, that new forms of professional agency need to be explored, shifting the current emphasis on professionalized interventions toward more facilitating roles aimed at community empowerment and participation. As the UK political agenda moves further in the direction of an ethos of localism and the right to build, this may well become one of urban design's most pressing issues if quality of life is to be achieved in future urban developments.

14.3 Conclusions

Urban design's greatest contribution to quality of life spans across scales, from the city-wide to the pedestrian and detailed one, through the distribution of basic services, the design of streets and blocks, and their combination, in terms of walkability, intended as a complex term, inclusive of spatial convenience (permeability), environmental quality (safety, appearance, interest, environmental comfort), and overall legibility. Moreover, the modulation of density and complexity (of activities) encourages exposure to diversity, the practice of social norms, the establishment of social networks, and engagement in civic activities (Berger 2013).

Urban design should be intended as a process that, especially at neighborhood and pedestrian scales, enables self-organization and modification through new forms of local space control. People-space relationships are, indeed, reciprocal. We need a substantial shift in how we see ourselves as part of the world, the city, and the neighborhood, in our personal, social and civic lives. Contextual pressures, from the environment, the climate and its resources, to the scale and pace of urbanization, require a change in how we make our choices. We might only just be seeing the end of a century in which choice was based on accumulation, individuality, and substitution, and we might just be at the dawn of a time of awareness of legacy and durability, and the convenience and affordability that they can offer. This requires learning how to move from compartmentalizing our activities and environments to blending them for efficiency, so that both efforts and effects contribute to more than their individual worth. Urban life is here to stay and indeed to grow at an unprecedented pace, so we need to understand that the synergies it can offer hold a large stake in our well-being. As the philosopher Berleant eloquently observed; *"What we need now is to reconceptualize our world in a way that comes to terms with this, for what we do in the environment we do to ourselves."* (Berleant 1997, p. 121).

It may be important, therefore, in moving forward to address contemporary challenges associated with the delivery of urban environments that actively benefit human quality of life, that we reconsider the concept of human-environment relationships that underpin our approaches to the practice of urban design. Recognizing, and then responding to, the mutually reciprocal relationship highlighted by Berleant (ibid.) may come to rest on two essential components of the urban design process: (i) the development of a better understanding of the aspects of spatial organization associated with the social dimensions of urban order; (ii) the nature of relationships between professional processes of urban place-making and the participation of urban occupants in these processes, to recognize the importance of achieving a better balance of top-down professionalized decision-making with community-led bottom-up, informal practices on the ground. This may be especially important as communities begin to explore further the implications of a more localized approach to service delivery and environmental management.

It seems clear from our investigations that getting the spatial arrangement right "by design" can only go so far in the delivery of quality of life, and this appears increasingly true as design attention reaches the human scale of urban place-making. The moment may have arrived to recognize that the quality that Alexander called "quality without a name" (Alexander 1979), which makes places lively and loved over time by their inhabitants and users, *does not come by design*. Acknowledging this means reconsidering the role of urban design in society, moving towards one whose task is to set the conditions, the spatial ones first, to enable such dynamics to flourish (Romice et al.

2016). It is about designing the structure, not the solution, so that the solution can emerge by itself and continue doing so over time, "without effort" (Wolfe 2013).

We have tried to highlight that, at some hard to define point in the delivery of spatial arrangement, a fusion needs to happen between what the professional fraternity does and what must be left to patterns of user occupation, appropriation and adaptability. Understanding this point means designing structurally for progressive adjustments and requires, first of all, an understanding of what belongs to the structure that we must design, and what does not, that is what we should *not* design; this is mainly a matter of research (Porta et al. 2014; Feliciotti et al. 2016). The development of new conceptual frameworks, for example the concepts of Masterplanning for Change (Romice et al. 2016) and Socially Restorative Urbanism, is beginning to set new agendas of thinking in this respect through the blending of new socio-spatial concepts of urban order and the role of urban inhabitants in how they become shaped, managed and adapted through time (Thwaites et al. 2013). Habraken (1998) has provided a particularly useful example by showing how the structure of the ordinary is often more a matter of control relationships, rather than external planning and design. Habraken demonstrates that social and spatial dimensions of urban order cannot be easily disentangled, and attempts to do so run the risk of producing planning and design solutions that are not necessarily conducive to human quality of life. This is one way of reflecting on the various lines of research and practice currently emerging that look at resilience, adaptability, plot-based urbanism (Porta and Romice 2014), smart urbanism, and a socially-responsive, time-conscious way of planning (Thwaites et al. 2007). With increasing international and national focus on localism, this kind of mental reorientation at the root of our approaches to urban design may become increasingly important in the determination of policy and, if this is to be effective in the long term, ways will need to be found to enable appropriate reorientation of professional practice and, by extension, the education of practitioners. As our investigation in this chapter highlights, this may involve a shift away from the large scale and rapid pace of delivery, characteristic of much contemporary urban regeneration and design, towards a longer term and more time-conscious approach, which will need to be informed by new avenues of research.

We hope to have made a contribution to beginning this process by asserting that fruitful lines of inquiry might focus on the relationship between social processes and spatial organization. Clearly, much has already been done in this respect, but it seems that whatever understanding we have acquired thus far is being hindered in its effective application, partly by sustaining disciplinary divisions and partly because of communication gaps and power imbalances, which continue to exist between professional specialists and those who live with the consequences of their decisions. Perhaps the further development of new readings of the environment and the relationship people have with it, in terms that can be accessed by all, may ultimately break down the professional-layperson divide to deliver alternative approaches to urban place-making and management that have explicit socio-spatial foundations.

Foundations for such an alternative approach might productively include wider consideration of the holistic nature of the human-environment relationship within research, teaching and practice to underpin a better understanding of the mutually transforming nature of our relationship with the settings we use. This essentially philosophical stance may well make it easier to frame new theoretical perspectives capable of recognizing the interdependency of urban morphology and social processes, and how this can then begin to shape approaches to research and practice better able to integrate professional, top-down processes with community-led bottom-up processes in urban design, management and adaptation. Accepting, embracing and delivering human quality of life within an urban design framework is necessarily cross-disciplinary, requiring a hitherto rare blend of psychology, sociology, architecture, landscape and urban design (and more besides). Nevertheless, this can enable research in environment-behavior studies help address urban problems (Marans 2012). It will require significant developments in accessible and inclusive forms of communication capable of addressing professional and community boundaries as well as discipline-specific boundaries. Inclusive communication may help to address better the territorial dimensions of urban quality of life, which are at the heart of its socio-spatial nature, emphasizing new readings of the urban realm more closely related to the need for a better balance between professional intervention and occupant self-organization and highlighting the importance of longitudinal, time-sensitive working partnerships. This alternative approach suggests a different kind of professional disciplinary position to that prevailing in the current mainstream, perhaps highlighting a need to re-think the relationship between professional interventions and the participation of urban inhabitants, starting with the reconsideration of the ultimate mission of design in society as advanced by Turner (1976), Turner and Fichter (1972), and Rudolfsky (1964), as well as a need for more effective cross-disciplinary relationships, ultimately to inform a renewed interest in the "right to build" well within advanced western planning systems (DCLG 2012; Wainwright 2014).

References

Akbar, J. (1988). *Crisis in the built environment: The case of the Muslim City*. Singapore: Concept Media.
Alexander, C. (1979). *The timeless way of building*. Oxford: Oxford University Press.
Alexander, C., Ishikawa, S., Silverstein, M., Jacobson, M., Fiksdahl-King, I., & Angel, S. (1977). *A pattern language*. New York: Oxford University Press.
Altman, I. (1975). *The environment and social behavior: Privacy, personal space, territory, and crowding*. Monterey: Brooks/Cole Publishing Company.
Angel, S., Parent, J., Civco, D. L., & Blei, A. M. (2011). *Making room for a planet of cities*. Cambridge: Lincoln Institute of Land Policy.
Appleton, J. (1996). *The experience of landscape* (2nd ed.). Chichester: Wiley.
Barton, H., Grant, M., & Guise, R. (2013). *Shaping neighbourhoods: For local health and global sustainability*. Abingdon: Routledge.
Bell, A. C., Ge, K., & Popkin, B. M. (2002). The road to obesity or the path to prevention: Motorized transportation and obesity in China. *Obesity Research, 10*(4), 277–283.
Bell, P. A., Green, T., Fisher, J. D., & Baum, A. (1996). *Environmental psychology* (4th ed.). Fort Worth: Harcourt Brace College Publishers.
Bentley, I., Alcock, A., Martin, P., McGlynn, S., & Smith, G. (1985). *Responsive environments*. London: The Architectural Press.
Berger. (2013). *Health and urbanism report*. Cambridge, MA: MIT Press.
Berleant, A. (1997). *Living in the landscape: Toward an aesthetics of environment*. Kansas: University Press of Kansas.
Bettencourt, L., & West, G. (2010). A unified theory of urban living. *Nature, 467*(7318), 912–913.
Bettencourt, L. M., Lobo, J., Strumsky, D., & West, G. B. (2010). Urban scaling and its deviations: Revealing the structure of wealth, innovation and crime across cities. *PLoS One, 5*(11), e13541.
Biddulph, M. (2007). *Introduction to residential layout*. London: Butterworth-Heinemann, Elsevier.
Bonaiuto, M., Aiello, A., Perugini, M., Bonnes, M., & Ercolani, A. P. (1999). Multidimensional perception of residential environment quality and neighbourhood attachment in the urban environment. *Journal of Environmental Psychology, 19*(4), 331–352.
Bonaiuto, M., Fornara, F., & Bonnes, M. (2003). Indexes of perceived residential environment quality and neighbourhood attachment in urban environments: A confirmation study on the city of Rome. *Landscape and Urban Planning, 65*(1), 41–52.
Bosselmann, P. (2008). *Urban transformation: Understanding city design and form*. Washington, DC: Island Press.
Bosselmann, P., & Cervero, R. (1994). *An evaluation of the market potential for transit oriented development using visual simulation techniques*. National Transit Access Center, Institute of Urban and Regional Development, University of California at Berkeley, Working Paper UCTC n.247.
Boyko, C. T., & Cooper, R. (2011). Clarifying and re-conceptualising density. *Progress in Planning, 76*(1), 1–61.
Brown, B., Perkins, D. D., & Brown, G. (2003). Place attachment in a revitalizing neighborhood: Individual and block levels of analysis. *Journal of Environmental Psychology, 23*(3), 259–271.
Buchanan, P. (1988). Facing up to facades: A report from the front. *Architects' Journal, 188*, 24–27.
Carmona, M. (2010). *Public places, urban spaces: The dimensions of urban design*. Oxford: Routledge.
Carmona, M. (2014). Investigating urban design. In M. Carmona (Ed.), *Explorations in urban design* (pp. 1–11). Surrey: Ashgate.
Carmona, M., & Tiesdell, S. (2007). *Urban design reader*. London: Routledge.
Carmona, M., Heath, T., Oc, T., & Tiesdell, S. (2003). *Public place-urban space: The dimensions of urban design*. Oxon: Architectural Press.
Carneiro, C., Morello, E., & Desthieux, G. (2009). Assessment of solar irradiance on the urban fabric for

the production of renewable energy using LIDAR data and image processing techniques. In *Advances in GIScience* (pp. 83–120). Berlin: Springer. doi:10.1007/978-3-642-00318-9.

Castells, M. (1983). *The city and the grassroots: A cross-cultural theory of urban social movements*. Berkeley: University of California Press CA.

Castells, M. (2000). *The information age: Economy, society and culture: the rise of the network society*. Oxford: Wiley-Blackwell.

Chalfont, G. (2005). Building edge: An ecological approach to research and design environments for people with dementia. *Alzheimer's Care Quarterly, 6*(4), 341–348.

Chermayeff, S., & Alexander, C. (1963). *Community and privacy: Toward a new architecture of humanism*. New York: Doubleday.

Churchman, A., & Ginosar, O. (1999). A theoretical basis for the post-occupancy evaluation of neighborhoods. *Journal of Environmental Psychology, 19*(3), 267–276.

Clifton, K., Ewing, R., Knaap, G. J., & Song, Y. (2008). Quantitative analysis of urban form: A multidisciplinary review. *Journal of Urbanism, 1*(1), 17–45.

Cooper-Marcus, C., & Francis, C. (1997). *People places: Design guidelines for urban open space* (2nd ed.). New York: London, Van Nostrand Reinhold.

Cooper-Marcus, C., & Sarkissian, W. (1986). *Housing as if people mattered: Site design guidelines for medium-density family housing*. Berkeley: University of California Press.

Cowan, R., Rogers, L., & Hall, P. G. (2005). *The dictionary of urbanism* (Vol. 67). Tisbury: Streetwise Press.

Cullen, G. (1971). *The concise townscape*. London: Routledge.

Cuthbert, A. R. (2007). Urban design: Requiem for an era—review and critique of the last 50 years. *Urban Design International, 12*(4), 177–223.

Day, C. (2002). *Spirit & place: Healing our environment: Healing environment*. Oxford: Architectural Press.

DCLG Department for Communities and Local Government. (2012). Giving people more power over what happens in their neighbourhood. https://www.gov.uk/government/policies/giving-people-more-power-over-what-happens-in-their-neighbourhood. Accesssed in Dec 2014.

Dee, C. (2011). *Form and fabric in landscape architecture: A visual introduction*. London: Spon.

Dempsey, N., & Burton, M. (2012). Defining place-keeping: The long-term management of public spaces. *Urban Forestry & Urban Greening, 11*(1), 11–20.

Dovey, K., & Polakit, K. (2010). Urban slippage: Smooth and striated streetscapes in Bangkok. In K. Dovey (2010c Ed.), *Becoming places: Urbanism/architecture/identity/power*. London: Routledge.

Dovey, K., & Rahaarjo, W. (2010). Becoming prosperous: Informal urbanism in Yogyakarta. In K. Dovey (2010c Ed.), *Becoming places: Urbanism/architecture/identity/power*. London: Routledge.

Ewing, R., & Cervero, R. (2001). Travel and the built environment: A synthesis. *Transportation Research Record: Journal of the Transportation Research Board, 1780*, 87–114.

Feliciotti, A., Romice, O., Porta, S. (forthcoming, accepted). Stepping stones towards a masterplan for change: Measuring resilience of the urban form. In *Open Housing International*.

Fernando, N. A. (2007). Urban streets in different cultural contexts. In K. A. Frank & Q. Stevens (Eds.), *Loose space: Possibility and diversity in urban life*. London: Routledge, Taylor & Francis Group.

Ferrai, C.P. (2014). *Urban agriculture as a socio-sustainable-ecological model*. Dissertation submitted for MA Landscape Architecture, University of Sheffield.

Foster, S., Giles-Corti, B., & Knuiman, M. (2010). Neighbourhood design and fear of crime: A social-ecological examination of the correlates of residents' fear in new suburban housing developments. *Health & Place, 16*(6), 1156–1165.

Francis, J., Giles-Corti, B., Wood, L., & Knuiman, M. (2012). Creating sense of community: The role of public space. *Journal of Environmental Psychology, 32*(4), 401–409.

Frank, K. A., & Stevens, Q. (Eds.). (2007). *Loose space: Possibility and diversity in urban life*. London: Routledge.

Frank, L. D., Schmid, T. L., Sallis, J. F., Chapman, J., & Saelens, B. E. (2005). Linking objectively measured physical activity with objectively measured urban form: Findings from SMARTRAQ. *American Journal of Preventive Medicine, 28*(2), 117–125.

Frank, L. D., Saelens, B. E., Powell, K. E., & Chapman, J. E. (2007). Stepping towards causation: Do built environments or neighborhood and travel preferences explain physical activity, driving, and obesity? *Social Science & Medicine, 65*(9), 1898–1914.

Freedman, J. L. (1975). *Crowding and behavior*. Oxford England: WH Freedman.

Gallimore, J. M., Brown, B. B., & Werner, C. M. (2011). Walking routes to school in new urban and suburban neighborhoods: An environmental walkability analysis of blocks and routes. *Journal of Environmental Psychology, 31*(2), 184–191.

Gardner, P. J. (2011). Natural neighborhood networks: Important social networks in the lives of older adults aging in place. *Journal of Aging Studies, 25*(3), 263–271.

Gehl, J. (1986). "Soft Edges" in residential streets. *Scandinavian Housing and Planning Research, 3*(2), 89–102.

Gehl, J. (1987). *Life between buildings: Using public space*. New York: Van Nostrand Reinhold.

Gehl, J. (2010). *Cities for people*. Washington, DC: Island Press.

Gehl, J., Brack, F., & Thornton, S. (1977). *The interface between public and private territories in residential areas*. Melbourne: Department of Architecture and Building.

Gehl, J., Kaefer, L. J., & Reigstad, S. (2006). Close encounters with buildings. *Urban Design International, 11*(1), 29–47.

Gifford, R. (2007). *Environmental psychology: Principles and practice*. Colville: Optimal Books.

Greenberg, M., & Crossney, K. (2007). Perceived neighborhood quality in the United States: Measuring outdoor, housing and jurisdictional influences. *Socio-Economic Planning Sciences, 41*(3), 181–194.

Greenberg, S. W., Rohe, W. M., & Williams, J. R. (1982). Safety in urban neighborhoods: A comparison of physical characteristics and informal territorial control in high and low crime neighborhoods. *Population and Environment, 5*(3), 141–165.

Greenbie, B. (1978). Social privacy in the community of diversity. In B. Greenbie & A. H. Esser (Eds.), *Design for communality and privacy*. New York: Plenum Press.

Habraken, N. J. (1998). *The structure of the ordinary: Form and control in the built environment*. Cambridge: MIT Press.

Hachem, C., Athienitis, A., & Fazio, P. (2011). Investigation of solar potential of housing units in different neighborhood designs. *Energy and Buildings, 43*(9), 2262–2273.

Hall, E. T. (1966). *The hidden dimension*. New York: Doubleday.

Hall, P. (1997). Modelling the post-industrial City. *Futures, 29*(415), 311–322.

Hampton, K. N., & Wellman, B. (2000). Examining community in the digital neighborhood: Early results from Canada's wired suburb. In *Digital cities* (pp. 194–208). Heidelberg: Springer.

Hanlon, P., Walsh, D., & Whyte, B. (2006). *Let glasgow flourish*. Glasgow: Glasgow Centre for Population Health.

Hanson, J. (2000). Urban transformations: A history of design ideas. *Urban Design International, 5*(2), 97–122.

Hartig, T. (2004). Restorative environments. In C. Spielberger (Ed.), *Encyclopedia of applied psychology* (Vol. 3, pp. 273–279). San Diego: Academic.

Hartig, T., Mang, M., & Evans, G. W. (1991). Restorative effects of natural environment experiences. *Environment and Behaviour, 23*, 3–26.

Hebbert, M. (2014). Historical exploration/explanation in urban design. In M. Carmona (Ed.), *Explorations in urban design* (pp. 287–297). Surrey: Ashgate.

Hillier, B. (1996). *Space is the machine: A configurational theory of architecture*. New York: Cambridge University Press.

Honneth, A. (1995). *The struggle for recognition: The moral grammar of social conflicts*. Cambridge: Polity Press.

Hoogland, C. (2000). *Semi-private zones as a facilitator of social cohesion*. Nijmegen: Katholieke Universiteit Nijmegen.

Hunter, A. (1978). Persistence of local sentiments in mass society. In *Handbook of contemporary urban life* (pp. 133–162).

Hur, M., Nasar, J. L., & Chun, B. (2010). Neighborhood satisfaction, physical and perceived naturalness and openness. *Journal of Environmental Psychology, 30*(1), 52–59.

IREC Institut de Recherche sur l'Environment Construit. (1990). *Le Temps des Rues: vers un Nuovel Aménagement de l'Espace Rue*. Lausanne: EPFL.

ITE Institue of Transportation Engineers. (1993). *Residential street design and traffic control*. Englewood Cliffs: Prentice Hall.

Jacobs, J. (1961). *The death and life of great American cities*. New York: Random House LLC.

Jacobs, A. B. (1993). *Great streets*. Cambridge: MIT Press.

Jacobs, A., & Appleyard, D. (1987). Toward an urban design manifesto. *Journal of the American Planning Association, 53*(1), 112–120.

Jonge, D. (1967). Applied hodology. *Landscape, 17*(2), 10–11.

Kaplan, R., & Kaplan, S. (1989). *The experience of nature: A psychological perspective*. New York: Cambridge University Press.

Katzmarzyk, P. T. (2010). Physical activity, sedentary behavior, and health: Paradigm paralysis or paradigm shift? *Diabetes, 59*(11), 2717–2725.

Koohsari, M. J., Kaczynski, A. T., Giles-Corti, B., & Karakiewicz, J. A. (2013). Effects of access to public open spaces on walking: Is proximity enough? *Landscape and Urban Planning, 117*, 92–99.

Korosec-Serfaty, P. (1985). Experience and the use of the dwelling. In I. Altman & C. M. Werner (Eds.), *Home environments*. New York: Plenum.

Krier, L. (2009). *The architecture of community*. Washington, DC: Island Press.

Kuyken, W., Orley, J., Power, M., Herrman, H., Schofield, H., & Murphy, B. (1995). The World Health Organization quality of life assessment (WHOQOL): Position paper from the World Health Organization. *Social Science and Medicine, 41*(10), 1403–1409.

Kyttä, M., Broberg, A., Tzoulas, T., & Snabb, K. (2013). Towards contextually sensitive urban densification: Location-based soft GIS knowledge revealing perceived residential environmental quality. *Landscape and Urban Planning, 113*, 30–46.

Latour, B. (2005). *Reassembling the social: An introduction to actor-network-theory*. Oxon: Oxford University Press.

Lehrer, J. (2010). A physicist [solves] the city. *New York Times*. Available at: http://www.nytimes.com/2010/12/19/magazine/19Urban_West-t.html?_r=2&adxnnl=1&pa. Accessed 30 May 2016.

Lewicka, M. (2010). What makes neighborhood different from home and city? Effects of place scale on place attachment. *Journal of Environmental Psychology, 30*(1), 35–51.

Llewelyn-Davies. (2000). *Urban design compendium 1*. London: English Partnerships.

Lopez, T. G. (2003). Influence of the public–private border configuration on pedestrian behaviour. The case of the city of Madrid. PhD. Spain, La Escuela Te'cnica Superior de Arquitectura de Madrid.

López, M., & van Nes, A. (2007). Space and crime in Dutch built environments. Macro and micro spatial conditions for residential burglaries and thefts from cars. In A. S. Kubat (Ed.), *Proceedings Space Syntax, 6th International Symposium*. Istanbul: Technological University Istanbul.

Lorenc, T., Clayton, S., Neary, D., Whitehead, M., Petticrew, M., Thomson, H., et al. (2012). Crime, fear of crime, environment, and mental health and wellbeing: Mapping review of theories and causal pathways. *Health & Place, 18*(4), 757–765.

Lynch, K. (1960). *The image of the city* (Vol. 11). Cambridge: MIT Press.

Macdonald, E. (2005). Street-facing dwelling units and liviability: The impacts of emerging building types in Vancouver's new high-density residential neighbourhoods. *Journal of Urban Design, 10*(1), 13–38.

Madanipour, A. (2003). *Public and private spaces of the city*. London: Routledge.

Mangiarotti, A., Paoletti, I., & Morello, E. (2008). A model for programming design interventions aimed at reducing thermal discomfort in urban open spaces. *Journal of Green Buildings, 3*, 119–129.

Marans, R. W. (2012). Quality of urban life studies: An overview and implications for environment-behaviour research. *Procedia-Social and Behavioral Sciences, 35*, 9–22.

Martin, M. (1996). Back-Alley as community landscape. *Landscape Journal, 15*, 138–153.

Matsuoka, R. H., & Kaplan, R. (2008). People needs in the urban landscape: Analysis of landscape and urban planning contributions. *Landscape and Urban Planning, 84*(1), 7–19.

McCullough, M. E., Fincham, F. D., & Tsang, J. (2003). Forgiveness, forbearance and time: The temporal unfolding of transgression-related interpersonal motivations. *Journal of Personality and Social Psychology, 84*(3), 540–557.

Mehaffy, M., Porta, S., Rofe, Y., et al. (2010). Urban nuclei and the geometry of streets: The 'emergent neighborhoods' model. *Urban Design International, 15*(1), 22–46.

Mehaffy, M., Porta, S., & Romice, O. (2014). The "neighborhood unit" on trial: A case study in the impacts of urban morphology. *Journal of Urbanism: International Research on Placemaking and Urban Sustainability* (ahead-of-print), 1–19.

Milne, E. M. (2012). A public health perspective on transport policy priorities. *Journal of Transport Geography, 21*, 62–69.

Milun, K. (2007). *Pathologies of modern space: Empty space, urban anxiety, and the recovery of the public self*. London: Routledge.

Montgomery, C. (2013). *Happy City: Transforming our lives through urban design*. Doubleday: Macmillan.

Moser, G. (2012). Cities. In S. D. Clayton (Ed.), *The Oxford handbook of environmental and conservation psychology*. Oxford: Oxford University Press.

Mumford, E. (2009). *Defining urban design: CIAM architects and the formation of a discipline, 1937–69*. New Heaven: Yale University Press.

Newman, O. (1972). *Defensible space: Crime prevention through environmental design*. New York: Macmillan.

Newman, O. (1976). *Design guidelines for creating defensible space*. Washington, DC: National Institute of Law Enforcement and Criminal Justice.

Newman, P., & Kenworthy, J. (1999). *Sustainability and cities: Overcoming automobile dependence*. Washington, DC: Island Press.

Nooraddin, H. (2002). In-between space: Towards establishing new methods. *Street Design, GBER, 2*(1), 50–57.

Pacione, M. (1986). Quality of life in Glasgow: An applied geographical analysis. *Environment and Planning, 18*(11), 1499–1520.

Pacione, M. (1990). Urban liveability: A review. *Urban Geography, 11*(1), 1–30.

Pagel, M. (2011). Cities as gardens. http://edge.org/conversation/cities-as-gardens. Accessed in Dec 2014.

Perry, C. A. (1929). *The neighborhood unit: A scheme of arrangement for the family life community, (Regional study of New York and its environs, VIII, neighborhood and community planning, Monograph 1, pp. 2–140)*. New York: Regional Plan of New York and its Environs.

Phillips, J., Walford, N., Hockey, A., Foreman, N., & Lewis, M. (2013). Older people and outdoor environments: Pedestrian anxieties and barriers in the use of familiar and unfamiliar spaces. *Geoforum, 47*, 113–124.

Porta, S., & Renne, J. L. (2005). Linking urban design to sustainability: Formal indicators of social urban sustainability field research in Perth, Western Australia. *Urban Design International., 10*, 51–64.

Porta, S., & Romice, O. (2014). Plot-based urbanism: Towards time-consciousness in place-making. In C. Mäckler & W. Sonne (Eds.), *New civic art: Dortmunder lectures on civic art* (Vol. 4, pp. 82–111). Sulgen: Niggli.

Porta, S., Latora, V., & Strano, E. (2010). Networks in urban design. Six years of research in multiple centrality assessment. In E. Estrada, M. Fox, & D. J. Higham (Eds.), *Network science* (pp. 107–129). London: Springer.

Porta, S., Latora, V., Wang, F., Rueda, S., Strano, E., Scellato, S., et al. (2012). Street centrality and the location of economic activities in Barcelona. *Urban Studies, 49*(7), 1471–1488.

Porta, S., Romice, O., Maxwell, J. A., Russell, P., & Baird, D. (2014). Alterations in scale: Patterns of change in main street networks across time and space. *Urban Studies, 5*(16), 3383–3400.

Punter, J. (2011). Urban design and the English urban renaissance 1999–2009: A review and preliminary evaluation. *Journal of Urban Design, 16*(1), 1–41.

Remali, A. M., Porta, S., & Romice, O. (2015). *Correlating street quality, street life and street centrality in Tripoli, Libya. The past, present and future of high streets* (pp. 104–130). London: UCL Eds.

Rioux, L., & Werner, C. (2011). Residential satisfaction among aging people living in place. *Journal of Environmental Psychology, 31*(2), 158–169.

Rogers, D. S., Duraiappah, A. K., Antons, D. C., Munoz, P., Bai, X., Fragkias, M., & Gutscher, H. (2012). A vision for human well-being: Transition to social sustainability. *Current Opinion in Environmental Sustainability, 4*(1), 61–73.

Romice, O., Porta, S., Feliciotti, A., & Barbour, G. (2016). Masterplanning for change: Design as a way to create the conditions for time-sensitive placemaking. In H. AlWaer & B. Illsley (Eds.), *Placemaking: Rethinking the masterplanning process*. London: ICE Publishing. ISBN 978-0-7277-6071-5.

Rudlin, D., & Falk, N. (1999). *Sustainable urban neighbourhood: Building the 21st century home*. Oxford: Architectural Press.

Rudolfsky, B. (1964). *Architecture without architects: A short introduction to non-pedigreed architecture*. Albuquerque: Doubleday.

Rydin, Y., Bleahu, A., Davies, M., Dávila, J. D., Friel, S., De Grandis, G., et al. (2012). Shaping cities for health: Complexity and the planning of urban environments in the 21st century. *Lancet, 379*(9831), 2079.

Saelens, B. E., Sallis, J. F., Black, J. B., & Chen, D. (2003). Neighborhood-based differences in physical activity: An environment scale evaluation. *American Journal of Public Health, 93*(9), 1552–1558.

Sallis, J. F., Frank, L. D., Saelens, B. E., & Kraft, M. K. (2004). Active transportation and physical activity: Opportunities for collaboration on transportation and public health research. *Transportation Research Part A: Policy and Practice, 38*(4), 249–268.

Schalock, R. L. (2004). The concept of quality of life: What we know and do not know. *Journal of Intellectual Disability Research, 48*(3), 203–216.

Schalock, R. L., Verdugo, M. A., et al. (2002). *Handbook on quality of life for human service practitioners*. Washington, DC: American Association on Mental Retardation.

Schetke, S., & Haase, D. (2008). Multi-criteria assessment of socio-environmental aspects in shrinking cities. Experiences from Eastern Germany. *Environmental Impact Assessment Review, 28*(7), 483–503.

Schwanen, T., & Mokhtarian, P. L. (2005). What affects commute mode choice: Neighborhood physical structure or preferences toward neighborhoods? *Journal of Transport Geography, 13*(1), 83–99.

Scott Brown, D. (1990). Urban concepts. *Architectural Design, 60*(1/2), 5–96.

Sennett, R. (1992a). *The conscience of the eye: The design and social life of cities*. London: WW Norton & Company.

Sennett, R. (1992b). *The fall of public man*. London: WW Norton & Company.

Speck, J. (2013). *Walkable city: How downtown can save America, one step at a time*. New York: Macmillan.

Spivak, M. (1973). *Archetypal place*. Architectural Forum. October, pp. 44–49. Chicago.

Sundstrom, E. (1977). *Theories in the impact of the physical working environment: analytical framework and selective review* (ARCC Workshop on The Impact of the Work Environment on Productivity). Washington, DC: ARCC.

Sundstrom, E., Bell, P. A., Busby, P. L., & Asmus, C. (1996). Environmental psychology 1989–1994. *Annual Review of Psychology, 47*(1), 485–512.

Taylor, R. (1978). Human territoriality-review and a model for future-research. *Cornell Journal of Social Relations, 13*(2), 125–151.

Thwaites, K., Porta, S., Romice, O., & Greaves, M. (2007). *Urban sustainability through environmental design: Approaches to time-people-place responsive urban design*. London: Routledge.

Thwaites, K., Mathers, A. R., & Simkins, I. M. (2013). *Socially restorative urbanism: The theory, process and practice of experiemics*. Abingdon: Routledge.

Turner, F. C. (1976). *Housing by people: Towards autonomy in building environments*. London: Marion Boyars.

Turner, F. C., & Fichter, R. (Eds.). (1972). *Freedom to build*. London: Macmillan.

Turrell, G., Haynes, M., Wilson, L.-A., & Giles-Corti, B. (2013). Can the built environment reduce health inequalities? A study of neighbourhood socioeconomic disadvantage and walking for transport. *Health & Place, 19*, 89–98.

Ulrich, R. S. (1979). Visual landscapes & psychological wellbeing. *Landscape Research, 4*, 17–23.

Urban Task Force. (1999). *Towards an urban renaissance*. London: English Partnership.

Van Cauwenberg, J., Van Holle, V., De Bourdeaudhuij, I., Clarys, P., Nasar, J., Salmon, J., et al. (2014). Physical environmental factors that invite older adults to walk for transportation. *Journal of Environmental Psychology, 38*, 65–70.

Vine, D., Buys, L., & Aird, R. (2012). The use of amenities in high density neighbourhoods by older urban Australian residents. *Landscape and Urban Planning, 107*(2), 159–171.

Wainwright, O. (2014, May 7). Right to build: Nick Boles tells councils to offer land for self-builds 'or be sued'. *The Guardian*. http://www.theguardian.com/artanddesign/architecture-design-blog/2014/may/07/right-to-nick-boles-councils-self-build-sued. Accessed in Dec 2014.

Weaver, W. (1948). Science and complexity. *American Scientist, 36*(4), 536–544.

Weeks, G. (2014). Objectively healthy cities. In E. Edgerton, O. Romice, & K. Thwaites (Eds.), *Bridging the boundaries: Human experience in the natural and built environment and implications for research, policy, and practice* (Vol. 5, pp. 19–33). Gottingen: Hogrefe Publishing.

Weiner, F. H. (1976). Altruism, ambiance, and action: The effects of rural and urban rearing on helping behavior. *Journal of Personality and Social Psychology, 34*(1), 112.

Werner, C. M., Brown, B. B., & Gallimore, J. (2010). Light rail use is more likely on "walkable" blocks: Further support for using micro-level environmental audit measures. *Journal of Environmental Psychology, 30*(2), 206–214.

Whyte, W. H. (1980). *The social life of small urban spaces*. New York: Project for Public Spaces.

Whyte, W. H. (1988). *City: Rediscovering the centre*. New York: Doubleday.

Wolch, J. R., Byrne, J., & Newell, J. P. (2014). Urban green space, public health, and environmental justice: The challenge of making cities 'just green enough'. *Landscape and Urban Planning, 125*, 234–244.

Wolfe, C. (2013). *Urbanism without effort*. Washington, DC: Island Press.

Wood, L., Shannon, T., Bulsara, M., Pikora, T., McCormack, G., & Giles-Corti, B. (2008). The anatomy of the safe and social suburb: An exploratory study of the built environment, social capital and residents' perceptions of safety. *Health & Place, 14*(1), 15–31.

Zimbardo, P. G. (1969). The human choice: Individuation, reason, and order versus deindividuation, impulse, and chaos. In *Nebraska symposium on motivation*. Lincoln: University of Nebraska Press.

The City as an Environment for Urban Experiences and the Learning of Cultural Practices

15

Pablo Páramo

15.1 The Quality of Urban Life

The concept of quality of life takes on a special meaning in the urban environment, not only because it is where the highest proportion of human settlements of the world's population are currently concentrated, but also because they create social conditions that require planning and management to ensure the well-being of people and the future of civilization. As is well known, cities now host nearly 50 % of the world's population and it is estimated that this will rise to 70 % by mid-century (http://www.un.org/en/ecosoc/integration/pdf/unesco.pdf). Cities have become more complex; the majority of national populations are concentrated there, having abandoned rural areas. Both domestic and foreign migrations have complicated the social and spatial relationships of the urban environment. Planners now have to design for crowds. Inaccessibility, the deterioration of the environment, social relationships, poverty, social insecurity, and the saturation of services are just some of the problems that are currently evident in cities, at least in South America. Hence, the quality of urban life is understood as the conditions governing a habitable space in terms of the comfort associated with the ecological, biological, economic-productive, socio-cultural, typological, technological and aesthetic elements in its spatial dimensions (Luengo 1998). Urban environmental quality is, by extension, a product of the interaction of these variables to create a healthy, comfortable habitat capable of satisfying the basic requirements of the sustainability of individual human life in social interaction within the urban environment (Moyano 2010). Zeisel (2006) says it is essential to meet five basic needs that individuals have in any type of built environment, whether it is housing, a neighborhood or a city: security, legibility, privacy, sociability and identity with that environment, to which we should add social support (as argued in Chap. 2 of this volume).

Therefore, projecting a good quality of life in the urban habitat means being able to respond to the demands for housing, utilities, equipment, ease of mobility and public space, while preserving the environmental quality and maintaining fairness among the inhabitants. In the year 2000, the Congress of Local Authorities of the Council of Europe adopted the European Declaration of the People's Right to the City (http://www.idhc.org/esp/documents/Biblio/DHE_7_esp.pdf). It defines the right to the city as: "the right to a collective space that belongs to all its inhabitants (who) have the right to find conditions for their political, social, and economic realization; assuming the duties of solidarity" (World Charter on the Right to

P. Páramo (✉)
Universidad Pedagógica Nacional, Bogotá, Colombia
e-mail: pdeparamo@gmail.com

the City www.hic-net.org). It also establishes the fundamental rights of citizens: the right to security, work, housing, mobility and coexistence between the various users of the public space, to health and to a healthy environment, sports and leisure, culture, and multicultural integration. It secures the right to a pleasant environment, built and stimulated as a result of a quality contemporary architecture produced by conservation and careful rehabilitation of heritage. In addition, it asserts that the city shall have a physical structure in which the various elements – houses, factories, and facilities – relate topologically to form a coherent whole, so that the phase of spontaneous growth is the subject of a planned order that transforms the real city into the desired city. The physical structure is a stage for urban life. Finally, the city is required to have a beautiful and attractive identifiable image, which is legible and able to remain in the memory of its inhabitants and of those who visit it. Apart from being an instrument of ethical education, it should also be aesthetic. Individual well-being should be an equal entitlement, resulting from the creation of an urban environment that helps personal fulfillment and the social, cultural, moral and spiritual development of each inhabitant and, finally, solidarity. It is in this context that it is stated in the document that the city must be seen as a collective space, a place for the political, economic, social and cultural development of the population, not only as a city but as Civitas and Polis. These aspects are of paramount importance as objects of study and for the application of environmental psychology.

The Monocle magazine publishes an annual list of the most livable cities and points out those that offer better living conditions. Among the criteria taken into account are: security, international connectivity, climate, the quality of the architecture, public transportation, tolerance, environmental conditions and access to nature, urban design, economic conditions, proactive policy and healthcare. According to this method, the best cities to live in are in Europe, Canada, and the United States. There are currently no Latin American cities included in this list.

The Gallup Poll on *Quality of Urban Life* established indicators of quality of life in the American region: the quality of public transportation, roads, the education system, quality and price of available homes, air and water quality, and security. Using these indicators for the Latin American region, public security stands out as the weakest point, based on the low percentage of people who feel safe walking around their city or neighborhood at night. As stated in the report of the Banco Interamericano de Desarrollo (BID 2008), nearly sixty percent of Latin American and Caribbean populations feel unsafe at night in the streets of their neighborhoods. The report notes that no other region in the world suffers from such a climate of insecurity. Hence, the importance of the role of environmental psychology in contributing to the improvement of the living conditions of the inhabitants of urban areas.

15.2 The Tradition in Research on the Relationships of Individuals with the Urban Environment

15.2.1 The Pessimistic View of Living in the City

For a long time, it was predominant in urban research to identify the negative aspects of living in big cities. The work of sociology and psychology for much of the twentieth century suggested that many of the conditions of urban life, such as high population density, noise, pollution, and the need to use mass transit, can generate stress. Academics, envisaging the population growth in cities during the twentieth century, wanted to warn of the risks to the health of their inhabitants. Therefore, they compared, mainly in terms of statistical correlations, living in the city and in the countryside with respect to crime, vandalism, prostitution, mental illness, addiction to drugs, etc. Several works identified some of these negative effects of the urban experience such as stress and the city (Roberts 1977; Webb and Collette 1977; Krupat 1985; Moser 1992); segregation of homosexuals (Aldrich 2004), overcrowding (Baldassare 1974); segregation of women (Drucker and Gumpert 1997; Franck 2002), vandalism, crime in the streets, noise and other types of pollution (Sternlieb and Hughes 1983). Such research sought to explain how living in the city was responsible for psychological stress, social problems, and a general decrease in the quality of life of the inhabitants.

It seems this pessimistic view was inherited from sociology and its most influential theorists, Park, Wirth, and Burgess, who formed the school of human ecology of Chicago. At the beginning of the twentieth century, they reflected a strong interest in studying the urban influences of overcrowding on aspects such as mental illness, isolation and conflicts between people. Carr and Lynch (1968) stated that the city is more than a set of buildings, streets, lights, and a transportation system; it is a state of mind, a set of traditions and customs transmitted through its tradition. Thus began the sociological study of the psychological consequences of living in an urban environment. More recently, other sociologists and geographers, such as Lefevre, Harvey, Borja and Castells, have shown a greater interest in demonstrating the various ways of appropriating space and have called attention to social inequalities in large cities. For Harvey (1973), for example, cities have been founded on the exploitation of the many by the few while Castells (1988) criticizes the concept of urban culture for how much it has homogenized the population, showing it as integrated and disguising the class divide.

One of the first psychologists to approach the urban experience as a whole was Milgram (1970) who described city life as an overload of stimuli that exceeds the capacity of the cognitive processing of its residents and that is reflected in social isolation, boredom and, in many cases, aggression as a protective shield against this stimulation overload. In this regard, the work of Milgram (1970), following the sociologist Simmel (1905), characterized the urban personality as rationalist and cold. Emphasizing the variables, Wirth (1938) identified the key features of the city as size, density and heterogeneity. These demographic variables, external to the individual, were responsible for what he called overload and for the inability to process the diversity of stimuli. Protection through psychological strategies minimized the time of attention to environmental stimuli, and led to inattention to stimuli considered less important, rethinking of how to relate to others, and cancellation of inputs and information that cannot be handled. As a psychologist, Milgram explained these forms of responding to the environment as internal processes and adaptive mechanisms to cope with complex environments.

To validate the assumptions or arguments about the risks of living in the city, several comparative studies between the city and the countryside were conducted on issues such as the incidence of mental illness, in which an emphasis on population density predominated as a determining factor for many of the problems faced by the urbanite. This led Milgram to define the city as a relatively important settlement, dense and permanent with socially heterogeneous individuals. These characteristics lead to impersonal, segmented, superficial and antisocial relationships, which also affect social responsibility, altruism, the seeking of anonymity, and the way space is represented by people.

Studies comparing rural life with city life showed a higher incidence of psychiatric disorders in the latter and a more solid social structure and clarity in social roles in the former. Some, like Faris and Dunham (1939) and Burgess (1925), proposed a spatial distribution for different types of disorders within the city limits.

Specifically, the high incidence of criminality, a deficiency in interpersonal behaviors, stress, and other social issues were attributed to population density (Baum and Paulus 1987; Evans and Cohen 1987). Alcoholism, drug addiction, and sexual deviations were also attributed to this population condition. Moser (1992), for example, examined the different urban stressors faced by the inhabitants of a big city as a result of the high population density, attitudes of aggression, personal relationships, memory, competences, emotions, and health. Stemming from Seligman's idea of learned helplessness (Seligman 1974), the environmental conditions of the city affect the potential for greater stress and its effects on the feeling of a loss of control to modify the consequences. For Moser, such reactions are mediated by several factors including motivation, habituation or adaptability, social support, and the characteristics of the situation (the possibility of exercising control over them). The stress of living in the city is induced by the threat posed by certain environmental stimuli and an inability to confront them or exercise control over complex situations generated by the city: time pressures, the need to follow many rules to engage in the space, etc.

However, most of these studies have been largely disproved by comparing the levels of these problems with highly populous countries in Asia, where the results contradict the initial findings. Apparently, the inconsistency in the results of studies on crowding and stress, for example, can be attributed to the fact that it is not the same to talk about population density as it is to talk about overcrowding, because density is principally concerned with the proportion of people in the space, while overcrowding is an eminently psychological topic that concerns the demand for space that exceeds the availability to the individual (Stokols 1972) or is a matter of perception of the possibility of exercising control over this (Responsible Urban Behavior, Páramo 2004).

In terms of crime, research on the perception of risk in cities associates them with a high rate of common crime, particularly involving young people (Dunn 1980; Fabrikant 1979). Other studies show that high crime in some parts of the city is associated with a lack of control by the authorities and a lack of informal social control, which leads the criminal to have a greater chance of success (Taylor and Gottfredson 1986; Taylor 1987). Criminals, according to these studies, prefer narrow, dark places with trees and out of sight of people, with escape routes. The work of Hunter (1978, 1987) identified environmental deterioration, the presence of the homeless, and workplaces that are seen by people as mysterious (Herzog and Smith 1988) as signs of a greater risk of being assaulted, a syndrome that is known as "broken windows". Brantingham and Brantingham (1991) developed a distribution model for crime, which included the distribution of criminals based on where they lived and their patterns of urban mobility; the distribution of potential victims; the spaces known to the offender and others; and factors that contribute to the estimation of the success or failure of the crime. Many studies were devoted, and still are, to exploring fear as a condition of urban life and to proposing theoretical models to decrease the risk of becoming victims, as in the proposal of Newman on defensible spaces (1972). This included designs giving the impression of always being under surveillance by installing false corridor windows and illuminating places where people walk without demonstrating that it is the urban environment as such that generates it. Similarly, the affirmation that the city generates social isolation, little altruism and distrust has also generated conflicting results. While some studies showed that there was less contact with neighbors and strangers in the city, others showed that helping behavior was more common in cities (Lofland 1978, 1985). Although there are real dangers, which are demonstrated objectively in the rates of homicides, robberies, rapes, etc., there are important subjective elements constructed by the psychological influence of the media. These contribute to increasing fear, giving rise to a fear built in the city and also a city built by fear (Fraile et al. 2010). For example, the decision to broadcast the news is based on the rating power of television channels, coupled with the interest that citizens show for sensationalism, tabloids, and violent news. It is presented day after day, and repeated morning, noon and night, relaying how dangerous the city is while helping to create a culture of fear (Altheide and de Gruyter 2002; Soyinka 2004; Macek 2006; Katz 2006; Linke and Smith 2009; Carro et al. 2010; Valera 2012). Thus, the pathology they are trying to show due to living in the city seems like a statement supported by the culture in which the study is carried out, and the role played by the media in their broadcasting, rather than a general law.

15.2.2 The Optimistic View of Living in the City

It is curious that the different strategies that researchers have used are viewed mainly from social pathology. By contrast, in modern cities, social situations give people the opportunity to develop new interests and new activities. Social groups develop new cognitive strategies that seem to lead to a relativism of values, at first glance, placing their own forms of resistance against the negative aspects of living a social behavior without rules. Although it is often assumed that overcrowding has negative effects on human behavior, the positive effects of stimulation diversity, informal social support and newness of unexpected stimuli that are offered by a great city should also be considered. Only recently have researchers paid attention to the vision of the city as a system of places with opportunities for learning and informal social support, with the exception of the work of Jacobs (1961, 1970a, b), Stone (1954) and Goffman (1983). Despite the difficulties already identified by urban planners and the risks of living in a big city, we can make use of great resources and benefits by living in large urban centers. Cities are the main areas of cultural, scientific and technological development, which affect the way of life of their inhabitants. They also have universities, which are major sources of employment and economic development and constitute an exciting atmosphere with a variety of options that apparently we are not willing to give up. This is why there have been important developments on a theoretical level and in intervention from an optimistic perspective to understand the urban experience. Such is the case in the following publications: *Image of the City* (Lynch 1965), *Where Learning Happens* (Carr and Lynch 1968), *Educating City* (Barcelona 1990), *The Sustainable City* (Lorenzo 1998), *The City of the Bambini* (Tonucci 1997), *The Conquered City* (Borja 2004), and *The Meaning of Public Places for the People of Bogotá* (Páramo 2004).

These approaches address topics ranging from the provision of public spaces that contribute to the meeting of people and participation to the utilization of municipal institutions as a resource for public education. The educating city movement, for example, has promoted the idea that the city as a whole can organize itself to provide educational opportunities. From this perspective, the city can plan a place designed for the purpose of learning and the personal growth of the citizen. Thus, cities constitute an educational resource, since they contain universities, museums, schools, etc. As an educational agent, the city provides diverse opportunities for socialization and for non-formal education. It offers a wide range of information, from informative signs to historic monuments. As an educational object, the city can be seen as an important element in itself, from which to learn its architecture, structure, and history (Trilla 1997). This is the starting point to collect studies that show other dimensions about the ways we interact with the city environment and with whom we relate.

15.3 Public Space

The public space is an element of importance for the assessment of the quality of life in the city. It is the suitable setting for the social and symbolic expression of different individuals and social groups and a vital element in the evocation of the collective historical memory of cities, commemorations, and demonstrations, which are part of the political, social, and cultural identity of its people. It is a scenario in which different aspects of urban life, such as economics, urban planning, the search for equity, gender, and relationships between the city and the natural environment, intersect. Perhaps this is what has caught the attention of researchers from different disciplines, and particularly that of city managers, in recent years. The city in its entirety is the way in which public life develops, by learning rules to coexist among strangers and how to mobilize and read symbols to orient ourselves spatially. For example, Jacobs mainly characterizes urban life in terms of meetings between strangers. Although superficial, they can still be satisfying (Jacobs 1961). This author also made contributions to the profound *Manual of Environmental Psychology* (Proshansky et al. 1978), where she explored life on the streets, showing the dynamics created by the interaction between neighbors, generating a feeling of support to the individual in the community and leading to confidence. This trust is formed over time from many small encounters between people who move along the sidewalks. She contemplated the value of sidewalks for these types of casual and informal encounters, which are not of the same warmth as those found in gated play areas or commercial interactions between people. Social life on the street offers positive aspects and quality relationships associated with an enjoyable experience and a break from different urban stressors.

On the other hand, Carr et al. (1992) focused their work on public space. They identified the forces that drive individuals to activities in open spaces and the rights of the people in them. Among these forces, climate and topography are often highlighted as conditions determining activities in open public scenarios, with cultural baggage making social activity a predominant feature in certain cultures, and also the function to meet the basic needs of a society (to navigate roads or places, to protect themselves from changes in the weather, to protect the members of the group, or to facilitate trade). Symbolic public life establishes another important force; the shared meanings that occur and are built in public, the spiritual and mystical experiences of a society, the celebration of past events as sacred or national days, and historical events that merit celebration. Other forces identified by the authors are the social, political, and economic systems. One of the first rights to be lost in a totalitarian government is the ability to meet and speak in public. For example, many tragic events for Latin American society, like the assassinations of political leaders, students or police, occurred in a public space. The economy also affects the availability and accessibility of the public space, focusing directly on the number of street vendors, unemployed people, displaced people, and the homeless on the streets. Due to our phylogenetic heritage, it is also possible to say that the attraction of people to the natural features of the environment contributes to sustaining public life. The vegetation, lakes, rivers, gardens and trees in the streets are valuable parts of cities. In many societies, parks invite people into the public space through designs that evoke natural qualities. These elements play a restorative function (for example, in the work of Kaplan: see Chaps. 7 and 8 in this volume) in offering opportunities for different groups to find themselves in positive ways. Public life – or being in the public space – gives relief from the stress of work by providing opportunities for entertainment, recreation, and social contact. In the same way, people use public spaces in search of stimulation. They satisfy their curiosity of the unknown by visiting sites that represent novelty, such as fairs, or making trips to places where the landscape is unknown or novelty can be found.

Despite the importance of social life, we favor the automobile, urban sprawl, the proliferation of shopping malls, and privatization (Berroeta and Vidal 2012). This displaces the sociability that formerly took place in the squares and streets of public places with a social vocation such as bars, cafes, discos, restaurants, social clubs, etc. These places led to the reconceptualization of public spaces with disappearing boundaries between the public and private, giving rise to the "semi" referring to the fuzzy, fluid, grayscale spaces (Monnet 2001, 2012), or the socioplaces (Páramo 2011) which, although remaining private, meet the need for a social life.

From the perspective of urban planning, New Urbanism has tried to find a compromise with social life on a neighborhood scale. It has provided the facilities people need, such as schools, banks, health centers and parks, which are accessible by foot, bicycle, or mass transit between one community and another (Duany and Speck 2009). There have been several important experiences with this objective, though not necessarily related to New Urbanism. One of the criticisms highlighted is the automobile. In Barcelona, for example, in the transformation planned by Bohigas in 1985 (Maragall et al. 2004), there was an effective policy for the transformation of the street and public space. The objective was to reverse gradually the dominant trend of the supremacy of private vehicles. Enhancing the social use of the street was an attempt to find a 50/50 balance between private and public use (Barcelona City Council 1987, 2012; Esteban 1998; FAD 2009). In Cúcuta (Colombia), they have recently introduced something called "The Mall with Open Sky", which tries to maintain the advantages of a shopping center in terms of security while ensuring the free access of everyone to these spaces. These experiences have led to the gradual transformation of the street, looking for a balance between the space for vehicles and sidewalks for citizens, while also seeking to reduce polluting emissions from vehicles. The reduction of circulatory spaces is used to deter the use of the private car in favor of public transport and to promote public spaces for pedestrians and the development of cultural activities. Hence, the discussion of the distinction between public and private spaces begins.

However, in some cities in the United States, and in the vast majority of Latin American cities, the public space has become a confrontational setting. Here, the social differences are marked and generate tension in the cities. They show the power relationships; not only by political demonstrations, but also in the way they have been designed and are appropriate. Public space design (or non-design) often results in gender inequalities brought about by segregation and may result in attacks on women (Valentine 1989). It excludes people for reasons of age, marginalizing children, adolescents and the elderly. It also disenfranchises via socio-economic status, excluding the poorest from access to transport to cultural sites and contact with nature, among other things. The majority of the population is affected by the conditions of insecurity seen by those living in these cities (Cuevas and Gómez 2014).

Although the space is often presented as belonging to everyone, in that we are equal, communicate, and exercise our right to citizenship, the reality shows a different scenario disputed by its characters – street vendors, social movement participants, artists, builders, businesspeople, and women and men of different conditions – all fighting for ownership (see Chaps. 4 and 5). Public space in Latin America is no longer the place for social meetings and much less for recreation. It is now the place where order is manifested and the activities carried out there tend to be criminalized. It is also the space used for economic purposes, mobility, and, in many cases, where crime is carried out. Low (2009), Irazábal (2008), Moser (2012), Berroeta and Vidal (2012), and Páramo (2014) have noted the tensions that arise today in these scenarios: the fight for it as a workplace, a stage for artistic expression or urban art, which some call graffiti or vandalism, street protests against government policies, and the defense of democracy. When these are forbidden, or certain officials are bribed to keep them out, the only resource that remains is the street. Protests are in turn criminalized when confrontations with municipal governments arise. Other scenarios that are fostered are the development of shopping malls that exclude the poor and young people (Jiménez 2014), closed communities (Low 2000), the enclosure of parks at night, which gives administrative power to neighboring communities that exclude others, etc.

Public space is a prerequisite for the expression of different manifestations and the essence of a society that claims to be truly democratic. In times of crisis, it becomes a unique stage to show the desires of the population to governments and the rest of society, demanding the recognition of certain groups of that population. In Latin America, there have been important instances when the public space, as a stage for democracy, has played a very significant role; for example, the major manifestations of the mothers in the Plaza de Mayo in Argentina, against Pinochet in Chile, opposing Banzer in Bolivia, against Rojas Pinilla in Colombia and, more recently, against the guerrilla group FARC.

The squares and main streets have become scenes of protest and the vindication of the rights of citizens; society taking to the squares and the streets has contributed to the change in policies and the change in totalitarian regimes. Not only in Latin America, but in the whole world, protest in the public space over time shows, without a doubt, the way in which the squares and streets have become the major scenarios for the protesting citizen. This can be seen in the mobilizations of the Arab spring that began in 2010 and in the social movements in China in Tiananmen Square in 1989; in the new social movements in Europe (15 M in the Puerta del Sol of Madrid, Plaza Cataluña in Barcelona, Plaza Syntagma in Athens and, in a more violent way, the racial-religious-ethnic conflicts but basically 'subtle' social exclusion on the streets of the suburbs of French cities).

Public space is also being seen as a resource for the education of the citizen and hence the importance of using it as a setting for study. This perspective recognizes that the city and public places may offer valuable educational opportunities to learn the rules of coexistence between strangers (Páramo 2013). Public cultural life, for example, is provided in the inventory of cultural history reflected in the monuments, streets, squares, and other elements of the public space. These elements have been erected to the memory of Juarez, Bolívar, San Martín, Lincoln, Cromwell, and Napoleon as well as more controversial ones in Latin America such as Colón. They may help us rediscover the lost stories of ordinary citizens (women, children, slaves) which evoke connections to past events while stimulating feelings of national pride and contributing to identity with the place or city. Based on the thesis that people convert spaces into places by endowing them with meaning, in the work of Hayden (1999), Donovan (2002), Páramo and Cuervo (2009, 2013), Remesar et al. (2012), and Salazar and Frechilla (2006), there have been historical studies showing how ordinary people use and experience public space with the purpose of increasing its meaning. This contributes to greater ownership of such places and strengthens urban identity. These studies propose spatial designs and the creation of places and monuments that contribute to this memory recall, through representation on the city walls (e.g. Vidal et al. 2012) of the everyday practices of the past, or by travelling exhibitions of historical photographs, which seek to strengthen the urban identity through contact with social history located in the public space in cities.

15.4 Women and Urban Space

Gender scholars have shown that, from infancy, women are taught differently from men in their behaviors and attitudes towards interacting with space. These differences vary historically and culturally according to age, social class, religion, and social role. In this context, some researchers have considered it important to investigate the way that women experience public space and their representations or forms of relationships with strangers in different scenarios or urban places.

The works in this field have identified a number of difficulties encountered by women in today's built environment, which are associated significantly with their age, sexual orientation, place of residence, and many other individual and cultural circumstances key to male and female identities, along with the division of jobs as Franck (2002) describes very well. The studies collected by Drucker and Gumpert (1997) and Burbano (2014a, b, c) try to explain gender differences in relation to uses of public space from the cultural and historical tradition, which shows that women's spaces have been matched with private spaces, and public spaces are still the space of men. In this sense, there is a symbolic opposition between the house and the rest of the world. The feminine sphere is opposite to the masculine one, which corresponds to public life, so that the role of women has mainly been associated with the house. Assuming the dwelling place of women is the house significantly affects the way space is designed, shaped, and used. This makes them meaningful and understandable from the social patterns imposed on women; therefore, public spaces are planned mainly from the perspective of male use (Duncan 1996).

This is the result of the tendency to split the asymmetric environments sexually between the private and the public, which seems to endure today in a variety of forms in Latin American countries. It is a division that contributes to restricting the mobility of women in the public space and to preventing them from participating fully as workers and as citizens.

The systematic observation of the everyday life of Latin American cities shows that the public space is not neutral. The use of public space by women in their daily lives, like taking public transport to work or moving about the street, shows that they are restricted in where, when, and how they can use these spaces. The harassment to which some men resort controls the presence of women in public spaces. Behaviors such as racy compliments, abuse on public transport, or theft of cell phones show that once women are in public, and not accompanied by men, they cannot claim their right to privacy or security as men do (Burbano 2014a). In Panama, a law is currently being studied that prohibits such harassing comments. Mexico City, Rio de Janeiro, and Medellin have exclusive subway cars for women. Harassing women in public places is evidence that women are still defined and perceived in terms of their sexuality and do not enjoy the right to privacy. The fear of this aggression is the main argument given for the distrust of strangers and an inability to circulate fearlessly through the streets at any time of day (Pineda 2007). The factors previously mentioned constitute the main reasons why women prefer shopping malls to streets (Burbano 2014b).

It seems that the Latin American discussion about gender still needs to be delimited thematically to make this topic more visible so that research continues to add to the theoretical contributions. In recent years, it has sought to provide evidence about the role of women in the public space, precisely highlighting inequalities by addressing the needs of women against those of men. The boundaries between public and private spatiality have been questioned, as in the latter women develop many activities of a public nature and achieve an overlapping of both spheres. Although there has been some progress, Veleda da Silva (2007) concludes that it is not sufficient to look at an axis separately from spatiality or as a result of the production of different geographical spaces.

15.5 Spatial Cognition

Spatial cognition is particularly relevant in urban studies because it is important to investigate the spatial orientation, navigation, and displacement that people make when they move through the public space of the city. The research explores the process by which we reach a spatial representation, known as a cognitive map, and its application to spatial orientation, to the image formed by individuals in the city, the way they solve the task of finding a route, the way people with spatial needs orient themselves (the blind, the deaf, the elderly and those with cognitive difficulties), and transcultural studies that examine social differences in urban representations (Gärling 2005; Navarro and Rodriguez 2015).

The information collected in this field of research has resulted in the improved management of public space, in the design of maps for tourists and for transport routes, signage in the city, naming towns, forms of teaching spatial orientation, the creation of a spatial grammar, etc.

15.6 Appropriation

A mechanism that helps to give meaning to spaces, to strengthen identity with the city and that has given rise to various studies in environmental psychology is the appropriation made by individuals of places, which transforms the environment through different forms of action (Korosec-Serfaty 1976; Pol 2002b; Rivlin 2007; Masso et al. Chap. 5 and Valera-Vidal Chap. 3 in this volume). Here, activities like sports, religious practices, music, protests or simply passing the time are observed. Recently, cities have organized concerts in parks, flash mobs, and festivities. They have encouraged gastronomy in the parks and the painting of graffiti as artistic expression on certain walls, generating activities that seek social cohesion, identity, attachment, and the appropriation of the city.

Groups of young people in Spain and several Latin American countries have been intervening, seeking to empower vulnerable population dynamics and inviting people to make things happen in places where they do not happen or temporarily build what is needed; for example, demarcated pedestrian zones or drawing attention to gaps in the streets that pose risks for both pedestrians and drivers. In the streets of Barcelona, Rio de Janeiro, Bogota and many other cities, there have been interventions of this type. Some people dress up as cones to encourage drivers to give way to pedestrians at dangerous crossroads. Sofas are placed in the streets so that pedestrians can rest. These are proposals that seek to promote activities for young people and enable them to take over the city.

However, some of these forms of appropriation generate tension. In recent years, the media have been registering the proliferation of graffiti in many Latin American cities. This has generated controversy about whether these demonstrations are acts of vandalism or urban art. At the same time, they have been promoting the recovery of monuments that are part of the tangible heritage of the cities. They emphasize the graffiti as an artistic cultural expression. Although this is seen today in many cities around the world, how can we understand the graffiti on monuments that make up the cultural heritage of the city? Does this reform the monuments, giving them new meaning, or is it simply vandalism? And what about the graffiti on stores, homes and parks? How do we differentiate artistic expression from a symbol that marks the appropriation of a territory by a group of young people who want to tell another group that they are not welcome? If works of art are only understood by a few artists and fail to communicate with the public in general, do we consider them artistic? How do we mediate between the general interests of the public, through the collective right to the public space, and those who want to seize it through such practices of appropriation?

15.7 The Design of Cultural Practices to Improve the Quality of Urban Life

Much of environmental psychological research has been aimed at finding solutions to environmental problems. Psychological and educational strategies have been directed at minimizing the impact of certain behaviors on the pollution or degradation of the environment. Such actions could include waste production, the consumption of non-renewable energy resources, and consumerism in general. Encouraging behavior that contributes to the protection of the environment, called environmentally responsible behavior (ERB), also includes the incentive of alternative transportation. The approach that has been followed has been that of informational campaigns and the application of the principles of experimental behavior analysis aimed at modifying behavior. These have achieved mixed results in terms of sustainability over time and when they have withdrawn the campaigns and the reinforcement contingencies, they have generated changes (Geller 2002; Schultz and Kaiser 2012).

When dealing with the particular problems of the urban environment, it is worth specifying that living in the city demands another kind of action for the sustainability of life. This includes all those other behaviors that facilitate social organization: the observance of traffic laws (signage, respecting pedestrians, giving way to ambulances and firefighters), making use of public transport and bicycles, taking care of the public space, including monuments, as a common good, participating in projects that affect the urban environment, following the rules of coexistence among strangers, no smoking in public places, acting in solidarity with whoever is in trouble, etc. The contributions of environmental psychology can help mitigate the impact that many of these actions have on the environment of the city. They improve the quality of life in cities by delegating much of the responsibility to the individual and by not only including environmentallyrelevant behaviors in the solution of environmental problems, but also all those that contribute to the quality of urban life; these are responsible urban behaviors (RUB) (Páramo 2010, 2013). URB are fundamental to the sustainability of urban life, and are mainly characterized by relationships between strangers (Lofland 1998), which are essentially agreements between individuals about certain behaviors to ensure peaceful coexistence and to improve the quality of life of the inhabitants of the city.

The notion of responsible urban behavior (RUB) assumes that it is sustainable over time and, by maximizing the achievements of informational campaigns and behavior modification programs, a metacontingency or functional relationship should establish a sustained collective action in an interdependent way between individuals who are part of a community. The result of this practice should produce greater benefits to the individual than their isolated action (Glenn 1991) so that when multiple interdependent behaviors act on social cohesion, the relationship between them and their similar consequences lead to a metacontingency. As a result, this leads to the selection of a group of behaviors, URB, which ensures its sustainability between generations and leads to it becoming a cultural practice.

Some examples of what could lead to a metacontingency are: the notice located on the back of some vehicles: "How am I driving? Call 757, toll free"; registering and advertising the total amount of taxes collected by the city on giant screens in the streets or transportation terminals, showing which works they are funding; posting vehicle accident rates accompanied by traffic laws that must be followed; displaying ordinary people demonstrating positive and negative RUB and sending them to television stations to be broadcast as indicators of coexistence. Another way is to ask for the participation of "Civil Servants" who call attention to pedestrians showing unsafe behavior when moving around the city. In Bogota, there is currently an application for smartphones, Waze, where citizens can report places where they have been the victim of a crime.

For people to be integrated into the metacontingencies that promote RUB, they must learn the rules, understood as verbal statements, which establish a relationship between what happens prior to the behavior, the said behavior, and its consequences (Catania et al. 1989). As a result of the verbal behavior, these rules describe the function and organization of the contingencies that are established with the environment. Rules usually indicate what to do, when to do it, and what should happen when it is done. They serve as a bridge between conduct and consequence, the latter usually being delayed. Rules, if they are taught, will be adopted depending on how successful and beneficial they have been in the past and what consequences are offered in the present by following them. Additionally, the individual will establish possible links between them to derive their own rules and act accordingly when the environment does not explicitly set the rules on how to deal with a situation that is new for the person. Examples of rules that are found in the city or that guide our behavior in urban spaces are: "Parking on the sidewalk will result in a fine of "x" amount". In this example, the rule also serves as a bridge between the expected conduct and the social or legal consequences when the latter occurs in a delayed manner (Hayes et al. 1989). Although the rules are taught to influence the way we relate to other people or objects, their main purpose is to teach self-regulation, hence their importance to teach and support desirable cultural practices for a good quality of life in the city.

What is being sought by attempting to use rules to guide the behavior of citizens is a lack of dependence on the physical presence of an external regulator. This initially requires the participation of the ordinary citizen to demand compliance with the rule in order to ensure the metacontingencies; the common benefit through interdependent action by individuals. Thus, rules present in the verbal repertoire of the individual mediate the relationship of the individual with society, even if it is only in one place (Guerin 2001). In many cities, for example, people already follow urban rules without external controls. Publictransport fares are paid even if the ticket does not have to be presented, vehicles give way to pedestrians, garbage is recycled, pedestrian bridges are used, streets are crossed at corners, dog droppings are collected, purchases are paid for at the supermarket, etc., all without external controls. Individuals regulate themselves following rules that contribute to social practices of coexistence. Thus, social behavior that involves following a rule does not necessarily require immediate or direct consequences or the presence of another individual in the workplace. People do not have to be controlled. Instead, they should be guided through rules that act as a mechanism for self-regulation.

In recent years, Bogota held a mass experiment during the administration of Mayor Mockus (1995–1998), which sought to educate citizens by spreading rules that contribute to the quality of urban life. It was called "Cultural Citizen". However, as with most behavior modification programs and advertising campaigns, their achievements have declined with the passage of time, as the advertising to publicize the rules and their consequences for citizens was removed.

Focusing more on a contingency design goal, Páramo developed this approach by way of field experiments in two cities in Colombia with promising results. In Bogota, signage was designed that advertised rules for the safety of pedestrians. Three places in the city were identified where pedestrians did not follow the rules, even when they were near the marked areas to cross a road at the signal light or by a pedestrian bridge. The 11 rules that were identified included: crossing underneath the pedestrian bridge, crossing the road between vehicles, running across, getting off the bus in the middle of the road. Each symbol that was part of the signage specified the rule to be followed in terms of the situation, the conduct, and the individual and social consequences. It was symbolically represented and invited citizens explicitly to point out those who were not following these rules and to accept being pointed out by fellow citizens. The signage was supported by the work of "Civil Servants", who applied social recognition for the observance of the rule or called attention to people for failing to meet it, thus guaranteeing the metacontingency. In Cúcuta, on the other hand, the signage was designed to take advantage of large advertising totems, which showed pedestrian safety rules and others on the management of waste in the streets. The achievements that were reached regarding fewer offending pedestrians, or which welcomed the observance of rules, were advertised through banners and a giant screen at one of the major intersections. The results have shown their effectiveness over time (Páramo and Páramo, Fundación MAPFRE 2015; Páramo and Contreras in press).

In this way, public spaces can be seen as political education scenarios, in which the rules of coexistence between strangers can be learned by arrays of metacontingencies that guarantee their sustainability over time.

15.8 An Interdisciplinary and Methodologically Mixed Approach to Research on Urban Affairs

An important trend in urban studies, which has been observed since their inception in environmental psychology, is the interdisciplinary and mixed character of the methodological approach to investigation. As noted by Uzzell and Romice (2003) and Romice and Uzzell (2005), it is very positive to break disciplinary barriers in the solution of environmental projects, and in their physical interventions. These are beneficial to communities though not always achieved (Churchman 2002). The studies on which this chapter is based use different methodological approaches, including the collection and analysis of qualitative and quantitative information: in-depth interviews, narratives, direct non-participating behavior, documentary reviews, georeferencing analysis (GIS), statistical analysis, SPSS, software systems and observations for qualitative analysis such as Atlas.ti. Thanks to this multi-method approach, a greater validity and greater communication between disciplines is achieved.

15.9 Conclusion

The central theme of this chapter is the theoretical development applied by environmental psychology to approach the urban experience as a whole. After a trend showing the negative impact of living in the city, sociological and psychological research now predominantly shows urban life as an enriching experience, which provides multiple opportunities for learning and interaction in the context of the city and contributes to shaping the identity of the citizen. The perspective of looking at the city as a stage for learning and the formation of citizen rights constitutes a starting point for a new theoretical development from environmental psychology, contributing to better planning of urban environments. By seeing the city as a learning environment and focusing on exploring how people understand, experience, and value the city, psychology contributes to urban studies in re-directing attention to the positive aspects of living in the city in this century and the importance of improving the quality of life in cities. As Moser affirmed (2012), it is not enough to live next to someone; it is necessary to live together sharing "urbanity" and creating the conditions to live in a community.

Several problems for research arise. Theoretical construction and social intervention should be addressed within the perspective of the psychological experience of the city compared to the conditions of currently living in cities. The identification and recovery of significant places, particularly for Latin American societies, beginning with their social history should also be addressed. Other issues are the search for spatial equity, the promotion of shared rules that can facilitate relationships between strangers in increasingly globalized cities, the impact of uprooting displaced populations and their identity in the place, etc. From the theoretical development, there should emerge a field of research. This should promote further development in the integration of theories through meta-concepts, especially in an interdisciplinary field such as environmental psychology. The developments of Pol (2002a) and Wiesenfeld and Sanchez (2002) represent a good guide to the role of environmental psychologists in urban interventions being carried out by public institutions in interdisciplinary teams of urban planners, architects, and, of course, the members of the community.

References

Aldrich, R. (2004). Homosexuality and the city: An historical overview. *Urban Studies, 41*(9), 1719–1737.

Altheide, D. L., & Gruyter, A. (2002). Creating fear: news and the construction of crisis. *Canadian Journal of Sociology Online.* http://www.cjsonline.ca/pdf/fear.pdf

Baldassare, M. (1974). *Crowding and huan behavior: Are cities behavioral sinks?* Monticello: University of California, Berkeley.

Barcelona. (1990). Ciudad Educadora. Congrés internacional de ciutats educadores, regidora d'Edicions. Publicacions, Ajuntament de Barcelona.

Barcelona City Council. (1987). *Programa d'Actuació Municipal 1988–1991* [Municipal Action Plan 1988–1991]. Technical Programming office. Barcelona City Council.

Barcelona City Council. (2012). *Plan de Movilidad Urbana de Barcelona 2013–2018* (Barcelona's Mobility Plan). http://prod-mobilitat.s3.amazonaws.com/PMU2013-2018IntroDiagnosiEscenaris_llarg_0.pdf. 9 Mar 2015.

Baum, A., & Paulus, P. (1987). Crowding. In D. Stokols & I. Altman (Eds.), *Handbook of environmental psychology* (pp. 533–570). New York: Wiley.

Berroeta, H., & Vidal, T. (2012). La noción de espacio público y la configuración de la ciudad: fundamentos para los relatos de pérdida, civilidad y disputa. *Polis, 31/2012.* http://polis.revues.org/3612; doi: 10.4000/polis.3612.

Bohigas, O. (1985). *Reconstrucció de Barcelona.* Barcelona: Edicions 62. Spanish version: Madrid, Publicaciones Ministerio de Obras Públicas y Urbanismo, 1986.

Borja, J. (2004). *La Ciudad Conquistada.* Madrid: Alianza.

Brantingham, P. J., & Brantingham, P. L. (1991). *Environmental criminology.* Prospect Heights: Waveland.

Burbano, A. M. (2014a). La investigación sobre el espacio público en Colombia: su importancia para la gestión urbana. *Territorios, 31*, 185–205. doi:dx.doi.org/10.12804/territ31.2014.08.

Burbano, A. M. (2014b). El espacio público urbano situado en la ciudad latinoamericana contemporánea: Una aproximación a su estudio desde la perspectiva del género. In E. Licona (Ed.), *Espacio y Espacio Público. Contribuciones para su estudio* (pp. 151–167). Puebla: Benemérita Universidad Autónoma de Puebla.

Burbano, A. M. (2014c). La investigación en espacio público desde la academia y las instituciones gubernamentales en Colombia. In A. M. Burbano & P. Páramo

(Eds.), *La ciudad habitable* (pp. 27–52). Bogotá: Universidad Piloto de Colombia.

Burgess, P. (1925). *The growth of the city: An introduction to a research project*. Chicago: The University of Chicago Press.

Carr, S., & Lynch, K. (1968). Where learning happens. *Daedalus, 97*, 1277–1291.

Carr, S., Francis, M., Rivlin, L., & Stone, A. (1992). *Public space*. New York: Cambridge University Press.

Carro, D., Valera, S., & Vidal, T. (2010). Perceived insecurity in the public space: Personal, social and environmental variables. *Quality and Quantity, 44*, 303–314.

Castells, M. (1988). *La cuestión urbana*. México: Siglo XXI. Edición original en francés, 1972.

Catania, A. C., Shimoff, E., & Matthews, B. A. (1989). An experimental analysis of rule-governed behavior. In S. C. Hayes (Ed.), *Rule-governed behavior: Cognition, contingencies, and instructional control* (pp. 119–150). New York: Plenum Press.

Churchman, A. (2002). Environmental psychology and urban planning: Where can the twain meet? In A. Churchman (Ed.), *Handbook of environmental psychology* (pp. 347–362). New York: Wiley.

Cuevas, E., & Gómez, S. Y. (2014). En busca de la seguridad en el espacio público: Aproximaciones urbanas de Xalapa, México. *Papeles de Coyuntura, 38*, 46–57. Universidad Piloto de Colombia.

Banco Interamericano de Desarrollo (2008). *Calidad de Vida: Más allá de los hechos*. http://idbdocs.iadb.org/wsdocs/getdocument.aspx?docnum=1775347

Donovan, M. G. (2002). *Space wars in Bogotá: The recovery of public space and its impact on street vendors. Urban planning*. Boston: MIT.

Drucker, S., & Gumpert, G. (Eds.). (1997). *Voices in the street: Explorations in gender, media, and public space*. Cresskill: Hampton Press.

Duany, A., & Speck, J. (2009). *The smart growth manual*. New York: McGraw-Hill.

Duncan, N. (1996). Renegotiating gender and sexuality in public and private spaces. In N. Duncan (Ed.), *Bodyspace: Destabilizing geographies of gender and sexuality* (pp. 127–145). London: Routledge.

Dunn, C. J. (1980). The social area structure of suburban crime. In D. E. Georges-Abeyie & K. D. Harries (Eds.), *Crime: A spatial perspective* (pp. 5–25). New York: Columbia University Press.

ECOSOC. (2014). *Intégration segment: Sustainable urbanization, 27–29 May 2014*. http://www.un.org/en/ecosoc/integration/pdf/unesco.pdf

Esteban, N. (1998). *Elementos de ordenación urbana*. Barcelona: UPC.

Evans, G., & Cohen, S. (1987). Environmental stress. In D. Stokols & I. Altman (Eds.), *Handbook of environmental psychology* (pp. 571–610). New York: Wiley-InterScience.

Fabrikant, R. (1979). The distribution of criminal offenses in an urban environment: A spatial analysis of crime spillovers and juvenile offenders. *American Journal of Economics and Sociology, 38*, 31–47.

FAD. (2009). *La U urbana. El libro blanco de las calles de Barcelona. Fomento de las Artes y el Diseño (FAD) y Ayuntamiento de Barcelona* [The Urban U. The White Book of the streets of Barcelona. Fostering Arts and Design (FAD) and Barcelona City Council]. http://issuu.com/ecourbano/docs/la-u-urbana. Accessed 9 Mar 2015

Faris, R. E., & Dunham, W. (1939). *Mental disorders in urban areas: An ecological study of schizophrenia and other psychoses*. Chicago: University of Chicago Press. Reprint 1965.

Fraile, P., Bonastra, Q., Rodríguez, G., & Arella, C. (2010). *Seguridad, temores y paisaje urbano*. Barcelona: Serbal.

Franck, K. A. (2002). Women and environment. In B. R. Bechtel & A. Churchman (Eds.), *Handbook of environmental psychology* (pp. 347–362). New York: Wiley.

Gärling, T. (2005). Spatial cognition in travel analysis. In Invited keynote address, 6th biennial conference on environmental psychology, Ruhr-University, Bochum, Germany.

Geller, E. S. (2002). The challenge of increasing proenvironmental behavior. In B. R. Bechtel & A. Churchman (Eds.), *Handbook of environmental psychology* (pp. 525–540). New York: Wiley.

Glenn, S. (1991). Contingencies and meta-contingencies: Relations among behavioral, cultural and biological evolution. In P. A. Lamal (Ed.), *Behavioral analysis of societies and cultural practices* (pp. 39–73). Washington: Hemisphere.

Goffman, E. (1983). The interaction order. *American Sociological Review, 48*(1), 1–17.

Guerin, B. (2001). Individuals as social relationships: 18 ways that acting alone can be thought of as social behavior. *Review of General Psychology, 5*(4), 406–428.

Harvey, D. (1973). *Social justice and the city*. Baltimore: John Hopkins University Press.

Hayden, D. (1999). *The power of place*. Cambridge: The MIT Press.

Hayes, S. C., Zettle, R., & Rosenfarb, I. (1989). Rule-following. In S. C. Hayes (Ed.), *Rule-governed behavior: Cognition, contingencies, and instructional control* (pp. 191–220). New York: Plenum Press.

Herzog, T. R., & Smith, G. A. (1988). Danger, mystery and environmental preference. *Environment and Behavior, 20*, 320–344.

Hunter, A. (1978). *Symbols of incivility: Social disorder and fear of crime in urban neighborhoods*. In Paper presented to the annual meeting of the American Society of Criminology, Dallas, USA.

Hunter, A. (1987). The symbolic ecology of suburbia. In I. Altman & A. Wandersman (Eds.), *Human behavior and environment* (Neighborhood and community environments, Vol. 9, pp. 191–219). New York: Plenum.

Irazábal, C. (2008). Ordinary places, extraordinary events in Latin America. In C. Irazábal (Ed.), *Ordinary places, extraordinary events*. New York: Routledge.

Jacobs, J. (1961). *The death and life of great American cities*. New York: Random House.

Jacobs, J. (1970a). The use of sidewalks. The contact. In H. M. Proshansky, W. H. Ittelson, & L. G. Rivlin (Eds.), *Environmental psychology: Man and his physical setting* (pp. 398–409). New York: Holt, Rinehart & Winston.

Jacobs, J. (1970b). The use of sidewalks. Children in assimilation process. In H. M. Proshansky, W. H. Ittelson, & L. G. Rivlin (Eds.), *Environmental psychology: Man and his physical setting* (pp. 497–501). New York: Holt, Rinehart & Winston.

Jiménez, B. (2014). Espacios públicos sustitutos por apropiación espacial juvenil en centros comerciales de Guadalajara y Puerto Vallarta. In A. M. Burbano & P. Páramo (Eds.), *Habitar la ciudad. Espacio Público y Sociedad*. Bogotá: U. Piloto.

Katz, C. (2006). Power, space and terror: Social reproduction and the public environment. In S. Low & N. Smith (Eds.), *The politics of public space* (pp. 105–121). New York: Routledge.

Korosec-Serfaty, P. (1976). Protection of urban sites and appropriation of public squares. In P. Korosec-Serfaty (Ed.), *Appropriation of space: Proceedings of the Strasbourg conference-IAPC-3* (pp. 46–61). Strasbourg-Louvain la Neuve: CIACO.

Krupat, E. (1985). *People in cities: The urban environment and its effects*. New York: Cambridge University Press.

Linke, U., & Smith, D. (2009). *Cultures of fear*. New York: Pluto Press.

Lofland, L. H. (1978). *The craft of dying: The modern face of death*. Thousand Oaks: Sage Publications.

Lofland, L. H. (1985). *A world of strangers: Order and action in urban public space*. New York: Waveland Press.

Lofland, L. H. (1998). *The public realm: Exploring the city's quintessential social territory*. New Brunswick: Transaction Publishers.

Lorenzo, R. (1998). *La Cita sostenible*. Milano: Eleuthera.

Low, S. (2000). *On the plaza: The politics of space and culture*. Austin: The University of Texas Press.

Low, S. (2009). Cerrando y reabriendo el espacio público en la ciudad latinoamericana. *Cuadernos de Antropología Social, 30*, 17–38.

Luengo, F. G. (1998). *Elementos para la definición y evaluación de la calidad amlbiental urbana: Una propuesta teórico-metodológica*. Ponencia presentada en el IV Seminario Latinoamericano de Calidad de Vida urbana, Septiembre de 1998, Tandil, Buenos Aires.

Lynch, K. (1965). The city as environment. *Scientific American, 213*, 209–219.

Macek, S. (2006). *Urban nightmares: The media, the right, and the moral panic over the city*. Minneapolis: Minnesota Press.

Maragall, P., Benach, N., & Bohigas, O. (2004). *Transforming Barcelona*. New York: Routledge.

Milgram, S. (1970). The experience of living in cities. *Scientific American, 167*, 1461–1468.

Monnet, J. (2001). Espacios públicos y lugares comunes en la Ciudad de México y Los Angeles: del modelo de sociedad nacional a las escenas metropolitanas. *Perfiles Latinoamericanos, 19–2001*, 131–151. http://redalyc.uaemex.mx/redalyc/pdf/115/11501907.pdf

Monnet, J. (2012). *El Espacio Público definido por sus usos. Una Propuesta Teórica*. http://virtual.unipiloto.edu.co/file.php/966/Lecturas_de_la_catedra/usos_del_espacio_publico-Monnet20121221_1_.pdf

Moser, G. (1992). *Le stress urbain*. Paris: Armand Colin.

Moser, G. (2012). Cities. In S. D. Clayton (Ed.), *The oxford handbook of environmental and conservation psychology* (pp. 203–220). New York: Oxford University Press.

Moyano, E. (2010). Globalización, crecimiento económico, salud y calidad de vida. In E. Moyano (Ed.), *Calidad de Vida y Psicología* (pp. 97–128). Talca: Universidad de Talca.

Navarro, O., & Rodríguez, U. (2015). Los mapas cognitivos o la adquisición de un saber espacial como método de investigación social. In P. Páramo (Ed.), *La investigación en ciencias sociales: la recolección de información* (pp. 381–398). Bogotá: Universidad Piloto de Colombia.

Newman, O. (1972). *Creating defensible space: Crime prevention through urban design*. New York: Macmillan.

Páramo, P. (2004). Algunos conceptos para una perspectiva optimista de vivir la ciudad. *Territorios, 10–11*, 91–109.

Páramo, P. (2010). Aprendizaje situado: Creación y modificación de prácticas culturales en el espacio público urbano. *Revista Psicología & Sociedade, 22*(1), 130–138.

Páramo, P. (2011). *Sociolugares*. Bogotá: Ediciones Universidad Piloto de Colombia.

Páramo, P. (2013). Comportamiento Urbano Responsible: las reglas de convivencia en el espacio público. *Revista Latinoamericana de Psicología, 45*(3), 473–485.

Páramo, P. (2014). El miedo a la ciudad. In A. Burbano & P. Páramo (Eds.), *La ciudad habitable: Espacio público y sociedad* (pp. 181–199). Bogotá: Ediciones Universidad Piloto de Colombia.

Páramo, P., & Contreras, M. (in press). Estrategias de Gestión Urbana para promover la formación ciudadana en Comportamientos Urbano Resonsables orientados a la convivencia ciudadana en el espacio public. *Revista Signo y Pensamiento, 34*(66).

Páramo, P., & Cuervo, M. (2009). *La experiencia urbana en el espacio público de Bogotá en el siglo XX*. Bogotá: Universidad Pedagógica Nacional-Santo Tomás-Piloto de Colombia.

Páramo, P., & Cuervo, M. (2013). *Historia social situada en el espacio público de Bogotá desde su fundación hasta el siglo XIX*. Bogotá: Universidad Pedagógica Nacional.

Páramo, P., & Páramo, J. J. (2015). *Los Comportamientos Urbano Responsables del Peatón*. Fundación MAPFRE. http://www.fundacionmapfre.org/

fundacion/es_es/seguridad-vial/investigacion/peatones-responsables-colombia.jsp

Pineda, J. (2007). Hacia una ciudad incluyente: género e indicadores sociales en Bogotá. *Revista Pre-til, 13*, 29–47.

Pol, E. (2002a). Environmental management: A perspective from environmental psychology. In B. R. Bechtel & A. Churchman (Eds.), *Handbook of environmental psychology* (pp. 55–84). New York: Wiley.

Pol, E. (2002b). El modelo dual de la apropiación del espacio. In R. Garcia-Mira, J. M. Sabucedo, & J. Romay (Eds.), *Psicología y Medio Ambiente. Aspectos psicosociales, educativos y metodológicos* (pp. 123–132). A Coruña: Publiedisa-Asociación Gallega de Estudios e Investigación Psicosocial. https://www.academia.edu/8036325/EL_MODELO_DUAL_DE_LA_APROPIACIÓN_DEL_ESPACIO. Accessed 9 Mar 2015.

Proshansky, H. M., Ittelson, W. H., & Rivlin, L. G. (Eds.). (1978). *Psicologia Ambiental*. México: Trillas.

Remesar, A., Salas, X., Padilla, S., & Esparza, D. (2012, October). Inclusion and empowerment in public art and urban design. *On the w@terfront, 24*. http://www.raco.cat/index.php/Waterfront/article/view/259235/346497

Rivlin, L. G. (2007). Found spaces: Freedom of choice in public life. In K. A. Franck & S. Tevens (Eds.), *Possibility and diversity in urban life* (pp. 38–53). New York: Routledge.

Roberts, C. (1977). *Stressful experiences in urban places: Some implications for design*. 8th EDRA Conference Proceedings.

Romice, O., & Uzzell, D. (2005). Community design studio: A collaboration of architects and psychologists. *Transactions, 2*(1), 73–88.

Salazar, R., & Frechilla, J. M. (2006). Orden, desorden y democracia en el espacio público de Caracas: Siglo XVIII desde el XXI. *Revista del Programa de Cooperació Interfacultades (PCI) Tharsis, Año 8, 5*(16), 39–52.

Schultz, P. W., & Kaiser, F. G. (2012). Promoting pro-environmental behavior. In S. D. Clayton (Ed.), *The oxford handbook of environmental and conservation psychology* (pp. 556–580). Oxford: Oxford University Press.

Seligman, M. E. P. (1974). Depression and learned helplessness. In R. J. Friedman & M. M. Katz (Eds.), *The psychology of depression: Contemporary theory and research* (pp. 83–113). Washington, DC: Winston.

Simmel, G. (1905). The metropolis and mental life. In P. K. Hatt & A. J. Reiss (Eds.), *Cities and society*. New York: Free Press.

Soyinka, W. (2004). *Climate of fear: The reith lectures 2004*. London: Profile Books Limited.

Sternlieb, G., & Hughes, W. (1983). The uncertain future of the central city. *Urban Affairs Quarterly, 18*, 455–472.

Stokols, D. (1972). On the distinction between density and crowding: Some implications for future research. *Psychological Review, 79*, 275–278.

Stone, L. J. (1954). A critique of studies of infant isolation. *Child Development, 25*(1), 9–20.

Taylor, R. B. (1987). Toward an environmental psychology of disorder: Delinquency, crime and fear of crime. In D. Stokols & I. Altman (Eds.), *Handbook of environmental psychology* (pp. 951–986). New York: Wiley.

Taylor, R. B., & Gottfredson, S. D. (1986). Environmental design, crime and prevention: An examination of community dynamics. In A. J. Reiss Jr. & M. Tonry (Eds.), *Communities and crime* (pp. 387–416). Chicago: University of Chicago Press.

Tonucci, F. (1997). *La ciudad de los niños*. Madrid: Fundación Germán Sánchez Ruipérez.

Trilla, J. (1997). *Animación sociocultural: teorías, programas y ámbitos*. Barcelona: Ariel.

Uzzell, D. L., & Romice, O. (2003). L'analyse des expériences environnementales. In G. Moser & K. Weiss (Eds.), *Espaces de vie: aspects de la relation homme-environnement* (pp. 49–84). Paris: A. Colin.

Valentine, G. (1989). The geography of women's fear. *Area, 21*(4), 385–390.

Valera, S. (2012). Reseña de "La ciudad y otros ensayos de ecología urbana" de Robert Erza Park. *Athenea Digital. Revista de Pensamiento e Investigación Social, 12*(1), 261–265.

Veleda da Silva, S. M. (2007). Estudios de geografía del género en América Latina: Brasil y Argentina. *Documents d'Anàlisi Geogràfica, 49*, 99–118.

Vidal, T., Salas, X., Viegas, I., Esparza, D., & Padilla, S. (2012). El mural de la memoria y la Rambla Ciutat d'Asunción del barrio de Baró de Viver (Barcelona): repensando la participación ciudadana en el diseño urbano [The BdV screen of memory and Ciutat d'Asunción boulevard at the Baró de Viver neighborhood (Barcelona): (re)thinking citizens' participation in urban design]. doi:http://dx.doi.org/10.5565/rev/athenead/v12n1.933. http://atheneadigital.net/article/view/Vidal

Webb, S. D., & Collette, J. (1977). Rural-urban differences in the use of stress-alleviative drugs. *American Journal of Sociology, 83*, 700–707.

Wiesenfeld, E., & Sánchez, E. (2002). Sustained participation: A community based approach to addressing environmental problems. In B. R. Bechtel & A. Churchman (Eds.), *Handbook of environmental psychology* (pp. 629–643). New York: Wiley.

Wirth, L. (1938). Urbanism as a way of life. American journal of sociology. In R. T. Le Gates & F. Stout (Eds.), *The city reader* (pp. 97–105). New York: Routledge.

Zeisel, J. (2006). *Inquiry by design: Environmental/behavior/neuroscience in architecture, interiors, landscape, and planning*. New York: W.W. Norton.

Adjustment to Geographical Space and Psychological Well-Being

Thierry Ramadier

In the introduction to his co-edited volume, Canter (1988 p. 1) argued that "one of the great discoveries that has emerged from the psychological study of the physical environment has been that the human experience of environment is essentially social". A year earlier, Stokols (1987) stressed the importance of the cultural dimension in environmental psychology; 20 years earlier, Lee (1968) sought to connect social relationships to relationships to the physical environment with a "socio-physical" approach. However, the social dimension in environmental psychology has yet to be properly defined; it appears indistinctly related to a social framework (of social relationships) and to a cultural one (of cultural differences in the relationships between signifiers and signified), when it is not reduced to different individual experiences based on different sets of shared material and economic conditions. This chapter discusses the role granted to the social dimension in the analysis of the person/environment relationship with a view towards reconsidering the concept of well-being in environmental psychology.

Translated from the French by Jean-Yves Bart
This chapter has received support from the Excellence Initiative of the University of Strasbourg funded by the French government's Future Investments program.

T. Ramadier (✉)
Laboratory SAGE, UMR 7363 CNRS/Université de Strasbourg, Strasbourg, France
e-mail: thierry.ramadier@misha.fr

Social psychology sees the environment as a historical social construct: "The standards of well-being and health have shifted alongside those of moral responsibility. This relativity of prevalent values is neither arbitrary nor predestined, neither erratic nor finalized: it has a meaning. It shows that environmental issues are always raised in a contextualized manner; in other words, they have no intrinsic legitimacy" (Rouquette 2006 p. 13). Environmental psychology, on the other hand, has developed an essentially physical approach of the social dimension (Moser 2009), laying emphasis on the physical presence of individuals and the physical markers of their social attributes (practices, socio-spatial interactions, density, spatial distribution, etc.) or, when absent, through the material traces of their practices (degradation, material appropriation, physical consequences of their practices, etc.), or through arrangements of space as legitimate traces of social practices (architectural and urban development codes and signs). This chapter will show that this physical approach of the social environment is a defining feature of environmental psychology, and has allowed all of its proponents to keep their distance from social psychology and community psychology (Moser 2009). As Bourdieu wrote (2000 p. 110), "technical competence and scientific knowledge function simultaneously as instruments of accumulation of symbolic capital; intellectual conflicts are always also power struggles, the polemics of reason are the contests of scientific rivalry, and so on." Yet there might be a scientific benefit in adopting a transdisciplinary approach (Ramadier 2004), bringing face to face advances in environmental psychology with those made in sociology and geography, especially when seeking to identify the environmental conditions of well-being (and the lack thereof).

Based on what is arguably the most solid finding in the history of environmental psychology – the congruence between the individual and the environment – this chapter sets out to revisit the place accorded to the sociological dimension in environmental psychology using a tridimensional conceptual model of person/environment congruence. It will end by presenting some examples of research providing illustrations of this model.

16.1 The Role of the Social Dimension in Environmental Psychology

In its earliest stages, environmental psychology initially focused on the physical dimension of the environment, so that the individual's social environment, a core dimension in social psychology that had already existed for 50 years, could be assigned a smaller role in research protocols and conclusions. The first decade of research following Lynch (1960) continued to evidence social variations in relationships to the same geographical space. These studies include works by Hall (1966), Lee (1968), Rapoport (1969) and Canter (1969). This is why Wohlwill (1970) argued that, on the one hand, too few psychologists deigned to support the construction of the new discipline of environmental psychology and, on the other hand, the concept of physical environment was too often perceived as a tool for the study of behavior, being as such of little interest to the psychologist. In the same year, the publication of the first book to include the designation "environmental psychology" in its title (Proshansky et al. 1970) also contributed to the construction of this research field. In the 1970s, it was the social dimension that became a secondary concern for environmental psychology. Research on the effects of the physical properties of geographical space (from the scale of a room to that of the world) on cognitive processes, emotional activities, social interactions and spatial behaviors were extensively investigated. This was made easier by the fact that the legacy of ecological perception theory (Gibson 1966), which was itself derived from Gestalt psychology (Köhler 1929; Koffka 1935), enabled an approach of the construction of environmental meanings (affordances), which was directly based on the physical environment, with no consideration given to the social environment.

The social dimension has never completely disappeared from environmental psychology research; moreover, the duality between the social and physical environment has become recurrent if not cyclical. It was through phenomenology (Lee 1968) that researchers in this field of psychology sought to resolve this tension. At the same time, sociologists, particularly Lefebvre (1968), posited a structural correspondence between physical and social space. The phenomenological approach of environmental psychology suggested emphasizing co-presence and social interactions (Chombart de Lauwe 1974) in order to understand changes induced by the social environment in an individual's relationship to his/her physical environment. In other words, the tension between the social and physical was ultimately resolved by approaching the social environment on the basis of the individual experience of its physical features, leading to an increasingly biological and subjective view of social relationships, to the extent that they became partly obscured.[1] In this approach, grasping what an individual feels in his/her relationship with others is crucial; in other words, greater emphasis is laid on the individual than on the context. When context is emphasized, social dimensions are often reduced to physical dimensions and/or the subjective dimension of individual reception. Analyses of population density are particularly telling in this respect; the topic was addressed in terms of physical features (over-stimulation of

---

[1] For instance, Holahan (1982 p. 3) described environmental psychology as "an area of psychology whose focus of investigation is the relationship between the physical environment and human behavior and experience."

density) and of what the individual feels (crowding). Thus, Stokols (1972) proposes a distinction between "density" and "crowding" so that the researcher can begin by ensuring the salience of population density for individuals. This is an important distinction, but, for instance, in a situation where a politician goes out in a crowd to mingle, it does not tell us the difference between the experience of the politician and that of the people who make up the crowd. In other words, the social positions of all involved are not considered.

From a structuralist perspective, Laborit (1971) proposed an even stronger biological connection between social and physical environment – one that is notably more functionalist than the previous phenomeno-physicalist model. For this author, the physical environment (the city) is only a means of retaining a social structure; physical context only has a functional finality. As a cultural product, the physical environment is an intermediate variable (effector) constructed by the social group (factor), which acts on the group structure (retroactive effect) that defines its existence. In contrast to Lefebvre's approach, this systematic approach does not envisage a direct correspondence between the physical and social environment. By considering the physical environment as one of the means for a social group to maintain its structure, the author lumps the individual in with the physical environment instead of looking at them as two connected entities. Laborit constructed a closed system whose dynamic ultimately lies in the search for an internal point of equilibrium, reached only when goals are matched by the means used to achieve them. In this system, well-being is merely the consequence of the sound functioning of the social structure as a whole. The social structure is perceived as an aggregation of individuals, more or less consciously concerned with the conservation of a structure that transcends them and in which they are participants. While, as shown below, well-being has something to do with the adjustment between social and physical structures, are we for that reason a component of a compact, entirely conservation-oriented social structure, or are we the product of a social structure that follows an indeterminate trajectory, encouraging us to reproduce actions, tastes and thoughts so that our acquired behaviors structure our future experiences?

As noted earlier, numerous theoretical models concerning the relationships between physical and social environment were constructed during the founding stages of environmental psychology. Two main streams of thought strongly and durably influenced environmental psychology. Derived from the Gestalt theory, the first is based on experimentation and a holistic approach. Works on environmental affordance (Gibson 1977) are the main theoretical basis for this research; they contributed to the emergence of the concept of person/environment congruence based on a physicalist concept of the environment. Kaplan (1983) then identified contemplation as an important component of congruence (in addition to knowledge and perceptions, constraints or required behaviors, and projects or intentional behaviors). He went on to argue that some environments (mainly so-called "natural" environments) enable "restorative" experiences by reducing environmental stress and cognitive fatigue, as they offer the opportunity to take a step back from cognitive and behavioral sources of stimulation (Kaplan 1992). Affordance and the restorative character of environments are now considered two dimensions of environmental well-being. The second stream of thought is derived from ecological psychology and relies on observation and experimentation in the field. It was inspired by Lewin's studies (1936) on the concept of field and by Barker's (1968) methodological efforts to take psychology out of laboratories with the concept of behavior setting. Bronfenbrenner (1979) built on this line of research to propose an ecological model of human development based on a set of interacting systems (onto-, micro-, meso-, exo-, macro-systems[2]).

---

[2]The ontosystem refers to intra-psychological processes and states; the microsystem pertains to the person's close environment; the mesosystem is the locus of interaction between different microsystems (for instance, family and school for a child); the exosystem relates to the broader environment beyond the individual's direct experience;

Lynch's work on the image of the city (1960) had a tremendous impact on environmental psychology, arguably because his research was situated at the intersection of these streams of thought: like affordances, the concepts of legibility and imageability are based on Werner's "signal objects" (Niveleau 2006) and like behavior settings, on fieldwork serving as a cognitive and behavioral force field. More broadly, these two streams converge on two key points in environmental psychology. The first is based on the progressive construction of a psychology that may be termed as functionalist, even if it is expressed differently in each stream. On the one hand, the Lewinian action-research approach is a core component of studies focused on changes in behavior or the evaluation of "acceptability", which are mostly concerned with finding means of managing a situation without undermining the social order (Stevens 2011). Consequently, neither the beliefs that shape the formulation of research questions nor the social relationships pertaining to the situation under study are questioned. For its part, the Gibsonian approach lays emphasis on the perception of the utility of objects and spaces; however, it also fails to investigate the social relationships that underlie the construction of the functional meanings perceived. For each of these streams of research, and for environmental psychology at large, variations in the relationship to an environmental context are considered primarily functional. The cognitive processes they discuss are in reference either to the singularity of an individual who projects (project, intention, self-fulfillment) or protects (control, rest, restorative withdrawal) him/herself, or to cultural differences (value and belief systems). Socio-cognitive processes resulting from social relationships (legitimacy, social distance, etc.) have only very recently started to be addressed by environmental psychology, and are not yet perceived as relevant in identifying the dimensions of individual well-being. The second intersecting point between these two streams of thought, and a core feature of environmental psychology, lies in the concept of congruence (fit, adaptation, compatibility, etc.). The Gibsonian approach is strongly influenced by the Gestaltist idea of a fit between cognitive and physical structures. The same applies to Barker's behavior setting; his works have also contributed to the emergence of environmental psychology.

16.2 Person/Environment Congruence and Well-Being

The concepts of congruence, and especially of adjustment, have played a crucial part in the theoretical development of environmental psychology and in work on well-being. Moser (2009) wrote that the goal ultimately pursued by environmental psychology consists in defining and identifying the conditions of individual well-being, insofar as the latter "cannot be independent from territorial roots" (p. 240) and as threats to well-being are "situations of concern for environmental psychology" (p. 132). In connection with quality of life, a condition "enabling the individual to fulfill his/her own needs" (Uzzel and Moser 2006), he defines well-being as an objective situation of weak exposure to a nuisance and a subjective situation of weak expression of discomfort, reflecting the congruence between individual and physical environment. Thus, "the state of well-being, of 'congruence' or of psychological adjustment between the individual and his/her environment, sets in when the qualities of the context concur with his/her goals and sensitivity to certain environmental qualities" (Moser 2009 p. 93). However, here "congruence" also hinges on a specific acceptation of social environment and its relationship with physical environment. A brief overview of the development of the concept will help understand its impact on environmental psychology. First, it is worth mentioning that the success of cognitive dissonance theory (Festinger 1957) in social psychology probably had an impact on this conceptualization of congruence in environmental psychology, and that the principle of lessening induced psychological tension implicitly relates to the concept of psychological

---

the macrosystem refers to the cultural context that influences all the other systems).

well-being. The concept of congruence appeared even earlier with the Stroop effect (1935) and the congruity or incongruity between the meaning and the shape of a word.

The first environmental sketches of the congruence concept came in the wake of studies in ecological psychology and of the behavior setting concept (Barker 1968). Edney (1976) then developed the idea of a "behavior/environment congruence" to analyze the relationship between a behavior and a given setting. The notion of behavior setting is relevant insofar as the behavioral standards of a place are effectively implemented. Yet this concept reflects a deterministic view of a place's impact on behavior. Wicker (1973) also used the concept of congruence but, unlike Edney (1976), he built on Barker's work and referred to a social determinism of behavior and environment. In other words, behaviors and surroundings are congruent because the two entities are socially determined: as places are constructed according to the behaviors projected in them, behaviors are effectively guided by standards that are objectified in space. However, not every individual is able to participate in the construction of such a site. This type of congruence between behavior and surroundings is also the result of a range of psychological mechanisms, such as operant learning, observational learning, environmental perception and social exchange (Wicker 1979). Congruence as adjustment between behavior and the physical features of the environment is also called "synomorphism" in ecological psychology. As with the concept of affordance, the early versions of congruence referred to what the environment evokes from the individual – they did not consider the future, but focused on past experiences. At the behavioral level, congruence refers to necessary actions prescribed by the environment on the basis of individual and collective experiences rather than to purposive actions coming from the individual (Kaplan 1983). At the social level, it pertains to the construction of norms reified by spatial categories. Those coming from the conceptors, and as such endowed with legitimacy, are rarely connected with those coming from the users; the latter are generally considered learners who have mastered – or failed to master – ways of behaving and thinking about their physical environment. As the social dimension is given the cultural status of a general framework that is automatically related to the physical configurations of space, the social impacts of the person/environment convergence are overlooked. As a result of this, psychological issues, such as the control and appropriation of the environment, are also studied independently of social issues (legitimacy, power, etc.).

Stokols (1978) argued that cognitive control of the environment underpins the concept of congruence, in the sense that it allows individuals to carry out planned actions and achieve goals that are important to them, whereas another concept developed by the author, salience, relates to the fact that the environment is associated with goals or psychological needs that matter to the individual (housing, work, leisure, etc.). These two concepts involve references to the quality of the environment and the individual's well-being. They are considered by the author as environmental dimensions defined by the subject. Both result from the interaction between the features of the individual and the environment; their combination enables an evaluation of the quality of that environment. In other words, a functional approach to congruence introduces the concept of need by giving it a bio-psychological character rather than a socio-psychological one, when the three dimensions taken simultaneously would yield useful findings. This utilitarian approach to congruence ultimately favors the concept of quality of life understood as a setting that allows individual needs to be fulfilled.

Michelson (1976), for his part, made a distinction between two forms of congruence. The first reflects the qualities of the environment that fit the individual's goals; the second concerns the individual's representation of this adjustment between environment and behavior. He also proposed a functionalist approach to congruence, defined as a cognitive, affective and behavioral state reached by the person/environment system when the characteristics of the environment allow individuals to fulfill their needs and carry out their activities (Michelson 1980). This state of congruence is achieved through a process of optimization. Yet, in this case, it would again be worth considering the person's social position or trajectory, their spatial position (socio-spatial category of the environment) or even their geographical trajectory (residential, occupational, consumer, etc.) in order to have a better grasp of this optimization process between individuals and surroundings.

Kaplan (1983) includes the concept of congruence in the broader person/environment compatibility model. This model describes the environment as supportive of all activities of individuals, cognitive or behavioral, when the person/environment congruence is appropriate. Conversely, the environment becomes a constraint when this congruence does not exist. It is not the absence of some features in an environment that is problematic, but their presence, which prevents individuals from finding support for their activities (noise, confusion due to the illegibility of the environment, etc.). Based on this model, Kaplan claims that the environment fulfills a specific function of support/constraint, which significantly affects the individual's relationship with his/her environment. In other words, he suggests analyzing the quality of the person/environment interface to describe and explain the individual/surroundings relationship better. At the behavioral level, he takes stock of both the approaches of ecological psychology and symbolic interactionism in psychology. To this end, he makes a distinction between behaviors required by the environment and deliberate behaviors – intentional or associated with goals – such as purposive actions and projects. However, some individual intentions, which he calls "inclinations", do not only have behavioral implications. The quality of the person/environment interface does not only depend on the behaviors envisioned by the individual; it also depends on the organization of information resulting from cognitive processes implemented to elaborate and structure the environment represented on the basis of the physical surroundings. The author mentions the perceptive and representative processes traditionally addressed in environmental psychology, but he also adds a new category of processes derived from the concept of inclination.

By granting the status of cognitive mechanism to this concept, he asserts that it encompasses a range of processes related to "contemplation" and "reflection". These processes, he argues, enable the organization of thought and feelings related to the environment by selecting and processing information stored in the memory: "Reflection is a means of extracting information from the past and anticipating possibilities in the future" (Kaplan 1983). This time the temporal dimension is a key element of congruence.

Kaplan notes that these processes have rarely been investigated, mostly because anthropologists concealed their importance by stressing the fact that evolution processes occurred under pressure. He suggests that in the event of incompatibility between the various activities described (required action, purposive action and inclination, environmental perception and knowledge, and reflection), an order of priorities occurs, with action and the environment favored over cognition and individual aspirations, in order to maintain a sufficient degree of adaptation and avoid inaction. Therefore, reflection is the last activity required in the case of incompatibility; in the case of compatibility, the four types of activity are pursued in parallel.

If one considers that, through an optimization process, the person/environment relationship leads towards a state of congruence or compatibility, one needs to look beyond the instrumental aspects of this relationship. According to Stokols (1990), who echoed Canter's introductory remark (1988), theoretical models in environmental psychology have made steps in that direction. In addition to the minimal approach, which neglects environment-induced changes to practices and social interactions, scholars have successively developed an instrumental and a "spiritual" approach to the person/surroundings relationship (Stokols 1990). In the latter, the environment is studied as a physical and social context laden with meanings: through the diversity and complexity of meanings, the environment becomes supportive for the individual's well-being or lack thereof. In the past two decades, emphasis has been laid on the socio-cultural and symbolic aspects of the environment and of the person/environment relationship, in which the instrumental approach saw the environment solely as a means of pursuing a wide array of individual activities. The role of surroundings in social relationships deserves further investigation.

16.3 Social Interaction, Cultural Contrasts or Social Relationships: Three Approaches to Social Facts

The teleological approach to well-being and quality of life has the benefit of being easily compatible with the econometric models of human behavior. The concept of utility in economics – or preference in environmental psychology and competence in ergonomic psychology – is part of the construction of a now multi-disciplinary (economics, psychology, sociology, geography, etc.) rational model of human behavior. Even if this rationality is limited (Simon 1947), including by the individual's subjective perceptions (Tversky and Kahneman 1981), it is chiefly an approach where the individual constructs him/herself based on his/her sensibilities and interactions with the world. This perspective rests on an internal/external duality; in other words, on the opposition between objectivity and subjectivity and their possible congruence. This is why, in rational models, the social environment is only seen as an outside element, like the physical environment. Approaches focused on needs overlook the fact that both individuals and geographical space are also the product of social relationships that do not only manifest themselves through cultural differences. In other words, differences in social relationships are not necessarily cultural; they are also shaped by structural asymmetries. For instance, more differences might be observed between an employee and a CEO of a corporation regarding their tastes and relationships with the environment than between a Thai and a French CEO or a Spanish and a German employee (though this does not mean we should neglect cultural differences, which can cause misunderstandings and discomfort – see, for instance, the cultural dimension of proxemics evidenced by Hall in 1966). At the individual level, social relationships are not stable; they are embedded within force fields (Lewin 1936) and fields of struggle (Bourdieu 1984). Due to these struggles, beliefs, norms and ideologies are not only constructed within groups, but also through the relationships and interactions between these social groups. For this reason, knowledge of the social, geographical and psychological positions of individuals is crucial to gaining an understanding of the person/environment congruence – and so is knowledge of social relationships. The social environment cannot be reduced to the quality of co-presence or the experiential quality of the individuals' relationship networks (whether their relationships are positive or not, fulfilling or not, conflict-laden or friendly), i.e. to immediate interactions or those that individuals may assess subjectively. Even in environmental psychology, the stakes of conflict and co-operation, and the engagement in the game that they encourage, matter just as much as conflict and co-operation themselves. For instance, can we truly understand why some people remain happy living in a neighborhood that has a bad reputation when others are unhappy, if we fail to see that the former do not have any hopes for better living conditions as they consider themselves to have reached the apex of an ever-improving residential trajectory, whereas the latter, who sometimes come from the same neighborhood and have achieved upward social mobility thanks to their degrees, aspire to live in a neighborhood where they can more easily self-identify as residents? Why should the prospect of a visit to the presidential palace be more anxiety-inducing for a teacher than for a doctor or a lawyer? Is this merely about being accustomed to such physical environments (even though a teacher may already have attended official receptions at city hall)? Is it about perceived hierarchical differences? Or is it because this situation appears to him/her to be very unlikely and unprecedented (this dimension may be objectivized by measuring the respective probabilities of a teacher, a doctor and a lawyer pursuing a political career)? These are all probably combining factors. However, by formulating it in this way, we are not referring to a difficulty relating to cultural differences or to a lack of familiarity with the physical and social environment (the first example shows that familiarity with the environment is not necessarily an asset), but rather *a disconnect between environmental expectations and the objective chances of finding oneself in the expected environmental setting*. This brings us back to the concept of congruence, but this time with less emphasis on the characteristics of the physical environment and on the individual's supposed needs. Likewise, the opposition between objectivity (or exteriority) and subjectivity (or interiority) is not as central here. This is a socio-cognitive form of familiarity rather than an experiential one. It depends on the relationship between expectations and chances, in which each entity is constructed on the basis of the other: environmental expectations (for instance, considering setting up an office in a room of a city center apartment) are adjusted to the individual's chances of meeting them (which are higher for a university professor than for a schoolteacher, even though both may live with a partner and children and need to work from home). In their study of the failure of prophecy, Festinger et al. (1956) also observed that expectations are adjusted to chances: expectations are remodeled, at least in their meanings, on the basis of the facts that shape the chances of them being met. Bourdieu (2000) added that when an individual's most fundamental expectations cannot be adjusted to the chances of meeting them, the hopes they formulate can become utterly unrealistic. For instance, people in very poor housing conditions often claim in interviews that they would like to live in the city's well-off neighborhoods. Likewise, overall assessments of residential situations always report high rates of satisfied individuals (around 80 %): in this case, everything happens as if expectations are adjusted to the objective chances of fulfilling them, and as if these chances ultimately reflect the current environmental situation.

This perspective appears well suited to the study of the environmental well-being of individuals. In this sense, I concur with Moser's claim (2009 p. 241) that "environmental psychology must move beyond an instantaneous vision of the man-environment relationship". However, this study will not use the concept of time as a dimension situated outside the individual, instead considering the individual's social and environmental trajectory. In other words, time is inscribed in practices (Bourdieu 2000) and it is "really experienced only when the quasi-automatic coincidence between expectations and chances [...] is broken" (Bourdieu 2000 p. 208). Likewise, emphasis will no longer be laid on the individual's tangible, direct social environment (though it will not be neglected, as research on attachment to places shows the importance of the networks formed there), but rather on the positions and socio-cognitive stakes for the person; in other words, social relationships that are not limited to social interaction and physical co-presence (Ramadier 2007).

16.4 A Tridimensional Theoretical Proposal

To sum up the state of the art on the role of the social dimension in environmental psychology, one might say that economic utility (needs, preferences and functionality) cannot be the only way of addressing the concept of individual environmental well-being. This is especially so as it has major drawbacks: either it assumes the needs of individuals (generally on the basis of needs considered universal) or, as is more frequent in psychology, when respondents are considered able to make their own informed decisions, researchers take the meanings of practices for the determinants of these practices (Beauvois 1994). Sociological (not social) and psychological utility, i.e., the relation between the conditions of possibility of a practice (what is possible for me to do?) and the way of experiencing the world in practice – action or representation- (what do I do or what can I consider doing?), provides a way out of this impasse. These two questions no longer relate to an opposition between subjectivity and objectivity: each can be addressed in a way that sheds light on how individuals construct themselves (subjectivation processes) and how practices and their context are constructed (objectification processes).

The social representations paradigm of environmental psychology has partly supported this functionalist approach to the person/environment relationship. While Milgram and Jodelet (1976) developed this psychosocial approach in a study of urban space, environmental objects of research have since become more diverse. However, all studies based on this paradigm also posit that representations are forged within the group through interaction between members. Relationships between groups are not perceived as relevant to the analysis of the social construction of an object. Everything happens as if social psychology applied to the environment is also subject to the fundamental attribution error (Ross et al. 1977) and the norm of internality (Jellison and Green 1981) that it enabled researchers to evidence. In fact, numerous experimental studies have shown that we more readily attribute the cause of an event to the characteristics of those involved in it than to the context (fundamental attribution error); as this explanation of facts is learned socially, it is also applied to individuals (norm of internality). In other words, both theoreticians in psychology and survey respondents have a very hard time moving beyond this opposition between internality and externality. Yet, a representation of an object can perfectly well be developed within a group, but according to its relationship to other groups rather than regardless of it. The first environmental psychology study linking social group and social positions dates back to the 1980s, when Deschamps and Doise (1988) investigated the representations of school among children born to Swiss and migrant parents. It reported no difference between the two groups of children regarding their representations of how school works and how they envisage what they will do after school; however, insofar as the socio-spatial setting of each group was different and as these differences were also correlated with the social positions of their parents, their representations of their mother tongues, regions of origin and of the places where they considered working after completing their studies all differed. While the two groups were sociologically similar in some aspects, the children of migrant parents had representations of the institution that matched those of Swiss parents; the parents' social positions most influenced the socio-spatial dimensions of their representations of school.

From a cognitive standpoint, not considering positional level in psychological explanation (Doise 1982) entails using the concepts of individual cognitive, affective and conative competences rather than that of dispositions to act and think. It means retaining the idea of individual self-construction, even if based on social materials, and ultimately thinking that expectations and chances can be similar when competences are similar. It means positing that, from a cognitive and affective point of view, individuals are always potentially interchangeable (which happens to reinforce the naturalistic idea that fundamental differences are ultimately of a biological nature). This universalist vision of human psychology masks the fact that from the beginning of social and geographical trajectories, the conditions for acquiring all sorts of "competences", including "environmental competences", differ; their appropriation – or even incorporation – make them lasting because with time and repetition, these "competences" are no longer conscious (for instance, we do not think about our movements when we walk, even though we have learned them, and it has become difficult to change them). Does the term "competence" still apply in this case? Other theoretical models, using econometric approaches, use the concept of "preference". Instead of learning, a core feature of their investigation of relationships with the environment is the rational and/or emotional ranking of environmental possibilities by individuals. In other words, in this model, individual expectations determine the chances of a given environmental context presenting itself. This theoretical approach is particularly common in environmental psychology, as it fits easily with approaches focused on individual sensibility, by placing more emphasis on expectations than on chances in the relationships with geographical space, and consequently by individualizing the concept of expectation. This theoretical approach also fits well with the "rational choice" concept of econometric models, insofar as preference is always the individual's ranking of a known set of possibilities, whose achievement, without this individual ranking, can be posited as equally probable.

In contrast, the dispositional model of sociology, whose most advanced version was proposed by Bourdieu (1984), posits that our ways of doing, thinking and being are distributed across social space; in other words, they depend on social positions (and trajectories) within that space. Unlike competence and preference, disposition is a socio-cognitive and socio-behavioral scheme, which strongly depends both on the social field to which the individual must adjust (for instance, physical strength is not an indispensible disposition in the academic field, but it is in sports), and on the individual's position within the field (while strength is a pre-requisite for the athlete, this is less the case for the trainer and arguably even less for the club manager). The importance of the individual's spatial position is key in environmental psychology, even if it is ultimately addressed in terms of location rather than position: "The issues faced by environmental psychologists lead researchers and practitioners to inscribe their analyses in spatial and temporal dimensions; concepts of space and place thus occupy a central position" (Moser 2009 p. 19). Position is also a geographical concept. Unlike geographical location, which refers to a single point within an isotopic physical space, geographical position contributes to the construction, structuring and even the analysis of the geographical space under study. Position is a spatial condition that produces differentiation. For instance, the city center cannot be defined without the suburbs and vice versa. Furthermore, this form of spatial categorization necessarily rests on a concentric concept of geographical space. Lastly, insofar as the same object may elicit different socio-cognitive representations, each of these representations can be envisaged as a socio-cognitive position. Indeed, on the one hand, a representation is social because, beyond the social dimensions of its construction, it is shared without being the only one possible: other representations co-exist, to the extent that there is a space of representations for each object. However, the concept of cognitive space or space of representations is conceptualized at the individual level. For instance, Slater (1976) proposed the concept of intrapersonal space, referring to a mental space of sorts, to understand the organization and relationships between the constitutive elements of representations.[3] Ehrlich (1979) then submitted the concept of cognitive mobility. However, we will see that an analysis of representations of space based on individual social trajectories shows that "cognitive mobility" comes with social mobility (Viaud 2003); in other words, the space of representations or the cognitive space of an object, in line with social representation theory, is not only an intrapersonal space.

Concerning the relationships between geographical, social and cognitive spaces, several studies on non-environmental objects have, however, shown a match between the person's representation and their social position (Tafani and Bellon 2001), and the adjustment of representation to the individual social trajectory (Viaud 2003). These studies are based on the idea that social representations are stances (or psychosocial anchorings) that relate to the positions occupied within the social system of production of values (Doise and Palmonari 1986). Likewise, there is ample evidence that an individual's geographical position has an impact on their representation of space; one has only to compare world maps produced on various continents to observe that each region of reference systematically places itself in the upper central section of the map; this geographical position thereby "displaces" the location of all other areas on the map. Lastly, numerous studies on social segregation in space have shown that spatial positions are associated with social positions. In other words, these different spaces (geographical, sociological and psychological) interact to form

---

[3]Portugali (1996), based on the concept of inter-representations network (IRN), refers to this intrapersonal space under the ambiguous term of "cognitive maps".

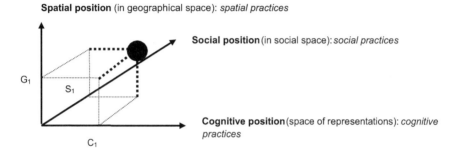

Fig. 16.1 Tridimensional model of person/environment congruence

a trijection of sorts, to use a mathematical model. Several pairwise correspondences have already been proposed. In sociology, Durkheim and Mauss (1903), then Bourdieu (Wacquant 1992) have long evidenced a correspondence between social and mental structures. In psychology, the "Geneva school" of social representations (Doise and Lorenzi-Cioldi 1989) has reached the same conclusions. Sociologists from the Chicago school were the first to demonstrate the correspondence between social and geographical space based on their studies on socio-spatial segregation. Examinations of the correspondence between mental and geographical structures, launched with a study by Lynch (1960) on the images of three US cities, were early landmarks in environmental psychology. Here I have devised a tridimensional model of the person/environment congruence (Fig. 16.1) that introduces social space into the correspondence between geographical and mental space; thus, the socio-cognitive space is no longer limited to an intrapersonal space; cognitive position is associated with a social and geographical position drawn from all possible socio-spatial representations.

16.5 A Few Examples Illustrating Tridimensional Congruence Applied to the Urban Environment

The tridimensional (geographical, sociological and cognitive) congruence model was progressively devised on the basis of studies that aimed to show that the legibility of urban space was not only physical (Lynch 1960), but also sociophysical (Ramadier and Moser 1998). An analysis of foreign students' representations of Paris (Ramadier 1997) clearly showed that differences observed between the European sample and the sub-Saharan African sample had nothing to do with individual cultural differences, being rather related to the congruence (or lack thereof) between the cultural characteristics of buildings (signs and architectural or urbanistic codes of geographical space) and the dispositions acquired by the student (signs and architectural or urbanistic codes internalized, making up the cultural capital of relationship to urban space). The difficulty in making Parisian space legible (constructing a spatially consistent and unified representation of the city) was only temporary for some of the less-adjusted students and more lasting for others. On the other hand, the better-adjusted socio-spatial relationships enabled individuals to transfer their ways of representing the city and to move within it immediately upon arriving. While from a cognitive standpoint, the non-adjustment between people and surroundings showed that the difficulty in representing urban space derived from a difficulty in making geographical space legible in its physical, social and functional dimensions and forced individuals to focus on their use of space (egocentric function), from an affective standpoint, many of those who struggled with this also reported difficulties in travelling, fear of getting lost and in some cases a form of "homesickness" – thus, such experiences of urban space were connected to a decrease in everyday mobility (Ramadier 2009). In sum, the

Table 16.1 Raw occurrences of spatio-temporal decisions for daily mobility patterns within a week ($n = 74$)

Type of decision in:		Space				
		Routine	Programmed	Contextual		
	Routine	568	7	27	602	**38.76 %**
Time	Programmed	324	226	47	597	**38.44 %**
	Contextual	196	41	117	354	**22.80 %**
		1088	274	191	**1553**	
		70.05 %	**17.64 %**	**12.30 %**		

relationship to urban space is not a simple matter of competences to acquire, but instead primarily relies on dispositions that can be transferred from one geographical space to another, thanks to the socio-cognitive proximity of the two places for the individual. This may be termed as socio-cognitive accessibility to the place of destination (Ramadier and Enaux 2012). The possible evolutions of these dispositions show that so-called "cultural" differences pertaining to cognitive mapping cannot be ascribed to different "cognitive styles" (Evans 1980) but to socio-cognitive processes that remain to be determined. This leads to the following research question: which social and cognitive conditions favor individual socio-cognitive adjustment to a new urban space when this adjustment cannot be made through the transfer of already internalized dispositions to read geographical space?[4]

More recent research (Dias and Ramadier 2015) has shown that representations of urban space depend on individual social trajectories. When social positions (managers) and areas of residence are equal, representations of urban space differ according to whether the individuals' parents were already managers (stable social mobility) or had intermediate socio-occupational positions (downward social mobility).[5] Likewise, people who occupy intermediate socio-occupational positions but whose parents were managers (downward social mobility) also have socio-cognitive representations of space that differ from the other two groups. In other words, inheriting, acquiring or losing the position of manager affects the spatial representation of the city. Primary socialization (in childhood) is therefore not singlehandedly responsible for forging social dispositions to read urban space. Secondary socialization (social integration through professional activity) is also important. Yet, these findings also confirm that socialization and individualization are not opposite processes; individuals vs. society is not as well-founded an opposition as is often thought (Elias 2010).

Other recorded geographical facts confirm the strong relationship between cognitive and geographical dimensions in relation to space. Spatial practices are more frequently investigated than spatial representations. A study, also conducted in a residential area of the nearby suburbs of Strasbourg, has shown that decisions pertaining to weekly mobility patterns are characterized by the dominance of spatial routines. For example, 70 % of all weekly mobility patterns[6] concern activities that no longer require decisions or choices regarding the place of destination – they are routine (Table 16.1).[7] This geographical stability is not specific to our research: the Canadian research protocol that inspired the aforementioned

---

[4] In the same way that someone who has learned to read may sometimes not be able to read on a regular basis or understand a text.

[5] It is worth remarking that, as these social trajectories are partly influenced by the social history of the residential area under study (which has experienced progressive gentrification since the 1990s), the ages of respondents vary according to the social trajectories observed.

[6] Seventy-four respondents from a residential area of Strasbourg built in the 1950s–1960s were asked to write down their mobility patterns during a 7-day period; information was collected from them every 48 h by phone.

[7] Programmed mobility patterns are those for which a choice is made between possible destinations accessible to the individual and for which decisions do not depend on the current context but on a more general spatio-temporal organization. Contextual mobility patterns are those that are neither routine nor programmed.

Table 16.2 Raw occurrences of spatio-temporal decisions for daily mobility patterns within a week for retired persons ($n = 22$)

Type of decision in:		Space				
		Routine	Programmed	Contextual		
	Routine	111	3	5	119	**28.40 %**
Time	Programmed	105	74	21	200	**47.73 %**
	Contextual	63	10	27	100	**23.87 %**
		279	87	53	**419**	
		66.59 %	**20.76 %**	**12.65 %**		

study (Ramadier et al. 2005) yielded similar findings about the spatio-temporal distribution of this daily mobility pattern.

These findings cannot be attributed to massive amounts of necessary mobility patterns like home/work commutes, as they are somewhat similar for pensioners in the same sample (Table 16.2).

These results show that our geographical positions are particularly stable. Yet this geographical stability is not only cognitive (decisional), it is socio-cognitive insofar as not all spatial routines have the same intensity for all practices. It is also spatio-cognitive insofar as, as Feldman (1990) demonstrated regarding residential mobility, daily mobility patterns are also based on the continuity between the geographical features of the places of departure and destination.

On the one hand, only certain activities are associated with routine mobility patterns in time and space. Unsurprisingly, for respondents who are employed, the bulk of home/work mobility patterns are spatio-temporal routines. Taking the children to school or to extra-curricular activities, taking walks, going to the bakery, to an art class or to a place of worship are the most routine activities both in space and in time.

As the spatio-temporal stability of mobility patterns depends on the activity, unsurprisingly we observe differences between social groups – for instance, between managers and employees. Within the same sample of Strasbourg residents, taking the children to school, taking walks, going to do sport, to a class or to a place of worship are spatio-temporal routines that are more specific to employees; going to the bakery, to the hairdresser, to an art class or taking the children to an extra-curricular activity are routines more often observed in managers. In other words, both the spatio-temporal relationships with activities and the activities themselves involve socialization.

On the other hand, the analysis of the Strasbourg household travel survey carried out in 1999 by the National Institute of Statistics and Economic Studies (INSEE) in 1999, with a representative sample of frequented places (Ramadier et al. 2011), showed that a significant part of daily mobility patterns are between geographically similar positions. In 1999, 36.29 % of the daily mobility patterns of the residents of Strasbourg's historic city center showed the destination to be another neighborhood[8] of the historic city center; 27.32 % of the mobility patterns of the residents of the part of the city center built during the so-called German period (1870–1914) were to another neighborhood in that sector; 21.49 % of the mobility patterns of the residents of the former villages that are now part of Strasbourg's urban area (the old towns of Schiltigheim, Cronenbourg, etc.) were to another one of these sectors; residents in large housing developments more often travelled to the other large housing developments than elsewhere (13.89 % of their daily mobility patterns): the mobility patterns of the other residents only proved to be between 2.19 % and 8.81 % in such places, even though a major shopping center is located in one of these neighborhoods. However, the more

---

[8] For the purposes of the study, the "neighborhoods" are the smallest spatial units in the nomenclature of the French National Institute for Statistics (INSEE) – called in French *Ilots Regroupés pour l'Information Statistique* (IRIS).

central the place of residence, the more similar to it are the destinations.

Overall, these findings suggest that these daily mobility patterns are in part *replacements* (Ramadier 2010), in that *individuals cross the functional borders of geographical space without crossing any social and cognitive borders*. This also shows that daily mobility might contribute to reinforcing social segregation in space, which is already strongly structured by residential mobility.

16.6 Conclusion

Since environmental psychology considers well-being, it cannot content itself with addressing the environment as an external constraint to which individuals adjust themselves, or as a means of adapting the material dimensions of physical and social environment to individual preferences and inclinations, or even to simply combining the two approaches by emphasizing the complexity of the regulations leading to the person/environment congruence. This approach, which is based on a dichotomy between inside and outside and between the objectivity and subjectivity of facts, systematically falls into the trap of the opposition between the individual and society. This entails that the adjustment of the individual's relationship to the environment is at the service of a general balance of human relationships on the scale of society (Elias 2010). This functionalist and homeostatic vision of the relationship to the environment and of social relationships precisely neglects social, cognitive and geographical structures. It describes relationships between individuals, groups and places as symmetrical. Either the objective properties of individuals, groups and places or the subjective perceptions of individuals (preferences) or groups (cultures) differ. As the individual is perceived as a self-constructing actor within a group and an environment, adjustments to the environment and of the environment are generally studied using the concept of control rather than of power. For instance, is it because an adolescent can close his bedroom door that he is free from his parents' authority and demands? Symmetrically, with the concept of affordance, places are perceived as regulators of human behavior – as controllers. Yet, for instance, in a classroom, the desk that faces all the others, which is conventionally attributed to the teacher, can be used by a student when the professor is absent, and the student may in turn profit from the symbolic value of this geographical position in the class. In other words, asymmetrical relationships of domination, of legitimacy of certain values or beliefs, or between certain places in a given socio-cognitive and geographical context, are rarely considered in environmental psychology. Social relationships are often reduced to the physical co-presence of individuals. Nevertheless, the environmental well-being of an individual is also constructed with social and spatial recognitions, feelings of social and spatial justice, more or less strong identifications with the groups and institutions that manage given places, etc. The trials of adjustment owing to the cost of entry to certain places, to incentives to visit or avoid them, etc. can be powerful barriers to psychological, social and physical well-being. For these reasons, I have proposed a (dis)positional model in which social, geographical and cognitive dimensions are based on the same concept of position (and consequently of social, cognitive and geographical space), in order to grasp processes of person/environment congruence involved in individual well-being jointly and simultaneously, without favoring any one of the three dimensions over the other.

While well-being cannot be limited to individual experience, it cannot either be reduced to rigid structures that automatically have effects. There is no way of knowing if statistical frameworks will apply to a given individual. However, the fact remains that individuals have a personal experience of these frameworks that can only be perceived on the basis of statistical probabilities by researchers; social realities can also be invested in an individualized form. Therefore, at the level of the individual, this statistical framework is a dimension of a living environment that is not as physical and tangible as that which environmental psychologists usually refer to as a "living environment". In fact, the adjustment of the individual's expectations to their chances of fulfilling these expectations constitutes a physical, social and cognitive framework that appears to have an impact on the place of residence, the way of inhabiting places, of representing the city and travelling, etc. This socio-cognitive set of references in the relationship to space is as evident to the individual as his/her family name or mother tongue. For instance, it is easier to speak a language when the chances of encountering other speakers of that language are adjusted to the expectations of being able to speak it freely. While the tridimensional model of well-being reveals that *entre-soi* can in some cases reflect a logic of person/environment congruence at the service of well-being, it also shows that it is a social, cognitive and geographical construction and not a natural human inclination (objectivity) or even a natural inclination of some individuals (subjectivity), and even less a necessity for the person's well-being. Socio-cognitive adjustment to space, as a construction that defines the individual's expectations according to the chances of fulfilling them and vice versa, is ultimately a model that opens up a great many possibilities. Indeed, the individual's trajectory in a field of forces and of struggles on the scale of society (the social space) redefines the framework of expectations and possibilities at each instant.

References

Barker, R. G. (1968). *Ecological psychology: Concepts and methods for studying the environment of human behavior*. Stanford: Stanford University Press.
Beauvois, J.-L. (1994). *Traité de servitude libérale. Analyse de la soumission*. Paris: Dunod.
Bourdieu, P. (1984). *Distinction: A social critique of the judgment of taste*. Cambridge: Harvard University Press.
Bourdieu, P. (2000). *Pascalian Meditations* (Richard Nice, trad.). Palo Alto: Stanford University Press.
Bronfenbrenner, U. (1979). *The ecology of human development: Experiments by nature and design*. Cambridge: Harvard University Press.
Canter, D. (1969). An intergroup comparison of connotative dimensions in architecture. *Environment and Behavior, 1*, 37–48.
Canter, D. (1988). Environmental (social) psychology: An emergin synthesis. In D. Canter, J. Correira Jesuino, L. Soscka, & G. Stephenson (Eds.), *Environmental social psychology*. London: Kluwer Academic Publishers.
Chombart de Lauwe, P.-H. (1974). Eth(n)ologie de l'espace. In *De l'espace corporel à l'espace écologique*. In Symposium de l'Association de Psychologie Scientifique de Langue Française. Paris: PUF.
Deschamps, J.-C., & Doise, W. (1988). Similar or different? Young immigrants in the Genevan educational context. In D. Canter, J. Correira Jesuino, L. Soscka, & G. Stephenson (Eds.), *Environmental social psychology*. London: Kluwer Academic Publishers.
Dias, P., & Ramadier, T. (2015). Social trajectory and socio-spatial representation of urban space: The relation between social and cognitive structure. *Journal of Environmental Psychology, 41*, 135–144.
Doise, W. (1982). *L'explication en psychologie sociale*. Paris: Presses universitaires de France.
Doise, W., & Lorenzi-Cioldi, F. (1989). Sociologues et psychologie sociale. *Revue Européenne des Sciences Sociales, 27*, 147–196.
Doise, W., & Palmonari, A. (1986). *L'étude des représentations sociales*. Neuchâtel: Delchaux & Niestlé.
Durkheim, E., & Mauss, M. (1903). De quelques formes de classification. Contribution à l'étude des représentations collectives. *Année Sociologique, 6*, 1–72.
Edney, J. J. (1976). Human territoriality. *Psychological Bulletin, 81*, 959–975.
Ehrlich, S. (1979). La mobilité cognitive. *Bulletin de Psychologie, no 340, 32*(8–14), 413–423.
Elias, N. (2010). *Au-delà de Freud: sociologie, psychologie, psychanalyse* (p. 2010). Paris: La Découverte.
Evans, G. W. (1980). Environmental cognition. *Psychological Bulletin, 88*(2), 259–287.
Feldman, R. M. (1990). Settlement identity: Psychological bonds with home places in a mobile society. *Environment and Behavior, 22*(2), 183–229.
Festinger, L. (1957). *A theory of cognitive dissonance*. Palo Alto: Stanford University Press.
Festinger, L., Riecken, H. W., & Schachter, S. (1956). *When prophecy fails: A social and psychological study of a modern group that predicted the destruction of the world*. New York: Harper-Torchbooks.
Gibson, J. J. (1966). *The senses considered as perceptual systems*. Boston: Houghton Mifflin.
Gibson, J. J. (1977). The theory of affordances. In R. Shaw & J. Bransford (Eds.), *Perceiving, acting, and knowing*. Hoboken: Wiley.
Hall, E. T. (1966). *The hidden dimension*. New York: Doubleday.
Holahan, C. (1982). *Environmental psychology*. New York: Random House.
Jellison, J. M., & Green, J. (1981). A self-presentation approach to the fundamental attribution error: The norm of internality. *Journal of Personality and Social Psychology, 40*(4), 643–649.
Kaplan, S. (1983). A model of person/environment compatibility. *Environment and Behavior, 15*(3), 311–332.

Kaplan, S. (1992). The restorative environment: Nature and human experience. In D. Relf (Ed.), *The role of horticulture in human well-being and social development* (pp. 134–142). Portland: Timber Press.

Koffka, K. (1935). *Principles of Gestalt psychology*. New York: Harcourt, Brace & World.

Köhler, W. (1929). *Gestalt psychology*. New York: Liveright.

Laborit, H. (1971). *L'homme et la ville*. Paris: Flammarion.

Lee, T. R. (1968). Urban neighborhood as a socio-spatial schema. *Human Relations, 21*, 241–268.

Lefebvre, H. (1968). *Le Droit à la Ville*. Paris: Éditions Anthropos.

Lewin, K. (1936). *Principles of topological psychology*. New-York: McGraw Hill.

Lynch, K. (1960). *The image of the city*. Cambridge: MIT Press.

Michelson, W. (1976). *Man and his urban environment: A sociological approach* (2nd ed.). Reading: Addison-Wesley.

Michelson, W. (1980). Long and short range criteria for housing choice and environment behavior. *Journal of Social Issues, 36*(3), 135–149.

Milgram, S., & Jodelet, D. (1976). Psychological maps of Paris. In H. Proshansky et al. (Eds.), *Environmental psychology* (2nd ed.). New York: Holt, Rinehart & Winston.

Moser, G. (2009). *Les relations homme-environnement. Psychologie environnementale*. Brussels: De Boeck.

Niveleau, C. (2006). Le concept gibsonien d'affordance: entre filiation, rupture et reconstruction conceptuelle. *Intellectica, 43*, 159–199.

Portugali, J. (1996). *The construction of cognitive maps*. London: Springer.

Proshansky, H. M., Ittelson, W. H., & Rivlin, L. G. (1970). *Environmental psychology: Man and his physical setting*. New York: Holt, Rinehart & Winston.

Ramadier, T. (1997). *Construction cognitive des images de la ville. Evolution de la représentation cognitive de Paris auprès d'étudiants étrangers*. Thèse de doctorat, Université Paris-Descartes.

Ramadier, T. (2004). Transdisciplinarity and its challenges: The case of urban studies. *Futures, 36*(4), 423–439.

Ramadier, T. (2007). Mobilité quotidienne et attachement au quartier: une question de position ? In J.-Y. Authier, M.-H. Bacqué, & F. Guerin-Pace (Eds.), *Le quartier: enjeux scientifiques, actions politiques et pratiques sociales* (pp. 127–138). Paris: La Découverte.

Ramadier, T. (2009). Capital culturel, lisibilité sociale de l'espace urbain et mobilité quotidienne. In F. Dureau & M.-A. Hily (Eds.), *Les mondes de la mobilité* (pp. 137–160). Rennes: Presses Universitaires de Rennes.

Ramadier, T. (2010). *La géométrie socio-cognitive de la mobilité quotidienne: distinction et continuité spatiale en milieu urbain*. Université de Nîmes. http://halshs.archives-ouvertes.fr/view_by_stamp.php?&halsid=kh3libj8lvv3c358lmqbgu7mc1&label=SHS&langue=fr&action_todo=view&id=tel-00564812&version=1

Ramadier, T., & Enaux, C. (2012). L'accessibilité socio-cognitive. In P. Frankhauser & D. Ansel (Eds.), *La décision d'habiter ici ou ailleurs* (pp. 76–101). Paris: Economica.

Ramadier, T., & Moser, G. (1998). Social legibility, the cognitive map and urban behavior. *Journal of Environmental Psychology, 18*(3), 307–319.

Ramadier, T., Lee-Gosselin, M., & Frenette, A. (2005). Conceptual perspective for explaining spatio-temporal behavior in urban areas. In M. E. H. Lee-Gosselin & S. T. Doherty (Eds.), *Integrated land-use and transportation models: Behavioral foundations* (pp. 87–100). Oxford: Elsevier.

Ramadier, T., Petropoulou, C., Haniotou, H., Bronner, A.-C., & Enaux, C. (2011). Daily mobility and urban form: Constancy in visited and represented places as indicators of environmental values. In M. Theriault & F. Desrosiers (Eds.), *Geographical information and urban dynamics: Mobility and urban form: From analysis to simulation* (pp. 129–157). London: ISTE-Wiley.

Rapoport, A. (1969). *House form and culture*. Englewood Cliffs: Prentice Hall.

Ross, L., Greene, D., & House, P. (1977). The "false consensus effect": An egocentric bias in social perception and attribution processes. *Journal of Experimental Social Psychology, 13*(3), 279–301.

Rouquette, M.-L. (2006). Introduction. In K. Weiss & D. Marchand (Eds.), *Psychologie sociale de l'environnement* (pp. 11–16). Rennes: Presses Universitaires de Rennes.

Simon, H. (1947). *Administrative behavior*. New York: Macmillan.

Slater, P. (Ed.). (1976). *Exploration of intrapersonal space* (Vol. 2). London: Wiley.

Stevens, H. (2011). De l'intervention psychosociologique au "développement personnel" dans l'entreprise. Esquisse d'une généalogie des relations entre management et psychologie en France. *Regards Sociologiques, 41*(42), 57–74.

Stokols, D. (1972). On the distinction between density and crowding: Some implications for future research. *Psychological Review, 79*(3), 275–277.

Stokols, D. (1978). Environmental psychology. *Annual Review of Psychology, 29*, 253–295.

Stokols, D. (1987). Conceptual strategies of environmental psychology. In D. Stokols & I. Altman (Eds.), *Handbook of environmental psychology* (pp. 41–70). New York: Wiley.

Stokols, D. (1990). Instrumental and spiritual views of people-environment relations. *American Psychologist, 45*(5), 641–646.

Stroop, J. R. (1935). Studies of interference in serial verbal reactions. *Journal of Experimental Psychology, 18*, 643–662.

Tafani, E., & Bellon, S. (2001). Principe d'homologie structurale et dynamique représentationnelle. In P. Moliner (Ed.), *La dynamique des représentations sociales* (pp. 163–193). Grenoble: Presses universitaires de Grenoble.

Tversky, A., & Kahneman, D. (1981). The framing of decisions and the psychology of choice. *Science, 211*, 453–458.

Uzzell, D., & Moser, G. (2006). On the quality of life of environment. *European Review of Applied Psychology, 56*(1), 1–4.

Viaud, J. (2003). Positions sociales et représentations: Contribution d'un cas de mobilité sociale ascendante à l'étude des dynamiques représentationnelles. *Cahiers Internationaux de Psychologie Sociale, 58*, 39–49.

Wacquant, L. (1992). Introduction. In P. Bourdieu & L. Wacquant (Eds.), *Réponses. Pour une anthropologie réflexive*. Paris: Seuil.

Wicker, A. W. (1973). Undermanning theory and research: Implications for the study of psychological and behavior effects of excess populations. *Representative Research in Social Psychology, 4*, 185–206.

Wicker, A. W. (1979). *An introduction to ecological psychology*. Monterey: Brooks and Cole Publishing Company.

Wohlwill, J. F. (1970). The emerging discipline of environmental psychology. *American Psychologist, 25*(4), 303–312.

Part V

Well-Being and Daily Environments – *Residential Environments*

Residential Satisfaction and Quality of Life

17

Juan Ignacio Aragonés, María Amérigo, and Raquel Pérez-López

Many theoretical reflections and empirical studies have been developed in relation to the concept of quality of life applied to residential quality. In the classic study by Campbell et al. (1976), there was already a consideration of residential satisfaction as one of the domains of satisfaction with life. It is not the objective of this chapter to describe what is understood by quality of life or well-being; however, before starting to develop the specific issues related to residential satisfaction, we should briefly review the concept of well-being with respect to its direct effect on the concept of residential satisfaction.

Within the psycho-social research that studies quality of life, terms are frequently found that are apparently synonyms and that, depending on the specific field in question or the background of the experts addressing the issue, refer to different terms such as quality of life, well-being, satisfaction or even happiness. These concepts are very much related and therefore require some precision, albeit briefly as other authors have already gone into detail on this topic (Amérigo 1993; Campbell et al. 1976; Marans and Stimson 2011).

Subjective quality of life becomes well-being, with this being understood as the satisfaction that the person or group experiences with their life or any aspect of the same. In the case of the latter, residential satisfaction could be considered an indicator of well-being or, the same thing, subjective quality of life. These social indicators, whether objective or subjective, can be used to analyze reality although obviously with the bias to paying attention to some and not others in each case.

Although it is certainly more difficult to specify subjective than objective aspects, because measuring feelings, perceptions, attitudes, etc. is more complex, it is essential to have subjective criteria for quality of life indicators in order to set the level of people's satisfaction with life. As previously indicated by Campbell et al. (1976), the mere possession of material resources is not directly linked to satisfaction itself. This had led, among other issues, to the proliferation of psycho-social studies related to the quality of life.

In relation to the topic of this chapter, it is worth remembering another classic study by Marans and Rodgers (1975) who indicated in their residential satisfaction model "that objective measures of environmental attributes are inadequate in themselves as indicators of life quality" (p. 343). This issue is repeatedly stated nowadays

J.I. Aragonés (✉)
University Complutense of Madrid, Madrid, Spain
e-mail: jiaragones@psi.ucm.es

M. Amérigo
University of Castilla-La Mancha, Ciudad Real, Spain
e-mail: Maria.Amerigo@uclm.es

R. Pérez-López
International University of La Rioja, La Rioja, Spain
e-mail: raquel.perez@unir.net

in empirical research, as demonstrated by the study of Hassine et al. (2014) who maintain that environmental quality of life must consider the interaction between people and their environment. This relationship reveals the manner in which the objective attributes of the environment are perceived and evaluated and, in this way, quality of life studies can serve as a guide for political decision-making.

17.1 Residential Satisfaction; Brief Review of This Concept Over Time

One of the topics being systematically studied in environmental psychology is residential satisfaction, as stated by Guiliani and Scopelliti (2009) in their bibliometric study. Gifford (2014), in a recent review of environmental psychology, also highlights the importance of studies on the home and the neighborhood, although in this case he also shows the growth that has occurred in recent years in favor of attachment over residential satisfaction. The importance of the first studies carried out on this topic in relation to the development of residential satisfaction merits their attention.[-45pt] Please confirm the identified section head levels are okay.This is correct

Although the beginnings of residential satisfaction studies mostly occurred in the 1980s, there are precedents such as those developed in the Chicago School, in which there are constant references to the living conditions of the residents of different neighborhoods. Among these, the study by Zorbaugh (1929), which provides a detailed description of the different social problems in adjacent urban spaces, should be highlighted. This study, considered by many to be ethnographic in nature, showed the enormous social distances in a single area of Chicago, Gold Coast, in the 1920s. Here, the residents of the social elite without links to the neighborhood were notable, with no feeling of a neighborly community, where the priorities for many were professional and economic interests located outside the neighborhood they lived in, and with the bond tying them together being their feeling of belonging to an elite group. In the slum, the identity symbol of the residents was the sub-culture created in this space itself, where the feeling of belonging to the community brought its members together. Here, loyalty and the control of some members over others led residents to develop specific attitudes of an identity nature.

At the beginning of the 1960s, studies were developed that included the notion of residential satisfaction. However, several years passed before this concept became a central part of environmental psychology. The studies from these years were carried out with the motivation of considering the social aspects of programs for planning and re-accommodating homes, particularly in the light of problems that had arisen in this field. At this time, the focus was on deprived areas of cities where the lowest classes of society lived.

Specifically, these studies responded to the need to find adequate criteria for an evaluation that would reveal why the residents of a neighborhood demonstrated strong attachment to the residential environment even when the living conditions in the homes were poor. They demonstrated the importance of social and psychological aspects of the constructed environment when explaining the attachment felt by the affected populations in the neighborhoods they lived in. The pioneering studies in this context were carried out in Boston's West End in relation to the renovation of deprived areas (Fried and Gleicher 1961; Gans 1959) and the effects of forced population movements from one area in the city to another (Fried 1963). Here, the role of the decisive psychosocial processes in the level of satisfaction can be observed, such as feelings of attachment to the area of residence, social networks between neighbors and the extension of the home to public residential areas. All of the above compensated, in many cases, the dire residential conditions.

In the same line was the work carried out by Hollinshead and Rogler (1963) in a low-status area of San Juan de Puerto Rico. These authors compared the reactions of people who lived in a public housing neighborhood with those who lived in a slum, concluding that the living conditions in the former were better, although the people in the slum were happier with their neighborhood.

One of the public housing interventions that is found in the majority of texts in the literature on residential satisfaction refers to the failed neighborhood of "Pruitt-Igoe" in the North American city of St. Louis (Missouri). This set of public housing constructed in the 1950s had to be destroyed in less than 20 years due to the lack of social control and security. The residents abandoned the apartments despite the fact that the majority of them considered them to be better than their previous homes. In the study carried out by Rainwater (1970), serious problems suffered by the majority of residents were identified. Some were related to the deterioration and vandalism of the shared equipment in the buildings, such as broken windows, lack of cleaning, poor operation of lifts, while others were related to security; for example, the women did not feel safe in community spaces, there was frequent theft and numerous anti-social behavior events in these spaces. This experience promoted concepts such as "defensible space" coined by Newman (1972). These are semi-private spaces that favor supervision by residents and therefore facilitate visitors recognizing them as private land, thus avoiding vandalism and criminal acts.

The disaster of Pruitt-Igoe was also discussed by Yancey (1971), who evaluated the deterioration of the neighborhood and also demonstrated the different ways in which the environment is perceived depending on social status. These variations are often not taken into account by planners and architects who, as they have a different social status from that of the residents, plan and design based on their own perceptions and not those of the future users. As stated by Bechtel (1997), "Pruitt-Igoe became the symbol of architectural arrogance and ignorance" (p. 322).

These series of studies have not only helped to configure models in order to discover how residential satisfaction works, but also their conclusions are currently applied when rehousing large families from poor conditions, such as avoiding breaking the social network before rehousing.

17.2 The Domains of the Residential Environment

As can easily be inferred from the above studies, the residential environment must be understood in wider terms. Thus, as well as the home itself in its exact dimensions, the nearby environment, both physical and social, where it is located and where the resident carries out a large proportion of their daily activities must also be considered. In the study carried out by Canter and Rees (1982), three elements were recognized – house, neighborhood and neighbors – that must be examined in any study related to the residential environment. After separately interviewing 902 couples in the United Kingdom through a questionnaire with 44 items, these authors showed, using a multi-dimensional analysis, that these three elements work independently when questions are asked in relation to the satisfaction experienced for each one of them.

Therefore, before beginning to discuss what is understood by residential satisfaction and from what perspective it is analyzed, it is necessary to review, albeit briefly, each of the three components of the residential environment. This is not about showing the numerous research projects for each one, but rather the way in which residential satisfaction has affected them.

17.2.1 The Home

The center upon which the residential environment is structured is, without doubt, the home. A large part of the daily life of most people is carried out within it and its surroundings, therefore it must be the key element in explaining residential satisfaction. However, as shown below, the literature on this topic has focused more heavily on the study of the neighborhood.

The majority of research on the home has considered it the place. Multi-dimensionality has also contributed to this concept, in relation to the development of environmental psychology, agreeing at this time on a definition for the home based on numerous studies such as those by Bachelard (1969), Rapoport (1969), Porteus (1976), Buttimer (1980), Saegert (1985), and Tognoli (1987), to name a few classics from the literature. In summary, the home could be considered the most immediate primary environment, stable, predictable and controllable; a central point around which human activities are organized. This is where the most significant experiences occur, where one feels comfortable, warm and isolated from the rest of the world. It is also important because it provides shelter, security, and protection; it gives meaning and identity; it structures social relationships, limits the space of family members and serves as a framework for developing domestic activities. All of this brings meaning to the home and transforms it into "home sweet home".

Together with this aspect that is largely personal and interpersonal, where the home has a relationship with the resident and the most immediate social network, there is an equally interesting perspective in which the socio-cultural aspects are highlighted. Authors such as Rapoport (1969) and Sanoff (1970) indicated that the house does not only provide a roof but also shows the way of life of a society, which converts it into a spatial unit with a clearly social character. In a similar sense, Altman and Chemers (1980) stated in the conclusion of their study that the home is a reflection of the relationship between culture and surroundings and, therefore, it adapts to the prevailing social values, becoming a physical manifestation of the culture it pertains to.

In fact, when a home is being constructed, although there are many physical possibilities, the real choices are limited by the cultural context where it is built. The constructed surroundings reflect socio-cultural strengths such as religious beliefs, family structure and social relationships. The solutions are varied and the homes will be the visible expression of the relative importance attributed to different aspects of life and the different ways of perceiving reality, thus acquiring a symbolic value. At the same time, the shape of the house is influenced by geographical contingencies such as climate, location and the availability of construction materials and techniques.

Both factors, geographical and particularly cultural, have led to homes evolving over time in order to adapt to the needs of residents, and to the specific cultural characteristics at each time in the history of humanity (Rybczynski 1986).

One of the fundamental factors that has favored change in the home has been the different concepts of family that have been held over the passage of time. Currently, the clear desire for intimacy and independence sought by the Western nuclear family has produced a parallel in the design of homes, where private rooms are created for family members.

The cultural factors shared by a large majority of Western populations produce a prototype home, in the terms of Altman and Chemers (1980), of a permanent nature as they tend to be permanent homes; differentiated by containing rooms with specific functions and not for communal use, as they are inhabited by a single nuclear family.

Clearly, it is necessary to address all these common factors that are derived from the cultural context and that influence the structural and functional aspects of the house, converting it into a home. The resident will perform actions at a cognitive level, such as comprehension of the space and their facilitator role in the performance of programs from daily life (Omata 1995); at an emotional level, such as a sense of belonging (Hidalgo and Hernández 2001; Pasca et al. 2015); and at a behavioral level, in relation to the activities that people perform in their daily lives (Ahrentzen et al. 1989; Aragonés 2002).

Although many of the topics about the relationship between people and their home have been addressed in an empirical manner (Shin 2014), there are still many others that have been treated at a purely speculative level or that infer certain relationships based on similar experiences in other contexts. The reasons for this lack are possibly due to an epistemological resistance similar to that indicated by Anzieu and Martin (1968) resulting from the study of groups. These authors suggest that groups are belatedly incorporated as subjects of research because their study implies the recognition of the individual's alignment in social life. In the same way, the study of the home involves invading the most intimate space and this is difficult to tolerate. However, as indicated by Lee (1968), the study of the home from a psychological point of view may be of great interest as this space is the main witness of the development of personality and of couple and parent relationships.

In summary, the house is not simply a place for living as mentioned by Cooper (1974), but rather it has a deep psychological significance beyond pure shelter and a place to carry out domestic behavior. With the home, or through it, the satisfaction of many aspirations, motivations and personal values that are related to the lifestyle of the person residing in it takes place.

17.2.2 The Neighborhood

As stated by Hur and Morrow-Jones (2008), the neighborhood can be considered the most basic unit of the urban medium, where numerous social relationships are produced, and therefore affects the residential satisfaction of its residents. It is apparently easy to share a definition of this urban element, however its meaning is more complex when we observe how it is treated in the psycho-environmental literature. Classic studies, such as that of Warren (1978), have already demonstrated the difficulty of defining what is understood by the neighborhood by confusing spatial and social issues. This inevitably implies difficulty when carrying out empirical research when it is necessary to operationalize a certain context, and even more so when the results from different research studies are to be compared.

The problem is exacerbated when the research refers to synonymous terms, such as community, district, and vicinity, among others, without providing an operative definition in each case or establishing geographical or social limits to delimit the urban or social reality in question. Among some of the few exceptions in which an effort is made to specify this element of the residential environment, we can highlight the work of Marans and Rodgers (1975), who differentiated the urban residential environment into the *macroneighborhood* and the *microneighborhood*. The former is equivalent to the officially recognized districts, or areas that are defined by the main public roads. The latter refers to the area immediately adjacent to the home and is formed by approximately six blocks. They also considered an intermediate level defined as "*community*" but, in the same study, the authors themselves showed that there were hardly any differences between *macroneighborhood* and *community*, and the residents found it difficult to differentiate these concepts.

The majority of research has focused on intermediate places between the *macroneighborhood* and the *microneighborhood*, commonly known as the neighborhood. This can be defined as an area of intermediate size next to the home where *facilities* are found and where social ties are established; in such a manner that the majority of studies are similar to what Lee (1968) defines as "socio-spatial schema". In his study carried out in the city of Cambridge, he established that both the physical dimensions (house elements, facilities, etc.) and social dimensions (relationships with neighbors) configure a specific schema for each resident, so that an overlap of the schemas of all residents configures the neighborhood or community. This is the concept referred to by all authors in their research, when they do not precisely define the limits or the type of social variables considered. A similar concept appears in Rapoport (1977) under the term "house-settlement system". For this author, this refers to the set of urban elements that are related to the home and that configure a system. In this way, the street is not a structure formed by a set of aligned buildings, but rather is based on the function of the elements that are carried out in them. Thus, for example, a cinema, restaurant or other functional place produces a manner of behavior in the street that gives it meaning.

As can be understood from the above, the neighborhood does not have a precise definition and each resident understands it in terms and magnitudes that are differentiated from those of their neighbors. However, in its physical dimension, we can talk about the area next to the home where the principal facilities and services are found for health (hospitals, health centers, pharmacies, etc.); recreation (sports hall, green areas, *clubs*, etc.); commerce, culture, religion, etc., in such a way that the resident can walk to the majority of them and where there is a relationship of mutual dependence, determined by the activities that are carried out in them.

Despite the difficulties in defining this urban element, in research it is easy to find that the physical characteristics of the neighborhood together with the social or economic characteristics have an influence on residential satisfaction (Sirgy and Cornwell 2002). The physical aspects have not received a great deal of attention in this type of study. Moreover, as indicated by Hur et al. (2010), when the physical nature of the neighborhood has been considered, the evaluations of the aesthetics (beauty or appearance) have been focused on more than the physical characteristics themselves. These authors showed how physical variables, such as quantity of vegetation and building density, together with their perception by the residents (naturalness and openness) are related to satisfaction with the neighborhood, demonstrating in one case that the true density is related to neighborhood satisfaction, while vegetation does so through the subjectivity of the resident relating to it.

In other cases, objective and perceived physical aspects produce certain feelings of social scope that affect neighborhood satisfaction. Thus, as Hur and Nasar (2014) have demonstrated, the deterioration and lack of maintenance of the neighborhood leads to feelings of insecurity that eventually influence neighborhood satisfaction.

Another variable that moderates the evaluation of the physical aspects of the neighborhood is the number of years of residency. It has been observed that newly-arrived residents give more importance to the physical characteristics of the neighborhood in relation to satisfaction than long-term residents. For the former, the adequacy of public services, the level of accessibility, the air quality and, above all, the general attractiveness of the new neighborhood are highlighted. However, the latter give more importance to stressful factors in relation to their neighbors and employment, improvements in health services, traffic and cleanliness, among others (Potter and Cantarero 2006).

Without a doubt, there is a lack of standardization in the operative and consensual definition of the neighborhood and objective measures of its physical characteristics, which enable more precise models to be formulated when simulating satisfaction with this element of the residential environment. For this, researchers must describe geographically the context in which their research projects are located. Moreover, it would improve the research if the residents, before being questioned about any physical or social aspect of their neighborhood, were informed of the definition of the area that is being researched, as on many, if not all, occasions, the different responses about equipment, facilities and social relationships that are mentioned in the interview are based on their own interpretation of "their neighborhood". Although the results obtained to date with these procedures can be interesting in each context, there is a loss of rigor and a difficulty in comparing results when it is not known which environments the person answering the questions has in mind. It can be concluded that the neighborhood, as a place where a great many social relationships take place, has been given little attention in comparison to the social component of these.

From a sociological or psychological point of view, the neighborhood is the environment that enables the establishment of social networks between its residents, providing a feeling of belonging to these, and thus becoming a type of home, as indicated by Fried (1986). Many studies have tried to show the characteristics of the neighborhood that facilitate social networks. The study by Warren (1978) established three dimensions to classify neighborhoods: intensity of interaction between residents, feeling of identity of residents due to living in the neighborhood, and number of connections with the outside world. In another study on neighborhood cohesion carried out by Weenig et al. (1990), its role among neighborhood residents was shown, highlighting two components: "neighboring", related to the number of communication or interaction links between neighbors; and "sense of community", linked to the quality of these links. Although Unger and Wandersman (1985) had tried to group both components, the empirical research has resisted this synthesis.

This second focus on the psycho-social character of the neighborhood has received the most attention in residential satisfaction studies, to the detriment of a focus on the neighborhood as a place. This means that, in a way, the essence of the place is lost and the neighborhood is largely seen as a space where many social relationships are established and thus becomes a diffuse place, sharing a large proportion of the issue with the component of neighbors in a residential environment.

17.2.3 The Neighbors

This is the component of the residential environment with a clearly social character, which overlies both the home and the neighborhood. Research projects in relation to this topic are mostly interested in the interactions with residents, and to what extent they are affected by the residential design. Two fields have been principally addressed in this area: the relationship between neighbors' proximity to each other and the feeling of community.

With respect to the former, we should highlight the classic psycho-social study carried out by Festinger et al. (1950) on propinquity, understood as the closeness between the places occupied by people. Two different types of proximity are identified in these studies. One is linked to the objective physical distance between two people as a facilitator for bonds of friendship, showing that those that live on the same floor or in the same block are more likely to become friends than those that live further apart. However, the physical distance is not the only indicator of friendship relationships between neighbors, as the second type of proximity is the "functional distance", understood as the probability of two people entering into frequent contact, which would promote friendly relationships. This type of distance leads to a greater probability of friendship than physical distance. Consequently, designs must consider both types of distance, physical and functional, if they wish to promote friendships. In any case, we must take into account the *similarity of the residents* in relation to their interests, attitudes, values, backgrounds, and/or personality as this promotes the development of relationships that are produced due to propinquity.

At an analytical level, Taylor et al. (1994) provided four explanations of why propinquity leads to friendship. First, because proximity usually increases familiarity as it facilitates meetings. Secondly, proximity is often related to similarity; those that work or live together share common elements. A third aspect is that people that are close-by are more available than those that are distant, and therefore it is easier to establish friendships with them. Finally, the theories of consistency state that people think well of each other when they are forced to be together.

The characteristics of the residential environment undoubtedly affect the relationships of neighbors and, indirectly, satisfaction. Although the residential environment has optimum qualities that favor friendship between neighbors, some may create annoyances (noise, dogs, naughty children, etc.) which may deteriorate the environment as noted in the environmental spoiling hypothesis of Ebbesen et al. (1976), which led Skjaeveland and Gärling (2002) to conclude that it is not clear whether a socially optimum environment leads to improved residential satisfaction and there is always a possibility of cognitively reconstructing the situation and even moving to a new home. In this sense of avoiding the annoyances caused by neighbors, Bell et al. (1996) highlighted the preference of many people for single-family houses in the suburbs, limiting the territory and seeking greater privacy from their neighbors and, therefore, regulating interaction with them.

Another sense, called the sense of community, is more successful and, in some way, transcends residential satisfaction, being understood as the neighborhood bonds that are established in a neighborhood or community. McMillan and Chavis (1986) established four elements to distinguish this concept; (1) membership, in relation to the sense of belonging to a group or sharing a sense of personal relationships, (2) mutual influence between the members, where one and all feel pressured to conform, (3) community resources that satisfy the needs of the members and (4) a shared emotional connection, resulting from the experiences of the community. In another contemporary study, Brown and Werner (1985) indicated that the studies referring to this topic must consider three different aspects: neighborhood cohesion, belonging to the place, and territoriality in the residential environment. In their research, they aimed to relate these three aspects. They compared the feeling of belonging to the neighborhood between two different housing dispositions; in one, houses were located in a street with no exit, a *cul-de-sac*, and in the other, they were positioned on a street with its corresponding exits. The results demonstrated the importance of the feeling of belonging in the physical design, as the residents in the *cul-de-sac* showed greater belonging and more intimate relationships between them.

In summary, we can conclude that the residential environment of the individual is characterized by both physical and social attributes that pertain to one of the three levels of analysis, home, neighborhood and neighbors, and that residential satisfaction as a whole depends on certain attributes of these three components.

17.3 Conceptual Approximations of Residential Satisfaction

Once the domains of residential environment that are taken into account at the time of evaluating residential satisfaction are known, we can consider what is understood by this concept. For this, it is worth noting which focuses and aspects are shared by experts in this topic.

Among the manuals for environmental psychology, it is easy to find a definition of residential satisfaction that can indicate what the authors of the manual understand; but, in a more systematic way, we could refer to some of the documents regarded as milestones for the discipline and that deal with this construct. In the first Environmental Psychology handbook, Schorr (1970) defined residential satisfaction as "the absence of complaint, when opportunity for complaint is provided, or as an explicit statement that the person likes his housing" (p. 323). Years later, Tognoli (1987) focused on the process of adaptation and adjustment of the resident to their environment in order to reach residential satisfaction. This author states that this construct is not a stable and permanent feeling, but rather a process contingent on the changes that may occur to the person or their environment, in such a way that they attempt to reach a balance with their residential surroundings. A plausible explanation for this search for balance can be found in the theories of cognitive consistency, either modifying the residential environment or the levels of aspiration faced with unsatisfied needs or even moving to a new environment. The adjustments and adaptations required to reach the desired residential satisfaction will depend on the personal characteristics of the resident and/or the opportunities offered by the residential environment.

From the point of view of Tognoli (1987) and those developed in the empirical research synthesized by Amérigo and Aragonés (1997) and Amérigo (2002), these processes of adjustment and adaptation do not focus on behavior for improving the residential environment only when the resident does not reach the level of satisfaction, but also when feelings towards the residential environment are positive. Thus, the resident will at least try to carry out maintenance tasks to continue or increase the status of well-being experienced. The response in both cases, satisfaction and dissatisfaction, will be that the behavior will be related to the physical, social and organizational environment, as indicated by Oktay et al. (2012).

Briefly summarizing the ways of approaching this construct, it is observed that the majority of the theory models respond to a systemic relationship between the different elements considered

(Aragonés et al. 2002), although mainly from two different points of view. Some understand the feeling of residential satisfaction as a product derived from a cognitive evaluation, which originates from the difference between what the resident possesses in their residential environment and what they would like to possess (Wu 2008). This form of studying residential satisfaction has led authors such as Marans and Rodgers (1975) to see it as a product of a series of previous events, thus, from a methodological perspective, it is a variable criterion for a set of variables that are predictive, either of an objective or subjective nature for the residential environment, or referring to the person's characteristics. The relationship suggested in this formulation takes into account the level of satisfaction or wellbeing of the resident at each time, but does not show the dynamic vision of residential satisfaction as this varies depending on the specific residential situation present in their lives. Another, less frequent, theoretical approach, such as that presented by Speare (1974), attempts to see residential satisfaction as a predictive variable. In this case, residential satisfaction is considered from an attitudinal perspective, as a prior variable that explains the adaptive behaviors performed by the resident to achieve a balance between what they have and what they desire.

Referring to the field literature, it is easy to find theoretical models that are included in the first group, such as those developed by Marans and Rodgers (1975), Gifford (1987), and Francecato et al. (1989), to mention some of the most well known. However, when residential satisfaction is considered a predictive variable, there are fewer models, as in this case it is normally observed to what extent the dissatisfaction or insufficient satisfaction leads residents to carry out adaptation or adjustment measures. In this sense, it is worth mentioning the model by Speare (1974), which includes the consideration of movement to another residential environment when the satisfaction is relative.

Following the models mentioned, it could be concluded that residential satisfaction is a positive feeling towards the residential environment that is derived from a process through which certain of its objective attributes are evaluated by the resident, leading to adjustment or adaptation, depending on the situation, of the environment through a set of behaviors that enable its congruence to be maintained or increased. These feelings may be influenced by the so-called "Pollyanna hypothesis", according to which there is a universal trend to use more positive words than negative words in the evaluations (Boucher and Osgood 1969).

Residential satisfaction, in both its predictive variable and criterion aspects, considers the result, that is to say the level of dissatisfaction, at a specific time and in a certain situation. However, it is easy to think that the feeling or cognition of a resident towards their residential environment passes through very different stages over the time that they live in that place, especially as people often stay in the same environment for a long portion of their life (Dieleman and Mulder 2002). Therefore, the models that understand residential satisfaction as a predictive variable or criterion pay more attention to the results than to the process completed by the residential satisfaction, and consequently exclude an integrated model that attempts to study the dynamics implied by residential satisfaction over the time that a person lives in the same place.

In this sense and in summary, it can be concluded that residential satisfaction is considered "an emotional result, an emotional response or a consequence of a positive nature that arises from establishing comparisons between the residential environment and the situation of the subject. All of this is considered in a cyclical and dynamic process where the subject adapts to each specific residential situation", as expressed by Amérigo (1995 p. 55). This approach of a systemic nature led the author to propose an integrated model that would enable analysis of residential satisfaction with the double function of predictive variable and criterion, and thus allow the process that this construct works on to be observed (Amérigo 2002) (see Fig. 17.1).

As a variable criterion, the objective attributes of the residential environment corresponding to

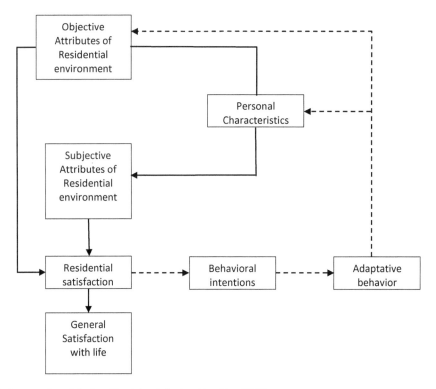

Fig. 17.1 Integrated model of residential satisfaction (Amérigo 1995)

the three levels (home, neighborhood and neighbors) directly affect the residential satisfaction experienced by the resident. Moreover, the subjective attributes, that is to say the objective attributes that the residents perceive and evaluate in their residential environment, directly influence their satisfaction with it. At the same time, this predicts part of their satisfaction with life, of which it is a relevant domain. On the other hand, residential satisfaction, understood as an attitude, explains the adjustment behavior of the resident to reach it when they are dissatisfied or wish to improve their level of satisfaction. These adjustments can be carried out both on the objective attributes of the residential environment, such as moving to a new home, renovating the home or intervening in the corresponding levels of the neighborhood and community if possible, and by modifying the levels of aspiration related to the residential situation. This would imply modifying the evaluation of the objective attributes that negatively influence residential satisfaction, referring to a "residential quality pattern" of an essentially regulatory character, where comparisons are carried out between the ideal and the true residential environment, thus generating new subjective attributes that conform more to the residential situation.

17.4 Environmental Quality Indicators and Residential Satisfaction Predictors

To begin this section, it is appropriate to refer to the model presented in the previous section and to differentiate between objective and subjective attributes of the residential environment. Both have their corresponding indicators of residential quality. In the first case, these are observable characteristics of environmental or social environments while in the second case, it refers to the ways in which people mentally represent certain outstanding categories of the environment. The impact of some indicators over others in relation to environmental issues was demonstrated in the study by Craik and Zube (1976), which observed the importance of the subjective over the objective at the time of evaluating the perception of environmental quality. The higher the residential quality indicators for a certain environment, the greater the residential quality. Not necessarily all of the residential quality indicators establish a link with satisfaction. Only those that, on each occasion, are capable of being related to residential satisfaction may be considered predictive.

It is now appropriate to make a methodological intervention. All the models proposed are based on the correlation between the residential quality indicators and the satisfaction experienced by people with their residential environment, whether through procedures of regression analysis or structural equations. In both cases, there is variability in the predictive variables and criteria, but if any of them were to offer great homogeneity, the correlation would be zero and, therefore, this indicator would not appear in the regression equation; therefore, it cannot be considered a predictor of residential satisfaction. Let us imagine an environment perceived by all residents as very insecure. In this case, the correlation with satisfaction would be zero or almost zero due to low variability, which would lead to the probably incorrect conclusion that the level of security perceived in a residential environment does not contribute to satisfaction with it. Consequently, the ad hoc studies carried out in a certain environment must consider this methodological slant.

By carefully reviewing the research on residential satisfaction, it is easy to find a framework that organizes the different satisfaction predictors. As stated by Amérigo and Aragonés (1997), there are four types of predictors to consider according to the combination of the dimensions of physical vs. social environment and subjective vs. objective environment. As shown in Fig. 17.2, the different residential satisfaction predictors can have: an subjective physical nature, such as maintenance of the neighborhood, appearance of the place, evaluation of the apartment, etc.; a subjective social nature, such as security, attachment to the residential area, overcrowding, etc.; an objective physical nature, such as noise levels, pedestrian areas of the neighborhood, type of home, etc.; or a objective social nature, such as lessor or owner, time of residence in the place, life cycle, etc.

In other studies, other groups of residential satisfaction predictors can be found. Thus, Francescato (2002) differentiates four domains – physical environment, psychological and social environment, organizational environment, and surrounding community – which include both subjective and objective predictors. Another classification related to satisfaction with the neighborhood was carried out by Sirgy and Cornwell (2002) who provide a long repertoire of predictors gathered from numerous research projects that create three features: physical, economic and social.

The majority of the research gives more weight to the subjective than the objective values, although the objective also directly affect residential satisfaction (e.g. Amérigo and Aragónes 1990; Francescato 2002). This idea is more than justified empirically if we consider that both the socio-demographic and personality characteristics of the residents and the environment in which they live lead to different realities and specific residential needs. An example of this can be found in the work by Galster and Hesser (1981), who concluded that, depending on certain socio-demographic characteristics such as age, marital status, race and number of children, it is possible to obtain different satisfaction scores for the same place. A similar statement was made by Vemuri et al. (2011), on finding that those who express greater satisfaction with the quality of life in their neighborhood have a higher level of education and belong to the White-Caucasian racial group. More recently, Robin et al. (2009) showed similar results when comparing evaluations of environmental annoyances perceived according to the socio-economical status and the level of education of the participants.

This situational character is applicable to the residential satisfaction models in which the resident typologies vary, as well as the environments where they live, which complicates the

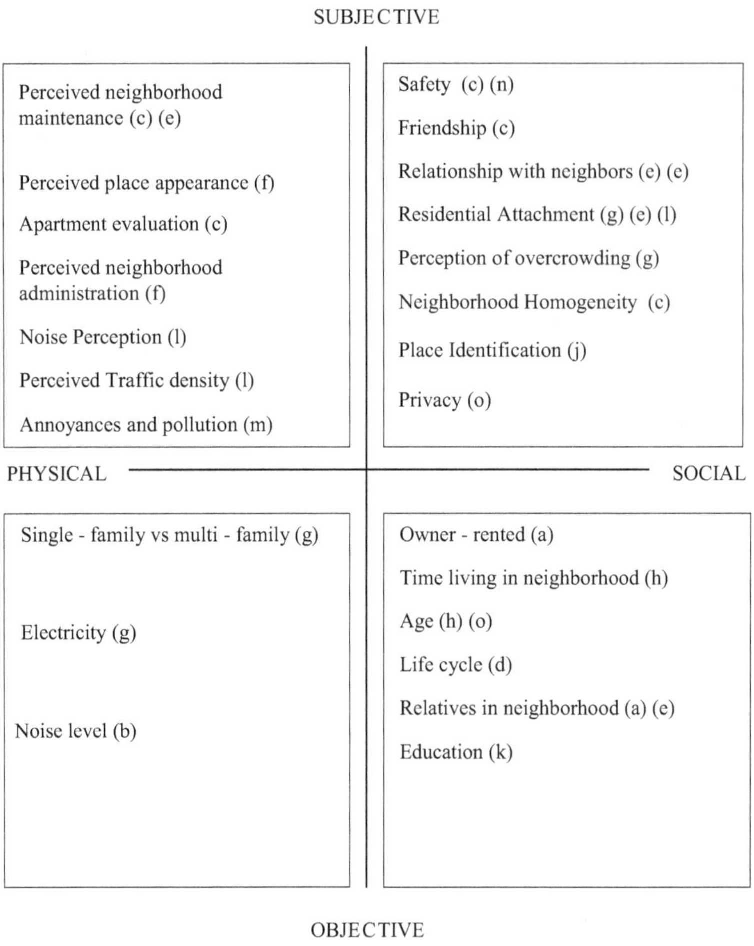

Fig. 17.2 Some predictors of residential satisfaction. (*a*) Rent and Rent (1978), (*b*) Miller et al. (1980), (*c*) Weidemann et al. (1982), (*d*) Hourihan (1984), (*e*) Amérigo and Aragonés (1990), (*f*) Antony et al. (1990), (*g*) Aragonés and Corraliza (1992), (*h*) Bonnes et al. (1991), (*i*) Rioux and Velasco (2004), (*j*) Fleury-Bahi et al. (2008), (*k*) Vemuri et al. (2011), (*l*) Oktay et al. (2012), (*m*) Hassine et al. (2014), (*n*) Hur and Nasar (2014), (*o*) Ahn and Lee (2015)

SUBJECTIVE

PHYSICAL	SOCIAL
Perceived neighborhood maintenance (c) (e)	Safety (c) (n)
Perceived place appearance (f)	Friendship (c)
Apartment evaluation (c)	Relationship with neighbors (e) (e)
Perceived neighborhood administration (f)	Residential Attachment (g) (e) (l)
Noise Perception (l)	Perception of overcrowding (g)
Perceived Traffic density (l)	Neighborhood Homogeneity (c)
Annoyances and pollution (m)	Place Identification (j)
	Privacy (o)
Single - family vs multi - family (g)	Owner - rented (a)
Electricity (g)	Time living in neighborhood (h)
	Age (h) (o)
	Life cycle (d)
Noise level (b)	Relatives in neighborhood (a) (e)
	Education (k)

OBJECTIVE

generalization of which are the best predictors. However, studies such as that by Francescato (2002) highlight that, in general, from pioneering studies until the present day, the attachment to a place, the social network or the feeling of community are concepts that are very linked to residential satisfaction in a positive way.

It has already been stated that sociodemographic variables constitute [objective] predictors arising from the social residential environment, and their influence on satisfaction has been demonstrated in earlier studies. For example, Onibokun (1976) observed in a study on public housing that the residents with higher levels of education, higher professional status and a higher level of income said that they were dissatisfied with their homes. This result, and many others, must be analyzed by taking into account the contextual aspect of where it was carried out.

For example, research carried out by Canter and Rees (1982), with a sample of almost a thousand citizens of the United Kingdom, must be considered in terms of its general character, recognizing the complication of extrapolating it to specific contexts. These authors asked residents to respond to a questionnaire where they were shown characteristics of the residential environment. A multidimensional analysis of the responses to the items revealed two different spaces: one for husbands and another for wives. However, these results were nuanced by a study carried out by Amérigo (1992) which demonstrated, through an analysis similar to that of Canter and Rees (1982), that in couples consisting of elderly people with a low socio-economic status and low level of education, both partners show an identical perception of the residential environment. On the other hand, in the same study, it was observed that when the residents are middle-aged couples, with an average socio-economic status and high level of education, both professional, the resulting space from the analysis does not allow the definition of a region for each category, nor an identical perception by both members of the couple. Instead, the participants occupy the two-dimensional space without obeying a law that would position them depending on their gender or the husband-wife relationship. These results modulate those obtained by Canter and Rees (1982), which could be accepted in reference to the general population, but would be nuanced on observing that they are moderated by variables of a socio-demographic nature.

Despite the variability of predictors that arise from empirical research, as shown in the studies of Fleury-Bahi et al. (2013) and Hassine et al. (2014) where different structures of predictors among French and Tunisian populations were obtained on the same scale, the experience obtained in this field can be useful as a guide for planning residential areas that will be inhabited by different types of populations.

In other words, both the predictors and the perceived quality indicators can facilitate the detection of the impact of housing intervention plans. Thus, Carp and Carp (1982) described some of the advantages of these indexes when applied to the study of a neighborhood: (1) they enable the perceived quality of the neighborhoods to be studied in terms of objective differences, (2) they can be used in known neighborhoods to see in what aspects the perception of the residents differs, and (3) they are useful for planners to evaluate the effects of intervention programs or designs in a neighborhood. As indicated by Francescato et al. (1989), the results of the research can not only predict the probability of success of the environment studied, but also be used in Post Occupational Evaluations (POEs) of the homes (Preiser 1989).

In summary, the knowledge of proven indicators of residential quality and the capacity to predict satisfaction have a clear application, guiding planners and architects in the decision-making processes involved in the construction and renovation of residential areas. Therefore, it would be very useful to know which variables of the residential environment, based on previous models, have predicted satisfaction in different contexts.

17.5 Attachment to Place

Attachment to place is one of the concepts that are becoming increasingly relevant in the literature related to residential satisfaction, principally focused on the neighborhood component (Hidalgo and Hernández 2001). Its importance is such that researchers often indicate attachment to the neighborhood as the most relevant variable when studying the residential environment, even above satisfaction with it (Bonaiuto et al. 1999). This concept appeared in some form in the first studies on residential satisfaction, principally in relation to the neighborhood and the feeling of community, as commented on above. A review of the literature on the subject reveals some research on this construct in later years, but it was in the 1980s when it was included in the models of residential satisfaction. Giulianni (2003) indicates in his review on place attachment that it was at the beginning of the twenty-first century when attachment acquired the promising future expected by Francescato (2002).

Further studies on attachment to place inevitably led to a significant reduction in those on residential satisfaction towards the end of the twentieth century. Perhaps one reason for this omission is the proximity of both concepts and the difficulty of operationalizing them in a differentiated way. Some authors, such as Giulianni (2003), suggested that the difference between attachment and satisfaction is more empirical than theoretical. However, this same author recognized that there were researchers such as Brown and Werner (1985), among others, for whom satisfaction is included within attachment, while Bonaiuto et al. (2015), for example, accept that there are positions in which attachment predicts satisfaction, or the contrary. In their study carried out with residents of Tabriz, they showed that satisfaction with the neighborhood is predicted by the attachment to it, which at the same time is predicted by the Perceived Residential Quality Indicator (PRQIs), of a cognitive nature. A similar position, although with certain differences, is developed by Oktay et al. (2012). These authors understand that residential satisfaction pays more attention to the cognitive components and efficiency in quality of life, as they focus on evaluating to what extent the objective attributes of the environment adjust to the needs and expectations of the residents. This reflection leads to the positioning of satisfaction as most linked to the cognitive component of well-being, while attachment, inevitably understood as a feeling, is linked to the emotional component of the quality perceived. However, in the empirical research, attachment to the neighborhood appears as a satisfaction predictor, although conditioned by the environmental role of the participants. Thus, within one single neighborhood, attachment to it was a strong satisfaction predictor for owner residents, whereas it was not for the international students who lived there occasionally. For these, the resulting predictors were attractiveness, accessibility and maintenance of the neighborhood, issues that are clearly linked to physical aspects. This difference between the results obtained for residents of one single neighborhood demonstrates that someone may feel satisfaction with their neighborhood but not feel attachment to it because they have not generated emotional links to the place or to its residents.

There are other aspects that occur in parallel between residential satisfaction and attachment to place. As commented on above, residential satisfaction predictors can be physical and/or social, as can the types of attachment, as indicated by Hidalgo and Hernández (2001). The former makes reference to the place itself, and the latter to the social relationships that are established in it. In the same study, the authors also differentiate between attachment to the neighborhood and attachment to the home. Numerous studies have been carried out that demonstrate, in general, that a high percentage of residents say that they feel attached to the neighborhood they live in. However, in relation to the home, there is only an occasional result in this sense (Pasca et al. 2015), which is surprising considering that the home is the primary place *par excellence* where feelings of belonging are produced (Serfaty-Garzon 2003) and a there is a greater personalization of the space (Pérez-López et al. 2013).

Although in both constructs the emotional component is clearly underlying, residential satisfaction is more related to cognitive processes. Thus, the level of satisfaction for an individual is determined by a series of physical and social attributes that can be changed without modifying the level of satisfaction, as these attributes are exchangeable in the regression equation. However, attachment to a place is a feeling that is not necessarily derived from changes in the residential environment (Giuliani 2003) and, therefore, has a different nature from satisfaction.

The empirical research has shown certain differences between both concepts. One of these, undoubtedly, is related to studies on the relevance of residential attachment for people with a low socio-economic status and its importance in re-housing the lowest social class. In these studies, there is occasionally contradiction between the residents. On the one hand, they are dissatisfied due to the impoverished status of their residential environment while maintaining a strong attachment to it, which causes problems for moving, even if this is to achieve a better residential situation. This discrepancy and its consequences led Fried (2000) to indicate that an attachment to place can often create dysfunctional situations.

A second variable that reveals differences or nuances between attachment and satisfaction can be found in the studies considering the length of time of residency in relation to both variables. The research carried out by Kasarda and Janowitz (1974) in the United Kingdom obtained a positive relationship between the time of residency, attachment and the feeling of belonging to the community. These results were confirmed by the work of Bonaiuto et al. (1999) in Rome, where it was observed that the time of residency in the neighborhood had a positive correlation with attachment to it. However, in the study carried out by Fleury-Bahi et al. (2008) in three French cities, identification with the neighborhood acted as a mediating variable between the time of residency in it and residential satisfaction. That is to say, in the first study, a direct relationship is found between time of residency and attachment, which is comprehensible as the number of links in the neighborhood easily increase over time. However, in the second study, the time of residency appears to be linked to satisfaction, through a variable that mediates the relationship.

Satisfaction and attachment follow a parallel course, but with numerous meeting points. The fact that researchers have focused on one construct or another and that they have been jointly treated as similar concepts in the majority of cases, without going deeper into the study of their differences and complementarity, leads to the conclusion that a great deal remains to be studied in relation to the joint course of these two constructs.

References

Ahn, M., &amp;amp; Lee, S. J. (2015, April). Housing satisfaction of older (55+) single-person householders in U.S. Rural Communities. *Journal of Applied Gerontology*, 1–34. doi:10.1177/0733464815577142.

Ahrentzen, S., Levine, D. W., & Michelson, W. (1989). Space, time and activity in the home: A gender analysis. *Journal of Environmental Psychology, 9*, 123–131. doi:10.1016/S0272-4944(89)80001-5.

Altman, A., & Chemers, M. M. (1980). *Culture and environment*. Monterey: Brooks/Cole Publishing Company.

Amérigo, M. (1992). Patrones perceptivos diferenciales en función del Rol Ambiental. *Psicothema, 4*, 123–131.

Amérigo, M. (1993). La calidad de vida. Juicios de satisfacción y felicidad como indicadores actitudinales de bienestar. *Revista de Psicología Social, 8*, 101–110.

Amérigo, M. (1995). *Satisfacción Residencial. Un análisis psicológico de la vivienda y su entorno*. Madrid: Alianza Editorial.

Amérigo, M. (2002). A psychological approach to the study of residential satisfaction. In J. I. Aragonés, G. Francescato, & T. Gärling (Eds.), *Residential environments. Choice, satisfaction, and behavior* (pp. 81–99). Westport: Greenwood Press.

Amérigo, M., & Aragonés, J. I. (1990). Residential satisfaction in council housing. *Journal of Environmental Psychology, 10*, 313–325. doi:10.1016/S0272-4944(05)80031-3.

Amérigo, M., & Aragonés, J. I. (1997). A theoretical and methodological approach to the study of residential satisfaction. *Journal of Environmental Psychology, 17*, 47–57. doi:10.1006/jevp.1996.0038.

Antony, K. H., Weidemann, S., & Chin, Y. (1990). Housing perceptions of low-income single parents. *Environment and Behavior, 22*, 147–182. doi:10.1177/001391659702900405.

Anzieu, D., & Martin, J. Y. (1968). *La dynamique des groupes restreints*. Paris: PUF.

Aragonés, J. I. (2002). The dwelling as place: Behaviors and symbolism. In J. I. Aragonés, G. Francescato, & T. Gärling (Eds.), *Residential environments. Choice, satisfaction, and behavior* (pp. 163–182). Westport: Greenwood Press.

Aragonés, J. I., & Corraliza, J. A. (1992). Satisfacción residencial. *Psicothema, 4*, 329–341.

Aragonés, J. I., Francescato, G., & Gärling, T. (2002). Evaluating residential environments. In J. I. Aragonés, G. Francescato, & T. Gärling (Eds.), *Residential environments. Choice, satisfaction, and behavior* (pp. 1–13). Westport: Greenwood Press.

Bachelard, G. (1969). La *poétique de l'espace*. Paris: PUF.

Bechtel, R. B. (1997). *Environment & behavior. An introduction*. Thousand Oaks: Sage.

Bell, P. A., Greene, T. C., Fisher, J. D., & Baum, A. (1996). *Environmental psychology*. Fort Worth: Harcourt Brace College Publishers.

Bonaiuto, M., Aiello, A., Perugini, M., Bonnes, M., & Ercolani, A. P. (1999). Multidimensional perception of residential environment quality and neighbourhood attachment in the urban environment. *Journal of Environmental Psychology, 19*, 331–352. doi:10.1006/jevp.1999.0138.

Bonaiuto, M., Fornara, F., Ariccio, S., Cancellieri, U. G., & Rahimi, L. (2015). Perceived Residential Environment Quality Indicators (PREQIs) relevance for UN-HABITAT City Prosperity Index (CPI). *Journal of Environmental Psychology, 45*, 53–63. doi:10.1016/j.habitatint.2014.06.015.

Bonnes, M., Bonaiuto, M., & Ercolani, A. P. (1991). Crowding and residential satisfaction in the urban environment. A contextual approach. *Environment and Behavior, 23*, 531–552.

Boucher, J., & Osgood, C. E. (1969). The Pollyanna hypothesis. *Journal of Verbal Learning and Verbal Behavior, 8*, 1–8. doi:10.1016/S0022-5371(69)80002-2.

Brown, B. B., & Werner, C. M. (1985). Social cohesiveness, territoriality, and holiday decorations: The influence of cul de sacs. *Environment and Behavior, 17*, 539–565.

Buttimer, A. (1980). Home, reach and sense of place. In A. Buttimer & D. Seamon (Eds.), *The human experience of space and place* (pp. 166–187). London: Croom Helm.

Campbell, A., Converse, P. E., & Rodgers, W. L. (1976). *The quality of life of America. Perceptions, evaluations, and satisfactions*. New York: Russell Sage.

Canter, D., & Rees, K. (1982). A multivariate model of housing satisfaction. *International Review of Applied Psychology, 31*, 185–208.

Carp, F. M., & Carp, A. (1982). Perceived environmental quality of neighborhoods: Development of assessment scales and their relation to age and gender. *Journal of Environmental Psychology, 2*, 295–312. doi:10.1016/S0272-4944(82)80029-7.

Cooper, C. (1974). The house as a symbol of self. In J. Lang (Ed.), *Architecture and human behavior* (pp. 130–146). Stroudsburg: Dowden.

Craik, K. H., & Zube, E. H. (1976). *Perceiving environmental quality*. New York: Plenum Press.

Dieleman, F. M., & Mukder, C. H. (2002). The geographical of residential choice. In J. I. Aragonés, G. Francescato, & T. Gärling (Eds.), *Residential environments. Choice, satisfaction, and behavior* (pp. 35–54). Westport: Greenwood Press.

Ebbesen, E. B., Kjos, G. L., & Konecni, V. J. (1976). Spatial ecology: Its effects on the choice of friends and enemies. *Journal of Experimental Social Psychology, 12*, 505–518.

Festinger, L., Schachter, S., & Back, K. (1950). *Social pressures in informal groups; a study of human factors in housing*. Oxford: Harper.

Fleury-Bahi, G., Félonneau, M.-L., & Marchand, D. (2008). Processes of place identification and residential satisfaction. *Environment and Behavior, 40*, 669–682. doi:10.1177/0013916507307461.

Fleury-Bahi, G., Marcouyeux, A., Préau, M., & Annabi-Attia, T. (2013). Development and validation of an environmental quality of life scale: Study of a French sample. *Social Indicators Research, 113*, 903–913. doi:10.1007/s11205-012-0119-4.

Francescato, G. (2002). Residential satisfaction research: The case for and against. In J. I. Aragonés, G. Francescato, & T. Gärling (Eds.), *Residential environments. Choice, satisfaction, and behavior* (pp. 15–34). Westport: Greenwood Press.

Francescato, G., Weidemann, S., & Anderson, J. R. (1989). Evaluating the built environment from the users' point of view: An attitudinal model of residential satisfaction. In W. F. E. Preiser (Ed.), *Building evaluation* (pp. 181–198). New York: Plenum Press.

Fried, M. (1963). Grieving for a lost home. In L. Duhl (Ed.), *The urban condition* (pp. 141–151). New York: Basic Books.

Fried, M. (1986). The neighborhood in metropolitan life: Its psychological significance. In R. B. Taylor (Ed.), *Urban neighborhoods. Research and policy* (pp. 331–363). New York: Praeger.

Fried, M. (2000). Continuities and discontinuities of place. *Journal of Environmental Psychology, 20*, 193–205. doi:10.1006/jevp.1999.0154.

Fried, M., & Gleicher, P. (1961). Some sources of residential satisfaction in an urban slum. *Journal of American Institute of Planners, 27*, 305–315.

Galster, G. C., & Hesser, G. W. (1981). Residential satisfaction. Compositional and contextual correlates. *Environment and Behavior, 13*, 735–758.

Gans, H. J. (1959). The human implications of current redevelopment and relocation planning. *Journal of American Institute of Planners, 25*, 15–25.

Gifford, R. (1987). *Environmental psychology. Principles and practice*. Newton: Allyn and Bacon.

Gifford, R. (2014). Environmental psychology matters. *Annual Review of Psychology, 65*, 541–579. doi:10.1146/annurev-psych-010213-115048.

Giuliani, M. V. (2003). Theory of attachment and place attachment. In M. Bonnes, T. Lee, & M. Bonaiuto (Eds.), *Psychological theories for environmental issues* (pp. 137–170). Aldershot: Ashgate.

Guiliani, M. V., & Scopelliti, M. (2009). Empirical research in environmental psychology: Past, present, and future. *Journal of Environmental Psychology, 29*, 375–386. doi:10.1016/j.jenvp.2008.11.008.

Hassine, K., Marcouyeux, A., Annabi-Attia, T., & Fleury-Bahi, G. (2014). Measuring quality of life in the neighborhood: The cases of air-polluted cities in Tunisia. *Social Indicators Research, 119*, 1603–1612. doi:10.1007/s11205-013-0565-7.

Hidalgo, M. C., & Hernández, B. (2001). Place attachment conceptual and empirical questions. *Journal of Environmental Psychology, 21*, 273–281. doi:10.1006/jevp.2001.0221.

Hollingshead, A., & Rogler, L. (1963). Attitudes toward slums and public housing in Puerto Rico. In L. Duhl (Ed.), *The urban condition* (pp. 229–245). New York: Basic Books.

Hourihan, K. (1984). Context-dependent models of residential satisfaction. An analysis of housing groups in Cork, Ireland. *Environment and Behavior, 16*, 369–393.

Hur, M., & Morrow-Jones, H. (2008). Factors that influence residents' satisfaction with neighborhoods. *Environment and Behavior, 40*, 619–635. doi:10.1177/0013916507307483.

Hur, M., & Nasar, J. L. (2014). Physical upkeep, perceived upkeep, fear of crime and neighborhood satisfaction. *Journal of Environmental Psychology, 38*, 186–194. doi:10.1016/j.jenvp.2014.02.001.

Hur, M., Nasar, J. L., & Chun, B. (2010). Neighborhood satisfaction, physical and perceived naturalness and openness. *Journal of Environmental Psychology, 30*, 52–59. doi:10.1016/j.jenvp.2009.05.005.

Kasarda, J. D., & Janowitz, M. (1974). Community attachment in mass society. *American Sociological Review, 39*, 328–339.

Lee, T. (1968). Urban neighbourhood as a sociospatial schema. *Human Relations, 21*, 241–268.

Marans, R. W., & Rodgers, W. (1975). Toward an understanding of community satisfaction. In A. H. Hawley & V. P. Rock (Eds.), *Metropolitan America in contemporary perspective* (pp. 299–352). New York: Wiley.

Marans, R. W., & Stimson, J. S. (Eds.). (2011). *Investigating quality of urban life: Theory, methods, and empirical research*. Dordrecht: Springer.

McMillan, D. W., & Chavis, D. M. (1986). Sense of community: A definition and theory. *Journal of Community Psychology, 14*, 6–23.

Miller, F. D., Tsemberis Malia, G. P., & Greca, D. (1980). Neighborhood satisfaction among urban dwellers. *Journal of Social Issues, 36*, 101–107.

Newman, O. (1972). *Defensible space*. New York: Macmillan.

Oktay, D., Rüstemli, A., & Marans, R. W. (2012). Determinants of neighborhood satisfaction among local residents and international students: A case study in Famagusta, N. Cyprus. *Journal of Architectural Planning and Research, 29*, 224–240.

Omata, K. (1995). Territoriality in the house and its relationship to the use of rooms and the psychological well-being of Japanese married women. *Journal of Environmental Psychology, 15*, 147–154. doi:10.1016/0272-4944(95)90022-5.

Onibokun, A. G. (1976). Social system correlates of residential satisfaction. *Environment and Behavior, 8*, 323–344.

Pasca, L., Aragonés, J. I., & Poggio, L. (2015). Dimensions that explain the attachment to housing. *Psyecology, 7*, 113–129. doi: 10.1080/21711976.2016.1138667.

Pérez-López, R., Aragonés, J. I., & Amérigo, M. (2013). Thin slices of competence and warmth via personalized primary spaces. *Psyecology, 4*, 267–286. doi:10.1174/217119713807749878.

Porteus, J. D. (1976). Home: The territorial core. *Geographical Review, 66*, 383–390.

Potter, J., & Cantarero, R. (2006). How does increasing population and diversity affect resident satisfaction? A small community case study. *Environment and Behavior, 38*, 605–625. doi:10.1177/0013916505284797.

Preiser, W. F. E. (1989). Towards a performance-based conceptual framework for systematic POEs. In W. F. E. Preiser (Ed.), *Building evaluation* (pp. 1–7). New York: Plenum Press.

Rainwater, L. (1970). *Behind ghetto walls. Black families in a federal slum*. Chicago: Aldine.

Rapoport, A. (1969). *House form and culture*. Englewood Cliffs: Prentice Hall.

Rapoport, A. (1977). *Human aspects of urban form. Towards a man-environment approach to urban form and design*. New York: Pergamon Press.

Rent, G. S., & Rent, C. S. (1978). Low income housing: Factors related to residential satisfaction. *Environment and Behavior, 10*, 459–488.

Rioux, L., & Velasco, L. (2004). Satisfacción de las mujeres con respecto al habitat o vivienda: Un análisis de la teoría de las facetas. *Revista Latinoamericana de Psicología, 36*, 85–95.

Robin, M., Matheau-Police, A., & Couty, C. (2009). Perception de la gêne environnementale en fonction de différents types d'unités urbaines françaises. *Revue Européenne de Psychologie Appliquée, 59*, 101–112.

Rybczynski, W. (1986). *Home. A short history of an idea*. New York: Viking Penguin.

Saegert, S. (1985). The role of housing in the experience of dwelling. In I. Altman & C. M. Werner (Eds.), *Home environments: Human behavior and the environment* (pp. 287–309). New York: Plenum Press.

Sanoff, H. (1970). House form and performance. In J. Archea & C. Erastman (Eds.), *Proceedings of the 2nd annual environmental design research association conference* (pp. 334–338). Stroudsburg: Dowden, Hutchinson & Ross.

Schorr, A. L. (1970). Housing and its effects. In H. M. Proshanky, W. H. Ittelson, & L. G. Rivlin (Eds.), *Environmental psychology. Man and his physical setting* (pp. 319–333). New York: Holt, Rinehart and Winston.

Serfaty-Garzon, P. (2003). *Chez soi. Les territoires de l'intimité*. Paris: Armand Colin.

Shin, J. (2014). Making home in the age of globalization: A comparative analysis of elderly homes in the U.S. and Korea. *Journal of Environmental Psychology, 37*, 80–93. doi:10.1016/j.jenvp.2013.12.001.

Sirgy, M. J., & Cornwell, T. (2002). How neighborhood features affect quality of life. *Social Indicators Research, 59*, 79–114.

Skjaeveland, O., & Gärling, T. (2002). Spatial-physical neighborhood attributes affecting social interactions among neighbors. In J. I. Aragonés, G. Francescato, & T. Gärling (Eds.), *Residential environments. Choice, satisfaction, and behavior* (pp. 182–203). Westport: Greenwood Press.

Speare, A. (1974). Residential satisfaction as an intervening variable in residential mobility. *Demography, 11*, 173–188.

Taylor, S. E., Pepalu, L. A., & Sears, D. O. (1994). *Social psychology* (8th ed.). Englewood Cliff: Prentice Hall.

Tognoli, J. (1987). Residential environments. In D. Stokols & I. Altman (Eds.), *Handbook of environmental psychology* (Vol. 1, pp. 655–690). New York: Wiley.

Unger, D. G., & Wandersman, A. (1985). The importance of neighbors: The social, cognitive, and affective components of neighboring. *American Journal of Community Psychology, 13*, 139–165.

Vemuri, A. W., Morgan Grove, J., Wilson, M. A., & Burch, W. R. (2011). A tale of two scales: Evaluating the relationship among life satisfaction, social capital, income, and the natural environment at individual and neighborhood levels in Metropolitan Baltimore. *Environment and Behavior, 43*, 3–25. doi:10.1177/0013916509338551.

Warren, D. I. (1978). Explorations in neighborhood differentiation. *The Sociological Quarterly, 19*, 310–331.

Weenig, M. W. H., Schmidt, T., & Midden, C. J. H. (1990). Social dimensions of neighborhoods and the effectiveness of information programs. *Environment and Behavior, 22*, 27–54.

Weidemann, S., Anderson, J. R., Butterfield, D. J., & O'Donell, P. M. (1982). Residents perception of satisfaction and safety. Basis for change in multifamily housing. *Environment and Behavior, 14*, 695–724.

Wu, C. (2008). The role of perceived discrepancy in satisfaction evaluation. *Social Indicators Research, 88*, 423–436. doi:10.1007/s11205-007-9200-9.

Yancey, W. L. (1971). Architecture, interaction and social control: The case of a large-scale public housing project. *Environment and Behavior, 3*, 3–21.

Zorbaugh, H. (1929). *The gold coast and the slum. A sociological study of Chicago's near north side*. Chicago: University of Chicago Press.

18. Spatial Inequalities, Geographically-Based Discrimination and Environmental Quality of Life

Ghozlane Fleury-Bahi* and André Ndobo*

18.1 Introduction

Many European and North American cities are characterized by their social and spatial disparities and by the existence of residential segregation. Within the same city, we never fail to encounter sharp differences in environmental quality of life, with some neighborhoods experiencing multiple problems, such as scarce shops and services, ill-adapted or non-existent urban transit networks, a degraded and polluted environment, precarious socio-economic circumstances, and security problems, which all build the neighborhood's poor reputation and contribute to its stigmatization. In France, the suburban areas or "*banlieues*" are hampered by all of these handicaps. In concrete terms, exclusion and social inequalities are concentrated in these areas, which fuel fears of insecurity. Moreover, the state of the "*banlieues*" is a major social and political issue since it highlights the spatial inscription of social inequalities in French society (Avenel 2009) – to which residential segregation should be added.

Massey and Denton (1988) identified five dimensions of residential segregation: evenness, exposure, concentration, clustering or spatial aggregation, and centralization. Evenness refers to the distribution of either one or several groups of population in a given space. Evenness indices measure either the over-representation or the under-representation of a group within the areal units: a group is considered segregated if it is unevenly distributed. Exposure is the degree of potential contact or of possible interaction between members of the same group or between members of two groups within a given space or areal unit. Concentration refers to the amount of space occupied by a group. Clustering implies that the more a group occupies adjoining geographical units, the more it is segregated. Conversely, a low level of clustering refers to distant geographical units. Lastly, centralization is "*the degree to which a group is spatially located near the center of an urban area*" (p. 291). This dimension is very relevant for many North American cities, in which pauperized ethnic minorities often live in city centers. However, it is less relevant for European cities, as well as several other North American cities, whose central districts have undergone gentrification over the last few decades.

Residential segregation is characterized by the differential distribution of groups within areal units and by the possibilities of various interactions between individuals who belong to similar or different groups. It can thus produce discrimination. In France, the importance of

---
*Authors contributed equally to this chapter.

G. Fleury-Bahi (✉) • A. Ndobo
University of Nantes, Nantes, France
e-mail: ghozlane.fleury@univ-nantes.fr;
Andre.ndobo@univ-nantes.fr

this type of discrimination is highlighted by the interest shown by political stakeholders, institutions, and researchers. Political stakeholders pay attention to discrimination for obvious reasons: they are responsible for public welfare and are mandated to make sure that their citizens receive fair and equal treatment, regardless of their sociodemographic specificity (gender, origin, age, place of residence, etc.). However, as well as its institutional dimension, discrimination is an interesting issue for non-specialists. Indeed, the promise of equality and justice, which underlies the mission of most democratic societies, raises expectations that should be met. Recent legislation, which was initiated by the French Senate, was aimed at punishing any attempt to discriminate or stigmatize anyone based on their place of residence. Discrimination is also a major scientific question that calls for a thorough study and analysis of its mechanisms.

As a political and scientific issue, residential discrimination is based on the social and spatial differences among urban neighborhoods or between urban and rural areas. If such a gap was caused by unequal public investment and state presence, its consequences are of a psychological nature. In fact, individuals can be assumed to feel included or excluded according to the condition of their places of residence. The nature and quality of such feelings can thus produce well-being or ill-being. Whenever a residential space generates some discomfort, the self-regulation and coping capacities of its inhabitants need to be addressed.

In the present chapter, we consider residential discrimination as an emerging question and take into account its institutional factors and psychological impacts. More precisely, our aim is to focus on the inhabitants of discriminated spaces in order to examine how they perceive this discrimination, and how they identify with and appropriate their place of residence. Our goal is also to study the psychological effects of this perceived discrimination, particularly in terms of self-esteem and well-being. Thus, we first show how identification with one's place of residence is a fundamental process in the individual's relationship with such a place. Relying on the psychosocial models of place identification, we then show how inhabitants experience social and spatial discrimination and how spatial inequalities can influence the levels of residential satisfaction and quality of life. Lastly, we describe some of the adaptive strategies that discriminated individuals resort to in order to maintain a certain level of self-esteem and, ultimately, of well-being.

18.2 Place and Identity

18.2.1 The Concept of Place-Identity

However differently, every dwelling place participates in the construction of both social identity and individual identity, by conveying particular representations and enabling various social interactions. In other words, individual identity is built on the individual's various experiences of places and residences (i.e., on the social interactions that occur within these places and on the individual's affective links with them). In order to examine the influence of the place of residence on the development and evolution of an individual's identity, various authors consider that the development of identity does not only rely on the comparison between the self and other individuals, but extends to comparisons with objects and locations (Proshansky 1978; Proshansky et al. 1983).

This is why place-identity is a relevant notion. Proshansky et al. (1983) defined it as a substructure of individual identity, which includes the cognitions related to the physical settings of the individual's dwelling place. These cognitions are memories, ideas, feelings, attitudes, values and preferences for varied and complex environments, which define the daily life of each human being (Proshansky 1978). For Proshansky et al. (1983), the individual dimensions of place identity are foregrounded, thereby leaving aside the social processes at stake in the individual-environment interaction. Yet, Proshansky et al. also suggested that the cognitions that constitute place identity do not only pertain to the physical environment but also involve the social insertion of an individual, more particularly through the social norms that apply to one's use and experience of a given space. All in all, Proshansky et al. (1983) considered place identity a distinct component of identity, along with social or individual identity. Conversely, through their references to identification processes, the psychosocial models of place identity consider places an indifferent identity referent. On this basis, each place can be considered a social category and, accordingly, place identification can be subject to the same operational principles as those of identification with a social group.

18.2.2 Social and Spatial Identification

Since the early 1990s, researchers (e.g., Twigger-Ross and Uzzell 1996; Uzzell et al. 2002) have evidenced the processes at stake in spatial identification within the paradigms of social psychology and, more particularly, with references to the notion of social categorization and the social identity theory (or SIT) (Tajfel 1978).

The SIT defines the self as a process that is influenced by the individual's membership of one or several specific groups. Tajfel holds that the social identity of an individual is *"his knowledge that he belongs to certain social groups together with some emotional and value significance to him of his membership"* (Tajfel 1981, p. 255). Thus, the identification processes do not solely rely on group membership but also involve the perception of several contrasts between the in-group and the out-group. In fact, the SIT also relies on the processes of categorization and social comparison. Cognitive research on categorization has shown that categories are determined by the perceived similarities between the members of a group and by the perceived differences between these members – hence the integration of the standards of comparison within the study of the processes of social identification. Added to this, the SIT also posits that identity dynamics directly involve individual and social dimensions. Individuals constantly oscillate between the two poles of an identity continuum: at one end of the continuum, individuals consider themselves members of a group (as opposed to members of another group); at the other end, they see themselves as singular and specific beings. Following on from the SIT, which focuses on the processes by which an individual defines him/herself in terms of social categories, the self-categorization theory (Turner et al. 1987) put greater emphasis on the cognitive dimension than the theories it drew on. In the social field, several types of self-categorization regulate perceptions and behaviors and make the self-concept dependent on context. Social identity thus depends on the salience of a given social category at a given time.

Both the SIT and the self-categorization theory have been used to study the comparisons conducted by groups that are spatially distinct. According to Twigger-Ross and Uzzell (1996), whenever we speak about place identification, the place is considered a social category that can be subject to identification processes similar to those that can be at stake with any other social group. The structuration of social identity thus involves several social identification processes that are more or less salient according to the context (Hogg and Abrams 1988). In fact, in the same way that social identification corresponds to the actualization of identity referents belonging to social categories (nationality, gender, job, etc.), place identification implies that the individual belongs to a group that defines itself by its spatial or geographic inscription. Thus, similarly to the way that the social categorization process leads to the definition of identity in terms of social category membership (e.g., gender, ethnic group, etc.), social and spatial categorization enables identification with a social and spatial category (e.g., urban identity, regional identity, etc.). Within this perspective, several authors have referred to the individuals' urban social identity (Valera and Guardia 2002; Valera and Pol 1994), or to their place-related social identity (Uzzell 2000).

In the environmental framework, the SIT also posits that whenever the image of a group that is defined by its spatial inscription is negative, its status is comparatively weak, which affects the group members' positive self-esteem. Several studies have shown that individuals tend to identify with places that help them foster a positive self-esteem. In 1989, Korpela even evidenced how specific locations could actually help maintain a high level of self-esteem in adolescents. Thus, when Bonaiuto et al. (1996) investigated the existence of a possible link between the degree of local and national identification in English students and the perception of pollution on English beaches, they showed that the more the participants identified with their city or country, the more likely they were to give a positive evaluation of the national beaches. According to them, these results confirm that an individual's environment and identification with that environment play a role in identity construction, along with the other dimensions of social identity. The strategies of denial of pollution that their participants resorted to were aimed at preserving the positive identity of their spatial in-group. The participants' positive perception bias echoes the strategies of self-protection that individuals resort to in the case of identity threat.

Similar results have been evidenced in middle-school students from either stigmatized or privileged neighborhoods (Fleury-Bahi and Marcouyeux-Deledalle 2010). The aim of the research was to test a model that associated the evaluation of the social image of a given location, place identification and self-esteem while positing the mediating status of the level of identification. The mediating role of place identification was confirmed since the effect of the location's social image on the degree of identification was observed, with identification maintaining and developing a positive self-esteem in participants. In other words, the perception of the socially-valued image of the students' middle-school contributed to their identification with their school, which, in turn, helped the students maintain a positive self-esteem. Conversely, a poor image is likely to hamper both the identification process and the attempt to maintain positive self-esteem. These results empirically support the use of the SIT in analyzing the relationships between place identification and self-esteem. The more a location is likely to enhance social and spatial identification (through the social image it conveys), the more an individual will easily maintain a positive self-esteem.

Maintaining a positive self-esteem is considered a fundamental identity motivation by various theorists of the Self (Abrams 1992; Hogg and Abrams 1988; Tajfel 1978). Within Tajfel's SIT, group evaluation and the emotional sphere play a central role. Accordingly, three components of social identity have emerged: a cognitive component (or self-categorization component, i.e., being aware of one's group membership), an evaluative component that corresponds to the self's esteem of the group (the positive or negative evaluation associated with one's group), and an emotional component (or involvement in one's group, i.e., how the individual feels about his/her group membership) (Ellemers et al. 1999).

For Tajfel, self-esteem is the main motivational principle of the SIT. Yet, according to the distinction between personal and social identity, the distinction between the individual's personal self-esteem and his/her social self-esteem should be highlighted. The SIT postulates that group members resort to intergroup differentiation strategies in order to obtain, maintain or strengthen the positive image of their social identity. In other words, group members are likely to reinforce their social self-esteem – a notion that thus also relies on the redefinition of one's self-esteem within a group, and not solely on an individual level.

18.2.3 Place as a Social Category

A study of the fluctuation of social self-esteem shows that it is mostly influenced by the individual's group status (Hogg and Abrams 1988; Ellemers et al. 1999). We should remind our readers that self-esteem is also one of the motivational principles described in Breakwell's Identity Process Theory. The explanation of intergroup conflict in terms of identity is functional in the perspective of social and environmental psychology. The notion of the place of residence refers more or less implicitly to the ways that individuals collectively appropriate a place, transfer values and significations and create the conditions of a collective and shared identity. Ecological inhabited areas should thus be thought of as a source of identity questioning for their residents. They can be considered social and spatial categories, with which one can associate entitative properties (Campbell 1958; for a research review of entitativity, see Lickel et al. 2000), which are the proximity, similitude, organization and interdependence of group members (i.e., goal similarity, group desirability and common fate).

The mechanisms that regulate the relationships between social and spatial groups are thus similar to those regulating the relationships between other social groups, which have been examined within the framework of the social psychology of intergroup relationships. Dwelling places can be assimilated to social categories, which can be distinguished by various criteria (e.g. ethnic, socio-economic and urban). These social categories are relevant because of (1) the common attachment that binds each group member to their dwelling place, (2) the identities and symbolic meanings that they associate with their dwelling place and (3) the group members' desire to evolve in an environment populated by individuals who are alike and share common values (Emerson et al. 2001). Such concepts can help us understand the problematic dimension of mixed residency initiatives, which have not questioned the symbolic value that inhabitants attach to their dwelling place, as well as their implicit fears of the modification of their ecosystem and the subsequent devalorization of facilities and services. In fact, since these initiatives are perceived as a threat to the locals' lifestyle, way of thinking and identity, they should entail a reinforcement of the locals' attachment to space and social and spatial discrimination.

In North America, for example, Krysan et al. (2009) compared African Americans and white Americans and demonstrated that the former preferred a mixed neighborhood while the latter preferred a homogeneous neighborhood. This was particularly true when these white Americans subscribed to racial and ethnic stereotypes. Other studies have shown that, for North American neighborhoods, the perception of insecurity in the place of residence may be influenced by the presence of minorities in the neighborhood and the development of stereotypes toward these minorities (Matei et al. 2001; Quillian and Pager 2001). In the same vein, in Spain, Di Masso et al. (2014) recently showed that discrimination opinions, and more precisely xeno-racist opinions, are linked with the perception of urban insecurity and crime-related experiences. Other studies demonstrated that perceived discrimination may be explained by cultural differences between the majority and minority groups in the neighborhood, the visibility of the minority group, and the duration of residence in the neighborhood (Jasinskaja-Lahti et al. 2006). Bourg and Castel (2011) also showed that, for a French urban population, the evaluation of one's neighborhood is more positive when the perceived proportion of French people increases; conversely, the evaluation becomes more negative when the perceived proportion of people of foreign origin increases. Residents tend to match a neighborhood's good reputation with the presence of French people, just as they associate a bad reputation with the presence of at least one category of people of foreign origin.

18.2.4 Place Identification and Discrimination

All the studies previously mentioned have shown that place identification and discrimination seem to go hand in hand. In fact, the more one is attached to a place, the more one identifies with it and tends to differentiate from individuals who do not live there. The SIT (Tajfel et al. 1971) enables us to understand better the proposed connection between identification with a social category and the tendency to discriminate against the out-group. In order to examine this link, Tajfel rejects both the utilitarian (Sherif 1966) and the psychopathological explanations (Adorno et al. 1950) of intergroup conflicts and their consequences. For Tajfel, conflicts and intergroup discrimination are the products of normal psychological functioning. One can indeed note that they also occur outside any situation of competition.

In concrete terms, Tajfel investigates both the cognitive and the motivational dimensions of discrimination. On the cognitive dimension, Tajfel (1969) assumes that the ways in which the in-group and out-group are perceived are determined by the system of information processing that people develop in social groups. He also refers to the notion of the cognitive categorization of objects as a determinant of object perception. In a seminal experiment, Tajfel and Wilkes (1963) tested and validated the hypothesis according to which the categorization of objects as distinct classes tends to yield an emphasis effect, i.e., the differences and contrasts between the classes of objects are emphasized, and so are the similarities between the objects that belong to a similar class. The paradigmatic value of the hypothesis linking categorization and perceptions was later confirmed for social judgments (McGarty and Penny 1988). In their experiment, McGarty and Penny (1988) showed that the emphasis placed on interclass differences and intraclass similarities occurred significantly. Yet, they also showed that the tendency to process information in that way was more common in participants who belonged to or identified with privileged social groups. This result enables us to consider the existence of a link between the phenomenon of cognitive emphasis and the SIT (Tajfel and Turner 1979) and, more particularly, to investigate the motivations of such an attitude.

On the motivational dimension, the experiment of the minimal group paradigm (Tajfel et al. 1971) cemented the relationship between discrimination and social identity. During the experiment, it appeared that whenever participants faced a category division (even when it was an artificial division), they tended to resort to strategies of maximal differentiation and maximal in-group profit at the expense of more cooperative strategies. This experiment clearly illustrates the role of social categorization in intergroup discrimination. Yet, the cognitive explanation of this phenomenon is insufficient as it does not highlight the reasons why people ignore the fairness strategies when allocating a small amount of money to other boys in the in-group and the out-group, and display in-group favoritism. Tajfel (1978, 1981) explains these attitudes in terms of social identity, according to which the discrimination biases are the consequence of the individuals' desire to give a positive value to their in-group. Through this bias, individuals actively maintain and improve their personal social identity. More generally, they lean towards a positive social identity and are motivated by the prospect of belonging to socially privileged groups. According to the theory of identity, discrimination is thus the consequence of the motivation of individuals to define themselves positively and to distinguish their social group from other groups in a given social context. However, this definition of the self is only possible when the bias introduced by membership of several social categories comes into play. Besides, one should note that developing a positive definition of the self from one's social categories is a universal tendency. We can thus consider that the biases of intergroup discrimination and favoritism towards the in-group bring together cognitive and motivational processes. Intergroup discrimination functions whatever the nature of the entities of social groups at stake.

18.2.5 What Is Discrimination?

Since one's dwelling place is subject to comparisons and identification, the question of residential discrimination needs to be addressed. Discrimination is a popular notion that is often used to refer to attitudes that pertain to either prejudices (on an affective level) or stereotypes (on a cognitive level), as well as to any form of injustice experienced by people or groups of people because of their social membership. Clearly, these are interrelated notions that rely on different aspects of the denigration that individuals and groups may undergo; however, if we wish to characterize the specificities of discriminatory practices per se, these notions should be distinguished. Generally speaking, discrimination refers to "*all types of distinction, exclusion or preference that is based on race, ethnicity, gender, religion, political opinion, national origin or social background. Its effects include destroying or altering the possibility of equal opportunity and equal treatment*" (Nys and Beauchesne 1992, p. 18). This definition follows up the 1949 Declaration of the UN, according to which "*discrimination includes any conduct based on a distinction made on grounds of natural or social categories, which have no relation either to individual capacities or merits, or to the concrete behavior of the individual person*". In these conditions, and independently of the intrinsic competences or qualities of an individual or a group, what matters and what justifies discrimination is the fact that the individual or the group that is being judged is not the source of the said judgment. In other words, unjust treatment is often caused by an individual's social group or by the possession of an arbitrary or stigmatizing trait (Dion 2003).

A triple-component perspective, taking into account the affective, cognitive and behavioral components of attitudes, would hold that discrimination is the concrete acting-out of the hostile feelings (prejudices) and negative beliefs (stereotypes) of individuals or groups towards other individuals or groups. Such a perspective involves a redefinition of the notion of discrimination as the negative behavior towards members of an outgroup shown by the members of an in-group, who are prejudiced against the said out-group (Dovidio and Gaertner 1986). Yet, such a conceptual connection certainly raises quite a few questions. For example, the link between prejudices and discrimination can be questioned, especially since individual actions are determined both by personal convictions and by extra-personal factors that the individual cannot control. Moreover, individuals without any prejudice towards a given social group can still discriminate against that group in a favorable environment (e.g., existence of discriminatory habits or laws) (Bourhis et al. 1999). However, we should also note that individuals do not necessarily transform their prejudices into discriminatory behavior (Allport 1954; Lapierre 1934).

Although all unequal treatments can be considered discriminatory practices, Allport (1954) warns against such generalizations and suggests that we distinguish between discrimination per se and examples of behavior that could arbitrarily be considered discriminatory. Allport (1954) also draws on the Declaration of the Commission of Human Rights (1949) according to which discrimination occurs whenever individuals or groups, because of their initiatives or behavior, are not given the equal rights to which they aspire. Accordingly, what distinguishes discriminatory practices from other types of behavior is the nature of the individual's motivation, i.e., the attempt to exclude or give unequal treatment to a target because of his/her membership of a particular social category. Similarly, other authors highlight the differences between legitimate and illegitimate discrimination (Crandall et al. 2002). The psychosocial literature evidences many examples of discrimination that are considered legitimate even by the targeted individuals (Crandall et al. 2002; Crocker and Major 1994; Oakes et al. 1994), e.g., giving unfair treatment to someone whom we do not like, refusing to grant a criminal the legal rights that a "normal" citizen would be allowed, or even preventing young people below a certain age from obtaining a driver's license are often considered examples of legitimate discrimination. Similarly, illegitimate discrimination is often defined as unfair treatment based on an individual's group membership, which violates non-discriminatory legislation. Residential discrimination, like ethnic and racial discrimination, is thus an illegitimate form of discrimination.

Unequal treatment is a universal criterion in the definition of discrimination. Yet, the production of discrimination as well as the way it is described varies according to the social issues at stake. For example, the production of discrimination can be favored by the resentment and identity crisis that members of a group can experience towards another group whom they consider intruders (Sanchez-Mazas and Licata 2005). In this particular case, treatment inequality and the rejection of other individuals can also involve a form of essentialization (Chulvi and Perez 2003). Threatening situations can also enhance discrimination towards members of the threatening out-group (Devos 2005; Stephan and Stephan 1996; Bobo and Hutchings 1996). Lastly, the identity motivations that stem from the positive image of one's in-group, compared to the competing out-groups (Branscombe et al. 1999a, b, 2002; Branscombe and Wann 1994; Tajfel et al. 1971; Tajfel and Turner 1979) can also be seen as sources of discriminatory behavior. Accordingly, discrimination gives some legitimacy to social categorization. As has been observed in the cases of racism and sexism, such categorization tends to add an intergroup hierarchization to the already existing intergroup differentiation (Wieviorka 1998).

Discrimination can have individual, group, and institutional origins and can take up different forms. Several types of discrimination are explicit and rely on a high level of belief or feeling endorsement. Other types are more subtle (e.g., being followed around or closely monitored by a shop assistant), especially when they take place in non-discriminatory legal contexts. Nevertheless, whatever its form, discrimination always has an impact on people. It is a systemic phenomenon that is rooted in the foundations of societies and the values that they convey.

18.3 Spatial Inequalities and Residential Satisfaction

18.3.1 Perceived Quality of the Neighborhood and Residential Satisfaction

The question of a link between place identification and the level of residential satisfaction arises when we investigate social and spatial inequalities, which develop on the differences in quality of neighborhoods (which in turn relies on either physical aspects, e.g., general maintenance, street lighting, etc. or social aspects, e.g., insecurity, reputation, etc.). In social sciences, residential satisfaction has been shown to be a good indicator of an individual's quality of life (Amerigo and Aragones 1997; Marans and Rodgers 1975) and of perceived well-being (Kahana et al. 2003).

The perceived quality of the residential environment is based on several factors that belong to its physical, spatial, social and functional characteristics. Three components of residential satisfaction or environmental quality of life have recurrently been highlighted, regardless of the country and type of population under scrutiny: satisfaction with the social environment, the physical environment and the amenities (references). For example, Fleury-Bahi et al. (2013) studied the influence of the place of residence on the quality of life in three French cities. Four components of environmental satisfaction were isolated: image of the neighborhood, shops and services, traffic and state of the public road network, and annoyance and pollution. In the same vein, Hassine et al. (2014) identified the indicators of neighborhood quality of life in a North African country. Exploratory and confirmatory factor analyses individualized three environmental quality of life dimensions in Tunisia: image of the neighborhood, shops and services, annoyances and pollution.

How do we go from objective environmental attributes to a level of satisfaction that is specific to each individual? Most of the models of environmental evaluation distinguish three successive phases in this process. The individual first perceives several objective environmental attributes, which are then evaluated (e.g., this space is clean/dirty; there is a high population density etc.) with psychological, sociological, and cultural factors. Thirdly, these attributes, whose valence is either positive or negative, produce a varying level of satisfaction (Amerigo and Aragones 1997; Marans 1976; Nasar 1998). Both the attribution of a valence to these different environmental attributes and the subsequent feeling of satisfaction or dissatisfaction directly depend on several factors, such as the sociodemographic and personality variables, the individual's residential experiences, expectations and needs, etc. However, psychosocial variables (such as the level of place identification) should also be taken into account (Lalli 1992; Bonaiuto et al. 1996; Fleury-Bahi et al. 2008).

18.3.2 Place Identification and Residential Satisfaction

The question of the link between place identification and the level of place satisfaction comes up whenever a transactional perspective (Amerigo and Aragones 1997; Altman and Rogoff 1987) is adopted in the study of the relationship between individuals and their environment. The transactional approach pays particular attention to the context and assumes change is a central notion. Such an approach seems particularly productive in the study of the relationship to the environment, as this relationship cannot be separated from the individual's spatial and socio-cultural anchoring.

The nature of the hypothesized link between the level of environmental satisfaction and place identification varies from author to author. Several works have maintained that satisfaction is a predisposition to identification (Mesch and Manor 1998); others have shown that satisfaction and identification are concomitant (Pol 2002; Uzzell et al. 2002) while a third group have argued that identification is a predictor of satisfaction (Bonaiuto et al. 1996; Fleury-Bahi et al. 2008). In their CIS model, Pol (2002) and Uzzell et al. (2002) considered residential satisfaction and place identification as two simultaneous processes that develop with social cohesion and that participate in the structuration of the place-related social identity.

Along with Lalli (1992) and Bonaiuto et al. (1996) hypothesized that place identification invariably produces a positive evaluation of the said place. In their research mentioned above, Bonaiuto et al. (1996) showed that young people with a strong local identity tend to minimize the impact of pollution on their city while young people with a strong national identity minimize the impact of pollution on their country, and even more so when the evaluation is asked by foreign groups. Lalli (1992) also showed that the more the inhabitants identify with their city, the more they evaluate it positively, even when the age and duration of residence are controlled. Félonneau (2004) argues that students with a high level of identification with their city tend to underestimate the frequency of incivility they experience while students with a low level of identification tend to overestimate these same difficulties.

In the same vein, Fleury-Bahi et al. (2008) tested a causal model that associates residential satisfaction, place identification and duration of residence on 278 participants from Paris, Nantes, and Bordeaux. The originality of this model consists in jointly considering the effect of the duration of residence on place identification and place satisfaction, while studying identification as a mediating variable. The model shows that identification directly depends on the duration of residence and that it works as a predictor of the level of place satisfaction. The results also indicate that the individual's inscription in his/her residential neighborhood does not modulate their satisfaction with the various components of the place of residence. More precisely, the social components of satisfaction (the social image of the place and the local relationships) play the most important role in the process. A rather high level of satisfaction can be analyzed as the outcome of the psychological phenomena of rationalization and the resolution of cognitive dissonance (Festinger 1957). Negatively evaluating a place that one identifies with would indeed produce cognitive inconsistency. Accordingly, studies in sociology have shown that even in degraded neighborhoods, inhabitants have a hard time expressing their dissatisfaction (Avenel 2005; Lepoutre 2000). If an individual declares that he/she is not satisfied with his/her neighborhood and its inhabitants while feeling identified with them, this means that he/she is endorsing a negative self-image. In other words, the individuals' awareness of belonging to a residential in-group in which they have invested affectively reinforces their positive evaluation of the social status of the neighborhood and their relational potential. These various works raise questions about the mechanisms of identification with and differentiation from neighboring communities, especially since these mechanisms are likely to modulate social and spatial identification, residential satisfaction, and, eventually, the general level of well-being.

18.4 Geographically-Based Discrimination and Psychological Well-Being

The questions of social and spatial segregation and place identification can be included in the issues raised by social psychology about intergroup relationships, social discrimination and the social integration of minority groups. In fact, the classic studies (Sherif and Sherif 1969; Avigdor 1953; Ferguson and Kelley 1964; Rabbie and Horwitz 1969) that show how an intergroup objective competition can be a source of intergroup bias also explain the mechanisms of in-group benevolence and intergroup differentiation. Referring to the minimal group paradigm, Tajfel investigated intergroup biases in terms of social identity (Tajfel 1978, 1981; Tajfel and Turner 1979) and argued that out-group discrimination would be caused by the individuals' inclination to define themselves positively and to differentiate their group positively from various relevant outgroups in a given social context. Thus, what is at stake is not so much the objective benefits that individuals can gain from the situation but rather the way that they see themselves in this situation. Moreover, as mentioned earlier, the inclination to define oneself positively and to define oneself according to one's social category is universal. The social identity theory developed by Tajfel enables us to understand the persistence of biases in intergroup evaluation, i.e., anti-out-group attitudes are supposed to protect the social identity of the individual's in-group against the threat represented by the out-group; according to Stephan and Stephan's typology (1996), the threat can be realist, symbolic, or affective. In each case, a positive self-definition performs both a defensive (intergroup differentiation) and an offensive (out-group promotion) function. Nevertheless, regardless of the type of threat, many works (Stephan and Stephan 1996, 2000; Stephan et al. 1998; Bizman and Yinon 2001) have evidenced the relationship between a perceived threat (of any type), the level of identification, and the level of intergroup discrimination.

Although discrimination is now a major topic in social science, most research has focused on the attitudes and motivations pertaining to it. However, such a narrow field of study cannot but raise a good many questions since it underestimates the experiences, reactions and retroactions of the victims of discrimination. Neither does it offer a full investigation of the variety of mechanisms at stake in a discriminatory process. Conversely, a dynamic and synoptic approach relies on the study of both the source and the target of discrimination, their attitudes and their consequences.

Addressing discrimination from the victims' point of view involves a study of (1) the emotional consequences of stigmatization (stress, depression, anxiety, aggressive attitudes and relational conflicts), (2) the individual differences in the experience of stigmatization, and (3) the strategies that victims develop in order to cope with stigmatization and to adapt to the social context. The theory of perceived discrimination, which refers to the intimate and subjective experience of stigmatization, enables us to investigate these questions. The subjective experience depends on the frequency of objective and subtle discrimination observed or faced by the victims (Clark et al. 1999; Liang and Borders 2012).

Discrimination produces a number of negative health outcomes, such as stress (McCoy and Major 2003) and anxiety, in members of discriminated groups. Research has shown that victims of discrimination are comparatively more depressed (William and Mohammed 2009), more aggressive (Smokowski and Bacallao 2006) and more prone to enter into interpersonal conflicts (Lee 2003) than non-discriminated individuals. Several authors refer to the social identity theory to confirm the link between the experience of discrimination and the emergence of negative psychological consequences in victims. For instance, Schmitt and Branscombe (2002) observed that the experience of discrimination affects the emotional well-being of victims and lowers their self-esteem. They identified several causes of the deterioration in the victims' self-esteem: (1) discriminated individuals have to suppose that a major part of their social identity is devalued by the stigmatizing group; (2) all threats aimed at the social identity of the victim's in-group are experienced as threats against the self; (3) members of stigmatized groups tend to perceive discrimination as a global, stable and widespread phenomenon, which exacerbates the negative implications of experienced and/or perceived discrimination for self-esteem. In the same vein, for geographically-based discrimination, some studies show a negative impact on mental health outcomes. For example in Japan, Tabushi et al. (2012) identified an association between perceived geographically-based discrimination and mental health, such as depressive symptoms and diagnosis of mental illness. Liebkind and Jasinskaja-Lahti (2000) also observed that the experience of discrimination and prejudice contribute to explain the level of well-being and psychological stress of immigrants in Finnish neighborhoods. As for African-American youth, Seaton and Yip (2009) argued that high levels of discrimination in the neighborhood were associated with lower life satisfaction.

Addressing discrimination from the victims' point of view also involves studying their coping strategies. Several works have indeed studied the range of coping strategies that individuals resort to in stressful and depreciatory circumstances. The seminal cognitive theory of stress and coping by Lazarus and Folkman (1984) enabled us to understand coping mechanisms in stressful situations. This theory posits that *"coping is a cognitive and behavioral effort to manage or reduce demands that are appraised as taxing or exceeding individuals' resources. From this theoretical perspective, coping is viewed as having two major functions: the management of the problem that is causing the distress (i.e., problem-focused coping) and the regulation of emotions or distress caused by the stressor (i.e., emotion-focused coping)"* (Outten et al. 2009; p. 149).

The cognitive theory of stress and coping built momentum in discrimination studies. Yet, *"within the stress and coping literature, the appraisal of coping options tends to be exclusively conceptualized as beliefs about what an individual can do as an individual to cope with their particular stress circumstances"* (Outten et al. 2009; p. 148). Other works were carried out within the framework of the social identity theory (Tajfel and Turner 1979) and self-categorization (Turner et al. 1994), *"according to which the cognitive appraisals process can be affected by peoples' group membership"* (Outten et al. 2009; p. 150). Therefore, the appraisal of one's stressful circumstances might also include more collective, or group-based, options for coping (Haslam and Reicher 2006; Outten et al. 2009). In other words, when the members of a stigmatized group feel devalued by a higher status group, psychologically investing more in one's disadvantaged group can buffer the psychological harm caused by feelings of unjust treatment. According to both the social identity theory (Tajfel 1978) and the self-categorization theory, people do not always define themselves as individuals, but rather as members of a collective – in which case they are motivated to achieve a positive intergroup identity based on favorable comparisons with relevant out-groups.

Since most forms of discrimination are considered intergroup phenomena that affect individuals and, as a result, their in-group, it is understandable that coping strategies should be considered at the intergroup level. This analysis is also relevant for groups that are defined by their social and spatial insertion; but what coping strategies do such groups resort to? Brondolo et al. (2009) have identified three strategies that help coping with stress induced by discrimination: identification with the in-group, social support, and confrontation with the source of stress. Several other studies have identified more types of coping strategies, while emphasizing that the victims of discrimination are not passive and that their use of specific means or strategies depends on their evaluation of the stressful situation they are in. Among these strategies, we can mention the discounting hypothesis, which refers to the propensity of members of discriminated or stigmatized groups to protect their well-being and self-esteem while jointly absolving themselves from any responsibility for the negative events that they experience and blaming prejudice as the cause of the events (Crocker and Major 1989; Crocker et al. 1998). We can also mention the self-protection strategy, which relies on the reject-identification theory (Branscombe et al. 1999a, b; Jetten et al. 1996, 2001; Schmitt and Branscombe 2002) and suggests that a high level of psychological identification with one's in-group reduces the negative effects of experienced and perceived discrimination on self-esteem. Yet, such over-investment mostly enables discriminated individuals to feel less alone when facing stigmatization (Bourguignon et al. 2006). A third strategy, the theory of the belief in a just world (Lerner 1980) which works as an antithesis of the belief in an unjust world theory (Lench and Chang 2007), maintains that individuals often see the world as unjust and arbitrary. It also assumes that individuals are victims of injustices and that they do not always deserve what happens to them. What is at stake is thus a rationalization of the negative experiences that were either witnessed or experienced. In a way, this is a cost-avoidance strategy, which enables the individual to face the situation.

We have mentioned social support as a possible coping strategy that is developed within an in-group defined by its social and spatial insertion. However, we should remind our readers that discrimination, particularly geographically-based discrimination, is not inevitable. For example, the contact hypothesis (Amir 1969; Brown 1995; Duckitt 1992) argues that intergroup conflicts tend to reduce when the duration of contact increases, which, in the case of geographically-based discrimination, amounts to the duration of residence. This can be explained by the individual's shared experiences with the out-group, and his/her subsequent growing familiarity with its social and cultural specificities.

18.5 Conclusion

In this chapter, we have tried to emphasize the links between three phenomena: residential identification, the discrimination that such identification produces between individuals who belong to different social and spatial groups, and the psychological consequences of such discrimination. Taking a psychosocial and environmental approach has enabled us to emphasize the emergence of geographically-based discrimination as a modern, social and political issue.

Indeed, as a political question, discrimination challenges the promise of equality and justice, which is at the heart of the democratic pact that binds most democratic societies, in which political institutions are expected to tackle the issue of discrimination. Their tools include corrective and preventive legislation as well as education programs and awareness campaigns. However, discrimination is also of interest for social and environmental psychology. This research topic enables us to understand the mechanisms of residential discrimination, and identify its differences from and shared points with more common criteria of discrimination (age, gender, race, etc.). Research also enables us to measure the impact of residential discrimination on individual well-being.

We have thus shown how the theory of social identity can help us understand better the mechanisms of residential discrimination. Drawing on Tajfel's model and the self-categorization theory, we have argued that residential categorization works in a similar way to other social categorizations, i.e., individuals identify with their place of residence and consider it more positively than they would consider other residential spaces. This identification process can rely on instrumental (the physical quality of the environment, the availability of services and facilities, etc.) or affective (the sociability of the group members) components. This is why individuals tend to stigmatize the members of other residential places. Through its focus on social and spatial discrimination, this analysis thus calls for political and public initiatives aimed at correcting the social inequalities of access to urban amenities in residential areas.

References

Abrams, D. (1992). Processes of social identification. In G. M. Breakwell (Ed.), *Social psychology of identity and the self-concept*. Surrey: Surrey University Press.

Adorno, T. W., Frenkel-Brunswik, E., Levinson, D. J., & Sanford, R. N. (1950). *The authoritarian personality*. New York: Harper & Row.

Allport, G. W. (1954). *The nature of prejudice*. Reading: Addison-Wesley.

Altman, I., & Rogoff, B. (1987). World views in psychology: Traits, interactional, organismic, and transactional perspectives. In D. Stokols & I. Altman (Eds.), *Handbook of environmental psychology* (Vol. 1, pp. 7–40). Malabar: Krieger Publishing Company.

Amerigo, M., & Aragones, J. I. (1997). A theoretical and methodological approach to the study of residential satisfaction. *Journal of Environmental Psychology, 17*(1), 47–57.

Amir, Y. (1969). Contact hypothesis in ethnic relations. *Psychological Bulletin, 71*, 319–342.

Avenel, C. (2005). *Sociologie des quartiers sensibles*. Paris: Armand Colin.

Avenel, C. (2009). La construction du problème des banlieues entre ségrégation et stigmatisation. *Journal Français de Psychiatrie, 3*(34), 36–44.

Avigdor, R. (1953). Etude expérimentale de la genèse des stéréotypes. *Cahiers Internationaux de Sociologie, 14*, 154–168.

Bizman, A., & Yinon, Y. (2001). Intergroup and interpersonal threats as determinants of prejudice: The moderating role of in-group identification. *Basic and Applied Social Psychology, 23*, 191–196.

Bobo, L., & Hutchings, V. L. (1996). Perceptions of racial group competition: Extending Blumer's theory of group position to a multiracial social context. *American Sociological Review, 61*(6), 951–972.

Bonaiuto, M., Breakwell, M. G., & Cano, I. (1996). Identity processes and environmental threat: The effects of nationalism and local identity upon perception of beach pollution. *Journal of Community & Applied Social Psychology, 6*, 157–175.

Bourg, G., & Castel, P. (2011). The relevance of psychosocial maps in the study of urban districts. *Journal of Environmental Psychology, 31*, 245–256.

Bourguignon, D., Seron, E., Yzerbyt, V., & Herman, G. (2006). Perceived group and personal discrimination: Differential effects on personal self-esteem. *European Journal of Social Psychology, 36*, 773–789.

Bourhis, R. Y., Gagnon, A., & Moïse, L. C. (1999). Discrimination et relations intergroupes. In R. Y. Bourhis & J.-P. Leyens (Eds.), *Stéréotypes, discriminations et relations intergroupes* (pp. 161–200). Liège: Mardaga.

Branscombe, N. R., & Wann, D. L. (1994). Collective self-esteem consequences of outgroup derogation when a valued social identity is on trial. *European Journal of Social Psychology, 24*, 641–657.

Branscombe, N. R., Ellemers, N., Spears, R., & Doosje, B. (1999a). The context and content of social identity threat. In N. Ellemers & R. Spears (Eds.), *Social identity: Contexts, commitment, content* (pp. 35–59). Oxford: Blackwell Science.

Branscombe, N. R., Schmitt, M. T., & Harvey, R. D. (1999b). Perceiving pervasive discrimination among African-Americans: Implications for group identification and well-being. *Journal of Personality and Social Psychology, 77*, 135–149.

Branscombe, N. R., Ellemers, N., Spears, R., & Doosje, B. (2002). Intragroup and intergroup evaluation effects on group behavior. *Personality and Social Psychology Bulletin, 28*, 744–753.

Brondolo, E., Brady ver Halen, N., Pencille, M., Beatty, D., & Contrada, R. J. (2009). Coping with racism: A selective review of the literature and a theoretical and methodological critique. *Journal of Behavioral Medicine, 32*, 64–88.

Brown, R. (1995). *Prejudice. Its social psychology*. Oxford: Blackwell.

Campbell, D. T. (1958). Comon fate, similarity and other indices of the status of aggregates of persons as social entities. *Behavorial Science, 3*, 14–25.

Chulvi, B., & Perez, J. A. (2003). Ontologisation vs. discrimination d'une minorité ethnique (les gitans). *Nouvelle Revue de Psychologie Sociale, 2*, 6–15.

Clark, R., Anderson, N. B., Clark, V. R., & Williams, D. R. (1999). Racism as a stressor for African Americans. A biopsychological model. *American Psychologist, 54*, 805–816.

Crandall, C. S., Eshleman, A., & O'Brien, L. (2002). Social norms and the expression and suppression of prejudice: The struggle for internalization. *Journal of Personality and Social Psychology, 82*, 359–378.

Crocker, J., & Major, B. (1989). Social stigma and self-esteem: The self-protective properties of stigma. *Psychological Review, 96*, 608–630.

Crocker, J., & Major, B. (1994). Reactions to stigma: The moderating role of justification. In M. P. Zanna & J. M. Olson (Eds.), *The psychology of prejudice: The Ontario symposium* (Vol. 7, pp. 289–314). Hillsdale: Lawrence Erlbaum Associates.

Crocker, J., Major, B., & Steele, C. M. (1998). Social stigma. In D. T. Gilbert, S. T. Fiske, & G. Lindzey (Eds.), *The handbook of social psychology* (4th ed., Vol. 2, pp. 504–553). Boston: McGraw-Hill.

Devos, T. (2005). Identité sociale et émotions intergroupes. *Les Cahiers Internationaux de Psychologie Sociale, 67–68*, 85–100.

Di Masso, A., Castrechini, A., & Valera, S. (2014). Displacing xeni-racism: The discursive legitimation of native supremacy through everyday accounts of "urban insecurity". *Discourse & Society, 25*(3), 341–361.

Dion, K. L. (2003). Prejudice, racism and discrimination. In T. Millon & M. Lerner (Eds.), *Handbook of psychology: Personality and social psychology* (Vol. 5, pp. 507–536). New York: Wiley.

Dovidio, J. F., & Gaertner, S. L. (1986). Prejudice, discrimination, and racism: Historical trends and contemporary approaches. In J. Dovidio & S. L. Gaertner (Eds.), *Prejudice, discrimination and racism* (pp. 1–34). New York: Academic.

Duckitt, J. (1992). *The social psychology of prejudice*. New York: Praeger.

Ellemers, N., Kortekaas, P., & Ouwerkerk, J. W. (1999). Self-categorisation, commitment to the group and group self-esteem as related but distinct aspects of social identity. *European Journal of Social Psychology, 29*(2–3), 371–389.

Emerson, M. O., Yancey, G., & Chai, K. J. (2001). Does race matter in residential segregation? Exploring the preferences of white Americans. *American Sociological Review, 66*, 922–932.

Félonneau, M. L. (2004). Love and loathing of the city: Urbanophilia and urbanophobia, topological identity and perceived incivilities. *Journal of Environmental Psychology, 24*, 43–52.

Fergusson, C. K., & Kelley, H. H. (1964). Significant factors in overevaluation of own-group's product. *Journal of Abnormal and Social Psychology, 69*, 223–228.

Festinger, L. (1957). *A theory of cognitive dissonance*. Stanford: Stanford University Press.

Fleury-Bahi, G., & Marcouyeux-Deledalle, A. (2010). Place evaluation and self-esteem at school: The mediated effect of place identification. *Educational Studies, 36*(1), 85–93.

Fleury-Bahi, G., Félonneau, M. L., & Marchand, D. (2008). Processes of place identification and residential satisfaction. *Environment & Behavior, 40*(5), 669–682.

Fleury-Bahi, G., Marcouyeux, A., Préau, M., & Annabi-Attia, T. (2013). Development and validation of an environmental quality of life scale: Study of a French sample. *Social Indicators Research, 113*(3), 903–913.

Haslam, S. A., & Reicher, S. (2006). Stressing the group: Social identity and the unfolding dynamics of responses to stress. *Journal of Applied Psychology, 91*, 1037–1052.

Hassine, K., Marcouyeux, A., Annabi-Attia, T., & Fleury-Bahi, G. (2014). Measuring quality of life in the neighborhood: The cases of air-polluted cities in Tunisia. *Social Indicators Research, 119*(3), 1603–1612.

Hogg, M. A., & Abrams, D. (1988). *Social identifications: A social psychology of intergroup relations and group processes*. London/New York: Taylor & Frances/Routledge.

Jasinskaja-Lahti, I., Liebkind, K., & Pehoniemi, R. (2006). Perceived discrimination and well-being: A victim study of different immigrant groups. *Journal of Community and Applied Social Psychology, 16*(4), 267–284.

Jetten, J., Spears, R., & Manstead, A. S. R. (1996). Intergroup norms and intergroup discrimination: Distinctive self-categorization and social identity effects. *Journal of Personality and Social Psychology, 71*, 1222–1233.

Jetten, J., Branscombe, N. R., Schmitt, M. T., & Spears, R. (2001). Rebels with cause: Group identification as a response to perceived discrimination from the mainstream. *Personality and Social Psychology Bulletin, 27*, 1204–1213.

Kahana, E., Lovegreen, L., Kahana, B., & Kahana, M. (2003). Person, environment, and person-environment fit as influences on residential satisfaction of elders. *Environment and Behavior, 35*(3), 434–453.

Krysan, M., Couper, M. P., Farley, R., & Forman, T. A. (2009). Does race matter in neighborhood preferences? Results from a video experiment. *American Journal of Sociology, 115*(2), 527–559.

Lalli, M. (1992). Urban related identity: Theory, measurement and empirical findings. *Journal of Environmental Psychology, 12*, 285–303.

LaPierre, R. T. (1934). Attitudes versus actions. *Social Forces, 13*, 230–237.

Lazarus, R. S., & Folkman, S. (1984). *Stress, appraisal, and coping*. New York: Springer.

Lee, R. M. (2003). Do ethnic identity and other-group orientation protect against discrimination for Asian Americans? *Journal of Counseling Psychology, 50*, 133–141.

Lench, H. C., & Chang, E. S. (2007). Belief in an unjust world: When beliefs in a just world fail. *Journal of Personality Assessment, 89*, 126–135.

Lepoutre, D. (2000). *Cœur de Banlieue. Codes, rites et langages*. Paris: Odile Jacob.

Lerner, M. J. (1980). *The belief in a just world: A fundamental delusion*. New York: Plenum Press.

Liang, C. T. H., & Borders, A. (2012). Belief in an unjust world mediates the association between perceived ethnic discrimination and psychological functioning. *Personality and Individual Differences, 53*, 528–533.

Lickel, B., Hamilton, D. L., Wieczorkowska, G., Lewis, A., Sherman, S. J., & Uhles, A. N. (2000). Varieties of groups and the perception of group entitativity. *Journal of Personality and Social Psychology, 78*, 223–246.

Liebkind, K., & Jasinskaja-Lahti, I. (2000). The influence of experiences of discrimination on psychological stress: A comparison of seven immigrant groups. *Journal of Community and Applied Social Psychology, 10*, 1–16.

Marans, R. W. (1976). Perceived quality of residential environments. Some methodological issues. In K. H. Craick & E. H. Zube (Eds.), *Perceiving environmental quality. Research and application*. New York: Plenum Press.

Marans, R. W., & Rodgers, W. (1975). Toward an understanding of community satisfaction. In A. Hawley & V. Rock (Eds.), *Metropolitan America in contemporary perspectives*. New York: Halsted.

Massey, D. S., & Denton, N. A. (1988). The dimensions of residential segregation. *Social Forces, 67*(2), 281–315.

Matei, S., Ball-Rokeach, S. J., & Linchuan Qiu, J. (2001). Fear and misperception of Los Angeles urban space: A spatial-statistical study of communication-shaped mental maps. *Communication Research, 28*(4), 429–463.

McCoy, S. K., & Major, B. (2003). Group identification moderates emotional responses to perceived prejudice. *Personality and Social Psychology Bulletin, 29*, 1005–1017.

McGarty, C., & Penny, R. E. C. (1988). Categorization, accentuation and social judgment. *British Journal of Social Psychology, 27*, 147–157.

Mesch, G. S., & Manor, O. (1998). Social ties, environment perception, and local attachment. *Environment and Behavior, 30*(4), 504–519.

Nasar, J. L. (1998). *The evaluative image of the city*. London: Sage.

Nys, M., & Beauchesne, M.M. (1992). La discrimination des travailleurs étrangers et d'origine étrangère dans l'entreprise. *Courrier Hebdomadaire du CRIPS* (Centre de Recherche et d'Information Socio-Politique), *1381–1382*, 1–78.

Oakes, P. J., Haslam, S. A., & Turner, J. C. (1994). *Stereotyping and social reality*. Maiden: Blackwell Publishing.

Outten, H., Schmitt, M. T., Garcia, D. M., & Branscombe, N. R. (2009). Coping options: Missing links between minority group identification and psychological well-being. *Applied Psychology: An International Review, 58*(1), 146–170.

Pol, E. (2002). The theoretical background of the city-identity-sustainability network. *Environment and Behavior, 34*(1), 8–25.

Proshansky, H. M. (1978). The city and self-identity. *Environment and Behavior, 10*(2), 147–169.

Proshansky, H. M., Fabian, A. K., & Kaminoff, R. (1983). Place-identity: Physical world socialization of the self. *Journal of Environmental Psychology, 3*, 57–83.

Quillian, L., & Pager, D. (2001). Black neighbors, higher crime? The role of racial stereotypes in evaluations of neighborhood crime. *American Journal of Sociology, 107*(3), 717–767.

Rabbie, J. M., & Horwitz, M. (1969). Arousal of ingroup-outgroup bias by a chance win or loss. *Journal of Personality and Social Psychology, 13*, 269–277.

Sanchez-Mazas, M., & Licata, L. (2005). *L'autre. Regards psychosociaux*. Grenoble: PUG.

Schmitt, M. T., & Branscombe, N. R. (2002). The meanings and consequences of perceived discrimination in disadvantaged and privileged social groups. *European Review of Social Psychology, 12*, 167–199.

Seaton, E. K., & Yip, T. (2009). School and neighborhood contexts, perceptions of racial discrimination, and psychological well-being among African American adolescents. *Journal of Youth and Adolescence, 38*, 153–163.

Sherif, M. (1966). *In common predicament. Social psychology of intergroup conflict and cooperation*. Boston: Houghton Mifflin.

Sherif, M., & Sherif, C. W. (1969). *Social psychology*. New York: Harper & Row.

Smokowski, P. R., & Bacallao, M. L. (2006). Acculturation and aggression in Latino adolescents: A structural model focusing on cultural risk factors and assets. *Journal of Abnormal Child Psychology, 34*, 657–671.

Stephan, W. G., & Stephan, C. W. (1996). Predicting prejudice. *International Journal of Intercultural Relations, 20*, 409–426.

Stephan, W. G., & Stephan, C. W. (2000). An integrated threat theory of prejudice. In S. Oskamp (Ed.), *Reducing prejudice and discrimination* (pp. 23–45). Mahwah: Lawrence Erlbaum.

Stephan, W. G., Ybarra, O. Y., Martinez, C. M., Schwarzwald, J., & Tur-Kaspa, M. (1998). Prejudice toward immigrants to Spain and Israel: An integrated threat theory analysis. *Journal of Cross-Cultural Psychology, 29*, 559–576.

Tabushi, T., Fukuhara, H., & Iso, H. (2012). Geographically-based discrimination is a social determinant of mental health in a deprived or stigmatized area in Japan: A cross-sectional study. *Social Science and Medicine, 75*, 1015–1021.

Tajfel, H. (1969). The cognitive aspect of prejudice. *Journal of Social Issues, 25*, 79–97.

Tajfel, H. (1978). *Differentiation between social groups*. London: Academic.

Tajfel, H. (1981). *Human groups and social categories*. Cambridge: Cambridge University Press.

Tajfel, H., & Turner, J. C. (1979). An integrative theory of intergroup conflict. In W. G. Austin & S. Worchel (Eds.), *The social psychology of intergroup relations* (pp. 33–47). Monterey: Brooks/Cole.

Tajfel, H., & Wilkes, A. L. (1963). Classification and quantitative judgement. *British Journal of Psychology, 54*, 101–114.

Tajfel, H., Billig, M.-G., Bundy, R.-P., & Flament, C. (1971). Social categorization and intergroup behaviour. *European Journal of Social Psychology, 1*, 149–178.

Turner, J. C., Hogg, M. A., Oakes, P. J., Reicher, S. D., & Wetherell, M. S. (1987). *Rediscovering the social group; a self categorisation theory*. Oxford: Blackwell.

Turner, J. C., Oakes, P. J., Haslam, S. A., & McGarty, C. A. (1994). Self and collective: Cognition and social context. *Personality and Social Psychology Bulletin, 20*, 454–463.

Twigger-Ross, C. L., & Uzzell, D. L. (1996). Place and identity processes. *Journal of Environmental Psychology, 16*, 205–220.

Uzzell, D. L. (2000). The psycho-spatial dimension of global environmental problems. *Journal of Environmental Psychology, 20*, 307–318.

Uzzell, D., Pol, E., & Badenas, D. (2002). Place identification, social cohesion and environmental sustainability. *Environment and Behavior, 34*, 26–53.

Valera, S., & Guardia, J. (2002). Urban social identity and sustainability: Barcelona's Olympic village. *Environment and Behavior, 34*(1), 54–66.

Valera, S., & Pol, E. (1994). El concepto de identidad social urbana: una aproximacion entre la psicologia social y la psicologia ambiental. *Anuario de Psicologia, 62*(3), 5–24.

Wieviorka, M. (1998). *Le racisme, une introduction*. Paris: La Découverte.

Williams, D. R., & Mohammed, S. A. (2009). Discrimination and racial disparities in health: Evidence and needed research. *Journal of Behavioral Medicine, 32*, 2–47.

Children in Cities: The Delicate Issue of Well-Being and Quality of Urban Life

19

Sandrine Depeau

19.1 Introduction

The urban changes that have accompanied the main technological and environmental transitions of the beginning of the twenty-first century have sparked a growing interest in the well-being of children. It is fast developing, because of the numerous international comparative studies conducted to help results appraisal of public policies. Within the context of urban planning, particularly when faced with questions of urban densification and urban sprawl control and with the rapid expansion of the "de-motorization of cities" paradigm (transition from a city with cars to one without), children's quality of life remains a health and social issue in the search for sustainable city models. Largely emphasized and extensively dealt with in Anglo-Saxon and Northern European literature, the issue of the child in the city is still underrepresented in French urban and social research. For instance, in the transportation sector, and particularly in national household travel surveys ("Enquêtes nationales Ménage-Déplacements"), children have for a long time only been considered in the analysis through the prism of the family, and especially the mother. Therefore, and because the principle of sustainability remains a fundamental element in the organization of cities (which are the main living environment of the majority of children), thinking today about future generations implies thinking with and about the generations of children who will be the adults of tomorrow. Improving their quality of life necessarily requires thinking about their place in the city, a place that usually depends on the world of adults, and which is sometimes normative and often unsuitable. The issue of children's quality of life is quite well represented and is a target of either direct or indirect surveys in the Anglo-Saxon literature. This is why this chapter can only be partial, at the risk of seeming biased. It is thus a matter of questioning, within the context of an ecological approach to human development, the notion of quality of life during a specific stage of childhood (admittedly the most studied): between the ages of 6 and 12. The concept of quality of life will be discussed mainly, but not only, within the environmental approach to psychology. It will be largely enriched by literature drawing sometimes on sociology, sometimes on geography and environmental planning, political science, architecture, etc. This approach will benefit from this accumulation of knowledge and from results derived from interdisciplinary research, which is, in fact, the very essence of the ecological approach to psychology.

Firstly, I refer to urban history and the history of education in order to show how childhood

S. Depeau (✉)
UMR ESO 6590 – CNRS, Université Rennes 2, Rennes, France
e-mail: sandrine.depeau@univ-rennes2.fr

has progressively been considered within a few urban utopias, and how, through principles of hygiene or safety, the quality of living spaces has been associated, from very early on, with spaces dedicated to childhood.

I then focus on developmental approaches, in order to point out that well-being and its conditions have clearly become important issues for research in this field. I concentrate on a few fundamental paradigms, and more specifically on the ecological approaches to development. This viewpoint leads me to show how child-environment relationships can be studied differently depending on whether well-being refers to objective conditions or has a more subjective meaning. The living environment of children can thus be outlined by making a distinction – depending on the developmental processes at stake – between what are called "proximal and distal environments". These two scales are used throughout the chapter to discuss the notion of quality of life through two main approaches: the first and more direct one, known as the psychometric approach, prioritizes well-being scales that are mostly used to evaluate international public policies (cf. OCDE), and a second, more holistic and indirect one, highlighting a few fundamental processes in child-environment relationships. With the first approach, I focus, on the one hand, on the different environmental dimensions that are studied to define children's living environment; and, on the other, on the fundamental role of housing, which is still a *sine qua non* condition for the good development of the child. Through the second approach, and without reviewing the primary condition for quality of life, meaning living in decent accommodation, I choose to focus on the environment outside the house. This is partly because it constitutes a fundamental context of development during the period of childhood concerned by this chapter (namely between the ages of 6 and 12, the period of first explorations and experiences of the environment); but also because the outer residential environment gives a priori greater guarantees of equal access – and is less dependent on the state of families' resources as it can be with home. I then discuss the quality of the residential environment through two key principles: attractiveness or "place-appeal" and accessibility.

19.2 The Place of Children in the Evolution of Urban Planning: Quality of Life Within Utopia

Interest in the quality of life has evolved throughout history as urbanization phenomena have developed, and with them, their consequences in terms of the population's public health (Tobelem-Zanin 1995). Amongst the most significant periods initiating numerous changes in favor of children's living conditions, there is one that stands out, connected to the urban utopias of the nineteenth century, which played a role in the major evolution of society (population growth, technological and scientific advances, particularly the institution of psychology, and urbanization trends). These ideologies (i.e. the utopian socialism of Fournier or Owen), whether philosophical or scientific, (Preyer), took on a fundamental role in the vision of society in which the place of children and pedagogy are more or less central (Becchi and Julia 1998). Nevertheless, regarding the design of spaces, childhood still retains different forms depending on which social group is concerned. In the upper class, under cover of educational concerns, the spaces that are granted to children are separated from those of adults (in some bourgeois milieus, children even have their own apartments). On the contrary, in the working classes, where living in the lack of privacy is still the norm, children rarely have any other options than the street, school (for those who are provided with schooling) or the workplace to acquire experience and basic skills alongside adults (Becchi and Julia 1998). However, in all social classes, the consideration of the well-being of children *"arises at the same time as the institutionalization of spaces for education and the privatization of family life."* (Depeau 2003 p. 10). Quality of life preoccupies philanthropists, who are also the advocates of social reforms and urban projects via public health committees, often focused on the working class. Blanqui's report (1849) can serve as a good illustration of this:

Whatever efforts are made by the government and the municipalities, whatever wisdom they generously try to spread; until children can be born and live in salubrious and tolerable dwellings, until they stop being prematurely bound to the glebe of the workshop instead of going to school, there is no hope to be had for the physical and moral improvement of the working classes... (quoted in Ragon 1986 p. 40).

The zoning principle and the creation of "familistères" (phalansteries, also called "social palaces") created some structures where "*children are taken charge of by the community, from "nursery" to elementary school to learning center.*" (Ragon 1986 p. 144). Despite the question of housing salubrity, education, schooling (hardly widespread) and the abolition of child labor (laws governing the minimum working age), constituted the main contexts of understanding children's living conditions. In these domains, a slow-paced evolution came up against the needs of low-income families (justifying child labor) and above all the difficulty of clearly and scientifically identifying the different ages of childhood – even though the fast rise of scientific psychology and the development of pedagogy helped (the Binet Institute in France, Claparède in Switzerland and Montessori in Italy (Becchi and Julia 1998).

Besides the education spaces integrated in houses or schools, children's quality of life also refers to the areas dedicated to their thriving and physical health. At the beginning of the nineteenth century, parks and gardens were laid out with these objectives in mind, and also to limit idleness. Pernoud (2010) shows how the depiction of childhood is intimately linked to the representation of public squares in painting.

> [...] *the playing child is a practically automatic attribute of the portrayal of public squares and public gardens of the capital – a compulsory figure denoting the formative role of the public square, an alternative playground with the fresh air and greenery qualities advocated by the psychology of the end of the century, influenced by foreign models such as the German and Anglo-Saxon* Kindergarten. (Pernoud 2010 p. 152).

Well-being, when associated with children's socialization, is largely grounded in urban spaces of relaxation such as public gardens. Later, some of which were widely criticized by Jacob in the 1960s because of the separation they marked from the rest of the city. Though associated with outdoor play equipment and nature, they remain key spaces in the city as they enable young people to be controlled within their railings. Later on, contrary to childhood that needed to be watched over, and with the development of psychology and sociological studies, childhood creativity and spontaneity also needed to blossom in adventure playgrounds. Originating in Europe and designed using vacant areas in the city, they emphasize children's creative and cooperative play. Progressively abandoned by public policies because they were unsuited to the sanitary norms of spaces for children, these playgrounds seem to have been regaining some popularity in recent years, in the wake of more health-related concerns linked to the risks of sedentary behavior for children in urban environments. Therefore, two sectors of environmental and city planning seem to be more seriously studied in current public policy: roads and footpaths – encouraging pedestrian traffic more generally – and recreational areas in cities. These two urban contexts are discussed in one part of this chapter, particularly because they demonstrate the on-going dual tensions between "the full", which consists of planning good spaces for children and "the empty", which refers to:

> *spaces in hallow allowing the tacit city reappropriation by children themselves, informal, less visible and recognized even illegitimate.* (Garnier 2015)

19.3 The Place of Children's Well-Being and Quality of Life in Ecological and Positive Development Approaches in Psychology

19.3.1 Ecology and Positive Development in Psychology

In the field of childhood, the nuances between quality of life and well-being remain inextricable, even though generally, in environmental and/or developmental psychology, studies are logically aimed at well-being issues. This is particularly the case in the ecological field of psychology; moreover, it is the explicit aim of research focused on the notion of "positive youth development". This field progresses mainly through action research; it is not exclusively carried forward by psychology, and covers a variety of disciplines (developmental psychology, sociology, medicine, education, public health, etc.) whose basic premise is to overcome the mere explanations of children's problem-behavior in order to find potential solutions thanks to key processes contributing to the good development of the individual. The emphasis is on all the processes observed in the broad field of everyday life (school, family, civil life, etc.) where children's well-being even becomes a key notion (Benson et al. 2006). Just like the ecological approach of development, it is anchored in a paradigm change regarding international children's rights, shifting from a need for protection (the child is primarily considered a vulnerable being) to the right to participate in the processes of space production and planning. Therefore, it entails a full recognition of the child's need for autonomy in order to grow up as a person. The understanding of factors linked to the positive development of the person is centered both on the individual and on whole life contexts (Benson et al. 2006). Bronfenbrenner's ecological approach (1979) is then used to understand better the connection between contexts in the implementation of support programs for children in difficulty. The goal is to overcome the state of irreversibility of risky behaviors, by implementing public programs to reveal, and even to consolidate the set of (individual and environmental) factors at play in the child's "well-becoming".

The ecological approach to development stems from the idea that the psychological functioning of an individual cannot be separated from the situations he or she encounters. The studies that fall within the scope of this approach, inherited from Lewin's research (1926/1931) and his definition of the "life space", define the context in terms of situations in which individuals react or are involved, and in which the levels of attention and interpretation of the experienced situations are still dependent on the physical and social characteristics of the context. There are two scale levels when considering contexts (also involving the levels of similar psychological processes): the proximal[1] level, which refers to the situation providing direct contact in the individual-environment relationship (microsystem or proximal environment) and the distal level, which may have similarities with the exosystem (distal environment) and refers to the cultural system in which the individual's experience situation is inscribed.

Along the proximity dimension, the total PE system to which an individual belongs forms a hierarchic system in which immediate situations, proximal environments and distal environments are integrated. (Magnusson and Stattin 2006 p. 406).

With the approach of positive development, the relationships between individual and environment are understood through a transactional model that prioritizes the action of the individual. Intentionality thus becomes a fundamental variable in the processes directing the action and investment in the environment (Brandtstädter 1998), and with it comes the notion of resiliency. This notion is integrated into the postulate of psychological plasticity, meaning that behavior or psychological states involved in difficulties can always potentially be ameliorated through what is called a "developmental asset". This is defined as follows:

The framework of developmental assets is a theory-based model linking features of ecologies (external assets) with personal skills and capacities (internal assets) guided by the hypothesis that external and internal assets are dynamically interconnected "building blocks" that, in combination, prevent high risk health behaviors and enhance many forms of developmental success (i.e. Thriving). (Benson et al. 2006 p. 906).

Consequently, the environment does not exist independently of the individual, but is situated within the scope of situations that initiate action. Some situations are also more fundamental than others, particularly exploration and play (Wohlwill 1987).

---

[1] Based on Vygostsky's notion of proximal development (1931/1978).

19.3.2 Some Developmental Assets and Fundamental Processes Involved in Exploration and Play

By adopting an ecological perspective on positive development, implying the active participation of children in the environment and the interlinked role of contexts, Benson et al. (2006) observed some developmental assets that they separated into two orders: external and internal. The first refers to the notions of *"support*, *"empowerment"*, *limitations, expectations and constructive use of time".* The second is about *"the commitment to learning, positive values, social skills, positive identity."* Both assets, under certain dimensions, trigger psychological processes, similar to those involved in the notion of environmental preference, which is rather more used for evaluating children's quality of life in a given environment. In the register of internal assets, the notion of commitment includes the whole body of active experiences in the environment that can have a positive impact on individual development. Understood as the interest children have in an activity or in a context, it contributes to self-esteem and to a stronger sense of optimism towards the future (Hunter and Csikszentmihalyi 2003). It is through the notion of children's autonomy that commitment is examined in this chapter in order to define their quality of life in the environment. Autonomy implies the active role of children, on the one hand, in the development of environmental knowledge and skills, contributing to increasing feelings of safety and self-efficacy when confronted with certain situations; on the other, in the appropriation of space and different types of attachments that contribute to the process of spatial identity. Lastly, it undergoes to the construction of environmental preferences, which are involved in the process of attachment and play a part in the notion of quality of life. Included in the register of external assets is the quality of relationship with life contexts (in particular family, friends, school) that is part of the child's well-being (Scales et al. 2000). The quality of the relationships with these different contexts, most often measured in terms of attachment and identification, plays a pivotal role in the socialization of children, in his/her involvement at school, and in his/her feeling of self-efficacy (Catalano et al. 2004; Bahi-Fleury 2009). Finally, the role played in development by the diversity of contexts can also be observed in its temporal forms and particularly in its principles of recurrence, naturally leading to the cumulative logic of milieu. It would then be very similar to the notion of capital as defined by Bourdieu (1979), and less similar to Coleman's definition of it (1988), as mentioned by Benson et al. (2006). This is particularly true because the cumulative dimension seems to refer rather to a skill resources logic, certainly contributing to enriching individual-environment relationships, but above all to supporting, encouraging and securing individual experiences in the diversity of contexts. Thus, all the principles in this double approach to a child's development justify a more specific interest in the residential environment (the neighborhood) as a plural context in order to approach children's well-being and access conditions of the spaces of proximal development.

19.4 Well-Being and Quality of Life: Psychometric Approaches

The definitions of well-being and quality of life are linked to the mutations of life spaces, the transformations of social structures and the evolution of the principles of Children's Rights enforcement: nowadays, they are undergoing a plethora of measurements that are sometimes confused (Koescher et al. 2014). Nevertheless, the two notions can be distinguished as they refer to different disciplinary fields:

> *The two came from different disciplines – 'well-being' more from psychology and 'quality of life' more from sociology and social policy – which have only gradually and incompletely become open to each other.* (Gasper 2010 p. 351).

Slight differences between these two notions become even clearer when domains of intervention in public policies are differentiated by scientific approaches (health, sociology, community studies, poverty studies, etc.) and more decisive depending on the values associated with society models (Gasper 2010). Moreover, as underlined by Kamerman et al. (2010), in terms of childhood studies, viewpoints are currently shifting *"from child-saving to child development"* or *"from child welfare to child well-being"*. In any case, and because the quality of life, depending on the various contexts, embraces some health and economic values, it is increasingly subject to measurement. Consequently, how are the context levels and the child's relationship to the environment understood, and what subjective dimensions are favored?

A non-exhaustive list of measurements of children's well-being will help to understand the place assigned to the context, particularly to the residential environment and to the psychological and social processes involved in the subjective construction of well-being. The bibliographical research carried out reveals that tools for measuring well-being in the field of childhood classically refer to either the medical field (public health or pediatrics) or the socio-economic field. These are two areas in which the requirement of optimal validity for the metrological conditions of the tools (reliability, validity of the tool) sometimes leads to more reluctance when it comes to allowing children to speak. Nonetheless, although for a long time it was studied through children's family contexts or it was directly provided by the adult care, well-being is gradually being defined not only from living conditions, (objective well-being) (Land et al. 2001), but also from the child's own answers (subjective well-being), in accordance with the rise of children's rights. It is only with Ben Arieh's studies (2001) that children have finally come to participate in studies measuring well-being. Furthermore, the contexts taken into account are varied but recurrent. Generally, they are micro-systems with some permanency in the child's daily life. These refer to some ecological niches of childhood: neighborhood, playground, school, street – proximal life spaces that give useful information on dimensions of accessibility, stress or pollution, as well as on material and economic resources.

19.4.1 Land et al.'s Well-Being Scale (2001)

In Land et al.'s measurement tool (2001), contextual conditions named by the authors "material well-being", are distributed within a category referring principally to economic resources (employment and income). Although the neighborhood is little taken into account, the material domain is still supplemented by social relationships and a mesosystemic part characterizing the school and other forms of contexts participating in the child's socializing, such as the "place in community" domain. The context is only an environment for living, socializing and personal relationships, the latter being understood only indirectly. The different scale dimensions cover states that have been recorded rather than children's relationships to their environment.

This is why Ben Arieh et al. (2001) suggested involving children in all stages of the survey in order to consider them as a specific population. That is to say that the notion of well-being must also be understood from a body of affective components (including both negative and positive emotions) and cognitive components characteristic of childhood. This justifies a multidimensional approach that includes objective and subjective measurements. As an example, the scale that has been tested in a European comparison can be mentioned.

19.4.2 Bradshaw et al.'s Well-Being Scale (2007): The Place of Fundamental Life Spaces: Home and Neighborhood

Using the ecological approach to development (Bronfenbrenner 1979) and considering well-being a process, Bradshaw et al. (2007) claimed that

> [Children] *are the result of the interplay between resources and risk factors concerning the personal situation of the child, his or her family, friends, situation at school and the wider society. These factors are constantly changing and children – with their evolving capacities – create their well-*

being actively by mediating these different factors. (Bradshaw et al. 2007 p. 136).

The notion of contexts is central here. It refers to two major dimensions, "economic situation" and "housing", documented from a body of studies or from national databases associating surveys conducted with children. The first level of context is defined from a group of economic resources (the family's economic situation), material resources (linked to the personal workspace at home and the family's mobility and communication equipment), and health and social resources (the number of vacation weeks of the family). The second level is built on the notion of "housing", which reveals, besides accommodation and its social density (overcrowding), the quality of the external living environment and the safety level perceived from criteria that are really indicators of incivility. The mesosytem that is mainly made up of the neighborhood and the school is also observed from the child's social relationships (intra-familial and between peers), which also constitutes its own field. It is not measured by quantitative criteria (which could be the time spent in a place or the number of friends, for instance), but rather by qualitative criteria allowing assessment of climate and feelings of support. Finally, subjective variables refer to well-being evaluation in fundamental domains of childhood (health, education[2], civic participation, social relationships, risks and safety).

The relationship to the life context is limited to questions of material properties, satisfaction levels or the implication levels regarding these areas. Well-being is thus associated with the possession of certain material conditions and with different states. These can very quickly be assimilated to personality traits, if the measured psychological state remains attached to a specific and non-contextualized situation – as it is the case for some questions related to personal well-being. However, well-being only makes sense if measured from cognitive and affective dimensions referring to the association of a present situation with a level of expectation (the future). Brought back to levels of satisfaction in certain domains of life, this implies measuring what is experienced relative to what was expected, that is to say, measuring the discrepancy between these two states. This appears to be more relevant when it concerns the development stage of children's autonomy within an environment, and some frustrations that can be triggered by a wide discrepancy between the expected and experienced levels.

Although the relationship to the environment is less studied here than in environmental psychology research, it is still worth noting that, in this tool, the socio-economic dimensions have a structuring effect. In fact, when applied to different European countries (cf. the OECD's report on well-being), the results show a strong link between socio-economic characteristics (family structure, poverty index) and well-being experienced by children. They also show that the level of personal dissatisfaction, the conditions of resource deprivation, and the level of social relationships between peers are related. Finally, some of the more implicit dimensions of well-being can also be understood as those related to the norms and cultural values of children's life contexts.

19.4.3 Middle Years Development Instrument (MDI): Taking into Account Free Spaces and Free Time

In order to avoid the side effects generated by the norms related to the age categories of institutional offers directed towards young people, some scales can also be associated with and built for a given period of childhood. This is the case of the MDI, which is applied to a turning point in children's development: the middle years of childhood (Schonert-Reichl et al. 2013).

This instrument consists of involving children and of better understanding some of the psychological characteristics specific to this developmental period, though without associating the children too arbitrarily with

---

[2]Because it will not be the focus of this chapter, let us note that the relationship with school is also a fundamental domain that is regularly exploited in numerous studies and measurements related to well-being.

fixed categories. It is thus used to observe behaviors, practices or potential responses read as signs of adaptation to the context. The scale covers five domains that include processes and behaviors linked to social and emotional development ("*positive and negative feelings, pro-social behaviors, social competences, self-efficacy, emotion regulations*", etc.), social development ("*connectedness, quality, range and support of relationships with adults and peers*"), health development ("*health habits, sleep, nutrition, hygiene and stress*"), and school-related development ("*climate, satisfaction, social relationships, support and motivation in school*"). Children's life context covers here not only the whole extracurricular world in its spatial, temporal and behavioral dimensions, but also activity desires and perception of barriers ("*Use of after-school time questioning (what, where, with whom) and undesired after-school activities, barriers to after-school time activities.*").

This extracurricular domain is all the more fundamental that nowadays it is tending to become more and more planned, modeled, indexed to school skills and, thus, it is leading to further inequalities. Free time, from a spatiotemporal approach close to methods used in time-geography, covers many different types of children's relationships with life contexts. It is notably linked to the notion of experience (dear to Wohlwill, cited by Heft 1998) that can be used to consider some fundamental activities such as exploration, play (more important for younger children), interpersonal relationships (whose role is crucial during this time of life) (Chu et al. 2010). They cover structured as well as unstructured activities. The latter, in children's daily lives, play a large part in their environmental preferences and relationships with natural spaces through play (Moore and Wong 1997). On this account, Burdette et al. remind us that:

> *among primary school-aged children, active free-play or unstructured physical activity that takes place outdoors in the child's free time may potentially be the major contributor to children's physical activity.* (Burdette et al. 2004 p. 354).

This part of unscheduled free time enjoyed by children but feared by parents also represents an important educational issue that deserves more attention (Chardonnel and Depeau 2014). Besides, the use of free time in the construction of children's well-being helps to address some health and social challenges; partly because studies tend to show that children's physical activity in urban contexts is decreasing (Maitland et al. 2013), but also because the occupation of free time for children who go to school is an important source of social inequality today, at least in France (Sue 2006), and of learning inequality. Free time, which represents 24 % of children's after-school activity time in France (Jacquemain 2003), has to be finely appreciated, particularly when facing the issues of school time reorganization that, in France, call into question the place of proximity spaces in neighborhoods.

It is also in this logic that numerous authors argue for the need to fine-tune the tools of understanding the child's well-being by taking into account the domains of life and daily situations that make sense to the child.

19.4.4 Well-Being Without an Instrument and Discussed Directly with Children

Fattore et al. (2007) studied the notion of children's well-being from a qualitative (ethnographic) approach, based upon researcher-child interaction. The objective was to have:

> *enabled children to describe their well-being in the present, as well as to discuss the way aspirations for the future impacted on a sense of current well-being and their ideas of future well-being.* (Fattore et al. 2007 p. 14).

By interviewing a population of children between 8 and 15 years old in two different contexts (urban and rural) in cities of New South Wales (Australia) using a variety of collection tools to help children build their own definition of well-being, they noticed two central developmental processes in the definition of well-being: autonomy and socialization. The formal expression of these processes illustrates situations where participation/involvement and real-life sensorial and social experiences are at the heart of children's narratives.

Autonomy and socialization are indeed the two fundamental processes of children's cognitive and social development (Depeau 2003). The results of Fattore et al.'s study are very similar to those observed in other environmental psychology studies. Spatial autonomy allows children to be active in their interactions with the environment, contributing to increasing play opportunities (Kytta 2004), sustaining relationships between peers (Prezza et al. 2001; Depeau 2003), and challenging self-efficacy by encouraging the development of spatial representations and thus the cognitive spatial familiarity that conditions part of the environmental accessibility (Depeau 2003). Therefore, it contributes to reducing feelings of unsafety by reducing feelings of loneliness. In this definition of well-being by children themselves, mobilized contexts share the same worlds as those developed by the MDI scales that take free time into account, or by Ben Arieh et al.'s measurements (2001).

These various measurements, not specific to the field of environmental psychology, have primarily demonstrated (besides the psycho-social processes at stake in the definition of well-being) which levels and characteristics of the contexts have to be taken into account. For instance – and this is mostly limited to home and neighborhood domains, i.e. to the proximal context –there is one fundamental living space in this system of contexts explaining the well-being of children, which is avoided, to say the least, in the indirect studies that are used in the next part. This living space is the home, approached here from wealth or resources indexes that can be used by families. Its role is all the more important that it contributes to factors of well-being inequality, especially when it comes to school life conditions that contribute to children's well-being at school, but also in the domain of socialization.

Moreover, what the controversies reveal about the measurement of well-being is the risk of fixing populations in levels of normality, thus preventing the meaning associated to well-being by children themselves from being taken into account, unless they are fully involved and observed in their everyday lives. It has been noted by Fattore et al. (2007):

Abstracting children from the social and economic contexts in which they live their lives, ignores the complexities of individual children's lives and thereby risks inappropriately simplistic policy responses, such as blaming parents for children's lack of coping skills or poor self-esteem. (Fattore et al. 2007, p. 9).

Well-being scales thus have the advantage of enabling populations coming from different contexts to be compared, provided that the meaning attributed to some dimensions are not only available (external collected data not necessarily available everywhere), but also understood in the same way. It is in this sense that White (2009) introduces her critical approach to the notion. She mentions that well-being takes on different meanings depending on the cultural models to which one is referring (individualist culture vs. collectivist culture). For White, it has to come from a variety of fields grouped around three dimensions: material, social and subjective. From this point of view, the definition of well-being needs to be part of a model with different domains of influence fitting together – not unlike Bronfenbrenner's model of ecological development (1979).

19.5 Holistic Approaches: From Attractiveness to Accessibility of Environment

Removed from psychometric approaches, holistic perspectives – which fall within a multidisciplinary framework – create a debate about questions related to quality of life in the domain of childhood. In environmental psychology, they play a part in the study of child-environment relationships. Within this framework and under the influence of significant urban mutations and the key principles that govern the redevelopment of cities (e.g. sustainability), the notion of quality of life is often associated with the principle of a Child Friendly City (CFC). This principle also refers to a charter adopted by many political actors in local governments, who endeavor not to reduce it to its mere territorial marketing function. The CFC makes it possible to find a "universal" way of determining the principles of urban quality of life for children, particularly because it is in line with the Convention on the Rights of the Child. Since the beginning of the 2000s, it has often been used and discussed in studies. By focusing on the improvement of children's living environments, it contributes to favoring living contexts that are bearable, accessible and even playable for children. However, although it has found some more or less consensual definitions at an international level, this notion remains very difficult to set up (Whitzman et al. 2010). Nevertheless, based on an extended literature review, Horelli (2007) puts forward a series of dimensions to define the notion of CFC, which have also been tested in other studies (Nordström 2010). These dimensions cited by Horelli cover spatial and social resources of children's life contexts, principles of environmental quality, as well as more psychological processes emblematic of child-environment relationships: (1) Housing and dwelling, (2) Basic services, (3) Participation, (4) Safety and security, (5) Family, peers and community, (6) Urban and environmental qualities, (7) Provision and distribution of resources and poverty reduction, (8) Ecology, (9) Sense of belonging and continuity, and (10) Good governance. This multidimensional definition denotes the need for a holistic approach covering both the individual and collective parts of children-environment relationships. This makes the notion difficult to use when considered as a whole, but it can be reexamined in other models, such as the "Bullerby" model (Broberg et al. 2013), discussed further in this chapter.

Based on some of the dimensions of the definition, two key principles applied to the definition of the quality of children's living environments, attractiveness and accessibility will be discussed. These two principles will be defined and described based on a selection of studies marking the field of childhood research: the selection is admittedly partial, but large and operational enough to cover the question of children's quality of life.

19.5.1 Attractiveness

The notion of attractiveness is here approached from external life contexts taken from children's daily life and studied on a proximal spatial scale.

19.5.1.1 Territories of Childhood and the Notion of Behavior-Setting: Between Structures and Informal Relationships to the Environment

Because of its spatiotemporal continuities and discontinuities, everyday life constitutes a fundamental observation variable to understanding child-environment relationships (Wapner 1998). Therefore, the concept of behavior-setting has been an essential contribution to describing not only the evolution of childhood territories, but also the variety of involvement of children in the environment outside the home (Barker and Wright 1955). It is measured by indexes such as the time spent outside or the penetration index, which is defined as:

> the index of individual involvement in a behavioral site and the degree from which the individual takes on the responsibility of staying within this behavioral model. (Barker and Wright 1955, cited by Rogoff and Lave 1984 p. 246).

Despite its reduced use in contemporary research, this concept is still interesting and deserves to be reexamined for the normative dimension and the notions of structure, routine, and rules it conjures up in order to understand the spaces frequented by children (Chardonnel and Depeau 2014). It is also useful to describe the diversity and environmental variability of the spaces frequented by children and, above all, the proportion of their active or passive involvement. This is what Barker and White have shown by comparing two contexts: a city center and urban outskirts. They observed that there is greater behavioral similarity in children in the same region if the same spaces are regularly frequented. Finally, as sites that are more or less spatially, socially and temporally structured, behavior-settings require activity levels from children that also summon up the register of individual perceptions and representations. Although this concept has the advantage of revealing some normative, or even cultural dimensions, it is not enough to predict the quality or congruence of the spaces. Besides, the related observation of unsupervised free time would reveal a whole field of activities that is still insufficiently exploited, related to the informal experiences of spaces (Chardonnel and Depeau 2014). In the urban area of a medium-sized city such as Rennes (France), very few children associate their regular extracurricular activities with their spatial preferences (Depeau 2012). Few refer to the spaces that are supervised and regulated by adults. On the contrary, natural amenities and little-structured spaces in the neighborhood area constitute elements of environmental preferences that are not part of the children's scheduled activities (Depeau 2012). As specified by Min and Lee (2006): "*the use of a setting may not necessarily lead to preference.*" (Min and Lee 2006 p. 53). It is at this critical point that the concept of affordance seems relevant to complement the notion of behavior-setting, as pointed out by Heft (1988).

19.5.1.2 The Attractiveness of Spaces and the Notion of Environmental Preferences

To reply to the principle of attractiveness, the notion of preference remains appropriate for studying child-environment relationships. Associated with the attachment process in the child's development, and stemming from needs, but also from abilities and action intentions in the environment, these preferences vary depending on the periods of development (Malinowski and Thurber 1996). Chawla (1992) sets out two important stages in the development of the child when it comes to preferences of relationships to the environment. The first stage, up to 11 years old, is that of intensive exploration of the environment, motor activities and construction of the first relationships. Then, as soon as children enter adolescence, it is the stage of consolidation of social relationships and a stronger need for "privacy" and non-programmed spaces. The gradual disconnection from familiar (therefore close to adults) worlds during preadolescence and the abandonment of routines in games started between peers (Pellegrini et al. 2004) also remain an adjustment variable in the choice of spaces (Chombart de Lauwe 1987; Chawla 1992; Depeau 2003) which can satisfy children's needs for both security and privacy (Korpela 2002). The notion of preference remains linked to the more general development of relationships to space: cognitive relationship to the surrounding world (Hart and Moore 1973), need for socialization (Depeau 2003, 2012), need for exploration and play, construction of place-attachment (Jack 2010), regulation of emotions (Korpela 1992), relationship to spatial forms (nature vs. virtual) (Malinowski and Thurber 1996). Preference also reveals, in some cases, the effect of parental prescriptions and their normative dimension through gender differences (Depeau 2003). For boys, preferences are more often directed towards outdoor and public spaces and physical activities, contrary to girls who tend to be more attracted to private spaces. Moreover, for Min and Lee (2006), this notion can dissociate behavioral reality (linked to the use of spaces) from psychological reality (which makes sense for and in the development of the child) in the definition of important sites that characterize children's living environments. Children's neighborhood settings include these two realities, summed up by the authors as settings (1) "*favored or important but not used well*", (2) "*used well but not favored or important*", (3) "*favored or important and used well*" and (4) "*not favored/important and not used well*" (Min and Lee 2006, p. 52). These different levels also reinforce for them the nuances between the notions of "place" and "space" so that the meaning of places completes or compensates their use. By working towards informing diverse dimensions useful for the qualification of spaces either in the behavioral domain (affordance), the emotional domain (restorative environment) or the social domain (social environment), the notion of "environmental preference" becomes an operational concept to define the quality of environments for children. Korpela (2002) notes that:

Place preferences during childhood and adolescence are assumed to provide support for the developing self-identity, the need for security, social attachments to caregivers and to the peer group, and the practice of social roles. (Korpela 2002 p. 365).

19.5.1.3 Attractiveness Through Play and Exploration: The Place of Playing Spaces

It is particularly through play that the representation of childhood in society is built. While being one of the favored activity of children, it still remains very unequally taken into account in cities today, even though efforts are made in its name to give a place to children in the city (Chombart de Lauwe 1987; Brougère 1991) – risking sometimes completely specializing the spaces whereas playful activities between children require flexibility. Without reexamining all of the literature about the interest of outdoor play for the development of children, a few points on the evolution of the planning and the fundamental characteristics contributing to the quality of spaces in the city should be underlined. Whether supported by structured spaces or not, play or recreational forms contribute to the social, cognitive and emotional development of the child. Moreover, the etymology of the term "recreation" refers to the reconstruction of the vital strengths of the individual. Appreciating the quality of spaces that enable children to play involves understanding not only the availability of these spaces in the city but also the properties or attractive qualities of the spaces, equipment and social context inviting the two forms of play and thereby contributing to the development of children: free play (play) and playing with rules (games).

This quality depends on several dimensions related to: (1) the flexibility of sites and equipment in order to make them match children's intentions or projects, which is partly related to the notion of affordance; (2) the offer of stimulation sought by children, which requires working on the balance between stimulation/novelty and safety (Ellaway et al. 2007); (3) opportunities for social relationships thereby contributing to social adjustments in general (Pellegrini et al. 2004). The availability of sites or, of spaces dedicated to play, depends on the size of cities, the weight of the dominating urban models (Chombart de Lauwe 1987) and the age of the equipment. Blinkert (2004 p. 106), notes:

Cities not only have grown but one also can observe a process of inner urban differentiation. There are residential areas, areas of production, of consumption and of recreation and certainly areas for children. The possibilities for playing in these designated children's places are to a high degree standardized by their equipment and order: slides, swings, specific equipment for wobbling and creeping, and expensive play systems.

Besides the closed structures dedicated to children, the question of outdoor play areas within the city remains fragile if these spaces trap the children in a too-rigid system of rules and order (Blinkert 2004) and if safety is questioned by parents (Ellaway et al. 2007). The planning of playgrounds requires a balance between simulation/safety and openness/tranquility, satisfying the two sub-categories of children that use playgrounds the most: "middle childhood" and adolescents, even though the latter group also uses parks less (Timperio et al. 2008). These two categories do not appreciate the same types of equipment and therefore of activities in urban parks. For example, Baran et al. (2014) showed that adolescents prefer sedentary activity structures whereas *"free play and unstructured activities appeal to younger children."* (p. 786).

Children's play and sociability areas are linked to health issues related to the increased sedentary lifestyles of children and to the mutation of families' housing modes. Today, they are also understood in terms of well-being from their relationship to nature and environment. Classically associated with restorative qualities, green spaces have an important sanitary and social role to play in neighborhoods, (Wells and Evans 2003) because they allow the regulation of interpersonal relationships (Flouri et al. 2014) and emotions (Faber Taylor et al. 2002). This is particularly true in some neighborhoods that are disadvantaged in terms of resources (Faber et al. 2001) and also in the "sense of commonality" (Kuo et al. 1998). The role of green spaces in stress-coping for children living in neighborhoods with fewer resources has also been observed (Wells and Evans 2003). Recent studies show that people have different needs depending on the structure of their neighborhoods, the resources of families (particularly whether they own a private garden or not) and the attractiveness of the equipment (Depeau 2003, 2012; Lehman-Frisch et al. 2012). However, the proportion of this green dimension in spatial attractiveness remains very variable from one study to the next. It depends on the metric distance to these spaces, on their form (surface, particular components), on the experience individuals have of them (visual experience vs. physical experience) and on the levels of maintenance and spatial amenities that play a fundamental role in the social and functional evaluation of spaces (Pitner and Astor 2008; Foster et al. 2015). Current research about the index of green impregnation[3] in children's living spaces suggests that the green spaces of neighborhoods condition and reinforce the variety of children's autonomous activities (Depeau and Quesseveur 2014). Approached from the daily travel routines of children as well as from their outside activities, this dimension can reveal (by combining these data with the social attractiveness of the place) which areas are used, preferred, or avoided by children. Lastly, the quality of spaces dedicated to play in the city has become a central issue in terms of planning because of the transformation of children's urban practices, and particularly the social variability of their occupation of free time, which generates different degrees of control and sedentary behavior depending on the milieu (Mahoney and Eccles 2008). Governed for a long time by safety requirements, the design of spaces for children seems to have become more audacious today as it tries to emphasize children's creativity and autonomy – even though the spaces are still insufficient and free play is diminishing in cities (Hart 2002). It is with this same goal in mind (nurturing children's creativity) that adventure playgrounds, established in Europe[4] after World War II, have been developing (Staempfli 2009). They consist of reinforcing the freedom of engagement in the activity, reducing the part of pre-set equipment, as noted by Staempfli (2009), and – on the contrary – selecting and shaping bits of nature. They favor a diversity of individual and collective learning processes, from the actual creation of play structures to the manipulation-experimentation of elements of the environment. They allow children to engage

> in social interactions across different age groups and with children from different neighborhoods. (Staempfli 2009 p. 272).

Adventure playgrounds cover a large part of the dimensions defining the quality of spaces in cities, including natural spaces. It is also because of their informal dimension, which leads to autonomy and diversity of social relationships that these adventure playgrounds may contribute to the development of the child.

19.5.1.4 Hidden Dimensions of Practiced Spaces or the Notion of Affordance

Strongly linked to play and to children's activities in outdoor spaces, and developed at first under the ecological paradigm favoring action, this notion of affordance is interesting because it emphasizes the functional particularities of the environment in accordance with children's goals for a given activity in the environment. The taxonomy that Heft first developed (1988) corresponds to a stage of childhood when relationships to the environment are mostly based on strong physical experiences. It is divided into ten categories: "*1. flat, relatively smooth surfaces, 2. relatively smooth slopes, 3. graspable/detached objects, 4. attached objects, 5. non-rigid attached objects, 6. climbable feature, 7. shelter, 8. moldable material (dirt, sand, snow), 9. water and 10. aperture (locomotion, looking and listening*

---

[3] Defined from the combination of data about the presence of hedges and green spaces close (20 m) to a path used by children during their trips recorded by GPS. The data is then refined according to the length of the travelled paths for each mobility sequence.

[4] "*Current estimations are that approximately 1000 adventure playgrounds exist in Europe, primarily in Denmark, Switzerland, France, Germany, the Netherlands and in England*" (Staempfli 2009 p. 270).

to an adjacent place." The relevance of space evaluation based on affordances also requires distinguishing between different registers of affordances depending on whether they are perceived, actually used, or wished for – even dreamed of – by children (Kyttä 2002). These different registers in the end refer to the distinction operated by Min and Lee (2006). Besides, the taxonomy of affordances, very much centered on children's motor behaviors, has been applied to the social domain in order to answer children's need for socialization during their various activities (Clark and Uzzell 2002; Kyttä 2002). Finally, the study of affordances makes it possible to apprehend intensity levels that children are looking for during their activities, and to characterize the diversity of their behaviors, which remains an important condition for learning (Prieske et al. 2015). In relation to this, Prieske et al. (2015) have shown how "*jumping is a kinetic marker of playing*" for children between 7 and 10 years old: how some behaviors are more appropriate than others to meet the challenges of behavioral intensity and diversity desired by children. Affordances then have the potential of contributing to the definition of children's "sense of place", as soon as it is about the spatial perception of action. Nevertheless, the quality of the sense of place remains dependent upon the conditions of children's autonomy in the environment. Tsoukala (2001) distinguishes the activities of children depending on their social dimension (autonomous character or not) and their orientation, whether it has a spatial characteristic or not (space looked for and used vs. space as support), and from that, defines four types of "socio-spatial activities": (1) Strategic activity, corresponding to activities initiated by children themselves. (2) Passive activity, i.e. not chosen by children – that is to say that they remain under the control and surveillance of other people. (3) Strategic spatial activity, which refers to the ones that are spatially oriented "*in conditions of social interactions that do not obey to the orders and rules of the adult*" (Tsoukala 2001 p. 60). It corresponds to the exploration of space, for instance strolling or skate-boarding in a structured area. (4) Passive spatial activity, supposing an activity that is "*directed towards spatial goals defined by other individuals*" (Tsoukala 2001 p. 60), like sports for instance. By focusing on a child's initiative, we can better understand which environmental situations contribute to spatial learning processes and the acquisition of autonomy. The "Bullerby model" distinguished four types of relationships to space, depending on the type of the child's independent mobility (which ultimately defines the level of access of the spaces) and on the level of affordance provided by a context. Therefore, it helps to understand the relationship between the sense of space in terms of affordances and the autonomy of children (Broberg et al. 2013) more appropriately. This model has also been applied to different neighborhoods in order to highlight the socio-spatial differences in accessibility and attractiveness of the residential environment.

19.5.1.5 Proximity of Other Children in the Neighborhood

The attractiveness of places in the neighborhood also depends on the presence of other children, who are potential companions or sometimes play observers. The proximity of other children is also an important dimension of children's well-being in cities when defined by parents (Depeau 2012). Veitch et al. (2006) also mention it: "*Social networks were frequently raised by parents as having a significant impact on their child's active freeplay.*" (Veitch et al. 2006 p. 388). All the more so, since the group of peers, through the bonds that are established, provides a basis for both safety and audacity in the process of mobility learning (Depeau 2003). It is a factor of encouragement for spatial exploration/discovery as well as for the social and spatial learning of the child (Bryant 1985; Depeau 2003; Prezza et al. 2001), for the broadening of the home-range (Matthews 1987; Depeau 2003), for the improvisation of games – particularly in peripheral towns with lower urban density (Depeau 2012). The number of friendships of children is strongly correlated with the scope of their activities (Torell and Biel 1985). Moreover, the lack of friendships in a neighborhood is perceived as an important social barrier to children's well-being (Depeau 2012), particularly when they start being allowed to play outside (Carbonara-Moscati 1985; Lehman-Frisch et al. 2012) and start exploring new spaces. Clark and Uzzell (2002) have shown how preference for certain places (downtown, house, neighborhood and school) is driven by two key principles, one of which is linked to social interactions. The presence of peers plays a role in children's social development (Depeau 2003) and in the construction of social and spatial identities based on a social comparison process – as suggested in Fattore et al.'s approach to well-being (2007). In the context of children's daily travel routines, and particularly for the walking-school-buses, the density of children in the neighborhood becomes a fundamental criterion that shapes the group of those walking to school and reinforces children's visibility. The inter-knowledge between children and families even constitutes a determining factor in the formation and perpetuity of those groups (Depeau 2012). Finally, as pointed out by Karsten (2015):

> The playground may function as a place – carved out of the adult world – to encounter other children and to build on social networks. Some children will develop intimate knowledge of their playground. These resident children are the ones who frequently come to the playground and who have a broad network among the other visitors. As a group they communicate in a specific way (backstage language), which makes their status difficult to contest. They are in a position to negotiate or even dictate the rules and may behave like private owners of the playground. (Karsten 2015 p. 459).

However, the existence of friendships is still dependent upon conditions of access to the spaces.

19.5.2 Environmental Accessibility

Attractiveness and "sense of place" cannot be dissociated from environmental accessibility because of the reduction in children's independent mobility (CIM) and uses in urban contexts (Hillman 1990/1997; Gaster 1991). CIM remains a popular and shared issue in diverse research fields, from social sciences to public health, from town planning to geography and psychology (Hillman et al. 1997; Kyttä 1997; Depeau 2003; Prezza et al. 2001; Prezza and Pacilli 2007). It has become a flourishing and even redundantly studied concept (Tranter and Whitelegg 1994; Collins and Kearns 2001; Page et al. 2010; Freeman and Tranter 2011; Rissotto and Tonucci 2002; Fotel and Thomson 2004; Kyttä 2004; Malone and Rudner 2011). Although such attention is explained in part because of an interest in development, it is also because it has triggered some studies linked to the urban paradigm of the "de-motorization of cities". Moreover, by focusing on children, their needs, their practices and what stops or encourages their autonomous movements, it is possible to contribute to improving neighborhoods in terms of active movements or physical activity. Besides the physical dimensions, it is worth considering first the proximal environment in order to define the levels of environmental accessibility, before taking a more social approach (referring to some of the dimensions of exo- and mesosystems of the environment) as embodied by the role of children's family backgrounds to extend the definition of environmental accessibility.

19.5.2.1 Urban Forms, Micro-Contexts and Risks

Research focused on the conditions of access to residential environments follows the evolution of societal challenges (town planning emphasizing safety; sustainable cities). Through its approach to urban forms and the variability of road-related risks, the first wave of studies described inequalities of access to the neighborhood (Sell 1985; Hillman et al. 1997; Blakely 1994; Lee and Row 1994). Through the prism of the paradigm of city de-motorization and of health objectives linked to the risks of children being sedentary, as well as the constant development of collecting (GPS) and finer analytical (FIS) methods, a substantial proportion of studies has focused on the search for more micro indexes (Islam et al. 2014), which mostly refer to principles of walkability (Gallimore et al. 2011). In the field of public health, and particularly the promotion of activities, a review focusing on residential factors used in health and psychology studies between 2005 and 2011 highlights the balance between in-house equipment vs. neighborhood equipment (Maitland et al. 2013).

The first wave of studies following the work of Hillman et al. (1990) showed that accessibility depends mostly on urban forms and the types of city planning that shape residential neighborhoods (city compacity), and therefore on the distance to school and to children's activity facilities (Depeau 2003; McMillan 2007; Loebach and Gilliland 2014). The distance between home and school is also the principal factor of children's independent travel (Depeau 2003; Loebach and Gilliland 2014) and of the time spent outside (Schlossberg et al. 2006; Loebach and Gilliland 2014); even more so in the context of a broadening of school zones triggered by school desectorization chosen by parents in France (Depeau 2003; Hillman 2006; Lehman-Frisch et al. 2012). Traffic and risks of accidents are also factors to be taken into account when it comes to the quality of living spaces. The highest number of accidents occurs in poorer social backgrounds and in dense urban residential areas (Thomson 1996). By reviewing the factors linked to accident risks (Kaufer-Christoffel et al. 1991), some areas may be listed as opposite to those that contribute to defining safe zones: the absence of one- or two-lane roads; residential areas controlling traffic and speed. Granié (2010) has also observed and explained the role of social categorization and road-rule categorization in children's road safety behaviors. The analysis of databases collected at three Parisian urban sites with different pedestrian planning (Depeau 2003) showed that the highest accident risk areas (in terms of gravity and affected children) were situated in the neighborhood of the new town, which was a priori the one with the most footpaths and playgrounds for children (compared to the city center of Paris). The urban form defined by buildings and social density, and sometimes by the volume of green spaces or the quality of pedestrian facilities, is a condition that still depends on planning periods, and its role varies from one study to another. Children living in semi-closed spaces (condominium yards) and close to green spaces were more autonomous (Prezza et al. 2001). Kyttä (2002) asserts that the diversity of environmental opportunities (affordances) in the inner city contributes to increasing children's independent mobility. In France, children living in the center of Paris traveled longer distances, and more often by themselves; whereas children living in the new town more frequently traveled in groups (Depeau 2003). The accessibility of neighborhoods, assessed from the number and type of street crossings on the way to school, is not so much linked to the density of traffic as to the speed of drivers, which is more variable in a neighborhood that is a priori more pedestrian-friendly. It increases the break effects on the children's trip to school and contradicts the adaptation to traffic if the child is less independent (Depeau 2003). As a matter of fact, nowadays, dense cities are tending to reassess their pedestrian plans. In this perspective, Gallimore et al. (2011) measured accessibility in terms of "micro- and macro-walkability" by comparing different urban forms, and particularly those related to the "new urban neighborhood", characterized by

> *well-connected streets [...], no cul-de-sacs, several small parks, protected open spaces, single family and town home residences and land designated for a commercial town center.* (Gallimore et al. 2011 p. 186).

By measuring micro-walkability designs (from "traffic safety, pleasantness, crime safety") and macro-walkability (based on the 3D principle: density, land-use diversity, pedestrian-friendly design), they showed that *"the new urban community was more walkable than the other forms"*. The notion of walkability applied to trips remains a principle often studied in urban planning, in the context of improving pedestrian-friendly streets. However, and despite the abundant diversity of planning factors taken into account in the observation of children's independent mobility, as noted by Larsen et al. (2012), the results are not consensual, especially when it comes to "the connectivity of streets" and "residential density".

The safety dimension of accessibility for travel and children's activities refers to the notion of legibility, conditioned by the imaginable property of the space (Lynch 1960), and by environmental familiarity, which is based on the cognitive and social construction of space (Depeau 2003). The observed relationship between the results of school itinerary spatial representations and and those of social representations of danger to the modalities of children's independent trips actually shows a few differences in qualification about levels of environmental accessibility. For instance, when experiences of children's autonomy are linked to the frequentation of open spaces in the neighborhood, the normative dimension of the structure of social representation of danger is associated with unknown people (Depeau 2003). This normative dimension varies between parents and children: the world of the road (accidents, vehicles and passengers) for parents and the more social and moral world (public spaces and crimes) for children. These worlds often do not refer to spaces dedicated to childhood, but rather to less specialized areas with more polymorphic uses and mixed publics: the street, for instance, and public spaces in general. The accessibility of these places is not necessarily obvious, and their uses often exclude children. In fact, because childhood has progressively found dedicated and always more specialized places in the city (Germanos 1995), the more informal spaces like the street or the empty spaces in the city remain marginal and the young populations that use them are marginalized (Valentine 1996). More than the space itself, it is its functions of openness, social heterogeneity, behavioral spontaneity or even serendipity that are fundamental qualities. These must also be studied in order to define the spatial accessibility of cities for children. The street has all of these qualities, which, as noted by Hart (2002), provide children with opportunities to diversify their interpersonal relationships in public spaces:

> Opportunities to interact with people of different social classes, cultures, ages, and to learn how to cooperate with them. (Hart 2002 p. 137).

Nevertheless, accessibility seems to be shaped by the whole socio-normative and technocratic system that today produces childhood spaces. Accessibility is inscribed in a strong social and functional legibility that can make children visible in the city but rather with controlling aims than to grant them behavioral ease and freedom (Hart 2002; Valentine 1996). However, when the street constitutes the vicinity of home, it remains a place favored by children for playing and socializing, especially when it offers distance from adults (cf. the woornerf). In public spaces, accessibility is thus provided by a balance between security and flexibility, social mixing, acceptable proximity of adults, and freedom of access. This freedom of access – mainly studied from the perspective of autonomy – remains fundamental in development since, as indicated by Deci and Ryan (2008):

> The first and most important component of self-determination theory is the basic need for autonomy, or feeling free to make one's own choices. (Deci and Ryan 2008)

Particularly because it also allows environmental skills to be developed (Chawla and Heft 2002). Entering through urban forms provides feeling of accessibility in terms of distances, urban amenities, physical and social density. Nevertheless, it only makes sense if taken in its historical context in terms of planning, as shown by the plethora of studies made at different periods (Carver et al. 2013). Accessibility goes beyond this mere morphological dimension and requires taking into consideration more sociocognitive processes based upon the relationships between individuals and the sense of place they create.

19.5.2.2 Accessibility Through the Prism of Parents and Their Relationship to the Neighborhood

The physical and material dimensions of the accessibility of children's proximal environments are only relevant if understood in the larger context of the everyday world, which includes, among other things, the family (Depeau 2003; Lewis and Torres 2010). In the context of children's independent mobility, the daily social context facilitates and/or constrains the internalization of values and the motivation for action. Control and support of autonomy (proactive encouragement) are the two modalities of social context that participate in the child's autonomy process (Joussemet et al. 2008). Therefore, accessibility still varies depending on the quality of educational attitudes, which, in order to be operative, are linked to the interpretation of situations to which children can/must be confronted, i.e. zones of proximal development[5] that Valsiner (1985) extended to the ecological domain of development in order to distinguish situations of current skills of children's "zone of free movement" and their "zone of encouraged movements" skills. These different zones remain variable depending on the child's age, the characteristics of the environment, and above all the parent's social category. They contribute, more or less directly, to producing parental styles that interfere with the levels of access and opening into the external physical or social environment (Kellerhalls and Montandon 1991). One of the opening modalities of environmental access is based on the perception of risks by parents, which still differs depending on the child's gender. In fact, restrictions on independent travel are stronger for girls than for boys (Prezza et al. 2001; Stone et al. 2014), particularly when it comes to traveling to school (McMillan et al. 2006). Moreover, linked people-related risks are perceived as more threatening for girls. Parental prescriptions imposed on girls for outings are often linked to those of "Little Red Riding Hood" (Depeau 2003). Primary guardians of child-environment relationships, parents really are an important filter in the translation and transmission of codes, values and meaning in relation to outdoor spaces and the environment (O'Neil et al. 2001), especially when the practices of the neighborhood differ from some educative attitudes. This is what Lehman-Frisch et al. (2012) demonstrate in terms of social mixing in gentrified neighborhoods. Neighborhood accessibility also and above all depends on the parent's own relationship to the environment. For instance, Coulton and Irwin (2009), in a study of low-income urban neighborhoods, showed that parents involved in neighborhood programs are more likely to encourage their children to participate in "out-of-school" programs. The fears, rules or control strategies of parents depend on their residential history, the conditions of urban amenities and resources (Foster et al. 2015) and the social housing characteristics of the neighborhood, which are sometimes qualified (Edwards and Bromfield 2009). In this regard, social mixing is an important issue of the living context and its effect on children's development and socializing experiences. By comparing the neighborhoods of two gentrified cities (Paris and San Francisco), Lehman-Frisch and Authier (2012) showed how the relationships to social mixing for families were operated by actions. They revealed that limits to social mixing (control of outings, choice of extracurricular activities, children's sociability) in children's social relationships do not occur in public spaces but rather in relation to home, particularly for more underprivileged children. When analyzed through social bonds, feelings of belonging, residential satisfaction and parents' insecurity, relationships with the neighborhood constitute a body of factors that define the type of anchoring in the neighborhood and the perception of children's quality of life in the neighborhood (Martinez et al. 2002).

Perception of risks, involvement in the neighborhood and relationships with the environment are also the bases of a new device that contributes to transforming the accessibility of neighborhoods in terms of traveling: walking-school-buses[6]. They have been developing for the last 15 years mainly in suburban towns around western cities. This system is often inscribed in a very strong form of children's travel institutionalization in neighborhoods (Depeau 2008). Presented as an alternative to cars, it is a way of reducing traffic around schools, as well as risks linked to cars and pollution. It also reinforces children's physical activity. Without being a very original form of organization, the walking-school-bus particularly develops parental dynamics, often very anchored in the neighborhood or in the school (Depeau 2012). Social inter-knowledge and the spatial proximity of families constitute the conditions of their being involved in the system. Nevertheless, the rather homogenous social com-

---

[5]first conceptualized by Vygotsky

[6]Group of children supervised by one or two parents to walk to and from school on a daily basis.

position of parent/child groups involved in all the towns makes the walking-school-bus a form of "entre-soi", sometimes consciously desired, sometimes only experienced, and makes the system an actual "in-bubble" traveling device (Depeau 2008). While encouraging an active mode of traveling for children, it is not an initiatory travel mode, furthering children's independence for other out-of-school travel routines. It is precisely what Hillman criticizes (2006), since the mean age for joining the walking-school-bus is 6 or 7. This transport device is only useful for very young children, who are learning how to travel (acquisition of road codes and co-presence in urban space) since they are free from car escorts, which allows them to obtain a place in the urban space. However, it loses its interest as soon as (older) children want to experiment with unauthorized traveling and to put their own autonomy capacity to the test. By restricting children's freedom more and more, not in functionally specialized spaces, but in socially and spatially adapted systems, children's relationships to public spaces cannot be built on a concept of frivolity and uncertainty (Roucous 2007) except sometimes in some residential areas. Isn't there a risk of crystallizing certain educational norms in the representation of the city and practices of the child?

19.6 Conclusion and Perspectives

Far from the ideas of Bion, who wanted children to go to the countryside in order to improve their living conditions, and because today a majority of children live in the city, the development of knowledge linked to child-environment relationships has to contribute to children's well-being and well-becoming. This is possible by participating in the construction of observation tools but, above all, by favoring a holistic and transactional approach. In fact, this approach enables us to go beyond the temptation of a simply spatial interpretation, which would reduce quality of life to the mere material and physical characteristics of the environment. In the end, conditions are rather relevant for current measurements, being developed particularly in the fields of geography and economy. Nevertheless, if (as shown by results collected with psychometric instruments of well-being) these material conditions have consequences for children-environment relationships (especially when it comes to housing), they are modulated by the social context in which children evolve – particularly the familial context and its own relationship to the environment, and by the ideological framework that produces models of spaces or living contexts. This relies on the delicate issue of planning for children where the difficulty (with the balance between the urban "full" and "empty") is to offer a good environment without making too many assigned spaces for children, which would be too normative and boring for them (Garnier 2015).

The overview of the main instruments of children's well-being with regard to more holistic and empirical approaches linked to environmental psychology highlights the interest of each of these approaches. In the first, with psychometric scales, the effort to consider well-being in a multidimensional approach is a way of making up for a few domains that are sometimes minimized in the other. This is particularly the case of the economic resources dimension, the housing domain and the availability of resources (equipment/space) at home, which are still insufficiently considered in the study of children's relationships to their neighborhoods in environmental psychology. Nevertheless, measuring inequalities in this field would help to understand better the differences in children's occupation of free time, the factors of independent mobility or sedentary behavior, and differences in children's socialization. The importance of the relationship with school could also be noted. It is a fundamental area conditioning children's development and has not been studied in great detail in this chapter.

In the second approach, well-being or rather children's quality of life is strongly defined by relationships to the environment, and also educational attitudes filtering meaning and perceived functions of the environment. Besides, although this second approach – discussing accessibility and attractiveness definitions – is more ecological, some dimensions still seem insufficiently studied and taken into account in environmental psychology. This is the case of the exo- and macro-systems, referring to the ideological model, and the notion of time, which is as fundamental as the notion of space in the field of childhood. In fact, while quality of life can be related to the characteristics of the spaces daily-used by children, it can also be defined through the daily times of childhood and its conditions of free-time occupation. This involves the notion of distance conditioning spatial access, which inevitably constrains children's travelling times and thus their autonomy, but also their free-time occupation. These free activities, often associated with children's idleness, are still an issue in the adults' and the city's organization and in planned surveillance.

Finally, questioning children's quality of life in an ecological approach to development requires broadening the definition of spaces in the digital domain. The rise of connected objects and the omnipresence of screens in the children's everyday life have to be further explored in environmental psychology. Clearly, these new digital uses question the role of new forms of socialization in everyday spaces and thus the development of social skills in the broadening of networks. They also question the self-regulation of identity, which seems to be akin to a quest for multiple identities. Besides, the use of cellphones, by allowing voluntary reactivation of proximal affective bonds, tests relationships between children and space when they are becoming autonomous, as well as their representation of space.

Acknowledgements This chapter was translated by Alice Verney (holder of the "aggregation"/qualified teacher in English, ENS, France)

References

Bahi-Fleury, G. (2009). Identification au lieu et aux pairs: quels effets sur la réussite scolaire? *Cahiers Internationaux de Psychologie Sociale, 81*(1), 97–113.

Baran, P. K., Smith, W., Moore, R. C., Floyd, M. F., Bocarro, J. N., Cosco, N. G., et al. (2014). Park use among youth and adults. Examination of individual, social, and urban form factors. *Environment and Behavior, 46*(6), 768–800.

Barker, R. G., & Wright, H. F. (1955). *Midwest and its children*. New York: Row, Peterson and Company.

Becchi, E., & Julia, D. (1998). *Histoire de l'enfance en Occident. Tome 2. Du XVIIIè siècle à nos jours*. Paris: Editions du Seuil.

Ben-Arieh, A., Hevener, K. N., Bowers, A. A., Goerge, R. M., Joo, L. B., & Lawrence, A. J. (2001). Measuring and monitoring children's well-being. *Social Indicator Research Series,* Vol 7. 188 pages. Springer Edition.

Benson, P. L., Scales, P. C., Hamilton, S. F., & Sesma, A., Jr. (2006). Positive youth development: Theory, research, and applications. In R. M. Lerner (Ed.), *Handbook of child psychology. Vol. one: Theoretical models of human development* (pp. 894–941). Hoboken: Wiley.

Blakely, K. S. (1994). Parents' conceptions of social dangers to children in the urban environment. *Children's Environments, 11*(1), 16–25.

Blanqui, A. (1849). Des classes ouvrières en France pendant l'année 1848. In l'Académie des sciences politiques et morales (Ed.), *Petits traités*. Paris: Pagnerre/Paulin & cie.

Blinkert, B. (2004). Quality of the city for children: Chaos and order. *Children, Youth and Environments, 14*(2), 100–112.

Bourdieu, P. (1979). *La distinction. Critique sociale du jugement*. Paris: Minuit.

Bradshaw, J., Hoelscher, P., & Richardson, D. (2007). An index of child well-being in the European union. *Social Indicators Research, 80*, 133–177.

Brandtstädter, J. (1998). Action perspectives on human development. In W. Damon (Editor in chief) & R. M. Lener (Vol. Ed.) *Handbook of child psychology: Vol. 1 theoretical models of human development* (5th ed,. pp. 807–863). New York: Wiley.

Broberg, A., Kyttä, M., & Fagerholm, N. (2013). Child-friendly urban structures: Bullerby revisited. *Journal of Environmental Psychology, 35*, 110–120.

Bronfenbrenner, U. (1979). *The ecology of human development*. Cambridge, MA: Harvard University Press.

Brougère, G. (1991). Espace de jeu et espace public. *Architecture & Comportement, 7*(2), 165–176.

Bryant, B. K. (1985). The neighborhood walk: Sources of support in middle childhood with commentary by Ross D. Parke. *Monographs of the Society for Research in Child Development, 50*(3), Serial N° 210.

Burdette, H., Whitaker, R., & Daniels, S. (2004). Parental report of outdoor playtime as a measure of physical activity in preschool-aged children. *Archives of Paediatric and Adolescent Medicine, 158*, 353–357.

Carbonara-Moscati, V. (1985). Barriers to play activities in the city environment: A study of children's perceptions. In T. Gärling & J. Valsiner (Eds.), *Children within environments: Toward a psychology of accident prevention* (pp. 119–126). New York: Plenum Press.

Carver, A., Watson, B., Shaw, B., & Hillman, M. (2013). A comparison study of children's independent mobility in England and Australia. *Children's Geographies, 11*(4), 461–475.

Catalano, R. F., Berglund, M. L., Ryan, J. A. M., Lonzak, H. S., & Hawkins, J. D. (2004). Positive youth development in the United States: Research findings on evaluations of positive youth development programs. *Annals of the American Academy of Political and Social Science, 591*, 98–124.

Chardonnel, S., & Depeau, S. (2014, March 26–27). Des routines quotidiennes familiales à l'imprévu dans l'apprentissage et l'accès à la ville des enfants: l'usage des "Behavior-Setting" et des "Pockets of local order" pour explorer les cultures éducatives urbaines. Colloque MSFS "Métro, boulot, dodo: quoi de neuf dans nos routines". Lille.

Chawla, L. (1992). Childhood place attachments. *Human Behavior and Environment: Advances in Theory and Research, 12*, 63–86.

Chawla, L., & Heft, H. (2002). Children's competence and the ecology of communities: A functional approach to the evaluation of participation. *Journal of Environmental Psychology, 22*(1–2), 201–216.

Chombart de Lauwe, M. J. (1987). *Espaces d'enfants: la relation enfants-environnement, ses conflits*. Fribourg: Del Val.

Chu, P. S., Saucier, D. A., & Hafner, E. (2010). Meta-analysis of the relationships between social support and well-being in children and adolescents. *Journal of Social and Clinical Psychology, 29*(6), 624–645.

Clark, C., & Uzzell, D. (2002). The affordances of the home, neighborhood, school and town centre for adolescents. *Journal of Environmental Psychology, 22*, 95–108.

Coleman, J. (1988). Social capital in the creation of human capital. *American Journal of Sociology, 94*, 95–120.

Collins, D., & Kearns, R. (2001). The safe journeys of an enterprising school: Negotiating landscapes of opportunity and risk. *Health and Place, 7*(4), 293–306.

Coulton, M., & Irwin, M. (2009). Parental and community level correlates of participation in out-of-school activities among children living in low income neighborhoods. *Children and Youth Services Review, 31*, 300–308.

Deci, E. L., & Ryan, R. M. (2008). Self-determination theory: A macrotheory of human motivation, development, and health. *Canadian Psychology, 49*, 182–185.

Depeau, S. (2003). *L'enfant en ville: autonomie de déplacement et accessibilité environnementale*. Thèse de doctorat (Phd).en Psychologie. Paris, René Descartes, Paris 5.

Depeau, S. (2008). Nouvelles façons de se déplacer vers l'école ou l'expérimentation du pédibus dans un quartier rennais. Quelles incidences sur l'apprentissage de l'autonomie de déplacement des enfants et leurs rapports à l'espace? *Revue Recherche Transport Sécurité, 101*, 253–271.

Depeau, S. (2012). Les bus pédestre dans les périphéries urbaines rennaises: dynamiques et fonctions dans les rapports à l'espace des familles. In M. Dumont (Ed.), *Trames de mobilités collectives: nouvelles expériences « publiques » du déplacement dans les périphéries urbaines. Analyse comparative dans les territoires de l'Ouest* (pp. 37–113). Rapport final, Programme « L'avenir des périphéries urbaines – la mobilité et le péri urbain », 2009–2012.

Depeau, S., & Quesseveur, E. (2014). A la recherche d'espaces invisibles de la mobilité: usages, apports et limites des techniques GPS dans l'étude des déplacements urbains à l'échelle pédestre. *Netcom, Networks and Communication Studies, 28*(1–2), 35–54.

Edwards, B., & Bromfield, L. M. (2009). Neighborhood influences on young children's conduct problems and pro-social behavior: Evidence from an Australian national sample. *Children and Youth Services Review, 31*, 317–324.

Ellaway, A., Kirk, A., Macintyre, S., & Mutrie, N. (2007). Nowhere to play? The relationship between the location of outdoor play areas and deprivation in Glasgow. *Health & Place, 13*, 557–561.

Faber Taylor, A., Kuo, F. E., & Sullivan, W. C. (2002). Views of nature and self-discipline: Evidence from inner-city children. *Journal of Environmental Psychology, 22*, 49–63.

Fattore, T., Mason, J., & Watson, E. (2007). Children's conceptualisation(s) of their well-being. *Social Indicators Research, 80*, 5–29.

Flouri, E., Midouhas, E., & Joshi, H. (2014). The role of urban neighborhood green space in children's emotional and behavioral resilience. *Journal of Environmental Psychology, 40*, 179–186.

Foster, S., Wood, L., Francis, J., Knuiman, M., Villanueva, K., & Giles-Corti, B. (2015). Suspicious minds: Can features of the local neighborhood ease parents' fears about stranger danger? *Journal of Environmental Psychology, 42*(June), 48–56.

Fotel, T., & Thomson, U. (2004). The surveillance of children's mobility. *Surveillance and Society, 1*(4), 535–554.

Freeman, C., & Tranter, P. (2011). *Children and their urban environment: Changing worlds*. London: Earthscan.

Gallimore, J. M., Brown, B. B., & Werner, C. M. (2011). Walking routes to school in new urban and suburban neighborhoods: An environmental walkability analysis of blocks and routes. *Journal of Environmental Psychology, 31*, 184–191.

Garnier, P. (2015, April). Une ville pour les enfants: entre ségrégation, réappropriation et participation. *Métropolitiques*. 10 avril 2015. Available, http://www.metropolitiques.eu/Une-ville-pour-les-enfants-entre.html

Gasper, D. (2010). Understanding the diversity of conceptions of well-being and quality of life. *The Journal of Socio-Economics, 39*, 351–360.

Gaster, S. (1991). Urban children's access to their neighborhood: Changes over three generations. *Environment and Behavior, 23*(1), 70–85.

Germanos, D. (1995). La relation de l'enfant à l'espace urbain. Perspectives éducatives et culturelles. *Architecture et Comportement, 11*(1), 55–61.

Granié, M. A. (2010). Socialisation au risque et construction sociale des comportements de l'enfant piéton: éléments de réflexion pour l'éducation routière. *Enfances, Familles et Générations, 12*, 88–110.

Hart, R. (2002). Containing children: Some lessons on planning for play from New York City. *Environment & Urbanization, 14*(2), 135–148.

Hart, R. A., & Moore, G. T. (1973). The development of spatial cognition: A review. In R. M. Downs & D. Stea (Eds.), *Image and environment: Cognitive mapping and spatial behavior*. Chicago: Adline.

Heft, H. (1988). Affordances of children's environments: A functional approach to environmental description. *Children's Environments Quarterly, 5*(3), 29–37.

Heft, H. (1998). Toward a functional ecology of behavior and development: The legacy of Joachim F. Wohlwill. In D. Görlitz, H. J. Harloff, G. Mey, & J. Valsiner (Eds.), *Children, cities and psychological theories: Developing relationships* (pp. 85–110). Berlin: Walter de Gruyter GmbH & Co.

Hillman, M. (2006). Children's rights and adults' wrongs. *Children's Geographies, 4*(1), 61–67.

Hillman, M., Adams, J., & Whitelegg, J. (1990). *One false move...a study of children's independent mobility*. London: PSI.

Hillman, M., Adams, J., & Whitelegg, J. (1997). *One false move...a study of children's independent mobility*. Dordrecht: Kluwer Academic Publishers.

Horelli, L. (2007). Constructing a theoretical framework for environmental child-friendliness. *Children, Youth and Environments, 17*(4), 267–292.

Hunter, J. P., & Csikszentmihalyi, M. (2003). The positive psychology of interest adolescents. *Journal of Youth and Adolescence, 32*, 27–35.

Islam, M., Moore, R., & Cosco, N. (2014). Child-friendly, active, healthy neighborhoods: Physical characteristics and children's time outdoors. *Environment and Behavior*, 1–26, Advance online publication. doi:10.1177/0013916514554694.

Jack, G. (2010). Place matters: The significance of place attachments for children's well being. *British Journal of Social Work, 40*, 755–771.

Jacquemain, D. (2003, October 2). Face à des rythmes sociaux qui ont évolué, et qui ont un impact sur la vie des enfants, les enjeux du temps libre sont majeurs. Quelques préconisations. In D. Gourland (Ed.), *Les temps de l'enfant. Les enfants sont-ils aux 35h?* Rapport du Forum "questions de temps".

Joussemet, M., Landry, R., & Koestner, R. (2008). A self-determination theory perspective on parenting. *Canadian Psychology, 49*(3), 194–200.

Kamerman, S. B., Phipps, S., & Ben-Arieh, A. (2010). *From child welfare to child well-being. An international perspective on knowledge in the service of policy making*. New York: Springer.

Karsten, L. (2015). Children's use of public space in the gendered world of the playground. *Childhood, 10*(4), 457–473.

Kaufer, C., Schofer, J. L., Lavigne, J. V., Tanz, R. R., Wills, K., White, B., Barthel, M., Mc Guire, P., Donovan, M., Buergo, F., Shawner, N., & Jenq, J. (1991). "Kids'n'Cars", an ongoing study of pedestrian injuries: Description and early findings. *Children's Environments Quarterly, 8*(2), 41–50.

Kellerhalls, J., & Montandon, C. (1991). *Les stratégies éducatives des familles: milieu familial, dynamique familiale et éducation des pré-adolescents*. Neuchâtel: Delachaux & Niestlé.

Koescher, A., Jiang, X., Ben Arieh, A., & Hebner, E. S. (2014). Advances in children's rights and children's well-being measurement: Implications for school psychologists. *School Psychology Quarterly, 29*(1), 7–20.

Korpela, K. M. (1992). Adolescents' favorite places and environmental self-regulation. *Journal of Environmental Psychology, 12*, 249–258.

Korpela, K. M. (2002). Children's environments. In R. B. Bechtel & A. Churchman (Eds.), *Handbook of environmental psychology* (pp. 363–373). New York: Wiley.

Kuo, F. E., Sullivan, W., Coley, R., & Brunson, L. (1998). Fertile ground for community: Inner-city neighborhood common spaces. *Journal of Community, 26*(6), 823–851.

Kyttä, M. (1997). Children's independent mobility in urban, small town, and rural environments. In R. Camstra (Ed.), *Growing up in a changing urban landscape* (pp. 41–52). Assen: Van Gorcum.

Kyttä, M. (2002). Affordances of children's environments in the context of cities, small towns, suburbs and rural villages in Finland and Belarus. *Journal of Environmental Psychology, 22*, 109–123.

Kyttä, M. (2004). The extent of children's independent mobility and the number of actualized affordances as criteria for child-friendly environments. *Journal of Environmental Psychology, 24*(2), 179–198.

Land, K. C., Lamb, V. L., & Mustillo, S. K. (2001). Child and youth well-being in the United States, 1975–1998: Some findings from a new index. *Social Indicators Research, 56*, 241–320.

Larsen, L., Gilliland, J., & Hess, P. M. (2012). Route-based analysis to capture the environmental influences on a child's mode of travel between home and school. *Annals of the Association of American Geographers, 102*(6), 1348–1365.

Lee, T., & Row, T. (1994). Parents' and children's perceived risks of the journey to school. Special issue: Children and the city. *Architecture and Behaviour, 10*(4), 379–389.

Lehman-Frisch, S., Authier, J. Y., & Dufaux, F. (2012). *Les enfants et la mixité sociale dans les quartiers gentrifiés à Paris, Londres et San Francisco*. Dossier d'Etudes. Paris: Caisse Allocations Familiales. N°153. 185 pages.

Lewin, K. (1926/1931). Environmental forces in child behavior and development. In C. Murchison (Ed.), *Handbook of child psychology* (pp. 590–625). Worcester: Clark University Press.

Lewis, P., & Torres, J. (2010). Les parents et les déplacements entre la maison et l'école primaire: quelle place pour l'enfant dans la ville? *Enfances, Familles et Générations, 12*, 44–65.

Loebach, J. E., & Gilliland, J. A. (2014, July). Free range kids? Using GPS-Derived activity spaces to examine children's neighborhood activity and mobility. *Environment and Behavior*, 1–33. Advance on-line publication doi:10.1177/0013916514543177

Lynch, K. (1960/1976). *L'image de la cité*. (M. F. Vénard & J. L. Vénard, traducteurs) Paris: Dunod. (Ouvrage original publié en 1960).

Magnusson, D., & Stattin, H. (2006). The person in context: A holistic-interactionistic approach. In R. M. Lerner (Ed.), *Handbook of child psychology. Vol. one: Theoretical models of human development* (pp. 894–941). Hoboken: Wiley.

Mahoney, J. L., & Eccles, J. S. (2008). Organized activity participation for children from low- and middle-income families. In A. Booth & A. C. Crouter (Eds.), *Disparities in school readiness* (pp. 207–222). New York: Lawrence Erlbaum Associates.

Maitland, C., Stratton, G., Foster, S., Braham, R., & Rosenberg, M. (2013). A place for play? The influence of the home physical environment on children's physical activity and sedentary behavior. *International Journal of Behavioral Nutrition and Physical Activity*, 99(10), 1–21.

Malinowski, J. C., & Thurber, C. A. (1996). Developmental shifts in the place preferences of boys aged 8–16 years. *Journal of Environmental Psychology*, 16, 45–54.

Malone, K., & Rudner, J. (2011). Global perspective on children's independent mobility: A socio-cultural comparison and theoretical discussion on children's lives in four countries in Asia and Africa. *Global Studies of Childhood*, 1(3), 243–259.

Martinez, M. L., Black, M., & Starr, R. H. (2002). Factorial structure of the perceived neighborhood scale (PNS): A test of longitudinal invariance. *Journal of Community Psychology*, 30(1), 23–43.

Matthews, M. H. (1987). Gender, home-range and environmental cognition. *Transactions of the Institute of British Geographers, New Series*, 12, 43–56.

McMillan, T. E. (2007). The relative influence of urban form on a child travel mode to school. *Transportation Research Part A: Policy and Practice*, 41, 69–79.

Min, B., & Lee, J. (2006). Children's neighborhood place as a psychological and behavioral domain. *Journal of Environmental Psychology*, 26, 51–71.

Moore, C., & Wong, H. (1997). *Natural learning. Creating environments for rediscovering nature's way of teaching*. Berkeley: MIG Communications.

Nordström, M. (2010, March). Children's views on child-friendly environments in different geographical, cultural and social neighborhoods. *Urban Studies*, 47(3), 514–528.

O'Neil, R., Parke, R. D., & McDowell, D. J. (2001). Objective and subjective features of children's neighborhoods: Relations to parental regulatory strategies and children's social competence. *Applied Developmental Psychology*, 22, 135–155.

Page, A. S., Cooper, A. R., Griew, P., & Jago, R. (2010). Independent mobility, perceptions of the built environment and children's participation in play, active travel and structured exercise and sport: The PEACH project. *International Journal of Behavioral Nutrition and Physical Activity*, 7, 17.

Pellegrini, A. D., Blatchford, P., Kato, K., & Baynes, E. (2004). A short-term longitudinal study of children's playground games in primary school: Implications for adjustment to school and social adjustment in the USA and the UK. *Social Development*, 13(1), 107–123.

Pernoud, E. (2010). L'enfant au square. *Enfance et Psy*, 46(1), 148–156.

Pitner, R. O., & Astor, R. A. (2008). Children's reasoning about poverty, physical deterioration, danger and retribution in neighborhood contexts. *Journal of Environmental Psychology*, 28, 327–338.

Prezza, M., & Pacilli, M. G. (2007). Current fear of crime and sense of community and loneliness in Italian adolescents: The role of autonomous mobility and play during childhood. *Journal of Community Psychology*, 35(2), 151–170.

Prezza, M., Pilloni, S., Morabito, C., Sersante, C., Aparone, F. R., & Guiliani, M. V. (2001). The influence of psycho-social and environmental factors on children's independent mobility and relationship to peer frequentation. *Journal of Community Applied Social Psychology*, 11, 435–450.

Prieske, B., Withagen, R., Smith, J., & Zaal, F. T. J. M. (2015). Affordances in a simple playscape: Are children attracted to challenging affordances. *Journal of Environmental Psychology*, 41, 101–111.

Ragon, M. (1986). *Histoire de l'architecture et de l'urbanisme modernes. 1. Idéologies et pionniers 1800–1910*. Paris: Casterman.

Rissotto, A., & Tonucci, F. (2002). Freedom of movement and environmental knowledge in elementary school children. *Journal of Environmental Psychology*, 22, 65–77.

Rogoff, B., & Lave, J. (1984). *Everyday cognition: Its development in social context*. Cambridge: Harvard University Press.

Roucous, N. (2007). Les loisirs de l'enfant ou le défi de l'éducation informelle. *Revue Française de Pédagogie*, 160(3), 63–73.

Scales, P. C., Benson, P. L., Leffert, N., & Blyth, D. A. (2000). Contribution of developmental assets to prediction of thriving among adolescents. *Applied Developmental Sciences*, 4(1), 27–46.

Schlossberg, M. J., Greene, P. P., Johnson, B., & Parker, B. (2006). School trips: Effects of urban form and distance on travel mode. *Journal of the American Planning Association*, 72, 337–346.

Schonert-Reichl, K. A., Guhn, M., Gadermann, A. M., Hymel, S., Sweiss, L., & Hertzman, C. (2013). Development and validation of the middle years development instrument (MDI): Assessing children's well-being and assets across multiple contexts. *Social Indicators Research*, 114, 345–369.

Sell, J. L. (1985). Territoriality and children's experience of neighborhoods. *Children's Environments Quarterly*, 2(2), 41–48.

Staempfli, M. B. (2009). Reintroducing adventure into children's outdoor play environments. *Environment and Behavior, 41*(2), 268–280.

Stone, M. R., Faulkner, G. E. J., Mitra, R., & Buliung, R. N. (2014). The freedom to explore: Examining the influence of independent mobility on weekday, weekend and after-school physical activity behavior in children's living in urban and inner-suburban neighborhoods of varying socio-economic status. *International Journal of Behavioral Nutrition and Physical Activity, 11*(5), 1–11.

Sue, R. (2006). Les temps nouveaux de l'éducation. *Revue du Mauss, 28*, 193–203.

Thomson, J. A. (1996). Increasing traffic competence in young children. In B. Gillham & J. A. Thomson (Eds.), *Child safety: Problem and prevention from preschool to adolescence* (pp. 86–112). New York: Routledge.

Timperio, A., Giles-Corti, B., Crawford, D., Andrianopoulos, N., Ball, K., Salmon, J., et al. (2008). Features of public open spaces and physical activity among children: Findings from the CLAN study. *Preventive Medicine, 47*(5), 514–518.

Tobelem-Zanin, C. (1995). *La qualité de la vie dans les villes françaises*. Rouen: Presses Universitaires de Rouen.

Torell, G., & Biel, A. (1985). Parental restrictions and children's acquisition of neighborhood knowledge. In T. Gärling & J. Valsiner (Eds.), *Children within environments: Toward a psychology of accident prevention* (pp. 107–117). New York: Plenum Press.

Tranter, P., & Whitelegg, J. (1994). Children's travel behaviors in Canberra: Car-dependent lifestyles in a low-density city. *Journal of Transport Geography, 2*(4), 265.

Tsoukala, K. (2001). *L'image de la ville chez l'enfant*. Paris: Anthropos.

Valentine, G. (1996). Children should be seen and not heard: The production and transgression of adults' public space. *Urban Geography, 17*(3), 205–220.

Valsiner, J. (1985). Theoretical issues of the child development and the problem of accident prevention. In T. Gärling & J. Valsiner (Eds.), *Children within environments: Toward a psychology of accident prevention* (pp. 13–36). New York: Plenum Press.

Veitch, J., Bagley, S., Ball, K., & Salmon, J. (2006). Where do children usually play? A qualitative study of parents' perception of influences on children's active free-play. *Health & Place, 12*(4), 383–393.

Vygotsky, L. (1931/1978). *Mind in society*. Harvard University Press, Cambridge, M.A.

Wapner, S. (1998). A holistic, developmental, systems-oriented perspective: Child-environment relations. In D. Görlitz, H. J. Harloff, G. Mey, & J. Valsiner (Eds.), *Children, cities and psychological theories: Developing relationships* (pp. 278–300). Berlin: Walter de Gruyter GmbH & Co.

Wells, N. M., & Evans, G. W. (2003). Nearby the nature: A buffer of life stress among rural children. *Environment and Behavior, 35*(3), 311–330.

White, S. (2009). *Bringing wellbeing into development practice*. University of Bath, WeD Working Paper 09/50.

Whitzman, C., Worthington, M., & Mizrachi, D. (2010). The journey and the destination matter: Child-friendly cities and children's right to the city. *Built Environment, 36*(4), 474–486.

Wohlwill, J. F. (1987). Varieties of exploratory activities in early childhood. In D. Görlitz & J. F. Wohlwill (Eds.), *Curiosity, imagination and play* (pp. 60–77). Hillsdale: Erlbaum.

Everyday Environments and Quality of Life: Positive School and Neighborhood Environments Influence the Health and Well-Being of Adolescents

20

Taciano L. Milfont and Simon J. Denny

20.1 Introduction

A key purpose that unites environmental psychology research is the quest to understand better the complex relationships between individuals and their built and natural surroundings (Gifford 2009). At the same time, Moser (2009) has suggested that a major role of environmental psychology research is to address people–environment congruity, which he defines as "the interrelation between the individual and his or her (especially: residential) environment, considering the match between individual life satisfaction and objective standards of living" (p. 351). In his view, it is this people–environment congruity that truly enables quality of life (QOL). Environmental psychology research is thus intrinsically linked to the understanding and promotion of objective (social welfare) and subjective (well-being and life satisfaction) indicators of QOL (see, for example, Uzzell and Moser 2006).

The present chapter focuses on two particular everyday environments—schools and neighborhoods—and on their role in promoting QOL. The influence of everyday environments on individuals' thinking, feeling, and actions has been extensively examined (for recent reviews, see Gifford 2014; Oishi 2014), and pioneers in environmental psychology have explored the influence of school and neighborhood environments (for historical reviews, see, for example, Bonnes and Secchiaroli 1995; Pol 2006, 2007).

One of the earliest environmental psychology studies of how school behavioral settings influence behavior was carried out by Barker and Gump (1964). These authors investigated the relations between high school size in terms of enrolment number (e.g., four high schools of 83–151 students compared to a high school of 2287 students), school settings and student participation in extracurricular activities (e.g., music festivals, journalistic competitions). They observed that the average number and kinds of extracurricular activities were twice as great for students in small schools than in large schools, suggesting that smaller groups are more satisfying for their members. These findings led the authors to conclude that "a school should be sufficiently small that all of its students are needed for its enterprises. A school should be small enough that students are not redundant" (p. 202).

Regarding neighborhood environments, Lee (1968) examined mental representations that inhabitants of Cambridge (UK) had of the physical-social space of their neighborhoods. He proposed an index called the 'neighborhood

T.L. Milfont (✉)
School of Psychology, Victoria University of Wellington, Wellington, New Zealand
e-mail: taciano.milfont@vuw.ac.nz

S.J. Denny
Department of Pediatrics, Child and Youth Health, University of Auckland, Auckland, New Zealand
e-mail: denny@auckland.ac.nz

quotient' expressing ratios of the physical properties of the environment that represent low-to-high socio-cognitive involvement and active participation with the neighborhood. Lee showed that the cognitive maps and socio-spatial schemas held by inhabitants are associated with the range and frequency of activities they engage in and their integration with the surrounding neighborhood. His observations have important urban design implications, particularly in terms of subdivision and size/density of urban spaces. Lee concluded by noting that planners in modern societies are "employed to fashion the environment of the future, and in this he [sic] idealistically includes the creation of *communities*. (...) Also there is accumulating research evidence that behavior and environment are interdependent, which implies that the planners' manipulations *can* influence behavior" (p. 265, emphasis in the original).

The work by Barker and Gump (1964) and Lee (1968) illustrates an early concern of environmental psychologists for person–environment relationships within these two important everyday environments, and more recent reviews of the literature describing the particularities of school and neighborhoods are also available (e.g., Clark and Uzzell 2002; Olivos 2010; Rivlin and Weinstein 1984). The present chapter provides a much more focused approach. First, data from individuals in a particular school or neighborhood context tend to be more similar than data from individuals from different school or neighborhood contexts; and the structure of such data is hierarchical, with individuals nested within the contextual unit of the particular school or neighborhood. This implies that a multilevel perspective is required for a proper examination of the influence of school and neighborhood environments on QOL indicators. Paraphrasing Afonzo (2005, p. 809), adopting a narrow approach to the multilevel problem of person–environment relationships might lead to a piecemeal understanding of the many characteristics of school and neighborhood environments affecting individuals' thinking, feeling and actions. We therefore focus our attention on studies that have used a multilevel approach. We also focus on studies that have examined the influence of school and neighborhood environments on youth.

The chapter is divided into three main sections. The first discusses the necessity of using multilevel models when examining the influence of macro factors, such as characteristics of school/neighborhood environments, on individual outcomes. This discussion on multilevel models is relevant in the present context because the chapter summarizes studies employing multilevel models. This section is also important because multilevel models should be further fostered in environmental psychology research since it centers on the relationships between individuals and their built and natural surroundings. Since the research reviewed focuses on youth, we also discuss strengths-based understanding of adolescent health and well-being in the first section of the chapter. The second section of the chapter provides a selective coverage of theoretical models and multilevel and review studies examining school and neighborhood environments. The third section of the chapter provides a more detailed discussion of a case study, offering a background of the Youth2000 Survey Series and a more detailed summary of selected studies that have used this survey dataset. The concluding section considers directions for future studies.

20.2 Multilevel and Strengths-Based Approaches

This section covers two related sub-topics. A multilevel approach is discussed first followed by a review of strengths-based understanding of adolescent health and well-being.

20.2.1 Studying School and Neighborhood Environments: A Multilevel Approach

One of the difficulties in studying school and neighborhood environments is understanding the distinction between the effects of individual-level variables and school/neighborhood-level variables on individual outcomes. This is due to the hierarchical structure of the dataset under scrutiny. Studies exploring the influence of school or neighborhood environments on adolescents have to deal with multilevel data. Such datasets have a hierarchical structure in which observations at the individual level of analysis (e.g., students' life satisfaction) are nested within observations at the school/neighborhood level of analysis. For example, in one setting students are embedded within classrooms, which are embedded within distinct schools, while in another setting students are embedded within households, which are embedded within distinct neighborhoods.

This issue has a long history in educational research. In 1976, Cronbach summarized some of the methodological pitfalls and pointed out that existing research did not distinguish within-school effects from between-school effects and was therefore drawing erroneous conclusions. In discussing this issue, Cronbach (1976) concluded: "The majority of studies of educational effects—whether classroom experiments, or evaluations of programs, or surveys—have collected and analyzed data in ways that conceal more than they reveal. The established methods have generated false conclusions in many studies" (p. 8).

This issue becomes even more problematic in observational and survey studies, as the direction of the associations cannot be ascertained. For example, there is an association between individual-level school connectedness and suicide attempts (Fleming et al. 2007). In longitudinal data, however, the relationship between school connectedness and mental health outcomes is far from clear. A recent systematic review of longitudinal studies found little evidence that school environments play a major role in the development of adolescent mental health concerns (Kidger et al. 2012). It is likely that the robust correlation between individual measures of school connectedness is due to the fact that students experiencing depression (or other mental health concerns) are more likely to report school factors more negatively than students without these mental health concerns.

To explore school/neighborhood-level phenomena, it is therefore important that researchers use school-level measures, which can be aggregated data from individual surveys, direct observation or third-party informants. This approach requires the use of multilevel models, where the effects of variables at each level of analysis on the outcome variable are estimated while controlling for the other variables, and cross-level interaction can also be estimated (see Luke 2004; Nezlek 2008; Raudenbush and Bryk 2002 for discussions on multilevel models).

One issue to consider when aggregating student information to the school-level or neighborhood-level is ensuring that the measure is reliable. Reliability in the context of multilevel studies refers to the number of students surveyed in each school/neighborhood and the correlation between students in the same school/neighborhood (Raudenbush 1999). In general, for intra-class correlations over 0.1, the reliabilities flatten out once there are more than 10–20 students. Thus, in the studies reviewed below that have examined neighborhood contexts, neighborhoods with fewer than ten students have been excluded.

Recent environmental psychology research has used multilevel models (see, for example, Kumar et al. 2008; Milfont and Markowitz 2016; Oshio and Urawaka 2012; Regoeczi 2003; Schultz et al. 2014) but, due to the nature of our field, multilevel models should be more widespread and it is an important future direction to foster in our research endeavors.

20.2.2 Strengths-Based Understanding of the Health and Well-Being of Adolescents

Besides multilevel modeling, another important issue refers to the conceptualization of everyday environments impacting youth. Over the last decades, our understanding of youth health and well-being has undergone significant shifts in conceptualization—perhaps due to the same paradigm shift that has led to the new approach of QOL research and positive psychology (see, for example, Corral-Verdugo 2012). Among the most important shifts has been the move towards strengths-based understanding of young people's health and well-being, in recognition that defining health and well-being as the 'absence of risk' was insufficient to capture fully the needs and lives of young people. This shift in focus reflected a renewed interest in identifying factors that promote competence or positive outcomes (for a recent review, see Masten 2014).

Initially, this emerged from studies looking at populations characterized by poor outcomes and identified protective factors, such as parental attachment or involvement with a church or community group, which promote good outcomes despite adversity (Werner and Smith 1992). These ideas spread to studies examining competence and positive outcomes among general populations of children and youth. Some of the 'protective factors' identified in this body of research included characteristics inherent in the individual, such as an easy-going temperament and at least average intellectual functioning (Fergusson and Lynskey 1996; Herrenkohl et al. 1994); characteristics within the family, such as parental attachment (Bradley et al. 1994; Gribble et al. 1993); and connections outside the family, such as external interests and/or identification with a non-related adult (Garmezy 1991; Jenkins and Smith 1990; Resnick et al. 1997). In fact, outcome evaluations across a wide range of youth health programs have shown that the successful programs used strategies to increase the competences of young people as well as to reduce their risk behaviors. This has led to the conceptualization of 'positive youth development' as central to efforts to improve the health and well-being of young people.

In a review of this body of work, Masten (2004) identified a 'short list' of protective factors or predictors of resilience in youth that have been consistently identified. As can be seen in the table below, effective schools and communities are among these predictors (Table 20.1).

Table 20.1 Summary of correlates and predictors of resilience in youth

One or more effective parents
Connections to other promoting and caring adults
Cognitive, attention, and problem-solving skills
Effective emotional and behavioral regulation
Positive self-perceptions (of efficacy, worth)
Beliefs that life has meaning; hopefulness
Religious faiths and affiliations
Aptitudes and characteristics valued by society (e.g., talent, attractiveness)
Pro-social friends
Socio-economic advantages
Effective schools, school bonding
Effective community (e.g., safe, with emergency services, recreation centers)

Source: Masten 2004, p. 315

Another shift has been the growing appreciation of the 'up-stream' determinants of young people's health and well-being, and attempts to understand the contextual, ecological and historical factors that place communities at risk and produce disparities between groups. Schools and neighborhoods are important contexts in this respect. For example, a study concluded that students who attended schools that deliberately set out to be health-promoting were not only healthier but also achieved better academically (Lee et al. 2006). Another illustration of this upstream approach is the Gatehouse project led by Patton and colleagues (2003). This was a large cluster randomized school intervention aimed at improving student health and emotional well-being through an innovative school-led program designed to promote a sense of social inclusion and connectedness among students. At the 4-year follow-up, the prevalence of substance use, antisocial behaviors and early initiation of sexual intercourse was approximately 20 % lower in the intervention schools than in the control schools (Patton et al. 2006). This study was remarkable in the efforts the research team took to work collaboratively with schools to change the school psycho-social environment. This up-stream approach is intrinsically linked to multilevel models as discussed above.

20.3 Selective Review of General Theoretical and Empirical Studies

After discussing the importance of multilevel and strengths-based approaches for a better understanding of how characteristics of school/neighborhood environments impact youth outcomes, this section provides a selective review of past research. First, there is a brief overview of useful theoretical models, and then past multilevel studies and summary studies using meta-analysis or critical reviews are described.

20.3.1 Review of Theoretical Models

Pikora et al. (2003) provided a framework of four main physical environmental determinants of walking and cycling in the local neighborhood. In this framework, the fundamental structural and physical attributes of the neighborhood comprise the *functional* feature (e.g., intersection and path design, direct route, and traffic volume). The *safety* feature incorporates personal and traffic elements of safety, such as the presence of lighting and availability of crossings. The presence of parks and private gardens, the level of pollution and architectural designs within the neighborhood are examples of factors in the *aesthetic* feature. Finally, factors in the *destination* feature comprise the availability of relevant facilities in the neighborhood, such as parks, schools, post offices, and train/bus stations.

A related approach was proposed by Alfonzo (2005), who developed an interesting socio-ecological model detailing the hierarchy of walking needs. Drawing on Maslow's well-known model, her model presents five hierarchically-organized levels of needs posited as antecedents of the walking decision-making process. These needs, with example factors, are: feasibility (mobility, time and responsibilities), accessibility (walking-related infrastructure), safety (physical incivilities and fear of crime), comfort (urban design amenities), and pleasurability (diversity and complexity, aesthetic appeal). Alfonzo argued that this hierarchy of walking needs could help explain the extent to which individual, group, regional and physical-environmental factors affect walking, by taking into account distinct stages of the behavioral decision-making process.

A model developed by the Royal Incorporation of Architects in Scotland describes quality indicators in the design of schools (or QIDS; Tombs 2005). It includes nine indicators in three main groups: functionality (uses and spaces, access, and external environment), build quality (engineered systems and performance, and construction), and impact (character and form, internal environment, social integration, and sustainability/ecology). The model provides a set of issues that stakeholders need to address when designing new school buildings but it is also a useful post-occupancy evaluation approach to assess school buildings. The indicators forming the impact grouping have clear implications for QOL of those using the school buildings.

20.3.2 Review of Selected Multilevel Studies

A multilevel study examined the influence of school environments on adolescent risk and health behavior across 29 Flemish schools in Belgium (Maes and Lievens 2003). A total of 31 independent school-level variables were considered, such as selling sweets and snacks at school, type of education offered, gender ratio of students, and both class and school size. Results indicated that, after accounting for individual-level variables, the school-level variables made a difference for three health-related behaviors: regular smoking, regular alcohol consumption, and teeth-brushing. Regular smoking was more likely in schools where the teachers' workload was higher than average, and in schools where the score on policy on rules for students was lower. Regular alcohol consumption was also more likely in schools where the score on policy on rules for students was lower, but less likely in schools where the administrator was female.

Finally, regular teeth-brushing was also more likely in schools where the administrator was female, and also in schools where the gender ratio of teachers was unbalanced, with male teachers outweighing female teachers.

According to the broken window theory, if a broken window is left unrepaired, all the windows in the building will soon be broken because disorder and crime are inseparably linked at the community level (Wilson and Kelling 1982). More recent environmental psychology research has indeed shown that when individuals observe others violating social norms (e.g., presence of graffiti, not returning shopping carts to the parking area, using illegal fire crackers), the individuals are more likely to violate other norms, causing disorder to spread (Keizer et al. 2008). In a multilevel study investigating predictions derived from the broken window theory, Brown and colleagues (2004) examined the extent to which objective and perceived physical incivilities in the neighborhood were related to crime. Objectively observed physical incivilities included amounts of litter and peeling paint, roofs and sidewalks in poor condition, the absence of flower or vegetable gardens, and the presence of broken windows or lights, graffiti and lawns. The multilevel findings indicated a cross-level interaction between observed physical incivilities and weak home attachment in the prediction of police-reported crime. In particular, houses in the neighborhood with more observed physical incivilities were more likely to be involved in crime occurrences, and this effect was intensified for householders who had fewer social contacts with their neighbors.

Flouri et al. (2014) conducted a multilevel study examining the effect of urban neighborhood green space on parent-reported emotional and behavioral adjustment and resilience in early-to-middle childhood (ages 3, 5 and 7) in the UK. Neighborhood green space was measured by the percentage of green space within a standard small area, but excluding domestic gardens. The findings indicated that neighborhood green space predicted emotional resilience. Specifically, poor children living in neighborhoods with a higher percentage of green space had fewer emotional problems from age 3 to 5 than children living in less green neighborhoods.

Another multilevel study examined the influence of the built environment on physical activity in residents of 14 multifamily housing complexes in the USA (Larco et al. 2012). The built environment characteristics examined included pedestrian-friendly features (i.e., external route directness from houses to a shopping area, presence of protected pedestrian paths, and presence of external streets that residents would have to cross and/or travel along), distance to a shopping area, and density of apartment complex. Physical activity (or lack thereof) was measured in terms of the number of bike and walking trips to a shopping area, number of driving trips to a shopping area, and percentage of all trips to a shopping area using an active mode (bike or walking). In line with other findings, residents in the more pedestrian-friendly and physically connected housing complexes were more likely to walk or bike (and less likely to drive) to the shopping area compared to residents in less connected and pedestrian-friendly areas.

20.3.3 Review of Selected Meta-analysis and Revision Studies

Durlak and Wells (1997) reported a meta-analysis of 177 primary programs aimed at preventing behavioral and social problems in youth. Among the environment-centered programs reviewed, 15 targeted school settings, modifying existing or creating settings for school-aged children. They focused on changing the psychosocial aspects of the existing classroom environment, improving classroom features including curricula, teacher-student relationships, and parent involvement in school activities, and creating a new child development center. Durlak and Wells observed that all reviewed environment-centered programs produced a significant positive effect (mean effect size = .35) in preventing behavioral and social problems in children and adolescents.

Another review of the literature focused on 150 studies examining the associations between environmental characteristics and physical activity in children and adolescents (Ferreira et al. 2007). Characteristics of the neighborhood physical environment included availability and accessibility of physical activity programs or facilities, and neighborhood safety and hazards such as no lit crossings, heavy traffic, physical disorder and pollution. All neighborhood physical environments considered were consistently unrelated to physical activity in children, and only low crime incidence in the neighborhood was associated with physical activity in adolescents. Regarding school characteristics, school policy environment, such as time allowed for free play, time spent outdoors and number of field trips, showed a positive association with physical activity in children, while type of school attended (high school versus vocational) had a positive association with physical activity in adolescents. The review also observed that some characteristics of home environments (i.e., father's level of physical activity with children, support from significant others, mother's education level, and family income in adolescents) were also positively associated with physical activity.

A similar review reported associations between objective measures of neighborhood environment and physical activity in youth (Ding et al. 2011). The review showed that walkability, access to recreation facilities and open space, presence of street trees, land-use mix, residential density, walking facilities such as sidewalks, and traffic speed/volume were associated with reported physical activity among children. Moreover, land-use mix and residential density were associated with reported physical activity among both children and adolescents, which led the authors to conclude that environment and policy change priorities could target these environmental conditions in neighborhoods to foster physical activity in youth.

Sellström and Bremberg (2006) conducted a review of 17 multilevel studies examining whether school environment characteristics influenced student outcomes. Having a health or anti-smoking policy, a good school climate, high average socio-economic status and an urban location were the four main school environment characteristics positively influencing student outcomes. Most of the outcomes considered (smoking habits, well-being, problem behavior, and school achievement) have implications for the QOL of students. This review suggests that the broad physical and socio-economic environment of schools has a positive contextual effect on the outcomes of students, after determinants operating at the individual level are taken into account.

Overall, the multilevel and review studies described so far strongly suggest that friendly and positive physical features of schools and neighborhoods can afford physical activity and other related factors, with clear implications for QOL indicators. The next section presents a case study investigating the influence of these everyday environments on the health and well-being of young people in New Zealand.

20.4 Case Study: The Youth2000 Survey Series

The Adolescent Health Research Group was established by researchers at The University of Auckland, New Zealand in 1997 with the aim of providing accurate and timely information on young people that communities, schools, parents, and policy-makers could use to improve the health status of young people. Under the Youth2000 project set up by this research group, large national youth surveys of over 8000 secondary school students were conducted in 2001, 2007 and 2012 (see Clark et al. 2013). The development of these national youth surveys was based on wide consultation and guidance from many steering groups. The overall aim was to improve the health and well-being of youth through the collection, analysis and dissemination of accurate and timely information about young people aged 12–18 years in New Zealand.

A two-stage random sample cluster design was employed to select schools and students for the Youth2000 Surveys, comprising a nationally representative sample of secondary school students in New Zealand. The first national survey carried out in 2001 used a self-report questionnaire administered on laptop computers with questions available by voiceover via headphones. A total of 133 high schools from those with more than 50 students in Year 9 through to Year 13 (i.e., ages from about 13 to 18) were randomly selected and invited to participate. At each of the 114 schools that agreed to participate, 15 % of students were randomly selected from the school roll. In total 9699 students took part in the 2001 survey. The second national survey conducted in 2007 used a development of the same questionnaire, but this time administered on internet tablets—essentially small handheld computers (see Denny et al. 2008). A total of 115 high schools were randomly selected and invited to participate and, in those schools agreeing to participate, 18 % of students were then randomly selected from the school roll. In total 9107 students from 96 secondary schools took part in the 2007 survey. The third national survey was administered in 2012 following similar procedures and with handheld internet tablets. A total of 125 high schools were randomly selected and invited to participate. In total 8500 students from 91 schools took part in the 2012 survey.

School principals gave consent for their own school to take part. A few weeks before the survey, information was sent to each school for distribution to parents and students. Parents were able to opt to have their child withdrawn from the study. On the day of the survey, invited students were given a verbal briefing and provided with an opportunity to ask questions, after which each student gave their own consent to participate at the beginning of the survey.

Findings from the Youth2000 Surveys highlight both the diversity and the common features of young people attending secondary schools in New Zealand. The data from the Youth2000 Surveys have been presented and published extensively and can easily be found online.

20.4.1 Selective Summary of Multilevel Youth2000 Studies

When designing the Youth2000 Surveys, the goal was to examine which factors keep young people well in New Zealand. In other words, what protects them from the risks inherent in growing up in a society where potentially negative influences, such as alcohol and drugs are readily available, cars can be bought relatively cheaply, and the media normalize violence and the initiation of sexual activity at an early age. Considering the 'positive youth development' framework employed by the project, the focus was on the protective factors (and not merely on risky factors) impacting the life of young people.

Here, we center our selective summary on published material that has used data from the second Youth2000 Survey conducted in 2007. In this survey, students completed the health and well-being survey as in the other waves, but data were also gathered from senior staff from participating schools who completed a senior management survey, and from teachers of these schools who completed a separate questionnaire. This triangulation of data sources enabled us to explore the influence of macro factors, such as the physical, interpersonal and social environments of each school and the surrounding neighborhood, on the health and well-being of students. Given the focus of the present chapter, our review will center on the multilevel models examining the influence of school/neighborhood factors on the outcome variable of interest. In all the multilevel models described, individual-level variables of the students, such as age, sex and ethnicity, are included as student-level covariates. The specific results regarding the influence of individual-level factors are not discussed here but can be obtained from the respective publications.

20.4.1.1 School Environments

An initial research report presented the overall findings regarding the social climate of New Zealand schools (Denny et al. 2009). As expected, results showed that the social climate usually varies between schools as a reflection of particular school characteristics, such as school type (public vs. private), school size (i.e., number of students attending school, or social density), and socio-economic make-up. Overall, smaller schools did better in terms of support for students and staff than larger schools. Compared to students from larger schools, students from smaller schools reported better school connection, better perception regarding their safety, were more likely to report feeling part of their school, and were less likely to report problems getting along with other students. These findings suggest that social density is a key variable in school environments. Past research has also shown that spatial density (i.e., square footage per school child) can influence student outcomes (Maxwell 2003), and that both household and neighborhood density are important in predicting aggression and withdrawal (Regoeczi 2003).

There were also distinctions between small and larger single-sex schools. Compared to students in other schools, students attending small boys-only schools were more likely to report that their school encouraged students of different ethnic groups to get along. However, teachers from the same small boys-only schools perceived their students to have poorer academic orientation, poorer teacher-student interactions, and more student disruptiveness than teachers from other schools. This lack of student-teacher congruity in perceptions of their school environment can have detrimental effects on the overall social climate. Interestingly, teachers from small girls-only schools had some of the best ratings of their school climate in terms of support for students, support for teachers and teachers' perceptions of their students (e.g., academic orientation, helpfulness and student interactions).

Another study linked data from the teacher survey conducted alongside the 2007 survey and supportive school environments for LGBT students, and identified a cross-level interaction (Denny et al. 2016). As mentioned above, multilevel models also allow the estimation of cross-level interactions examining possible associations between contextual-level variables and individual-level variables (for other interesting examples in the environmental domain, see Liu and Sibley 2013; Milfont and Markowitz 2016). Denny et al. showed that schools where teachers reported more supportive environments were associated with fewer depression symptoms among both-sex-attracted students. Among opposite-sex-attracted students, there was no relationship between school environments and their risk of depressive symptoms (Fig. 20.1). This suggests that for the majority of students who are opposite-sex attracted, supportive school environments are normative and therefore of less consequence than for sexual minority students where supportive school environments may matter a great deal.

Another recent study examined the impact of school-level characteristics on bullying behaviors and victimization (Denny et al. 2015). The five characteristics considered were: (1) school size, in terms of number of enrolled students, (2) school type, i.e., girls-only, boys-only or co-educational schools, (3) school funding, in terms of public, integrated or private schools, (4) school-level socio-economic status, indicating the extent to which students attending the school are from low socio-economic communities, and (5) school-level student and teacher behavior with respect to bullying (i.e., student or teacher takes action to stop bullying). The outcome variables considered were being bullied (i.e., students who reported being bullied in school about once a week, several times a week, or most days) and bullying others (i.e., students who reported bullying other students in their school about once a week, several times a week, or most days).

Results showed that approximately 6% of students reported being frequently bullied and 5% reported bullying others, and these levels are comparable to other New Zealand and international studies. Results from the multilevel models showed that the most important school-level characteristic to impact being bullied was the variable representing student action to stop bullying. Schools where students often take action when they know a student is being bullied have less bullying. The only other school-level characteristics to impact students bullying others was school funding, with public and integrated schools having fewer students bullying others

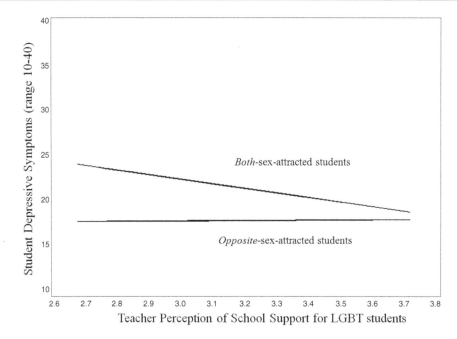

Fig. 20.1 Significant cross-level interaction between students' depressive symptoms and teachers' perceptions of school support for LGBT in the context of students' sexual orientation (Adapted from Denny et al. 2016)

compared to privately-funded schools. These results suggest that one aspect of the school interpersonal environment (students taking action) influences students being bullied, while one aspect of the school economic environment (funding) influences students bullying others. The findings also indicate that students have an important role in preventing bullying since encouraging students to take action to stop bullying may decrease the prevalence of bullying in schools.

Another study examined whether student risk-taking behaviors and depressive symptoms were influenced by a larger number of school-level characteristics (Denny et al. 2011). Six risk-taking behaviors were considered: (1) attempted suicide, measured as one or more suicide attempts in the last 12 months; (2) six questions related to motor vehicle risk behavior; (3) four questions on violence-related behaviors; (4) two questions on smoking cigarettes; (5) ten questions related to alcohol use risk; and (6) three questions related to unsafe sexual behaviors. Depressive symptomatology was measured with a standard depression measure for adolescents (Milfont et al. 2008; Szabo et al. 2014). The school-level characteristics included: (1) rural vs. urban schools based on where students lived, (2) school social climate as perceived by students, (3) student participation at school, (4) student representation at school, (5) staff work environment as perceived by teachers, (6) teacher well-being, (7) teacher burnout, (8) health and welfare services available to students, and (9) health-promoting school organization.

The results from the multilevel models showed that some school-level characteristics impacted student risk-taking behaviors and depressive symptoms. Compared to other schools, those where students reported a more positive school climate had fewer students with alcohol use problems, engaging in both violent behaviors and in risky motor vehicle use. These results suggest that increasing students' sense of belonging, supportive relationships with adults at school, expectations from people at school, and perceptions of safety can serve as protective factors for these particular risk-taking behaviors. Additionally, schools where teachers had a higher average well-being had lower rates of depressive symptoms among students. Compared to other schools, those where teachers and senior management reported better levels of health and welfare services for students had fewer students engaging in unsafe sexual health behaviors. (Unexpectedly, a higher level of unsafe sexual health behaviors among students was observed in schools with greater student participation in the running of their school, which may be attributed to a spurious effect.)

Overall, the results showed that only a few school-level characteristics influenced student-level outcomes, and this influence was often small in terms of effect size. As discussed elsewhere (Denny et al. 2011), this might reflect the homogeneity of high schools in New Zealand, which might limit the variance between and within schools. Notwithstanding the similarities among New Zealand schools, the results suggest an important contribution of school environments to the health and well-being indicators of adolescents, in line with previous findings (Kumar et al. 2008).

20.4.1.2 Neighborhood Environments

One study was carried out to examine the influence of school-level as well as neighborhood-level characteristics on self-reported days of physical activity of students in New Zealand (Utter et al. 2011). The three school-level measures accessed the perception of students regarding how much their school encouraged physical activity, if students had attended a physical exercise class in the past week, and if students belonged to any school sports teams. The four neighborhood-level indicators included: (1) six questions on community cohesion, (2) seven questions on neighborhood physical disintegration, (3) one question on safety perception of the neighborhood, and (4) one question assessing availability of recreational facilities in the area where students lived that they could walk to from home.

The results from the multilevel models showed that opportunities for sports participation at school and strong social connections in the neighborhood were associated with physical activity. Greater levels of physical activity among students were observed in schools where a high number of students belonged to sports teams, and in neighborhoods perceived as having strong social cohesion (e.g., students liked their neighborhood, trusted and liked the people in their neighborhood, and felt they belonged). These results suggest that increasing school opportunities for adolescents to be physically active, by means of sports teams, as well as fostering positive social aspects of neighborhoods can increase the amount of physical activity of adolescents. Interestingly, another recent multilevel study conducted in two cities in France and Spain observed no impact of perceived physical activity opportunities in the neighborhood on adolescents' daily moderate-to-vigorous physical activity (Aibar et al. 2015)—only warmer weather, lower levels of precipitation and walking/biking to school had an impact. It seems that perceived physical activity opportunities at school are more likely to foster subsequent physical activity in adolescents than perceived opportunities in the neighborhood.

A second study examined the extent to which neighborhood social capital impacts self-reported well-being of adolescents (Aminzadeh et al. 2013). Well-being was indexed with three measures accessing general mood, life satisfaction, and the well-being measure of the World Health Organization. Neighborhood social capital was indexed with seven indicators, including some used in the Utter et al. study reviewed above: neighborhood social cohesion accessed by measures of reciprocity (e.g., 'Do the people in your neighborhood help each other?'), psychological sense of community (e.g., 'Do you trust the people in your neighborhood?') and safety perception of the neighborhood (i.e., 'Do you feel safe in your neighborhood?'); membership of community organizations (i.e., 'Do you belong to a group, club or team which is not run by your school?'); availability of recreational neighborhood facilities; neighborhood physical disintegration; and residential stability (i.e., 'In the past year, how many times have you moved home?').

Multilevel model results showed that a higher level of self-reported well-being was positively associated with greater levels of social cohesion and membership of community organizations. These findings closely support the idea of people–environment congruity discussed by Moser (2009). Greater levels of subjective QOL, here indexed by positive mood, being very happy/satisfied with life and perceived well-being, matched greater levels of social capital both in terms of subjective indicators (a general perception of mutual trust, reciprocity, safety and sense of community) and structural indicators (youth membership of community organizations). The importance of neighborhood social connection for adolescents and, in particular, in influencing greater adolescent involvement in civic activities, has also been observed in other studies (e.g., Lenzi et al. 2013).

The study by Aminzadeh et al. (2013) also identified a cross-level interaction between the socio-economic status of students, their self-reported well-being, and membership of community organizations. This significant cross-level interaction indicates that the positive association between well-being and community organization membership is more pronounced for students who are most socio-economically deprived (Fig. 20.2). In other words, membership of community organizations may buffer some of the negative effects of poverty for young people.

20.5 Conclusions

Everyday environments are clearly important in shaping our QOL. This is because the objective physical, interpersonal, economic and political environments with which we interact daily influence our thinking, feeling and action. In the present chapter, we have provided a selective summary of research examining how two particular everyday environments (school and neighborhood) influence indicators related to QOL. Based on theoretical and methodological grounds, we have focused on studies using multilevel and strengths-based approaches, and have presented as a case study a more detailed review of research examining the health and well-being of adolescents attending high schools in New Zealand.

The multilevel results reviewed show that friendly and positive physical and social features of schools and neighborhoods are related to student-level QOL indicators. Overall, these findings support the view that both school and neighborhood environments can afford social interaction and retreat to adolescents (Clark and Uzzell 2002). School-level variables related to economic environments (e.g., funding) and interpersonal environments (e.g., positive social

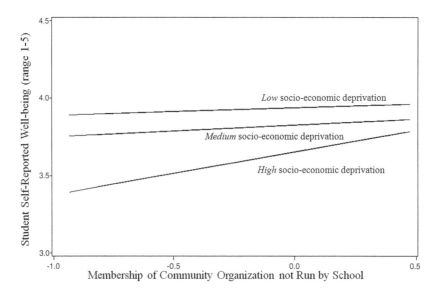

Fig. 20.2 Significant cross-level interaction between students' self-reported well-being and membership of community organizations in the context of their socio-economic deprivation (Adapted from Aminzadeh et al. 2013)

climate) influence being bullied, bullying others, risk-taking behaviors and depressive symptoms. In the same vein, neighborhood-level variables related to interpersonal environments (e.g., social cohesion, membership of community organizations) influence engagement in physical activity and well-being. In contrast, objective physical environments (e.g., school size) have minimal associations with the student-level outcomes considered. In general, it seems that socio-economic characteristics of both schools and neighborhoods have a greater influence on QOL indicators compared to the physical characteristics of these everyday environments. Further systematic and multilevel research is necessary to examine and confirm this observation.

There are other issues that future research could take into account. One missing key environment, which has an essential influence on the QOL of adolescents, is their families. The Youth2000 Surveys asked students a number of questions regarding their families (e.g., Clark et al. 2011; Utter et al. 2012), but there was no parent/guardian survey. Collecting data from students, parents/guardians and teachers would provide a much richer perspective on the health and well-being of adolescents, and how particular everyday environments (family, school and neighborhood) impact these indicators. The student-level QOL outcomes, as well as most of the variables considered, often rely on self-reports. A combination of self-reports plus more objective indicators of the same student-level outcomes would provide stronger evidence for the observed associations between school/neighborhood environments and the given outcome. The research summarized gives a number of indications regarding the association between particular characteristics of schools and neighborhoods and QOL indicators, but no directional effect can be assumed. Longitudinal designs are suitable for examining the direction of the associations.

There are methodological aspects worth noting. To estimate the unbiased association between characteristics of schools or neighborhoods and health outcomes, the treatment assignment must be 'strongly ignorable' (Rosenbaum and Rubin 1983). Experimental studies achieve this through randomization of people to treatments. In observational studies, treatment assignment is 'strongly ignorable' when outcomes are independent of treatment assignment given a set of covariates. In the multilevel studies reviewed, treatment assignment is either the school the student attends or the neighborhood of residence of the student's family. School selection is mostly driven by neighborhood of residence which, in turn, is mostly influenced by household income and ethnicity (Grbic et al. 2010). However, there are other factors over and above ethnicity and socio-economic status that may influence school and/or neighborhood selection, and it remains debatable whether school and neighborhood effects can be estimated from observational models, such as those reviewed in the present chapter. Moreover, Oakes (2004) criticizes current neighborhood effects research and identifies intractable identification problems inherent in observational studies of neighborhoods: their social stratification, endogenous group phenomena, and lack of exchangeability of people between neighborhoods. Another methodological limitation of neighbourhood research is the modifiable area unit problem (Openshaw and Taylor 1979), which highlights the difficulty in defining the relevant geographical scale especially for young people. Future research needs to consider randomized intervention studies at the school or neighborhood level, and also consider these methodological issues.

Lastly, Oishi (2014) makes a useful distinction between studies investigating the adaptation of humans to their surroundings. He proposes three types: (1) *association* studies focusing on the link between social ecology (e.g., a hot climate) and a given psychological response (e.g., aggression); (2) *process* studies focusing on the psychological process that might clarify the social ecology-psychology link (i.e., what mediates the association between the social ecology and the psychological response; e.g., a hot climate → hostility/crankiness → aggression); and (3) *niche construction* studies focusing on how psychological states might create and maintain a particular social ecology (e.g., hostility/crankiness → self-defense and permissive gun ownership laws). His review of the literature led to the conclusion that most studies conducted so far are either association or process studies, and that niche studies have been rare. Clearly, the studies we have reviewed above are examples of association studies. Future research on the QOL of adolescents in particular and QOL research in general should attempt to identify psychological mediators of the associations between everyday environments and QOL indicators, and to investigate how individuals intentionally modify their surroundings to increase their QOL.

The multilevel studies reviewed in this chapter illustrate a broad line of research designed to investigate the complex interrelations between adolescents and their school/neighborhood surroundings, and the implications these interactions might have on their QOL. Overall, the results suggest that QOL increases as a function of the congruity individuals experience with their social ecology, which supports the general argument put forward by Moser (2009). Multilevel models are best suited for examining nested data (e.g., students within classrooms within schools) and research questions derived from the key purpose of environmental psychology of understanding the relationships between individuals and their built/natural surroundings (Gifford 2009) often imply such data. Therefore, we believe multilevel models should be used further in environmental psychology research. Multilevel models are also useful in QOL research interested in fostering people–environment congruity.

Acknowledgements We thank all researchers, staff and funding agencies involved in the running of the Youth2000 Survey Series. Special thanks go to all the students and school staff who participated in the surveys. Without their work and help this project would not have been possible.

References

Aibar, A., Bois, J. E., Generelo, E., Bengoechea, E. G., Paillard, T., & Zaragoza, J. (2015). Effect of weather, school transport, and perceived neighborhood characteristics on moderate to vigorous physical activity levels of adolescents from two European cities. *Environment and Behavior, 47*, 395–417.

Alfonzo, M. A. (2005). To walk or not to walk? The hierarchy of walking needs. *Environment and Behavior, 37*, 808–836.

Aminzadeh, K., Denny, S. M., Utter, J., Milfont, T. L., Ameratunga, S. N., Teevale, T., & Clark, T. (2013). Neighbourhood social capital and adolescent self-reported wellbeing in New Zealand: A multilevel analysis. *Social Science and Medicine, 84*, 13–21.

Barker, R. G., & Gump, P. V. (1964). *Big school, small school: High school size and student behavior*. Stanford: Stanford University Press.

Bonnes, M., & Secchiaroli, G. (1995). *Environmental psychology: A psycho-social introduction*. London: Sage.

Bradley, R. H., Whiteside, L., Mundfrom, D. J., Casey, P. H., Kelleher, K. J., & Pope, S. K. (1994). Early indications of resilience and their relation to experiences in the home environments of low birthweight, premature children living in poverty. *Child Development, 65*, 346–360.

Brown, B. B., Perkins, D. D., & Brown, G. (2004). Incivilities, place attachment and crime: Block and individual effects. *Journal of Environmental Psychology, 24*, 359–371.

Clark, C., & Uzzell, D. L. (2002). The affordances of the home, neighbourhood, school and town centre for adolescents. *Journal of Environmental Psychology, 22*, 95–208.

Clark, T., Robinson, E., Crengle, S., Fleming, T., Ameratunga, S. N., Denny, S. M., Bearinger, L., Sieving, R., & Saewyc, E. (2011). Risk and protective factors for suicide attempt among indigenous Māori youth in New Zealand: The role of family connection as a moderating variable. *Journal of Aboriginal Health, 7*, 16–31.

Clark, T., Fleming, T., Bullen, P., Crengle, S., Denny, S. M., Dyson, B., Peiris-John, R., Robinson, E., Rossen, F., Sheridan, J., Teevale, T., Utter, J., & Lewycka, S. (2013). Health and well-being of secondary school students in New Zealand: Trends between 2001, 2007 and 2012. *Journal of Paediatrics and Child Health, 49*, 925–934.

Corral-Verdugo, V. (2012). The positive psychology of sustainability. *Environment, Development and Sustainability, 14*, 651–666.

Cronbach, L. J. (1976). *Research on classrooms and schools: Formulation of questions, design, and analysis*. Stanford: Stanford Evaluation Consortium.

Denny, S. J., Milfont, T. L., Utter, J., Robinson, E. M., Ameratunga, S. N., Merry, S. N., Fleming, T., & Watson, P. D. (2008). Hand-held internet tablets for school-based data collection. *BMC Research Notes, 1*, 52.

Denny, S., Robinson, E. M., Milfont, T. L., & Grant, S. (2009). *Youth'07: The social climate of secondary schools in New Zealand*. Auckland: School of Population Health, The University of Auckland.

Denny, S. J., Robinson, E. M., Utter, J., Fleming, T. M., Grant, S., Milfont, T. L., Crengle, S., Ameratunga, S. N., & Clark, T. (2011). Do schools influence student risk behaviors and emotional health symptoms? *Journal of Adolescent Health, 48*, 259–267.

Denny, S. J., Peterson, E. R., Stuart, J., Utter, J., Bullen, P., Fleming, T., Ameratunga, S. N., Clark, T., & Milfont, T. L. (2015). Bystander intervention, bullying, and victimization: Multilevel analysis of New Zealand high schools. *Journal of School Violence, 14*, 245–272.

Denny, S. J., Lucassen, M. F., Stuart, J., Fleming, T., Bullen, P., Peiris-John, R., Rossen, F. V., & Utter, J. (2016). The association between supportive high school environments and depressive symptoms and suicidality among sexual minority students. *Journal of Clinical Child and Adolescent Psychology, 45*, 248–261.

Ding, D., Sallis, J. F., Kerr, J., Lee, S., & Rosenberg, D. E. (2011). Neighborhood environment and physical activity among youth: A review. *American Journal of Preventive Medicine, 41*, 442–455.

Durlak, J. A., & Wells, A. M. (1997). Primary prevention mental health programs for children and adolescents: A meta-analytic review. *American Journal of Community Psychology, 25*, 115–182.

Fergusson, D. M., & Lynskey, M. T. (1996). Adolescent resiliency to family adversity. *Journal of Child Psychology and Psychiatry, and Allied Disciplines, 37*, 281–292.

Ferreira, I., Van der Horst, K., Wendel-Vos, W., Kremers, S., Van Lenthe, F. J., & Brug, J. (2007). Environmental correlates of physical activity in youth – A review and update. *Obesity Reviews, 8*, 129–154.

Fleming, T. M., Merry, S. N., Robinson, E. M., Denny, S. J., & Watson, P. D. (2007). Self-reported suicide attempts and associated risk and protective factors among secondary school students in New Zealand. *Australasian Psychiatry, 41*, 213–221.

Flouri, E., Midouhas, E., & Joshi, H. (2014). The role of urban neighbourhood green space in children's emotional and behavioural resilience. *Journal of Environmental Psychology, 40*, 179–186.

Garmezy, N. (1991). Resilience and vulnerability to adverse developmental outcomes associated with poverty. *American Behavioral Scientist, 34*, 416–430.

Gifford, R. (2009). Environmental psychology: Manifold visions, unity of purpose. *Journal of Environmental Psychology, 29*, 387–389.

Gifford, R. (2014). Environmental psychology matters. *Annual Review of Psychology, 65*, 541–579.

Grbic, D., Ishizawa, H., & Crothers, C. (2010). Ethnic residential segregation in New Zealand, 1991–2006. *Social Science Research, 39*, 25–38.

Gribble, P. A., Cowen, E. L., Wyman, P. A., Work, W. C., Wannon, M., & Raoof, A. (1993). Parent and child views of parent-child relationship qualities and resilient outcomes among urban children. *Journal of Child Psychology and Psychiatry, and Allied Disciplines, 34*, 507–519.

Herrenkohl, E. C., Herrenkohl, R. C., & Egolf, B. (1994). Resilient early school-age children from maltreating homes: Outcomes in late adolescence. *American Journal of Orthopsychiatry, 64*, 301–309.

Jenkins, J. M., & Smith, M. A. (1990). Factors protecting children living in disharmonious homes: Maternal reports. *Journal of the American Academy of Child & Adolescent Psychiatry, 29*, 60–69.

Keizer, K., Lindenberg, S., & Steg, L. (2008). The spreading of disorder. *Science, 322*, 1681–1685.

Kidger, J., Araya, R., Donovan, J., & Gunnell, D. (2012). The effect of the school environment on the emotional health of adolescents: A systematic review. *Pediatrics, 129*, 925–949.

Kumar, R., O'Malley, P. M., & Johnston, L. D. (2008). Association between physical environment of secondary schools and student problem behaviour: A national study, 2000–2003. *Environment and Behavior, 40*, 455–486.

Larco, N., Steiner, B., Stockard, J., & West, A. (2012). Pedestrian-friendly environments and active travel for residents of multifamily housing: The role of preferences and perceptions. *Environment and Behavior, 44*, 303–333.

Lee, T. (1968). Urban neighbourhood as a social-spatial schema. *Human Relations, 21*, 241–267.

Lee, A., Cheng, F. F., & Fung, Y. (2006). Can health promoting schools contribute to the better health and well-being of young people? The Hong Kong experience. *Journal of Epidemiology and Community Health, 60*, 530–536.

Lenzi, M., Vieno, A., Pastore, M., & Santinello, M. (2013). Neighbourhood social connectedness and adolescent civic engagement: An integrative model. *Journal of Environmental Psychology, 34*, 45–54.

Liu, J. H., & Sibley, C. H. (2013). Hope for the future? Understanding self-sacrifice among young citizens of the world in the face of global warming. *Analyses of Social Issues and Public Policy, 12*, 190–203.

Luke, D. (2004). *Multilevel modeling* (Quantitative applications in the social sciences, 143). Thousand Oaks: Sage.

Maes, L., & Lievens, J. (2003). Can the school make a difference? A multilevel analysis of adolescent risk and health behaviour. *Social Science and Medicine, 56*, 517–529.

Masten, A. S. (2004). Regulatory processes, risk, and resilience in adolescent development. *Annals of the New York Academy of Sciences, 1021*, 310–319.

Masten, A. S. (2014). *Ordinary magic: Resilience in development*. New York: Guilford Press.

Maxwell, L. E. (2003). Home and school density effects on elementary school children: The role of spatial density. *Environment and Behavior, 35*, 566–578.

Milfont, T. L., & Markowitz, E. (2016). Sustainable consumer behavior: A multilevel perspective. *Current Opinion in Psychology, 10*, 112–117.

Milfont, T. L., Merry, S. N., Robinson, E. M., Denny, S., Crengle, S., & Ameratunga, S. N. (2008). Evaluating the short form of the Reynolds adolescent depression scale in New Zealand Adolescents. *Australian and New Zealand Journal of Psychiatry, 42*, 950–954.

Moser, G. (2009). Quality of life and sustainability: Toward person-environment congruity. *Journal of Environmental Psychology, 29*, 351–357.

Nezlek, J. B. (2008). An introduction to multilevel modeling for social and personality psychology. *Social and Personality Psychology Compass, 2*, 842–860.

Oakes, J. M. (2004). The (mis)estimation of neighborhood effects: Causal inference for a practicable social epidemiology. *Social Science and Medicine, 58*, 1929–1952.

Oishi, S. (2014). Socioecological psychology. *Annual Review of Psychology, 65*, 581–609.

Olivos, P. (2010). Ambientes escolares. In J. I. Aragonés & M. Américo (Eds.), *Psicología ambiental* (3rd ed., pp. 205–224). Madrid: Ediciones Pirámide.

Openshaw, S., & Taylor, P. J. (1979). A million or so correlation coefficients: Three experiments on the modifiable areal unit problem. In N. Wrigley (Ed.), *Statistical methods in the spatial sciences* (pp. 127–144). London: Pion.

Oshio, T., & Urawaka, K. (2012). Neighbourhood satisfaction, self-rated health, and psychological attributes: A multilevel analysis in Japan. *Journal of Environmental Psychology, 32*, 410–417.

Patton, G., Bond, L., Butler, H., & Glover, S. (2003). Changing schools, changing health? Design and implementation of the Gatehouse Project. *Journal of Adolescent Health, 33*, 231–239.

Patton, G. C., Bond, L., Carlin, J. B., Thomas, L., Butler, H., Glover, S., et al. (2006). Promoting social inclusion in schools: A group-randomized trial of effects on student health risk behavior and well-being. *American Journal of Public Health, 96*, 1582–1587.

Pikora, T., Giles-Corti, B., Bull, F., Jamrozik, K., & Donovan, R. (2003). Developing a framework for assessment of the environmental determinants of walking and cycling. *Social Science and Medicine, 56*, 1693–1703.

Pol, E. (2006). Blueprints for a history of environmental psychology (I): From first birth to American transition. *Medio Ambiente y Comportamiento Humano, 7*, 95–113.

Pol, E. (2007). Blueprints for a history of environmental psychology (II): From architectural psychology to the challenge of sustainability. *Medio Ambiente y Comportamiento Humano, 8*, 1–28.

Raudenbush, S. W. (1999). Ecometrics: Towards a science of assessing ecological settings, with application to the systematic social observation of neighborhoods. *Sociological Methodology, 29*, 1–41.

Raudenbush, S. W., & Bryk, A. S. (2002). *Hierarchical linear models: Applications to data analysis methods* (2nd ed.). Thousand Oaks: Sage.

Regoeczi, W. C. (2003). When context matters: A multilevel analysis of household and neighbourhood crowding on aggression and withdrawal. *Journal of Environmental Psychology, 23*, 457–470.

Resnick, M. D., Bearman, P. S., Blum, R. W., Bauman, K. E., Harris, K. M., Jones, J., ... Udry, R. (1997). Protecting adolescents from harm: Findings from the National Longitudinal Study on Adolescent Health. *Journal of the American Medical Association, 278*, 823–832.

Rivlin, L. G., & Weinstein, C. S. (1984). Educational issues, school settings, and environmental psychology. *Journal of Environmental Psychology, 4*, 347–364.

Rosenbaum, P. R., & Rubin, D. B. (1983). The central role of the propensity score in observational studies for causal effects. *Biometrika, 70*, 41–55.

Schultz, P. W., Milfont, T. L., Chance, R. C., Tronu, G., Luís, S., Ando, K., Rasool, F., Roose, P. L., Ogunbode, C. A., Castro, J., & Gouveia, V. V. (2014). Cross-cultural evidence for spatial bias in beliefs about the severity of environmental problems. *Environment and Behavior, 46*, 267–302.

Sellström, E., & Bremberg, S. (2006). Is there a "school effect" on pupil outcomes? A review of multilevel studies. *Journal of Epidemiology and Community Health, 60*, 149–155.

Szabo, A., Milfont, T. L., Merry, S. N., Robinson, E. M., Crengle, S., Ameratunga, S. N., & Denny, S. J. (2014). Equivalence of the short form of the Reynolds adolescent depression scale across groups. *Journal of Clinical Child and Adolescent Psychology, 43*, 592–600.

Tombs, S. (2005). Quality indicators in the design of schools (QIDS): A tool for assessing school design? *OECD/PEB Report: Evaluating Quality in Educational Facilities*. http://www.oecd.org/edu/innovation-education/37905277.pdf

Utter, J., Denny, S. J., Robinson, E. M., Ameratunga, S. N., & Milfont, T. L. (2011). Social and physical contexts of schools and neighborhoods: Associations with physical activity among young people in New Zealand. *American Journal of Public Health, 101*, 1690–1695.

Utter, J., Denny, S. J., Robinson, E. M., Fleming, T. M., Ameratunga, S. N., & Grant, S. (2012). Family meals among New Zealand young people: Relationships with eating behaviours and body mass index. *Journal of Nutrition Education and Behavior, 45*, 3–11.

Uzzell, D., & Moser, G. (2006). On the quality of life of environments. *European Review of Applied Psychology, 56*, 1–4.

Werner, E. E., & Smith, R. S. (1992). *Overcoming the odds: High risk children from birth to adulthood*. Ithaca: Cornell University Press.

Wilson, G. L., & Kelling, J. Q. (1982). Broken windows: The police and neighbourhood safety. *The Atlantic, 249*(3), 29–38.

Part VI

Well-Being and Daily Environments – *Work and Institutional Environments*

The Effect of Workplace Design on Quality of Life at Work

21

Jacqueline C. Vischer and Mariam Wifi

21.1 Why Quality Matters

This chapter argues that people's Quality of Life (QoL) is directly affected by their quality of work life (QWL), which is in turn influenced by the quality of the work environment. First, we explore some definitions of these terms and demonstrate how environmental psychology plays a key role in both QWL and QoL. Then, we outline a model that combines several well-established theories of the relationship between workers and their physical environment in order to show how the quality of workspace as perceived by users, contributes to and predicts their QWL and consequently their QoL. In conclusion, we offer an approach to collecting empirical data from workers that can be used to diagnose the quality of the physical work environment. The results of data analysis are applied to the 'treatment' of environmental problems and barriers so as to improve users' experience and make them more effective at work, as well as more satisfied with their QWL.

The concept of Quality of Life (QoL) has been defined as "the degree to which the experience of an individual's life satisfies that individual's wants and needs (both physical and psychological)" (Rice 1984). The World Health Organization defines QoL as the "individual's perception of their position in life in the context of the culture and value system in which they live and in relation to their goals, expectations, standards and concerns" (WHO 1994). Several definitions of QoL in this volume make reference to the natural environment (see Chaps. 8 and 10), and some consider the larger scale of urban and cultural environments (see Chaps. 12 and 27). In this chapter, we apply QoL concepts to environments for work.

Evidence suggests that measured QoL outcomes are both predictors of and result from QoL: "Happiness and a feeling of well-being will also result from QoL. When one rates his or her life as having quality, one will concurrently have a sense of self-esteem and pride regarding his or her life. It must be noted that a confounding scenario seems to be apparent with each of these consequences of quality of life in that each can contribute to, as well as result from, quality of life" (Meeberg 1993). The ideological importance of Quality of Life is that it promotes the idea of supporting people to live in ways that are best for them in the environments they occupy. Individual assessment of QoL varies according to perceptions, personal needs, individual differences, preferences, culture and expectations. Lack of quality, however, may be perceived in more uniform ways.

J.C. Vischer, Ph.D. (✉) • M. Wifi, M.Arch.
Faculté de l'aménagement, University of Montreal, P.O. Box 6128 succursale Centreville, Montreal, H3C 3J7 QC, Canada
e-mail: jacqueline.vischer@umontreal.ca

Quality of life is a holistic concept composed of the cumulative contributions of a range of different life domains such as work, family, housing, neighborhood, religion, and social networks (Rice et al. 1985). The quality of each life domain can be assessed separately and will vary according to activity, place, social role and human relations as well as cultural values and individual expectations. Perceived quality of life results from an infinite number of cumulative life experiences; the degree to which people's wants and needs are satisfied in each domain determines the distribution of their QoL (Rice 1984).

One domain of life experience is the built environment that people occupy and the series of interior and exterior environments in which they behave, interact, perform activities and react. In western cultures, it is estimated that people spend 90 % of their time indoors and, consequently, "Beyond their biological effects, [places] make us feel uncomfortable and ill-at-ease, energetic and stimulated or relaxed and at peace... They can work so deeply into our being that they affect our state of health" (Day 2002). The field of environmental psychology has spent many decades studying the effects of various types of built and natural environments on occupants – on their health, comfort, safety, attachments, behavior and attitudes. By showing the degree to which people are affected by the environment, the quality of each environment can be said to have a direct impact on people's QoL.

Notions of QoL depend largely on an understanding of human needs. Much environmental psychology research examines how aspects of their physical environment succeed or fail in meeting people's needs. Human needs have been classified in many different ways, beginning with Maslow's (1954) hierarchy. The categories in this hierarchy form a pyramid ranging from that which is most basic to survival to the less basic but nonetheless essential; they include physiological needs, safety, esteem, love, and self-actualization. The more people's needs are met, the better their QoL. The debate on needs also applies to the physical environment and how the environment affects human behavior, using a similar premise: the more a specific built environment meets the needs of its occupants, the more effective or successful it is considered to be.[1]

21.2 Quality of Work Life

"Quality of Work Life is that part of overall quality of life that is influenced by work. It is more than just job satisfaction or work happiness, but the widest context in which an employee would evaluate their work environment" (Varghese and Jayan 2013). QWL is a sub-category of QoL research that has been studied independently since the late 1970s (Davis and Cherns 1975; Hackman and Suttle 1977; Lawler 1982). For QoL researchers, "Efforts to improve the quality of work life of employees may also affect their sense of quality of life" (Elizur and Shye 1990). As with QoL, people's QWL is affected by their work-related goals, desires, expectations and needs, and how well these are fulfilled. The concept of QWL evolved from a concern for the negative impacts of work on employees' health and well-being and the urge to improve the quality of the work domain by making changes in the design and conditions of work.

Quality of Work Life is a dynamic multidimensional construct focusing on worker well-being. It is concerned with workers' productivity, yet also addresses their emotional need to feel satisfied with their experience of work. However, QWL is not the same as job satisfaction (Lawler 1982). QWL is a philosophy or a set of principles based on a view of employees as the most important and meaningful resource in the organization, who should be treated with dignity and respect (Straw and Heckscher 1984). QWL combines factors related to the job itself – such as job satisfaction, salary, and relationships with colleagues – with intangibles, such as overall life satisfaction and feelings of well-being (Danna and Griffin 1999). The eight factors that affect workers' QWL are fair compensation, health and

---

[1] It should be noted that the needs model has been criticized on the grounds that it is premised on the human user as a passive recipient of environmental stimuli, rather than on the human user as an active agent with a reciprocal effect on his/her environment (Vischer 1985).

safety, self-development, growth and security, social integration, constitutionalism, life space and social relevance (Walton 1991). A model of needs in the work domain includes job requirements, work environment, supervisory behavior, ancillary programs, and organizational commitment. Work domain needs can be fulfilled through resources, activities, and outcomes resulting from participation in the workplace (Sirgy et al. 2001). Later, the physical workspace was added to QWL as a factor affecting job satisfaction and productivity (Cummings and Worley 2005).

Space-related needs in the work environment have been identified by concepts such as Preiser's (1983) habitability framework and Vischer's (1989, 1996, 2005) functional comfort pyramid according to which different workspace qualities can be ranked. Functional comfort is discussed later in this chapter; it is based on the habitability framework, which connects buildings and settings with users, and occupants' needs with the work environment. Habitability is a relative concept that may differ from one culture to another: "Habitability defines the degree of fit between individuals or groups and their environment, both natural and man-made, in terms of an ecologically sound and humane, built environment" (p. 87, Preiser op.cit.). Habitability requires that the physical environment meet three categories of users' needs: health and safety, functional and task performance, and psychological comfort. Improving habitability through a better fit between the occupant and the workspace means a better quality work environment and improved QWL. As QWL is considered a key factor in the sustainability and viability of organizations, finding ways of improving employees' QWL is an investment in human capital and in the viability of the organization (Sheel et al. 2012). Aspects of the work environment that have been found to affect QWL include the job or task, physical conditions, such as the building design, materials and technology, as well as economic and social aspects, such as administrative policies and the work-life relationship (Cunningham and Eberle 1990; Elizur and Shye 1990).

A poor QWL often means increased stress at work. Workers in North America spend at least 50 % of their indoor time in the workplace, and reducing work stress is a concern shared by managers, designers, environment-behavior researchers and environmental psychologists (Bagnara et al. 2001). The term "workspace stress" has been coined to distinguish the stress caused by functionally uncomfortable (unsupportive) workspace (Vischer 2007). Higher levels of stress at work are related to increased insomnia, anxiety, depression, job dissatisfaction, decreased organizational commitment, reduced job performance, and absenteeism (Woo and Postolache 2008). Studies indicate that workspace stress levels can be reduced by according more environmental control to occupants (Csikszentmihalyi 1990; Karasek and Theorell 1990; Kaplan 1983; Walton 1980; Lawler 1975). Environmental control can take mechanical and instrumental forms, such as light switches, furniture adjustments, and thermostats, or social-psychological forms, such as access to information about workspace decisions and participation in workspace design and planning. Offering occupants greater environmental control is known as environmental empowerment and contributes to employee well-being (Vischer 2005; Vischer and Malkoski 2015).

There is significant evidence that workers waste time and energy coping with poorly designed workspace, which reduces the time and energy they invest in work (Vischer 2008). In addition to the negative impact of absenteeism due to illnesses such as respiratory infections, eye strain, and back and neck pains, the stress of functioning in an unsupportive or adverse physical environment has behavioral effects. These include low morale, reduced motivation, employee turnover, and inadequate work performance as a result of performing tasks slower and making more errors – all factors that affect organizational productivity (Vischer 1989, 2003, 2008; Haynes 2007; Damian 2004; Heerwagen et al. 2004; Sundstrom et al. 1994).

In contrast, numerous studies have demonstrated the positive effects on both worker morale and productivity of occupying workspace where ambient conditions such as lighting, temperature and sound levels, as well as furniture comfort, aesthetics and architectonic details, are managed appropriately for the tasks being performed (McCoy and Evans 2005; Vischer and Fischer 2005; Damian 2004; Brill and Weideman 2001; Fisk 2000; Monk 1997). Studies have found that lighting quality, ventilation rates, access to natural light and the acoustic environment are significantly related to workers' satisfaction and productivity (Humphreys 2005; Veitch et al. 2004; Becker 1981). In addition, workers' attitudes and behaviors are affected by ambient conditions such as indoor air quality, illumination, temperature, and views (Larsen et al. 1998; Veitch and Gifford 1996). Natural elements, such as views of nature and indoor plants, have a positive influence on mental fatigue and restorative value (Kaplan 1995; Kaplan et al. 1988; Haber 1977).

Well-being and emotional health are also affected by ambient conditions at work. For example, natural light in the workspace makes people happier and more motivated (Hameed and Amjad 2009; Heschong et al. 2002). Improved light and temperature, indoor air quality, ergonomic furniture and lighting are also linked to worker health (Dilani 2004; Milton et al. 2000; Veitch and Newsham 2000). It has been suggested that 16–37 million cases of colds and flu per year could be avoided by improving indoor environmental quality in US office buildings (Fisk et al. 2002; Milton et al. 2000).

Space layouts and furniture also affect QWL. Numerous studies, as well as popular journalism, have reported on the (usually adverse) effects of what are called 'open-plan' layouts (Konnikova 2014; Rashid and Zimring 2008). In the 1990s, the populart cartoonist Scott Adams made fun of office cubicles in his 'Dilbert' cartoons (Adams 1997). However, with few exceptions, little effort has been made in these critiques to identify exactly what is meant by 'open plan' – a term that was first used in the 1940s to denote workspace layouts that were widely spaced and screened by plants, tall furniture, and, later, moveable partitions. Contemporary workspace design is characterized by a rich variety of open and partially enclosed layouts, depending on the culture and values of the organization as well as on the type of work people are doing. Increasingly, the private individual office is a space-consuming anachronism as managers prefer to work collaboratively with their groups and teams, and co-workers prefer to co-locate and move around as needed during the workday. From a design and facilities management perspective, some form of open-plan configuration is essential in order to keep pace with the constant moves and changes of the modern office. As a result of changing technology, updated business practices, and project-based communication and collaboration, contemporary furniture products and office design concepts aim to accommodate the changing business environment (Tarricone and Luca 2002).

Open-plan workspace supports some important QWL values, such as egalitarian space allocation, communication, and collaboration opportunities. Studies have shown that an open work environment encourages mutual support behaviors, fostering cooperation and engagement with co-workers (Mubex 2010; O'Neil 2008). Rapid and reliable communication is critical in today's business environments and the speed and accuracy of task performance has a direct impact on productivity (Quilan 2001; Fleming and Larder 1999). However, dense open-plan configurations cause distractions due to noise, and task performance can be affected by poor visual and sound privacy (Chu and Warnock 2002; Evans and Johnson 2000). Workplace distractions may reduce employee productivity by up to 40 % and increase errors by 27 % (Bruce 2008).

Users' ratings of quality in green or sustainable buildings show little variation from these priorities in conventional buildings with regard to the performance of work (Lee and Kim 2008; Paul and Taylor 2008; Abbaszadeh et al. 2006). While users working in buildings certified as green or sustainable generally report better indoor air quality, thermal comfort, and overall satisfaction levels, there are few significant differences between sustainable and conventional buildings when it comes to ratings of interior layouts, noise, and interior lighting. Generally, green-certified and green-intent buildings tend to be perceived more positively by occupants than conventional buildings in general ways, whereas for specific conditions the differences are not clear-cut. Occupants' quality ratings may be influenced by their knowledge and expectations of how green buildings are intended to improve occupants' health, comfort and productivity (Leaman and Bordass 2007).

21.3 Models of Environmental Quality at Work

Environmental Quality (EQ) is a general concept applied to assessing user experiences of the physical environments they occupy (Khattab 1993; Rapoport 1990). It has frequently been applied to research on users' needs in work environments (Elzeyadi 2001). EQ is a key element in QWL and therefore an important determinant of workers' QoL. One approach to assessing EQ is post-occupancy evaluation (POE). POE is a general term for a range of types of study aimed at collecting and analyzing data from building occupants in order to measure EQ and the user impact of various building features. 'Objective' POE measures building performance in relation to air handling systems (ventilation and temperature) and lighting, and may include energy use, workspace reconfiguration of interiors, and preventive maintenance. Study results are typically compared to published standards and building code requirements (Kerce 1992). 'Subjective' POE measures rely on data on occupant perceptions, which are related to users' needs and therefore to perceived quality (Blishen and Atkinson 1980). Like POE, EQ is assessed in objective and subjective terms. Rapoport (1977) distinguishes between physical and perceived qualities of built environments, where 'physical' includes material aspects that are measurable using physical instrumentation, such as indoor air quality, light, sound and temperature. Material and immaterial qualities as perceived by occupants are measured using social and behavioral measurement tools such as surveys and interviews. According to Rapoport (1988), the quality of a built environment can be determined in terms of its instrumental, latent, and symbolic levels of meaning. At the instrumental level are the physical properties and functional qualities of the environment, enabling occupants to perform their tasks. The latent level refers to psychological, socio-cultural, and socio-psychological qualities, such as privacy, safety, territoriality, way-finding, and personalization (Fischer 1997; Ornstein 1992; Becker 1990). The symbolic level comprises the meanings and values of spatial elements in terms of users' traditions, beliefs, historical values, pride, and culture (Rapoport 1983; Doxtater 1994; Turner 1990). As these and other theoretical models suggest, user perception is a crucial factor in the person-environment relationship and essential to defining quality: "Most managers know that environmental quality does not exist outside the context of users' perceptions" (Vischer 1989).

Building Performance Evaluation (BPE) evolved from POE to take a more comprehensive approach (Preiser and Schramm 1997, 2012; Preiser and Vischer 2004). According to the BPE model, POE has a role in each stage of the life cycle of a building because each stage affects building performance and occupants' experience of quality. The BPE process model of building delivery and life cycle integrates the perspective of all parties involved at each stage – including owner or tenant, space programmer and designers, contractor and construction team – and includes commissioning and eventual occupancy.

The goals of measuring EQ using POE and related efforts are overall quality improvement: that is to say, a better managed and more cost-effective process resulting in a better quality building and more effective and satisfied users. This broad-brush approach to quality assessment has emerged out of several decades of post-occupancy studies, which have traced the causes of building performance and occupancy problems to decisions made early on in the process, often in the erroneous assumption of reducing short-term costs. The same intention can be found in the Leadership in Energy and Environmental Design (LEED) category 'Integrative process' in which points can be gained from involving designers, builders and specialists in different ways to solve

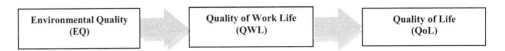

Fig. 21.1 Links between EQ, QWL and QoL

problems early on in the process, leading to more cost-efficient and sustainable outcomes (USGBC 2014). Figure 21.1 illustrates how EQ – as measured using POE and BPE research – has a direct effect on QWL and thereby on QoL.

In keeping with the concept of users' needs described above, a common approach to measuring EQ uses occupants' satisfaction ratings of different building features. It can also be argued that an overarching need for people at work is to be able to perform their tasks as effectively as possible. Asking workers to assess their environment in terms of its functionality and the ease or comfort of task performance provides both a user-based diagnostic measure of workspace quality and a direct indicator of EQ. The concept of functional comfort addresses how well users perceive their tasks and activities to be supported (or unsupported) by the physical environment in which they work. Collecting and analyzing functional comfort feedback from building occupants provides an empirical measure of how well people can get work done and is thus a more focused indicator of EQ than a general satisfaction rating. Vischer (1989) defines environmental quality as "the combination of environmental elements that interact with users of the environment to enable that environment to be the best possible one for the activities that go on in it". Assessing EQ in this way provides an indicator of both user satisfaction (likes and preferences) and user effectiveness (productivity) that affect overall well-being and QoL.

While users' functional comfort is a useful and practical empirical indicator of EQ, it is not the only one. However, unlike satisfaction and other measures of human behavior, functional comfort ratings provide a diagnosis of workspace quality that is based on occupants' task performance and can therefore be used to identify and correct practical problems of workspace design and management. In the next section, we will explain more about functional comfort and how it is measured, as well as how study results can be applied to problem-solving and better workspace design. We present the concept of functional comfort in the broader context of the environmental comfort model, in which users' physical comfort and their psychological comfort play an important part.

21.4 Functional Comfort and Environmental Quality

Defining EQ in terms of the functional comfort ratings that occupants attribute to and experience in their workspace facilitates the assessment of this aspect of QWL and ultimately of workers' QoL. Functional comfort is one measure of environmental quality, indicating that people who can do their work efficiently and effectively, without stress and with a sense of environmental support, occupy a good quality environment and experience it as comfortable. While workers' comfort experiences are filtered through cognitive processes, emotions, expectations, personality traits, learned behaviors and past experiences, their perceptions of comfort are based both on how they use the physical features around them and on their deep and detailed knowledge of their work and the requirements of their tasks.

The workspace comfort or workability pyramid referred to previously is shown in Fig. 21.2, below.

The diagram indicates that physical comfort is the basis of occupants' workspace experience and sets the minimum standard for basic habitability. If physical comfort is not adequate, people feel that their health and well-being might be in danger and, in some cases such as indoor air contamination, they cannot or will not perform their work. Most modern office buildings meet basic health and safety standards and rarely threaten occupants' physical comfort unless there is a system malfunction or a threat of danger, such

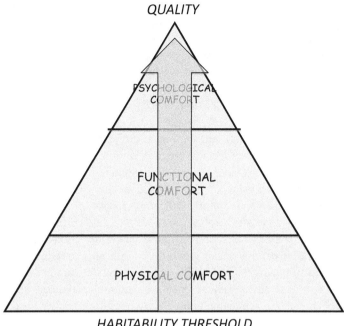

Fig. 21.2 Workspace comfort pyramid (First published in Vischer 1989)

as a fire. While most building standards protect occupant safety and health – and even comfort – they do not ensure functionally comfortable workspace, that is, an environment supportive of the tasks that people are performing, whether these are computer-based, interactive, focused, collaborative, or specialized.

The diagram indicates that workspace that is designed to be functionally comfortable supports people's tasks, whereas uncomfortable workspace causes workers to expend their energy on overcoming environmental barriers to task performance, thus causing stress. All work environments can be classified somewhere on the functional comfort continuum, ranging from very supportive and comfortable at one extreme to slowing down work and stressful at the other. The degree to which tasks are supported is measured through systematic feedback obtained from building users. Extensive research on functional comfort has shown that a limited number of environmental dimensions directly affect task performance. These include:

- Thermal comfort, ventilation and indoor air quality
- Lighting and the illumination environment
- Windows and daylighting
- Acoustic comfort and noise management
- Access to privacy for concentration and confidentiality
- Workstation dimensions, storage, enclosure and lay-outs
- Access to collaborative and shared spaces
- Cleaning and maintenance
- Safety and security.

Each dimension can be evaluated in more than one way. For example, a wide range of different types of lighting are available for work, most of which perform more than adequately in modern buildings. However, whether lighting is comfortable or not depends on whether the lighting provided is appropriate to the tasks people are performing. Long hours of screen-based work require low background levels of lighting that is not too bright or direct, whereas judging visual displays, graphic tasks, and selecting colors and materials all require direct and color-sensitive light fixtures. As experts on their tasks, the workers themselves assess whether their work is supported or not by the lighting they have, thereby providing data on the functional comfort and EQ of their work environment.

While the exact nature and number of functional comfort dimensions varies slightly among organizations and types of work, the adequacy and convenience of building support spaces such as coffee areas, elevators and washrooms also have an impact on functional comfort. When these and other critical environmental supports are insufficient or absent, workers experience fatigue and discomfort and may have difficulty communicating. They spend more time completing their tasks and there is a higher risk of making mistakes; their quality of life is reduced. Sustained functional discomfort, indicating a poor QWL, leads to stress (Vischer 2007).

At the peak of the comfort pyramid shown in Fig. 21.2 is psychological comfort. As with all notions of quality, including EQ, office workers' perceptions of ambient environmental conditions, such as lighting and temperature, have a psychological component. For example, if indoor temperatures are too warm, workers start to suspect indoor air quality and become concerned about pollution. Occupants have strong feelings about windows even though they do not necessarily use natural light to see. Background noise levels that are distracting to some go unnoticed by others. The symbolic meanings identified by Rapoport (described above) also play a role in psychological comfort. From a behavioral perspective, the three measurable determinants of psychological comfort are territoriality, privacy, and control. A sense of territory is associated with a feeling of belonging and ownership; privacy is best understood as the need to exercise control over one's accessibility to others; while environmental control exists on at least two levels, mechanical or instrumental control – access to tools that enable users to control conditions, such as a thermostat to control temperature – and empowerment – participation in decision-making by those who occupy and use workspace. Workers need to feel they have control over intrusions and distractions. Depending on how the need for privacy is defined, physical enclosure is a less influential factor than behavioral norms, such as managing interactions with co-workers and social contact. Where technology enables people to work from a variety of locations, territorial appropriation is no longer limited to physical space and traditional definitions of place: employees find other ways of taking ownership.

Nevertheless, what workers want and like is not always the same as what they need to get their work done. For example, most people will say they want and like private individual places to work with a minimum of distraction, whereas what teams need to work effectively are open, flexible, collaborative environments. Consequently, the ways in which data on functional comfort are sought from building occupants must be tested and validated as measures of functional comfort – or quality – rather than measuring individual wants and preferences. Measuring functional comfort through occupants' ratings of key environmental elements serves as a diagnostic tool of workspace quality and should not be confused with subjective satisfaction ratings, which are better indicators of users' psychological states.

21.5 Using Functional Comfort to Measure Environmental Quality

Environmental quality (EQ) can be measured through carefully acquired feedback data using a questionnaire survey designed to measure functional comfort. Results enable researchers, managers and designers to assess the quality of various environmental attributes in terms of the degree to which they support the tasks being performed. This means that a comparative quantitative indicator is available for key dimensions of the work environment; it can be used to ascertain how well workspace helps occupants work as well as to set priorities for environmental problem-solving and workspace change.

The best way to find out how comfortable or uncomfortable workers are is to ask them, as they are the experts on the performance and requirements of their tasks. However, it is important to ask them the right questions, and in such a way as to know exactly what their answers mean. The functional comfort survey is a short, standardized questionnaire that has been used to measure various types of workspace since the early 1990s. It is both reliable (can be used in different kinds of building) and valid (collects data that measure users' experiences of actual building conditions). The questions invite respondents to rate how comfortable they are on critical dimensions of functional comfort. Survey results help to identify both positive and supportive building conditions as well as those that might cause workspace stress. The data from the questionnaire survey provide constructive feedback that diagnoses workspace functionality, comfort and quality in precise locations. The results are designed to measure workspace as a tool for work, and feedback has been collected from buildings in North America, Europe and Australia.[2] Measuring functional comfort is one of many tools available to assess EQ. Using an existing standardized survey system means that planners can benchmark data to use for comparing and understanding results.

Once survey data have been collected, they must be analyzed, the results interpreted, and useful conclusions developed to aid in correcting problems and guiding workspace design and planning. Many surveys collect data on demographic differences, such as age, gender and job type. However, in using occupant feedback as a diagnostic tool, demographic distinctions are less relevant than locational differences such as proximity to windows, type of desk or furniture, and temperature and ventilation conditions. Assessing functional comfort requires user ratings of environmental conditions that affect task performance rather than data on who users are and what they like and dislike.

Functional comfort assessments and the results of other evidence-gathering activities have been used to measure EQ in a range of workspace design and planning situations, including:

- Setting priorities for building maintenance and renovation,
- Responding to long-term employee complaints,
- Promoting continuous improvement,
- Comparing user assessments from different buildings, floors, areas, and pre- and post-change,
- Establishing benchmarks for workspace quality over time.

The results of such assessments provide a valid and reliable indicator of environmental quality and facilitate the evaluation of workers' quality of life. Obtaining reliable feedback from occupants who are knowledgeable about their tasks offers a firm basis for new workspace design and a constructive opportunity to engage and involve occupants in situations of workspace change.

21.6 Designing a Better Workspace to Improve Users' QoL at Work

The theoretical model outlined in this chapter posits that identifiable elements of the physical environments in which people work can be assessed in terms of EQ (through measuring occupants' functional comfort levels) and thereby affect QWL and ultimately contribute to QoL. Implicit in this argument is that there is a direct link between environmental design and QoL. Studies of workspace that has been strategically designed to support workers' tasks and activities demonstrate increases in all levels of workspace comfort, especially where the planning and design process has met psychological comfort needs by including opportunities for feedback and by empowering users' through participation in the planning process. Improved functional and psychological comfort gives rise to a sense of ownership. Feeling supported leads not only to better work performance and improved client and co-worker relations, but also to more commitment and loyalty to the organization and better QWL, which results in reduced staff turnover and higher morale. Measuring occupants' levels of functional comfort produces information that can be applied to space changes and building renovations as well as to the design of new space.

---

[2]Previously published in Vischer (1996, 2005).

As a critical component of QWL, EQ is also a key predictor of QoL. Collecting feedback from users that allows a diagnostic approach to be taken to building performance also generates information that can be applied to improving people's QWL and QoL. Diagnostic information on workspace quality is needed to shape the interior environment to provide improved support to what users are actually doing. This means that once physical comfort is assured in terms of occupants' health and safety, workers should occupy an adaptable workspace where lighting levels are adjustable, various configurations of desks and chairs and meeting spaces are possible, and where environments support constructive and flowing collaboration as well as focused concentration and privacy. As the workspace comfort pyramid (Fig. 21.2) indicates, once work tasks and activities are functionally supported, there are opportunities for increasing psychological comfort, such as more individual and team control over workspace through easily reconfigurable territorial definition, a sense of security, and aesthetic advantages. As indicated by the arrow in the diagram, the vector propels towards quality improvement – that is, more EQ – as more needs at the three levels are met. The inclusion of psychological comfort makes the connection to less tangible components of the work environment which are related to human needs and affect workers' QWL. These may include opportunities for autonomy and responsibility, employer-employee relations, organizational culture and values, opportunities for reward and advancement, and a social support network (Herbst 1962; Lawler 1975; Walton 1980).

As this chapter makes clear, QoL includes health, comfort, and satisfaction of needs. It is seen as a "systematic framework through which to view work aimed toward improving the lives of individuals" (p. 54, Keith 2001). QoL can be assessed in three ways: as a "sensitizing notion that provides reference and guidance", as a "social construct", and as an "organizing concept" or "unifying theme" (Keith and Schalock 2000). QoL is commonly used as an indicator of total well-being based on how people feel about different aspects of their lives. Objective measures of QoL are preferred by public agencies and policy-makers, as they translate into codes and standards that can be applied and their effects assessed. Subjective QoL measures are preferred by students of social and behavioral psychology, who use self-report data from validated research instruments. Subjective or perceived QoL typically contains 'cognitive' and 'affective' components; cognitive feedback is often based on individual judgments and can be subsumed as an element in need satisfaction, whereas affective feedback is typically more emotional and can be considered an element of human happiness (Kerce 1992). While satisfaction means comparing external conditions to one's internal standards, happiness is an appraisal of an emotional experience (Cheng 1988). Andrews and Withey (1976) believe that it is "only when both types of measures (subjective and objective) are concurrently measured will it be possible to know how demonstrable changes in living conditions are affecting peoples' sense of life quality and, conversely, whether changes in people's sense of life quality can be attributed to changes in external conditions."

Numerous POE, BPE and other studies of user-environment fit have focused on need satisfaction through users' own ratings of how well the occupied space meets their expectations and internalized standards. The implicit assumption is that if users state that their needs are met and that they are satisfied, then the built space they occupy is a success (e.g. Marans and Yan 1989; Humphreys 2005; Veitch et al. 2007; Schakib-Ekbatan et al. 2010). In this paradigm, meeting needs as a criterion for building quality is conflated with meeting needs as an indicator of quality of life, although the former measures building performance and the latter measures user experience. While other types of study use objective or instrumental measures of EQ – such as measuring indoor air contaminants, humidity levels, luminance and illuminance, and other ambient conditions – results still have to be interpreted in the context of the real or assumed comfort experience of occupants in order to have meaning.

In this chapter, we have argued for measuring QWL – and therefore QoL – by testing levels of functional comfort as rated by occupants. This can be construed as a subjective measure of EQ because measuring tools are designed to elicit occupants' feedback on their experiences of the environment. However, the functional comfort paradigm is designed to apply the results of data analysis to assessing a building's EQ and diagnosing building problems with a view to improvement, so can also be considered objective.

Field studies of EQ in work environments often highlight differences between instrument measurements of building performance or quality and occupants' judgments and perceptions (Leaman and Bordass 1999; Sekhar et al. 2003). For example, a workplace where a thermometer measures temperature at 20 °C may be rated 'cold' by occupants – perhaps because the majority are seated near windows or under ceiling diffusers or simply dressed in light clothing. Hence, there may be variation in what is objectively satisfactory and what is satisfactory from the users' perspective in measuring both EQ and QoL. However, by considering the rating as a diagnosis rather than a judgment and seeking out the reason why users have assigned a 'cold' rating to an objectively 'warm' environment, it becomes possible to identify the cause of the problem and therefore its solution.

"From the occupant's point of view, the ideal situation is an indoor environment that satisfies all occupants (i.e. they have no complaints) and does not unnecessarily increase the risk or severity of illness or injury" (Bluyssen et al. 2003). However, the many parallels between QoL and EQ research – and specifically the emphasis on satisfaction of human needs – support the argument that EQ is a valid and measurable indicator of QoL and that, in this context, measuring people's satisfaction is as valid an indicator of EQ as functional comfort ratings. While functional comfort is a diagnostic tool for assessing building performance through specific outcomes such as occupants' task performance and effectiveness, eliciting occupants' satisfaction ratings is a complementary approach to predicting QWL and QoL at work.

In finding out more about the complex relationships between people at work and the environmental tools they need, a more rational and substantive basis is being developed for informing design decisions that lead to more supportive workspace. In parallel, more is being discovered about how people experience QoL at work and the degree to which they feel their needs are being met. These are two interestingly complementary approaches for future research.

References

Abbaszadeh, S., Zagreus, L., Lehrer, D., & Huizenga, C. (2006). Occupant satisfaction with indoor environmental quality in green buildings. In *Proceedings, healthy buildings 2006, Vol. III* (pp. 365–370). Lisbon, Portugal, 4–6 June.

Adams, S. (1997). *The Dilbert principle: A cubicle's-eye view of bosses, meetings, management fads & other workplace afflictions*. New York: HarperBusiness; Reprint edition.

Andrews, F. M., & Withey, S. B. (1976). *Social indicators of well-being: The development and measurement of perpetual indicators*. New York: Plenum.

Bagnara, S., Mariani, M., & Parlangeli, O. (2001). Ergonomics for quality of working life in services. In G. Bradley (Ed.), *Humans on the net – Information and Communication Technology (ICT) work organization and human beings* (pp. 139–154). Stockholm: SE Prevent.

Becker, F. (1981). *Workspace: Creating environments in organizations*. New York: Praeger.

Becker, F. (1990). *The total workplace*. New York: Van Nostrand Reinhold.

Blishen, B., & Atkinson, T. (1980). Anglophone and francophone differences in perceptions of the quality of life in Canada. In A. Szalai & F. M. Andrews (Eds.), *The quality of life: Comparative studies* (pp. 25–39). London: Sage.

Bluyssen, P. M., Cox, C., Seppänen, O., Oliveira Fernandes, E., Clausen, G., Müller, B., & Roulet, C. A. (2003). Why, when and how do HVAC-systems pollute the indoor environment and what to do about it? The European AIRLESS project. *Building and Environment, 38*(2), 209–225.

Brill, M., & Weideman, S. (2001). *Disproving widespread myths about workplace design*. Jasper: Kimball International.

Bruce, D. (2008). *How much can noise affect your worker's productivity*. http://www.office-Soundmasking.com

Cheng, S. (1988). Subjective quality of life in the planning and evaluation of programs. *Evaluation and Program Planning, 11*, 123–134.

Chu, W. T., & Warnock, A. C. (2002). *Measurement of sound propagation in open plan offices*. Ottawa: Institute for Research in Construction, National Research Council of Canada.

Csikszentmihalyi, M. (1990). *Flow: The psychology of optimal experience*. New York: Harper Collins.

Cummings, T. G., & Worley, C. G. (2005). *Organization development and change*. Mason: Thomson/South-Western.

Cunningham, J. B., & Eberle, T. (1990). A guide to job enrichment and redesign. *Personnel, 67*(2), 56–61.

Damian, A. (2004). *Airy offices create 15 % work boost*. http://www.building.co.uk/airy-offices-create-15-work-boost/3038316.article. Accessed 15 July 2014.

Danna, K., & Griffin, R. W. (1999). Health and well-being in the workplace: A review and synthesis of the literature. *Journal of Management, 25*, 357–384.

Davis, L. E., & Cherns, A. B. (1975). *The quality of working life* (Vol. 1). New York: The Free Press.

Day, C. (2002). *Spirit & place: Healing our environments*. London: Architectural Press.

Dilani, A. (2004). *Design and health III: Health promotion through environmental design*. Stockholm: International Academy for Design and Health.

Doxtater, D. (1994). *Architecture, ritual practice and co-determination in the Swedish office*. Aldershot: Avebury.

Elizur, D., & Shye, S. (1990). Quality of work life and its relation to quality of life. *Applied Psychology, 39*(3), 275–291.

Elzeyadi, I. (2001). *Ten palaces tell their stories: Environmental quality assessment of offices inside adaptively re-used historical palaces in Cairo, Egypt*. Ph.D. Dissertation, University of Wisconsin, Milwaukee.

Evans, G. W., & Johnson, D. (2000). Stress and open-office noise. *Journal of Applied Psychology, 85*(5), 779–783.

Fischer, G.-N. (1997). *Individuals and environment. A psychosocial approach to workspace*. New York: Walter de Gruyter.

Fisk, W. J. (2000). Health and productivity gains from better indoor environments and their implications for the U.S. Department of Energy. In *Proceedings of E-Vision 2000 conference, indoor environment department*, Washington, DC, USA, 11–13 October.

Fisk, W. J., Price, P. N., Faulkner, D., Sullivan, D. P., Dibartolomeo, D. L., Federspiel, C. C., et al. (2002). Worker performance and ventilation: Analyses of time-series data for a group of call-center workers. In *Proceedings of the Indoor Air 2002 conference, Vol. 1* (pp. 790–795). Monterey: ISIAQ.

Fleming, M., & Larder, R. (1999). When is a risk not a risk. *Professional Safety, 69*(3), 30–38.

US Green Building Council. (2014). *LEED v4 integrative design process*. http://www.usgbc-ncc.org/calendar/event/1451. Accessed 10 Oct 2014.

Haber, G. M. (1977). The impact of tall buildings on users and neighbors. In D. Conway (Ed.), *Human response to tall buildings* (pp. 45–55). Stroudsburg: Dowden, Hutchinson, & Ross.

Hackman, J. R., & Suttle, J. L. (1977). *Improving life at work*. Glenview: Scott, Foresman.

Hameed, A., & Amjad, S. (2009). Impact of office design on employees' productivity: A case study of banking organisations of Abbottabad, Pakistan. *Journal of Public Affairs, Administration and Management, 3*(1), 1–13.

Haynes, B. P. (2007). The impact of the behavioural environment on office productivity. *Journal of Facilities Management, 5*(3), 158–171.

Heerwagen, J., Kampschroer, K., Powell, K., & Loftness, V. (2004). Collaborative knowledge work environments. *Building Research & Information, 32*(5), 510–528.

Herbst, P. G. (1962). *Autonomous group functioning*. London: Tavistock.

Heschong, L., Wright, R. L., & Okura, S. (2002). Daylighting impacts on human performance in school. *Journal of the Illuminating Engineering Society, 31*(2), 101–112.

Humphreys, M. (2005). Quantifying occupant comfort: Are combined indices of the indoor environment practicable? *Building Research and Information, 33*(4), 317–325.

Kaplan, R. D. (1983). Person-environment fit, past, present and future. In C. L. Cooper (Ed.), *Stress research: Issues for the eighties* (pp. 35–78). Hoboken: Wiley.

Kaplan, S. (1995). The restorative benefits of nature: Toward an integrative framework. *Journal of Environmental Psychology, 15*(3), 169–182.

Kaplan, S., Talbot, J. F., & Kaplan, R. (1988). *Coping with daily hassles: The impact of nearby nature on the work environment*. Project Report, US Forest Service. Urban Forest Unit Cooperative Agreement 23-85-08. St. Paul: North Central Forest Experiment Station.

Karasek, R., & Theorell, T. (1990). *Health work: Stress, productivity and the reconstruction of working life*. New York: Basic Books.

Keith, K. D. (2001). International quality of life: Current conceptual, measurement and implementation issues. *International Review of Research in Mental Retardation, 24*, 49–74.

Keith, K. D., & Schalock, R. L. (2000). Cross-cultural perspectives on quality of life: Trends and themes. In K. D. Keith & R. L. Schalock (Eds.), *Cross-cultural perspectives on quality of life* (pp. 363–380). Washington, DC: American Association on Mental Retardation.

Kerce, E. W. (1992). *Quality of life: Meaning, measurement, and models (NPRDC-TN-92-15)*. San Diego: Navy Personnel Research and Development Center.

Khattab, O. (1993). Environmental quality assessment, an attempt to evaluate government housing projects. *Open House, 18*(4), 41–47.

Konnikova, M. (2014). The open-office Trap. *The New Yorker*, (blog) January 7.

Larsen, L., Adams, J., Deal, B., Kweon, B., & Tyler, E. (1998). Plants in the workplace: The effect of plant density on productivity, attitude and perceptions. *Environment and Behavior, 30*(3), 261–281.

Lawler, E. E. (1975). Measuring the psychological quality of working life: The why and how of it. In L. E. Davis & A. B. Cherns (Eds.), *The quality of working life* (Vol. 1, pp. 123–133). New York: Free Press.

Lawler, E. E. (1982). Strategies for improving the quality of work life. *American Psychologist, 37*(5), 486.

Leaman, A., & Bordass, B. (1999). Productivity in buildings: The 'killer' variables. *Building Research & Information, 27*(1), 4–19.

Leaman, A., & Bordass, B. (2007). Are users more tolerant of green buildings? *Building Research and Information, 35*(6), 662–673.

Lee, Y. S., & Kim, S.-K. (2008). Indoor environmental quality in LEED-certified buildings in the US. *Journal of Asian Architecture and Building Engineering, 7*(2), 293–300.

Marans, R. W., & Yan, X. (1989). Lighting quality and environmental satisfaction in open and enclosed offices. *Journal of Architectural and Planning Research, 6*, 118–131.

Maslow, A. H. (1954). *Motivation and personality*. New York: Harper and Row.

McCoy, J. M., & Evans, G. W. (2005). Physical work environment. In J. Barling, E. K. Kelloway, & R. M. Frone (Eds.), *Handbook of work stress* (pp. 219–245). London: Sage.

Meeberg, G. A. (1993). Quality of life: A concept analysis. *Journal of Advanced Nursing, 18*(1), 32–38.

Milton, D. K., Glencross, P. M., & Walters, M. D. (2000). Risk of sick leave associated with outdoor air supply rate, humidification and occupant complaints. *Indoor Air, 10*(4), 212–221.

Monk, R. (1997). The impact of open-plan offices on organizational performance. *International Journal of Management, 14*(3), 345–349.

Mubex, C. M. (2010). *Closed offices versus open plan layout*. http://www.mubex.com/sme/closed-vs-open-plan-officers.htm. Accessed 3 Aug 2012.

O'Neill, M. (2008). Open plan and enclosed private offices: Research review and recommendations, Knoll Workplace Research. http://www.aminteriors.com/wp-content/uploads/2011/03/OpenClosed_Offices_wp.pdf. Accessed 1 June 2013.

Ornstein, S. (1992). First impressions of the symbolic meanings connoted by reception area design. *Environment and Behavior, 24*(1), 85–110.

Paul, W. L., & Taylor, P. A. (2008). A comparison of occupant comfort and satisfaction between a green building and a conventional building. *Building and Environment, 43*(11), 1858–1870.

Preiser, W. F. E. (1983). The habitability framework: A conceptual approach towards linking human behavior and the built environment. *Design Studies, 4*(2), 84–91.

Preiser, W. F. E., & Schramm, U. (1997). Building performance evaluation. In D. Watson, M. J. Crosbie, & J. H. Callender (Eds.), *Time-saver standards: Architectural design data* (pp. 233–238). New York: McGraw-Hill.

Preiser, W. F. E., & Schramm, U. (2012). A process model for Building Performance Evaluation (BPE). In S. Mallory-Hill, W. F. E. Preiser, & C. Watson (Eds.), *Enhancing building performance* (pp. 19–31). Oxford: Wiley-Blackwell.

Preiser, W. F. E., & Vischer, J. C. (2004). *Assessing building performance*. Oxford: Elsevier Scientific Publishing.

Quilan, M. (2001). Precarious employment: Work reorganization and the factoring of OHS management. *International Journal of Systematic Occupational Health and Safety Management, 24*(10), 175–178.

Rapoport, A. (1977). *Human aspects of urban form: Toward a man-environment approach to urban form and design*. Oxford: Pergamon.

Rapoport, A. (1983). Environmental quality, metropolitan areas and traditional settlements. *Habitat International, 7*(3/4), 37–63.

Rapoport, A. (1988). Levels of meaning in the built environment. In F. Poyatos (Ed.), *Perspectives on non verabl communication* (pp. 317–336). Toronto: C.J. Hogrefe.

Rapoport, A. (1990). System of activities and system of settings. In S. Kent (Ed.), *Domestic architecture and the use of space* (pp. 171–232). Cambridge: Cambridge University Press.

Rashid, M., & Zimring, C. (2008). A review of the empirical literature on the relationships between indoor environment and stress in health care and office settings: Problems and prospects of sharing evidence. *Environment and Behavior, 40*(2), 151–190.

Rice, R. W. (1984). Work and the quality of life. In S. Oskamp (Ed.), *Applied social psychology annual. 5: Applications in organizational settings* (pp. 155–177). Beverly Hills: Sage.

Rice, R. W., McFarlin, D. B., Hunt, R. G., & Near, J. P. (1985). Organizational work and the perceived quality of life: Toward a conceptual model. *Academy of Management Review, 10*(2), 296–310.

Schakib-Ekbatan, K., Wagner, A., & Lussac, C. (2010). Occupant satisfaction as an indicator for the sociocultural dimension of sustainable office buildings – development of an overall building index. In *Proceedings of conference on adapting to change: New thinking on comfort*, Cumberland Lodge, Windsor, UK, 9–11 April.

Sekhar, S. C., Tham, K. W., & Cheong, K. W. (2003). Indoor air quality and energy performance of air-conditioned office buildings in Singapore. *Indoor Air, 13*, 315–331.

Sheel, S., Sindhwani, D., Goel, S., & Pathak, S. (2012). Quality of work life, employee performance and career growth opportunities: A literature review. *International Journal of Multidisciplinary Research, 2*(2), 291–300.

Sirgy, M. J., Efraty, D., Siegel, P., & Lee, D. (2001). A new measure of quality of work life (QoWL) based on need satisfaction and spillover theories. *Social Indicators Research, 55*, 241–302.

Straw, R. J., & Heckscher, C. C. (1984). QWL: New working relationships in the communication industry. *Labor Studies Journal, 9*, 261–274.

Sundstrom, E., Town, J. P., Rice, R. W., Osborn, D. P., & Brill, M. (1994). Office noise, satisfaction, and performance. *Environment and Behavior, 26*(2), 195–222.

Tarricone, P., & Luca, J. (2002). Employees, teamwork and social interdependence – A formula for successful business? *Team Performance Management: An International Journal, 8*(3/4), 54–59.

Turner, B. (1990). *Organizational symbolism*. Berlin: Walter de Gruyter.

Varghese, S., & Jayan, C. (2013). Quality of work life: A dynamic multidimensional construct at work place–part II. *Guru Journal of Behavioral and Social Science, 1*(2), 91–104.

Veitch, J., & Gifford, R. (1996). Choice, perceived control and performance decrement in the physical environment. *Journal of Environmental Psychology, 16*, 269–276.

Veitch, J. A., & Newsham, G. R. (2000). Exercised control, lighting choices and energy use: An office simulation experiment. *Journal of Environmental Psychology, 20*(3), 219–237.

Veitch, J. A., Charles, K. E., Newsham, G. R., Marquardt, C. J. G., & Geerts, J. (2004). *Workstation characteristics and environmental satisfaction in open-plan offices: COPE Field Findings (NRCC-47629)*. Ottawa: National Research Council.

Veitch, J. A., Charles, K. E., Farley, K. M. J., & Newsham, G. R. (2007). A model of satisfaction with open-plan office conditions: Cope field findings. *Journal of Environmental Psychology, 27*(3), 177–189.

Vischer, J. C. (1985). The adaptation and control model of user needs: A new direction for housing research. *Journal of Environmental Psychology, 5*(4), 239–255.

Vischer, J. C. (1989). *Environmental quality in offices*. New York: Van Nostrand Reinhold.

Vischer, J. C. (1996). *Workspace strategies: Environment as a tool for work*. New York: Chapman and Hall.

Vischer, J. C. (2003, June). Designing the work environment for worker health and productivity. In *Proceedings of the 3rd international conference on design and health* (pp. 85–93). Montreal, Canada.

Vischer, J. C. (2005). *Space meets status: Designing workplace performance*. New York: Routledge.

Vischer, J. C. (2007). The effects of the physical environment on job performance: Towards a theoretical model of workspace stress. *Stress & Health: Journal of the International Society for the Investigation of Stress, 23*(3), 175–184.

Vischer, J. C. (2008). Towards an environmental psychology of workspace: How people are affected by environments for work. *Architectural Science Review, 51*(2), 97–108.

Vischer, J. C., & Fischer, G. N. (2005). Issues in user evaluation of the work environment: Recent research. *Le Travail Humain, 68*(1), 73–81.

Vischer, J. C., & Malkoski, K. (2015, in press). *The power of workspace for business and people*. Melbourne: Schiavello.

Walton, E. R. (1980). Quality of work life activities: A research agenda. *Professional Psychology, 11*(3), 393–484.

Walton, E. R. (1991). From control to commitment in the workplace. In R. Steers & L. Porter (Eds.), *Motivation and work behavior* (pp. 15–29). New York: McGraw-Hill.

Woo, J. M., & Postolache, T. T. (2008). The impact of work environment on mood disorders and suicide: Evidence and implications. *International Journal on Disability and Human Development, 7*(2), 185–200.

World Health Organization. (1994). Development of the WHOQOL: Rationale and current status. *International Journal of Mental Health, 23*, 24–56.

Comfort at Work: An Indicator of Quality of Life at Work

22

Liliane Rioux

22.1 Introduction

According to Martel and Dupuis (2006), "Quality of Work Life, at a given time, corresponds to a condition experienced by the individual in his or her dynamic pursuit of his or her hierarchically organized goals within work domains where the reduction of the gap separating the individual from these goals is reflected by a positive impact on the individual's general quality of life, organizational performance, and consequently the overall functioning of society" (p. 355). By focusing on the way employees evaluate their current situation in relation to their expectations and to the speed with which they think they can meet their objectives, this definition of the quality of work life gives the work environment a central role. This role was disputed by Mayo (Roethlisberger and Dickson 1939), who concluded that psychosocial factors had a greater impact on productivity than changes in physical conditions (lighting, temperature, etc.) (see Pol 2006). Although comfort is a central concern of our modern period and a clear feature of our daily life and our history, it is only during the last 20 years that researchers in the Human Sciences (psychologists, sociologists, managers, architects, etc.) have taken it up and applied it to work contexts, investigating the way in which employees experience their workplace. From this standpoint, environmental psychology clearly has a key role, on account of its interest in the psychological processes that help maintain the dynamics between physical and social spaces.

After a brief historical overview of the concept of comfort, we look at the current state of research. We then present the theoretical model of comfort at work proposed by Rioux et al. (2013), together with its physical, functional and psychological dimensions. Finally, we look at the role of comfort in Human Resource Management, notably in relation to the WHO guidelines for the primary prevention of occupational hazards.

22.2 Comfort, an Evolving Composite Concept

In Europe, one of the first written traces of the word "comfort" ("confort" in French) dates back to the eleventh century in the "Chanson de Roland", an epic poem and "chanson de geste" that gives a highly romanticized account of the defeat of Charlemagne's troops by the Saracens. At that time, "confort" meant aid and assistance, but rapidly took on the meaning of solace or support ("réconfort"). This need for comfort-solace (confort-réconfort) in the

L. Rioux (✉)
Department of psychology, Laboratoire Parisien de Psychologie Sociale-LAPPS, EA 4386, Université Paris Ouest Nanterre La Défense, Nanterre, France
e-mail: lrioux@u-paris10.fr

medieval nobility led to the creation of places that people could call their own; private spaces that were restful and reassuring, as opposed to external spaces seen as threatening or hostile. Thus, dwelling places, initially built to receive people, gradually became private places too. Two types of space thus emerged, corresponding to two types of comfort: reception places (e.g. dining rooms) that accentuated comfort-luxury, and more intimate places (bedrooms, toilets) that symbolized comfort-sensation.

The meaning of the word gradually weakened over time, and in England at the end of the eighteenth century it became synonymous with well-being. As a result of rapid technological progress, European industrial societies made it a specific cultural form, a symbol of the bourgeois life-style of the period. For example, in France at the end of the nineteenth century, "les confortables" were slippers or armchairs. This comfort of luxury, restricted to an elite, enabled the bourgeoisie to maintain a certain distance from the "common" people, and hence a sense of recognition.

The technological advances of the twentieth century led to the idea of comfort for all. It is seen as a factor of progress and has become accessible to everyone, even if certain comforts still depend on the individual's place in society. This user comfort is thus considered an indicator of well-being, or even of happiness. Since 1945, the main aspect of comfort in the great majority of member countries of the Organization for Economic Cooperation and Development (OECD) has been housing. For example, in France, surveys by INSEE (French National Institute of Statistics and Economic Studies) have made it possible to monitor how individuals' comfort has changed over time and to categorize each housing unit by calculating a minimum level of comfort. Norms were established, notably with regard to sanitary facilities, which probably explains the strong links that emerged between comfort and hygiene and between comfort and health. This resulted in new practices, and comfort consequently became subject to not only technical but also social norms. Between 1945 and 1973, the period known as "Les Trente Glorieuses", it was used as the reference value reflecting the growth of a society (Le Goff 1994).

In Europe, following the economic crisis of 1974, comfort took on a more qualitative dimension; it was no longer a question of offering more comfort, but rather of targeting areas of discomfort, notably related to life outside the home. The French expression "metro, boulot, dodo" (metro, work, sleep) (inspired by a poem by Pierre Béarn (1951)) illustrates this desire to relieve the monotony of daily life and the lack of collective comfort. The demand for comfort thus extended across the whole public domain, from public transport to hospitals, and including residential housing and the workplace. Over and above personal comfort in the private sphere, the notion of a more collective comfort developed, in harmony with the immediate environment, related to a certain idea of the quality of life.

Finally, a new type of comfort has recently emerged, which we will call comfort-sustainable development. This corresponds to the search for a way of life in keeping with technological advances that helps preserve the resources of the planet (Cole et al. 2008) and in harmony with pro-environmental attitudes (Becker and Félonneau 2011). It is widely disseminated via social networks, fostering the adoption of environmentally-friendly behaviors in order to ensure the comfort of future generations, while also calling for vigilance regarding countries and companies that do not respect this sustainable comfort.

From the "comfort-solace" (confort-réconfort) of the Middle Ages to the "comfort-sustainable development" of recent years, via the "comfort of luxury" restricted to the bourgeois elite of the nineteenth century and the "user comfort for all" that symbolized the technical and social progress of the twentieth century, it is clear that "comfort" can have multiple meanings, covering spatial, temporal, physical and psychological aspects. Looking at its history, notably in Europe, shows that it is complex and multi-faceted. Nevertheless, whichever definition is adopted, it always relates to space, to an environment in which the person feels at ease. As such, environmental psychology is probably the discipline that is best placed to understand comfort in all its diversity.

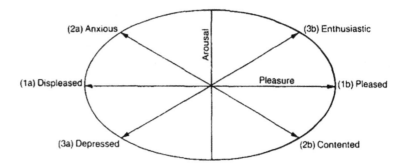

Fig. 22.1 Three principal axes for the measurement of affective well-being (Warr 1990)

22.2.1 Models for Studying Comfort at Work

22.2.1.1 Warr's Model

It is difficult to discuss comfort at work without looking at Warr's (1990) vision of affective or psychological well-being. From a review of scientific articles dealing with well-being and a study of more than 1500 job-holders in various occupational sectors in the UK, Warr developed a model based on two independent dimensions that he called "pleasure" and "arousal". He thus identified three bipolar factors to measure the affective well-being of an individual (Fig. 22.1).

These factors constitute groups of affects that can be located within a two-dimensional space and can be differentiated qualitatively along three axes: (a) displeasure–pleasure, (b) depression–enthusiasm, (c) anxiety–comfort. This model was extended by Daniels (2000) who proposed a five-factor structure of bipolar affects: angry-placid, anxiety-comfort, depression-pleasure, bored-enthusiastic, and tiredness-vigor. However, in both models, comfort is opposed to anxiety. While anxiety corresponds to a low level of pleasure associated with a very high level of arousal, comfort refers to a high level of pleasure without mental arousal.

In an organizational setting, this state of low arousal associated with a degree of pleasure at work could indicate a state of "resigned job satisfaction", in which the employee does not complain about his/her work, but shows relative apathy or lack of interest (Warr 1994). In a very large study with a sample of 14,127 employees in 1177 different workplaces, Wood et al. (2012) observed that there was no significant relationship between the anxiety-comfort dimension and the four measures of organizational performance (financial performance, productivity, quality of work and absenteeism), whereas the latter were all significantly related to job satisfaction. According to these authors, these results could indicate that performance is influenced more by the pleasure experienced at work than by the level of activation of the cirumflex model of emotions (Wood et al. 2012). The link between comfort and performance would be made via pleasure at work.

22.2.1.2 Vischer's Model

The studies on comfort carried out by Vischer and the Groupe de recherche sur les environnements de travail (GRET; Work Environments Research Group) fall within the framework of organizational/environmental psychology. The researchers postulate that a workspace can enhance or hinder the comfort of workers and can be situated on a stress-to-comfort continuum, independently of the characteristics of the building, its internal layout or the jobs carried out there. From this standpoint, comfort is defined independently of the arousal dimension, and contrasts with stress.

Vischer (2005; 2006) drew up a model of comfort with a hierarchical structure involving three levels of comfort: physical, functional and psychological. This model is presented in greater detail in Chap. 21.

It postulates that to experience a degree of comfort at work, employees should not just work in a safe and healthy environment, but also need an environment that supports them in the activities they have to carry out (Vischer 1996). This comfort, which Vischer calls functional comfort, is conditioned not only by a viable environment that ensures the physical health of workers but also by an environment that creates "workability". To this end, the workspace must be perceived as having both good sensorial qualities, including visual (e.g. non-glare lighting), auditory (e.g. prevention of disturbing noise), thermal and hydrometric (e.g. temperature and humidity levels can be regulated), and spatial qualities with a layout that provides easy access and use of work tools. The study of functional comfort thus requires an evaluation of the building's performance. Optimization of this performance involves the interior design, based on the extent to which managers adapt workspaces to the professional activity of the people working in them, and the behavior of the workers themselves, and thus their preferences in terms of the sensory qualities of the building and the layout of the workspace. As such, functional comfort is at the heart of the "dialogue" between architecture and the environment, creating living and work spaces that evolve over time (Moser and Weiss 2003).

One of the main interests of this model lies in its associated diagnostic method and its eminently applied character. The evaluation tool takes into account the specific features of the building and the expectations and aspirations of the employees of a given organization. In other words, it must be adapted to each organization. For further details, the reader can consult the excellent review published in Le Travail Humain (Human Work) (Vischer and Fischer 2005).

22.2.1.3 Rioux's Model (2012)

Rioux's model is based on those of both Warr and Vischer, breaking down comfort at work into three dimensions: physical, evaluative and psychological.

The first dimension is evaluated by physical measures of the workspace. For example, light can be measured using a light meter or a luminance meter, and temperature and hygrometry using a thermohygrometer. The second dimension can be measured by an evaluation of environmental satisfaction at work (satisfaction overall and with different places within the organization) (Ittelson et al. 1974; Moffat et al. in press). The third dimension, the psychological aspect, refers, for instance, to attachment to the workplace, defined as "a global relationship of people's identity and affective evaluations with their [...] environment" (Bonaiuto et al. 1999 p. 333). Comfort at work could thus be evaluated using a three-dimensional tool, which could be used to identify the psycho-environmental risks incurred by employees.

22.2.2 Physical Dimension of Comfort at Work

Before discussing the physical dimension of comfort at work in more detail, it is appropriate to make a detour via the texts that impose obligations on company managers, namely those concerning the regulations of the country where the company is situated (e.g. the Code du Travail in France or the Codice del lavoro in Italy) and those involving the recommendations of international bodies, which enable companies to obtain certification. The best-known and most dynamic organization is probably the International Organization for Standardization (ISO), which is composed of representatives of the national standardization organizations of 157 countries. Its objective is to produce norms, called ISO standards, in industrial and commercial sectors. These standards are only guidelines and recommendations that countries are encouraged to adopt, taking into account their specific social, cultural, political and legislative characteristics.

ISO 26000 can be cited as an example. In 2005, at the request of consumers, ISO members agreed to work on a social responsibility standard. It was finally published in 2010. It provides guidance to organizations for initiating procedures in seven core subjects related to social responsibility (employment relationships and work conditions, governance, human rights, fair operating practices, consumer issues, the environment and social investment). ISO 26000 states that working conditions have profound repercussions on workers' quality of life and that organizations should thus not only respect the legislation and regulations in force (action area 3) but also promote and maintain the highest degree of physical, mental and social well-being of employees and prevent the negative effects of work conditions on health (action area 4). It is up to each state to draw up their own normative text dealing with social responsibility and sustainable development. In France, the NF X50-135 standard, published on 16th July 2011 and dealing with responsible purchases, applies the concept of social responsibility, as defined by ISO 26000, to French organizations. To ensure that employees are in a physically comfortable situation, there must be no potential nuisance factor in their risk zone that could have an impact on their health. The term "nuisance" is taken here in its broadest sense and refers to known and listed physical agents or, more generally, to the atmosphere and layout of the workspace.

22.2.2.1 Physical Agents

Evaluation of a nuisance involves taking into account its duration, frequency, and the level of exposure. In particular, exposure must be estimated and compared to a reference level, taking the form either of a norm on a legal basis (e.g. Code du Travail in France), or of a recommendation (e.g. ISO standard). For example, the European directive 2003/10/CE of 6th February 2003 regarding noise at work is applied in French law via decree no. 2006-892 of 19th July 2006; article R. 231-127 of this decree stipulates that an exposure value of over 70 dB must trigger a preventative action by the employer (provision of personal protection equipment, sound-proofed walls, absorbent materials, enclosure of machines, etc.).

Recommendations for measuring and evaluating moderate thermal atmospheres are provided by ISO 7730. It provides a method of predicting the sensation and degree of discomfort of people exposed to moderate thermal environments, notably within an organizational setting. It also recommends standards of acceptability of thermal environment conditions and is applicable in most regions of the world, even though adjustments are required in order to take into account geographical and ethnic differences.

Respecting current standards and recommendations in terms of occupational health should, in principle, guarantee acceptable health and safety conditions and thus the physical comfort of employees by protecting them from environmental nuisances.

It should be recalled that establishing a standard of exposure is based on the principle that there is a relationship between the amount of exposure and the frequency of the emergence of a harmful effect on workers' health. When the curve showing the dose-response relationship reveals the existence of a threshold, as in the case of noise, it is relatively easy to fix a standard. However, when the dose-response relationship does not indicate a threshold, as is the case with carcinogenic substances and ionizing radiation, it is not possible to determine a risk-free level. In this case, levels of risk can be defined but it is usually recommended that exposure be reduced as much as possible.

While a number of organizations work on updating these standards annually (e.g. the American Conference of Governmental Industrial Hygienists (ACGIH) in the USA and the Deutsche Forschungsgemeinschaft in Germany), the rapid development of knowledge, the number and diversity of national and/or international recommendations, together with the ever-increasing specialization of work stations, make it relatively complex for companies to apply them. This is particularly true given that the standards are established on the basis of pure exposure to a single nuisance factor and cannot take into account the effect of exposure to several different factors, which is however very common and constitutes what is called the work atmosphere.

22.2.2.2 The Work Atmosphere

In a work environment, employees are subject to several types of nuisance factors or physical agents that together constitute the work atmosphere. This is particularly the case of noise, which frequently interacts with heat and vibrations in industrial environments, producing a more intense physical discomfort than that created by each factor alone (Pellerin and Candas 2004). One example is people working on motorbikes, such as the police or dispatch riders. Paradoxically, because these people work in extreme environments, they benefit from specific preventative measures and are often less exposed than those who are subject to noise of lower intensity (Rabinowitz et al. 2007). However, most companies generate background noise of low intensity that can affect workers' health. In office environments, this can be linked to computer cooling fans, desktop devices, or even the chatter of colleagues. A large number of studies have been carried out on the effects of auditory distraction on performing office tasks (Banbury and Berry 1998; Hughes and Jones 2003), and it has been clearly shown that background noise is not always synonymous with auditory distraction. For example, the recall of a conversation against a noisy background is dependent on age, with older people having poorer recall (Tun et al. 2009) even though they do not necessarily show a greater effect of auditory distraction than younger adults (Beaman 2005; Bell and Buchner 2007). Furthermore, a certain level of background noise is required to ensure the physical comfort of employees and a "sense of place", while the absence of noise can produce a distraction effect (e.g. a trading room without a noisy background). Thus, the physical comfort of employees does not involve eliminating all sounds, and every acoustic environment should be thoroughly assessed to identify relevant and important sounds before taking any measures aimed at modifying the sounds that an employee may hear.

Many sources of contaminants can impair the quality of the atmosphere in a work building; for example, the occupants themselves (e.g. allergens brought in on clothing), building and equipment material (e.g. volatile organic compounds), the contamination of certain parts of the building (e.g. microbial agents), human activities (e.g. smoking), or outside air (e.g. micro-organisms). As most of these contaminants are transmitted through the air, particularly indoor air, it is not surprising that the Grenelle 2 Environment forum identified air-quality control as a priority. In organizational settings, it is recommended that indoor air should be at least as healthy as outdoor air to ensure the physical comfort of employees.

With regard to the impact of indoor air quality on health, a general distinction is made between (a) building-related diseases (BRD), which are disorders with objective clinical signs and which have been proven to be related to the occupation of a given building (e.g. Legionnaires' disease), and (b) building-related illness (BRI) or building-related symptoms (BRS), which refer to clusters of symptoms perceived by employees working in the same building. The best-known example is "sick-building syndrome", most commonly involving eye, nose and throat irritations, headaches, chronic fatigue, or even dizziness and nausea (Frontczak and Wargocki 2011; Pejtersen et al. 2006; Witterseh et al. 2004). Low levels of humidity could also increase the overall intensity of these symptoms (Reinikainen and Jaakkola 2001).

Whether it is a question of a disease or symptoms, a deterioration of the work atmosphere is involved; however, symptoms, over which doctors are often powerless, are of particular interest to environmental psychologists because they can indicate a state of organizational malaise that can generate psychosocial risks. It thus seems crucial to consider employees who feel threatened by their work environment and to provide recommendations in order to improve their work atmosphere (real and/or perceived).

22.2.2.3 Layout of the Workspace

A large number of international standards have been drawn up and a review would be tedious if not impossible. As more than 50 % of the working population is estimated to work in offices, a percentage that continues to rise (Brill et al. 2001; Duffy 1999), we will focus on the work environment of office employees. With regard to work stations, ISO 9241, entitled "Ergonomic requirements for office work with visual display terminals (VDTs)", includes a section (Part 5) on work-station layout and postural requirements. ISO 9241-410 concerns the design criteria for physical input devices, including the keyboard, mouse and trackball, recommending that they should be designed such that the fingers can reach and activate the keys without excessive deviation from the neutral posture (Aptel et al. 2011 p. 95).

AFNOR NF X35-102, issued in 1998, can also be cited, which provides recommendations to companies with regard to office layout in order to reduce occupational hazards. In particular, it recommends a minimum office space of 10 m^2 per person, 12 m^2 for two people, etc. Temperature should be maintained between 22 and 26 °C; natural lighting should be given priority, and when this is insufficient, ambient and individual lighting of 250–500 lux should be provided. Rooms should have a minimum height of 2.5 m, and the air renewal rate should be 25 m^3 per person per hour. More recently, ISO 3382-3:2012 recommended levels of acoustic performance for different types of space, notably individual offices, shared offices, open-plan offices and open multi-purpose spaces. The clearly stated aim of these standardization documents is to provide guidelines to evaluate the physical comfort and safety of office workers.

It should be noted that some academic studies have examined the introduction of indoor plants into work environments (e.g. Bringslimark et al. 2009), notably with regard to the beneficial role of plants in the absorption of pollutants, dust and CO_2 (e.g. Waku et al. 1995), in air humidification, temperature control and acoustic improvement (e.g. Van den Berg 2005). While further research is required, it would not be surprising if these studies gave rise to future international recommendations.

22.2.2.4 The Physical Environment: A Factor Influencing Attitudes and Behavior at Work

There is agreement among researchers that the physical environment has an influence on job satisfaction (e.g. Probst et al. 2010), productivity (e.g. Haka et al. 2009) and stress (e.g. Bodin-Danielsson and Bodin 2010). Noise has been widely studied in research on office space, because it is considered a major factor of environmental stress with an impact on job satisfaction (e.g. Raffaello and Maas 2002) and on employee performance (Haka et al. 2009). For example, Haka et al. carried out an experiment in a laboratory resembling an open-plan office and observed that in an operation span task, serial recall and activation of knowledge deteriorated when participants were exposed to speech that was irrelevant but intelligible at 0.65 STI (Speech Transmission Index). By contrast, Sundstrom et al. (1994) did not observe any significant relationship, while a continuous noise during short work sessions was associated with improved performance and could be seen as a positive distraction (Sundstrom 1986). This raises the question of whether this experiment did in fact involve noise, defined as an undesirable sound, rather than a positively perceived acoustic environment; indeed, the sounds made by colleagues could be stimulating, strengthen group cohesion and give workers the impression of being less isolated. This is particularly the case in trading rooms where sound intensity is part of the atmosphere of the workplace and can be beneficial for employee productivity (Beaman and Holt 2013). These authors showed that while noise can be a source of discomfort, the absence of noise is not necessarily a source of comfort. Thus, there is no linear correlation between acoustic comfort and sound level. This result is in line with research by Servais (2011) who considered that the opposite of noise is not silence but a set of sounds perceived as pleasant or harmonious.

Some studies have focused on the layout of workspaces, notably the impact of the type of office (closed vs. open-plan) on employee collaboration (Hua et al. 2011) or job satisfaction (Bodin-Danielsson 2013; De Croon et al. 2005; Veitch et al. 2007). Bodin-Danielsson and Bodin (2008) carried out a survey through a questionnaire sent to 469 Swedish office workers and found that employee satisfaction was influenced by the type of office (individual office, shared office, small, medium or large open-plan office, flexible office space, combi-office). They demonstrated that employees working in cell offices, flexible offices and shared-room offices had the highest job satisfaction. By contrast, job satisfaction was lowest in combi-offices and medium-sized open-plan offices. They observed that employees' health, as evaluated by sick leave, was also influenced by the type of office; people working in medium and small open-plan offices had a low level of health, while those in individual and flexible offices had good health. Moreover, studies by Knight and Haslam (2010) found that "lean", undecorated office spaces led to lower productivity than those designed by architects with artistic or old objects and plants (enriched condition) and those decorated by the workers themselves (empowered condition).

Plants and greenery reduce the level of real stress, measured using physiological techniques, by allowing physiological and psychological recovery in stressful situations (Ulrich 2002). This result is in line with a laboratory study by Ulrich et al. (1991) who observed that recovery from stress, assessed by physiological measures, was better following visual exposure to video-recorded scenes of nature. Moreover, contact with nature has been shown to have a positive effect on work performance (El-Zeiny 2012; Raanaas et al. 2011), notably by reducing perceived stress (Leather et al. 1998; Uzzell 2013). However, these positive impacts on workers depend largely on the characteristics of the context in which these indoor plants are found and of the people involved (Bringslimark et al. 2009). This also raises the question of the effect of fashion and/or of the Hawthorne effect on these impacts.

In line with some of these studies, a large body of research suggests that physical comfort based on objective parameters is different from perceived physical comfort, and that both have an influence on workers' evaluation of their work environment (Fisher 2004; Fornara et al. 2006). Thus, the physical characteristics of the environment interact with psychosocial and affective aspects, and this interaction not only gives meaning to the place and space but is also fundamental to the way we evaluate a place (Moser and Uzzell 2003). The environmental context is essential to understand evaluative (e.g. environmental satisfaction at work) and psychological (e.g. attachment to the workplace) processes.

22.2.3 Evaluative Dimension: Satisfaction with the Work Environment

Environmental satisfaction, defined as an affective process giving rise to the expression of a level of satisfaction with a place or a part of this place (Russell and Mehrabian 1978), can be seen as an evaluation of the environment (Ittelson et al. 1974) based on the sense of pleasure experienced by a person when occupying a place (Bonaiuto 2004). Like overall satisfaction, it relies on both affective and cognitive elements (Fleury-Bahi 1997) and on the physical, social and human characteristics of the space (Bonnes et al. 1997). Environmental satisfaction thus describes the set of processes whereby users experience and judge their physical environment (Craik 1966).

22.2.3.1 Recent Studies

Environmental satisfaction has been the subject of a large number of studies involving different types of environment: residential (Bonaiuto et al. 2003; Mogenet and Rioux 2014), educational (Lannegrande-Willems and Bosma 2006; Deledalle-Marcouyeux et al. 2009; Nabli-Bouzid 2014) and work (Veitch et al. 2007; Fleury-Bahi and Marcouyeux 2011; Slama 2014).

Satisfaction with the work environment can be conceived as the result of a person's evaluation of the space in which he/she works, based on personal characteristics (sociodemographic and cultural aspects or environmental preferences) and on a comparison between the perceived and the ideal workplace (Fleury-Bahi and Marcouyeux 2011). While this has been relatively little investigated in organizational settings, it can be considered an indicator of the evaluative dimension of comfort at work (Rioux et al. 2013).

Over the last 20 years, there has been a large amount of research focusing on satisfaction with the work environment (Carlopio 1996; Fornara et al. 2006; Aries et al. 2010). Although some studies have evaluated overall levels of satisfaction (e.g. Carlopio and Gardner 1992; Veitch et al. 2011), the majority have aimed to identify its different components (Wagner et al. 2007;

Fischer et al. 2004). Amongst the latter, the study by Fischer et al. (2004) defined the workplace according to its physical, functional and human characteristics and studied satisfaction with each of these three components. In a study carried out with 165 French civil service employees, Rioux and Fouquereau (2004) evaluated the workers' satisfaction with different places in the office and observed that this varied in each place. They showed that environmental satisfaction was greatest in the social spaces not directly related to work (library, union office, staff room, medical room) and that satisfaction with places giving employees a degree of control over their work (cafeteria, archive room, meeting room, and particularly the computer room) provided predictors of job satisfaction.

Some studies looking at environmental satisfaction at work have focused on single components of the work environment, such as the environmental atmosphere (lighting, acoustics and air quality). For example, Sundstrom (1986) found that employees were more satisfied with their work environment when their office had windows providing daylight. Wagner et al. (2007) investigated thermal comfort in office buildings in Germany and found that, at similar temperatures, satisfaction with the work environment was greater in winter and corresponded to a sensation of being too cold, while in summer it corresponded to a feeling of being too hot, associated with dissatisfaction linked to an impression of mediocre indoor air quality.

22.2.3.2 Measuring Environmental Satisfaction

There are relatively few tools for measuring satisfaction with the work environment. Some are generalist and thus applicable to all work environments, while others are specific to a given type of environment. An example of the former is the scale of satisfaction with the workspace by Fleury-Bahi and Marcouyeux (2011), which can be applied to various workspaces (office environments, production units, institutions, etc.) and used with all classes of workers (p. 379). This tool is composed of 14 items, divided into two dimensions of seven items each: comfort/functionality (e.g. lighting quality) and control/privacy (e.g. the possibility of visual privacy). Its structure has been validated with people working in very different sectors of activity: automobile construction, transport, health, telecommunications (Fleury-Bahi and Marcouyeux 2011), commerce, research, defense (Moffat et al. in press) and banking (Pasquier and Rioux 2014).

With regard to more specific tools, an example is Carlopio's (1996) pioneering Physical Work Environment Satisfaction Questionnaire (PWESQ) comprising five factors (spatial layout, task organization, health and safety, recreation areas, facilities), validated with a sample of factory workers; however, to our knowledge, is has never really been used. Fischer and Vischer's (1997) functional comfort questionnaire was aimed more specifically at the office environment, based on data collected from 2500 office workers in Canadian government departments, and comprised seven factors of satisfaction: air quality, comfort linked to noise produced by staff, thermal comfort, spatial comfort, privacy, lighting quality, and comfort linked to noise not produced by staff. More recently, Veitch et al. (2007) drew up a scale for people working in open-plan offices, based on an enriched version of the tool used by Stokols and Scharf (1990). The structure of their scale highlights three dimensions of satisfaction with the office space: acoustic privacy, luminosity/lighting, and air quality.

Another reference is the Perceived Hospital Environment Quality Indicators-PHEQIs (Fornara et al. 2006; Andrade et al. 2013), adapted from Bonaiuto et al.'s (1999) Perceived Residential Environment Quality (PREQ) scale. It has three dimensions, two concerning physical environments ("External spaces" and "Care unit and in-/out-patient (waiting) area") and one on the social environment ("Social-functional features"). It has been used notably by Fornara and Andrade (2012) and Fornara et al. (2013) and shows that hospital staff are both less satisfied with their environment and exhibit a lower increase in satisfaction in "humanized" hospital environments than patients. Details of this inventory can be found in Chaps. 23

(Devlin and Andrade) and 24 (Fornara and Manca) of this Handbook.

To our knowledge, only one scale takes into account the different spaces in employees' work environments. This is Moffat's (2014) scale of satisfaction with the work environment (Echelle de Satisfaction Environnementale au Travail – ESET), which looks at satisfaction with the work station, the office space, the workspace and the external environment. It has 58 items divided into ten dimensions: layout (8 items), safety (8 items), peace and quiet (9 items), transport (5 items), sociability (7 items), shops (4 items), alienation (5 items), pleasantness of the neighborhood (5 items), green spaces (3 items), and pleasantness of the firm (4 items). It was validated with 180 people working in the primary, secondary or tertiary sectors. Used with people working in administrative, industrial or institutional work environments, this tool showed very satisfactory psychometric properties (Cronbach's Alpha of 0.70–0.91 on the dimensions, test-retest reliability coefficient of 0.75 at a 3-week interval). Moreover, there was a significant positive correlation of 0.001 between scores on the overall factor of ESET and those of the overall factor of Fleury-Bahi and Marcouyeux's (2011) generalist questionnaire of environmental satisfaction at work on the one hand, and between the ten dimensions of the ESET and the two sub-scales (comfort/functionality and control/privacy) of Fleury-Bahi's scale on the other (Moffat et al. in press). The ESET was recently validated in Italian by Scrima et al. (2014a, b).

22.2.3.3 Variables Linked to Satisfaction with the Work Environment

Satisfaction with the work environment is linked to perceived productivity (Leaman and Bordass 2007), perceived well-being (Wells 2000), organizational commitment (Carlopio 1996) and intention to leave the organization (Carlopio 1996). More specifically, Moffat et al. (2014) demonstrated that the "environmental alienation" dimension of satisfaction with the work environment is a predictor of the intention to leave the organization, mediated moreover through job satisfaction. Thus, the more routine becomes established in the work environment, the less satisfied workers are with their job and the greater their intention to leave the organization.

In environmental psychology applied to work, the links between satisfaction with the work environment and job satisfaction have been the subject of a large number of studies, and we therefore believe it is important to examine more closely the concept of job satisfaction as defined by occupational psychology. Meysonnier (2006) carried out a survey of definitions found in the literature, in which job satisfaction is variously described as a dynamic process (Barbash 1974), an emotional state (Locke 1976) or an evaluation (Roussel 1996). Researchers currently seem to agree that it is an attitude (Brief 1998; Wright 2006) and thus an overall evaluation of objects (Ajzen 2001). It would thus be the result of an evaluation of one's job or of the work situation, and the affective response to the object would be linked to evaluation of the object and hence to job satisfaction, but these would be two distinct constructs (Mignonac 2004). The concept of satisfaction, whether applied to the work object or to the work environment, thus seems to refer to a similar conceptualization.

These studies have revealed a positive link between environmental satisfaction and job satisfaction, which is either direct (e.g. Carlopio 1996; Wells 2000; Sundstrom et al. 1994; Pasquier and Rioux 2014; Veitch et al. 2007; Wineman 1982), or indirect (Newsham et al. 2009). We refer again to the classic study by Carlopio (1996) who found a significant correlation between the Physical Work Environment Satisfaction Questionnaire-PWESQ and job satisfaction (MSQ-short version). Wells (2000) found that the ability to personalize one's work area was positively related to environmental satisfaction, which in turn positively influenced job satisfaction. For their part, Pasquier and Rioux (2014) found that, unlike the comfort/functionality dimension of Fleury-Bahi and Marcouyeux's environmental satisfaction scale, the control/privacy dimension was a significant predictor of the level of job satisfaction, highlighting the importance of the control that employees believe they have over their environment when evaluating it.

An indirect link was identified by Newsham et al. (2009) in a study carried out in an open-plan office building in Michigan. They found a strong correlation between satisfaction with the work environment and job satisfaction, mediated by satisfaction with management and pay.

22.2.4 Psychological Dimensions: Attachment to the Workplace

Place attachment is currently a major topic of environmental psychology research. However, the relative lack of conceptual coherence surrounding the concept of place (Morgan 2010) has probably led to the persistence of some terminological confusion with its related processes (sense of belonging, community attachment, sense of community, place dependence, etc.) as well as a plethora of definitions of the concept of place attachment. For example, in their review of 1993, Giuliani and Feldman identified 11 different definitions. For further details, readers can also refer to the article by Kyle et al. (2004). Since 1985–1990, place attachment has been the concept used to designate the links held by people with places that are dear to them (Giuliani 2003; Manzo 2003; Pretty et al. 2003).

Today, attachment is seen as an attitude and thus a three-dimensional concept including affective, cognitive and behavioral dimensions (e.g. Aronson et al. 2005). Studies have not explored all three dimensions but focused mainly on just one or two (Scannell and Gifford 2010). This is notably the case for studies that concentrate on the affective component of the tie between a person and a given place (Altman and Low 1992; Lewicka 2005) and thus measure place attachment using one-dimensional scales (e.g. Lalli 1992; Austin and Baba 1990; Bonaiuto et al. 2006).

The concept of place attachment has been considered in relation to the type of place, particularly the home (e.g. Giuliani 1991; Windsong 2010), the neighborhood (e.g. Bonaiuto et al. 1999; Rioux and Mokounkolo 2005), the town (e.g. Hidalgo and Hernández 2001; Giuliani et al. 2003), and natural or wild areas (e.g. Vittersø et al. 2001; Halpenny 2007). Readers can refer to the excellent summary by Lewicka (2011). Somewhat curiously, studies investigating attachment to the workplace are fewer and above all more recent. However, Inalhan (2006, 2009, 2013) and Inalhan and Finch (2004) examined the processes of place attachment among workers going through a period of socio-spatial transition. For example, taking a longitudinal approach, Inalhan (2006, 2009) studied the development of attachment of employees to their former (vs. their new) workplace during the process of change in the physical work environment. She observed that place disruptions or move experiences interrupted the processes that bind people to their socio-environments, thereby significantly hindering their integration into their new workplace. Moreover, she showed that loyalty to an organization was increasingly determined by place attachment (Inalhan and Finch 2004). In an article focusing more specifically on the social dimension of attachment to the workplace and based on semi-structured interviews, Inalhan (2013) observed that employees working in dedicated offices did not feel very disturbed by their new physical environment but mentioned changes in their social environment, particularly in their relationships within their team, while mobile teleworkers who worked in shared offices when they were in the organization experienced little difference at either a spatial or a social level. By contrast, the employees who worked in dedicated offices in their former company and who opted for mobile work in their new company were the most disturbed; they expressed the most negative comments, particularly regretting the loss of community feeling. This change, both spatial and organizational, affected the way that these workers used the space, which in turn influenced the creation of social groups. It thus had an impact on the spatial and social dimensions of workplace attachment, reminiscent of the effects observed by Fried (1963) when the residents of a Boston suburb were forced to move.

Rioux has also explored workplace attachment, which she defines as the positive affective bond between workers and the organizational space, and has thus focused more specifically on the affective dimension of the concept. In 2006, she constructed a scale of attachment to the workplace (Echelle d'Attachement au Lieu de Travail – EALT) (Rioux 2006), based on an analysis of the content of semi-structured interviews and on the translation and adaptation of Bonnes et al.'s (1997) Neighborhood Attachment Scale (NAS). This one-dimensional scale is composed of seven items, five concerning the consequences of attachment and two referring to the process itself. Its structure has been confirmed among various groups of French workers: civil service employees (Rioux 2005), hospital workers (Velasco and Rioux 2010), high-school teachers (Rioux and Pignault 2013a) and people working in large retail outlets (Rioux and Pignault 2013b). It has demonstrated thoroughly acceptable psychometric properties when used with Romanian (Pavalache-Ilie and Rioux 2014), Gabonese (Bakita and Rioux 2011), Italian and English (Scrima et al. 2014a, b) samples.

In these studies, the organizational space constitutes a "place," which Canter (1986) defined as a unit of environmental experience that is the result of interrelationships between the person's past, present, and future actions and the physical or representational characteristics of the place. From this perspective, it can be conceived both as an entity and as a mosaic of specific spaces each offering a basis for attachment. Rioux (2005) carried out an empirical study of women working in a French government department, revealing notably that all the participants were attached to their workplace overall, but with varying degrees of attachment to different places within it. The entrance hall was perceived as a transitional space between outside and inside and as a day-to-day meeting place, and was given the highest attachment score. By contrast, the participants expressed moderate attachment to the places that were central to their work (work stations). Similar results have been observed among hospital workers (Velasco and Rioux 2010) and high-school teachers (Rioux and Pignault 2013a), the central places of their work being the "patient's bedside" and the classroom, respectively. Two types of predictor of workplace attachment were identified in these studies: (a) those that constitute the key aspect of the job and give meaning to work, and (b) those that enable workers to manage their level of organizational stress by withdrawing physically or psychologically to relieve the stress related to organizational pressure. It should be noted that this result was found irrespective of the type of profession, the level of responsibility or interpersonal issues.

From this series of studies, it can be concluded that the places that provide more or less informal opportunities for socializing give rise to a greater sense of attachment than those that are more directly work-related, the latter proving to be explanatory variables of workplace attachment.

During the last 10 years, the links between workplace attachment and other psycho-environmental or psycho-organizational variables have been investigated. Examples are the behaviors of pro-environmental organizational citizenship (Ajdukovic et al. 2012) and the perceived quality of the workspace (Dinç 2010; Ajdukovic et al. 2014). More specifically, workplace attachment constitutes a predictive factor of job satisfaction (Rioux and Angel 2014), affective organizational attachment (Scrima 2014), the intention to leave the organization (Rioux 2007, 2011), and organizational citizenship behaviors (Le Roy and Rioux 2012). Among these variables, we feel that particular importance should be given to personalization. Defined by Wells as "the deliberate decoration or modification of an environment by its occupants to reflect their identities" (Wells 2000 p. 239), personalization has been investigated specifically in an organizational environment. There is general agreement that personalizing one's work environment according to one's own tastes and needs enhances emotional attachment to the work environment (Brown 1987; Goodrich 1986; Heidmets 1994; Wells 2000). In a recent study of people working in the same tertiary building, Ajdukovic et al. (2014) found a significant link between personalization of the workspace and workplace attachment, whatever the type of office (open-plan, individual or shared).

22.2.5 Towards a Study Design for Comfort at Work

Following a number of studies that have demonstrated links between (a) physical indicators of comfort and satisfaction with the work environment (Slama 2014; Kim and de Dear 2013; Gou et al. 2013), (b) satisfaction with the work environment and workplace attachment (Moffat and Rioux 2013; Leconte et al. 2013) and (c) physical indicators of comfort and workplace attachment (Pasquier and Rioux 2013), Scrima (2015) tested the model of comfort at work proposed by Rioux et al. (2013) in a sample of 127 French government office workers. The results confirmed that the variables identified by Rioux are good indicators of the three levels of comfort. Moreover, the strong correlation between these three levels supports the existence of a general "Comfort" factor.

22.3 Applications in the Management Domain

There is currently agreement amongst researchers that a well-designed environment can produce job satisfaction and/or comfort and can increase productivity. This can be due not only to physical factors (removing or attenuating nuisance factors, suitable layout of space) but also to psychological processes that can be enhanced by the organization (control over the environment, possibility of personalizing the work station, workplace attachment, etc.).

With regard to nuisance factors, a noisy environment can be improved by dealing with loud, intense or unpredictable sounds. It should be recalled that acoustic comfort does not mean silence but rather a set of sounds perceived as pleasant or harmonious. Ensuring acoustic comfort at work should involve transforming disturbing noises into harmonious sounds rather than removing them.

The layout of the work environment can also have a clear impact on satisfaction, comfort and productivity, as shown by the studies of Bodin-Danielsson (2013) who recommended that priority be given to individual, flexible or shared offices.

Studies have also shown that in order to feel "comfortable", workers should be able to make their space their own. From this standpoint, the organization has an important role to play, aiding or hindering this process, regarding both the immediate work environment (in the company or at home) and the overall organizational space. Marking one's space with pictures, flowers, or pot plants provided by the organization or personalizing it with personal objects (photos, diplomas, children's drawings, etc.) gives workers a sense of control over their space and empowers them with regard to their work environment. This sense of control can also come from being free to adjust features such as lighting, temperature, and sound level. It can produce satisfaction and comfort, and it improves productivity.

Recent studies have also shown that workplace attachment is an important lever for generating job satisfaction and comfort at work. To increase this attachment, managers can act on the places that enable workers to withdraw physically or psychologically in order to relieve the stress linked to organizational pressure. Rather than designing spaces for this purpose, the organization needs to identify those that exist and make them available to their employees.

The studies presented in this chapter demonstrate that the concept of comfort is firmly rooted in the field of positive psychology and work, a sub-discipline concerned with the well-being of workers and ways of preserving it in organizational settings. It is thus involved in finding a new balance between factors that have a positive and a negative influence on workers, groups and the organization as a whole; it focuses on ways of preventing rather than dealing with conflicts, on the work atmosphere rather than on nuisance factors, on comfort at work rather than on stress. From this perspective, ensuring the comfort of workers can be seen as one of the responses to psychosocial occupational risks, coming within the framework of the primary prevention of occupational hazards as defined by the WHO. More specifically, we believe that our tool could be a first step in the construction of an inventory of psycho-environmental risks, as part of a more general consideration of employees' work conditions.

References

Ajdukovic, I., Girandola, F., & Weiss, K. (2012). *Bâtiment de travail durable et éco- responsabilité. Le rôle du bâtiment et de l'identification au lieu dans le changement éco-comportemental.* [Sustainable work building and eco-responsibility: The role of the building and of place identification in changing eco-behavior]. Paper presented at the 9th CIPSLF, Porto, Portugal.

Ajdukovic, I., Gilibert, D., & Labbouz-Henry, D. (2014). Confort au travail. Le rôle de l'attachement et de la personnalisation dans la perception de la qualité de l'espace de travail. [Comfort at work: The role of attachment and personalization in the perception of the quality of the workspace]. *Psychologie du Travail et des Organisations, Special Edition on Comfort at Work, 20*(3), 311–327.

Ajzen, I. (2001). Nature and operation of attitudes. *Annual Review of Psychology, 52*, 27–58.

Altman, I., & Low, S. M. (1992). *Place attachment.* New York: Springer.

Andrade, C. C., Lima, L., Pereira, C. R., Fornara, F., & Bonaiuto, M. (2013). Inpatients' and outpatients' satisfaction: The mediating role of perceived quality of physical and social environment. *Health & Place, 21*, 122–132.

Aptel, M., Cail, F., & Aublet-Cuvelier, A. (2011). *Les troubles des muscles squelettiques des membres supérieurs* [Musculoskeletal disorders of the upper limbs]. INRS, ED 957, 97 p.

Aries, M. B. C., Veitch, J. A., & Newsham, G. R. (2010). Windows, view, and office characteristics predict physical and psychological discomfort. *Journal of Environmental Psychology, 30*(4), 533–541.

Aronson, E., Wilson, T. D., & Akert, R. M. (2005). *Social psychology* (5th ed.). Upper Saddle River: Pearson Education International.

Austin, D. M., & Baba, Y. (1990). Social determinants of neighborhood attachment. *Sociological Spectrum, 10*, 59–78.

Bakita, M., & Rioux, L. (2011, September). *Valeurs et comportements éthiques* [Values and ethical behavior]. Paper presented at the 53rd Congress of the Société Française de Psychologie (SFP), Metz, France.

Banbury, S., & Berry, D. (1998). Disruption of office-related tasks by speech and office noise. *British Journal of Psychology, 89*, 499–517.

Barbash, J. (1974). Enquête sur les attitudes concernant la satisfaction au travail. [Survey of attitudes concerning job satisfaction.]. OECD Document, MS/IR/7431, 20 p.

Beaman, P. (2005). Auditory distraction from low-intensity noise: A review of the consequences for learning and workplace environments. *Applied Cognitive Psychology, 19*, 1041–1064.

Beaman, C. P., & Holt, N. J. (2013). L'environnement sonore au travail [The acoustic environment at work]. In L. Rioux, J. Le Roy, L. Rubens, & J. Le Conte (Eds.), *Le confort au travail. Que nous apprend la psychologie environnementale ? [Comfort at work. What does environmental psychology teach us?]* (pp. 68–88). Quebec: Presses Universitaires de Laval.

Béarn, P. (1951). *Couleur d'usine.* Paris: Seghers.

Becker, M., & Félonneau, M. L. (2011). Pourquoi être pro-environnemental ? Une approche socio-normative des liens entre valeurs et pro-environnementalisme [What are the reasons for being pro-environmental? Adopting a socio-normative perspective on the relationship between values and "pro-environmentalism"]. *Pratiques Psychologiques, Numéro Spécial "Psychologie Sociale Appliquée à l'Environnement", 17*(3), 219–236.

Bell, R., & Buchner, A. (2007). Equivalent irrelevant-sound effects for young and old adults. *Memory & Cognition, 35*, 352–364.

Bodin, D. C., & Bodin, L. (2010). *Office design's influence on employees' stress levels.* Paper presented at the ARCC/EAAE 2010 International Conference on Architectural Research, June 2010, Washington, USA.

Bodin-Danielsson, C. (2013). L'impact de la conception architecturale des bureaux sur le confort et le bien-être des employés. [The impact of architectural office design on workers' comfort and well-being]. In L. Rioux, J. Le Roy, L. Rubens, & J. Le Conte (Eds.), *Le confort au travail. Que nous apprend la psychologie environnementale ? [Comfort at work. What does environmental psychology teach us?]* (pp. 32–64). Quebec: Presses Universitaires de Laval.

Bodin-Danielsson, C., & Bodin, L. (2008). Office-type in relation to health, well-being and job satisfaction among employees. *Environment & Behavior, 40*(5), 636–668.

Bonaiuto, M. (2004). Residential satisfaction and perceived urban quality. In C. Spielberger (Ed.), *Encyclopedia of applied psychology* (Vol. 3, pp. 267–270). Rome: University of Rome la Sapienza.

Bonaiuto, M., Aiello, A., Perugini, M., Bonnes, M., & Ercolani, A. P. (1999). Multidimensional perception of residential environment quality and neighbourhood attachment in the urban environment. *Journal of Environmental Psychology, 19*, 331–352.

Bonaiuto, M., Fornara, F., & Bonnes, M. (2003). Indexes of perceived residential environment quality and neighbourhood attachment in urban environments: A confirmation study on the city of Rome. *Landscape and Urban Planning, 65*, 41–52.

Bonaiuto, M., Fornara, F., & Bonnes, M. (2006). Perception de la qualité résidentielle dans les villes italiennes de moyenne et petite étendues [Perceived residential environment quality in middle- and low-extension Italian cities]. *Revue Européenne de Psychologie Appliquée/European Journal of Applied Psychology, 56*, 23–34.

Bonnes, M., Bonaiuto, M., Aiello, A., Perugini, M., & Ercolani, A. P. (1997). A transactional perspective on residential satisfaction. In C. Despres & D. Piché (Eds.), *Housing surveys. Advances in theory and methods* (pp. 99–135). Saint Foy: Crad.

Brief, A. P. (1998). *Attitudes in and around organizations*. Thousand Oaks: Sage.

Brill, M., Weidemann, S., Allard, L., Olson, J., & Keable, E. (2001). *Disproving widespread myths about workplace design*. Jasper: Kimball International.

Bringslimark, T., Hartig, T., & Patil, G. (2009). The psychological benefits of indoor plants: A critical review of the experimental literature. *Journal of Environmental Psychology, 29*, 422–433.

Brown, B. (1987). Territoriality. In D. Stokols & I. Altman (Eds.), *Handbook and environmental psychology* (Vol. 1, pp. 505–531). New York: Wiley.

Canter, D. (1986). Putting situations in their place: Foundations for a bridge between social and environmental psychology. In A. Furnham (Ed.), *Social behaviour in context* (pp. 208–239). London: Allyon and Bacon.

Carlopio, J. (1996). Construct validity of a physical work environment satisfaction questionnaire. *Journal of Occupational Health Psychology, 1*(3), 330–344.

Carlopio, J., & Gardner, D. (1992). Direct and interactive effects of the physical work environment on attitudes. *Environment & Behavior, 24*, 579–601.

Cole, R., Robinson, J., Brown, Z., & O'Shea, M. (2008). Re-contextualizing the notion of comfort. *Building Research & Information, 36*(4), 323–336.

Craik, K. J. W. (1966). Brightness discrimination, borders, and subjective brightness. In S. L. Sherwood (Ed.), *The nature of psychology* (pp. 94–97). Cambridge: Cambridge University Press.

Daniels, K. (2000). Measures of five aspects of affective well-being at work. *Human Relations, 53*(2), 275–294.

De Croon, E., Sluiter, J., Kuijer, P., & Frings-Dresen, M. (2005). The effect of office concepts on worker health and performance: A systematic review of the literature. *Ergonomics, 48*(2), 119–134.

Deledalle-Marcouyeux, A., Fleury-Bahi, G., & Florin, A. (2009). Construction et validation d'une échelle de satisfaction envers le lycée. [Construction and validation of a scale of satisfaction with the high school]. *Revue Européenne de Psychologie Appliquée/European Journal of Applied Psychology, 59*(2), 91–100.

Dinç, P. (2010). Spatial and behavioral variables that affect â emotional attachment â of users: A multidimensional approach for private offices. *Gazi University Journal of Science, 20*(2), 41–50.

Duffy, F. (1999). *The new office* (2nd ed.). London: Conran Octopus Limited.

El-Zeiny, R. M. A. (2012). The interior design of workplace and its impact on employees' performance: A case study of the private sector corporations in Egypt. *Procedia – Social and Behavioral Sciences, 35*, 746–756.

Fischer, G.-N., & Vischer, J. (1997). *L'évaluation des environnements de travail [The evaluation of work environments]*. Paris: De Boeck Université.

Fischer, G.-N., Tarquinio, C., & Vischer, J. (2004). Effects of the self-schema on perception of space at work. *Journal of Environmental Psychology, 24*, 131–140.

Fisher, G.-N. (2004). Les environnements de travail [Work environments]. In E. Brangier, A. Lancry, & C. Louche (Eds.), *Les dimensions humaines du travail. Théories et pratiques en psychologie du travail et des organisations [The human dimensions of work. Theories and practices in the psychology of work and organizations]* (pp. 161–181). Nancy: PUN.

Fleury-Bahi, G. (1997). Histoire, identité résidentielle et attachement au quartier actuel: étude sur les habitants de la ville de Paris [History, residential identity and attachment to the residential environment: Study of the inhabitants of Paris]. *Psychologie Française, 42*(2), 183–184.

Fleury-Bahi, G., & Marcouyeux, A. (2011). Évaluer la satisfaction envers l'espace de travail: développement d'une échelle et première validation [Evaluating satisfaction with the workspace: Development and first validation of a scale]. *Revue de Psychologie du Travail et des Organisations, 17*, 376–392.

Fornara, F., & Andrade, C. C. (2012). Healthcare environments. In S. Clayton (Ed.), *The Oxford handbook of environmental and conservation psychology* (pp. 295–315). New York: Oxford University Press.

Fornara, F., Bonaiuto, M., & Bonnes, M. (2006). Perceived hospital environment quality indicators: A study of orthopaedic units. *Journal of Environmental Psychology, 26*, 321–334.

Fornara, F., Bonaiuto, M., & Bonnes, M. (2013). Les attentes du personnel hospitalier envers son lieu de travail [The expectations of hospital staff regarding their workplace]. In L. Rioux, J. Le Roy, L. Rubens, & J. Le Conte (Eds.), *Le confort au travail. Que nous apprend la psychologie environnementale ?* (pp. 225–244). Laval: Presses Universitaires de Laval.

Fried, M. (1963). Grieving for a lost home. In L. J. Duhl (Ed.), *The urban condition: People and policy in the metropolis* (pp. 124–152). New York: Simon & Schuster.

Frontczak, M., & Wargocki, P. (2011). Literature survey on how different factors influence human comfort in indoor environments. *Building and Environment, 46*(4), 922–937.

Giuliani, M.-V. (1991). Toward an analysis of mental representations of attachment to the home. *The Journal of Architectural and Planning Research, 8*(2), 133–146.

Giuliani, M.-V. (2003). Theory of attachment and place attachment. In M. Bonnes, T. Lee, & M. Bonaiuto (Eds.),

Psychological theories for environmental issues (pp. 137–172). London: Ashgate.
Giuliani, M. V., & Feldman, R. (1993). Place attachment in developmental and cultural context. *Journal of Environmental Psychology, 13*, 267–274.
Giuliani, M.-V., Ferrara, F., & Barabotti, S. (2003). One attachment or more? In G. Moser, E. Pol, Y. Bernard, M. Bonnes, J. Corraliza, & M.-V. Giuliani (Eds.), *People, places, and sustainability: 21st century metropolis* (pp. 111–122). Göttingen: Hogrefe & Huber.
Goodrich, R. (1986). The perceived office: The office environment as experienced by its users. In J. D. Wineman (Ed.), *Behavioral issues in office design* (pp. 109–133). New York: Van Nostrand Reinhold.
Gou, Z., Deo Prasad, D., & Stephen Siu-Yu Lau, S. (2013). Are green buildings more satisfactory and comfortable? *Habitat International, 39*, 156–161.
Haka, M., Haapakanga, A., Keränen, J., Hakala, J., Keskinen, E., & Hongisto, V. (2009). Performance effects and subjective disturbance of speech in acoustically different office types – laboratory experiment. *Indoor Air, 15*, 1–14.
Halpenny, E. (2007). Examining the relationship of place attachment with pro-environmental intentions. In R. Burns & K. Robinson (Eds.), *Proceedings of the 2006 northeastern recreation research symposium* (pp. 63–66). Newtown Square: Northern Research Station.
Heidmets, M. (1994). The phenomenon of personalization of the environment: A theoretical analysis. *Journal of Russian and East European Psychology, 32*(3), 41–85.
Hidalgo, M. C., & Hernandez, B. (2001). Place attachment: Conceptual and empirical questions. *Journal of Environmental Psychology, 21*(3), 273–281.
Hua, Y., Loftness, V., Heerwagen, J., & Powell, K. M. (2011). Relationship between workplace spatial settings and occupant-perceived support for collaboration. *Environment and Behavior, 43*(6), 807–826.
Hughes, R., & Jones, D. (2003). Benefits and unavoidable costs of unattended sound for cognitive functioning. *Noise and Health, 6*, 3–76.
Inalhan, G. (2006). *The role of place attachment on employees' resistance to change in workplace accommodation projects*. PhD thesis, University of Reading, Reading, UK.
Inalhan, G. (2009). Attachments: The unrecognised link between employees and their workplace (in change management projects). *Journal of Corporate Real Estate, 11*(1), 17–37.
Inalhan, G. (2013). Quitter le confort du familier. L'attachement social au lieu de travail [Leaving the comfort of what is familiar. Social attachment to the workplace]. In L. Rioux, J. Le Roy, L. Rubens, & J. Le Conte (Eds.), *Le confort au travail. Que nous apprend la psychologie environnementale ?* (pp. 174–216). Laval: Presses Universitaires de Laval.
Inalhan, G., & Finch, E. (2004). Place attachment and sense of belonging. *Facilities, 22*, 5–6.
Ittelson, W. H., Proshansky, H. M., Rivlin, L. G., & Winkel, G. H. (1974). *An introduction to environmental psychology*. New York: Holt, Rinehart & Winston.

Kim, J., & de Dear, R. (2013). Workspace satisfaction: The privacy-communication trade-off in open-plan offices. *Journal of Environmental Psychology, 36*, 18–26.
Knight, C., & Haslam, A. (2010). The relative merits of lean, enriched, and empowered offices: An experimental examination of the impact of workspace management strategies on well-being and productivity. *Journal of Experimental Psychology, 16*, 158–172.
Kyle, G., Graefe, A., Manning, R., & Bacon, J. (2004). Effect of activity involvement and place attachment on recreationists' perceptions of setting density. *Journal of Leisure Research, 36*(2), 209–231.
Lalli, M. (1992). Urban-related identity: Theory, measurement, and empirical findings. *Journal of Environmental Psychology, 12*, 285–303.
Lannegrand-Willems, L., & Bosma, H. (2006). Identity development-in-context. The school as an important context for identity development. *Identity: an International Journal of Theory and Research, 6*(1), 85–113.
Le Goff, O. (1994). *L'invention du confort. Naissance d'une forme sociale*. Lyon: Presses Universitaires de Lyon.
Le Roy, J., & Rioux, L. (2012). The mediating role of workplace attachment in the relationship between organizational commitment and organizational citizenship behavior. *Revue Internationale de Psychologie Sociale [International Journal of Social Psychology], 3*, 211–233.
Leaman, A., & Bordass, B. (2007). Are users more tolerant of 'green' buildings? *Building Research & Information, 35*(6), 662–673.
Leather, P., Pygras, M., Beale, D., & Lawrence, C. (1998). Windows in the work-place: Sunlight, view, and occupational stress. *Environment and Behavior, 30*(6), 739–762.
Leconte, J., Rubens, L., Le Roy, J., & Rioux, L. (2013, July). *Perception of comfort in training places*. Poster presented at the 13th European Congress of Psychology (ECP 2013), Stockholm, Sweden.
Lewicka, M. (2005). Ways to make people active: Role of place attachment, cultural capital and neighborhood ties. *Journal of Environmental Psychology, 4*, 381–395.
Lewicka, M. (2011). Place attachment: How far have we come in the last 40 years? *Journal of Environmental Psychology, 31*(3), 207–230.
Locke, E. A. (1976). The nature and causes of job satisfaction. In M. D. Dunnette (Ed.), *Handbook of industrial and organizational psychology* (pp. 1297–1349). Chicago: Rand McNally.
Manzo, L. (2003). Beyond home and haven: Toward a reconceptualization of place attachment. *Journal of Environmental Psychology, 23*(1), 47–61.
Martel, J.-P., & Dupuis, G. (2006). Quality of work life: Theoretical and methodological problems, and presentation of a new model and measuring instrument. *Social Indicators Research, 77*, 333–368.
Meyssonier, R. (2006). Les sources d'attachement à l'entreprise, du point de vue des salariés: l'exemple des

ingénieurs [The sources of attachment to the company from the employees' perspective]. *Revue de Gestion des Ressources Humaines, 60,* 48–70.

Mignonac, K. (2004). Que mesure-t-on réellement lorsque l'on invoque le concept de satisfaction au travail ? [What is really being measured when referring to the concept of job satisfaction?]. *Revue de Gestion des Ressources Humaines, 53,* 80–93.

Moffat, E. (2014, July). *Echelle de Satisfaction Environnementale au Travail (ESET): Validation d'un outil d'évaluation de la qualité de vie professionnelle [Scale of Environmental Satisfaction at Work (ESET): Validation of a tool for evaluating the quality of professional life].* Poster presented at the 3rd study day of the Association pour la Recherche en Psychologie Environnementale-ARPEnv "Bien-être, mieux-être, où en sommes-nous avec ces notion en psychologie environnementale ?", Université Paris Sorbonne, Paris, France.

Moffat, E., & Rioux, L. (2013, September). *La satisfaction environnementale dans le cadre organisationnel [Environmental satisfaction in an organizational setting].* Paper presented at the 55th National Congress of French Psychology, "Psychologie et conscience" organized by the Société Française de Psychologie (SFP), Lyon, France.

Moffat, E., Scrima, F., Rioux, L., & Mogenet, J. L. (2014, August). *Satisfaction au travail, satisfaction environnementale au travail et intention de retrait. Une étude menée auprès d'une population d'employés de bureau [Job satisfaction, environmental satisfaction at work and intention to leave the organization. A study of office workers].* Poster presented at the 18th Congress of the Association Internationale de Psychologie du Travail de Langue Française (AIPTLF), Florence, Italy.

Moffat, E., Rioux, L., & Mogenet, J. L. (in press). Développement et validation d une Echelle de Satisfaction Environnementale au Travail [*Development and validation of a Scale of Environmental Satisfaction at Work*]. *Psychologie française.*

Mogenet, J. L., & Rioux, L. (2014). Students' satisfaction with their university accommodation. *Nordic Psychology, 66*(4), 303–320.

Morgan, P. (2010). Towards a developmental theory of place attachment. *Journal of Environmental Psychology, 30*(1), 11–22.

Moser, G., & Uzzell, D. (2003). Environmental psychology. In T. Millon & M. Lerner (Eds.), *Comprehensive handbook of psychology, volume 5: Personality and social psychology* (pp. 419–445). New York: Wiley.

Moser, G., & Weiss, K. (2003). *Espaces de vie. Aspects de la relation homme-environnement. [Living spaces. Aspects of the people-environment relationship].* Paris: Armand Colin.

Nabli-Bouzid, D. (2014). Confort des élèves et performance du bâtiment scolaire [Comfort of students and school building performance]. *Psychologie du Travail et des Organisations, Special Edition on Comfort at Work, 20*(3), 241–258.

Newsham, G. R., Brand, J. L., Donnelly, C. L., Veitch, J. A., Aries, M. B. C., & Charles, K. E. (2009). Linking indoor environment conditions to job satisfaction: A field study. *Building Research & Information, 37*(2), 129–147.

Pasquier, D., & Rioux, L. (2013, November). *Satisfaction envers la formation et confort au travail des étudiants. Approche corrélationnelle et approche implicative [Satisfaction with training and students' comfort at work. Correlational and implicative approach].* Paper presented at the international conference: "Le confort au travail. Regards croisés entre la psychologie environnementale et les sciences de gestion", Nanterre, France.

Pasquier, D., & Rioux, L. (2014). Satisfaction et confort au travail. L'apport de la démarche implicative [Satisfaction and comfort at work. The contribution of the implicative procedure]. *Psychologie du Travail et des Organisations, Special Edition on Comfort at Work, 20*(3), 275–293.

Pavalache-Ilie, M., & Rioux, L. (2014). Résilience organisationnelle et attachement au lieu de travail [Organizational resilience and attachment to the workplace]. In M. Tomita & S. Cace (Eds.), *Proceedings of the second world congress on resilience: From person to society* (pp. 817–822). Bologna: Medimond.

Pejtersen, J., Allermann, L., Kristensen, T. S., & Poulsen, O. M. (2006). Indoor climate, psychosocial work environment and symptoms in open-plan offices. *Indoor Air, 16,* 392–401.

Pellerin, N., & Candas, V. (2004). Effects of steady-state noise and temperature conditions on environmental acceptability. *Indoor Air, 14,* 129–136.

Pol, E. (2006). Blueprints for a history of environmental psychology (I): From first birth to American transition. *Medio Ambiente y Comportamiento Humano, 7*(2), 95–113.

Pretty, G. H., Chipuer, H. M., & Bramston, P. (2003). Sense of place amongst adolescents and adults in two rural Australian towns: The discriminating features of place attachment, sense of place community and place dependence in relation to place identity. *Journal of Environmental Psychology, 23,* 273–287.

Probst, J. C., Baek, J. D., & Laditka, S. B. (2010). The impact of workplace environment on job satisfaction among nursing assistants: Finding from a national survey. *Journal of the American Medical Directors Association, 11*(4), 246–252.

Raanaas, R., Evensen, K., Rich, D., Sjøstrøm, G., & Patil, G. (2011). Benefits of indoor plants on attention capacity in an office setting. *Journal of Environmental Psychology, 31*(1), 99–105.

Rabinowitz, P. M., Galusha, D., Dixon-Ernst, C., Slade, M. D., & Cullen, M. (2007). Do ambient noise exposure levels predict hearing loss in a modern industrial cohort. *Occupational and Environmental Medicine, 64,* 53–59.

Raffaello, M., & Maas, A. (2002). Chronic exposure to noise in industry. *Environment and Behavior, 34*(5), 651–671.

Reinikainen, L., & Jaakkola, J. (2001). Effects of temperature and humidification in the office environment. *Archives of Environmental Health, 56*, 365–368.

Rioux, L. (2005). L'attachement au lieu de travail. Etude dans une administration française [Attachment to the workplace. A study in a French public services department.] [CD-Rom]. In A. Battistelli, M. Depolo, & F. Fraccaroli (Eds.), *La qualité de la vie au travail dans les années 2000 [The quality of work life in the 2000s]*. Bologna: CLUEB.

Rioux, L. (2006). Construction d'une échelle d'attachement au lieu de travail. Une démarche exploratoire [Construction of a workplace attachment scale. An exploratory study]. *Canadian Journal of Behaviour Science, 38*(4), 325–336.

Rioux, L. (2007, August). *Attachement au lieu de travail et demande de mutation professionnelle dans l'Education Nationale [Workplace attachment and application for transfer by State Education personnel]*. Paper presented at the 8th International Congress of Applied Social Psychology (CIPSA), Besançon, France.

Rioux, L. (2011). Workplace attachment and request for professional transfer. Study of a population of French employees. *Bulletin of the Transylvania University of Braşov, 4*(53), 91–96.

Rioux, L. (2012). *Modéliser le confort au travail [Modeling comfort at work]*. Nanterre: Research seminar, GREPON (Groupe de Recherches Environnementales de l'université Paris Ouest Nanterre).

Rioux, L., & Angel, V. (2014, in press). La satisfaction au travail. L'impact de quelques variables psycho-environnementales [Job satisfaction. The impact of some psycho-environmental variables]. In R. Kouabenan, M. Dubois, M. E. Bobillier Chaumon, P. Sarnin, & J. Vacherand-Revel (Eds.), *Conditions de travail, évaluation des risques, résilience et management de la sécurité*. Paris: L'Harmattan.

Rioux, L., & Fouquereau, E. (2004). Satisfaction de vie professionnelle et satisfaction liée aux espaces de travail [Satisfaction with professional life and satisfaction linked to workspaces]. In B. Gangloff (Ed.), *La personne et ses rapports au travail* (pp. 43–50). Paris: L'Harmattan.

Rioux, L., & Mokounkolo, R. (2005). Attachement au quartier et adolescence. Etude comparative dans deux banlieues à forte diversité culturelle [Neighborhood attachment and adolescence. A comparative study carried out in two neighborhoods presenting a great cultural diversity]. *Bulletin de Psychologie, 57*(6), 611–620.

Rioux, L., & Pignault, A. (2013a). Workplace attachment and meaning of work in a French secondary school. *The Spanish Journal of Psychology, 16*(e23), 1–14.

Rioux, L., & Pignault, A. (2013b). Workplace attachment, workspace appropriation, and job satisfaction/Apego al lugar de trabajo, apropiación del lugar de trabajo, y satisfacción laboral. *Psyecology: Revista Bilingüe de Psicología Ambiental – Bilingual Journal of Environmental Psychology, 4*(1), 39–65.

Rioux, L., Le Roy, J., Rubens, L., & Le Conte, J. (2013). *Le confort au travail. Que nous apprend la psychologie environnementale ? [Comfort at work. What does environmental psychology teach us?]*. Laval: Presses Universitaires de Laval.

Roethlisberger, F. J., & Dickson, W. J. (1939). *Management and the worker*. Cambridge: Harvard University Press.

Roussel, P. (1996). *Rémunération, motivation et satisfaction au travail [Pay, motivation and job satisfaction]*. Paris: Economica.

Russell, J. A., & Mehrabian, A. (1978). Approach-avoidance and affiliation as functions of the emotion-eliciting quality of an environment. *Environment and Behavior, 10*(3), 355–387.

Scannell, L., & Gifford, R. (2010). Defining place attachment: A tripartite organizing framework. *Journal of Environmental Psychology, 30*, 1–10.

Scrima, F. (2014). Comprendre l'attachement au travail pour agir sur le confort au travail [Understanding attachment to work in order to act on comfort at work]. *Psychologie du Travail et des Organisations, Special Edition on Comfort at Work, 20*(3), 295–310.

Scrima, F. (2015). *A three-factor structure of comfort in the workplace. An exploratory investigation*. Paper presented at GREPON (Groupe de Psychologie Environnementale de Paris Ouest Nanterre), University of Paris Ouest Nanterre La Défense, France.

Scrima, F., Rioux, L., & Lorito, L. (2014a). Three-factor structure of adult attachment in the workplace: Comparison of British, French, and Italian samples. *Psychological Reports, 115*(2), 327–342.

Scrima, F., Moffat, E., Rioux, L. (2014b, August). *Validation italienne de l'Echelle de Satisfaction Environnementale au Travail (ESET) [Italian validation of the scale of Environmental Satisfaction at Work (ESET)]*. Paper presented at the 8th AIPTLF Congress (Association Internationale de Psychologie du Travail en Langue Française), Florence, Italy.

Servais, M. (2011). *Nuisances sonores et open space. [Sound pollution and open space*. Master's thesis. http://mathieuservais.com/docs/Les_Nuisances_sonores_en_open_space-Mathieu_servais.pdf. Accessed 10 Oct 2014.

Slama, I. (2014). Vécu et confort des usagers des immeubles de bureaux tunisiens contemporains: de la normalization à la notion d'ambiance [Experience and comfort of people using contemporary offices in Tunisia: From standardization to the notion of atmosphere]. *Psychologie du Travail et des Organisations, Special Edition on Comfort at Work, 20*(3), 259–274.

Stokols, D., & Scharf, F. (1990). Developing standardised tools for assessing employees' ratings of facility performance. In G. Davis & F. T. Ventre (Eds.), *Performance of buildings and serviceability of facilities* (pp. 55–68). Philadelphia: American Society for Testing and Materials.

Sundstrom, E. (1986). *Work places: The psychology of the physical environment in offices and factories*. Cambridge: Cambridge University Press.

Sundstrom, E., Town, J. P., Rice, R. W., Osborn, D. P., & Brill, M. (1994). Office noise and satisfaction, and performance. *Environment and Behavior, 26*, 195–222.

Tun, P., McCoy, S., & Wingfield, A. (2009). Hearing acuity and the attentional costs of effortful listening. *Psychology and Aging, 24*, 761–766.

Ulrich, R. (2002). Health benefits of gardens in hospitals. Paper presented at the "Plants for People" conference, Florida, USA.

Ulrich, R., Simons, R., Losito, B., Fiorito, E., Miles, M., & Zelson, M. (1991). Stress recovery during exposure to natural and urban environments. *Journal of Environmental Psychology, 11*, 201–230.

Uzzell, D. (2013). Les plantes vertes, vecteurs de la satisfaction au travail [Indoor plants, vectors of job satisfaction]. In L. Rioux, J. Le Roy, L. Rubens, & J. Le Conte (Eds.), *Le confort au travail. Que nous apprend la psychologie environnementale?* (pp. 93–125). Laval: Presses Universitaires de Laval.

Van den Berg, A. (2005). *Health impacts of healing environments: A review of evidence of benefits of nature, daylight, fresh air and quiet in healthcare settings.* Groninge: Foundation 200 Years, University Hospital Groninge.

Veitch, J. A., Charles, K. E., Farley, K. M. J., & Newsham, G. R. (2007). A model of satisfaction with open-plan office conditions: COPE field findings. *Journal of Environmental Psychology, 27*, 177–189.

Veitch, J., Stokkermans, M., & Newsham, G. R. (2011). Linking lighting appraisals to work behaviors. *Environment and Behavior, 45*(2), 198–214.

Velasco, L., & Rioux, L. (2010). Enfoque psicosocial del "apego al lugar de trabajo". Estudio realizado con personal hospitalario [Psychosocial approach to workplace attachment: A study carried out among hospital staff]. *Estudios de Psicología, 31*(3), 309–323.

Vischer, J. C. (1996). *Workspace strategies: Environment as a tool for work.* New York: Chapman and Hall.

Vischer, J. C. (2005). *Space meets status: Designing workplace performance.* Oxford: Taylor and Francis/Routledge.

Vischer, J. C. (2006). The concept of workplace performance and its value to managers. *California Management Review, 49*(2), 62–79.

Vischer, J. C., & Fischer, G. N. (2005). User evaluation of the work environment: A diagnostic approach. *Le Travail Humain/Human Work, 68*(1), 73–96.

Vittersø, J., Vorkinn, M., & Vistad, O. I. (2001). Congruence between recreational mode and actual behavior. A prerequisite for optimal experience. *Journal of Leisure Research, 33*(2), 137–159.

Wagner, A., Gossauer, E., Moosmann, C., Gropp, T., & Leonhart, R. (2007). Thermal comfort and workplace occupant satisfaction: Results of field studies in German low energy office buildings. *Energy and Buildings, 39*(7), 758–769.

Waku, H., Tamura, I., Inoue, M., & Akai, M. (1995). Life cycle analysis of fossil power plant with CO_2 recovery and sequestering system. *Energy Conversion and Management, 36*(6–9), 877–880.

Warr, P. (1990). The measurement of well-being and other aspects of mental health. *Journal of Occupational Psychology, 63*, 193–210.

Warr, P. (1994). A conceptual framework for the study of work and mental health. *Work and Stress, 8*(2), 84–97.

Wells, M. M. (2000). Office clutter or meaningful personal displays: The role of office personalization in employee and organizational well-being. *Journal of Environmental Psychology, 20*(3), 239–255.

Windsong, E. A. (2010). There is no place like home: Complexities in exploring home and place attachment. *The Social Science Journal, 47*(1), 205–214.

Wineman, J. (1982). The office environment as a source of stress. In G. W. Evans (Ed.), *Environmental stress* (pp. 256–285). Cambridge: Cambridge University Press.

Witterseh, T., Wyon, D. P., & Clausen, G. (2004). The effects of moderate heat stress and open-plan office noise distraction on SBS symptoms and on the performance of office work. *Indoor Air, 14*(s8), 30–40.

Wood, S., van Veldhoven, M., Croon, M., & de Menezes, L. M. (2012). Enriched job design, high involvement management and organizational performance: The mediating roles of job satisfaction and well-being. *Human Relations, 65*(4), 419–444.

Wright, T. A. (2006). The emergence of job satisfaction in organizational behavior: A historical overview of the dawn of job attitude research. *Journal of Management History, 12*(3), 262–277.

Quality of the Hospital Experience: Impact of the Physical Environment

23

Ann Sloan Devlin and Cláudia Campos Andrade

23.1 Introduction

This chapter will focus on the quality of life experienced in one particular institutional setting: the hospital. Hospitals are in some sense the bookends of our lives; they may be the first physical environment we encounter and the last. In the interim, we may interact with the hospital as patients or visitors on numerous occasions.

With the possible exception of childbirth, most visits to the hospital are not welcome and activate high levels of stress. When patients are placed within the social arena of healthcare, they are required, among other things, to relate to physicians, healthcare providers, administrative staff, other patients, their illness, the diagnostic procedures, and the treatments (González-Santos 2011), which is psychologically demanding. Psychoneuroimmunology has a long history linking stress and health. People under stress might experience changes in perception, attention, memory processes, and decision-making, as well as feelings of distress, anxiety, fear, and depression. Stress responses involve cognitive, emotional, behavioral, and physiological effects (Steptoe and Ayers 2005), such as changes in the activity of immune, endocrine, cardiovascular, gastrointestinal, and other bodily systems. All of these stress-related changes may create susceptibility to disease, affect disease progression, or retard the speed of recovery (Dougall and Baum 2001). For example, studies have shown that enduring chronic stressors is associated with greater susceptibility to colds (Cohen et al. 1998), and with impairing cutaneous wound healing (Ebrecht et al. 2004).

This research indicates that the (unnecessary) stress patients experience in the hospital should be reduced as much as possible. Although some of the stressors patients face are unavoidable, such as illness and treatments, others are not (Powell and Johnston 2007). The physical environment, one important avenue to both create and reduce stress, is the focus of this chapter. The relationship between the quality of the physical environment of healthcare settings and users' well-being has been extensively documented, and a sizable body of empirical literature has been generated over the last three decades. For example, it has been repeatedly found that hospital design has an impact on patient satisfaction with the service, affective state, and speed of recovery (Devlin and Arneill 2003; Ulrich et al. 2008).

A.S. Devlin (✉)
Connecticut College, New London, Connecticut, USA
e-mail: asdev@conncoll.edu

C.C. Andrade
Centro de Investigação e Intervenção Social (CIS-IUL), Instituto Universitário de Lisboa (ISCTE-IUL), Lisbon, Portugal
e-mail: claudia.andrade@iscte.pt

In this chapter, we will examine what qualities and characteristics of hospital environments result in positive outcomes, with a particular focus on the patient. Specifically, we will explore what fundamental psychological needs are addressed by the hospital's physical environment. Meeting these needs may impact the patient's quality of life both subjectively and objectively. Subjective indicators may include satisfaction with hospital care, whereas objective indicators may include length of recovery.

Ulrich's (1991) theory of supportive design provides one useful theoretical framework to view the quality of the hospital experience (QHE). This theory proposes that healthcare physical environments will promote well-being if they are designed to foster a sense of control over physical-social surroundings, access to social support, and access to positive distractions. All of these needs have been found to be important. For example, the role of control was demonstrated in a study of individuals waiting to donate blood whose stress levels were actually higher when the television, not under their control, was turned on than when it was off (Ulrich et al. 2003). Research on single-occupancy rooms in hospitals has also suggested that patients may welcome the privacy (and control) such rooms provide (Chaudhury et al. 2005); specific categories of patients, such as adolescents, are particularly sensitive to the issue of privacy. Research on adolescents also emphasizes the need for social support, particularly from friends, as they recover in hospital settings (Blumberg and Devlin 2006). Increasingly, inpatient rooms are being designed to accommodate overnight stays by family members (e.g., Rabner et al. 2013). In addition, the importance of positive distraction in hospital settings, whether through views out of windows of nature, art on the walls, music provided through headphones, or self-paced activities such as reading, is thoroughly documented (see, for example, Arneill and Devlin 2002; Hathorn and Nanda 2008; Mazer and Smith 1999; Ulrich 1984).

We begin the chapter with an overview of two principles that guide research on hospitals and quality of life. We then consider specific evidence on the topics from the Theory of Supportive Design: positive distraction, perceived control, and social support.

23.2 Guiding Principles: Patient-Centered Care and Evidence-Based Design

Recent work on the environmental psychology of healthcare facility design has been guided by an emphasis on patient-centered care and evidence-based design.

23.2.1 Patient-Centered Care

With regard to quality of life, the emphasis on patient-centered care or patient-focused care is particularly relevant. This movement can be traced to the pioneering work of Angelica Thieriot. Her unhappy experience as a patient in what might be called the high-tech low-touch approach of the US healthcare system led to the creation of the Planetree model of healthcare (named after the tree in Greece under which Hippocrates taught his students). The model focuses on the needs of the patient (see www.planetree.org) including patient education, patient choice, and a variety of approaches to healing that concentrate on the senses (e.g., aromatherapy). Much of the research on the ambient environment of healthcare (e.g., sights, sounds) can be connected to the Planetree movement. Stichler (2008, 2011) similarly uses the model of care provided by Planetree as a framework to discuss how the needs of patients and staff can be addressed and comments that the concept of patient-centered care "has been discussed in nursing since the inception of the profession" (2011, p. 503).

23.2.2 Evidence-Based Design

The second initiative, evidence-based design, relates to the idea that the design of healthcare facilities should be based on evidence: "Evidence-Based Design (EBD) is the process of basing decisions about the built environment on credible research to achieve the best possible outcomes" (www.healthdesign.org/edac/about) and has a parallel in evidence-based medicine: "...the conscientious, explicit, and judicious use of current best evidence in making decisions about the care of individual patients" (Sackett et al. 1996, p. 71). Evidence-based Design Accreditation and Certification (EDAC) is offered by The Center for Health Design (CHD). The CHD was founded in 1988 with the goal of improving the facilities in which healthcare takes place. Among its numerous initiatives is The Pebble Project, in which collaborative research documents the ways in which the built environment is related to healthcare outcomes. The Center also offers a Knowledge Repository of whitepapers and reports (at no cost) on healthcare research. The Robert Wood Johnson Foundation also supports research in the area of patient-centered care and patient safety, among other topics (www.rwjf.org/en/grants/what-we-fund.html). Moreover, the American Institute of Architects has also formed a group emphasizing the importance of healthcare: the AIA Academy of Architecture for Health (AIA/AAH).

Ulrich et al. (2010) present a nine-faceted framework for evidence-based design: (1) audio, (2) visual, (3) safety enhancement, (4) wayfinding system, (5) sustainability, (6) patient room, (7) family support spaces, (8) staff support spaces, (9) physician support spaces. These dimensions impact the outcomes for participants (patients, families, physicians, nurses, and other staff), which in turn impact outcomes for the organization. Explicit within this nine-faceted framework are the topics of positive distractions and social support. The role of perceived control is implicit through such facets as opportunities for control in the patient's room and family support spaces, among others.

23.3 Patients' Needs for Quality of Life: Applying Ulrich's Theory of Supportive Design

23.3.1 Need for Positive Distractions

The ambient environment, or the environment that surrounds us and impinges on our senses, provides important sources of information that may affect the well-being of patients, positively or negatively. A number of the dimensions in the Ulrich et al. (2010) framework for evidence-based design dovetail with Ulrich's Theory of Supportive Design and reflect aspects of the ambient environment, including how the visual and auditory environment affect patient quality of life in hospital settings. What we see and what we hear may help divert our attention from aspects of the hospital environment that are stress-inducing or painful.

We will concentrate first on the concept of visual positive distraction and then move to auditory positive distraction. One of the most widely cited studies in environmental psychology (791 times since 1996, according to the Scopus database on June 24 2014) is Ulrich's 1984 publication in *Science* demonstrating that a view of everyday nature from a hospital window resulted in shorter post-operative stays and fewer doses of potent analgesics for patients compared to matched patients with a view of a brick wall. Since that early research, the role of what we see in the hospital environment has received the greatest research emphasis, followed by what we hear. Other aspects, such as what we smell (e.g., aromatherapy), have received relatively little attention in comparison.

23.3.1.1 Views of Nature, Healing Gardens, and Live Plants

Following Ulrich's 1984 work, other research has confirmed the potential of nature to affect well-being positively in a hospital setting, although not always with such unequivocal results. This research includes the availability of views of nature, healing gardens, and the presence of live plants.

In the environmental literature, the benefit of nature to quality of life has been addressed beyond the healthcare setting, primarily in the research of Stephen and Rachel Kaplan (e.g., Berman et al. 2008; Kaplan and Kaplan 1989; Kaplan 1995). The Kaplans developed Attention Restoration Theory (ART) to describe the benefits that nature has for human attention. Nature has the ability to create ideal restorative environments, which help alleviate stress by reducing directed-attention fatigue. Restoration relates to some level of recovery at a physical or psychological level, such as cognitive capacity, focus, vitality, stress reduction, rest and relaxation, etc. Restorative environments describe particular settings that promote "restoration" when one spends time in them or perceives elements of them. Although nature offers ideal environments for a restorative experience, other kinds of environments can provide opportunities for the restoration of "directed attention" (e.g., through artwork, photographs, window views, etc.) (e.g., museums; Kaplan et al. 1993).

The four interrelated factors thought to work in a restorative experience are: being away, extent, fascination, and compatibility. In healthcare facilities, it is possible that aspects of ART, such as soft fascination and the sense of being transported to another world, are responsible for the well-being that nature provides: nature is assumed to attract involuntary attention due to its fascinating qualities, and to provide escape from unwanted distractions in the surroundings. This possibility is important so that patients can be distracted from aversive situations, procedures, and thoughts. For example, burn patients who were assigned to watch nature videos had less pain and anxiety during dressing changes than did patients in a control group (Miller et al. 1992). Similarly, Diette et al. (2003) conducted a study of patients undergoing flexible bronchoscopy procedures at the Johns Hopkins Hospital. During the procedure, one group was exposed to a combination of a nature mural plus nature sounds whereas the second group received treatment as usual. The treatment group who received the nature intervention reported less pain than did the group that was not exposed to nature.

To promote a better quality of hospital experience, nature and other kinds of distracting environmental elements can also be part of the surroundings.

Views of Nature

There are significant challenges in providing views of nature in healthcare settings, ranging from the location itself to the proliferation of adjacent buildings on a medical campus. Often views are at least partially blocked by other buildings. Examining this issue using coronary and pulmonary patients, Raanaas et al. (2012) looked at the effects of occupying a room with a panoramic view vs. a view obstructed (partially or fully) by other buildings. As an overview of their results, the panoramic view had more positive benefits on physical and mental health than did the blocked view.

While most of the emphasis has been on the potential benefit of views of nature for patients, staff members may also benefit from this exposure. The value of exposure to nature was revealed in a study of pediatric nurses whose alertness remained the same or increased and whose stress levels remained the same or decreased with a view of nature, compared to nurses whose stress levels increased, and alertness decreased with either no view, or a view without nature (Pati et al. 2008).

Views of and access to nature are often among the positive variables emerging from a comprehensive summary of the literature on healthcare facilities (e.g., Salonen et al. 2013), but not all research presents clear findings about their role. For example, research by Shepley et al. (2012), comparing light levels and views in a pre-post comparison of two intensive care units (ICUs), failed to show a statistically significant reduction in pain perception and reduced errors in the ICU with views of nature (rooftop garden or newly planted gardens), but there were statistically significant reductions in staff absenteeism and staff vacancies.

Healing Gardens

Over the past two decades, healthcare facilities have increasingly embraced the idea of healing gardens, or garden areas in the healthcare setting that have therapeutic and restorative qualities. Gardens may be walked in, explored, or even just viewed from within the facility. Perhaps most well-known for advancing this concept are Clare Cooper Marcus and Marni Barnes, who have written two important books on this topic (1995, 1999) and contributed to a number of articles (e.g., Francis and Cooper Marcus 1992; Whitehouse et al. 2001). The garden areas described by both Cooper Marcus and Barnes (1999) and Carpman et al. (1986) may be welcoming not only to patients and visitors but also to staff (see also Naderi and Shin 2008). Cervinka et al. (2014) evaluated the qualities of hospital gardens that were rated appealing in an online survey completed by staff and other participants. Among the qualities that emerged from the factor analyses of semantic differential rating scales were tranquility and sense of touch. The researchers went on to incorporate these variables in recommendations to enhance hospital gardens. Other researchers suggest providing a variety of elements that support a range of activities from passive to active; different categories of users and different age groups are attracted to different elements in the garden (Sherman et al. 2005).

Aspects such as the amount of shade and the quality of seating are correlated with the use of garden areas by staff (Pasha 2013); accessibility and maintenance must also be considered in planning gardens (Davis 2011), as should the provision of private and semi-private spaces (Asano 2008). In addition, management must advertise or otherwise make known the existence of the garden (Whitehouse et al. 2001); simply creating the garden does not guarantee its use. Gardens may also provide volunteer opportunities for neighbors (Asano 2008). Beyond the hospital environment, the benefits of gardens have also been identified in special care units for individuals with dementia (Hernandez 2007), and horticultural therapy has been used for those undergoing rehabilitation for brain damage (Söderback et al. 2004).

Live Plants

Not all hospital settings provide views of nature. This problem may occur because the hospital is in an urban setting where other buildings are the only objects in view, because rooms are off interior corridors without windows, or because the technology being used requires high levels of shielding for radiation. Furthermore, live potted plants and flowers (both fresh and dried) may not be considered appropriate for some hospital populations (Centers for Disease Control and Prevention 2007). In these kinds of situations, researchers have investigated whether plants placed in a patient's room might produce benefits.

For example, in a randomized clinical trial involving post-appendectomy patients, potted plants and flowers were added to an inpatient room (with an identical room – same floor, same side of the building, but no plants – as the control) (Park and Mattson 2008). For the patients with plants and flowers, a number of positive benefits emerged including lower systolic blood pressure, fewer doses of postoperative analgesics, lower pain ratings and lower anxiety, more positive feelings about the room and satisfaction, compared to those in the control room. In a similar experimental approach, Park and Mattson (2009) used female patients recovering from thyroidectomy to assess the effect of adding plants to an inpatient room. Similar positive results emerged, in that the patients exposed to the plants (12 in this study) took fewer analgesics and had more positive vital signs, as well as more positive subjective responses, than did patients in the control rooms (same floor, same side of the building, but without plants). Research in a simulated hospital room environment (Park et al. 2004) had previously shown a decrease in pain sensitivity for those in the hospital room with plants. Even in a situation where nature is already available in a scenic mountainous area, indoor plants were able to influence the subjective well-being of patients in a Norwegian rehabilitation center (Raanaas et al. 2010).

Is the Effect of Nature Mediated?

Research is beginning to include psychological variables as mediators between the physical environment, on the one hand, and health-related outcomes, on the other (e.g., Andrade et al. 2013). That is, psychological processing (what we think about the environment) may mediate this relationship between environment and outcome. For example, Tanja-Dijkstra et al. (2008) added plants to a photograph of a hospital room (participants viewed it online) to investigate whether the perceived attractiveness of the room mediated the relationship between plants and stress. Measures included a stress arousal checklist with a ten-item bipolar adjective scale. Those exposed to the hospital room with plants perceived less stress, and the attractiveness of the room was found to be a mediator. This study suggests that one of the ways in which nature contributes to well-being in the hospital setting is by making the environment more attractive.

Nevertheless, caution is required before concluding that adding plants is necessarily beneficial. A review of 21 experimental studies from peer-reviewed journals suggests some evidence for the positive benefits of indoor plants (e.g., pain tolerance and stress reduction), but mixed findings as well (Bringslimark et al. 2009). Of these studies, only six involved healthcare facilities or simulations of such facilities, and of those six, only three involved patients. The authors conclude that more research is needed to understand more fully the potential of indoor plants for enhancing well-being and quality of life in the hospital setting.

23.3.1.2 Art and Simulated Nature

Not all settings, and not all functions within settings (e.g., computed tomography or CT scans), can provide a view of nature. Considerable research has examined substitutes for nature that might influence health outcomes. Among these alternatives, art has received a good deal of attention and recommendations have emerged about what kinds of art might be appropriate. A solid research base showing the benefits of art for patient well-being (in the inpatient room, the waiting area, and even in the operating theater) has accumulated (see Cusack et al. 2010; Diette et al. 2003; Eisen et al. 2008; Hathorn 1993; Miller et al. 1992; Nanda et al. 2010). A number of authors have provided review articles looking at the use of art in healthcare facilities (e.g., Hathorn and Nanda 2008; Rollins 2011; Ulrich 2009). Furthermore, companies such as American Art Resources have developed materials to be used in healthcare settings (see www.americanartresources.com). Early recommendations came from Ulrich and Gilpin (2003) who suggested using scenes of representational nature and avoiding abstract art. Not all nature scenes are deemed suitable, however, as those that depict threats (e.g., storms) are not recommended. The research of Ulrich et al. (1993) also points to the fact that "not just any art will do". Their study of patients recovering from open-heart surgery showed the beneficial effects (on anxiety and level of pain medication) of patients exposed to nature images, in contrast to those exposed to abstract art, a blank panel, or no panel.

Nanda and her colleagues have conducted a significant body of research on the use of art in healthcare settings. As an example of their field research, Nanda et al. (2012a) examined the effects on patients of using art in a waiting area. In a pre-post design, the researchers provided a continuous loop of nature images, including floral, water, and landscapes, on a plasma monitor as well as still photographs (the largest was $132 \times 40''$). The most notable change was a significant decrease in what were defined as restless activities, such as getting out of their seats and asking questions at the front desk, which the authors argue could be due to the presence of the positive distraction provided by their intervention. No subjective emotional measures were collected, and consequently patients' evaluations of and thoughts about the art installation are unknown. Other research from Nanda and colleagues (Nanda et al. 2011) looked at the effect of art in the lounge of a psychiatric facility where patients spent 3–4 days being evaluated. Three art conditions and a control were involved. The art conditions were: abstract ["Convergence" by Jackson Pollock]; abstract representational ["The Fields" by Van Gogh]; and realistic nature stock photography [image of the savannah]. These posters were large ($3 \times 4'$) and mounted on the wall of the lounge of the psychiatric unit. The dependent variable was the distribution of PRN (pro re nata or as needed) medication, which was significantly lower for the days on which the image of the savannah was displayed, in contrast to the control condition. At the same time, no significant differences were found between the control and the other two art conditions, which points to the particular power of nature, and not simply any visual distraction, to affect well-being.

23.3.1.3 Television

Television is considered a source of positive distraction, but its use in healthcare settings is debated because not all research shows positive benefits. One of the most compelling arguments against its use comes from research by Ulrich et al. (2003) in which blood donors had lower blood pressure when the television was off than when it was turned on, without the opportunity to adjust the channel or otherwise have control over it.

Examining waiting behavior in Dutch hospital polyclinics (Pruyn and Smidts 1998), patients with appointments either watched television or were exposed to the waiting area with the television turned off. Behavior was observed in the waiting area, and patients were intercepted to fill out a questionnaire after their appointment. Of interest were (1) the objective and perceived time spent waiting and (2) satisfaction with the service. Patients were actually good at estimating their waiting time (correlation between subjective and objective indices of 0.73). On average, patients spent 23 min waiting. Results indicated that having a television per se would not necessarily produce positive distraction, and that watching it was related to the length of wait, with longer waits associated with television watching due to boredom. Because non-television viewers underestimated the perceived time spent and television viewers overestimated this, "patients who have to wait longer and are thus more annoyed, seem to be more eager for distraction and start watching TV" (p. 327). Thus, the authors argue, the conditions within which people wait matter. Moreover, if television is to be used, they recommend longer and less segmented presentations, because the "more a time interval is segmented, the longer its perceived duration" (p. 332).

23.3.1.4 Music

A large body of research has examined the role of music in enhancing the well-being of patients, particularly cancer patients and those receiving palliative care (e.g., Renz et al. 2005). In general, the findings point to the use of music to reduce anxiety and tolerate pain better (Cooke et al. 2005; Dileo and Bradt 2005; Guétin et al. 2012; Lee et al. 2011, 2012; Thorgaard et al. 2004; Williamson 1992). The Continuous Ambient Relaxation Environment™ (CARE) Channel, a specialized set of musical selections, has even been developed for use in healthcare settings (see http://www.healinghealth.com; Mazer and Smith 1999).

Music therapy has been a recommended mode of treatment for cancer patients to help reduce the considerable anxiety they feel following their diagnosis, yet when strict criteria for experimental control are applied, the benefits are equivocal (Nightingale et al. 2013). Of an original list of 606 studies, 566 were excluded when rigorous methodology was used as a prerequisite while of the remaining 40, only 13 were included in a systematic review. Of these, only four were appropriate for the meta-analysis in terms of evaluating the same theoretical construct with measures that were sufficiently similar and with raw data that were accessible. While the systematic review including 13 studies showed that music had a positive effect on anxiety for cancer patients, the meta-analysis of just 4 did not. Similarly, in a review, Hilliard (2005) encountered problems with internal validity in that only 2 of the 11 studies controlled for diagnostic category, and only 3 of them met the criteria for randomized controlled research. In the end, in only one of these was there a significant difference suggesting that music therapy had a positive effect.

Given all the methodological differences reported in the Nightingale et al. (2013) article (e.g., when the musical intervention occurs, with whom, for how long, and so on), it would make sense to have a consortium of hospitals undertake an identical procedure across hospitals to ascertain the impact of music intervention on cancer patients. To advance the research, the authors themselves argue for more standardized approaches and more homogeneous populations of diagnosis and modality of treatment.

Although we might typically think of music produced by human intention (i.e., composed and performed), sounds of nature have been used in the healthcare setting. As an example, patients being gradually taken off mechanical ventilation after undergoing coronary bypass surgery listened through headphones to nature-based sounds or no sounds. Those in the experimental group had a choice of 36 selections from 6 general categories including birdsong, rain, water-course sounds, and sounds associated with walking in a forest. Those exposed to their selection of such sounds demonstrated positive benefits in terms of reduced anxiety and agitation, and also in terms of hemodynamic measures (i.e., systolic and diastolic blood pressure, heart and respiration rate, mean arterial pressure) in contrast to patients who wore headphones, but were not exposed to the nature-based sounds (Aghaie et al. 2014). According to the authors, this was perhaps a reflection of increased relaxation. In addition to the significant differences between the experimental and control groups, those who served as controls also saw positive benefits (although not as dramatic) in terms of reduced anxiety; the authors speculate this outcome may be related to shutting out background noise by wearing headphones. Another consideration in this research was that patients in the experimental group were able to choose what they wanted to hear, which may have contributed to their sense of well-being.

Beyond this use of music and sound as an intervention to enhance well-being, researchers have begun to consider the totality of the sounds that patients and staff might hear in a healthcare setting. Mackrill et al. (2013) used a semi-structured interview to understand the "soundscape" on a cardiothoracic ward of a public university hospital in the United Kingdom. A "soundscape" was defined as the "auditory version of the landscape" (p. 2) and included positive as well as negative aspects of sound. Responses were grouped into 11 categories (p. 3): perception of sounds, sound sources, emotional response, temporal factors, restoration, other physical attributes, future design, behavior of people, analogy, job duties, and patient interaction. The conceptual model they propose points to the fact that interventions could test the (1) physical sound sources (e.g., effect of manipulating sound such as a burbling brook); (2) cognitive intervention (e.g., by providing information about sound sources and effects to try to create a more positive perception). The authors argue that simply reducing the level of sound in a hospital environment should not be the sole goal, given the complexities of the sounds that occur and their potential positive as well as negative impact. For example, they point out that the patterns of the sounds over the course of the day can be reassuring to patients.

Nevertheless, it should be remembered that noise typically has negative effects on patients. For example, Hagerman and colleagues (2005) focused on the effects of room acoustics on patients with coronary artery disease. They compared patients who were in a unit with ceiling tiles that provided poor acoustics with those who were in the unit after a renovation had supplied sound-absorbing tiles with good acoustics. Patients with acute myocardial infarction and unstable angina showed lower pulse amplitude during the night in the presence of good acoustics. In addition to these physiological effects, patients in the good acoustics group considered the staff attitude to be much better and had a lower incidence of rehospitalization than did patients treated when the poor acoustics were present.

Under the umbrella of positive distractions and the ambient environment, there are a number of other topics that could be discussed, including the roles of light, color, and odor (aromatherapy) in patient well-being. Unfortunately, these topics are beyond the scope of this chapter. Readers are referred to the work of Malkin (2014) and Devlin (2015), among others, for a discussion of such issues. As previously mentioned, recent theories of pain highlight the role that sensory stimuli from the environment can play in influencing the particular experience of pain. A literature review can be found in Malenbaum et al. (2008).

23.3.1.5 Issues of Ecological Validity in Methodology

Virtual reality (VR) is being increasingly used as a medium for research, presumably because it provides more flexibility than still images and approximates more closely the experience of "being there". Despite these advantages, a question for researchers (and ultimately practitioners) is the degree to which such simulations mirror results that would be found in the actual settings. Given the difficulty of experimental control, a good deal of the research about healthcare settings uses simulated environments. On one end of the continuum, the research uses patients in the healthcare setting; on the other, there are non-patients (often students), who are exposed to simulated environments through digital images, videos, virtual reality (VR), or some combination. In the middle are combinations of these approaches, such as patients viewing controlled simulations, or non-patients lying in a hospital bed while being exposed to various conditions.

Tanja-Dijkstra et al. (2014) combined a mock dental office setting with virtual reality during a simulated dental procedure. This approach was taken to manage the vividness of (presumably aversive) dental memories for dental patients with higher anxiety than for those with lower anxiety. A lab was decorated to look like a dental office waiting area and procedure room, with appropriate posters of dental procedures. Participants listened to a tape of a dentist with the running dialogue one would commonly hear (e.g., "Please open your mouth..."). The VR environment, which the participant viewed while sitting in a dental chair, consisted of a path along the coast with the beach, water, and grassy areas visible. Those in the VR active condition could explore this scene; those in the VR passive condition could only view it (and in fact were yoked-controls, viewing the "walk" that the previous active VR participant had created). A control group wore the head-mounted device but saw only a black screen. There was a greater reduction in the vividness of memories for patients with higher dental anxiety than for those with lower dental anxiety over a week. The researchers argue these findings have implications for reducing dental anxiety. If memories become less vivid, dental anxiety may be reduced, which in turn may disrupt what is known as the cycle of dental anxiety.

Use of VR suggests that the hospital experience might be improved by exposure to more restorative environments. For example, VR viewing of a hospital room with a sea view has been shown to reduce laser pain for patients with chronic migraines (DeTommaso et al. 2013). In another example, Dinis et al. (2013) used VR to gauge the effect of interior design elements in a hospital room on university students' emotional responses. The rooms varied in the number of elements in the room (potted plant, landscape poster, Mondrian painting, or chair). Like those of Andrade and Devlin (2015), these results showed a correlation between the number of elements in the hospital room and a positive emotional response.

Virtual reality is gaining traction as a research tool as it becomes more available, but photographs are still used. The ability of images to reduce discomfort (reflected in mood and vital sign measures) was tested by Vincent et al. (2010a, see also b). The images in the study were based on categories from Appleton's (1975) prospect-refuge theory. This theory posits that we like to see out (i.e., prospect) but at the same time be sheltered from being seen by others (i.e., refuge). This research tested Appleton's theory with four photos: prospect, refuge, hazard, and a mixture of prospect and refuge. A control group who viewed a black screen was also included. To simulate the experience of being patients, participants lay in a hospital bed facing the screen (36″ long × 21″ high). The picture with a hazard (forest fire) had the greatest ability to distract patients (i.e., influence their thoughts) during a cold pressor test, but produced the lowest mood ratings (total mood disturbance) and therefore is not considered appropriate for therapeutic purposes. The cold pressor test is commonly used to induce stress; participants immerse one hand in a container of iced water for up to 2 min.

Simulated nature in the form of posters may be as effective as live plants, and certainly more effective in lowering stress than having no plants at all, based on research done by Beukeboom et al. (2012). The size of what is displayed may also have an impact, as other research showed that displaying a 10-min nature film on a larger screen (72″) was more effective than displaying it on a smaller screen (31″) in terms of positively affecting physiological responses related to stress (deKort et al. 2006).

23.3.1.6 Neuroarchitecture

Recently, mirroring the trend in psychology more broadly, designers have begun to ask questions about the neuroscience of the phenomena they are investigating, and a special organization, the Academy of Neuroscience for Architecture (ANFA; http://www.anfarch.org) has emerged from the AIA. *Intelligent Buildings International* published a special issue focusing on the first conference of the Academy of Neuroscience for Architecture (ANFA). In that publication, Nanda et al. (2013) reported beginning to investigate the role of contours, a specific visual feature, in the experience of emotion. The authors do this by delving into brain function and discussing using fMRI technology to investigate participants' responses to visual images. This approach has been described as "neuroarchitecture", which looks at the relationship between brain function in areas such as the amygdala, hippocampus, and cerebral cortex, on the one hand, and the physical environment, on the other, as a way of understanding more fully our emotional responses to the designed environment. A helpful review of this topic is provided by Nanda et al. (2012b).

As an example of this approach, Pati et al. (2014) used blood oxygen level dependent (BOLD) measures to track brain activation while participants looked at different images on a ceiling. These images fell into the categories of sky, positive (birds, flowers, pets), negative (burning cars, storms, trash) and neutral (blank wall, wood grain, white tiles, similar to those on a ceiling). The motivation for concentrating on the "sky" was the parallel it presents to patients in bed who are able to look out at the sky but cannot see any other natural elements because hospital buildings obscure their view. Results showed that the brain areas activated by the sky images were different from those activated by traditional ceiling images, but the findings are so preliminary that the authors' major contribution is to point out the need for more definitive studies.

23.3.2 Need for Control

The hospital environment is a major negative contributor with regard to privacy, personal control, and noise. These factors increase the levels of stress experienced by patients and affect their well-being (Devlin and Arneill 2003; Ulrich 1991). Perceived control, one of the three elements in Ulrich's Theory of Supportive Design, has received little attention, perhaps because it is difficult to isolate from other responses to the hospital environment.

In particular, research on the topic of environmental control is usually found in the workplace literature (e.g., Lee and Brand 2005). Environmental control is defined as the degree to which people perceive they have "control over various characteristics of their environment" (Lee and Brand 2010, p. 326). In the healthcare setting, most research on this topic comes indirectly, for example through research on patients' reactions to single-occupancy rooms, which give them privacy and presumably a sense of control over their surroundings.

As an illustration, in a survey featuring color and lighting in hospital design, Dalke et al. (2006) commented that "patients should have some potential control of their immediate environment; local lighting control and a customized ambience will be appreciated by the patient" (p. 361). Control can emerge in a variety of ways; even the opportunity to select one's own music can help with the tolerance of pain (Mitchell and MacDonald 2006). As was discussed earlier in this chapter, lack of control also has an impact on well-being. When people cannot exert control, as was the case in the study of blood donors when the television was not under their control (Ulrich et al. 2003), their well-being is negatively impacted.

Perceived control is seen as a protective cognitive factor for psychiatric patients (Kim 2014) and has been linked to emotional well-being for hospital workers where there is workplace violence and aggression (Schat and Kelloway 2000). Not feeling in control of one's self, and hearing alarms and buzzers (stimuli not under one's control), were primary sources of stress for patients in critical care units in Jordan (Hweidi 2007).

Patients seem to relinquish their sense of control when they enter the hospital and thus have relatively little awareness that they might be able to control aspects of their hospital room. Hospitals are frightening places. Patients feel overwhelmed by technology. Moreover, they do not want to be reminded of their situation by viewing medical equipment. There is now some effort by design practitioners to screen elements such as soiled linen and equipment in patient rooms (Malkin 2011). The idea of screening also relates to the body of the patient, where drapes, doors, and opaque glass can be used to provide visual and auditory privacy. When patients could overhear their health information being shared, they were less likely to reveal it to providers (Olsen and Sabin 2003).

To explore the effect of screening medical equipment, Tanja-Dijkstra (2011) exposed people to a photograph of a hospital room with or without medical equipment visible. Her goal was to examine the effect of the visibility of medical equipment on feelings of stress, hypothesizing that an emotional state (pleasure), present when there is no equipment visible, mediates the relationship between the presence/absence of medical equipment and stress reactions. In fact, there was less stress in the situation that represented "out of sight, out of mind." In mediational analysis, feelings of pleasure mediated the stress-reducing effects of placing the equipment out of sight.

In addition to the importance of being in control of the physical environment, some researchers construe control in terms of the decision-making process. Using a framework from the services industry (specifically Hui and Bateson 1991), Gotlieb (2000) argues the degree of participation in the healthcare decision-making process influences patients' perceived control. He defines perception of control as "the extent to which patients believe that they can influence the care that they receive from the hospital" (p. 6). He cites research by (1) Haskell and Brown (1998) that patients seek more control over the care they receive, (2) the work of Affleck et al. (1987) that perceptions of greater control are associated with more positive mood in patients, (3) the work of Mahler and Kulik (1990) that perceptions of control can influence rates of recovery and (4) the work of Thompson et al. (1993) in cancer patients showing that perceptions of control are associated with less depression and anxiety. In Gotlieb's research, patients' perceptions of the healthcare process and of their hospital rooms, together affected their perceptions of nurses (Gotlieb 2000); however, perception of control did not affect perception of hospital quality.

As previously indicated, the role of perceived control in the healthcare setting may appear indirectly from other aspects being studied, rather than being addressed head-on. As an example, lack of control over lighting and television emerged as a problem in a study of patients and families at the Cleveland Clinic (Trochelman et al. 2012). Trochelman et al. assessed the design features that were related to satisfaction and dissatisfaction. Using a pre-post design in the heart and vascular department of the Cleveland Clinic, the researchers combined interviews and data related to the physical environment from the Hospital Consumer Assessment of Health Providers and Systems (HCAHPS) and Press Ganey surveys. Patients were moved from an old (semi-private rooms) to a new (single-occupancy rooms) unit and thus were able to compare the two. Among the variables linked to satisfaction were privacy (single-occupancy room), spaciousness, and large windows. Dissatisfying aspects involved issues of control, including lighting and television, which allowed patients to scroll through channels in only one direction. Thus, when patients do not have control over aspects that are familiar to them (here, scrolling through television channels), there may be dissatisfaction.

23.3.2.1 Perceived Control and Single-Occupancy Rooms

Over the last 20 years, one of the biggest changes in the inpatient's experience of healthcare, at least in the United States, has been the widespread adoption of the single-occupancy room model. In 2006, the American Institute of Architects recommended that single-occupancy rooms become the standard in acute care hospitals (American Institute of Architects Media Relations 2006, July 19).

The use of single-occupancy rooms vs. multiple-occupancy rooms has received support in the literature (Chaudhury et al. 2005). Benefits include lowered possibility of infection, fewer medication errors, and better workflow for healthcare workers. However, there are also contradictory findings in terms of patient safety. For example, compared with controls, patients isolated for infection control precautions experienced more preventable adverse events and had less documented care (Stelfox et al. 2003). Moreover, some authors argue that private rooms by themselves may not be very effective in infection control without proper precautions, such as hand washing and use of protective gear (Chaudhury et al. 2005; Van de Glind et al. 2007). Furthermore, Lorenz and Dreher (2011) used a retrospective comparative design to look at outcomes related to room design (single vs. multiple occupancy). Results related to safety (i.e., risk of falling) led the authors to suggest that offering an array of room choices may meet the needs of patients better than offering only single-occupancy rooms. The risk of falling in older patients increases in such single-occupancy rooms because they may attempt to do things (e.g., use the bathroom) by themselves.

As these comments about drawbacks to single-occupancy rooms suggest, some authors believe that evidence is scarce and more research is needed to ensure that patients benefit from single rooms. In a literature review by Van de Glind et al. (2007), the authors make the point, which is echoed in the literature about single-occupancy rooms (e.g., Verderber and Todd 2012), that there are too few high quality studies on the built environment in healthcare to reach definitive conclusions on this issue. In particular, the number of randomized controlled trials is inadequate. Their review included only 25 studies after exclusions and targeted 6 outcome measures: "privacy and dignity of patients, noise and quality of sleep, patient satisfaction with care, hospital infection rates of methicillin-resistant Staphylococcus aureus (MRSA), patient safety (fall accidents, medication errors, patient recovery rates, complications) and length of stay" (p. 155). Only four of the studies qualified as randomized controlled trials (two on patient satisfaction with care, two on hospital infection rates). The review by Van de Glind et al. reports conflicting findings about whether single rooms reduce infection, and insufficient information about recovery rates and patient safety to reach conclusions about these potential benefits. The most consistent finding is higher satisfaction with care in single-bed rooms.

In terms of the quality of the patient experience – the focus of this chapter – results are also contradictory. Jongerden et al. (2013) showed advantages of single-occupancy rooms in a pre-post study dealing with an intensive care unit in a hospital in The Netherlands. Moving from a multiple-occupancy to a single-occupancy setting, benefits in the new unit included less noise, more daylight, and better facilities for families (more space; windows). Responses to the new unit showed significantly higher patient and family satisfaction, compared to the old unit. Research on a maternity unit of single-occupancy rooms (Janssen et al. 2000) also reflected higher satisfaction on a number of dependent variables, including continuity of nursing care and stability of room assignment (which is related to a feeling of control over what will happen). It should be noted that this was not a randomized controlled trial.

Although benefits of single rooms may be less noise and more space for family members (Chaudhury et al. 2005), isolated patients, compared with controls, expressed greater dissatisfaction with their treatment (Stelfox et al. 2003), and studies have found that some patients prefer multiple rooms (e.g., Rowlands and Noble 2008). Regarding personal control, single rooms offer more privacy for patients, families and caregivers, and improved control of physical environmental conditions (noise, lighting, etc.). However, more research is needed to understand whether single rooms are really the only or the best way to satisfy these needs, and if other needs not highlighted in the research may be affected.

Beyond the impact on the patient, there is research on the impact of single and multiple rooms on nurses. Chaudhury et al. (2006) used surveys to look at the reactions of nurses ($N = 77$) to single-occupancy vs. double-occupancy rooms in acute care settings in four hospitals in the northwestern US. Their research documents the positive perception of nurses toward single-occupancy rooms in terms of flexibility of use (which provides control); accommodation of family members (with implications for control); appropriateness for examination of patients (i.e., privacy provided), which also relates to control; reduced likelihood of medication and dietary errors; and lowered risk of acquiring an infection. Thus, the role of perceived control figures prominently in the single-occupancy inpatient room. Perceived control is a variable that deserves additional study and has implications for the way the environment is designed, with attention to possible cross-cultural differences (Devlin et al. 2014). In Devlin et al.'s research with patients on orthopedic units, patients in the US valued the potential for perceived control in the hospital room more than did patients in Portugal. The authors suggest this difference may be related to cultural differences in locus of control and individualism vs. collectivism, among other possibilities.

23.3.3 Need for Social Support

Social support refers to the provision by a social network of psychological and material resources intended to benefit an individual's ability to cope with stress (Cohen 2004). People facing a stressful event, such as going to hospital, often turn to those who care for them, seeking information, consolation, and reassurance (Taylor et al. 2007), and such social support is a valuable and effective means by which a person can reduce the negative mental and physical health effects of stress.

There is a long research history linking the quality and quantity of social ties not only to physical and mental health, but also to morbidity and mortality (see Holt-Lunstad et al. 2010; Uchino 2006). Social support promotes health by reducing the physical reactivity to stress. For example, college students who gave a speech in the presence of a supportive confederate had smaller cardiovascular reactivity in comparison with their counterparts who gave the speech alone (Lepore et al. 1993). Furthermore, increased sociability is associated with a decreased probability of developing a cold (Cohen et al. 2003), and low emotional support from the social network is associated with high anxiety of patients waiting for coronary artery bypass grafting (Koivula et al. 2002). In sum, social support appears to "lighten the load" that individuals physically incur when facing challenging situations (Schnall et al. 2008).

Social support may alleviate the impact of stress by providing a solution to the problem, reducing the perceived importance of the problem, or providing a distraction from the problem. Beyond actual support, perceptions that others (even one reliable source) will provide appropriate aid and resources may bolster one's perceived ability to cope with demands, thus changing the appraisal of the situation and lowering its effective stress (Cohen 2004). Evidence has been found that social support even shapes people's perception of the physical world. For example, a steep hill appeared less steep in both an in vivo study when a friend was physically present versus not present, and in a study that involved the mental recall of a supportive other versus a neutral or non-supportive other (Schnall et al. 2008). Thus, there is ample evidence that the presence of supportive others can enhance quality of life.

In such a potentially unfamiliar and stressful environment as the hospital setting, and with depleted physical and psychological resources, having the social support of others will ameliorate patient stress (Bolger and Amarel 2007). In his theory, Ulrich states that (1) in a medical setting, patients need social support to feel less stress, and that (2) the physical environment has a role in promoting opportunities for that to happen. A paradigmatic example of this last remark is the study by Sommer and Ross (1958) to test Osmond's theory (Osmond 1957) of the existence of "sociofugal" and "sociopetal" spatial settings, aimed at discouraging or encouraging, respectively, social interaction. These researchers studied the effects of furniture arrangement on social interaction in a geriatric ward. The furniture of a day-room was rearranged from shoulder-to-shoulder seating (against the wall) to chairs grouped around tables. As a result, communication among elderly woman increased more than 50 % with the table grouping.

González-Santos (2011) observed that the spatial factors of two hospital fertility clinics shaped both patient-physician and patient-patient relationships and social dynamics. A public clinic with a constrained physical setting (and a bureaucratic health system) led to a weakened patient-doctor relationship yet fostered a strong patient-patient relationship. The constricted spaces, the lack of consultation rooms, and the thin walls allowed patients to listen to each other's results and progress – which could be stressful, particularly if the feedback was not positive. Consequently, women sometimes held back from asking questions or giving doctors more information than that which was explicitly prompted; in doing so, they hindered the possibility of establishing a better patient–physician relationship. Interestingly, being alone in an area exclusively designated for patients promoted patient interaction and, as a result, spontaneous informal support groups were created. In contrast, in the private clinic, the patient's companion was welcome to accompany the patient all the time, the general waiting area was spacious, and there were independent private waiting areas, which provided both individual physical privacy as well as privacy of information. This study illustrates that – along with the characteristics of the system structure of the service – the physical elements of the clinical setting can delimit and constrain interaction, establish certain types and degrees of accessibility to the service, and enable or obstruct privacy and intimacy between patients, and between patients and their physicians. As a result of these configurations of physical elements, a particular type of social dynamic is established between patients, physicians, and the treatment.

For inpatients, receiving visits and calls are key elements in ensuring they have social support, and the environment clearly plays a role in these interactions. Beyond hospital visiting policies, the physical environment can offer more or fewer opportunities for engagement, for example, through providing a bedside phone, visual and auditory privacy, comfortable seating, or overnight accommodation. Single- and multiple-bed rooms could be considered another form of Osmond's "sociofugal" and "sociopetal" settings. Staying in a single-bed room or in a multiple-bed room will certainly create substantially different experiences in terms of social support. As previously described in this chapter, some research suggests that single rooms enable better control of infections, and more privacy and quality for patients (e.g., Chaudhury et al. 2005). However, shared rooms seem to meet patient needs for social support more effectively. Larsen et al. (2013) observed and interviewed patients in two-, three-, or four-bed hospital rooms. The environment contributed to the loss of privacy and personal control, noise, the enforced company of strangers, and the withholding of information from healthcare professionals. However, despite the challenges of managing privacy in shared hospital rooms, and managing the hospital environment in general, the company of fellow patients provided both essential support and information for the patients. In fact, a total of 18 of the 20 patients interviewed preferred to be hospitalized in a multiple-bed room unless they were too ill to interact. In the study of Rowlands and Noble (2008), patients also reported the importance of interacting with other patients for mutual support, and they preferred to be in the company of others when they were able to interact.

Potentially, multi-bed rooms could meet patient needs if the environment still made it easy to find refuge from fellow patients during hospitalization, both inside and outside these rooms (Larsen et al. 2013), although there may be cultural differences in preferences for such rooms (Andrade and Devlin 2015; Kim et al. 2008). Moreover, although the physical environment plays a role, it is the responsibility of healthcare workers to protect the patients and ensure that their privacy needs are met (Matiti and Trorey 2008). Single and multiple rooms have both advantages and disadvantages, and research pointing to a single configuration is still inconclusive. More research is needed, for example to explore whether the severity of the illness (being too ill to interact), and the quantity and quality of the patient's social network are important moderators of the impact of these two different kinds of room. A good solution may be to have both single- and multiple-bed rooms available and use them according to the situation.

Social support can be provided by other patients and by healthcare professionals, but the hospital environment should also create the best conditions for family and friends to keep in contact, to visit, and to stay with the patient. For example, the Mayo Clinic recommends creating these conditions as a way of fostering a sense of belonging, increasing a sense of self-worth, and increasing feelings of security (Mayo Clinic Staff n.d.). Beyond providing space and chairs/sofas, and overnight accommodation (e.g., a window seat wide enough for sleeping), the Internet and a bedside phone may also contribute to social support. When friends and family are not present, a loved one's picture on the bedside table can be an acceptable substitute: Master et al. (2009) showed that simply viewing a photograph of one's romantic partner could reduce the experience of physical pain.

23.4 Conclusion

The research discussed here shows how central the physical environment is to quality of life during a patient's hospitalization. Using Ulrich's (1991) Theory of Supportive Design, we suggest that positive distraction, perceived control, and social support affect important subjective and objective outcomes, such as reported stress and heart rate. The emergence of organizations such as the American Institute of Architect's Academy of Architecture for Health (AIA/AAH) points to a growing recognition that architecture has the capacity to affect health. Moreover, the Academy of Neuroscience for Architecture (ANFA) also shows the integration of architecture into what is becoming a dominant paradigm in psychology, neuroscience.

Within Ulrich's three-part model, positive distraction has received the greatest emphasis and has yielded the most practical benefits (e.g., the use of art and music for patients). This result may reflect issues of cost and ease of implementation because many of these applications (such as the installation of art) are relatively inexpensive.

The literature reviewed here shows the need to continue research of high methodological quality, particularly in terms of randomized controlled trials. The relative paucity of research about how the hospital environment can foster perceived control and different kinds of social support also points to the need to expand research on these topics. Furthermore, cultural differences in terms of these three psychological constructs (positive distraction, perceived control, and social support) deserve more attention.

References

Affleck, G., Tennen, H., Pfeiffer, C., & Fifeld, J. (1987). Appraisal of control and predictability in adapting to a chronic disease. *Journal of Personality and Social Psychology, 53*, 273–279. doi:10.1037/0022-3514.53.2.273.

Aghaie, B., Rejeh, N., Heravi-Karimooi, M., Ebadi, A., Moradian, S. T., Vaismoradi, M., & Jasper, M. (2014). Effect of nature-based sound therapy on agitation and anxiety in coronary artery bypass graft patients during the weaning of mechanical ventilation: A randomized clinical trial. *International Journal of Nursing Studies, 51*, 526–538. doi:10.1016/j.ijnurstu.2013.08.003.

American Institute of Architects Media Relations. (2006, July 19). *Private rooms are new standard in updated guidelines for design and construction of health care facilities.* http://www.speechprivacy.org/download/AIAPrivateHospitalRoomAnnouncement.pdf

Andrade, C. C., Lima, M. L., Pereira, C. R., Fornara, F., & Bonaiuto, M. (2013). Inpatients' and outpatients' satisfaction: The mediating role of perceived quality of physical and social environment. *Health & Place, 21*, 122–132. doi:10.1016/j.healthplace.2013.01.013.

Andrade, C. C., & Devlin, A. S. (2015). Stress reduction in the hospital room: Applying Ulrich's theory of supportive design. *Journal of Environmental Psychology, 41*, 125–134. doi:10.1016/j.jenvp.2014.12.001.

Appleton, J. (1975). *The experience of landscape.* London: Wiley.

Arneill, A., & Devlin, A. S. (2002). Perceived quality of care: The influence of the waiting room environment. *Journal of Environmental Psychology, 22*, 345–360. doi:10.1006/jevp.2002.0274.

Asano, F. (2008). Healing at a hospital garden: Integration of physical and non-physical aspects. *Acta Horticulturae, 775*, 13–22.

Berman, M. G., Jonides, J., & Kaplan, S. (2008). The cognitive benefits of interacting with nature. *Psychological Science, 19*, 1207–1212. doi:10.1111/j.1467-9280.2008.02225.x.

Beukeboom, C. J., Langeveld, D., & Tanja-Dijkstra, K. (2012). Stress-reducing effects of real and artificial nature in a hospital waiting room. *The Journal of Alternative and Complementary Medicine, 18*(4), 329–333. doi:10.1089/acm.2011.0488.

Blumberg, R., & Devlin, A. S. (2006). Design issues in hospitals: The adolescent client. *Environment and Behavior, 38*, 293–317. doi:10.1177/0013916505281575.

Bolger, N., & Amarel, D. (2007). Effects of social support visibility on adjustment to stress. Experimental evidence. *Journal of Personality and Social Psychology, 92*, 458–475. doi:10.1037/0022-3514.92.3.458.

Bringslimark, T., Hartig, T., & Patil, G. G. (2009). The psychological benefits of indoor plants: A critical review of the experimental literature. *Journal of Environmental Psychology, 29*, 422–433. doi:10.1016/j.jenvp.2009.05.001.

CARE Channel. Resources. http://healinghealth.com/hhs/sote/page/articles

Carpman, J. R., Grant, M. A., & Simmons, D. A. (1986). *Design that cares: Planning health facilities for patients and visitors*. Chicago: American Hospital Publishing.

Centers for Disease Control and Prevention (2007). *2007 Guidelines for isolation precautions: Preventing transmission of infectious agents in healthcare settings*. Washington: CDC. http://www.cdc.gov/hicpac/2007ip/2007ip_part4.html.

Cervinka, R., Röderer, K., & Hämmerle, I. (2014). Evaluation of hospital gardens and implications for design: Benefits from environmental psychology for architecture and landscape planning. *Journal of Architectural and Planning Research, 31*(1), 43–56.

Chaudhury, H., Mahmood, A., & Valente, M. (2005). Advantages and disadvantages of single versus multiple-occupancy rooms in acute care environments: A review and analysis of the literature. *Environment and Behavior, 37*, 760–786. doi:10.1177/0013916504272658.

Chaudhury, H., Mahmood, A., & Valente, M. (2006). Nurses' perception of single-occupancy versus multi-occupancy rooms in acute care environments: An exploratory comparative assessment. *Applied Nursing Research, 19*, 118–125. doi:10.1016/j.apnr.2005.06.002.

Cohen, S. (2004). Social relationships and health. *The American Psychologist, 59*(8), 676–684. doi:10.1037/0003-066X.59.8.676.

Cohen, S., Doyle, W. J., Turner, R., Alper, C. M., & Skoner, D. P. (2003). Sociability and susceptibility to the common cold. *Psychological Science, 14*(5), 389–395.

Cohen, S., Frank, E., Doyle, W. J., Skoner, D. P., Rabin, B. S., & Gwaltney, J. M. (1998). Types of stressors that increase susceptibility to the common cold in healthy adults. *Health Psychology, 17*, 214–223.

Cooke, M., Chaboyer, W., & Hiratos, M. A. (2005). Music and its effect on anxiety in short waiting periods: A critical appraisal. *Journal of Clinical Nursing, 14*(2), 145–155. doi:10.1111/j.1365-2702.2004.01033.x.

Cooper Marcus, C., & Barnes, M. (1995). *Gardens in health care facilities: Uses, therapeutic benefits, and design considerations*. Martinez: Center for Health Design.

Cooper Marcus, C., & Barnes, M. (Eds.). (1999). *Healing gardens: Therapeutic benefits and design recommendations*. New York: Wiley.

Cusack, P., Lankston, L., & Isles, C. (2010). Impact of visual art in patient waiting rooms. Survey of patients attending a transplant clinic in Dumfries. *Journal of the Royal Society of Medicine Short Reports, 1*(6), 52. doi:10.1258/shorts.2010.010077.

Dalke, H., Little, J., Niemann, E., Camgoz, N., Steadman, G., Hill, S., & Stott, L. (2006). Colour and lighting in hospital design. *Optics & Laser Technology, 38*, 343–365. doi:10.1016/j.optlastec.2005.06.040.

Davis, B. E. (2011). Rooftop hospital gardens for physical therapy: A post-occupancy evaluation. *Health Environments Research and Design Journal, 4*(3), 14–43.

deKort, Y. A. W., Meijnders, A. L., Sponselee, A. A. G., & Ijsselsteijn, W. A. (2006). What's wrong with virtual trees? Restoring from stress in a mediated environment. *Journal of Environmental Psychology, 26*(4), 309–320. doi:10.1016/j.jenvp.2006.09.001.

DeTommaso, M., Ricci, K., Laneve, L., Savino, N., Antonaci, V., & Livrea, P. (2013). Virtual visual effect of hospital waiting room on pain modulation in healthy subjects and patients with chronic migraine. *Pain Research and Treatment, 2013*, 8 pages. doi:10.1155/2013/515730.

Devlin, A. S. (2015). *Transforming the doctor's office: Principles from evidence-based design*. New York: Routledge.

Devlin, A. S., Andrade, C., & Lima, M. L. (2014, August). *Hospital rooms and patients' well-being: Exploring modeling variables*. Grant report to the Academy of Architecture for Health Foundation.

Devlin, A. S., & Arneill, A. (2003). Health care environments and patient outcomes: A review of the literature. *Environment & Behavior, 35*, 665–694. doi:10.1177/0013916503255102.

Diette, G. B., Lechtzin, N., Haponik, E., Devrotes, A., & Rubin, H. R. (2003). Distraction therapy with nature sights and sounds reduces pain during flexible bronchoscopy: A complementary approach to routine analgesia. *Chest, 123*(3), 941–948. doi:10.1378/chest.123.3.941.

Dileo, C., & Bradt, J. (2005). *Medical music therapy: A meta-analysis & agenda for future research*. Cherry Hill: Jeffrey Books.

Dinis, S., Duarte, E., Noriega, P., Teixeira, L., Vilar, E., & Rebelo, F. (2013). Evaluating emotional responses to the interior design of a hospital room: A study using virtual reality. *Lecture Notes in Computer Science, 8014 LNCS* (Part 3), 475–483.

Dougall, A. L., & Baum, A. (2001). Stress, health and illness. In A. Baum, T. A. Revenson, & J. E. Singer

(Eds.), *Handbook of health psychology* (pp. 321–327). Mahwah: Lawrence Erlbaum.

Ebrecht, M., Hextall, J., Kirtley, L.-G., Taylor, A., Dyson, M., & Weinman, J. (2004). Perceived stress and cortisol levels predict speed of wound healing in healthy male adults. *Psychoneuroendocrinology, 29*, 798–809. doi:10.1016/S0306-4530(03)00144-6.

Eisen, S. L., Ulrich, R. S., Shepley, M. M., Varni, J. W., & Sherman, S. (2008). The stress-reducing effects of art in pediatric health care: Art preferences of healthy children and hospitalized children. *Journal of Child Health Care, 12*(3), 173–190. doi:10.1177/1367493508092507.

Evidence-based design accreditation and certification (EDAC). Center for Health Design. http://www.healthdesign.org/edac/about.

Francis, C., & Cooper Marcus, C. (1992). Restorative places: Environment and emotional well-being. In E. G. Arias & M. D. Gross (Eds.), *Proceedings of the 23rd annual environmental design research association conference* (p. 274). Oklahoma City: EDRA.

González-Santos, S. P. (2011). Space, structure and social dynamics within the clinical setting: Two case studies of assisted reproduction in Mexico City. *Health & Place, 17*, 166–174. doi:10.1016/j.healthplace.2010.09.013.

Gotlieb, J. B. (2000). Understanding the effects of nurses, patients' hospital rooms, and patients' perceptions of control on the perceived quality of a hospital. *Health Marketing Quarterly, 12*(1–2), 1–14.

Guétin, S., Giniès, P., Siou, D. K. A., Picot, M.-C., Pommié, C., Guldner, E., ... Touchon, J. (2012). The effects of music intervention in the management of chronic pain: A single-blind, randomized, controlled trial. *Clinical Journal of Pain, 28*, 329–337. doi:10.1097/AJP.0b013e31822be973.

Hagerman, I., Rasmanis, G., Blomkvist, V., Ulrich, R. S., Eriksen, C. A., & Theorell, T. (2005). Influence of intensive coronary care acoustics on the quality of care and physiological state of patients. *International Journal of Cardiology, 98*(2), 267–270. doi:10.1016/j.ijcard.2003.11.006.

Haskell, D. M., & Brown, H. N. (1998). Calling all patients. *Nursing Management, 29*(9), 39–40.

Hathorn, K. (1993). The use of art in a health care setting. In P. L. Blyth (Ed.), *Health care interior finishes: Problems and solutions – an environmental services perspective* (pp. 1–11). Washington, DC: American Hospital Association.

Hathorn, K., & Nanda, U. (2008). *A guide to evidence-based art* (White paper for the Environmental Standards Council, Center for Health Design). Concord: Center for Health Design. http://www.healthdesign.org/sites/default/files/Hathorn_Nanda_Mar08.pdf.

Hernandez, R. O. (2007). Effects of therapeutic gardens in special care units for people with dementia: Two case studies. *Journal of Housing for the Elderly, 21*(1–2), 117–152. doi:10.1300/J081v21n01_07.

Hilliard, R. E. (2005). Music therapy in hospice and palliative care: A review of the empirical data. *eCAM, 2*(2), 173–178. doi:10.1093/ecam/neh076.

Holt-Lunstad, J., Smith, T. B., & Layton, J. B. (2010). Social relationships and mortality risk: A meta-analytic review. *PLoS Medicine, 7*(7), e1000316. doi:10.1371/journal.pmed.1000316.

Hui, M. K., & Bateson, J. E. G. (1991). Perceived control and the effects of crowding and consumer choice on the service experience. *Journal of Consumer Research, 18*, 174–185. doi:10.1086/209250.

Hweidi, I. M. (2007). Jordanian patients' perceptions of stressors on critical care units: A questionnaire survey. *International Journal of Nursing Studies, 44*(2), 227–235. doi:10.1016/j.ijnurstu.2005.11.025.

Janssen, P. A., Klein, M. C., Harris, S. J., Soolsma, J., & Seymour, L. C. (2000). Single room maternity care and client satisfaction. *Birth: Issues in Perinatal Care, 27*, 235–243.

Jongerden, I. P., Slooter, A. J., Peelen, L. M., Wessels, H., Ram, C. M., Kesecioglu, J., ... van Dijk, D. (2013). Effect of intensive care environment on family and patient satisfaction: A before-after study. *Intensive Care Medicine, 39*, 1626–1634. doi:10.1007/s00134-013-2966-0.

Kaplan, S. (1995). The restorative benefits of nature: Toward an integrative framework. *Journal of Environmental Psychology, 15*, 169–182. doi:10.1016/0272-4944(95)90001-2.

Kaplan, S., Bardwell, L. V., & Slakter, D. B. (1993). The museum as a restorative environment. *Environment and Behavior, 25*(6), 725–742. doi:10.1177/0013916593256004.

Kaplan, R., & Kaplan, S. (1989). *The experience of nature: A psychological perspective*. Cambridge: Cambridge University Press.

Kim, H. S., Sherman, D. K., & Taylor, S. E. (2008). Culture and social support. *The American Psychologist, 63*, 518–526. doi:10.1037/0003-066X.

Kim, Y. J. (2014, January). *The role of perception of control in predicting depression among individuals with mental illness after psychiatric hospital discharge*. Presentation at the 2014 annual conference of the society for social work and research, San Antonio, TX.

Koivula, M., Paunonen-Ilmonen, M., Tarkka, M.-T., Tarkka, M., & Laippala, P. (2002). Social support and its relation to fear and anxiety in patients awaiting coronary artery bypass grafting. *Journal of Clinical Nursing, 11*, 622–633.

Larsen, L. S., Larsen, B. H., & Birkelund, R. (2013). A companionship between strangers – the hospital environment as a challenge in patient-patient interaction in oncology wards. *Journal of Advanced Nursing, 70*(2), 395–404. doi:10.1111/jan.12204.

Lee, K.-C., Chao, Y.-H., Yiin, J.-J., Chiang, P.-Y., & Chao, Y.-F. (2011). Effectiveness of different music-playing devices for reducing preoperative anxiety: A clinical control study. *International Journal of Nursing Stud-*

ies, *48*(10), 1180–1187. doi:10.1016/j.ijnurstu.2011.04.001.

Lee, K.-C., Chao, Y.-H., Yiin, J.-J., Hsieh, H.-Y., Dai, W.-J., & Chao, Y.-F. (2012). Evidence that music listening reduces preoperative patients' anxiety. *Biological Research for Nursing, 14*(1), 78–84. doi:10.1177/1099800410396704.

Lee, S., & Brand, J. (2005). Effects of control over workspace on perceptions of the work environment and work outcomes. *Journal of Environmental Psychology, 25*, 323–333. doi:10.1016/j.jenvp.2005.08.001.

Lee, S. Y., & Brand, J. L. (2010). Can personal control over the physical environment ease distractions in office workplaces? *Ergonomics, 53*, 324–335. doi:10.1080/00140130903389019.

Lepore, S. J., Allen, K. A., & Evans, G. W. (1993). Social support lowers cardiovascular reactivity to an acute stressor. *Psychosomatic Medicine, 55*(6), 518–524.

Lorenz, S. G., & Dreher, H. M. (2011). Hospital room design and health outcomes of the aging adult. *Health Environments Research & Design Journal, 4*(2), 23–35.

Mackrill, J., Cain, R., & Jennings, P. (2013). Experiencing the hospital ward soundscape: Towards a model. *Journal of Environmental Psychology, 36*, 1–8. doi:10.1016/j.jenvp.2013.06.004.

Mahler, H. I., & Kulik, J. A. (1990). Preferences for health care involvement, perceived control and surgical recovery: A prospective study. *Social Science and Medicine, 31*, 743–751.

Malenbaum, S., Keefe, F. J., Williams, A. C. D. C., Ulrich, R. S., & Somers, T. J. (2008). Pain in its environmental context: Implications for designing environments to enhance pain control. *Pain, 134*, 241–244. doi:10.1016/j.pain.2007.12.002.

Malkin, J. (2011, May). *Evidence-based design that reflects patient-centered care: A journey from Father Knows Best to Have It Your Way. Part of the intensive: Implementing person-centered design in healthcare: Building connections*. Presentation at the 42nd annual environmental design research association conference, Chicago, Illinois.

Malkin, J. (2014). *Medical and dental space planning: A comprehensive guide to design, equipment, and clinical procedures* (4th ed.). New York: Wiley.

Master, S. L., Eisenberger, N. I., Taylor, S. E., Naliboff, B. D., Shirinyan, D., & Lieberman, M. D. (2009). A picture' s worth: Partner photographs reduce experimentally induced pain. *Psychological Science, 20*, 1316–1318. doi:10.1111/j.1467-9280.2009.02444.x.

Matiti, M. R., & Trorey, G. M. (2008). Patients' expectations of the maintenance of their dignity. *Journal of Clinical Nursing, 17*, 2709–2717. doi:10.1111/j.1365-2702.2008.02365.x.

Mazer, S., & Smith, D. (1999). *Sound choices: Using music to design the environments in which you live, work, and heal*. Carlsbad: Hay House.

Miller, A. C., Hickman, L. C., & Lemasters, G. K. (1992). A distraction technique for control of burn pain. *Journal of Burn Care Rehabilitation, 13*(5), 576–580.

Mitchell, L. A., & MacDonald, R. A. R. (2006). An experimental investigation of the effects of preferred and relaxing music listening on pain perception. *Journal of Music Therapy, 43*, 295–316.

Naderi, J. R., & Shin, W. H. (2008). Humane design for hospital landscapes: A case study in landscape architecture of a healing garden for nurses. *Health Environments Research & Design Journal, 2*(1), 82–119.

Nanda, U., Chanaud, C., Nelson, M., Zhu, X., Bajema, R., & Jansen, B. H. (2012a). Impact of visual art on patient behavior in the emergency department waiting room. *Journal of Emergency Medicine, 43*, 172–181. doi:10.1016/j.jemermed.2011.06.138.

Nanda, U., Eisen, S. L., Zadeh, R. S., & Owen, D. (2011). Effect of visual art on patient anxiety and agitation in a mental health facility and implications for the business case. *Journal of Psychiatry and Mental Health Nursing, 18*, 386–393. doi:10.1111/j.1365-2850.2010.01682.x.

Nanda, U., Gaydos, H. L. B., Hathorn, K., & Watkins, N. (2010). Art and posttraumatic stress: A review of the literature on the therapeutic implications of artwork for war veterans with posttraumatic stress disorder. *Environment and Behavior, 42*(3), 376–390. doi:10.1177/0013916510361874.

Nanda, U., Pati, D., Ghamari, H., & Bajema, R. (2013). Lessons from neuroscience: Form follows function, emotions follow form. *Intelligent Buildings International, 5*(Supplement 1), 61–78. doi:10.1080/17508975.2013.807767.

Nanda, U., Zhu, X., & Jaansen, B. H. (2012b). Image and emotion: From outcomes to brain behavior. *Health Environments Research & Design Journal, 5*(4), 40–59.

Nightingale, C., Rodriguez, C., & Carnaby, G. (2013). The impact of music interventions on anxiety for adult cancer patients: A meta-analysis and systematic review. *Integrative Cancer Therapies, 12*, 393–403. doi:10.1177/1534735413485817.

Olsen, J. C., & Sabin, B. R. (2003). Emergency department patient perceptions of privacy and confidentiality. *Journal of Emergency Medicine, 25*, 329–333. doi:10.1016/S0736-4679(03)00216-6.

Osmond, H. (1957). Function as the basis of psychiatric ward design. *Mental Hospital, 8*, 23–30.

Park, S.-H., & Mattson, R. H. (2008). Effects of flowering and foliage plants in hospital rooms on patients recovering from abdominal surgery. *HortTechnology, 18*, 563–568.

Park, S.-H., & Mattson, R. H. (2009). Therapeutic influences of plants in hospital rooms on surgical recovery. *HortScience, 44*, 1–4.

Park, S.-H., Mattson, R. H., & Kim, E. (2004). Pain tolerance effects of ornamental plants in a simulated

hospital patient room. *Acta Horticulturae, 639*, 241–247.
Pasha, S. (2013). Barriers to garden visitation in children's hospitals. *Health Environments Research and Design Journal, 6*(4), 76–96.
Pati, D., Harvey, T. E., Jr., & Barach, P. (2008). Relationships between exterior views and nurse stress: An exploratory examination. *Health Environments Research & Design Journal, 1*(2), 27–38.
Pati, D., O'Boyle, M., Armor, C., Hou, J., Valipoor, S., & Fang, D. (2014). Neural correlates of nature stimuli: An fMRI study. *Health Environments Research and Design Journal, 7*(2), 9–28.
Powell, R., & Johnston, M. (2007). Hospitalisation in adults. In S. Ayers, A. Baum, C. McManus, S. Newman, K. Wallston, J. Weinman, & R. West (Eds.), *Cambridge handbook of psychology, health & medicine* (2nd ed., pp. 109–113). Cambridge: Cambridge University Press.
Pruyn, A., & Smidts, A. (1998). Effects of waiting on the satisfaction with service: Beyond objective time measures. *International Journal of Research in Marketing, 15*, 321–334. doi:10.1016/S0167-8116(98)00008-1.
Raanaas, R. K., Patil, G. G., & Hartig, T. (2010). Effect of an indoor foliage plant intervention on patient well-being during a residential rehabilitation program. *HortScience, 45*(3), 387–392.
Raanaas, R. K., Patil, G. G., & Hartig, T. (2012). Health benefits of a view of nature through the window: A quasi-experimental study of patients in a residential rehabilitation center. *Clinical Rehabilitation, 26*(1), 21–32. doi:10.1177/0269215511412800.
Rabner, B. S., Lorenz, S., Peavey, E., & Somarriba, J. (2013, October). *A model for success: The University Medical Center of Princeton at Plainsboro*. Presentation at the Edra Inaugural Fall Symposium: The Landscape of Accountable Care: How a Patient Focus is Changing the Industry, New York.
Renz, M., Schütt Mao, M., & Cerny, T. (2005). Spirituality, psychotherapy and music in palliative cancer care: Research projects in psycho-oncology at an oncology center in Switzerland. *Supportive Care International, 12*, 961–966. doi:10.1007/s00520-005-0873-9.
Rollins, J. A. (2011). Arousing curiosity: When hospital art transcends. *Health Environments Research & Design Journal, 4*(3), 72–94.
Rowlands, J., & Noble, S. (2008). How does the environment impact on the quality of life of advanced cancer patients? A qualitative study with implications for ward design. *Palliative Medicine, 22*, 768–774. doi:10.1177/0269216308093839.
Sackett, D. L., Rosenberg, W. M. C., Gray, J. A. M., Haynes, R. B., & Richardson, W. S. (1996). Evidence based medicine: What it is and what it isn't. *BMJ, 312*, 71–72. doi:10.1136/bmj.312.7023.71.
Salonen, H., Lahtinen, M., Lappalainen, S., Nevala, N., Knibbs, L. D., Morawska, L., & Reijula, K. (2013). Design approaches for promoting beneficial indoor environments in healthcare facilities: A review. *Intelligent Buildings International, 5*(1), 26–50. doi:10.1080/17508975.2013.764839.
Schat, A. C., & Kelloway, E. K. (2000). Effects of perceived control on the outcomes of workplace aggression and violence. *Journal of Occupational Health Psychology, 5*, 386–402.
Schnall, S., Harber, K. D., Stefanucci, J. K., & Proffitt, D. R. (2008). Social Support and the perception of geographical slant. *Journal of Experimental Social Psychology, 44*, 1246–1255. doi:10.1016/j.jesp.2008.04.011.
Shepley, M. M., Gerbi, R. P., Watson, A. E., Imgrund, S., & Sagha-Zadeh, R. (2012). The impact of daylight and views on ICU patients and staff. *Health Environments Research & Design Journal, 5*(2), 46–60.
Sherman, S. A., Varni, J. W., Ulrich, R. S., & Malcarne, V. L. (2005). Post-occupancy evaluation of healing gardens in a pediatric cancer center. *Landscape and Urban Planning, 73*(2–3), 167–183. doi:10.1016/j.landurbplan.2004.11.013.
Söderback, I., Söderström, M., & Schälander, E. (2004). Horticultural therapy: The 'healing garden' and gardening in rehabilitation measures at Danderyd Hospital Rehabilitation Clinic, Sweden. *Pediatric Rehabilitation, 7*, 245–260. doi:10.1080/13638490410001711416.
Sommer, R., & Ross, H. (1958). Social interaction on a geriatrics ward. *International Journal of Social Psychiatry, 4*(2), 128–133. doi:10.1177/002076405800400207.
Stelfox, H. T., Bates, D. W., & Redelmeier, D. A. (2003). Safety of patients isolated for infection control. *Journal of the American Medical Association, 290*(14), 1899–1905. doi:10.1001/jama.290.14.1899.
Steptoe, A., & Ayers, S. (2005). Stress, health and illness. In S. R. Sutton, A. S. Baum, & M. Johnston (Eds.), *The SAGE handbook of health psychology* (pp. 169–196). London: Sage.
Stichler, J. F. (2008). Healing by design. *Journal of Nursing Administration, 38*, 505–509. doi:10.1097/NNA.0b013e31818ebfa6.
Stichler, J. F. (2011). Patient-centered healthcare design. *Journal of Nursing Administration, 41*, 503–506. doi:10.1097/NNA.0b013e31823278a3b.
Tanja-Dijkstra, K. (2011). The impact of bedside technology on patients' well-being. *Health Environments Research & Design Journal, 5*(1), 43–51.
Tanja-Dijkstra, K., Pahl, S. P., White, M. P., Andrade, J., Qian, C., Bruce, M., ... Moles, D. R. (2014). Improving dental experiences by using virtual reality distraction: A simulation study. *PLoS ONE, 9*(3), e91276. doi:10.1371/journal.pone.0091276.
Tanja-Dijkstra, K., Pieterse, M. E., & Pruyn, A. (2008). Stress-reducing effects of indoor plants in the built healthcare environment: The mediating role of perceived attractiveness. *Preventive Medicine, 47*, 279–283. doi:10.1016/j.ypmed.2008,01.013.

Taylor, S. E., Welch, W. T., Kim, H. S., & Sherman, D. K. (2007). Cultural differences in the impact of social support on psychological and biological stress responses. *Psychological Science, 18*(9), 831–837. doi:10.1111/j.1467-9280.2007.01987.x.

Thompson, S. C., Sobolew-Shubin, A., Galbraith, M. E., Schwankovsky, L., & Cruzen, D. (1993). Maintaining perceptions of control: Finding perceived control in low-control circumstances. *Journal of Personality and Social Psychology, 64*, 293–304. doi:10.1037/0022-3514.64.2.293.

Thorgaard, B., Henriksen, B. B., Pedersbæck, G., & Thomsen, I. (2004). Specially selected music in the cardiac laboratory. An important tool for improvement of wellbeing of patients. *European Journal of Cardiovascular Nursing, 3*(1), 21–26. doi:10.1016/j.ejcnurse.2003.10.001.

Trochelman, K., Albert, N., Spence, J., Murray, T., & Slifcak, E. (2012). Patients and families weigh in on evidence-based hospital design. *Critical Care Nurse, 32*(1), e1–e10.

Uchino, B. N. (2006). Social support and health: A review of physiological processes potentially underlying links to disease outcomes. *Journal of Behavioral Medicine, 29*, 377–387. doi:10.1007/s10865-006-9056-5.

Ulrich, R. S. (1984). View through a window may influence recovery from surgery. *Science, 224*, 420–421.

Ulrich, R. S. (1991). Effects of interior design on wellness: Theory and recent scientific research. *Journal of Healthcare Interior Design, 3*, 97–109.

Ulrich, R. S. (2009). Effects of viewing art on health outcomes. In S. B. Frampton & P. A. Charmel (Eds.), *Putting patients first: Best practices in patient-centered care* (2nd ed., pp. 129–149). San Francisco: Jossey-Bass.

Ulrich, R. S., Berry, L. L., Quan, X., & Parish, J. T. (2010). A conceptual framework for the domain of evidence-based design. *Health Environments Research & Design Journal, 4*(1), 95–114.

Ulrich, R. S., & Gilpin, L. (2003). Healing arts: Nutrition for the soul. In S. B. Frampton, L. Gilpin,& P. A. Charmel (Eds.), *Putting patients first: Designing and practicing patient-centered care* (pp. 117–146). San Francisco: Jossey-Bass.

Ulrich, R. S., Lunden, O., & Etinge, J. L. (1993, October). *Effects of exposure to nature and abstract pictures on patients' recovery from heart surgery*. Society for psychophysiological research, 33rd annual meeting, Rottach-Egern, Germany, S1–S7.

Ulrich, R. S., Simons, R. F., & Miles, M. A. (2003). Effects of environmental simulations and television on blood donor stress. *Journal of Architectural and Planning Research, 20*(1), 38–47.

Ulrich, R. S., Zimring, C. M., Zhu, X., DuBose, J., Seo, H.-B., Choi, Y.-S., ... Joseph, A. (2008). A review of the research literature on evidence-based healthcare design. *Health Environments Research and Design Journal, 1*, 61–125.

Van de Glind, I., de Roode, S., & Goossensen, A. (2007). Do patients in hospitals benefit from single rooms? A literature review. *Health Policy, 84*(2), 153–161. doi:10.1016/j.healthpol.2007.06.002.

Verderber, S., & Todd, L. G. (2012). Reconsidering the semiprivate inpatient room in U.S. hospitals. *Health Environments Research & Design Journal, 5*(2), 7–23.

Vincent, E., Battisto, D., & Grimes, L. (2010a). The effects of presence and influence in nature images in a simulated hospital patient room. *Health Environments Research & Design Journal, 3*(3), 56–69.

Vincent, E., Battisto, D., Grimes, L., & McCubbin, J. (2010b). The effects of nature images on pain in a simulated hospital patient room. *Health Environments Research & Design Journal, 3*(3), 42–55.

Whitehouse, S., Varni, J. W., Seid, M., Cooper-Marcus, C., Ensberg, M. J., Jacobs, J. R., & Mehlenbeck, R. S. (2001). Evaluating a children's hospital garden environment: Utilization and consumer satisfaction. *Journal of Environmental Psychology, 21*, 301–314. doi:10.1006/jevp.2001.0224.

Williamson, J. S. (1992). The effects of ocean sounds on sleep after coronary artery bypass graft surgery. *American Journal of Critical Care, 1*(1), 91–97.

Healthy Residential Environments for the Elderly

24

Ferdinando Fornara and Sara Manca

24.1 The Person-Environment Fit in the Elderly Population

This chapter focuses on the role of the spatial-physical dimensions of the residential environment in influencing the psychological well-being and the overall quality of life of the elderly population.

The first part of the chapter outlines theoretical approaches and constructs that have been applied (or could be useful) for exploring the relationship between the older adult and her/his residential place. In particular, the Person-Environment fit approach (Carp and Carp 1984) is used to organize this section.

The second part of the chapter reports the findings of recent research on the topic, and is structured along the dimension moving from the (private) home environment to the (community) residential care environment.

F. Fornara (✉)
Dipartimento di Pedagogia, Psicologia, Filosofia, Università degli Studi di Cagliari, CIRPA – Centro Interuniversitario di Ricerca in Psicologia Ambientale, Cagliari, Italy
e-mail: ffornara@unica.it

S. Manca
Dipartimento di Psicologia dei Processi di Sviluppo e Socializzazione, Sapienza Università di Roma, Rome, Italy
e-mail: sara.manca@uniroma1.it

The rapid aging of the global population is reflected in the current number of 868 million people over 60, nearly 12 % of the population worldwide, and this figure is expected to more than double by 2050 (Global AgeWatch Index 2014). Average life expectancy has dramatically increased the risk of health problems in the elderly, with a tendency to impaired hearing, vision, mobility, and mental function (The World Health Report 2013).

According to Birren (2006), the growth of elderly populations in both developed and developing countries has promoted an increasing interest in research findings about aging issues, so that understanding aging-related processes has become "one of the major challenges facing science in the 21st century" (Birren 2006 p. 15).

The aging process is a multidimensional pattern of change that includes biological, behavioral, social, and environmental aspects. The progressive and rapid aging of the world population (particularly in Western countries) makes it crucial to identify those factors that may reduce the impact of the negative consequences of aging (e.g., a possible decline in physical and cognitive functions) and improve the elderly's resilience and coping strategies in order to promote "successful aging" (Rowe and Kahn 1997) or "optimal aging" (Baltes 1996).

According to Rowe and Kahn (1997), "successful aging" is a multidimensional outcome due to factors such as absence of disease, good physical function, intact cognition, and active engagement with life. As stated by Aldwin and colleagues (Aldwin et al. 2006), this approach highlights the prominent role of physical versus psychological health in defining healthy aging, as reflected by the focus on disease, which in turn recalls the "bio-medical" approach to health issues (Stroebe 1995). Similarly, there has been an emphasis on loss reduction rather than on gain achievement. Nevertheless, the primacy of physical health in influencing elderly well-being and quality of life has been empirically contradicted. For example, Strawbridge et al. (2002) found relevant discrepancies between investigator-rated and self-rated successful aging; in fact, a third of respondents who had chronic physical illnesses rated themselves as aging successfully, whereas a third of those who were aging successfully by "objective" physical criteria did not consider themselves successful. Similarly, Snowdon (2001) found that successful aging is characterized by positive psychological features (such as happiness, intellectual curiosity, gratitude, spirituality, and a sense of community) somewhat independently of individuals' physical conditions. Moreover, large proportions of disabled older adults typically express high levels of quality of life, whereas a significant number of disability-free adults report low levels of quality of life (George 2006).

According to Lawton (1999), the identification of "successful" aging is too related to the comparison of the older adult situation with typical functioning in midlife. Following a life span developmental perspective, Aldwin and colleagues (2006) argued that what is missing from the definitions of "successful" aging is that losses are often balanced by those gains that (if achieved) are a distinctive feature of the final part of our lives. In this regard, Baltes (1996) proposed the term "optimal" aging, implying that individuals may choose to optimize different facets of their lives, depending on their current goal structures (Rothermund and Brandtstädter 2003). Thus, this approach recognizes more the importance of "subjective" psychological responses in determining the older adults' well-being and quality of life.

This discrepancy between "objective" and "subjective" assessments of well-being recalls a similar incongruence that may appear between "expert" (or objective) and "lay" (or subjective) evaluation of environmental quality (Gifford 2002). This dichotomy has been studied and detected mostly for design issues, where the contrast between formal satisfaction with technical requirements and users' satisfaction is rooted in the differences of opinion between designers and users, i.e. between experts, whose judgment is mostly influenced by their professional background, and lay people, whose representations and images are socially and culturally built up through the processes of knowledge and goal-directed action, which build on the systems of practices that characterize the specific place (Bonnes and Secchiaroli 1995; Fornara and Andrade 2012).

As stated by many scholars (e.g., Lawrence 2002; Bonaiuto and Alves 2012), the residential environment has a strict connection with quality of life, since the latter may be considered an overarching construct within which the former plays a role. This link is particularly prominent in old age for at least 2, partly related, reasons: (i) elderly people usually spend most of the day in their residential environment (their home and their neighborhood) (e.g., Bonaiuto et al. 2004); and (ii) the residential environment has a special meaning for older adults, given that it may help in providing a sense of continuity with the past (Korpela 2012), maintaining a positive self-image (Rubinstein and Parmalee 1992), and promoting identity, independence, and well-being (Eyles and Williams 2008).

In addressing the issue of care facilities for the elderly population, most gerontology literature refers to a continuum of care, identifying institutional settings for long-term care as the most restrictive and one's own home as the least restrictive. However, one's own home can be as restrictive as an institutional setting, if the older adult is confined at home and is not using (or does not have access to) those services that would increase her/his independence. On the other hand, a "home-like" environment may also be provided in an institutional setting (Stone 2006).

One important conceptual lens that can be useful for understanding how, and to what extent, the residential environment plays a role in the elderly's well-being and quality of life is the notion of Person-Environment (P-E) fit (Lawton and Nahemow 1973; Kahana 1982) and related models, e.g. the Complementary-Congruence Model of Person-Environment fit (Carp and Carp 1984; Carp 1987). According to this model, the fit between individual features (i.e., physical and psychological needs, lifestyle, and other personal characteristics) and characteristics of the environment is a predictor of several outcomes, such as physical and psychological well-being, satisfaction, autonomy, and so on. Outcomes may be moderated or mediated by "subjective" psychological patterns (such as a sense of personal competence, coping style, health attitudes) and more "objective" dimensions (such as resources, social support, and life events).

Similar theoretical constructs, which partially overlap with the P-E fit, are person-environment compatibility, in which the greater the compatibility, the better the well-being (Kaplan 1983); environmental support, i.e. the extent to which the environment facilitates or inhibits one's own goals (Bonaiuto and Alves 2012); environmental accessibility, i.e. the fit between the individual's functional capacity to perform activities and environmental demands (Iwarsson and Stahl 2003); and "affordances" (Gibson 1979), i.e. those properties of the environment that suggest its appropriate use to (the perceptual system of) the individual.

More generally, the P-E fit models and constructs are particularly pertinent within an environmental psychology perspective, since they reflect a transactional approach, in which Person and Environment are not conceived as independent units, but rather as interdependent aspects of the same unit (Bonnes and Secchiaroli 1995). This perspective, defined as "interactional-transactional" (Holahan 1978) or "transactional-contextual" (Altman and Rogoff 1987), has been one of the theoretical guidelines in the environmental psychology domain.

The application of the P-E fit approach to aging research stresses that behavioral and health outcomes vary as a function of personal competence and environmental press (Scheidt and Windley 2006). This has been demonstrated as particularly pertinent in studies on long-term care settings, since a design choice should necessarily take into account the complex interplay between physical setting, organization, staff, and the users' needs (Geboy and Diaz Moore 2005).

In line with all these considerations, we have chosen the P-E fit approach – which has also been specifically used to study residential satisfaction in the elderly population (Kahana et al. 2003) – to structure the first part of this chapter. One section focuses on the Person pole of the P-E unit, while another is dedicated to the Environment pole.

24.2 Person-Focused Environment-Related Features

In the scientific literature, an array of psychosocial factors has been studied as possible predictors of healthy aging. They include perceived control, everyday competence, autonomy, coping strategies, proactivity (vs. reactivity), assimilation (vs. accommodation), and social support. Psychosocial characteristics are hypothesized to have direct effects and to mediate partially the effects of objective life conditions on elderly well-being (George 2006).

Perceived control refers to the extent to which individuals believe that they can control their lives (George 2006) and represents one of the most studied psychological patterns in relation to elderly well-being (Aldwin et al. 2006). There are other well-known constructs partially overlapping with perceived control, such as mastery (Pearlin and Schooler 1978), self-efficacy, locus of control, and outcome expectancy (Rodin 1990). There is strong evidence that perceived control (or related patterns) is a significant mediator of the effects of objective life conditions on elderly well-being (George 2006) as well as a buffer of the detrimental effects of life stressors on their physical and mental health (Gadalla 2009). For example, Bisconti and Bergeman (1999) found that perceived control mediates the effects of social support on life satisfaction. Similarly, Windle and Woods (2004) found that mastery mediates the effects of functional status and social support on well-being.

The concept of "everyday competence" (Lawton 1982; Willis 1996) refers to one's ability to perform a broad array of activities that are considered essential for independent living. In general, lower everyday competence is associated with both lower self-esteem and lower life satisfaction (Kuriansky et al. 1976), greater use of home healthcare services (Wolinsky et al. 1983), greater risk of hospitalization and institutionalization (Branch and Jette 1982), and higher mortality (Keller and Potter 1994). Lawton (1982) suggested that higher competence is associated with greater independence from the behavioral effects of environmental press. This has been termed the "docility hypothesis", inferring that the lower the individual's competence, the lower their ability to adapt to environmental demands. The study of competence in the elderly population has often been related to the individual's ability to perform basic Activities of Daily Living (ADL – such as eating and drinking, toileting, bathing, and dressing) and Instrumental Activities of Daily Living (such as food preparation, house cleaning, medication use, and money management), which have been widely studied in gerontological research and are related to the ability to live independently and in good health (Fried et al. 2001). In this regard, maintaining independence is a key factor that should be encouraged in old age (Baltes and Carstensen 1996); in fact, there is strong evidence that quality of life is reduced by limitations in mobility and difficulties with everyday activities (Netuveli et al. 2006). Exposure to open and green spaces has emerged as a key environmental factor that can improve the older adult's independence (Lord et al. 2011), since it was found to be associated with an increased social network, increased physical activity, increased capacity to deal with major life events, and recovery of attentional resources, which are depleted in conditions of stress.

Lawton (1998) reported that older adults with lower competence are likely to experience a wider range of environmental demands as aversive, whereas higher levels of competence favor the consideration of resources and opportunities provided by the environment, thus allowing a greater sense of control over it.

Concerning the coping strategies used by the elderly for maintaining or improving their quality of life, Lawton (1989) indicated two classes of strategy to eliminate the perceived discrepancies between the actual and the desired course of development. The first involves environmental "proactivity", i.e. the tendency to adjust life circumstances to personal preferences; conversely, the second refers to environmental "reactivity", i.e. the tendency to adjust personal preferences to situational constraints. Brandtstädter and Renner (1990) distinguished two general coping strategies to maintain life satisfaction; assimilation, involving active modification of the environment in order to reach personal goals, and accommodation, involving a more passive acceptance of life circumstances and obstacles. Following this distinction, adaptive strategies can be placed along a continuum from the most assimilative to the most accommodative. Some studies (Wister 1989; Brandtstädter and Renner 1990; Brandtstädter and Rothermund 2002) have shown that older people tend to shift from assimilative to accommodative strategies as they become older. In any case, the use of both strategies is positively related to life satisfaction.

Slangen-de Kort and colleagues (1998) focused on categorizing the object or activity that was adapted. Referring to daily activities related to one's own home, these authors made a distinction between adaptation of: (a) the physical environment (modification of the home, use of assistive devices), (b) the social environment (divided into formal help, such as paid housekeeping, and informal help, such as help from friends), and (c) the person him/herself (changes in behavior, the "give-up" reaction). Strategies of adaptation of the physical environment are considered the most assimilative and proactive, whereas strategies of personal adaptation (particularly the "give-up" reaction) are categorized as the most accommodative and reactive. Adaptations of the social environment can be seen as most accommodative, since they imply giving up a goal that is relevant to most people, i.e. independence or autonomy. Requiring formal help, such as public, volunteer or paid assistance, represents a more active and goal-directed behavior aimed at modifying the situation, whereas relying on friends and relatives mirrors a more dependent and accommodative choice. Focusing on resources provided by the physical environment, some researchers (Slangen-de Kort et al. 1998; Scopelliti et al. 2005) investigated how the home environment can afford assimilative and proactive coping strategies in the elderly. Results showed that people who judged their home as more "adaptable" were more likely to choose an assimilative than an accommodative strategy.

A further psychosocial dimension that has been considered for predicting healthy aging is social support, which may include both an objective component (e.g., number of significant others) and a subjective component (in terms of perceived social support). In fact, a possible need for assistance may be satisfied by an informal network of friends, family, neighbors, and members of the local community. Overall, social relationships and social support provided by family and friends have emerged as strong predictors of perceived well-being in old age (George 2006). In a meta-analysis, Pinquart and Sörensen (2000) found that both the number of significant others and the perceived quality of relationships with significant others were predictors of perceived well-being. It is interesting to note that the latter (the subjective side) weighed more than the former (the objective side). In this meta-analysis, it was also found that relationships with friends and with children had positive, independent effects on life satisfaction.

To sum up, psychosocial characteristics have been shown to play either a direct or a mediating role in influencing elderly well-being and quality of life. As stated by George (2006), the appropriate interpretation of research findings is that objective life conditions are significant predictors of psychosocial characteristics, which, in turn, affect health outcomes. In this sense, the influence of psychosocial patterns on older people's quality of life seems always direct. Nevertheless, in line with the P-E fit perspective, a healthy or unhealthy outcome is assumed to be the result of how, and to what extent, the Environment side supports or impedes the Person goals. This is the focus of the next section.

24.3 Environment-Focused Person-Related Features

Scheidt and Windley (2006) suggested implementing the Lewinian "action research" (Lewin 1946) in environmental gerontology in order to "improve the quality of life of older people through collaborative knowledge building that informs design and environmental modification of the social and physical living arrangements" (Scheidt and Windley 2006 p. 114). According to these authors, action research would be beneficial due to its focus on "place" rather than the traditional emphasis put on the individual as the unit of analysis (Windley and Weisman 2004). Such a concept of "place", which includes the psychological, social, and architectural attributes of a setting, recalls the Place Theory (Canter 1977), one of the main theoretical underpinnings of environmental psychology, in which place is conceived as an interface of physical attributes, setting-related behaviors, and people's beliefs about such behaviors. Thus, the theoretical lens provided by environmental psychology theories and conceptualizations, in underlining the transactional view of the Person-Environment unit that is congruent with the Lewinian thought (Bonnes and Secchiaroli 1995), seems particularly appropriate for moving attention from the characteristics of the aged individual (the Person) to the environmental (spatial-physical) properties, as they are configured in the interplay with the individual her/himself.

A focus on those environmental properties that can support the elderly population is evident in some programs of urban regeneration. For instance, in US urban contexts (e.g., Boston, see Pol 2009), the attention paid to elderly residents is witnessed by the provision of more compact and multifunctional residential neighborhoods, which should promote access to services without using a private car.

24.3.1 The Home Environment

A specific focus on the spatial-physical features of the elderly home environment was included in the Enable-Age Project (Iwarsson et al. 2005), which was a cross-cultural and multidisciplinary research project funded by the EU, with the participation of research partners from Sweden, Germany, the United Kingdom, Hungary, and Latvia. The project aimed to verify the impact of both "objective" and "subjective" aspects of the home setting on the prediction of healthy aging, defined by behavioral (performing activities of daily living), cognitive-evaluative (life satisfaction), affective (signs of depression), and social participation facets (Iwarsson 2004). The transactional view reflected by this research project is witnessed by its core concept, i.e. environmental accessibility (Iwarsson and Stahl 2003), which is fully embedded in the P-E fit perspective (see the introduction of this chapter).

In order to assess the environmental accessibility of the home setting, a psychometric tool, labeled "Housing Enabler" (Iwarsson et al. 2005), was developed and validated. It includes interview and observational measures that tap the accessibility construct, i.e. the interface between an individual's functional capacity to carry out activities of daily living and the spatial-physical barriers of the home environment. Overall, the data collected in different countries showed that the level of accessibility decreases as age increases, and is positively associated with perceived health.

The focus on the home environment reflects the general preference of the elderly to remain in the familiar context of their own home (Callahan 1992; Challis and Davies 1985; Scanlon and Devine 2001). This is expressed by the so-called "aging-in-place" construct, which is defined as "a transaction between an aging individual and his or her environment that is characterized by changes in both person and environment over time, with the physical location of the person being the only constant" (Lawton 1990 p. 288). The older adults' preference for aging-in-place is related to feelings of freedom to do what they want, when and how they want to, and of control over their environment and daily lives (Leith 2006). Moreover, the experience of one's own home "integrates memories and images, desires and fears, the past and the present" (Pallasmaa 1995 p. 133). For this reason, the home and the surrounding environment represent the main target of the place attachment pattern, which taps those feelings, bonds, thoughts, and behavioral intentions that people develop over time with reference to their social-physical environment (Brown and Perkins 1992). Home is one of the most meaningful social-physical environments that we experience over time, since it includes physical, social, and biographical meanings (Fogel 1992), so that it can be encompassed as part of our self-identity (Moore 2000). Though a relocation experience usually provokes the breaking of attachment ties with the home place, the experience of "feeling at home" may also develop in places that are different from one's own home (Leith 2006).

Studies led by Küller (1988a, 1991) in geriatric hospitals showed that the creation of a "home-like" environment could counteract the institutionalization process. In particular, the decoration of the dining room in a similar vein to their previous house, and changes in layout, furnishing, lighting, and colors were all features eliciting positive responses from patients. It is thus likely that the shift from an "institutional" residential setting to a more "home-like" place would promote a parallel (positive) modification, both of the schema concerning such an environment (Imamoglu 2007) and of the patterns of spatial appropriation that characterize the relationship between the elderly and their living environment (Anderson 2011). This issue should be carefully considered above all in the design of residential care facilities within a "user-centered" and "architectural" humanization perspective, as addressed in the next sections.

24.3.2 Evidence-Based and User-Centered Design

Residential environments designed for older people can differ in many aspects such as the level of care, financial cost, social and caregiving support, location, and accessibility. Altman et al. (1984) identified two main types of residential environment for the elderly living in North America: retirement communities and supportive residential settings.

A retirement community, also known as an independent living community or village, is a housing complex or condominium for older people that enables them to remain in their own homes with access to specialized services, such as home health assistance, transportation, social activities, and help with housework (Marans et al. 1984). The living place is easier to navigate, focused on the older residents, and includes common areas for recreational or social activities, and exterior maintenance. Care and support services may be bought when required by the elderly. This specific residential environment ensures an independent life without the supportive help of family and friends.

A Supportive Residential Setting (SRS) is designed for older adults who need daily personal assistance, such as washing, dressing, and toileting, with staff available 24 h a day (Moos and Lemke 1984). The accommodation is provided in a single or shared bedroom, although a significant number of people move in with their relatives. A specific type of SRS is the Nursing Home (NH), which is characterized by the highest level of care outside the hospital; in fact, there is always a medical professional on site (and skilled nursing care is also available) to monitor the residents. NHs play an important role in supporting older people who cannot live independently in their own homes, due to either a recent hospitalization or a chronic illness.

In her review of the emerging issues in long-term care for the elderly, Stone (2006) stresses the importance of a system of services that responds to the users' needs, in order to reach the two "quality-oriented" goals of quality of (delivered) care and (users') quality of life. Schwarz (1997) suggested reversing the institutional character, derived from the medical model, of both the design and management of residential facilities for the elderly, since it fails to promote user well-being. Thus, such facilities should include more appropriate configurations and processes to increase the residents' quality of life. These goals could be reached by moving to a more "user-centered" design approach that is "evidence-based".

"Evidence-based" design is defined as "a deliberate attempt to base design decisions on the best available research findings" (Hamilton 2003 p. 19), thus the more empirical research evidence is used to guide design, the better the outcomes in terms of quality outputs (Becker et al. 2011).

This perspective has been developed for healthcare environments, but can be generalized across environments, particularly for similar ones like residential care facilities. It is by definition "user-centered", since it is based not only on designers' technical knowledge, but also on what users prefer, in the light of their quality of life (Fornara and Andrade 2012).

The key point of a "user-centered" (Gifford 2002), or "social" (Sommer 1983), design perspective is the essential consideration of the point of view of current and potential users in the design work. Thus, the underlying worldview expressed in this approach focuses on the main social actor for whom a given built space is intended, i.e. its user (Fornara and Andrade 2012). Applying this view to residential care facilities for the elderly means that they should be projected on the basis of specific information about the preferences and expectations of the current or potential older adults who will use the place. Such a view is reflected in the aging domain by a US-based movement named the Pioneer Network (www.pioneernetwork.net), which is composed of projects and associations such as the Eden Alternative, Deep Culture Change, Regenerative Care, and the Greenhouse Project (Stone 2006). This movement should witness a cultural change in residential care facilities like nursing homes, since it focuses on resident-centered living spaces where older adults' demands, preferences, and lifestyles drive the design of the physical, social, and organizational environments (Ronch 2003). As reported by Stone (2006), the Pioneer Network includes the following principles: (a) residents receive individualized care to nurture the human spirit; (b) residents make their own decisions; (c) the facility belongs to the residents; (d) the staff follow the resident's routine; (e) the staff have personal relationships with the residents; (f) residents and families are an integral part of the service team. This set of principles closely recalls the "humanization" perspective that was developed for healthcare environments.

24.3.3 Environmental Humanization

A user-centered design, which is evidence-based, should promote a "more humane" (Nagasawa 2000) care setting. The "humanization" construct covers a range of aspects, including organizational, social, relational, functional, and spatial-physical, which characterize a healing place. Within this perspective, organizational and management features play a significant role in improving both the quality of the delivered services and the comfort provided by the place. Thus, the latter point highlights the spatial-physical environment as a key factor influencing how the place is experienced by its occupants (Fornara and Andrade 2012).

Spatial-physical humanization is achieved through paying attention to a set of design attributes – such as layout and spatial configuration, colors and materials of furniture, walls, and floors, artwork, type, quantity, and focus of natural and artificial light, views inside and outside, size of windows, cleanliness, and climate – that should be provided in order to satisfy the fundamental needs of users (Devlin and Arneill 2003; Fornara et al. 2006). Such needs, as reported by Fornara and Andrade (2012), concern spatial and sensorial comfort (in visual terms, i.e., adequate lighting and panoramic views; in auditory terms, i.e., avoidance of annoying noises; and in climatic terms, i.e., adequacy of temperature and humidity), orientation, sense of welcome, privacy, social interaction, perceptual consistency, control over space, clear affordance, and restorativeness.

Though these considerations and guidelines have been put forward and developed for a specific category of places, i.e. healthcare environments, they are also relevant to the residential care environments for older adults, since they have many elements in common: both places share a care mission, include the presence of caregivers and visitors as the place's main social actors, and are characterized by facility users (healthcare patients or elderly residents) who typically experience reduced personal autonomy, helplessness, emotional distress, and anxiety in these places (this is particularly true for older adults during relocation).

The "home-like" and "hotel-like" concepts (Verderber and Fine 2000) that characterize a humanized care setting are particularly congruent with the goal of "de-institutionalization" of residential care environments for the elderly, which should be reflected by features such as the removal of barriers or other visual obstructions, the introduction of carpets in hallways, the possibility of personalizing room spaces with familiar objects, the preparation of areas for carrying out activities such as cooking or art, an increase in perceptual clues that facilitate orientation, the continuity of paths, the access to outdoor spaces such as gardens and other natural elements, and the design of areas for socialization and recreational activities (Ulrich et al. 2004). About the latter, the design of these kinds of spaces seems particularly relevant for the elderly, since significant relationships have been found between "active" aging (where the elderly are engaged in leisure activities) and positive outcomes in health, self-fulfillment, self-esteem, feelings of freedom, having social contacts, and personal well-being (Booth et al. 2000; Coleman and Iso-Ahola 1993). Frequently, an improvement in older adults' quality of life is not (only) related to the involvement of the cognitive area, but is rather reflected by an autonomy in daily routine behaviors, a reduction in the time spent in one's own room, and an increase in the time spent in walking and social interactions (Melin and Gotestam 1981). The importance of considering a variety of inner and outer spaces in the light of older people's well-being supports the view of the residential environment as a multi-place system.

24.3.4 The Residential Environment as a Multi-place System

As postulated by Canter (1977) in his Place Theory, which is one of the most cited in environmental psychology, the "place" is the result of the relationship between the setting's physical attributes, the behaviors that typically occur or are expected to occur in such a setting, and the descriptions or conceptions people hold of that behavior in that setting (Canter 1977 p. 159).

Merging the Place Theory with the systemic view of places expressed by the ecological approach of Bronfenbrenner (1979) – see also Fornara and Andrade (2012), for its application to healthcare environments, and Bonaiuto and Alves (2012), for its application to urban environments – Bonnes and Secchiaroli (1995) developed the "multi-place" perspective, where each place can be seen as a system of subplaces whose relationship is expressed by the main criteria of inclusion versus exclusion and nearness versus farness. From this perspective, the different subplaces are more or less connected with reference to the users' goals, activities, representations, and opportunities. Applying this view to a residential facility for the elderly, it can be analyzed as a place that includes different subplaces (such as residents' private rooms, common leisure spaces, a dining room, outside green spaces, staff areas, etc.), each of them in a relationship of nearness or farness with another subplace, and included in that specific facility, which in turn is included in a specific broader context (e.g., a specific neighborhood of a specific city) and excluded by other contexts (e.g., other neighborhoods of the same city). Thus, subplaces should not be seen as separated, since each place is part of a broader "place system," that is, within a complex of other places that are connected with it in users' representations and behaviors (Fornara and Andrade 2012). As stated by Bonnes and Secchiaroli (1995), the individual's place experience is organized in a multi-place sense, where the various levels of the different place systems appear as integrated in the action in the socio-physical environment. For example, the high or low frequency of use, and the kind of use of a specific leisure space in a given residential facility can be related to the frequency and kind of use of another space, which in turn may influence the use of a third area, and so on. Of course, each behavioral choice (what to do where and how) is assumed to be related to both the physical attributes of these spaces and the conceptions about a given behavior in a given space, as postulated by Canter's (1977) Place Theory. Empirical evidence of the multi-place organization of activities in the elderly population is provided by the study led by Bonaiuto and colleagues (2004) on the activities performed by the residents of a big city like Rome at different levels of scale (i.e., the home, the neighborhood of residence, the city center, and the periphery). Among the results of this study, it emerged that daily activities were organized by following a multi-place perspective, where the residential environment (home and neighborhood) had a key role (e.g., the opportunity or the choice of performing a given activity – such as shopping or playing sport – in the residential neighborhood implied avoiding doing that in other parts of the city). In particular, different ways of organizing daily activities were detected, from the most "open" and articulated, i.e. using different parts of the city for different activities and goals, to the most "closed" and poorly articulated, which reflected a substantial confinement to home and neighborhood. The latter pattern was expressed above all by older adults.

Analogously to other places, residential environments for the elderly are composed of various subplaces that include specific physical-spatial characteristics, specific systems of action (which in turn depend on place functions, types of social actors and roles), and the specific needs, expectations, and objectives of the different place-users involved (Fornara and Andrade 2012). For example, in a residential facility for the elderly, the elements that characterize a nursing space differ from those of a dining room in terms of function, social actors, staff tasks, typical behaviors, layout, physical features, residents' psychological state and responses, and so on.

Thus, the characteristics of subplaces should be viewed with reference to their different functional goals, also bearing in mind the differences between kinds of residential facilities, which are occupied by different kinds of elderly users (in terms of general health state and other related issues). In other words, it is likely that the needs expressed by residents in diverse kinds of residential facilities for the elderly (e.g., nursing home vs. residential care facility in the US context, see Lemke and Moos 1986) are quite different.

The systemic multi-place view of residential environments for older adults also takes into account the different social actors who occupy each place – and the interconnection of such places in their daily activity, as seen before – in terms of their role in that place. Canter (1977) proposed the construct of "environmental role" to identify "that aspect of a person's role which is related to his dealing with his physical environment" (Canter 1977 p. 128). In this sense, the individual's environmental role is shaped by specific functions, goals, motivations, meanings, expectancies, behaviors, and uses of place. Thus, the main distinction in a residential care environment is between care-recipients and caregivers, where the former express a passive and dependent role, and the latter typically play an active role, which involves a series of responsibilities (Fornara and Andrade 2012). The active role of caregivers is related to providing support not only to the elderly residents, but also to their families. It has been evidenced that communication between the staff of a residential care environment and the older adults' family can promote the involvement of family members in residents' care and provide social support to all the social actors involved, reducing the level of anxiety (Laitinen and Isola 1996). Clearly, it is also necessary to pay attention to the quality of caregivers' experiences, including the various facets of their job (how they are organized and managed, compensated and rewarded, etc.), since this significantly influences the residents' health and well-being (Stone et al. 2003).

24.4 Recent Research Trends in the Relationship Between the Elderly and Their Residential Environment

In this section, recent trends in the scientific literature concerning the relationship between quality of life and residential environment in the elderly population are reported. This process began with a preliminary screening of articles of the last decade included in the electronic databases of PsycINFO, PsycARTICLES, ScienceDirect, and Scopus. Moreover, empirical studies included in some book chapters were considered. About 8,000 papers were identified on the basis of different combinations of the following keywords: "elderly", "older people", "home", "physical environment", "well-being", and "health".

Firstly, a preliminary selection of potentially relevant articles was made through an analysis of the title and the abstract. Then, the full text was assessed in order to include only papers focusing on the impact of residential features on elderly well-being and quality of life. This resulted in a set of about 40 articles.

The analysis of the selected literature showed that a limited number of papers have been published in the two main journals addressing environmental psychology topics; the Journal of Environmental Psychology and Environment and Behavior. This reflects a marked absence of the issue of healthy residential environments in the environmental psychology domain and the consequent need for future development.

On the basis of the content of the recent literature on residential environments and elders' quality of life, this section is structured by focusing on aging-in-place, assistive domotics, relocation issues, residential care environments, and, finally, a special group of residents, i.e. older people affected by dementia, who have received particular attention in the literature. This thematic structure follows a sequential conceptual line, beginning with the consideration of home features (including the broader neighborhood level), then moving to a specific class of home aid features that are represented by assistive technological devices, then addressing the issue of "necessary" home abandon, or relocation, followed by typical relocation destinations, i.e. residential care facilities, and, finally, focusing on a specific class of residential care facilities, designed for the elderly affected by dementia and related pathologies.

24.4.1 Aging-in-Place

Oswald and Wahl (2004) identified four categories that define the meaning of home for the elderly: physical (spatial-physical features, furnishing, and access), emotional (perceived safety, intimacy, and sense of belonging), cognitive (home attachment), and social (relationships and interactions with relatives, neighbors, and visitors). The relationship between these dimensions has been verified in various studies. Starting from the development of the "Adaptable Design housing policy guidelines" for supporting the independent living of older adults in Canada, Danziger and Chaudhury (2009) carried out a Post Occupancy Evaluation study consisting of the assessment of these design guidelines by the older residents of five buildings built on the basis of them. A set of interviews was conducted in order to detect design preferences and perceptions of possible environmental modifications. The outcomes showed an overall satisfaction with all those features that were associated with a high level of independence (e.g., nonskid flooring, wide corridors, grab bars, many light switches, and a shower area to avoid the fear of falling due to the presence of a bath tub).

Regarding physical features, they were put under the lens of the above-mentioned Enable-Age Project (Iwarsson et al. 2006), a cross-national and multidisciplinary study aimed at analyzing environmental accessibility issues in the homes of the elderly. Although dwellings were characterized by features that differed across countries, environmental barriers and problems related to design issues showed some important similarities. In fact, the results pointed out the marked presence of environmental barriers in houses as well as accessibility problems, even though the latter varied between countries (for example, they mostly concerned the too-high placement of wall-mounted cupboards and shelves in Sweden, the overall lack of seating places in Germany, and the absence of grab bars in the shower or bath in Latvia).

In order to investigate the role of house physical features in influencing the older adults' choice of "aging-in-place", Safran-Norton (2010) carried out a longitudinal study with households living in a couple or alone. It emerged that the reasons for remaining in their present home were related to interior home modifications, such as the installation of ramps, railings, wheelchair accessibility equipment, and shower seats. These outcomes substantially confirmed those found in other studies, in which elders who were proactive or used an assimilation strategy (Slangen-de Kort et al. 1998; Scopelliti et al. 2005) in modifying their house were more inclined to stay longer in their current dwelling (Hutchings et al. 2008; Gitlin et al. 1999). Hwang and colleagues (2011) found that the spatial-physical aspects of one's own home play a crucial role in reinforcing personal autonomy and social interaction in older adults.

Another study led with elderly residents in Hong Kong (Phillips et al. 2005) verified the mediating role of residential satisfaction in the relationship between home conditions and psychological well-being. The impact of the interior and the open-space environment was also examined separately in order to detect possible differences in the association with specific dwelling features and psychological well-being. An overall influence of physical factors emerged, while the interior environment features (lighting, temperature, crowdedness, ventilation, presence of a lift, special facilities, and security devices) showed a greater impact than open-space features on explaining psychological well-being via residential satisfaction.

The daily use of common spaces was the analytical focus of an observational study led in assisted living units located in Gothenburg (Andersson et al. 2011). Interestingly, incongruence was found between the actual use of the spaces and their original function. In particular, the main difficulties in usability expressed by the older users concerned the configuration of the physical environment, the conflicts between different users for the use of the spaces, or a combination of both.

Although an extensive literature has focused on the home as a crucial setting for determining beneficial (or negative) consequences on older adults' quality of life, the importance of a broader level of environmental scale for coping successfully with factors such as isolation, loneliness, and physical decline has also been recognized (Windle et al. 2006; Rojo-Perez et al. 2007; Kellett et al. 2005). Several studies have identified the neighborhood as a specific environmental unit that may promote or discourage a healthy lifestyle through spatial-physical characteristics such as the presence of green spaces, good lighting and pavement quality, and proximity to shops and services (Stock and Ellaway 2013). These factors are strictly connected with social dimensions (such as perceived safety, social interactions, and fear of crime) that can influence the possibility of establishing social contacts with neighbors, particularly for older people who live alone (Macintyre et al. 2002). An extensive survey on almost 10,000 older Australian women investigated the influence of the sense of belonging to the residential neighborhood on their health and well-being (Young et al. 2004). The results showed a significant association between high neighborhood belongingness and overall satisfaction, high level of physical and mental health, lower stress, better social support, and a physically active life. A focus on the spatial-physical features of neighborhoods was made in two recent studies that reported similar outcomes. Nathan, Wood and Giles-Corti (2012) ran a survey study with elders residing in several Australian retirement villages, in which some spatial-physical features of the neighborhood (such as fewer physical barriers, services related to aesthetics, safety, and services) were found relevant in promoting healthy activities such as walking. In another study, Sugiyama, Ward Thompson and Alves (2009) explored the association between the neighborhood open space and quality of life in a British sample of individuals aged over 65. They found that features such as pleasantness, quality, and safety of open spaces were associated with a higher degree of health, physical activity, and life satisfaction.

A further aspect that has received attention in the research literature concerns green spaces and natural views. In this regard, the Environmental Standards Council of the Center for Health Design (American Institute of Architects) recommended access to nature as a guideline for both hospitals and long-term care facilities. This acknowledges the role of natural daylight and views of nature in the positive health benefits of both residents and caregivers. The influence of these aspects on health outcomes has been proven within the restorativeness literature (Staats 2012), with specific reference to residential settings (Wells and Rollings 2012). This effect also concerns the elderly population, since a positive impact was found on the longevity and residential satisfaction of older people who could access natural settings in their surrounding residential environment (De Vries et al. 2003; Kweon et al. 1998). Kellett et al. (2005) reported that trees and gardens emerged in the elders' narratives as promoters of pleasantness and privacy. Furthermore, green spaces and private gardens were described as part of one's own home as well as reflecting personal identity and recalling memories.

24.4.2 Assistive Domotics

When mental and physical changes undermine the possibility for older adults to live independently in their private home, assistive technology may play an important role in maintaining their autonomy, safety, and quality of life, by keeping the elderly in their living environments.

Video-monitoring, health monitors and electronic sensors (e.g., fall detectors, door monitors, bed alerts, pressure mats and smoke and heat alarms) are some examples of assistive technology that can improve older people's safety, security and ability to cope at home (Miskelly 2001).

Löfqvist et al. (2014) investigated the perceived unmet need for assistive technology related to elders' overall health, daily independence, and environmental barriers. A set of self-report data, health indicators, and observations of the living environment was collected from a sample of older people living in Sweden. Outcomes highlighted that assistive technology played a key role in providing adaption, whereas it seemed less important for issues related to communication support.

A 3-year Italian project (named "Robocare Domestic Environment", Cesta et al. 2007) aimed at developing a technological cognitive support for the elderly was led by research groups with different backgrounds (IT and AI experts, engineers, social psychologists, etc.).

Within this project, in a survey study comparing different age categories (Scopelliti et al. 2005), elderly people were the most frightened by the idea of having a robot at home, and they tried to ward off their anxiety by attributing to it those features that could reduce its (negative) impact, i.e. small size, slow motion, feminine voice, and executing collaborative tasks. More generally, though assessing technology as useful, older people showed a substantial mistrust towards machines that they expected to be unsafe. This effect was more pronounced in less educated elders, suggesting that the possibility of controlling technological devices is an essential requirement for their acceptability.

In a simulation study within the same project, Cesta et al. (2007) examined the complex relationship between the cognitive, affective and emotional components of the elders' images of robots with a focus on the interaction between the user, the robot, and the perceived usefulness of the latter's support role. In the simulation, the robotic mediator was equipped with sensors for continuous monitoring and intelligent software devices, and operated as the main communication channel between the older adult and the domestic environment. The physical aspect of the robot was important for its acceptability by the older people. In particular, robots with a no-face version were preferred and perceived as better integrated in the living environment than those with a human resemblance. Furthermore, the robot was considered useful in supporting various activities, thus witnessing the key role of compensation strategies in managing the loss of personal resources.

The role of assistive domotics has received particular attention for a specific category of older adults, i.e. those affected by dementia and related pathologies. For example, some studies (Bjørneby et al. 2004; Orpwood et al. 2004) have identified a set of guidelines for elders with initial or moderate dementia. In particular, the special needs of this group of people require technological help that is reassuring, controllable, user-friendly, and familiar (Orpwood et al. 2004). The general difficulty found in the comprehension and usability of assistive technology devices (Scopelliti et al. 2005) suggests a moderate introduction of these elements into those environments designed for people with Alzheimer's disease (van Hoof and Kort 2009). On the contrary, Marquardt and colleagues (2011) reported a beneficial role of such devices in the early stages of dementia, in order to reduce the symptoms of this kind of disease. These authors carried out a descriptive study focusing on the physical limitation related to the barriers in a community dwelling. The main obstacles observed were the absence of handrails, the lack of adequate spaces to maneuver wheelchairs, and stairs inside and outside the living environment. Furthermore, the study detected the insufficient adoption of assistive technology, such as automatic additional lights, alarm systems, signs and labels, and stove security sensors. Nevertheless, home modifications could engender confusion and irritation in the care recipient even if this negative feeling seemed to decrease with the progression of dementia.

To summarize, assistive domotics seems to play a role in increasing the level of safety and mobility of older adults within their home, therefore the design of a supportive technological environment is expected to increase significantly the quality of life of elderly people. Nevertheless, designers and producers of assistive domotics should consider that both the ease of use of technological devices and the adequate training of users are important requirements for this new technology to be accepted by the elderly population (Giuliani et al. 2005).

24.4.3 Relocation Issues

Although most elders show a general preference for aging-in-place, the natural health decline during aging makes home an unsuitable place, thus relocation may be preferable or necessary (Rubinstein and De Medeiros 2004; Cerina and Fornara 2011). Doubtless, a residential move is a complex experience, especially for older adults, which may influence individual well-being and quality of life. It is thus important to shed light on the motivations, attitudes, and all the psychosocial and environmental dimensions related to the relocation event.

Although the general preference of the elderly to remain in the familiar context of their own home has often been recorded (Callahan 1992; Challis and Davies 1985; Scanlon and Devine 2001), there is also some evidence of possible positive effects of relocation on the older adults' quality of life (Oswald et al. 2002).

An 11-year longitudinal study was carried out by Nygren and Iwarsson (2009) in order to explore the motivations underlying the decision to relocate. The relationship between housing and health was analyzed through a mixed-methods approach, including qualitative interviews with 13 very old participants and quantitative survey data collected in three waves. On the basis of the outcomes, the relocation process was defined as a non-linear procedure consisting of five phases: in the first phase, the approach towards relocation was ambivalent, i.e. including both positive and negative feelings. Later, when health-related issues (e.g., loss of independence, mobility problems, sudden illness) emerged, the relationship with home changed, until the decision to move was taken. The period of change provoked psychological responses such as frustration, doubts and expectations, but finally, in the last phase, positive experiences, subjective health enhancement, and increased social contacts were reported.

A similar study, which used a qualitative longitudinal perspective, included in-depth interviews with 16 older people from Sweden and Germany living in their private home, 8 years after the first collection (Granbom et al. 2014a). The purpose of the study was to investigate the reasoning processes concerning the choice between aging-in-place and relocation. Findings revealed several changes over time, related to environmental experience, the development or decline of the attachment to place, and the maintenance of "residential normalcy" (Golant 2011) during the period of declining health and loss of independence. In particular, elders who lived in a new dwelling showed a reduction in stress in the case of attachment to their former house. Furthermore, living out of comfort zones can activate relocation even if this process can generate a drop in residential normalcy due to a health decline or a growing awareness of impending death.

Another longitudinal study led by Granbom et al. (2014b) aimed at investigating which aspects of housing and health predict the relocation choice in both ordinary and special residences. The sample study included 384 older people living alone in ordinary houses in Sweden. A relevant role was found for cleaning, perceived functional independence, and living in a one-family house in predicting relocation to ordinary housing. On the other hand, dependence on cooking, cognitive deficits, and accessibility problems were identified as predictors of relocation to special housing. These outcomes demonstrate the different impact of house characteristics and health problems on orienting elders in their personal choice of a new residential environment.

Amedeo et al. (2009) focused their attention on the relocation choice of retirees in Australia. In particular, the analytical context was a retirement village, i.e. a community that includes private housing with or without direct care provision. Consistent with the push-pull approach (Bogue 1969), motivations underlying such a choice were identified in four push factors (or stressors, which were worries about home maintenance, declining health and complicated access to facilities, social innovation, and changes in lifestyle) and three pull factors (or attractors, which were maintaining existing lifestyle and familiarity, location, and built environment and affordability).

The search for predictors of the elderly's attitudes toward relocation was the focus of a study carried out with older Italians living in their own homes (Cerina and Fornara 2011). The findings highlighted the prominence of environmental competence as an antecedent of attitudes toward relocation and perceived well-being. Place attachment and perceived social network were also significant predictors in the negative sense of the decision to move to another residential setting. A further result concerned the difference between urban and rural residents: in particular, elders living in rural places evidenced a higher degree of psychological well-being and collective identity than those living in urban places.

In order to examine reflections and emotions related to relocation and aging-in-place, Löfqvist et al. (2013) ran a cross-national qualitative study with a sample of very old people residing in two European countries, Sweden and Germany. Similar outcomes were observed in both countries, suggesting some degree of generalizability of the relocation process. Older people were inclined to remain in place, especially in the presence of a strong attachment to their home and neighborhood, economic problems, and fear of losing the continuity of routine activities. On the contrary, determinants of moving to another living environment were the desire to maintain independence and the avoidance of loneliness. These findings are similar to those of the longitudinal study of Pope and Kang (2010), which showed that older people were more inclined to relocate for "reactive" reasons, such as a stressful event, life change, or crisis, rather than in the condition of relocation choices not related to triggering events.

24.4.4 Residential Care Environments

Older people may occasionally spend some periods in hospital rooms, which become their temporary home, or may relocate to residential care facilities or nursing homes permanently or for long periods.

In all these cases, some issues should be considered. How can a "home-like" and/or "hotel-like" environment, i.e. a comfortable setting, be ensured in line with the humanization framework? How can those feelings of identity, security, belongingness, and control (Pastalan and Barnes 1999), which are typically developed in residential environments like one's own home and one's own neighborhood, be promoted? In order to respond to these questions, specific attention has been paid to elders' needs and preferences in healthcare facilities (Arneill and Devlin 2002; Wakamura and Tokura 2001; Kearney and Winterbottom 2005; Wilmott 1986). Nevertheless, there is still little systematic investigation of quality issues in residential care facilities (Aud et al. 2004).

One important attempt to address the multidimensional nature of the quality of residential care environments is the Multiphasic Environmental Assessment Procedure (MEAP, Lemke and Moos 1986), which taps four quality domains; physical features, policies and program, human aspects, and social climate. Consistently, the MEAP tool consists of four instruments; the Physical and Architectural Features Checklist (filled in by an external observer), the Policy and Program Information Form (deduced by staff, observation, and archives), the Resident and Staff Information Form (concerning objective information about occupants) and, finally, the Sheltered Care Environment Scale (based on assessment by residents and staff members). Note that only the last tool reflects the subjective evaluation of users, whereas the others concern more or less objective information about the facility.

In a recent experimental study, adults over 65-years-old living in their private home were assigned to different conditions consisting of exposure to a scenario composed of two images, depicting the facade and the surrounding external spaces of a residential facility for the elderly, respectively (Cerina et al. 2015). The manipulation concerned the architectural style (home-like vs. hotel-like vs. standard) and the green areas (presence vs. absence). The outcomes evidenced both the importance of green areas and the positive impact of home-like and hotel-like residences in psychosocial responses, such as residential satisfaction, (lower) feelings of broken place attachment, and attitudes toward relocation.

The relevant role of the physical environment in the elders' quality of life was investigated in three residential care facilities, two refurbished and one not refurbished, in Gothenburg (Sweden), with a focus on the impact of refurbishment on residents' aesthetic impression, orientation, mobility, and social interactions in common spaces (entrance, corridors, kitchen, and living room: Falk et al. 2009). The results showed no significant increase in the elders' mood, behaviors, and social contacts, even though the refurbishment was appreciated in terms of its enhancement of the overall aesthetic impression. Surprisingly, the residents who experienced the refurbishment expressed a level of perceived quality of life that was slightly lower than in the residents who did not experience it. This seems to confirm the general low adaptation level of the elderly population toward physical environment changes. Another similar study examined the impact of the refurbishment of two communal spaces (the recreation room and the lobby room) on the subjective well-being of older adults living in a residential care facility (Weenig and Staats 2010). The purpose of the refurbishment was to improve the aesthetic aspect of both spaces (e.g., by inserting furniture made with natural materials, plants, warmer colors, and rearranging the position of the tables) and to facilitate social contacts. Analogously to that found by Falk and colleagues (2009), the physical changes concerning residential and aesthetic quality were positively evaluated but, in this case, the results also indicated a significant increase in the residents' subjective well-being and perceived support of social interactions.

With regard to the relationship between space and quality of life, Barnes (2006) conducted a cross-sectional study with elders living in 38 different residential care facilities. The purpose was to investigate the impact of the use of different spaces on well-being, active behavior, and environmental control. A strong association emerged between the articulation of space and quality of life; in particular, the richness and variety of spaces were associated with a high level of activity, well-being and environmental control.

The positive influence of the physical environment on older people's well-being and quality of life also emerged in a study led in community residences that differed in their degree of architectural humanization (Cerina and Fornara 2012). Results showed that the higher the degree of "objective" architectural humanization, the higher the users' perceived quality of life and overall satisfaction, thus confirming the findings of other studies focusing on healthcare humanization (Fornara et al. 2006; Andrade et al. 2012, 2013).

The impact of residential satisfaction and sense of belongingness on elders' perceived loneliness was the core of the study conducted by Prieto-Flores and colleagues (2011) in two different residential environments, a community home and a residential care facility located in Spain. As hypothesized, the authors found a negative relationship between residential satisfaction and loneliness in both residential contexts, with the sense of belongingness playing a mediating role. Furthermore, the relevant role of social contacts emerged (e.g., gathering with family, friends and neighbors) in increasing both the sense of belongingness of people living in the residential care facility and the residential satisfaction of the residents in the community home.

24.4.5 Special Residential Care Environments for Dementia Patients

Within the research on residential environments and healthy aging, a specific group that has received special attention is older people

affected by Alzheimer's disease or other age-related dementia. The global increase in the average age has resulted in a growing number of elderly people suffering from this kind of illness; in fact, the most recent World Alzheimer Report (2014) estimates that 44 million people worldwide are afflicted by these pathologies, and this figure is predicted to double by 2030, and more than triple by 2050.

The syndrome is characterized by a progressive decline in functioning, related to various domains (such as language, memory, visual or spatial abilities, and judgment), which interferes with daily life. In order to lower the impact of dementia on individuals and society, the role of the living environment has become increasingly important.

One of the key issues concerns familiarity: Küller (1988b) found that collective housing in small units decorated in an old and familiar style were more positively activating than conventional geriatric residences for the elderly affected by dementia. This is due to the fact that a familiar environment activates old habits established through lifelong experience, thus increasing the functional and social competence of these patients.

Several studies have underlined the importance of modifications in architecture, technology, and indoor design in reducing confusion, anxiety, and negative emotions (Migita et al. 2005), and in promoting and improving wayfinding and social interaction among elders with dementia (Fleming et al. 2003). According to Zeisel (2009), the provision of memory cues in their living environment, which can encourage independence and reduce frustration, is a key factor for improving the quality of life of these older people.

In their literature review integrated with the outcomes of focus groups, van Hoof and Kort (2009) showed how all the features of the living environment could be designed following a holistic approach (Diaz Moore et al. 2006). They identified several physical aspects that can influence and enhance the well-being of the care recipient. In particular, the dementia dwelling should be as open as possible and corridors should be removed in order to make older people affected by Alzheimer's disease more relaxed (Cohen-Mansfield et al. 1990). Concerning the interior design, pastel colors are preferable for painting the walls (Marx et al. 2002) and wallpaper print should be avoided, since it may cause confusion, anxiety, and fear (Cohen-Mansfield et al. 1990). About the layout of the circulation system, a consistent outcome of empirical studies (Marquardt and Schmieg 2009; Marquardt 2014; Passini et al. 2000) was that direct visual access is helpful for orientation.

An adequate design of a residential environment for dementia patients should consider other factors, such as lighting and thermal comfort, which can play an important role in residents' well-being. In fact, older people with Alzheimer's disease are characterized by behaviors such as nocturnal restlessness and wandering. Thus, in addition to large windows that provide good lighting, a high-intensity bluish light is suggested to influence these types of behavior, by enhancing sleep quality and reducing depression states (Abbott 2003; van Hoof et al. 2009). On the other hand, people with dementia, especially in the later stages of Alzheimer's disease, have a different perception of the thermal environment (van Hoof 2008) and thus a thermally comfortable place is important to provide efficient support. It is worth mentioning that people with dementia may perform behaviors such as undressing themselves, thus an optimal thermal comfort must be guaranteed in bathrooms and bedrooms. Moreover, the temperature between the various rooms should be kept as similar as possible through the implementation of heating, ventilation, and air-conditioning systems (van Hoof and Kort 2009).

A cross-sectional survey on residents living in special care units showed the role of personalization of bedrooms in reducing aggression and anxiety (Zeisel et al. 2003), confirming the results of previous research (Charras et al. 2010; Garcia et al. 2012; Morgan and Stewart 1998) on the influence of the availability of private rooms in reducing irritability and enhancing the quality of sleep.

Several studies have also pointed out the importance of green spaces and exposure to daylight for the health of older people with

Alzheimer's disease (van Hoof and Kort 2009). Green spaces, as evidenced by Ulrich (1984), may have therapeutic properties, encouraging walking activity (Joseph et al. 2005; Mooney and Nicell 1992), and reducing wandering and anxiety (Cohen-Mansfield and Werner 1998; Namazi and Johnson 1991).

The scientific literature underlines the rapid decline of some cognitive processes during the early period of dementia (Diaz Moore 2007). Attention fatigue is suggested as an antecedent of stress (Kaplan 2001), thus its control and a general attention reactivation may contribute to reducing the negative outcomes related to stress in the elderly with a cognitive disease. Based on this assumption, Diaz Moore (2007) investigated the role of the four "restorative" properties of environments (i.e. being away, fascination, extent, compatibility), identified by Kaplan's Attention Restoration Theory (1995), in designing restorative gardens for this specific group of users. An interpretative analysis of qualitative data was based on the observations of several experts regarding a set of five therapeutic gardens recognized for their design quality. Results pointed out the importance of low walls, thresholds, and trees to ensure a sense of enclosure and safety. Furthermore, some details such as fountains, ornamental grates and a canopy may guarantee a special visual intrigue able to capture the attention effortlessly. Changes occurring seasonally and in daylight hours can provide several restorative and fascinating views of the landscape, supported by the presence of plants and flowers. The variety of sensorial stimulation (including touch, sound, sense of smell, movement, sight, and the memory of previous experiences) was also identified as an essential feature of therapeutic gardens, which should have characteristics of richness and coherence able to engage the haptic system. Finally, the study also identified a series of physical components (plants, paving, planters, benches, raised beds, hedges, and movable chairs), sensorial factors (glare and contrast, scented flowers, tactile stimulation through contact with dirt, plants, and ornamental grasses) and spatial properties (simple configuration, presence of different places for social interaction, active participation as well as places for solitude or meditation), which involve a rich variety of senses overall, thus improving those abilities related to accessibility and wayfinding. Indeed, it is well known that people afflicted by dementia often reveal disorientation problems. Referring to this issue, Brawley (2007) underlined the importance of visible and recognizable clues linking people to familiar places. Moreover, gardening is a familiar and daily activity, involving planting and harvesting fruits and vegetables, with a positive influence on self-esteem and overall satisfaction.

Green spaces may also have an important role in the quality of sleep, extending their positive influence to night-time. Calkins et al. (2007) led an exploratory study with 17 individuals affected by dementia residing in three nursing homes. A repeated measures design assigned participants to four conditions: winter/no activity, winter/internal activity, summer/no activity, summer/outside activity. Through direct observations and the use of an actigraphy (a validated proxy measure of sleep and agitation), a modest improvement was found in the quality of sleep among older people who spent their time on outdoor activities.

24.5 Conclusion

It is widely acknowledged that the spatial-physical attributes of places affect people's health, both by producing well-being or distress feelings and by conveying positive or negative information for people's self-esteem, security, and identity (Evans and McCoy 1998). Designing and arranging spaces based on the needs and expectations of current and potential users is particularly important for residential settings destined for older adults, considering the progressive growth of the elderly population and bearing in mind that the residential place is typically the main environment(and often the only one,Bonaiuto et al. 2004) experienced by aged people. Spatial-physical features can be either facilitators or inhibitors of the older person's goals, thus triggering either satisfaction or

frustration. **Consequently, an appropriate design can foster healthy residential environments, thus influencing the quality of life of a population that is typically experiencing a state of physical and/or mental decline.**

This aim is reflected by the claim for "more humane" (Nagasawa 2000) environments to reduce the stress level of people who are frequently requested to cope with situations such as disease, pain, and death. In this chapter, we have shed light on the concept of environmental humanization, which has been developed with reference to healthcare environments but appears equally relevant for residential places for the elderly. Thus, designing residential environments that follow "user-centered" and "evidence-based" guidelines should improve the quality of environmental properties and, consequently, increase the likelihood of congruence of the Environment supply related to the Person's requests within a P-E fit perspective (Carp and Carp 1984).

Although the recent research trends presented in the second part of this chapter seem to indicate a growing interest in the topic of healthy residences for older people, there is still a prevalent concern for the Person side of the dichotomy, i.e. the psychological patterns that help cope with the environmental demands, and a minor focus on the Environment, or on the transactional unit Person-Environment, i.e. which place attributes fit better with the older adult's goals and needs. This statement is corroborated by the limited number of papers that have been published in the two main journals of environmental psychology; the Journal of Environmental Psychology and Environment and Behavior. Thus, there is ample room for the development of rigorous and systematic research proving the importance of "humanized" design aspects for elderly people's well-being and quality of life.

References

Abbott, A. (2003). Restless nights, listless days. *Nature, 245,* 896–898.

Aldwin, C. M., Spiro, A., III, & Park, C. L. (2006). Health, behavior, and optimal aging: A life span developmental perspective. In J. E. Birren, K. W. Schaie, R. P. Abeles, M. Gatz, & T. A. Salthouse (Eds.), *Handbook of the psychology of aging* (6th ed., pp. 85–104). New York: Academic.

Altman, I., & Rogoff, B. (1987). World views in psychology: Trait, interactional, organismic, and transactional perspectives. In D. Stokols &·I. Altman (Eds.), *Handbook of environmental psychology* (Vol. 1, pp. 7–40). New York: Wiley.

Altman, A., Powell Lawton, M., & Wohlwill, J. F. (1984). *Human behavior and environment: Advances in theory and research* (Elderly people and the environment, Vol. 7). New York: Plenum Press.

Amedeo, D., Golledge, R. G., & Stimson, R. J. (2009). *Person environment behavior research: Investigating activities and experiences in spaces and environments.* New York: The Guildford Press.

Anderson, J. E. (2011). *Appropriating space in an assisted living residence: On architecture and elderly frail people's spatial use.* Proceedings of the 2011 ARCC (Architectural Research Centers Consortium) Conference "Reflecting upon Current Themes in Architectural Research" (pp. 1–19). Detroit: Lawrence Technological University, 20–23 April 2011.

Andersson, M., Lindahl, G., & Malmqvist, I. (2011). Use and usability of assisted living facilities for the elderly: An observation study in Gothenburg Sweden. *Journal of Housing for the Elderly, 25,* 380–400.

Andrade, C., Lima, L., Fornara, F., & Bonaiuto, M. (2012). Users' views of hospital environmental quality: Validation of the Perceived Hospital Environment Quality Indicators (PHEQIs). *Journal of Environmental Psychology, 32,* 97–111.

Andrade, C., Lima, L., Pereira, C. R., Fornara, F., & Bonaiuto, M. (2013). Inpatients' and outpatients' satisfaction: The mediating role of perceived quality of physical and social environment. *Health & Place, 21,* 122–132.

Arneill, A. B., & Devlin, A. S. (2002). Perceived quality of care: The influence of the waiting room environment. *Journal of Environmental Psychology, 22,* 345–360.

Aud, M. A., Rantz, M. J., Zwygart-Stauffacher, M., & Manion, P. (2004). Developing a residential care facility version of the observable indicators of nursing home care quality instrument. *Journal of Nursing Care Quality, 19*(1), 48–57.

Baltes, M. M. (1996). *The many faces of dependency in old age.* New York: Cambridge University Press.

Baltes, M. M., & Carstensen, L. L. (1996). The process of successful ageing. *Ageing and Society, 16,* 397–422.

Barnes, S. (2006). Space, choice and control, and quality of life in care settings for older people. *Environment and Behavior, 38,* 589–604.

Becker, F., Bonaiuto, M., Bilotta, E., & Bonnes, M. (2011). Integrated healthscape strategies: An ecological approach to evidence-based design. *HERD Health Environments Research & Design Journal, 4,* 114–129.

Birren, J. E. (2006). Foreword. In J. E. Birren, K. W. Schaie, R. P. Abeles, M. Gatz, & T. A. Salthouse

(Eds.), *Handbook of the psychology of aging* (6th ed., pp. 15–16). New York: Academic.

Bisconti, T. L., & Bergeman, C. S. (1999). Perceived social control as a mediator of the relationships among social support, psychological well-being, and perceived health. *The Gerontologist, 39*, 94–103.

Bjørneby, S., Topo, P., Cahill, S., Begley, E., Jones, K., Hagen, I., Macijauskiene, J., & Holthe, T. (2004). Ethical considerations in the ENABLE project. *Dementia, 3*, 297–312.

Bogue, D. J. (1969). *Principles of demography*. New York: Wiley.

Bonaiuto, M., & Alves, S. (2012). Residential places and neighbourhoods: Toward healthy life, social integration, and reputable residence. In S. Clayton (Ed.), *The Oxford handbook of environmental and conservation psychology* (pp. 221–247). New York: Oxford University Press.

Bonaiuto, M., Bonnes, M., & Continisio, M. (2004). Neighborhood evaluation within a multi-place perspective on urban activities. *Environment and Behavior, 36*, 41–69.

Bonnes, M., & Secchiaroli, G. (1995). *Environmental psychology. A psycho-social introduction*. London: Sage.

Booth, M. L., Owen, N., Bauman, A., Clavisi, O., & Leslie, E. (2000). Social-cognitive and perceived environment influences associated with physical activity in older Australians. *Preventive Medicine, 31*, 15–22.

Branch, L. C., & Jette, A. M. (1982). A prospective study of long-term care institutionalization among the aged. *American Journal of Public Health, 72*, 1373–1379.

Brandtstädter, J., & Renner, G. (1990). Tenacious goal pursuit and flexible goal adjustment: Explication and age-related analysis of assimilative and accommodative strategies of coping. *Psychology & Aging, 5*, 58–67.

Brandtstädter, J., & Rothermund, K. (2002). The life course dynamics of goal pursuit and goal adjustment: A two-process framework. *Developmental Review, 22*, 117–150.

Brawley, E. C. (2007). Designing successful gardens and outdoor spaces for individuals with Alzheimer's disease. *Journal of Housing for the Elderly, 21*, 265–283.

Bronfenbrenner, U. (1979). *The ecology of human development: Experiments by nature and design*. Cambridge: Harvard University Press.

Brown, B., & Perkins, D. (1992). Disruption in place attachment. In I. Altman & S. Low (Eds.), *Place attachment, human behavior and environment* (pp. 279–304). New York: Plenum Press.

Calkins, M., Szmerekovsky, J. G., & Biddle, S. (2007). Effect of increased time spent outdoors on individual with dementia residing in nursing homes. *Journal of Housing for the Elderly, 21*, 211–228.

Callahan, J. J. (1992). Aging in place. *Generations, 16*, 5–6.

Canter, D. (1977). *The psychology of place*. London: Architectural Press.

Carp, F. M. (1987). The impact of planned housing. A longitudinal study. In V. Regnier & J. Pynoos (Eds.), *Housing the aged, design directives and policy considerations*. New York: Elsevier.

Carp, F. M., & Carp, A. (1984). A complementary/congruence model of well-being or mental health for the community older. In I. Altman, M. P. Lawton, & J. F. Wohlwill (Eds.), *Human behavior and environment* (Older People and the Environment, Vol. 7, pp. 279–336). New York: Plenum Press.

Cerina, V., & Fornara, F. (2011). The psychological determinants of attitudes toward relocation in the elderly: A survey study in rural and urban environments. *Psyecology, 2*(3), 335–348.

Cerina, V., & Fornara, F. (2012). *Promoting design quality in the community residences for the elderly: A comparison between high- vs. low-humanization structures in Sardinia (Italy)*. Proceedings of the International Conference ARCH12 "Architecture/Research/Care – Health perspectives on Nordic welfare environments" (pp. 1–15). Gothenburg: Chalmers University of Technology, 12–14 Nov 2012.

Cerina, V., Fornara, F., & Manca, S. (2015). *A pleasant residential facility for the older adults: The influence of architectural style and green space on attitudes toward relocation and expected wellbeing*. Manuscript under review.

Cesta, A., Cortellessa, G., Giuliani, M. V., Pecora, F., Scopelliti, M., & Tiberio, L. (2007). Psychological implications of domestic assistive technology for the elderly. *PsychNology Journal, 5*, 229–252.

Challis, D., & Davies, B. (1985). Long term care for the elderly: The community care scheme. *British Journal of Social Work, 15*, 563–579.

Charras, K., Zeisel, J., Belmin, J., Drunat, O., Sebbagh, M., Gridel, G., & Bahon, F. (2010). Effect of personalization of private spaces in special care units on institutionalized elderly with dementia of the Alzheimer type. *Non-Pharmacological Therapies in Dementia, 1*, 121–137.

Cohen-Mansfield, J., & Werner, P. (1998). The effects of an enhanced environment on nursing home residents who pace. *Gerontologist, 38*, 199–208.

Cohen-Mansfield, J., Werner, P., & Marx, M. S. (1990). The spatial distribution of agitation in agitated nursing home residents. *Environment and Behavior, 22*, 408–419.

Coleman, D., & Iso-Ahola, S. E. (1993). Leisure and health: The role of social support and self-determination. *Journal of Leisure Research, 25*, 111–128.

Danziger, S., & Chaudhury, H. (2009). Older adults' use of adaptable design features in housing units: An exploratory study. *Journal of Housing for the Elderly, 23*, 134–148.

de Vries, S., Verheij, R. A., Groenewegen, P. P., & Spreeuwenberg, P. (2003). Natural environments – healthy environments? An exploratory analysis of the relationship between green space and health. *Environment and Planning A, 35*, 1717–1731.

Devlin, A. S., & Arneill, A. B. (2003). Healthcare environments and patient outcomes. *Environment and Behavior, 35*, 665–694.

Diaz Moore, K. (2007). Restorative dementia gardens: Exploring how design may ameliorate attention fatigue. *Journal of Housing for the Elderly, 21*, 73–88.

Diaz Moore, K., Geboy, L. D., & Weisman, G. D. (2006). *Designing a better Day. Guidelines for adults and dementia day services centres*. Baltimore: The Johns Hopkins University Press.

Evans, G. W., & McCoy, J. M. (1998). Why buildings don't work: The role of architecture in human health. *Journal of Environmental Psychology, 18*, 85–94.

Eyles, J., & Williams, A. (2008). *Sense of place, health and quality of life*. Aldershot: Ashgate.

Falk, H., Wijk, H., & Persson, L. O. (2009). The effects of refurbishment on residents' quality of life and wellbeing in two Swedish residential care facilities. *Health & Place, 15*, 717–724.

Fleming, R., Forbes, I., & Bennett, K. (2003). *Adapting the ward for people with dementia*. Sidney: The Hammond Care Group.

Fogel, B. S. (1992). Psychological aspects of staying at home. *Generations, 16*, 15–19.

Fornara, F., & Andrade, C. (2012). Healthcare environments. In S. Clayton (Ed.), *The Oxford handbook of environmental and conservation psychology* (pp. 295–315). New York: Oxford University Press.

Fornara, F., Bonaiuto, M., & Bonnes, M. (2006). Perceived hospital environment quality indicators: A study of orthopaedic units. *Journal of Environmental Psychology, 26*, 321–334.

Fried, T. R., Bradley, E. H., Williams, C. S., & Tinetti, M. E. (2001). Functional disability and health care expenditures for older persons. *Archives of Internal Medicine, 161*, 2602–2607.

Gadalla, T. (2009). Sense of mastery, social support, and health in elderly Canadians. *Journal of Aging and Health, 21*, 581–595.

Garcia, L. J., Hébert, M., Kozak, J., Sénécal, I., Slaughter, S. E., Aminzadeh, F., Dalziel, W., Charles, J., & Eliasziw, M. (2012). Perceptions of family and staff on the role of the environment in longterm care homes for people with dementia. *International Psychogeriatrics, 34*, 753–765.

Geboy, L. D., & Diaz Moore, K. (2005). Considering organizational competence: A theoretical extension of Lawton and Nahemow's competence-press model. In H. Chaudhury & A. Mahmood (Eds.), *Proceedings of the environmental design research association* (Vol. 1). Edmond: EDRA.

George, L. K. (2006). Perceived quality of life. In L. H. Binstock & L. K. George (Eds.), *Handbook of aging and the social sciences* (6th ed., pp. 320–336). New York: Academic.

Gibson, J. J. (1979). *The ecological approach to visual perception*. Boston: Houghton-Mifflin.

Gifford, R. (2002). *Environmental psychology: Principles and practice*. Boston: Allyn & Bacon.

Gitlin, L. N., Corcoran, M. A., Winter, L., Boyce, A., & Marcus, S. (1999). Predicting participation and adherence to a home environmental intervention among family caregivers of persons with dementia. *Family Relations, 48*, 363–372.

Giuliani, M. V., Scopelliti, M., Fornara, F. (2005). *Elderly people at home: Technological help in everyday activities*. Proceedings of 14th IEEE international workshop on robot and human interactive communication (pp. 365–370). Nashville: Omnipress.

Global AgeWatch Index. (2014). *Insight report*. London: HelpAge International.

Golant, S. M. (2011). The quest for residential normalcy by older adults: Relocation but one pathway. *Journal of Aging Studies, 25*, 193–205.

Granbom, M., Himmelsbach, I., Haak, M., Löfqvist, C., Oswald, F., & Iwarsson, S. (2014a). Residential normalcy and environmental experiences of very old people: Changes in residential reasoning over time. *Journal of Aging Studies, 29*, 9–19.

Granbom, M., Löfqvist, C., Horstmann, V., Haak, M., & Iwarsson, S. (2014b). Relocation to ordinary or special housing in very old age: Aspects of housing and health. *European Journal of Ageing, 11*, 55–65.

Hamilton, K. (2003). The four levels of evidence-based design practice. *Healthcare Design, 3*, 18–26.

Holahan, C. J. (1978). *Environment and behavior: A dynamic perspective*. New York: Plenum Press.

Hutchings, B. L., Olsen, R. V., & Moulton, H. J. (2008). Environmental evaluations and modifications to support aging at home with a developmental disability. *Journal of Housing for the Elderly, 22*, 286–310.

Hwang, E., Cummings, L., Sixsmith, A., & Sixsmith, J. (2011). Impacts of home modifications on aging-in-place. *Journal of Housing for the Elderly, 25*, 246–257.

Imamoglu, C. (2007). Assisted living as a new place schema: A comparison with homes and nursing homes. *Environment and Behaviour, 39*(2), 246–268.

Iwarsson, S. (2004). Assessing the fit between older people and their physical environments: An occupational therapy research perspective. *Annual Review of Gerontology and Geriatrics, 23*, 85–109.

Iwarsson, S., & Stahl, A. (2003). Accessibility, usability, and universal design: Positioning and definition of concepts describing person–environment relationships. *Disability and Rehabilitation, 25*, 57–66.

Iwarsson, S., Nygren, C., & Slaug, B. (2005). Cross-national and multi-professional inter-rater reliability of the Housing Enabler. *Scandinavian Journal of Occupational Therapy, 12*, 29–39.

Iwarsson, S., Nygren, C., Oswald, F., Wahl, H., & Tomsone, S. (2006). Environmental barriers and housing accessibility problems over a one-year period in later life in three european countries. *Journal of Housing for the Elderly, 20*, 23–43.

Joseph, A., Zimring, C., Harris-Kojetin, L., & Kiefer, K. (2005). Presence and visibility of outdoor and indoor physical activity features and participation in physical activity among older adults in retirement communities. *Journal of Housing for the Elderly, 19*, 141–165.

Kahana, E. (1982). A congruence model of person-environment interaction. In M. P. Lawton, P. G. Windley, & T. O. Byerts (Eds.), *Aging the environment theoretical approaches* (pp. 97–120). New York: Sage Publications.

Kahana, E., Lovegreen, L., Kahana, B., & Kahana, M. (2003). Person, environment, and person-environment fit as influences on residential satisfaction of elders. *Environment and Behavior, 35*, 434–453.

Kaplan, S. (1983). A model of person environment compatibility. *Environment and Behavior, 15*, 311–332.

Kaplan, S. (1995). The restorative benefits of nature: Towards an integrative framework. *Journal of Environmental Psychology, 15*, 169–182.

Kaplan, S. (2001). Meditation, restoration, and the management of mental fatigue. *Environment and Behavior, 33*, 480–506.

Kearney, A. R., & Winterbottom, D. (2005). Nearby nature and long-term care facility residents: Benefits and design recommendations. *Journal of Housing for the Elderly, 19*, 7–28.

Keller, B. K., & Potter, J. F. (1994). Predictors of mortality in outpatient geriatric evaluation and management clinic patients. *Journal of Gerontology, 49*, 246–251.

Kellett, P., Gilroy, R., & Jackson, S. (2005). Space, identity and choice: Exploring the housing arrangements of older people. In B. Martens & A. G. Keul (Eds.), *Designing social innovation: Planning, building, evaluating* (pp. 291–299). Göttingen: Hogrefe & Huber Publishers.

Korpela, K. M. (2012). Place attachment. In S. Clayton (Ed.), *The Oxford handbook of environmental and conservation psychology* (pp. 148–163). New York: Oxford University Press.

Küller, R. (1988a). Housing for the elderly in Sweden. In D. Canter, M. Krampen, & D. Stea (Eds.), *Ethnoscapes: Environmental policy, assessment and communication* (pp. 199–224). Newcastle upon Tyne: Athenaeum.

Küller, R. (1988b). *Environmental activation of old persons suffering from senile dementia.* Abstracts of the X IAPS Conference (p. 76). Delft: Delft University Press.

Küller, R. (1991). Familiar design helps dementia patients cope. In W. F. E. Preiser, J. C. Vischer, & E. T. White (Eds.), *Design intervention: Toward a more humane architecture* (pp. 255–267). New York: Van Nostrand Reinhold.

Kuriansky, J., Gurland, B., Fleiss, J. L., & Cowan, D. (1976). The assessment of self-care capacity in geriatric psychiatric patients by objective and subjective methods. *Journal of Clinical Psychology, 32*, 95–102.

Kweon, B., Sullivan, W. C., & Wiley, A. R. (1998). Green common spaces and the social integration of inner city older adults. *Environment and Behaviour, 30*, 832–858.

Laitinen, P., & Isola, A. (1996). Promoting participation of informal caregivers in the hospital care of the elderly patient: Informal caregivers' perceptions. *Journal of Advanced Nursing, 23*, 942–947.

Lawrence, R. J. (2002). Healthy residential environments. In R. Bechtel & A. Churchman (Eds.), *Handbook of environmental psychology* (pp. 394–412). New York: Wiley.

Lawton, M. P. (1982). Time budgets of older people: A window on four lifestyles. *Journal of Gerontology, 37*, 115–123.

Lawton, M. P. (1989). Environmental proactivity in older people. In V. L. Bengston & K. W. Schaie (Eds.), *The course of later life: Research and reflections* (pp. 15–23). New York: Springer.

Lawton, M. P. (1990). Knowledge resources and gaps in housing for the aged. In D. Tilson (Ed.), *Aging in place: Supporting the frail elderly in residential environments* (pp. 287–309). Glenview: Scott, Foresman, and Company.

Lawton, M. P. (1998). Environment and aging: Theory revisited. In R. J. Scheidt & P. G. Windley (Eds.), *Environment and aging theory: A focus on housing* (pp. 1–32). Westport: Greenwood Press.

Lawton, M. P. (1999). Quality of life in chronic illness. *Gerontology, 45*, 181–183.

Lawton, M. P., & Nahemow, L. (1973). Ecology and the aging process. In C. Eisdorfer & M. P. Lawton (Eds.), *The psychology of adult development and aging* (pp. 619–674). Washington: American Psychological Association.

Leith, K. H. (2006). "Home is where the heart is … or is it?" A phenomenological exploration of the meaning of home for older women in congregate housing. *Journal of Aging Studies, 20*, 317–333.

Lemke, S., & Moos, R. H. (1986). Quality of residential settings for elderly adults. *Journal of Gerontology, 41*(2), 268–276.

Lewin, K. (1946). Action research and minority problems. *Journal of Social Issues, 2*, 34–36.

Löfqvist, L., Granbom, M., Himmelsbach, I., Iwarsson, S., Oswald, F., & Haak, M. (2013). Voices on relocation and aging in place in very old age: A complex and ambivalent matter. *The Gerontologist, 53*, 919–927.

Löfqvist, L., Slaug, B., Ekström, H., Kylberg, M., & Haak, M. (2014). Use, non-use and perceived unmet needs of assistive technology among Swedish people in the third age. *Disability and Rehabilitation. Assistive Technology*. doi:10.3109/17483107.2014.961180.

Lord, S., Despres, C., & Ramadier, T. (2011). When mobility makes sense: A qualitative and longitudinal study of the daily mobility of the elderly. *Journal of Environmental Psychology, 31*(1), 52–61.

Macintyre, S., Ellaway, A., & Cummins, S. (2002). Place effects on health: How can we conceptualise, operationalise and measure them? *Social Science & Medicine, 55*, 125–139.

Marans, R. W., Hunt, M. E., & Vakalo, K. L. (1984). Retirement communities. In I. Altman, M. P. Lawton, & J. F. Wohlwill (Eds.), *Human behavior and environment: Advances in theory and research* (Elderly people and the environment, Vol. 7, pp. 57–93). New York: Plenum Press.

Marquardt, G. (2014). Dementia-friendly architecture: Integrating evidence in architectural design. In E. Edgerton, O. Romice, & K. Thwaites (Eds.), *Bridging the boundaries: Human experience in the natural and built environment and implications for research, policy, and practice* (pp. 33–45). Göttingen: Hogrefe Publishing.

Marquardt, G., & Schmieg, P. (2009). Dementia-friendly architecture: Environments that facilitate wayfinding in nursing homes. *American Journal of Alzheimer's Disease & Other Dementias, 24*, 333–340.

Marquardt, G., Johnston, D., Black, B. S., Morrison, A., Rosenblatt, A., Lyketsos, C. G., & Samus, Q. M. (2011). A descriptive study of home modification for people with dementia and barriers to implementation. *Journal of Housing for the Elderly, 25*, 258–273.

Marx, L., Haschka, B., & Schnur, P. (2002). Mehr Lux – mehr Wohlbefinden. Die richtige Beleuchtung hat positiven Einfluss auf demente Bewohner. *Altenheim, 41*, 57–61.

Melin, L., & Gotestam, K. G. (1981). The effects of rearranging ward routines on communication and eating behaviors of psychogeriatric patients. *Journal of Applied Behavior Analysis, 14*, 47–51.

Migita, R., Yanagi, H., & Tomura, S. (2005). Factors affecting the mental health of residents in a communal-housing project for seniors in Japan. *Archives of Gerontology and Geriatrics, 41*, 1–14.

Miskelly, F. G. (2001). Assistive technology in elderly care. *Age and Ageing, 30*, 455–458.

Mooney, P., & Nicell, P. L. (1992). The importance of exterior environment for Alzheimer residents: Effective care and risk management. *Healthcare Management Forum, 5*, 23–29.

Moore, J. (2000). Placing home in context. *Journal of Environmental Psychology, 20*, 207–217.

Moos, R. H., & Lemke, S. (1984). Supportive residential settings for older people. In I. Altman, M. Powell Lawton, & J. F. Wohlwill (Eds.), *Human behavior and environment: Advances in theory and research* (Elderly people and the environment, Vol. 7, pp. 150–190). New York: Plenum Press.

Morgan, D., & Stewart, N. J. (1998). Multiple occupancy versus private rooms on dementia care units. *Environment and Behavior, 30*, 487–503.

Nagasawa, Y. (2000). The geography of hospitals. In S. Wapner, J. Demick, T. Yamamoto, & H. Minani (Eds.), *Theoretical perspectives in environment-behavior research* (pp. 217–227). New York: Kluwer.

Namazi, K. H., & Johnson, B. D. (1991). Physical environment cues to reduce the problems of incontinence in Alzheimer's disease units. *American Journal of Alzheimer's Disease & Other Dementias, 6*, 22–28.

Nathan, A. G., Wood, L. J., & Giles-Corti, B. (2012). Perceptions of the built environment and associations with walking among retirement village residents. *Environment and Behavior, 17*, 1–24.

Netuveli, G., Wiggins, R. D., Hildon, Z., Montgomery, S. M., & Blane, D. (2006). Quality of life at older ages: Evidence from the English longitudinal study of aging (wave 1). *Journal of Epidemiological Community Health, 60*(4), 357–363.

Nygren, C., & Iwarsson, S. (2009). Negotiating and effectuating relocation to sheltered housing in old age: A Swedish study over 11 years. *European Journal of Ageing, 6*, 177–189.

Orpwood, R., Bjørneby, S., Hagen, I., Mäki, O., Faulkner, R., & Topo, P. (2004). User involvement in dementia product development. *Dementia, 3*, 263–279.

Oswald, F., & Wahl, H. W. (2004). Housing and health in later life. *Reviews on Environmental Health, 19*(3–4), 223–252.

Oswald, F., Schilling, O., Wahl, H. W., & Gäng, K. (2002). Trouble in paradise? Reasons to relocate and objective environmental changes among well-off older adults. *Journal of Environmental Psychology, 22*, 273–288.

Pallasmaa, J. (1995). Identity, intimacy, and domicile: Notes on the phenomenology of home. In D. N. Benjamin, D. Stea, & D. Saile (Eds.), *The home: Words, interpretations, meanings, and environments* (pp. 131–147). Aldershot: Avebury.

Passini, R., Pigot, H., Rainville, C., & Tétreault, M. H. (2000). Wayfinding in a nursing home for advanced dementia of the Alzheimer's type. *Environment and Behavior, 32*, 684–710.

Pastalan, L. A., & Barnes, J. E. (1999). Personal rituals: Identity, attachment to place, and community solidarity. In S. Benyamin & B. Ruth (Eds.), *Ageing, autonomy, and architecture: Advances in assisted living* (pp. 81–89). Baltimore: John Hopkins University Press.

Pearlin, L. I., & Schooler, C. (1978). The structure of coping. *Journal of Health and Social Behavior, 19*, 2–21.

Phillips, D. R., Siu, O. L., Yeh, A. G. O., & Cheng, K. H. C. (2005). The impacts of dwelling conditions on older persons' psychological well-being in Hong Kong: The mediating role of residential satisfaction. *Social Science & Medicine, 60*, 2785–2797.

Pinquart, M., & Sörensen, S. (2000). Influences of socioeconomic status, social network, and competence on subjective well-being in later life: A meta-analysis. *Psychology and Aging, 15*, 187–224.

Pol, E. (2009). Sostenibilidad, ciudad y medio ambiente. Dinámicas urbanas y construcción de valores ambientales [Sustainability, city and environment. Urban dynamics and construction of environmental values]. In R. Garcia Mira & P. Vega Marcote (Eds.), *Sostenibilidad, valores y cultura ambiental [Sustainability, values and environmental culture]* (pp. 183–209). Madrid: Pirámide.

Pope, N., & Kang, B. (2010). Residential relocation in later life: A comparison of proactive and reactive moves. *Journal of Housing for the Elderly, 24*, 193–207.

Prieto-Flores, M. E., Fernandez-Mayoralas, G., Forjaz, M. J., Rojo-Perez, F., & Martinez-Martin, P. (2011). Residential satisfaction, sense of belonging and loneliness among older adults living in the community and in care facilities. *Health & Place, 17*, 1183–1190.

Rodin, J. (1990). Control by any other name: Definitions, concepts, and processes. In J. Rodin, C. Schooler, & K. W. Schaie (Eds.), *Selfdirectedness: Cause and effects throughout the life course* (pp. 1–17). Hillsdale: Erlbaum.

Rojo-Perez, F., Fernandez-Mayoralas, G., Rodriguez-Rodriguez, V., & Rojo-Abuin, J. M. (2007). The environments of ageing in the context of the global quality of life among older people living in family housing. In H. Mollenkopf & A. Walker (Eds.), *Quality of life in old age. International and multidisciplinary perspectives* (pp. 123–150). Dordrecht: Springer.

Ronch, J. L. (2003). Leading culture change in long-term care: A map for the road ahead. In A. Weiner & J. L. Ronch (Eds.), *Culture change in long-term care* (pp. 65–80). Binghamton: Haworth Press.

Rothermund, K., & Brandtstädter, J. (2003). Coping with deficits and losses in later life: From compensatory action to accommodation. *Psychology & Aging, 18*, 896–905.

Rowe, J. W., & Kahn, R. L. (1997). Successful aging. *The Gerontologist, 37*, 433–440.

Rubinstein, R. L., & De Medeiros, K. (2004). Ecology and the aging self. In H. W. Wahl, R. J. Scheidt, & P. G. Windley (Eds.), *Annual review of gerontology and geriatrics. Aging in context: Socio-physical environments* (pp. 59–84). New York: Springer.

Rubinstein, R. L., & Parmalee, P. A. (1992). Attachment to place and the representation of the life course by the elderly. In I. Altman & S. M. Low (Eds.), *Human behavior and environment: Advances in theory and research* (Place attachment, Vol. 12, pp. 139–163). New York: Plenum Press.

Safran-Norton, C. E. (2010). Physical home environment as a determinant of aging in place for different types of elderly households. *Journal of Housing for the Elderly, 24*, 208–231.

Scanlon, E., & Devine, K. (2001). Residential mobility and youth wellbeing: Research, policy, and practice issues. *Journal of Sociology and Social Welfare, 28*, 119–138.

Scheidt, R. J., & Windley, P. G. (2006). Environmental gerontology: Progress in the Post-Lawton Era. In J. E. Birren, K. W. Schaie, R. P. Abeles, M. Gatz, & T. A. Salthouse (Eds.), *Handbook of the psychology of aging* (6th ed., pp. 105–125). New York: Academic.

Schwarz, B. (1997). Nursing home design: A misguided architectural model. *Journal of Architectural and Planning Research, 14*, 343–357.

Scopelliti, M., Giuliani, M. V., & Fornara, F. (2005). Robots in a domestic setting: A psychological approach. *Universal Access in the Information Society, 4*, 146–155.

Slangen-de Kort, Y. A. W., Midden, C. J. H., & van Wagenberg, A. F. (1998). Predictors of the adaptive problem-solving of older persons in their homes. *Journal of Environmental Psychology, 18*, 187–197.

Snowdon, D. A. (2001). *Aging with grace: What the Nun Study teaches us about leading longer, healthier, and more meaningful lives*. New York: Bantam.

Sommer, R. (1983). *Social design*. Englewood Cliffs: Prentice-Hall.

Staats, H. (2012). Restorative environments. In S. Clayton (Ed.), *The Oxford handbook of environmental and conservation psychology* (pp. 445–458). New York: Oxford University Press.

Stock, C., & Ellaway, A. (2013). *Neighbourhood structure and health promotion: An introduction*. New York: Springer.

Stone, R. I. (2006). Emerging issues in long-term care. In L. H. Binstock & L. K. George (Eds.), *Handbook of aging and the social sciences* (6th ed., pp. 397–418). New York: Academic.

Stone, P., Ream, E., Richardson, A., Thomas, H., Andrews, P., Campbell, P., Dawson, T., Edwards, J., Goldie, T., Hammick, M., Kearney, N., Lean, M., Rapley, D., Smith, A. G., Teague, C., & Young, A. (2003). Cancer related fatigue – a difference of opinion? Results of a multicentre survey of healthcare professionals, patients and caregivers. *European Journal of Cancer Care, 12*, 20–27.

Strawbridge, W. J., Wallhagen, M. I., & Cohen, R. D. (2002). Successful aging and wellbeing: Self-report compared with Rowe and Kahn. *Gerontologist, 42*, 727–733.

Stroebe, W. (1995). *Social psychology and health*. Maidenhead: Open University Press.

Sugiyama, T., Ward Thompson, C., & Alves, S. (2009). Associations between neighborhood open space attributes and quality of life for older people in Britain. *Environment and Behavior, 41*, 3–21.

Ulrich, R. S. (1984). View through a window may influence recovery from surgery. *Science, 224*, 420–421.

Ulrich, R. S., Zimring, C., Quan, X., Joseph, A., & Choudhary, R. (2004). *The role of the physical environment in the hospital of the 21st century: A once-in-a-lifetime opportunity*. Report to the Center for Health Design for the Designing the 21st Century Hospital Project.

van Hoof, J. (2008). Forty years of Fanger's model of thermal comfort: Comfort for all? *Indoor Air, 18*, 182–201.

van Hoof, J., & Kort, H. S. M. (2009). Supportive living environments: A first concept of a dwelling designed for older adults with dementia. *Dementia, 8*, 293–316.

van Hoof, J., Aarts, M. P. J., Rense, C. G., & Schoutens, A. M. C. (2009). Ambient bright light in dementia: Effects on behaviour and circadian rhythmicity. *Building and Environment, 44*, 146–155.

Verderber, S., & Fine, D. J. (2000). *Healthcare architecture in an era of radical transformation*. New Haven: Yale University Press.

Wakamura, T., & Tokura, H. (2001). Influence of bright light during daytime on sleep parameters in hospitalized elderly patients. *Journal of Physiological Anthropology and Applied Human Science, 20*, 345–351.

Weenig, M. W. H., & Staats, H. (2010). The impact of a refurbishment of two communal spaces in a care home on residents' subjective well-being. *Journal of Environmental Psychology, 30*, 542–552.

Wells, N. M., & Rollings, K. A. (2012). The natural environment: Influences on human health and function. In S. Clayton (Ed.), *The Oxford handbook of environmental and conservation psychology* (pp. 509–523). New York: Oxford University Press.

Willis, S. L. (1996). Everyday cognitive competence in elderly persons: Conceptual issues and empirical findings. *The Gerontologist, 36*, 595–601.

Wilmott, M. (1986). The effect of a vinyl floor surface and carpeted floor surface upon walking in elderly hospital in patients. *Age and Aging, 15*, 119–120.

Windle, G., & Woods, R. T. (2004). Variations in subjective well-being: The mediating role of a psychological resource. *Aging and Society, 24*, 583–602.

Windle, G. S., Burholt, V., & Edwards, R. T. (2006). Housing related difficulties, housing tenure and variations in health status: Evidence from older people in Wales. *Health & Place, 12*, 267–278.

Windley, P. G., & Weisman, G. (2004). Environmental gerontology research and practice: The challenge of application. *Annual Review of Gerontology and Geriatrics, 23*, 334–365.

Wister, A. V. (1989). Environmental adaptation by persons in their later life. *Research on Aging, 11*, 267–291.

Wolinsky, F. D., Coe, R. M., Miller, D. K., Prendergast, J. M., Creel, M. J., & Chavez, M. N. (1983). Health services utilization among the non-institutionalized elderly. *Journal of Health and Social Behavior, 24*, 325–337.

World Alzheimer Report. (2014). *Dementia and risk reduction. An analysis of protective and modifiable factors*. London: Alzheimer's Disease International (ADI).

World Health Organization. (2013). *The world health report 2013: Research for universal health coverage*. Geneva: World Health Organization.

Young, A. F., Russell, A., & Powers, J. R. (2004). The sense of belonging to a neighbourhood: Can it be measured and is it related to health and well being in older women? *Social Science & Medicine, 59*, 2627–2637.

Zeisel, J. (2009). *I'm still here: A breakthrough approach to understanding someone living with Alzheimer's*. New York: Penguin.

Zeisel, J., Silverstein, N. M., Hyde, J., Levkoff, S., Lawton, M. P., & Holmes, W. (2003). Environmental correlates to behavioral health outcomes in Alzheimer's special care units. *The Gerontologist, 43*, 697–711.

Part VII

Quality of Life and Environmental Threats – *Environmental Stressors and Risks*

Environmental Stress

25

Birgitta Gatersleben and Isabelle Griffin

25.1 What Is Environmental Stress?

Environmental stress can be defined as the emotional, cognitive and behavioral responses to an environmental stimulus (or stressor). Much of the research on environmental stress focuses on examining how different environmental stimuli affect such psychological consequences. However, there are different theoretical perspectives on the mechanisms that underlie stress responses. These different models will be outlined below and they can be distinguished in two different types of models. The Arousal Theory and the Cognitive Load theory have a stronger basis in physiological arousal theories. The Environmental Stress Theory and the Behavior Constraint Theory, on the other hand, have a stronger basis in psychological models and focus on subjective appraisals of environmental stimuli. Individual differences are important in both types of perspectives and it is generally agreed that environmental stress is a function of individual and physical factors. As such, Bilotta and Evans (2013) define environmental stress as an "imbalance between environmental demands and human response capabilities" (p. 28).

It is important to distinguish chronic and acute stress. Cannon (1932) describes acute stress as a fight-flight reaction to a potential stressor, which is associated with an activation of the human sympathetic nervous system. For instance, the increase in heart rate that may be experienced on hearing a sudden very loud noise. Such a stress reaction is short-lived and the human body quickly returns to homeostasis. The General Adaptation Syndrome (GAS) proposed by Seyle (1956) explains how chronic exposure to such stressors can cause significant damage by suggesting that recovery from acute stress requires energy and is costly for the human body, thus long-term exposure and repeated responses to a stressor can result in wear and tear and exhaustion (inability to cope any longer). Similarly, the Behavior Constraint Model (Proshansky et al. 1970) suggests that continued frustration to regain control over an environmental stressor can result in learned helplessness and the Cognitive Load perspective suggests that continued engagement of directed attention can result in mental fatigue (Kaplan and Kaplan 1989).

Although environmental stress is usually understood as a negative experience this may not always be the case. Sometimes exposure to an acute stressor (e.g. a roller-coaster ride) can be fun. Moreover, too little environmental stimula-

B. Gatersleben (✉) • I. Griffin
Environmental Psychology Research Centre, School of Psychology, University of Surrey Guildford, Surrey GU2 7XH, UK
e-mail: B.Gatersleben@surrey.ac.uk

tion can be boring and cause drowsiness (Berlyne 1960; Yerkes and Dodson 1908). However, exposure to chronic stress is always negative. Whereas exposure to acute stress is commonly understood to result in short-term psychological and physiological responses that disappear rather quickly (e.g. increased heart rate after leaving the rollercoaster ride), long-term or frequent exposure to stressors will result in wear and tear, learned helplessness and exhaustion.

25.1.1 Environmental Stress Theory

An important model of stress in psychology, developed by Lazarus (1966) and known as the *environmental stress theory*, suggests that stress is a product of an external stimulus and an individual's appraisal of their ability to cope with this stimulus. This helps to explain why not all environmental stimuli will cause stress for everybody all of the time; whether stress occurs is dependent on individual and contextual factors. Two types of appraisal are important: "primary" occurs when evaluating the stressor from personal and situational factors, and "secondary" denotes appraisal of the individual's own coping mechanisms. Environmental stress theory thus has two key elements: an environmental stressor and a subjective cognitive appraisal of that stressor.

25.1.2 Behavior Constraint Model

People's subjective assessment of their ability to control a stressor is essential in the *behavior constraint model* (Proshansky et al. 1970; Bechtel and Churchman 2003). This model suggests that when people experience a loss of control in the face of environmental threats, they initially try to regain control. An example of this might be encountering a crowded tube station at rush hour and, instead of joining the throng and accepting a crowded and stressful commute home, turning around and finding an alternative way home or raising the issue with a station official. Another way to cope with such environmental stressors may be to reduce eye contact with others in an attempt to withdraw from the situation and regain a sense of privacy (Cave 1998). There are similarities here with Seyle's (1956) General Adaptation Syndrome, which suggests that the stress process has three stages – alarm, resistance, and finally exhaustion. According to the behavior constraint model, if efforts to regain control fail, this can result in learned helplessness.

The perception of control is extremely important. This does not have to be actual behavioral control over the stressor, stimulus or constraint but can also take the form of cognitive control (understanding the threat) or decisional control (feeling able to choose). For instance, hearing someone else's music while you are trying to write a report can be less stressful if you believe that your neighbor will turn down the music if you ask, if you understand why he/she is playing this music, or if you feel that you can go to another room, even if you do not do any of these things. The importance of a sense of control was demonstrated nicely in a study by Rodin et al. (1978), which found that people experience less crowding in a lift when they are positioned next to the lift buttons.

25.1.3 Conditional Model of Stress

Moser (1994) proposed a model of stress specifically related to the effect of the urban environment and the issue of control, which has clear links to the behavior constraint model (Proshansky et al. 1970). He suggests that stress arises from an interaction between a person and their environment, but the effects of a stressor depend on the individual's control and perception of control. Moser states that control can take four forms: directly dealing with the source of the stress in the situation (turning loud music down), adjusting your own behavior to deal with the stressor (leaving the room where the loud music is playing), or simply believing you are capable of doing any of these actions. The effects of a stressor are therefore moderated by a feeling of control over its source (Moser 1994, p. 152). Living in an urban environment means that people are exposed to a constant level of stress, from traffic noise to crowding, which country-dwellers do not necessarily experience. This leads to a process of adaptation and habituation. Stress increases when acute stressors (with a different level of stimulation than the individual is used to) arise. As such, city dwellers have a different reference point in relation to environmental stress, and thus Moser's model is based on the intensity of the environmental stimuli and cognitive operations. If the amount of activation does not differ significantly from their (urban) baseline level or if it is very weak, automatic reactions ensue (such as narrowing of attention or decrease in eye contact), and hence individuals often do not notice any change. If the activation is too intense, "automatic disorganized reactions" occur, such as aggressiveness or intense emotion. If activation is at a medium level, cognitive processes are employed to appraise the situation and attempts are made to cope with the stressor. However, if coping attempts do not work, "automatic disorganized reactions" occur, which can increase due to failing to cope with the situation in the first place (i.e. a lack of control) resulting in feelings of confusion and helplessness.

25.1.4 Adaptation Level Theory

Moser's model has links with other work that suggests that responses to environmental stimuli may depend on what we are used to. Adaptation level theory, as it is now known, was suggested by Wohlwill (1974) and stems from Helson's (1964) theory of adaptation level relating to sensation and perception. Before this, Fiske and Maddi (1961) referred to an adaptation level of stimulation where the individual's optimal level of stimulation could be influenced by past experiences of stimulation. This can be used to explain stressors such as noise and air pollution, which city dwellers may become used to resulting in a different optimal level of stimulation from those living in the country where noise and air pollution levels are much lower. More recent models of stress, however, suggest that long-term exposure to stressors may not simply result in adaptation to those stressors but may have longer-term negative effects on the body due to the constant state of adjustment (McEwen 1998).

25.1.5 Arousal Theory

In the models above, coping and control appraisals play a central role. There are, however, other models that do not specifically focus on such subjective appraisals but have stronger links to physiological stress models. *Arousal theory* proposes that there is an optimum level of arousal under which people perform best. Environmental stimulation from stressors outside of this optimum level can have physiological effects, such as increased heart rate and blood pressure (Cave 1998), and psychological effects. Importantly, both levels that are too high and those that are too low are suboptimal. Berlyne (1960), for instance, suggests that hedonic tone results from an optimum level of stimulation. When arousal is too low (boredom) or too high (stress), aesthetic preferences will suffer. Similarly, the Yerkes-Dodson law suggests that individual performance will suffer when arousal is too high or too low (Moch 1989; Yerkes and Dodson 1908). What is too low or too high will depend on individual and situational factors. This fits with the understimulation perspective, also known as restricted environmental stimulation, which may have originally been considered by Parr (1966) who discussed the monotonous effect of the urban environment on our behavior, thoughts and feelings. Other research related to this approach involves sensory deprivation of which Zubek (1969) gives a comprehensive overview.

25.1.6 Environmental Load Theory

Environmental load or overload theory has a basis in cognitive theory. Originally considered in his paper on the evolution of urban norms, Milgram (1970) states that overload derives from system analysis, and denotes "a system's inability to process inputs from the environment because there are too many inputs for the system to cope with, or because successive inputs come so fast that input A cannot be processed when input B is presented" (p. 1462). Cohen (1978) subsequently developed the theory to relate to individual responses by suggesting that total available attentional capacity is not fixed but shrinks when subjected to prolonged demands. Cohen's (1978) theory is linked to Kahneman's (1973) analysis of attention, which states that there is a limit to an individual's attention capacity to carry out tasks. Prolonged engagement of directed attention results in directed attention fatigue, which can lead to tunnel vision. However, new stimuli, especially those that are intense and unpredictable, still need attention that we cannot then cope with. Moser (1998a) called this a multitude of stimulation, where the individual's efforts to cope with different stressors can result in short- and long-term negative consequences. Depending on how good we are at screening out irrelevant information, how many stimuli need attention and how depleted our directed attention is, overload can result in irritability, intolerance, frustration and errors. Living in urban environments makes many demands on our directed attention. In urban environments (or at work or school), people are surrounded by many stimuli that demand directed attention (traffic lights, traffic noise, phones ringing, people talking, fire alarms, etc.). Spending time in such environments can result in directed attention fatigue from which people need to recover to resume optimum functioning. Getting away, and in particular spending time in natural environments, can promote such recovery (Kaplan and Kaplan 1989).

25.2 Individual Differences

Although some of the models described above focus on it more specifically, all of them allow for individual differences. Stress responses are a function of certain aspects of the stimulus as well as individual and contextual elements. As noted before, environmental stress is usually conceptualized as a psychological response to an external stimulus. Only when these stimuli outweigh a person's (perceived) ability to cope with them will stress occur. There are many individual factors that affect an individual's ability to cope with environmental demands, including personality factors and mental fatigue (Bell et al. 2001).

Although the arousal and environmental load perspectives do not specifically focus on individual subjective appraisals, they do allow for individual differences as well. For instance, the environmental load theory suggests that environmental stimuli are more likely to be stressful when someone is mentally fatigued than when they are not. Consider, for example, walking down a busy shopping street when you are not really looking for anything and are not in a hurry compared to walking down the same street when you have had a busy day at work, the shops are about to close, and you cannot find that important birthday present that you need for tonight. Your experiences in such a situation can be explained well by environmental load theory.

Moreover, for someone who is normally extremely conscientious and hardworking, being underprepared for an exam may cause much more stress than for someone else who has a more carefree and laidback attitude to passing the exam. As such, if the exam conditions become more difficult, for example due to extreme heat or noise, the more carefree student may cope better than the student who is already concerned about their lack of revision. This might also be explained by environmental load theory, since the underprepared student has already encountered other stressors and so has fewer resources left to deal with the exam, or perhaps the environmental stress perspective, since they may think that they cannot cope with the noise in the exam room and concentrate on the exam for which they feel already insufficiently prepared.

25.3 Environmental Stressors

Evans and Cohen (1987) distinguished four types of environmental stressors: cataclysmic events, stressful life events, daily hassles and ambient stressors. Cataclysmic events refer to infrequent events that have a major impact on people and their environment, such as natural disasters. They tend to affect larger groups of people. Stressful life events denote more personal events that people may experience on a daily basis, such as illness or family problems. Daily hassles refer to things that we experience every day, such as crowding and stressful commutes. Ambient stressors are also referred to as background stressors, such as air pollution or noise but, unlike other stressors, they tend to be tolerated for short periods. Lawrence (2002), in his recommendations for designing healthy residential environments, refers to the importance of avoiding certain stressors, such as extreme temperatures, air pollution and crowded housing conditions, highlighting the important role environmental stress can play in the design of our environment. It is also worth noting that, although different types of stressors have been distinguished and are often studied in isolation, in many situations people are exposed to more than one stressor at a time. Busy cities, for instance, are more likely to be noisy, crowded, and smelly. Equally, it is also important to acknowledge that not all outcomes from environmental stressors are negative. Glass and Singer (1972) note that a little stress in everyday life can be helpful or beneficial, challenging us to cope with the situation.

There are different characteristics of environmental stressors that affect whether they may cause stress or not. The arousal perspective, outlined earlier, refers to an optimum level of arousal, which has been proposed to be a function of complexity, novelty, incongruity and surprise (Berlyne 1960; Kaplan and Kaplan 1989). In terms of sounds, for instance, this means that sounds that are too monotonous or too complex and changeable are more likely to cause stress. Clearly, what is "too" depends on a range of individual and contextual factors.

In the environmental stress literature, the most commonly studied environmental stressors are ambient stressors, such as noise, and social-environmental stressors, which include daily hassles like crowding and personal space invasion. Some of these are briefly discussed below, including examples of research where stressors have been studied in isolation and with other stressors.

25.3.1 Ambient Environmental Stressors

Light is often researched in relation to health, for instance in the study of sleep or seasonal affective disorder. There is plenty of evidence that humans function better in optimum light conditions. Unlike nocturnal animals our arousal levels tend to be higher when we are exposed to daylight. The importance of daylight was shown in a recent study by Smolders et al. (2013) who found that daylight improved feelings of vitality. Whilst bright light is often perceived to be better for us, some recent research (Steidle and Werth 2013) demonstrated that dim light and priming darkness improved creativity by releasing social inhibitions. Moreover, other studies have shown that participants perform better on cognitive tasks in "warm" white lighting compared to "cool" or artificial "daylight" lighting (Knez 2001; Knez and Hygge 2002).

Whilst much research has focused on the brightness and warmth of light, there is also evidence that the frequency of subliminal flicker can have an impact on mood and performance in cognitive tasks (with low frequency being more beneficial than high), suggesting that subliminal perception of our environment can seemingly have an impact on our mood and behavior (Knez 2014). A recent special issue gives an overview of the latest research in environmental psychology examining the role of light in perceptions, behaviors and cognitions (de Kort and Veitch 2014).

Color it is an aspect of environmental stress that is often considered to be under-researched (Cassidy 1997; Bell et al. 2001). However, various studies on color have been carried out to investigate a number of different factors, including individual differences and its effect on other stressors. Much of the existing work has focused on color in relation to autonomous arousal, accompanied by the belief that warm colors such as red and yellow are more "arousing" in terms of psychological (e.g. anxiety; Jacob and Suess 1975) and physiological outcomes than cooler colors like green or blue (Gerard 1958; Wilson

1966). Some of the early work was contradictory, with some researchers finding red to be more physiologically arousing (Gerard 1958; Wilson 1966), and others not (Pressey 1921). However, Pressey (1921) did suggest that the brightness of a color can influence performance, which was a factor not accounted for by Gerard (1958) or Wilson (1966). Later research that controlled for brightness seems to support the finding that colors such as red can have an effect on arousal levels (Jacobs and Hustmyer 1974). More recent work (Mehta and Zhu 2009) has also shown that red (versus blue) induces primarily an avoidance (versus approach) motivation and that red enhances performance on a detail-oriented task, whereas blue enhances performance on a creative task. These effects occur outside of an individual's consciousness.

Despite the work on color being fairly limited, research has also suggested that it can have an effect on other stressors. For instance, rooms with a darker tone of the same color were perceived to be more crowded than their lighter-toned counterparts (Baum and Davis 1976), while red rooms have also been perceived as more closed than blue ones (Küller et al. 2009). Lighter rooms are also considered more open and spacious (Acking and Küller 1972), which suggests that colors may influence our perception of our environment and other people.

Research has also emphasized the importance of individual differences in our perception of color (Dijkstra et al. 2008), which may explain why some studies find significant results (Kwallek and Lewis 1990) and others do not (Ainsworth et al. 1993). For instance, it has been suggested that screening ability, the ability to ignore irrelevant stimuli in an environment (Mehrabian 1977), may have an influence on the way we perceive our environment. Dijkstra et al. (2008) found that the color of a hospital room appeared to have a greater influence on stress reduction (in a green room) and arousal induction (in an orange room) for participants with low stimulus screening ability, i.e. those who were less able to ignore irrelevant stimuli. Although these authors used photographs of colorful rooms, Küller et al. (2009) found similar results when entire rooms were repainted for the purpose of the study; participants in the red room experienced greater arousal that those in the blue room. Interestingly, they also found that those with personality traits such as introversion, as well as those experiencing a negative mood, were more affected by the color of the room than others.

Noise has received a great deal of attention in the literature, with most studies tending to measure and examine the effects of the physical properties of the stimulus as well as subjective perceptions of it. Noise has been defined as unwanted sound (Cohen 1981), a subjective appraisal of a sound, which can be generated in many different ways, such as by transport or other people (Benfield et al. 2012). It is important to note the various forms of noise that have been investigated, and equally the role of individual differences in this form of environmental stress. As Moch (1989) suggests, some people work better whilst listening to music whilst others do not, simply due to the individual differences in their required level of stimulation. Noise has been demonstrated to have a significant impact on human behavior; for instance, a study investigating the effect on altruism of environmental overload and roadworks with or without noise found that noise was the important factor impacting upon whether participants would help another person (Moser 1988).

It is interesting that different types of noise can have a similarly detrimental influence on our behavior. Hygge et al. (2003) demonstrated that performance on a semantic memory task was just as bad under the influence of traffic noise as meaningful irrelevant speech, suggesting that our attention can be impaired by many different types of noise. Studies have also investigated the negative effect of other sounds on our behavior, including a ringing mobile phone (Shelton et al. 2009), certain types of background music (Dobbs et al. 2011; Schlittmeier and Hellbrück 2009) and aircraft noise (Hygge 2003), showing how both indoor and outdoor environmental stressors can affect us. The detrimental effects of noise have been found for different groups, including adults (for a review, Beaman 2005) and children (for a review, see Klatte et al. 2013). For example, Rouleau and Belleville (1996) demonstrated that elderly people and students were both negatively affected by familiar and non-familiar irrelevant speech during a digit-recall task, but neither group was affected by white noise. Similarly, Enmarker (2004) found a detrimental effect of meaningful irrelevant speech and road traffic noise on teachers' recall abilities, but no effect of age. Interestingly, this effect does not extend to visually impaired people – Kattner and Ellermeier (2014) suggest that irrelevant speech does not have the same effect on performance in a word recall task for visually impaired people compared to sighted people due to their improved selective attention abilities. In some ways, this might be linked back to the effects of color and the idea that it has fewer effects on individuals who have a high ability to screen out certain environmental stimuli (Dijkstra et al. 2008), given that both groups are able to screen out the stimuli they deem irrelevant to the task at hand.

Whether noise results in annoyance (negative feelings about the noise) depends on many different factors such as volume, exposure time, predictability, the source (valuable), attitudes towards those who generate noise, beliefs about its consequences, satisfaction with other aspects of the environment, attitudes towards the noise, and sensitivity to it (Bell et al. 2001). Individual differences play an important part in the way we perceive noise and whether this results in noise annoyance; bird sounds may be perceived differently by a night shift worker trying to fall asleep compared to another who works regular hours (Benfield et al. 2012). Moreover, some recent work compared the effects of noise from traffic and an industrial site on perceived noise annoyance by mapping the noise in the survey area and distributing a questionnaire to evaluate noise annoyance (Pierrette et al. 2012). They demonstrated the importance of satisfaction with other environmental aspects when understanding noise annoyance in urban environments, given that a fear of the industrial site was positively correlated with residents' annoyance levels. Our perception of noise has also been linked to personality traits; for instance, it is suggested that neuroticism is related to noise annoyance, and neurotics tend to perform worse on certain cognitive tasks when exposed to noise (von Wright and Vauras 1980; Nurmi and von Wright 1983).

Because noise is easy to perceive it is often used to make judgments about other environmental stressors, such as air pollution caused by urban traffic. On the other hand, visual impact factors can influence noise annoyance as well, as demonstrated by Pedersen and Larsman (2008) in a study on the noise annoyance associated with wind turbines. Noise can also impact upon our perceptions of natural environments; a study comparing the effects of different types of noise on participants' responses to national parks found that motorized noise impacted negatively on the individual's assessment of the landscape quality, with motorcycle noise having the most detrimental effect (Weinzimmer et al. 2014). Although noise is usually studied using linear models (i.e. more noise is worse), researchers have also suggested that people may function best under moderate noise levels. For instance, Mehta et al. (2012) showed that a moderate (70 dB) versus low (50 dB) level of ambient noise enhance performance on creative tasks, whereas a high level of noise (85 dB) impaired creativity. This finding could be used as support for the arousal theory, which refers to an optimum level of stimulation.

Temperature While noise is usually studied in relation to human activity, temperature has more often been studied in relation to climate. Rising temperatures have been associated with aggressive behaviors such as assault (Bell and Fusco 1989) and car horn honking (Baron 1976). Rotton and Cohn (2002) have highlighted the link between temperature and crime, which they demonstrate can partly be explained by the Negative Affect Escape model (Baron 1972, 1978). This suggests that there is an inverted U-shaped relationship between the temperature and aggression; as temperature increases, so does the likelihood of aggressive behavior, but only up to a certain point when increased temperature is linked to decreased aggression. As Cassidy (1997) indicates, this relationship is very similar to that outlined in the Yerkes-Dodson law and the Arousal Theory.

An additional area of research related to temperature is altruism or helping behavior. However, as is the case with some of the other environmental stressors outlined in this chapter, the literature has yet to draw a clear conclusion about the influence of temperature on such behavior. For instance, altruism has been found to decrease over the summer as temperatures increase, but increase over the winter as temperatures increase (Cunningham 1979), suggesting there may be other factors, such as seasonal changes, which also play a role. Conversely, other studies have found no relationship between heat and helping behaviors (Schneider et al. 1980; Bell and Doyle 1983).

The effects of extreme cold have also been discussed in the literature; for example, Hinkle (1961) considered the effects that low temperature may have on brain function. Moreover, low temperature has also been connected to the understimulation theory; there is evidence to suggest that in environments such as the Antarctic, individuals often experience sensory deprivation as well as cognitive impairment (Mullin 1960). A study investigating the aggressive behavior that participants exhibited towards a confederate after receiving negative feedback about a passage they had written offers support for Baron's Negative Affect Escape model (1972). Participants were less likely to act aggressively (by administering electric shocks) in extremely cold (and hot) temperatures (Bell and Baron 1977), which supports the notion of an inverted U-shaped curve. However, it is important to note the contextual factors of temperature. As Suedfeld (1991) notes, often communities living in extreme environments like the Arctic do not experience the adverse effects generally associated with freezing temperatures and do not perceive it to be stressful.

25.3.2 Social-Environmental Stressors

The presence of other people in an environment can be a major source of environmental stress, not only because they cause ambient stressors, such as noise or smell, but also because they are simply there. This is related to a range of concepts in environmental psychology including privacy, personal space, territoriality and crowding.

Privacy is not the same as withdrawal but refers to a process of regulating social interactions. It is related to many of the other social-environmental stressors discussed below, including personal space, territoriality and crowding (Bell et al. 2001), but it is also linked to ambient stressors related to auditory and visual privacy. Pedersen (1999) distinguishes six types of privacy with different functions: solitude, isolation, anonymity, reserve, intimacy with friends, and intimacy with family. Research has demonstrated that we can differ in our preferences for these types of privacy, with women tending to prefer intimacy with family and men preferring intimacy with friends (Demirbas and Demirkan 2000). Altman (1975) defines privacy as "the selective control of access to the self or to one's group" (p. 18). The idea of control is key, since we only feel distressed when we no longer have control over what is known about us (Cassidy 1997; Evans and Cohen 1987). As such, it is only really when we lose control over our privacy that we notice it at all (Cassidy 1997). Privacy regulation is often understood in terms of an equilibrium model, which assumes that people seek a balance between the need to withdraw from others and the need to communicate with others. In some ways, this might be linked to the Arousal Theory, given the links to an optimal level of stimulation and the idea that this might vary from person to person. However, some recent work suggests that these two needs (to withdraw and to communicate) are not necessarily negatively correlated, but may be quite distinct from each other (Haans et al. 2007).

The concept of privacy is often researched with regard to design, since certain design features can facilitate feelings of privacy. This idea has been investigated in office environments, where research has demonstrated that employees moving from enclosed offices, i.e. with physical partitions, to more open-plan environments reported decreased privacy (Sundstrom et al. 1982). More recent research showed that workers in high-walled cubicles experienced less satisfaction in terms of privacy than those in shared enclosed offices or completely open-plan offices without partitions (Lee 2010). Whilst visual privacy is important, auditory privacy is considered, by some, to be more important (Kupritz 1998). Lee (2010) also found that workers were more satisfied with sound privacy when in open-plan offices or enclosed shared offices than offices with high partitions. This may suggest that the lack of control about who can listen to your conversation can impact upon the feeling of auditory privacy, since you may be more aware of who is listening in on your conversation in an open-plan office than in a high-partitioned cubicle.

As well as office design, the importance of privacy in healthcare has also been highlighted (for a review, see Barnes 2002). For instance, Duffy et al. (1986) found that, when considering designs for a new nursing home, staff selected those which enabled social interaction between residents, whereas the residents themselves preferred designs which afforded privacy. Equally, affording privacy through the use of private rooms has been shown to have beneficial effects for nursing home residents suffering from dementia (Morgan and Stewart 1998), and is considered to be a particularly important issue for elderly people (Morgan and Stewart 1999).

Although much of the literature focuses on physical privacy, the growth of the internet has led to the study of online privacy, an important factor in our increasingly technologically-centered lives. This brings into the debate the control we have over what people know about us on the internet, particularly in terms of social media, as well as other influences that might increase our lack of control, such as hackers, targeted advertisements and, more generally, the concept of a "Big Brother" society. Moreover, there is the suggestion that for young people today, the internet has become part of normal social life and that they have "no sense of privacy" (Livingstone 2008, p. 395). As Livingstone (2008) notes, social media sites display as the norm information about the individual that past generations considered private, such as age or religion. However, it has also been found, through qualitative research, that young people are particularly concerned about privacy and highlight the importance of control in determining who can know information about them (Livingstone 2008). In addition, "profile management" (Madden 2012, p. 2) is increasing with more social network users tracking their online privacy than before; for example, by removing their names from tagged photographs or deleting "friends". Another issue with online privacy seems to be the difficulty of using privacy controls, with almost half of users experiencing problems (Madden 2012), which can only contribute to the feeling of a lack of control. Interestingly, if online use is considered in terms of the equilibrium model mentioned above, in some ways it affords privacy, since the user can usually choose to withdraw from or engage with others using the internet.

Personal space has been defined as a bubble around a person's body (Katz 1937) – it is the distance maintained between individuals in different situations (Moser 1998b). Sommer (1959) defined it as an emotionally tinged zone around the human body that people feel is "their space". Personal space involves the expectation that, as the individual respects another's personal space, the other person will do the same for them, and this interaction relies very much on eye contact (Moser 1998b). The size of personal space, or proxemics, varies (Hall 1959, 1966) between an intimate distance (<1.5 ft), a personal distance (1.5–4 ft), a social distance (4–12 ft), and a public distance (12–25 ft).

Much work was carried out in the late 1970s and early 1980s that showed that the optimum amount of personal space varied with attraction, similarity, cultural determinants, gender, age and room shape and size (Bell et al. 2001; for a review, see Altman and Vinsel 1977). For instance, belonging to the same group reduces personal space (Novelli et al. 2010). Although earlier research suggested personal space may vary with gender, recent work (Uzzell and Horne 2006) demonstrated that gender role rather than gender is important for interpersonal distance, with people with more feminine gender roles maintaining less distance. Earlier work has studied the importance of optimal personal space for optimum human functioning; for instance, in learning and professional interactions. This work shows that we not like to have our personal space invaded and that we feel uncomfortable invading other people's personal space (Bell et al. 2001). Moreover, personal space invasion can result in flight behavior (walking away), compensatory reactions, and perceptual withdrawal (Bell et al. 2001; Gifford 2002). Although the majority of personal space research has been conducted with samples of students (Patterson et al. 1971; Krail and Leventhal 1976), other groups have also been investigated (for a review, see Suedfeld 1991) including people suffering from mental health problems (Felipe and Sommer 1966), the armed forces (Dean et al. 1975) and prisoners (Dabbs et al. 1973).

One avenue of research investigates personal space for people with disabilities, which is particularly interesting in terms of visual impairment since proxemics is said to rely on eye contact (Hall 1966). It has been suggested that when one individual closes their eyes, the other will stand closer to them (Argyle and Dean 1965). Although the research in this area is limited, some has considered personal space for elderly people who may experience visual and hearing impairments (Webb and Weber 2003). Hayduk and Mainprize (1980) conducted a study with blind and sighted samples in which the researcher would walk towards the participants (counting, so that completely blind participants could identify their distance through sound) and participants indicated when the distance made them feel slightly uncomfortable. The researchers found that the personal space of blind and sighted individuals was effectively the same. Interestingly, sighted individuals with their eyes closed tended to enlarge their personal space whereas the blind sample (partially and completely blind) mostly did not, suggesting that using auditory perception posed more of a problem for the sighted sample than for the partially and completely blind sample.

Although Hayduk and Mainprize (1980) provided important findings, theoretically it would also be interesting to consider whether visually impaired individuals rely more heavily on the other person to maintain an appropriate personal space than a sighted individual might in a real-life situation. Similarly, with the introduction of other stressors such as crowding, does personal space for visually impaired individuals actually increase without them knowing, i.e. do other people see a sign of their visual impairment, such as a guide dog or a white stick, and purposely allow them greater personal space? This has been investigated by some researchers who found that people who are "stigmatized" by society are afforded more personal space by others (Kleck et al. 1968; Worthington 1974; Rumsey et al. 1982; Davis and Lennon 1983). There have been technological advances to aid blind individuals to identify interpersonal distances, such as using rhythm and vibration (McDaniel et al. 2009), and future research could consider the impacts of technology in relation to personal space.

With advances in technology come changes in the way we perceive and investigate environmental stressors such as personal space. For example, Sardar et al. (2012) considered the differential effect of fast- and slow-moving humans and robots on human perception of personal space. Participants trusted the faster robot more and the faster human less when the "confederates" invaded their personal space, suggesting that what or who is invading is an important factor. Similarly, it is possible we are more forgiving of robots when they invade our personal space, which is supported by other psychological robotics research (Weiss et al. 2014).

Crowding was first investigated in animal studies carried out by Calhoun (1962) who found many behavioral and hormonal disruptions in rats living in crowded situations. It is important to distinguish density from crowding; density is the objective number of people whereas crowding is a subjective experience (Rapoport 1975). Crowding has been explained using many of the stress theories we have already mentioned, including overload and arousal theories, but the role of control must also be emphasized, as in the case of personal space. Crowding becomes stressful when the individual no longer feels as if they have control over the situation, resulting in negative affective and behavioral outcomes (Baum and Paulus 1991).

Most research on crowding focuses on examining its effects on a range of psychological factors, such as affect, physiological arousal, illness, attraction, withdrawal, helping behavior, aggression and performance (see Bell et al. (2001) and Gifford (2002) for good overviews). Crowding in urban environments has been shown to have serious negative consequences on human beings, and can lead people to adopt coping mechanisms such as social withdrawal (Evans et al. 2000). For instance, Evans et al. (2001, 2002) found that the mental health of children in low-income areas was associated with the number of people with whom they lived.

One interesting aspect of crowding research is that of chronic exposure to crowding, where the individual is in a crowded environment for an extended period of time. A good example is a prison, where the number of inmates can exceed the capacity of the prison, resulting in crowding. The effects of crowding in prisons have been investigated by numerous researchers and the evidence suggests that it may have an effect on prisoners' health (McCain et al. 1976; Cox et al. 1984; Haney 2012; Walker et al. 2014), negative affect (Paulus et al. 1975) and elevated blood pressure (D'Atri 1975). The subjective experience of crowding has also been associated with a greater likelihood of interpreting another's behavior as aggressive or violent (Lawrence and Andrews 2004). Crowding in prisons may also be linked to lower psychological well-being and an increased suicide rate, although this may be due to the fact that overcrowding means less opportunity for engagement in activities (Wooldredge 1999; Huey and McNulty 2005; Leese et al. 2006).

Territoriality is the final concept that is important to discuss. Territorial behavior can be understood as behavior that helps regulate social interactions and thus a way of dealing with, coping with or preventing social environmental stress. Altman (1975) defines territorial behavior as "a self-other boundary regulation mechanism that involves personalization of or marking of a place or object and communication that it is "owned" by a person or group" (p. 107). Personalization and ownership function to regulate social interaction. When territorial boundaries are violated defense response occur. The difference between territoriality and personal space is that a territory tends to be visible whereas personal space is not. A territory is larger, it is owned and controlled, and demarcation is an essential aspect of it. Three different types of territories are distinguished: primary (home), secondary (classroom) and public (station platform). These all vary in the extent to which they are personalized (from high to low) and their perceived ownership (permanent vs. temporary). If a primary territory is invaded, the consequences are much more severe compared to a secondary or public territory.

Due to its very nature, it is difficult to study territoriality in a laboratory and so research with humans tends to take the form of observation, interviews and field studies. As is the case with some of the other stressors we have discussed, the first researchers studied animals, particularly because of the prevalence of territorial behavior in many different species (Ruwet 1998). In the case of humans, however, territoriality is a key example of how environmental stress research has considered both indoor and outdoor settings. For instance, researchers have looked at indoor environments such as offices, where the territory may often take the form of an individual's own desk, or even their team's section of the larger office. In such settings threats to territory can cause a range of reactions (Brown and Robinson 2011). For example, prohibiting personalization in offices leads workers to look for other ways to assert territoriality (Brunia and Hartjes-Gosselink 2009). This is a growing area of interest in the field as open-plan design becomes more popular and the notion of "hot-desking", i.e. not having your own fixed desk, threatens territorial behavior for workers.

Public places have also received attention in the literature. For example, it has been demonstrated that if intrusion occurred whilst the participant was using a public telephone, they were more likely to stay on the phone for a longer period of time than if the intrusion had not occurred (Ruback et al. 1989). Interestingly, this is also true of car drivers leaving a car park, since drivers leaving took longer to leave if they saw a car waiting (i.e. intruding) or if the waiting car driver honked at them (Ruback and Juieng 1997). These two examples could be linked back to control, since the stress of an intrusion into their "territory" leads individuals to exercise control in any way they can, such as staying on the phone for longer or taking longer to leave the car park.

In terms of outdoor settings and territoriality, much research has been conducted to investigate the notion of a "home advantage" in sports fixtures. Field studies have examined the difference in the home advantage before and after teams have relocated to a new stadium, finding that the so-called "home advantage" decreased with time (Pollard 2002; Loughead et al. 2003), which has been attributed to territoriality, or a lack there-of (Allen and Jones 2014).

In other outdoor contexts, crime and territoriality attract much attention in the literature; for instance, researchers report that gang violence is more prevalent in the United States compared to Europe due, in part, to the greater prevalence of gang territoriality in the States (Klein et al. 2006). We can see here that territoriality may be more important to some gangs, and intrusion into their territory by other gangs may have criminal and violent consequences.

25.3.3 Multiple Stressors

Relatively little is known about how different environmental stressors may interact. Yet many environments contain a multitude of stressors. For instance, people living in low-income housing areas are more likely to experience crowding as well as noise, visual pollution, and neighborhood social and physical disorder (Bilotta and Evans 2013; Evans et al. 1998; Dupéré and Perkins 2007).

Environmental stressors can affect people in many ways. Evans (2003) suggested that stressors in the built environment affect people directly as well as indirectly by altering psycho-social processes associated with mental health, such as control, social support and restoration. For instance, multiple stressors affect residential satisfaction (Honold Beyer et al. 2012) and can have a negative impact on people's persistence in performing challenging tasks (Evans et al. 1998) and cause learned helplessness (Evans and Stecker 2004). Such consequences can make people more vulnerable to further stress, for instance by undermining their sense of control or perceived ability to cope.

Social support is known to help people cope with stress and yet, when people are in a situation where they are in need of such support, social networks tend to suffer due to social withdrawal. Dupéré and Perkins (2007) compared different neighborhoods with high and low levels of stressors and found that when few stressors are around, strong social ties have a positive effect on mental health. They did not find communities with many stressors and such strong ties. However, they did find that when many stressors are around, withdrawal may have a positive effect on mental health.

There is relatively little research on the effect of multiple stressors in urban environments. One of the few studies that looked into this (Honold et al. 2012) found that multiple stressors were associated with residential satisfaction but that health symptoms were only associated with perceived air pollution. Some experimental research has also investigated the effect of a combination of stressors. For instance, research shows that personal space invasion is more likely to occur in a crowded environment (Kaya and Erkip 1999; Evans and Wener 2007). Equally, lighting appears to affect the personal space preferences of female students, with a preference for a greater distance between themselves and the experimenter in dim lighting compared to bright lighting (Adams and Zuckerman 1991). Future research may wish to examine potential interactive or additive effects of multiple stressors in more detail in order to better understand human experiences in the built environment and help design and manage healthy built environments.

25.4 Conclusion

The physical environment is full of potential stressors such as noise, pollution and crowding. Environmental psychology research has examined the consequences of ambient and social environmental stressors in indoor and outdoor environments on individual affect, cognition and emotion. Much research has examined the role of one particular type of stressor and there is plenty of evidence that noise and crowding, in particular, have negative effects on people's health, wellbeing and behavior. However, in many built environments, people tend to be exposed to multiple stressors and the effects of these are not yet well understood. Moreover, although there is general agreement that stress is a function of both objectively measurable environmental demands and individual subjective responses, there are different theoretical perspectives on the psychological processes that underlie stress responses. Future research may wish to focus on developing and testing theories of stress in order to generate a better understanding of when and how environmental stress may occur and for whom.

References

Acking, C. A., & Küller, H. (1972). The perception of an interior as a function of its colour. *Ergonomics, 15*(6), 645–654. Published online: 24 October 2007.

Adams, L., & Zuckerman, D. (1991). The effect of lighting conditions on personal space requirements. *The Journal of General Psychology, 118*(4), 335–340.

Ainsworth, R. A., Simpson, L., & Cassell, D. (1993). Effects of three colors in an office interior on mood and performance. *Perceptual and Motor Skills, 76*, 235–241.

Allen, M. S., & Jones, M. V. (2014). The 'home advantage' in athletic competitions. *Current Directions in Psychological Science, 23*(1), 48–53.

Altman, I. (1975). *The environment and social behavior: Privacy, personal space, territory, crowding.* Monterey: Brooks/Cole.

Altman, I., & Vinsel, A. M. (1977). Personal space: An analysis of E. T. Hall's proxemics framework. In I. Altman & J. F. Wohlwill (Eds.), *Human behaviour and environment* (Vol. 2, pp. 181–259). New York: Plenum Press.

Argyle, M., & Dean, J. (1965). Eye-contact, distance and affiliation. *Sociometry, 28*, 289–304.

Barnes, S. (2002). The design of caring environments and the quality of life of older people. *Ageing and Society, 22*(6), 775–789.

Baron, R. A. (1972). Aggression as a function of ambient temperature and prior anger arousal. *Journal of Personality and Social Psychology, 21*, 183–189.

Baron, R. A. (1976). The reduction of human aggression: A field study of the influence of incompatible reactions. *Journal of Applied Social Psychology, 6*, 260–274.

Baron, R. A. (1978). Aggression and heat: The "long hot summer" revisited. In A. Baum, S. Valins, & J. E. Singer (Eds.), *Advances in environmental research* (Vol. 1, pp. 186–207). Hillsdale: Lawrence Erlbaum Associates.

Baum, A., & Davis, G. E. (1976). Spatial and social aspects of crowding perception. *Environment and Behavior, 8*(4), 527–544.

Baum, A., & Paulus, P. B. (1991). Crowding. In D. Stokols & I. Altman (Eds.), *Handbook of environmental psychology* (pp. 863–887). Malabar: Krieger Publishing Company.

Beaman, C. P. (2005). Auditory distraction from low-intensity noise: A review of the consequences for learning and workplace environments. *Applied Cognitive Psychology, 19*, 1041–1064.

Bechtel, R., & Churchman, A. (2003). *Handbook of environmental psychology.* New York: Wiley.

Bell, P. A., & Baron, R. A. (1977). Aggression and ambient temperature: The facilitating and inhibiting effects of hot and cold environments. *Bulletin of the Psychonomic Society, 9*, 443–445.

Bell, P. A., & Doyle, D. P. (1983). Effects of heat and noise on helping behaviour. *Psychological Reports, 53*, 955–959.

Bell, P. A., & Fusco, M. E. (1989). Heat and violence in the Dallas field data: Linearity, curvilinearity and heteroscedasticity. *Journal of Applied Social Psychology, 19*(17), 1479–1482.

Bell, P. A., Greene, T. C., Fisher, J. D., & Baum, A. (2001). *Environmental psychology* (5th ed.). Belmont: Wadsworth/Thomson Learning.

Benfield, J. A., Nurse, G. A., Jakubowski, R., Gibson, A. W., Taff, B. D., Newman, P., & Bell, P. A. (2012). Testing noise in the field: A brief measure of individual noise sensitivity. *Environment and Behavior, 20*(10), 1–20.

Berlyne, D. E. (1960). *Conflict, arousal, and curiosity.* New York: McGraw-Hill.

Bilotta, E., & Evans, G. (2013). Environmental stress. In L. Steg, A. E. van den Berg, & J. I. M. de Groot (Eds.), *Environmental psychology: An introduction* (pp. 27–35). Chichester: Wiley/Blackwell.

Brown, G., & Robinson, S. L. (2011). Reactions to territorial infringement. *Organization Science, 22*(1), 210–224.

Brunia, S., & Hartjes-Gosselink, A. (2009). Personalization in non-territorial offices: A study of a human need. *Journal of Corporate Real Estate, 11*(3), 169–182.

Calhoun, J. B. (1962). Population density and social pathology. *Scientific American, 206*(2), 139–150.

Cannon, W. B. (1932). *The wisdom of the body*. New York: W. W. Norton.

Cassidy, T. (1997). *Environmental psychology: Behaviour and experience in context*. Hove: Psychology Press.

Cave, S. (1998). The environment and stress. In *Applying psychology to the environment* (pp. 137–169). London: Hodder & Stoughton Education.

Cohen, S. (1978). Environmental load and the allocation of attention. In A. Baum, J. E. Singer, & S. Valins (Eds.), *Advances in environmental psychology* (The Urban Environment, Vol. 1, pp. 1–29). Hillsdale: Lawrence Erlbaum.

Cohen, S. (1981). Sound effects on behavior. *Psychology Today, 15*, 38–49.

Cox, V. C., Paulus, P. B., & McCain, G. (1984). Prison crowding research: The relevance for prison housing standards and a general approach regarding crowding phenomena. *American Psychologist, 39*(10), 1148–1160.

Cunningham, M. R. (1979). Weather, mood, and helping behavior: Quasi experiments with the sunshine samaritan. *Journal of Personality and Social Psychology, 37*(11), 1947–1956.

D'Atri, D. A. (1975). Psychophysiological responses to crowding. *Environment and Behavior, 7*(2), 237–252.

Dabbs, J. M., Fuller, J. P., & Carr, T. P. (1973). *Personal space when "cornered": College students and prison inmates*. In Proceedings of the annual convention of the American Psychological Association (pp. 213–214). Canada: Montreal.

Davis, L. L., & Lennon, S. J. (1983). Social stigma of pregnancy: Further evidence. *Psychological Reports, 53*, 997–998.

de Kort, Y. A., & Veitch, J. A. (2014). From blind spot into the spotlight: Introduction to the special issue 'Light, lighting, and human behaviour'. *Journal of Environmental Psychology, 39*, 1–4.

Dean, L. M., Willis, F. N., & Hewitt, J. (1975). Initial interaction distance among individuals equal and unequal in military rank. *Journal of Personality and Social Psychology, 32*(2), 294–299.

Demirbas, O. O., & Demirkan, H. (2000). Privacy dimensions: A case study in the interior architecture design studio. *Journal of Environmental Psychology, 20*(1), 53–64.

Dijkstra, K., Pieterse, M. E., & Pruyn, A. T. H. (2008). Individual differences in reactions towards color in simulated healthcare environments: The role of stimulus screening ability. *Journal of Environmental Psychology, 28*, 268–277.

Dobbs, S., Furnham, A., & McClelland, A. (2011). The effect of background music and noise on the cognitive test performance of introverts and extroverts. *Applied Cognitive Psychology, 25*, 307–313.

Duffy, M., Bailey, S., Beck, B., & Barker, D. G. (1986). Preferences in nursing home design: A comparison of residents, administrators and designers. *Environment and Behavior, 18*, 246–257.

Dupéré, V., & Perkins, D. D. (2007). Community types and mental health: A multilevel study of local environmental stress and coping. *American Journal of Community Psychology, 39*, 107–119.

Enmarker, I. (2004). The effects of meaningful irrelevant speech and road traffic noise on teachers' attention, episodic and semantic memory. *Scandinavian Journal of Psychology, 45*(5), 393–405.

Evans, G. W. (2003). The built environment and mental health. *Journal of Urban Health, 80*(4), 536–555.

Evans, G. W., & Cohen, S. A. (1987). Environmental stress. In D. Stokols & I. Altman (Eds.), *Handbook of environmental psychology* (Vol. 1). New York: Wiley.

Evans, G. W., & Stecker, R. (2004). Motivational consequences of environmental stress. *Journal of Environmental Psychology, 24*, 143–165.

Evans, G. W., & Wener, R. (2007). Crowding and personal space invasion on the train. *Journal of Environmental Psychology, 27*, 90–94.

Evans, G. W., Bullinger, M., & Hygge, S. (1998). Chronic noise exposure and physiological response: A prospective study of children living under environmental stress. *Psychological Science, 9*(1), 75–77.

Evans, G. W., Rhee, E., Forbes, C., Mata Allen, K., & Lepore, S. J. (2000). The meaning and efficacy of social withdrawal as a strategy for coping with chronic residential crowding. *Journal of Environmental Psychology, 20*(4), 335–342.

Evans, G. W., Saegert, S., & Harris, R. (2001). Residential density and psychological health among children in low-income families. *Environment and Behaviour, 33*(2), 165–180.

Evans, G. W., Lercher, P., & Kofler, W. (2002). Crowding and children's mental health: The role of house type. *Journal of Environmental Psychology, 22*(3), 221–231.

Felipe, N. J., & Sommer, R. (1966). Invasions of personal space. *Social Problems, 14*(2), 206–214.

Fiske, D. W., & Maddi, S. R. (1961). *Functions of varied experience*. Homewood: Dorsey Press.

Gerard, R. M. (1958). *Differential effects of coloured lights on psycho-physiological functions*. Unpublished doctoral dissertation. University of California, Los Angeles.

Gifford, R. (2002). *Environmental psychology: Principles and practice*. Colville: Optimal Books.

Glass, D. C., & Singer, J. E. (1972). *Urban stress: Experiments on noise and social stressors*. New York: Academic.

Haans, A., Kaiser, F., & de Kort, Y. (2007). Privacy needs in office environments development of two behavior-based scales. *European Psychologist, 12*(2), 93–102.

Hall, E. T. (1959). *The silent language*. New York: Doubleday.

Hall, E. T. (1966). *The hidden dimension*. New York: Doubleday.

Haney, C. (2012). Prison effects of in the age of mass incarceration. *The Prison Journal, 20*(10), 1–24.

Hayduk, L. A., & Mainprize, S. (1980). Personal space of the blind. *Social Psychology Quarterly, 43*(2), 216–223.

Helson, H. (1964). *Adaptation-level theory: An experimental and systematic approach to behavior*. New York: Harper & Row.

Hinkle, L. E. (1961). The physiological state of the interrogation subject as it affects brain function. In A. D. Biderman & H. Zimmer (Eds.), *The manipulation of human behaviour* (pp. 18–50). New York: Wiley.

Honold, J., Beyer, R., Lakes, T., & van der Meer, E. (2012). Multiple environmental burdens and neighbourhood-related health of city residents. *Journal of Environmental Psychology, 32*, 305–317.

Huey, M. P., & McNulty, T. L. (2005). Institutional conditions and prison suicide: Conditional effects of deprivation and overcrowding. *The Prison Journal, 85*(4), 490–514.

Hygge, S. (2003). Classroom experiments on the effects of different noise sources and sound levels on long-term recall and recognition in children. *Applied Cognitive Psychology, 17*, 895–914.

Hygge, S., Boman, E., & Enmarker, I. (2003). The effects of road traffic noise and meaningful irrelevant speech on different memory systems. *Scandinavian Journal of Psychology, 44*, 13–21.

Jacobs, K. W., & Hustmyer, F. E. (1974). Effects of four psychological primary colours on GSR, heart rate, and respiration rate. *Perceptual Motor Skills, 38*(3), 763–766.

Jacobs, K. W., & Suess, J. F. (1975). Effects of four psychological primary colors on anxiety state. *Perceptual and Motor Skills, 41*, 207–210.

Kahneman, D. (1973). *Attention and effort*. Englewood Cliffs: Prentice-Hall.

Kaplan, R., & Kaplan, S. (1989). *The experience of nature: A psychological perspective*. Cambridge: Cambridge University Press.

Kattner, F., & Ellermeier, Q. (2014). Irrelevant speech does not interfere with serial recall in early blind listeners. *The Quarterly Journal of Experimental Psychology, 67*(11), 2207–2217.

Katz, D. (1937). *Animals and men*. New York: Longmans, Green.

Kaya, N., & Erkip, F. (1999). Invasion of personal space under the condition of short-term crowding: A case study on an automatic teller machine. *Journal of Environmental Psychology, 19*(2), 183–189.

Klatte, M., Bergström, K., & Lachmann, T. (2013). Does noise affect learning? A short review on noise effects on cognitive performance in children. *Frontiers in Psychology, 4*, 1–6.

Kleck, R., Buck, P. L., Goller, W. L., London, R. S., Pfieiffer, J. R., & Vukcevic, D. P. (1968). Effect of stigmatizing conditions on the use of personal space. *Psychological Reports, 23*, 111–118.

Klein, M. W., Weerman, F. M., & Thornberry, T. P. (2006). Street gang violence in Europe. *European Journal of Criminology, 3*(4), 413–437.

Knez, I. (2001). Effects of colour of light on nonvisual psychological processes. *Journal of Environmental Psychology, 21*, 201–208.

Knez, I. (2014). Affective and cognitive reactions to subliminal flicker from fluorescent lighting. *Consciousness and Cognition, 26*, 97–104.

Knez, I., & Hygge, S. (2002). Irrelevant speech and indoor lighting: Effects on cognitive performance and self-reported affect. *Applied Cognitive Psychology, 16*(6), 709–718.

Krail, K. A., & Leventhal, G. (1976). The sex variable in the intrusion of personal space. *Sociometry, 39*(2), 170–173.

Küller, R., Mikellides, B., & Janssens, J. (2009). Color, arousal, and performance: A comparison of three experiments. *Color Research & Application, 34*(2), 141–152.

Kupritz, V. W. (1998). Privacy in the work place: The impact of building design. *Journal of Environmental Psychology, 18*, 341–356.

Kwallek, N., & Lewis, C. M. (1990). Effects of environmental colour on males and females: A red or white or green office. *Applied Ergonomics, 21*, 275–278.

Lawrence, R. J. (2002). Healthy residential environments. In R. B. Bechtel & A. Churchman (Eds.), *Handbook of environmental psychology* (pp. 394–412). New York: Wiley.

Lawrence, C., & Andrews, K. (2004). The influence of perceived prison crowding on male inmates' perception of aggressive events. *Aggressive Behavior, 30*, 273–283.

Lazarus, R. (1966). *Psychological stress and the coping process*. New York: McGraw Hill.

Lee, Y. S. (2010). Office layout affecting privacy, interaction, and acoustic quality in LEED-certified buildings. *Building and Environment, 45*, 1594–1600.

Leese, M., Thomas, S., & Snow, L. (2006). An ecological study of factors associated with rates of self-inflicted death in prisons in England and Wales. *International Journal of Law and Psychiatry, 29*, 355–360.

Livingstone, S. (2008). Taking risky opportunities in youthful content creation: Teenagers' use of social networking sites for intimacy, privacy and self-expression. *New Media & Society, 10*(3), 393–411.

Loughead, T. M., Carron, A. V., Bray, S. R., & Kim, A. J. (2003). Facility familiarity and the home advantage in professional sports. *International Journal of Sport and Exercise Psychology, 1*, 264–274.

Madden, M. (2012). Privacy management on social media sites. *Pew Internet Report*. http://www.isaca.org/Groups/Professional-English/privacy-data-protection/GroupDocuments/PIP_Privacy%20mgt%20on%20social%20media%20sites%20Feb%202012.pdf.

McCain, G., Cox, V. C., & Paulus, P. B. (1976). The relationship between illness complaints and degree of crowding in a prison environment. *Environment and Behavior, 8*(2), 283–290.

McDaniel, T., Krishna, S., Colbry, D., & Panchanathan, S. (2009). Using tactile rhythm to convey interpersonal distance to individuals who are blind. In *CHI'09 extended abstracts on human factors in computing systems* (pp. 4669–4674). Boston: ACM.

McEwen, B. S. (1998). Stress, adaptation, and disease: Allostasis and allostatic load. *Annals of the New York Academy of Sciences, 840*, 33–44.

Mehrabian, A. (1977). Individual differences in stimulus screening and arousability. *Journal of Personality, 45*(2), 237–250.

Mehta, R., & Zhu, R. (2009). Blue or red? Exploring the effect of color on cognitive task performances. *Science, 323*(5918), 1226–1229.

Mehta, R., Zhu, R., & Cheema, A. (2012). Is noise always bad? Exploring the effects of ambient noise on creative cognition. *Journal of Consumer Research, 39*(4), 784–799.

Milgram, S. (1970). The experience of living in cities. *Science, 167*(3924), 1461–1468.

Moch, A. (1989). *Les stress de l'environnement*. Paris: Presses Universitaires de Vincennes.

Morgan, D. G., & Stewart, N. J. (1998). Multiple occupancy versus private rooms on dementia care units. *Environment and Behavior, 30*(4), 487–503.

Morgan, D. G., & Stewart, N. J. (1999). The physical environment of special care units: Needs of residents with dementia from the perspective of staff and family caregivers. *Quantitative Health Research, 9*(1), 105–118.

Moser, G. (1988). Urban stress and helping behaviour: Effects of environmental overload and noise on behaviour. *Journal of Experimental Psychology, 8*, 287–298.

Moser, G. (1994). Environmental stress and urban behaviour. *European Review of Applied Psychology, 44*(2), 149–154.

Moser, G. (1998a). Stress environnementaux. In R. Doron & F. Parot (Eds.), *Dictionnaire de Psychologie* (2nd ed.). Paris: Presses Universitaires de France.

Moser, G. (1998b). Espace personnel. In R. Doron & F. Parot (Eds.), *Dictionnaire de Psychologie* (2nd ed.). Paris: Presses Universitaires de France.

Mullin, C. S. (1960). Some psychological aspects of isolated Antarctic living. *American Journal of Psychiatry, 117*, 323–325.

Novelli, D., Drury, J., & Reicher, S. (2010). Come together: Two studies concerning the impact of group relations on personal space. *British Journal of Social Psychology, 49*(2), 223–236.

Nurmi, J.-E., & von Wright, J. (1983). Interactive effects of noise, neuroticism and state anxiety in the learning and recall of a textbook passage. *Human Learning: Journal of Practical Research & Applications, 2*(2), 119–125.

Parr, A. E. (1966). Psychological aspects of urbanology. *Journal of Social Issues, 12*(4), 39–45.

Patterson, M. L., Mullens, S., & Romano, J. (1971). Compensatory reactions to spatial intrusion. *Sociometry, 34*(1), 114–121.

Paulus, P., Cox, V., McCain, G., & Chandler, J. (1975). Some effects of crowding in a prison environment. *Journal of Applied Social Psychology, 5*(1), 86–91.

Pedersen, D. M. (1999). Model for types of privacy by privacy functions. *Journal of Environmental Psychology, 19*(4), 397–405.

Pedersen, E., & Larsman, P. (2008). The impact of visual factors on noise annoyance among people living in the vicinity of wind turbines. *Journal of Environmental Psychology, 28*(4), 379–389.

Pierrette, M., Marquis-Favre, C., Morel, J., Rioux, L., Vallet, M., Viollon, S., & Moch, A. (2012). Noise annoyance from industrial and road traffic combined noises: A survey and a total annoyance model comparison. *Journal of Environmental Psychology, 32*(2), 178–186.

Pollard, R. (2002). Evidence of a reduced home advantage when a team moves to a new stadium. *Journal of Sports Sciences, 20*(12), 969–973.

Pressey, S. L. (1921). The influence of color upon mental and motor efficiency. *The American Journal of Psychology, 32*(3), 326–356.

Proshansky, H. M., Ittelson, W. H., & Rivlin, L. G. (1970). *Environmental psychology: Man and his physical setting*. New York: Holt, Rinehart & Winston.

Rapoport, A. (1975). Towards a redefinition of density. *Environment and Behavior, 7*(2), 133–158.

Rodin, J., Solomon, S. K., & Metcalf, J. (1978). Role of control in mediating perceptions of density. *Journal of Personality and Social Psychology, 36*, 988–999.

Rotton, J., & Cohn, E. G. (2002). Climate, weather & crime. In R. B. Bechtel & A. Churchman (Eds.), *Handbook of environmental psychology* (pp. 481–498). New York: Wiley.

Rouleau, N., & Belleville, S. (1996). Irrelevant speech effect in aging: An assessment of inhibitory processes in working memory. *Journal of Gerontology Series B: Psychological Sciences, 51*(6), 356–363.

Ruback, R. B., & Juieng, D. (1997). Territorial defence in parking lots: Retaliation against waiting drivers. *Journal of Applied Social Psychology, 27*(9), 821–834.

Ruback, R. B., Pape, K. D., & Doriot, P. (1989). Waiting for a phone: Intrusion on callers leads to territorial defense. *Social Psychology Quarterly, 52*(3), 232–241.

Rumsey, N., Bull, R., & Gahagan, D. (1982). The effect of facial disfigurement on the proxemics behaviour of the general public. *Journal of Applied Social Psychology, 12*(2), 137–150.

Ruwet, J.-C. (1998). Territorialité. In R. Doron & F. Parot (Eds.), *Dictionnaire de Psychologie* (2nd ed.). Paris: Presses Universitaires de France.

Sardar, A., Joosse, M., Weiss, A., & Evers, V. (2012). *Don't stand so close to me: Users' attitudinal and behavioural responses to personal space invasion by robots*. In Proceedings of the seventh annual ACM/IEEE international conference on Human-Robot Interaction (pp. 229–230). Boston: ACM.

Schlittmeier, S. J., & Hellbrück, J. (2009). Background music as noise abatement in open-plan offices: A laboratory study on performance effects and subjective preferences. *Applied Cognitive Psychology, 23*, 684–697.

Schneider, F. W., Lesko, W. A., & Garrett, W. A. (1980). Helping behavior in hot, comfortable, and cold temperatures. *Environment and Behavior, 12*, 231–240.

Seyle, H. (1956). *The stress of life*. New York: McGraw Hill.

Shelton, J. T., Elliott, E. M., Eaves, S. D., & Exner, A. L. (2009). The distracting effects of a ringing cell phone: An investigation of the laboratory and the classroom setting. *Journal of Environmental Psychology, 29*, 513–521.

Smolders, K. C. H. J., de Kort, Y. A. W., & van den Berg, S. M. (2013). Daytime light exposure and feelings of vitality: Results of a field study during regular weekdays. *Journal of Environmental Psychology, 36*, 270–279.

Sommer, R. (1959). Studies in personal space. *Sociometry, 22*(3), 247–260.

Steidle, A., & Werth, L. (2013). Freedom from constraints: Darkness and dim illumination promote creativity. *Journal of Environmental Psychology, 35*, 67–80.

Suedfeld, P. (1991). Extreme and unusual environments. In D. Stokols & I. Altman (Eds.), *Handbook of environmental psychology* (pp. 863–887). Malabar: Krieger Publishing Company.

Sundstrom, E., Herbert, R. K., & Brown, D. W. (1982). Privacy and communication in an open-plan office a case study. *Environment and Behavior, 14*(3), 379–392.

Uzzell, D., & Horne, N. (2006). The influence of biological sex, sexuality and gender role on interpersonal distance. *British Journal of Social Psychology, 45*(3), 579–597.

von Wright, J., & Vauras, M. (1980). Interactive effects of noise and neuroticism on recall from semantic memory. *Scandinavian Journal of Psychology, 21*(1), 97–101.

Walker, J., Illingworth, C., Canning, A., Garner, E., Woolley, J., Taylor, P., & Amos, T. (2014). *Acta Psychiatrica Scandinavica, 129*, 427–436.

Webb, J. D., & Weber, M. J. (2003). Influence of sensory abilities on the interpersonal distance of the elderly. *Environment and Behavior, 35*(5), 695–711.

Weinzimmer, D., Newman, P., Taff, D., Benfield, J., Lynch, E., & Bell, P. (2014). Human responses to simulated motorized noise in national parks. *Leisure Sciences: An Interdisciplinary Journal, 36*, 251–267.

Weiss, A., Vincze, M., Panek, P., & Mayer, P. (2014). *Don't bother me: Users' reactions to different robot disturbing behaviors*. In Proceedings of the 2014 ACD/IEEE international conference on Human-robot interaction (pp. 320–321). Bielefeld: ACM.

Wilson, G. D. (1966). Arousal properties of red versus green. *Perceptual and Motor Skills, 23*, 947–949.

Wohlwill, J. F. (1974). Human adaptations to levels of environmental stimulation. *Human Ecology, 2*(2), 127–147.

Wooldredge, J. D. (1999). Inmate experiences and psychological well-being. *Criminal Justice and Behavior, 26*(2), 235–250.

Worthington, M. E. (1974). Personal space as a function of the stigma effect. *Environment and Behavior, 6*(3), 289–294.

Yerkes, R. M., & Dodson, J. D. (1908). The relation of strength of stimulus to rapidity of habit-formation. *Journal of Comparative Neurology and Psychology, 18*(5), 459–482.

Zubek, J. P. (1969). *Sensory deprivation. Fifteen years of research*. New York: Appleton-Century-Crofts.

Living in an "At Risk" Environment: The Example of "Costal Risks"

26

Élisabeth Michel-Guillou and Catherine Meur-Ferec

26.1 Introduction

In the Western world, risk has permeated our daily lives. It has become a topical issue that cannot be ignored within societies concerned by safety and prevention issues. Modern, industrial societies are described by sociologists as "risk societies" because "In advanced modernity the social production of *wealth* is systematically accompanied by the social production of *risk*" (Beck 1992 p. 19). These "risk-producing" societies are characterized by the multiplication of anthropogenic hazards and by increasingly low thresholds of risk tolerance (Beck 1992; Giddens 1990; Peretti-Watel 2000).

É. Michel-Guillou (✉)
Environmental Psychology/Centre de recherches en psychologie, cognition et communication (CRPCC EA 1285), Institut Brestois des Sciences de l'Homme et de la Société (IBSHS)/Université de Bretagne Occidentale (UBO)/COMUE Université Bretagne Loire (UBL), CRPCC 20 rue Duquesne, CS 93837, 29238, Brest cedex 3, France
e-mail: Elisabeth.Michel-Guillou@univ-brest.fr

C. Meur-Ferec
Geography/Littoral, Environnement, Télédétection, Géomatique (LETG-Géomer, UMR 6554 CNRS)/Institut Universitaire Européen de la Mer (IUEM), Université de Bretagne Occidentale (UBO)/COMUE Université Bretagne Loire (UBL), IUEM, Rue Dumont D'Urville, Technopole Brest Iroise, 29280 Plouzané, France
e-mail: meurferec@univbrest.fr

Manifestations of risk are broad and innumerable, affecting daily activities (domestic risks, food risks, transport-related risks, etc.) and more infrequent activities (health risks, technological risks, so-called "natural" risks, etc.). Furthermore, although risk is often perceived as harmful, it can also be seen as an opportunity for gain (gambling, stock exchange, entrepreneurship, etc.). In this chapter, we discuss the damaging and socially constructed nature of risks, construed as a danger that individuals seek to avoid. We present the work carried out within the framework of an interdisciplinary approach in psychology and geography. Notably in this latter discipline, the concept of risk has been widely studied, specifically in relation to the issue of vulnerability. This interdisciplinary reflection lies at the core of the very foundations of environmental psychology. In fact, the research conducted in this field often requires the association of researchers from various backgrounds in order to understand complex phenomena (Altman and Rogoff 1987; Ittelson et al. 1974; Proshansky 1976).

In social psychology, two main approaches to risk can be identified. The first distinguishes the "layperson" from the "expert". The perception of laypeople, considered non-experts of risk, is then compared with the assessment made by one or more experts. This theoretical point of view underlines the differences between the two forms of thought (layperson vs. expert) in the way risk is conceived, and consequently, the different ways of adapting to it. The second theoretical approach, in terms of the *social representation* of risk, highlights the socio-cognitive construction of risks. This is addressed by taking into account particularly socio-cultural factors, or more generally the context in which the individuals, who are likely to be concerned by a given risk, live. The theory of social representations proposes, on the one hand, to abandon the hierarchy or dichotomy between forms of thought (expert/layperson) and, on the other hand, to link them with other concepts such as identity and sense of place.

This chapter focuses more specifically on the issue of so-called "natural" risks (D'Ercole and Pigeon 1999). The term "natural" is usually used in contrast to "technological" risks and describes risks associated with hazards considered of natural origin (earthquakes, storms, landslides, etc.). Nevertheless, although some hazards are indisputably of natural origin (volcanoes and earthquakes), others have an anthropogenic component. The risk is a result of the issues involved (individuals, personal belongings, etc.) in a given territory.

These natural risks have some characteristics that distinguish them from others for several reasons (Navarro-Carrascal and Michel-Guillou 2014). Firstly, it is difficult to grant them any "usefulness", in contrast to food (nutrition) or industrial risks (access to technological development) for example. In addition, as already mentioned, even though they may not be entirely natural, it still seems difficult to assign total and direct responsibility to human action for largely natural hazards (erosion, flooding, etc.). Of course, the increase in intensity or frequency of storms, which might result in damage affecting humans, can be attributed to climate changes, themselves generated by higher concentrations of greenhouse gases, which are partly man-made (Intergovernmental Panel on Climate Change 2007, 2013). Nevertheless, humans do not create storms. On the other hand, compared to industrial risks, natural risks are not as tangible or noticeable in the landscape as a factory, for instance. Finally, since it is not always easy to locate them precisely in space or to predict the probability of their occurrence, it is sometimes difficult to propose consensual adaptation strategies to deal with them. These strategies are based on identifying "vulnerable" territories, an approach to risk that has been extensively developed, particularly in geography (November 2002; Meur-Ferec et al. 2011).

From a human point of view, living in areas exposed to natural risks generates contrasting attitudes and behaviors depending on the vision of risk, varying between the absence of representations (or perceptions) of the risk, and minimizing or taking action against it. The psychological processes used to develop these visions of risk tend to create a balance, or at least a justification, for life in such areas, providing residential satisfaction and maintaining a sense of well-being and quality of life.

Risk prevention concerns a multitude of people and groups, who are locally or indirectly involved (inhabitants, managers at local, regional, national or even international levels, scientists, etc.) and who may have different visions of the risk. This diversity reflects a different use of the concept of risk that is particularly linked to knowledge and the personal and social experience of the place itself, taking into account the possible experience of risk(s). Thus, understanding these diverse viewpoints and whether they are interlinked is a major challenge in environmental psychology. This analysis enables a better comprehension of the vulnerability of territories defined as being "at risk" and thus contributes to a more adapted management.

This chapter therefore addresses the issue of natural risks in four parts. In the first, we define risks while in the second, we develop the main theoretical approaches used in social and environmental psychology to understand them. At the end of this theoretical presentation, we present an example of natural risks: coastal risks (marine erosion and submersion). This chapter, aimed at studying the quality of life, will then lead to the final part in which we discuss the value of a comprehensive approach to the study of complex phenomena.

26.2 From a Dichotomous Hazard/Vulnerability Approach to a Systemic Approach to Risks

The twentieth century was marked by the hope of eliminating risks through technological advances. Thus, the 1980s culminated with risk studies in which the unique "scientific" approach predominated (engineering, statistics, probabilistic modeling, mapping, etc.). With regard to natural risks, before the 1980s, research involved mainly the study of hazards, in particular those based on engineering science and geoscience approaches (geology, geomorphology, hydrology, meteorology, etc.) (Morel et al. 2006). The role of human societies in catastrophic events was not always developed; in this "naturalist" approach, humans are subject to the hazards of nature (Pigeon 2002).

However, at the end of the 1980s, in the face of the immensity of uncertainties, the so-called "hard" sciences reached their own limits (the Chernobyl accident, explosion of the Challenger shuttle, etc.). "A concept of risk as a danger to be eliminated through scientific development and normative action was succeeded by a concept of risk as a hazard to be managed" (Lascoumes 1991 p. 79). "The somewhat utopian project of the conquest of safety, through the eradication of risk, has reached an impasse and gives way to pragmatic management" (Peretti-Watel 2000 p. 51). We must now accept the obvious: "zero risk does not exist". Considering ways to limit vulnerability seemed the only realistic objective. The necessary involvement of the social sciences was then gradually recognized (Theys and Kalaora 1992). Vulnerability[1] became the fundamental and decisive element in the assessment and management of collective risks, notably natural risks (Becerra and Peltier 2009; Clarke et al. 1989; Cutter 1993; Gilbert 2009).[2] In the social sciences, this polysemous notion refers to the social conditions that existed before the event, which gave rise to a greater or lesser propensity to suffer damage.

Nevertheless, although the study of vulnerability greatly enriched the approach to natural risks, there often remained a dichotomy between natural hazards on the one hand and social vulnerability on the other, which did not satisfy many geographers. From the mid-1980s, some researchers, such as Bertrand (1986), tried to refocus the discussions on nature-society relationships and build a framework for the process of risk emergence, opening up the field of environmental geography. This involved going beyond a "Manicheaist and naive reading of the hazard and vulnerability pair" (D'Ercole and Pigeon 2000). Indeed, as D'Ercole and Metzger explained "this dichotomy between hazard and vulnerability clearly effects a division of work between the "hard" sciences that consider the hazard and the social sciences that analyze the vulnerability, although everyone claims the necessary multidisciplinarity to address the risks" (2011 p. 6). In contrast, adopting a broader, multidimensional concept of vulnerability means that "the fragility of a system as a whole" (D'Ercole and Pigeon 1999) can be estimated. This systemic view, bringing together nature and societies, is the result of advances linked to increased interdisciplinarity in the study of natural risks (Hellequin et al. 2013).

The concept of systemic vulnerability combines the natural and anthropogenic elements in a single system (Meur-Ferec 2008). Based on this principle, and drawing on the work of D'Ercole (1994), the notion of systemic vulnerability applied to coastal risks was developed in the early 2000s, as part of the National Coastal Environment Program (PNEC). It was the result of a long process of broad interdis-

---

[1] We use the term vulnerability here to mean "social" vulnerability, which differs from the physical, or intrinsic, vulnerability used by the engineering and earth sciences to talk about the fragility of a building for example.

[2] In general, vulnerability refers to the individual (vulnerable populations or territories — implying inhabited ones), while risk refers to a danger (hazard). However, these two concepts cannot be considered independently of the human presence. On the other hand, vulnerability is sometimes defined as a component of risk perception (e.g., Acuna-Rivera et al. 2014).

ciplinary research[3] (Jollivet 1992), enabling approaches in the natural and social sciences to be integrated into a common research subject: the risks of marine erosion and submersion. Thus, we consider that systemic vulnerability has four main components: (1) hazards (here, the natural phenomena, sometimes influenced by human action, such as cliff erosion, the breakdown of dune systems, submersion, etc.), (2) issues (people and property exposed to hazards), (3) management (public policies for prevention, protection and crisis management, sea defense equipment, etc.) and (4) representations (the way in which local actors, managers of coastal areas (e.g. elected representatives, government services) and users of the space (e.g. inhabitants, maritime professionals) perceive their environment and the risks.

The main originality of this approach is that it integrates hazards into the "vulnerability system" and thus allows an overall understanding of the issue of risks. Although research on coastal hazards (e.g. modeling of water levels, geomorphic processes) and ways to reduce them (e.g. coastline protection, beach nourishment) remains an essential element in the study of these risks, it is not sufficient on its own. A detailed and integrated study of the four components of systemic vulnerability is necessary to address the question of long-term coastal management strategies. This systemic approach echoes the transactional perspective developed in environmental psychology (Altman and Rogoff 1987; Bonnes and Secchiaroli 1995; Ittelson 1976; Moser and Uzzell 2003; Wapner 1981).

Within this systemic vulnerability, the representation dimension aims to identify the physical, psychological, social and cultural factors that can affect the representations of risk (Moser and Uzzell 2003; Navarro-Carascal and Michel-Guillou 2014). It focuses on the context, the living conditions and the relationship to place (proximity and experience of risks) in the construction of risk representations (Krien and Michel-Guillou 2014; Michel-Guillou et al. 2015). These factors contribute to understanding systemic vulnerability by increasing our knowledge of perceived vulnerability (Moser 1998), which depends on the interaction of individuals with their environment. This approach is characteristic of the work carried out in the context of social and environmental psychology.

26.3 Contributions of Social Psychology and Environmental Psychology

The notion of risk refers to the idea of a threat of damage that everyone seeks to avoid (Douglas 1994; Joffe 2003; Slovic 2000). From the social psychology point of view, two main approaches to risk can be distinguished: one interested mainly in individual processes and cognitive functioning (the *perception* of risk) and another more focused on social and contextual processes (the *social representation* of risk).

26.3.1 Social Psychology and the Perception of Risk

The perception of risk refers to the attitudes and judgments that people have with respect to risk (Slovic 2000). It is the result of a compromise between the risks and benefits arising from a given situation. Research results demonstrate that risk is perceived differently by different groups: the experts who study the risk based, for example, on the estimation of the number of deaths, compared to the laypeople who live with the risk and who will instead take into account its catastrophic potential, its controllability, the threat to future generations, its familiarity, etc. This comparison of different groups usually reveals a gap between the thinking of experts and of laypeople, a discrepancy that is generally ex-

---

[3]Geomorphology, human geography, economy and sociology in the National Coastal Environment Program (PNEC) [*Programme National d'Environnement Côtier*] (2003–2004), then the addition of physics and geology in the MISEEVA research programme (2001–2011) and finally psychology, anthropology and law in the COCORISCO research programme (2011–2014), both funded by ANR, the French National Research Agency.

plained by evidence of the cognitive dysfunction of the layperson's perception. Early studies thus showed that individuals, since they cannot deal with all the information from their environment (complex and uncertain), make errors of judgment and are victims of biases that can have important consequences (Gardner and Stern 1996; Tversky and Kahneman 1974). Whether these dysfunctions are described as biases, illusions or beliefs (e.g. Kouabenan et al. 2006; Slovic 2000), they refer to the idea of a hierarchy of two forms of thought, partly reproducing social hierarchies. Western societies' confidence in science (Doise 1982) confers a high degree of credibility on the expert's word or knowledge, even if it is sometimes doubted (Beck 1992): the expert is the one who knows. Accordingly, any misunderstanding or non-recognition of the risks as defined by the experts is considered a form of denial. There is thus an "objective" reality defined by the experts and when the individual deviates from this reality, he/she becomes "irrational" or "fallible" (Moscovici 1987). The psychometric perspective (Slovic 1987, 2000) is characteristic of this approach in the United States, but cognitive biases are also widely studied in France (Kouabenan et al. 2006). Furthermore, although Slovic (2000) recognizes the importance of contextual variables in the perception of risk by laypeople: "research has also demonstrated that there are important social, political and cultural factors that play important roles in the story" (p. 32), ultimately, these variables are rarely taken into account in the interpretation of the results. However, this is not the goal. This approach essentially focuses on the identification and understanding of cognitive processes. "The word *perception* is used here and in the literature to refer to various kinds of attitudes and judgments" (Slovic 2000 p. 37).

In contrast, another approach in terms of representation proposes abandoning the hierarchy between the forms of thought, since it is no longer interested in individuals only, instead focusing on the contextual, social and cultural factors at stake in the development of the risk representation (Joffe 2003). This approach is rooted in the theory of social representations (Moscovici 1976, 2001; Moscovici and Duveen 2000).

26.3.2 Social Representation of Risk

Social representations are a social construction of reality, developed and shared by a social group and with a practical purpose (Jodelet 1999) that aims to make this reality meaningful (Abric 2001). They are highly contextualized and depend on the groups' social anchoring. They enable individuals to understand their environment and provide them with a vision of the world. This theory addresses the question of "why" and assumes that a phenomenon known as a risk is linked to a mental construction, which is said to be social insofar as it is the work of a socially integrated individual. Doise and colleagues (Clémence et al. 1994; Doise 1993) define social representations as principles that generate positions and organize individual differences. Through this approach, the interest is no longer to demonstrate a discrepancy, or even a hierarchy, of forms of thought, but to understand their sense or meanings.

Moscovici (1976) was interested in how a scientific discipline becomes an object of common sense, and demonstrated two processes at work in the genesis and development of a social representation: *objectification* and *anchoring*. The goal of *objectification* is to materialize abstract concepts through images, so that they can be communicated throughout society. For example, "climate change" and "sea level rise" in scientific language become "warming" and "storms" in common sense (Cabecinhas et al. 2008; Kempton 1997; Lorenzoni et al. 2005; Michel-Guillou 2015). The scientific phenomenon that is difficult to perceive is *objectified*, or symbolized, through concrete, current and familiar marine-weather elements. *Anchoring* occurs: it roots the abstract scientific concept in familiar categories of thought or language, thus enabling it to be used in daily life and social exchanges because it is "translated" into a language shared by all, a common sense language (Doise 2005).

In terms of risks, social representations of a given risk may therefore be different, not only because of the person's status (expert/layperson on a particular subject) but also because of their emergence in different contexts. Social representations are strongly dependent on the context in which they are developed and evolve (Doise 1990, 1993; Moscovici 1976, 2001). As a guide for action, the representations that exposed populations have of risks, of their dangerousness and of their own susceptibility to suffer damage affect their individual or collective response capacity and their adaptation strategies. Consequently, it is important to take this into account in the construction of institutional preventive actions. Furthermore, as we mentioned and as Moscovici recalled (1976), these contextual elements naturally include the living space of individuals, their environment. As a result, the relationship to a living space is also important in the construction of a social representation of a given risk.

26.3.3 Environmental Psychology and the Study of the Relationship to Place

Environmental psychology is the study of the interrelations or transactions between the individual and his/her socio-physical environment (Bonnes and Secchiaroli 1995; Canter and Craik 1981; Moser and Uzzell 2003; Stokols and Altman 1987). The environment is understood in the broad sense; it refers to the *context* (Wapner and Demick 2002). For these authors, *context* takes into account aspects linked to the person (physical aspects like fitness, health, etc.; psychological aspects like emotional state, etc.; and socio-cultural aspects like status, etc.) and the surrounding aspects (physical aspects, either natural or constructed; interpersonal aspects like the entourage; and socio-cultural aspects such as rules, standards of the couple, community, or culture, etc.). Thus, the influence of the physical characteristics of the space is taken into account, as well as that of individual and social factors. More specifically, environmental psychology is particularly interested in the study of the environmental conditions of individuals' well-being and quality of life; in other words, their life environment. The characteristics of this life environment and the way in which it is occupied contribute to shaping the identity of the individual and give meaning to his/her behavior (Proshansky et al. 1983; Twigger-Ross and Uzzell 1996; Wester-Herber 2004). Any study about the life environment cannot exclude the social and cultural milieu considered (Ittelson et al. 1974; Lévy-Leboyer 1980) and the time (Moser and Uzzell 2003). The temporal dimension is omnipresent and determines the way in which the environment is perceived and assessed and the practices that are deployed there. Taking into account all of these properties means recognizing that the environment is not a neutral concept, without values, and admitting that it conveys meanings or sense.

The people-place relationship is one of the bases of geography, as can be seen in the notion of *ecumene*, i.e., geographicity (human beings' relationship to the earth) and inhabitation (Dardel 1952; Berque 2000; Lussault 2007). Early geographers highlighted how geography is the science of place, in which place is the link between nature and society (Gold 1980; Pinchemel and Pinchemel 1997; Reclus 1866; Tuan 1979). "It is not the human fact which is geography, any more than it is the environmental fact, but rather the relation which may exist between the two. Geography is a science of relationships" (Barrows 1923 p. 12).

In psychology, many studies have focused on concepts related to this concept of place, such as *sense of place* (Hay 1998; Jørgensen and Stedman 2001, 2006; Stedman 2002), *place attachment* (Bonaiuto et al. 1999; Giulliani 2003; Hidalgo and Hernandez 2001; Lewicka 2011), *place identity* (Bonaiuto et al. 2002; Proshansky et al. 1983; Twigger-Ross and Uzzell 1996) *place dependence* (Stokols and Shumaker 1981), and many others. The link between these concepts is not always clear (Giuliani and Feldman 1993; Hidalgo and Hernandez 2001; Lewicka 2011; Stedman 2002) due to the many theoretical perspectives and methodologies involved (Hernandez et al. 2007; Lewicka 2011; Stedman 2002). For example, Hernandez et al. (2007) showed that the concepts of *place attachment* and *place identity* are sometimes regarded as interchangeable or as a component of one or the other. The concept of *sense of place* tends to include all the other notions. It comprises cognitive, affective and conative dimensions (Jørgensen and Stedman 2001, 2006) and includes *place attachment, place identity, sense of community, rooting* and *membership*. The affective dimension deals with the emotional link that a person or a group has to a particular place (Altman and Low 1992) and is always present in *place attachment*, generally defined as "an affective bond or link between people and specific places" (Hidalgo and Hernandez 2001 p. 274). *Rooting* is a type of spatial anchoring that is often expressed and reinforced by length of residence, and also by memories, intergenerational transmission and heritage (Kelly and Hosking 2008). Temporality (e.g. through place of residence) is therefore a fundamental concept of this dimension that is frequently mentioned in the literature. *Sense of community* reflects social ties and memberships of groups (social networks, neighborhoods, association memberships, etc.) (Raymond et al. 2010), which can also lead to collective commitments and play a role in *place identity*. Twigger-Ross and Uzzell (1996) defined four dimensions for place identity: *distinctive character, continuity, self-esteem* and *self-efficacy*. *Distinctive character* refers to the characteristics of the place where the individual belongs that he/she uses to differentiate it from others. *Continuity* refers to the place with respect to past relationships (actions, experiences, etc.) that are maintained and developed in the present. According to the authors, this link provides a sense of community identity. This dimension can be compared to the theory of social representations to the extent that they are partly made up of images that enable a link with the past. "They [the images] keep alive the traces of the past, take up space in our memories in order to protect them from the hurly-burly of change and reinforce the feeling that there is a continuity to the environment and to individual and collective experiences" (Moscovici 1976 p. 45). *Self-esteem* not only corresponds to a positive assessment of place, but also indicates the capacity or ability to use the qualities of the place to improve this self-esteem. Finally, *self-efficacy* is defined by researchers as "an individual's belief in his or her capacities to cope with situational demands" (p. 208). It refers to the functions of the place and the capacity of individuals to manage them. Furthermore, the whole concept of identity can be compared to that of social representation. Social representations notably have an identity function insofar as (i) they give the group its specificity in its social environment, and (ii) they offer the individual a set of standards and values in keeping with his/her system of thought. They participate in the construction of social identity (Abric 2001). Thus, *place identity* and *place attachment* are strongly linked (Bonaiuto et al. 2002) and both play a role in the representations of risk (Michel-Guillou et al. 2015).

In connection with the above-mentioned concepts, several studies have shown a link between the representations associated with place, the relationship to the life environment and the representations/perceptions of risk. They demonstrate the importance of the representation of place to individuals faced with these risks, as well as the role played by proximity (physical and social) in the social representation of risk (Baggio and Rouqette 2006). Studies tend to show that the more individuals are attached to a space, the more they feel safe and the less they perceive the risk as a threat, although they may be fully aware of its existence (interpersonal communications, media, etc.) (Billig 2006). "When attachment is very strong, any environment, even if known to be very dangerous, may still be perceived as being good enough to live in, making one's home one's castle" (Billig 2006 p. 263–264). Individuals are aware that they live in an area that is exposed to risk, but they do not feel insecure. This representation of their environment and their living space enables them to face the risk.

Research studies, particularly on flood risk, have reached the same conclusions. Weiss et al. (2006) showed that when people report high levels of attachment to their living space, they accept the risk linked to their environment. These people, who are very attached to their house and who have lived there for many years, are also informed about and involved in the management of their life environment. These results are in agreement with those of Bonaiuto et al. (2011) who showed that, concerning flood risks in Italy, people with high levels of attachment to their living space also have a high level of risk perception and are concerned by this issue. This risk representation enables people who are strongly attached to their environment to maintain a sense of control over events (Weiss et al. 2011) by minimizing the changes. Thus, adaptation strategies are possible as long as they do not lead to stigmatizing changes in the environment or a forced relocation (Wester-Herber 2004) and thus do not interfere with place identity (Twigger-Ross and Uzzell 1996). In the context of coastal risks, we find similar results. An application of these concepts to this type of risk is provided in the following section.

26.4 Application to Coastal Risks

Along with climate change and sea level rise, the question of coastal risks (erosion and submersion) is a topical issue. This phenomenon obviously concerns coastal communities and includes the users (inhabitants, professionals, tourists, etc.) who may be affected and the local managers who are responsible for the local management of this risk (elected representatives, technicians, resident associations, etc.). In France, concerns about this type of natural risk have increased since 2010, the year the Xynthia storm struck the Vendée and Charente coasts and took a heavy toll in human lives. This storm left its mark and represents a turning point in the management of areas exposed to risks of submersion and erosion.

The concepts discussed previously can be applied to the study of coastal risks of erosion and submersion. They were particularly used as the basis for interdisciplinary research involving numerous disciplines including geography and environmental psychology.[4] The objective of this research was to understand systemic vulnerability in the face of coastal risks and to move towards sustainable management strategies for these risks. Among the four components previously presented, namely hazards, issues, risk management and social and individual representations, it is within the latter that the concepts of social representations of risk and sense of place were used. The goal was to study the environment and the quality of life of people living in coastal communities, who were concerned by the risk of erosion or submersion (Michel-Guillou et al. 2015; Michel-Guillou and Meur-Ferec 2014) and, more specifically, how these populations represent the risks and the link that these representations may have with attachment to their place of residence and their adaptation preferences. This "representation" component of the systemic model seems essential to understand the links between practices and representations and thus to inform public risk management policies.

The study focused on people living in communities exposed to coastal risks. It was based on a survey that combined semi-structured interviews and questionnaires. It asked how people concerned by coastal risks (local stakeholders such as local representatives, technicians, residents or cultural associations) or potentially concerned (users of "at risk" areas, such as people with main or secondary residences, land or home owners, maritime professionals, users of the maritime space) represented their life environment and whether they integrated the issue of coastal risks in this representation. What position did these people adopt faced with this issue? How did they integrate it, either in their daily lives, or in the management of their community?

The survey showed that user comments about their life environment were structured around two main themes: aspects linked to the life environment (humans, their ways and quality of life, community, issues and concerns), and aspects linked to the natural environment (hazards and

---

[4]COCORISCO (Coastal Risks: Hazards, Issues, Representations, Management): Programme funded by ANR (the French National Research Agency) in the framework of the call for proposals for Global Environmental Change [*Changements Environnementaux Planétaires*], March 2011–February 2015.

the risk of marine submersion, among others). It should be noted that nearly two thirds of respondents did not spontaneously cite sea-related risks. In fact, they placed more emphasis on the risk linked to tourism and the population drain experienced by the municipalities in winter, which threatened regular economic activities (fishing, agriculture, local businesses, etc.). This risk of devitalization was also associated with population ageing. Pollution risks were also mentioned and were agriculture-related (nitrates, green algae, water potability, etc.). Regulation policies for urbanization were also cited as a risk by the inhabitants because they limit building permits. Other people minimized the risks; risk was absent from their comments, or trivialized and associated with the minor risks of everyday life.

Particular attention was then paid to the third of respondents who spontaneously cited coastal risks (submersion, flooding, erosion, storm, etc.). It appeared that these people had a social representation of coastal risk (Moscovici 1976). They could define it and characterize it. It was subject to issues, particularly those related to group cohesion and identity (Krien and Michel-Guillou 2014). However, coastal risk was not mentioned locally; it was either situated outside the community or place of residence (elsewhere), moved to a time scale beyond the life of the people: *"I'll no longer be around"*,[5] or downplayed: *"I don't see any danger, except for the water but, well, ... we manage"*. A distancing from the object occurred. However, the larger the distance perceived by the individual between him/herself and the object from a geographical, temporal, social or presumed social point of view, the more the object is perceived as abstract (Liberman and Trope 2008; Trope and Liberman 2003, 2010). This abstraction, or distance from the risk, may contribute to increased feelings of well-being insofar as it places the dangerous phenomenon further away. It thus protects the individuals from uncomfortable feelings and, in a way, avoids them having to consider changes in their environment.

In summary, it appears that the vast majority of respondents did not say that they were concerned about coastal risks. The phenomenon was usually experienced as occasional or exceptional. *"There are common risks that are greater than the exceptional risk of submersion on Ile-Tudy"*. The respondents *"do not fret"* or *"are not at all worried"*, they *"don't think that it will happen"* or *"don't necessarily live with it on their minds"*. They declared that they *"don't risk much"* or *"live with it"*, *"don't see it"*, etc. Living in a coastal community was seen primarily as an advantage or even a privilege, defined by the beauty of the landscape and the view, the proximity of the beach and the sea, etc. This living environment gives access to leisure activities and provides a certain amount of peace, which was affirmed systematically, and is the opposite of the pace of urban life. Amongst the benefits, conviviality through the community dimension was also highlighted. The social bond was considered an important element of the life environment. *"I get on quite well with the population so it's nice because even though we are isolated, we still see people. (...) I am with someone all the time, I'm never alone"*. Other benefits were also mentioned: being close to shops, being at home or close to family. The family (relative) dimension highlighted a living space attachment or even a rooting, which reflects a deep attachment, an anchoring. It is asserted through memories, transmission between generations, and heritage (Kelly and Hosking 2008). *"My home is here. My children come here, my son has a house, my daughter just bought hers. So you see our roots are here"*. *"So my origins are here. My roots are here. And I've always liked this place"*. *"I've always been very close to the sea for various reasons: holiday, family, education, roots"*.

Thus, place attachment, as the affective bond that encourages people to stay close to a place, is important (Hidalgo and Hernandez 2001). It is linked to the community aspect, the social bond (Raymond et al. 2010), rooting, and the feeling of being privileged to live in this place. This latter dimension contributes to enhancing people's

---

[5]The quotations in italics are translations of interview excerpts.

identity insofar as the place, with which they identify, allows them to distinguish themselves from others (Twigger-Ross and Uzzell 1996). The place thus contributes to maintaining or enhancing self-esteem and the sense of self-efficacy (Twigger-Ross and Uzzell 1996; Wester-Herber 2004). In addition, these inhabitants felt very privileged to live in this place, which contributed to the distinctive character of place identity enabling them to differentiate their place positively from other places. Living in a coastal community was considered a privilege because of the beauty of the landscapes, access to sea-related activities, tranquility, etc. On the other hand, as in other studies on the risks of river flooding (Weiss et al. 2011; Bonaiuto et al. 2011), the recognition of the existence of erosion and submersion phenomena gave respondents a sense of control. This feeling was expressed notably by the belief that public authorities manage the risks. Here again, the recognition of risk and the sense of control enabled individuals to maintain a positive place identity.

Furthermore, the place of attachment varies, since it may refer to the home, the community or, more generally, to the environment. Thus, the attachment is not only to the place of habitation, but also to the object that characterizes it: the sea, the object of socially valued representations, or the object of attractiveness (Kelly and Hosking 2008). Most respondents reported a proximity to the sea, be it physical (perceived distance), psychological (affective link) and/or social or functional (increased activities, alone or in groups, sea- and coastal-related). Their image of the community was one of a maritime and sea-related space. All these characteristics demonstrated a strong proximity and familiarity with the sea.

Thus, several research studies (for example, Billig 2006) have shown that people who are emotionally close to their living place tend to minimize the risks. The place of attachment is a place of security. In addition, as previously mentioned, since the life environment contributes to the construction of people's identities (Twigger-Ross and Uzzell 1996; Wester-Herber 2004), it does not tend to be conceived as a potential danger zone or a source of anxiety. Nevertheless, all the respondents acknowledged the existence of the phenomena of *erosion* and *submersion*; there was no denial of these phenomena, but they were not regarded as a "risk" according to the common meaning. The sea and its dangers were elements with which they lived or had chosen to live, without thinking of them in terms of risk. By this process of *objectification*, submersion becomes a "flood" or "wet-feet", the storm becomes a "strong gust of wind", etc. Thus, the coastal risk was not present in the respondents' minds; it was not "re-presented". For the majority of people, this risk was probably not an object of social representation (Moscovici 1976), or only for a small number of individuals directly concerned, for whom it has become the object of issues. Since the Xynthia storm in 2010 in France, the concept of this risk and its management have taken a new turn. It is not so much the fear of suffering damage that has emerged but rather that of being penalized by building permit restrictions (Krien and Michel-Guillou 2014).

Finally, it appears that the representation of coastal risks presents a specific aspect with respect to other risks, such as industrial ones. This specificity is linked to the ambivalence of the sea, an element considered potentially dangerous but which also has a very positive image in contemporary Western societies (Corbin 1988). It is seen above all as a source of pleasure rather than of risk. This ambivalence explains how people can be both sensitive to coastal risks and feel very privileged to live in a place because of its proximity to the sea. Indeed, in modern societies, seaside communities have become an object of attractiveness (Miossec 1998) and the sea plays an important role in the motivation to live there (Kelly and Hosking 2008). As for preferences in terms of adaptation strategies, this specific representation also explains the strong desire to remain in a privileged position by rejecting strategies that involve changes (demolition and relocation of properties) and by preferring or even demanding strategies that could strengthen their position (e.g., adapting the architecture of dwellings, strengthening sea protection works). Once more, this choice can be explained by the wish to protect one's personal identity. Considering that the living place plays a role in the construction of identity (Bonaiuto et al. 2002), it is understandable that changes in the configuration of a place to which one is attached can be perceived as an identity threat (Wester-Herber 2004).

26.5 Conclusion on the Systemic Approach in the Context of Natural Risks

This chapter focuses on natural risks and highlights the links between social representations and sense of place (mainly place attachment and identity), using the example of coastal risks. These reflections and results emphasize the need to take into account the human and social dimension in understanding natural risks and the vulnerability of a system in its entirety. This is the relevance of a systemic approach, or transactional approach in environmental psychology terms, for understanding complex phenomena.

From this point of view, humans and the environment are an integral part of a system within which they are interdependent, they evolve and form a unity. This acceptance of the environment reflects that of *ecumene*, cherished by geographers and defined by Berque (1996) as being "both the Earth and humanity, but it is not the Earth plus humanity, nor the inverse, it is the Earth as it is inhabited by humanity and it is also humanity as it inhabits the Earth". (...) "The *ecumene* is the relationship of humanity to the terrestrial area" (p. 78). It also relates to the concept of *mediance*, defined by Berque as "the sense of milieu, the sense of the relationship of a society to the terrestrial area". For this author, *mediance* is the translation of the Japanese term "*fudo-sei*", a neologism of Watsuji (1935) designating "the engagement of the human in his or her milieu". This word comes from the term "*fudo*", which means milieu in the sense of "founding condition (of human existence) where the relationships of men among themselves and men to things are intertwined" (p. 115). In other words, it refers to the *environment*, in the sense given to it by the transactional approach in psychology. Thus, for the psychologist Ittelson (1973 p. 12–13), "one cannot be a subject of an environment; one can only be a participant. The very distinction between self and nonself breaks down: the environment surrounds, enfolds, engulfs, and nothing and no one can be isolated and identified as standing outside of, and apart from it". The individual and the environment are thus considered elements of a larger system, the characteristics of one being inevitably linked to the characteristics of the other, as well as those of the system as a whole (Weiss et al. 2014). Any modification of a part of this system causes changes to the system as a whole (Wapner 1981). In this context, understanding complex phenomena relates to the analysis of the totality of the circumstances in which they occur; in other words, taking into account the psychological processes and the sociophysical context. The situations are composed of actors engaged in psychological processes in physical, social and temporal contexts (Altman and Rogoff 1987) and it is not so much the "objective" conditions of this context that are taken into account, but the "represented" conditions of these contexts. Thus, the behavior of the individual in relation to the environment is interpreted depending on the representation that he/she has of it. In this transactional perspective, the various dimensions that may intervene in the human-environment transaction are simultaneously taken into account: the aspects of the person (as an individual in a societal context), the attributes of the environment (the physical, cultural, and social characteristics, etc.) and the temporal dimension (Legendre 2005). This transactional approach in psychology resembles the notion of *trajectivity* developed by Berque (1996), which formulated the incessant exchanges that occur within the *ecumene* to produce a *milieu*, dated, located and endowed with meaning (Lussault 1997). In the same way and due to these characteristics, a transactional or systemic analysis refers to a dynamic study of the phenomena. These are taken into account at a given moment, in a given context that is constantly evolving. Therefore, the configuration of this system and the way in which it evolves is at the heart of this approach (Altman and Rogoff 1987). However, the undeniable heuristic value of this approach in the way it treats complex phenomena makes its implementation complex by taking into account a multitude of variables (Weiss et al. 2014). This is why a considerable amount of research addresses these issues in a decontextualized manner.

With respect to the study of natural risks, this approach seems particularly appropriate. As mentioned above, it is often difficult to attribute a direct responsibility to human action in this type of risk. In this sense, an explanation of the risk perception uniquely in terms of cognitive functioning and centered on the individual is not enough (Michel-Guillou and Weiss 2007). The way in which individuals represent this type of risk is inevitably conditioned by the context (physical, social, cultural, temporal) within which they live. Although cognitive processes can be identified as an optimism bias (Slovic 2000), they can only be understood through an overall approach to the system. This can highlight optimism biases, resistance, and diverse reasoning modes that can only be understood in relation to current cultural and societal standards (e.g., promotion of the sea in Western societies), at the time in which they are situated (in our contemporary societies, since the twentieth century, cf. Corbin 1988), and in relation to group memberships (resident associations, local managers, etc.). The temporal dimension is particularly important (Moser and Uzzell 2003) since the phenomena, anchored in time, are by definition dynamic and evolving. The transactional approach (Altman and Rogoff 1987), like the social representation approach (Moscovici 1976), is mainly based on the study of these dynamics inherent to the system. In addition, as seen above, it contributes to the construction of the identity of individuals, notably in the "continuity" dimension of identity (Twigger-Ross and Uzzell 1996).

Furthermore, this type of approach can help to understand the phenomena of local resistance, which are expressed at both the individual and collective levels. Thus, diverse forms of resistance lead individuals to approve national measures in general, while pointing out the impossibility of their local application (Castro and Michel-Guillou 2010). For example, in terms of coastal risk management, it is increasingly agreed that urbanization of coastal areas should be limited. However, when it comes to applying national risk management policies at the local level, other factors hinder these applications. Beyond personal interests, the priorities in terms of local development and risk management in general (devitalization of communities, environmental management, etc.) are not necessarily the same as those at the national scale. For example, the application to coastal risks presented in this chapter may explain the disparity between the national management strategy for coastlines in France, which advocates a consideration of the "relocation of issues and activities" (Ministère de l'Écologie, du Développement durable et de l'Énergie[6] 2012), and the difficulties of its local application with respect to the resistance of the inhabitants (Meur-Ferec and Rabuteau 2014).

In this sense, social representations, constituted of beliefs and attitudes, are especially appropriate for understanding the way in which individuals represent themselves, position themselves vis-à-vis the risks and take action to deal with them. To the extent that they clearly establish a distinction between the beliefs of an individual about an object and the attitudes about it (Castro 2006; Moscovici 1976), social representations take into account the paradoxes that occur within individual thought. They also provide a link between contextualized local issues and overall thought systems (Castro 2006). Concerning natural risks, and coastal risks in particular, the theory of social representations makes it conceivable that people can identify submergence phenomena as important at the global level without admitting their existence at the local level, and thus prepare practices aimed at solving it. Faced with this subject, and depending on their preoccupations, individuals will activate their own representations based on their social anchorage (Doise 1992).

---

[6]French Ministry of Ecology, Energy and Sustainable Development.

References

Abric, J.-C. (2001). A structural approach to social representations. In K. Deaux & G. Philogène (Eds.), *Representations of the social: Bridging theoretical traditions* (pp. 42–47). Malden: Blackwell Publishing.

Acuña-Rivera, M., Uzzell, D., & Brown, J. (2014). The mediating role of risk perception in neighbourhood disorder and perceptions of safety about victimization. *Journal of Environmental Psychology, 40*, 64–75.

Altman, I., & Low, S. M. (Eds.). (1992). *Place attachment*. New York: Plenum Press.

Altman, I., & Rogoff, B. (1987). World-views in psychology: Trait, interactional, organismic and transactional perspectives. In D. Stokols & I. Altman (Eds.), *Handbook of environmental psychology* (pp. 7–40). New York: Wiley.

Baggio, S., & Rouquette, M. L. (2006). La représentation sociale de l'inondation: Influence croisée de la proximité au risque et de l'importance de l'enjeu. *Bulletin de Psychologie, 59*, 103–117.

Barrows, H. H. (1923). Geography as human ecology. *Annals of the Association of American Geographer, XIII*, 1–13.

Becerra, S., & Peltier, A. (2009). *Risques et environnement: recherches interdisciplinaires sur la vulnérabilité des sociétés*. Paris: L'Harmattan.

Beck, U. (1992). *Risk society. Towards a new modernity*. London: Sage Publications.

Berque, A. (1996). *Etre humains sur la terre: principes d'éthique de l'écoumène*. Paris: Gallimard.

Berque, A. (2000). *Écoumène. Introduction à l'étude des milieux humains*. Paris: Belin.

Bertrand, G. (1986). Le système et l'élément. *Revue Géographique des Pyrénées et du Sud-Ouest, 57*, 283–290.

Billig, M. (2006). Is my home my castle? Place attachment, risk perception, and religious faith. *Environment and Behavior, 38*, 248–265.

Bonaiuto, M., Aiello, A., Perugini, M., Bonnes, M., & Ercolani, A. P. (1999). Multidimensional perception of residential environmental quality and neighbourhood attachment in the urban environment. *Journal of Environmental Psychology, 19*, 331–352.

Bonaiuto, M., Carrus, G., Martorella, H., & Bonnes, M. (2002). Local identity processes and environmental attitudes in land use changes: The case of natural protected areas. *Journal of Economic Psychology, 23*, 631–653.

Bonaiuto, M., De Dominicis, S., Fornara, F., Ganucci Cancellieri, U., & Mosco, B. (2011). Flood risk, the role of neighbourhood attachment. In G. Zenz & R. Hornich (Eds.), *Proceedings of the international symposium on urban flood risk management: Approaches to enhance resilience of communities* (pp. 547–552). Graz: Verlag der Technischen Universität.

Bonnes, M., & Secchiaroli, G. (1995). *Environmental psychology: A psycho-social introduction*. London: Sage Publications.

Cabecinhas, R., Lázaro, A., & Carvalho, A. (2008). Media uses and social representations of climate change. In A. Carvalho (Ed.), *Communicating climate change: Discourses, mediations and perceptions* (pp. 170–189). Braga: Centro de Estudos de Comunicação e Sociedade, Universidade do Minho.

Canter, D. V., & Craik, K. H. (1981). Environmental psychology. *Journal of Environmental Psychology, 1*, 1–11.

Castro, P. (2006). Applying social psychology to the study of environmental concern and environmental worldviews: Contributions from the social representations approach. *Journal of Community & Applied Social Psychology, 16*, 247–266.

Castro, P., & Michel-Guillou, E. (2010). Le développement durable et l'innovation législative: de la construction des idées à la compréhension des pratiques. In K. Weiss & F. Girandola (Eds.), *Psychologie du développement durable* (pp. 141–155). Paris: In Press Editions.

Clarke, J. I., Curson, P., Kayastha, S. L., & Prithvish, N. (Eds.). (1989). *Population and disaster*. London: Basil Blackwell.

Clémence, A., Doise, W., & Lorenzi-Cioldi, F. (1994). Prises de position et principes organisateurs des représentations sociales. In C. Guimelli (Ed.), *Structures et transformations des représentations sociales* (pp. 119–152). Paris: Delachaux et Niestlé.

Corbin, A. (1988). *Le territoire du vide, l'Occident et le désir de rivage 1750–1840*. Paris: Flammarion.

Cutter, S. L. (1993). *Living with risks*. London: E. Arnold.

D'Ercole, R. (1994). Les vulnérabilités des sociétés et des espaces urbains, concepts, typologie, mode d'analyse. *Revue de Géographie Alpine (Numéro Spécial), 4*, 87–96.

D'Ercole, R., & Metzger P. (2011). Les risques en milieu urbain: éléments de réflexion, *EchoGéo [En ligne], 18*. http://echogeo.revues.org/12640

D'Ercole, R., & Pigeon, P. (1999). L'expertise internationale des risques dits naturels: intérêt géographique. *Annales de Géographie, 608*, 339–357.

D'Ercole, R., & Pigeon, P. (2000). L'évaluation du risque à l'échelle internationale. Méthodologie et application aux diagnostics préalables aux actions de préparation et de prévention des catastrophes. *Cahiers Savoisiens de Géographie, 1*, 29–36.

Dardel, E. (1952). *L'Homme et la Terre. Nature de la réalité géographique*. Paris: CTHS.

Doise, W. (1982). *L'explication en psychologie sociale*. Paris: Presses Universitaires de France.

Doise, W. (1990). Les représentations sociales. In R. Ghiglione, C. Bonnet, & J.-F. Richard (Eds.), *Traité de psychologie cognitive. Cognition, représentation, communication* (Vol. 3, pp. 111–174). Paris: Dunod.

Doise, W. (1992). L'ancrage dans les études sur les représentations sociales. *Bulletin de Psychologie, 45*, 189–195.

Doise, W. (1993). Debating social representations. In G. M. Breakwell & D. V. Canter (Eds.), *Empirical approaches to social representations* (pp. 157–170). Oxford: Clarendon Press.

Doise, W. (2005). Les représentations sociales. In N. Dubois (Ed.), *Psychologie sociale de la cognition* (pp. 152–207). Paris: Dunod.

Douglas, M. (1994). *Risk and blame: Essays in cultural theory*. London: Routledge.

Gardner, G. T., & Stern, P. C. (1996). *Environmental problems and human behavior*. Boston: Allyn & Bacon.

Giddens, A. (1990). *The consequences of modernity*. Cambridge: Polity.

Gilbert, C. (2009). La vulnérabilité: une notion vulnérable ? À propos des risques naturels. In S. Becerra & A. Peltier (Eds.), *Risques et Environnement: recherches interdisciplinaires sur la Vulnérabilité des sociétés* (pp. 23–40). Paris: L'Harmattan.

Giuliani, M. V. (2003). Theory of attachment and place attachment. In M. Bonnes, T. Lee, & M. Bonaiuto (Eds.), *Psychological theories for environmental issues* (pp. 137–170). Aldershot: Ashgate.

Giuliani, M. V., & Feldman, R. (1993). Place attachment in a developmental and cultural context. *Journal of Environmental Psychology, 13*, 267–274.

Gold, J. R. (1980). *An introduction to behavioural geography*. Oxford: University Press.

Hay, R. (1998). Sense of place in developmental context. *Journal of Environmental Psychology, 18*, 5–29.

Hellequin, A.-P., Flanquart, H., Meur-Ferec, C., & Rulleau, B. (2013). Perceptions du risque de submersion marine par la population du littoral languedocien: contribution à l'analyse de la vulnérabilité côtière. *Nature, Science, Société, 21*, 385–399.

Hernandez, B., Hidalgo, C., Salazar-Laplacea, M. E., & Hess, S. (2007). Place attachment and place identity in natives and non-natives. *Journal of Environmental Psychology, 27*, 310–319.

Hidalgo, M. C., & Hernández, B. (2001). Place attachment: Conceptual and empirical questions. *Journal of Environmental Psychology, 21*, 273–281.

IPCC. (2007). *Climate change 2007: Synthesis report. Contribution of working groups I, II and III to the fourth assessment report of the intergovernmental panel on climate change*. Geneva: IPCC.

IPCC. (2013). *Climate change 2013. The physical science basis*. New York: Cambridge University Press.

Ittelson, W. H. (1973). Environment perception and contemporary perceptual theory. In W. H. Ittelson (Ed.), *Environment and cognition* (pp. 1–19). New York: Seminar Press.

Ittelson, W. H. (1976). Some issues facing a theory of environment and behavior. In H. M. Proshansky, W. H. Ittelson, & L. G. Rivlin (Eds.), *Environmental psychology: People and their physical settings* (2nd ed., pp. 51–59). New York: Holt, Rinehart and Winston.

Ittelson, W. H., Proshansky, H. M., Rivlin, L. G., & Winkel, G. H. (1974). *An introduction to environmental psychology*. New York: Holt, Rinehart and Winston.

Jodelet, D. (1999). Représentations sociales: un domaine en expansion. In D. Jodelet (Ed.), *Les représentations sociales* (pp. 47–78). Paris: Presses Universitaires de France.

Joffe, H. (2003). Risk: From perception to social representation. *British Journal of Social Psychology, 42*, 55–73.

Jollivet, M. (1992). *Sciences de la nature, sciences de la société: les passeurs de frontière*. Paris: CNRS Editions.

Jørgensen, B. S., & Stedman, R. (2001). Sense of place as an attitude: Lakeshore property owners' attitudes toward their properties. *Journal of Environmental Psychology, 21*, 233–248.

Jørgensen, B., & Stedman, R. (2006). A comparative analysis of predictors of sense of place dimensions: Attachment to, dependence on, and identification with lakeshore properties. *Journal of Environmental Management, 79*, 316–327.

Kelly, G., & Hosking, K. (2008). Nonpermanent residents, place attachment and "sea change" communities. *Environment and Behavior, 40*, 575–594.

Kempton, W. (1997). How the public views climate change. *Environment, 39*, 12–21.

Kouabenan, D. R., Cadet, B., Hermand, D., & Munoz Sastre, M. T. (Eds.). (2006). *Psychologie du risque: identifier, évaluer, prévenir*. Brussels: De Boeck Université.

Krien, N., & Michel-Guillou, E. (2014). Place des risques côtiers dans les représentations sociales du cadre de vie d'habitants de communes littorales. *Les Cahiers Internationaux de Psychologie Sociale, 101*, 101–122.

Lascoumes, P. (1991). De l'atteinte à la prévention des risques industriels. In C. Dourlens, J.-P. Galland, J. Theys, & P.-A. Vidal-Naquet (Eds.), *Conquête de la sécurité, gestion des risques. Ouvrage collectif* (pp. 65–86). Paris: L'Harmattan.

Legendre, A. (Ed.) (2005). Enjeux environnementaux et urbains: Contribution de la psychologie environnementale. *Psychologie et Société, 8*, 7–24.

Lévy-Leboyer, C. (1980). *Psychologie et environnement*. Paris: Presses Universitaires de France.

Lewicka, M. (2011). Place attachment: How far have we come in the last 40 years? *Journal of Environmental Psychology, 31*, 207–230.

Liberman, N., & Trope, Y. (2008). The psychology of transcending here and now. *Science, 322*, 1201–1205.

Lorenzoni, I., Pidgeon, N. F., & O'Connor, R. E. (2005). Dangerous climate change: The role for risk research. *Risk Analysis, 25*, 1387–1398.

Lussault, M. (1997). Espace, société, nature. In R. Knafou (Ed.), *L'état de la géographie, autoscopie d'une science* (pp. 408–419). Paris: Belin, Mappemonde.

Lussault, M. (2007). *L'homme spatial. La construction sociale de l'espace humain*. Paris: Seuil.

MEDDE. (2012). *Stratégie nationale de gestion intégrée du trait de côte: vers la relocalisation des activités et des biens*. La Défense: Ministère de l'Écologie, du Développement durable et de l'Énergie/Direction générale de l'Aménagement, du Logement et de la Nature.

Meur-Ferec, C. (2008). *De la dynamique naturelle à la gestion intégrée de l'espace littoral: un itinéraire de géographe*. Paris: Edilivre.

Meur-Ferec C., & Rabuteau Y. (2014). Un territoire fictif pour souligner les dilemmes des élus locaux face à la gestion des risques côtiers *L'Espace Géographique, 43*(1), 18–34.

Meur-Ferec, C., Flanquart, H., Hellequin, A.-P., & Rulleau, B. (2011). Risk perception, a key component of systemic vulnerability of coastal zones to erosion-submersion. Case study on the French Mediterranean coast. *Actes du colloque Littoral 2010 – Adapting to Global Change at the Coast*. Article Number 10003. doi: http://dx.doi.org/10.1051/litt/201110003.

Michel-Guillou, E. (2015). Water resources and climate change: Water managers' perceptions of these related environmental issues. *Journal of Water and Climate Change, 6*(1), 111–123. doi:10.2166/wcc.2014.098.

Michel-Guillou, E., & Weiss, K. (2007). Representations and behaviours of farmers with regard to sustainable development: A psycho-environmental approach. In B. A. Larson (Ed.), *Sustainable development research advances* (pp. 207–221). New York: Nova Science Publishers.

Michel-Guillou, E., Lalanne, P. A., & Krien, N. (2015). Hommes et aléas : appréhension des risques côtiers par des usagers et des gestionnaires de communes littorales. *Pratiques Psychologiques, 21*(1), 35–53. doi : 10.1016/j.prps.2014.12.001.

Miossec, A. (1998). *Les littoraux, entre nature et aménagement*. Paris: SEDES.

Morel, V., Deboudt, P., Hellequin, A.-P., Herbert, V., & Meur-Ferec, C. (2006). Regard rétrospectif sur l'étude des risques en géographie à partir des publications universitaires (1980–2004). *L'Information Géographique, 1*, 6–24.

Moscovici, S. (1976). *La psychanalyse, son image et son public*. Paris: Presses Universitaires de France.

Moscovici, S. (1987). Le déni. L'influence cachée des minorités. In S. Moscovici & G. Mugny (Eds.), *Psychologie de la conversion. Études sur l'influence inconsciente* (pp. 251–263). Delval: Cousset.

Moscovici, S. (2001). Why a theory of social representations? In K. Deaux & G. Philogène (Eds.), *Representations of the social: Bridging theoretical traditions* (pp. 8–35). Malden: Blackwell Publishing.

Moscovici, S., & Duveen, G. (Eds.). (2000). *Social representations. Explorations in social psychology*. Cambridge: Polity Press.

Moser, G. (1998). Attribution causale et sentiment d'insécurité de victimes de différents types de vols. *Les Cahiers Internationaux de Psychologie Sociale, 39*, 43–52.

Moser, G., & Uzzell, D. (2003). Environmental psychology. In T. Millon & M. J. Lerner (Eds.), *Comprehensive handbook of psychology* (pp. 419–445). New York: Wiley.

Navarro-Carascal, O., & Michel-Guillou, E. (2014). Analyse des risques et menaces environnementales: un regard psycho-socio-environnemental. In D. Marchand, S. Depeau, & K. Weiss (Eds.), *L'individu au risque de l'environnement* (pp. 271–297). Paris: Editions In Press.

November, V. (2002). *Les territoires du risque: le risque comme objet de réflexion géographique*. Berne: Peter Lang.

Peretti-Watel, P. (2000). *Sociologie du risque*. Paris: Armand Colin.

Pigeon, P. (2002). Réflexions sur les notions et les méthodes en géographie des risques dits naturels. *Annales de Géographie, 627–628*, 452–470.

Pinchemel, P., & Pinchemel, G. (1997). *La face de la terre*. Paris: Armand Colin.

Proshansky, H. M. (1976). Environmental psychology and the real world. *American Psychologist, 4*, 303–310.

Proshansky, H. M., Fabian, A. K., & Kamino, R. (1983). Place-identity: Physical world socialization of the self. *Journal of Environmental Psychology, 3*, 57–83.

Raymond, C., Brown, G., & Weber, D. (2010). The measurement of place attachment: Personal, community, and environmental connections. *Journal of Environmental Psychology, 30*, 422–434.

Reclus, E. (1866). Du sentiment de la nature dans les sociétés modernes. *La Revue des Deux Mondes, 63*, 352–381.

Slovic, P. (1987). Perception of risk. *Science, 236*, 280–285.

Slovic, P. (Ed.). (2000). *The perception of risk*. London: Earthscan.

Stedman, R. C. (2002). Toward a social psychology of place: Predicting behavior from place-based cognitions, attitude, and identity. *Environment and Behavior, 34*, 405–425.

Stokols, D., & Altman, I. (1987). Introduction. In D. Stokols & I. Altman (Eds.), *Handbook of environmental psychology* (pp. 1–4). New York: Wiley.

Stokols, D., & Shumaker, S. A. (1981). People in places: A transactional view of settings. In J. H. Harvey (Ed.), *Cognition, social behavior, and the environment* (pp. 441–488). Hillsdale: Lawrence Erlbaum Associates.

Theys, J., & Kalaora, B. (Eds.). (1992). *La terre outragée, les experts sont formels*. Paris: Autrement.

Trope, Y., & Liberman, N. (2003). Temporal construal. *Psychological Review, 110*, 403–421.

Trope, Y., & Liberman, N. (2010). Construal level theory of psychological distance. *Psychological Review, 117*, 440–463.

Tuan, Y.-F. (1979). Space and place: Humanistic perspective. In S. Gale & G. Olsson (Eds.), *Philosophy in geography* (pp. 387–427). Boston: D. Reidel Pub.

Tversky, A., & Kahneman, D. (1974). Judgement under uncertainty: Heuristics and biaises. *Science, 185*, 1124–1131.

Twigger-Ross, C. L., & Uzzell, D. L. (1996). Place and identity processes. *Journal of Environmental Psychology, 16*, 205–220.

Wapner, S. (1981). Transactions of person-in-environments: Some critical transitions. *Journal of Environmental Psychology, 1*, 223–239.

Wapner, S., & Demick, J. (2002). The increasing contexts of context in the study of environment behavior relations. In R. B. Bechtel & A. Churchman (Eds.), *Handbook of environmental psychology* (pp. 3–14). New York: Wiley.

Watsuji, T. (1935). *Fûdo, ningengakuteki*. Tôkyô: Iwanami.

Weiss, K., Colbeau-Justin, L., & Marchand, D. (2006). Entre connaissance, mémoire et oublis: représentations de l'environnement et réactions face à une catastrophe naturelle. In K. Weiss & D. Marchand (Eds.), *Psychologie sociale de l'environnement* (pp. 145–156). Rennes: Presses Universitaires de Rennes.

Weiss, K., Girandola, F., & Colbeau-Justin, L. (2011). Les comportements de protection face au risque naturel: de la résistance à l'engagement. *Pratiques Psychologiques, 17*, 251–262.

Weiss, K., Michel-Guillou, E., & Navarro-Carascal, O. (2014). Psychologie sociale et psychologie environnementale: confrontations et complémentarités. In D. Marchand, S. Depeau, & K. Weiss (Eds.), *L'individu au risque de l'environnement* (pp. 87–112). Paris: In Press.

Wester-Herber, M. (2004). Underlying concerns in land-use conflicts: The role of place-identity in risk perception. *Environmental Science & Policy, 7*, 109–116.

Social Inequality and Environmental Risk Perception

Oscar Navarro

For several decades, environmental risks have generated increasing scientific, political and social interest, particularly in terms of their potential impact on people's quality of life. Currently, the undeniable reality of climate change and its consequences, evidenced in "extraordinary" events (tsunamis, cyclones, storms) that strongly impact on the life of human societies, are major issues on the international political, media and scientific scene. Faced with the huge human cost that such extraordinary events generate, the inevitable question of the unequal ability of institutions and populations to deal with these risks is raised. Some populations live in areas characterized by especially marked environmental degradation. Industrial activities, in particular, may provoke significant ecological changes. They can be the source of various types of pollution on a local scale and these pollutants can lead to public health problems. They can also be the origin of technological accidents. These negative conditions disrupting the environment and impacting negatively on the health and well-being of populations are called environmental risks. Whether they concern global changes and major disasters or local pollution, or even a combination of the two, these risks are defined in terms of their impact on the quality of life (QOL) of the populations exposed to them to varying degrees, and who do not all have the same capability of dealing with them. In so far as QOL relates to an individual's subjective well-being and satisfaction with life (Moser 2009), environmental risks have a direct impact on QOL since they disrupt, or even preclude, the possibility of living a peaceful daily life and of fulfilling one's expectations and realizing one's projects. Moser (2009) distinguishes between two types of approach to identify the conditions of well-being: the first concerns the facilities and services to which individuals have access (health, education, leisure activities, transport), while the second relates to the degree to which an individual values and is satisfied with objective aspects of QOL. The probability of an event becoming threatening and disrupting QOL therefore depends on the characteristics of the event itself and, above all, the way in which the individual perceives and evaluates them. As regards the characteristics of the event, this evaluation of the dangerousness of a risk of environmental pollution is possible only when an individual or population is effectively exposed to it. The variability of judgment faced with risks depends on the variability of exposure, which differs according to social and ecological conditions. Individuals' perception and identification of an event as being likely to harm their well-being depends on their assessment of this event. This assessment takes into account a set

O. Navarro (✉)
Faculty of Psychology, University of Nantes,
Nantes, France
e-mail: oscar.navarro@univ-nantes.fr

of conditions and information, and leads to an evaluation of the dangerousness of the event in question. Moser (op.cit.) identifies three types of evaluation in the definition of a threat: by comparison with previous harm (Lazarus and Launier 1978); by anticipation of the potential dangers; and based on the perceived possibilities of dealing with the threat. The two aspects of risk evaluation and social and environmental inequality are examined in this chapter through a literature review in which I seek to give meaning to an approach that does not yet enjoy consensus and is treated differently, depending on the cultural context.

A starting point for my reflection is the fact that environmental and social inequalities cannot be separated, as the former relate to the sources of risks, whereas the latter relate essentially to the characteristics of the individuals and populations concerned (Roussel 2009). The relationship between the two affords insight into the strategies that individuals use to adapt to their environments, and could constitute a means for improving QOL.

27.1 Social Inequality and Environmental Risks

Ecological crises and environmental disasters spawn new issues that link inequalities to environmental situations (climate refugees, energy precariousness, diverse forms of pollution, environmental health, etc.) (Faburel 2010). The aspect that enjoys the greatest consensus in this approach to inequality is the analysis of poor populations' exposure to major environmental risks. However, the fact that different terms are used with different meanings in the literature makes it difficult for authors to understand one another and to reach agreement – which also reveals diversity stemming from cultural contexts. For example, it is often difficult to understand the connection between inequality, injustice and disparity, and the terms environment and ecology are frequently confused. The fundamental question here pertains to the relevance of differentiating between social and environmental inequalities (Chaumel and La Branche 2008; Emelianoff 2008). An approach focusing on environmental inequalities enables us to complete, understand and even reveal the historical and social conditions fostering social divisions, characterized by the domination of certain social groups or categories. In Europe, for instance, studies have confirmed that the impact of multiple and cumulative risks may be two or more times greater on people living in disadvantaged socio-economic conditions than on their richer neighbors (Kohlhuber et al. 2006). Likewise, inequalities in terms of exposure to environmental threats are more frequent for the most vulnerable groups such as children, the elderly, households with little education, the unemployed, migrants and certain ethnic groups (WHO Europe 2010).

Three aspects or manifestations of environmental and social inequality can be identified. The first is the unequal exposure of certain groups to environmental risks and hazards, caused by historically entrenched structural social inequalities. The second is unequal access to environmental resources and wealth, which are mainly enjoyed by certain groups that hold power, at the expense of others that are disadvantaged. For example, it has been shown that populations with the greatest exposure to a natural environment and green spaces are often those with the highest income and the best health (Mitchell and Popham 2008). Finally, the third aspect relates to certain groups' unequal ability to adapt, take action and react to environmental risks. We consider that a level of informational and educational precariousness, characteristic of certain economically disadvantaged groups, tends to go hand in hand with a form of passivity or acceptance of poor environmental conditions (Roussel 2009). This raises the question of the need to find a form of social justice and structural equity, or at least a reduction of inequality. In this respect, the social sciences and environmental sciences are currently expected to identify and reduce the territorial, social and individual vulnerability of populations, and to identify and reduce inequalities (exposure, access to resources, possibility of acting), in order to promote social and environmental justice. Promoting physical environments that favor good health and encouraging access to cultural and educational resources could be important ways in which these socio-environmental inequalities and their impact on QOL may be reduced.

From a historical perspective, in this framework of reflection and action, the first approach was environmental justice, which appeared in the US in 1970 (Fol and Pflieger 2010), as part of the civic rights and anti-discrimination movement. The main objective was to defend ethnic minorities (African-American, Latin-American and Amerindian) and socio-economic groups (poor urban populations) considered to be victims of abuse by those in power. The high levels of exposure to sanitary risks and lower life expectancy rates of these populations informed court cases with extensive media coverage (Bullard 1994; Bullard and Johnson 2000). In this approach, a strong correlation between environmental and public health inequalities was firmly emphasized (Brulle and Pellow 2006).

A similar approach has recently been developing in other countries, especially in Europe, but is veering away from its ethnic focus to cover all social vulnerability with regard to the environment (Faburel 2010). Unequal exposure to sanitary risks is highlighted, along with exposure to industrial pollution and notably poor air quality (Kohlhuber et al. 2006). Here too, the epidemiological aspect is emphasized.

Another approach has been developing alongside the preceding one: a more macro-social, even planetary perspective of social and environmental inequalities, focusing on geopolitical "North-South" relations. It examines countries' levels of economic and industrial development to analyze the ecological consequences of inequalities. It also seeks to show the close correlation between social inequalities, poverty gaps and the global ecological crisis. The main theme is climate change and its effects. For example, a report of the Intergovernmental Panel on Climate Change (2007) showed that, in 2004, the poorest countries accounted for 37 % of the world's population but were responsible for only 7 % of CO_2 emissions, as opposed to the inhabitants of the richest countries who accounted for 15 % of the world's population but were responsible for 45 % of CO_2 emissions. Another topic of interest in this approach is the significant inequality of the impacts of natural disasters: for the same number of disasters, the number of victims in terms of human life is always much higher in poor countries (Faburel 2010).

By highlighting environmental inequality, we point out that an aspect of the environment is distributed unevenly amongst different social groups (differentiated by social class, ethnicity, gender, age, location, etc.). Degrees of inequality can differ, depending on how skewed an environmental parameter is towards or away from the social groups concerned (Stephens et al. 2007). Lima (2008) notes that, to understand the role of social inequality and exclusion in the study of perception, one has to focus on the analysis of unequal exposure to risks. People who live close to sources of pollution are generally the poorest. They live in areas with high environmental risks, have the fewest means to protect themselves against exposure, are the least informed on the subject, and have the least power to influence environmental decisions. It has also been shown that the magnitude of the impact of disasters differs depending on a country's level of development (Lima et al. 2005). For example, for every death caused by a natural disaster in Europe, there are 128 in Africa (Lima op.cit.). Likewise, income level is one of the factors studied by the psychological approach to risk perception, which shows that the poorest individuals have the acutest perceptions of risk (Chauvin and Hermand 2008). This category of the population is the most worried about technological risks (Pilisuk and Acredolo 1988) and is more afraid of personal exposure than are people in higher income groups (Savage 1993). Possible explanations for this phenomenon are the loss of commercial value of land situated in high-risk areas (both natural and industrial risks) and the weak capacity of these residents for social mobilization. This leads to the emergence of socially and ecologically stigmatized territories.

To sum up, it seems that the most salient characteristic of poverty impacting on health and QOL is exposure to the multiple factors of environmental risks. Inverse relationships between income level and environmental risks (toxic waste, toxins, air pollution, water quality, noise, residential density, poor housing, poor quality of the working environment and the neighborhood) bear witness to the importance of this aspect of social and environmental inequality (Evans and Kantrowitz 2002) – one that has become a classic in the literature and on which there is consensus. Yet, studying other conditions affecting individuals, groups or contexts can also further our understanding of this very particular relationship between the perception of environmental risks and social and environmental inequalities.

27.2 Social Inequalities as a Source of Vulnerability Faced with Environmental Risks

There is now consensus in the literature on the definition of environmental risk as the result of a combination of threat and vulnerability. In other words, risk corresponds to the probability of a danger materializing, and to the gravity of its consequences (Leplat 2006; Fleury-Bahi 2010). A threat is thus seen as the probability or even the possibility of a dangerous event occurring, while vulnerability refers to the fact that a person or group is likely to be "affected" by this dangerous event. It *"relates to a certain fragility, a certain propensity to suffer damages, and applies to all the objects of the social and natural world"* (Metzger et al. 2010: 240). Risk is thus seen as the product of the extent and characteristics of the threat, and of the degree of vulnerability of the victims.

In the general field of collective risk evaluation and management, the tendency in recent decades has been to focus on the dimension of vulnerability rather than that of unpredictability. Vulnerability relates to the conditions preceding the event, which influence the degree of propensity to incur damages. Building a framework of interpretation based not on the unknown but on stakes and vulnerability is becoming the main objective of social sciences in risk analysis (Gleyze and Reghezza 2007; Metzger et al. 2010). With regard to risk, vulnerability has gradually become a key concept, the importance of which is now recognized (Metzger and D'Ercole 2011; Gilbert 2009). A general definition of vulnerability is: *"the propensity of a given society to suffer damages when a natural or anthropic phenomenon arises. This propensity varies, depending on the weight of certain factors that have to be identified and analysed because they induce a certain type of social response"* (D'Ercole 1994: 88). This dimension is an integral part of the very definition of risk, and is understood as the structural condition of the populations concerned. In general, authors allude to material vulnerability or to functional vulnerability, seen as the structural living conditions that undermine QOL.

In this field of risk evaluation studies, researchers focus on social vulnerability insofar as it is seen to be the product of social inequalities, since social factors shape groups' susceptibility to risk, as well as their ability to react (Brooks 2003). Vulnerability also includes the spatial inequalities of certain communities, as well as the characteristics of their buildings, and the quality or level of their urbanization, growth rates and economic vitality (Cutter et al. 2003). The social and the biophysical thus interact to produce the vulnerability of a place and its inhabitants. Likewise, the key role of socio-economic conditions in judgments and beliefs concerning proven sanitary risks has also been recognized (Vaughan 1995). Vulnerability is considered to be the combination of three components: differing exposure to stress factors; differing sensitivity to risks (susceptibility, predisposition, exposure); and differing ability to respond to, adapt to, and recover from the impacts thereof (Downs et al. 2011; Rossignol et al. 2015).

The fact of being or of considering themselves vulnerable or not, faced with a threat, indicates the individuals' or populations' unequal relationship to it. This is a result of the objective and assessed conditions of social belonging, as well as environmental conditions, which make certain categories of people more vulnerable than others. Socio-environmental inequalities in social life emerge as the result of basic underlying historical conditions. Of concern is the fact that millions of people with a low income, of diverse ethnic origins, live in areas harboring highly stressful industrial activities. These conditions in themselves are a form of basic vulnerability in which psychosocial stress is present. Studies on the health conditions and perceived QOL of these populations show that there is a relationship between gender and race, on the one hand, and health disparities, on the other. Women, in particular, show higher morbidity rates than men, and women of color more than white women, especially when the individual claims to be depressed (Downs et al. 2011).

In environmental psychology, an approach focused on vulnerability looks at perceived and experienced vulnerability; in other words, at individuals' evaluation of their own vulnerability. This is the product of objective living conditions, but also takes into account other factors that need to be identified. The level of perceived vulnerability depends on five key factors (Navarro and Michel-Guillou 2014): degree of exposure to a significant risk; feelings of loss of control due a lack of protective measures, the means to defend oneself effectively and/or possibilities of avoiding the threat; anticipation of significant negative consequences (Moser 1998); causal attribution of the threatening situation, that is, beliefs and forms of explanation of threatening events; and familiarity with the environment and/or the threat, that is, physical and/or social proximity with it (through collective memory).

In particular, the perception of a loss or absence of control over the threatening situation can increase the feeling of fear and psychological suffering (Henslin 1967; Ward and Jenkins 1965). Having a strong feeling of control over an event means believing that one has a behavioral response that can alter the probability of its occurrence. Otherwise, the person experiences a feeling of powerlessness or even despair (Seligman 1975). In this sense, and connected to perceived vulnerability, the feeling of powerlessness experienced when faced with risk is produced by the feeling of a loss of control in the situation, and by the sense of having no possibility of preventing its effects or of protecting oneself from the event, or of recovering. The subject therefore assesses his/her own resources and the possibility of having some control. In short, perceived control relates to the way in which individuals assess the degree of influence that they may have on their environment (Nuissier 1994).

As noted above, a person or group is more likely to feel vulnerable in the face of a risk when their living conditions are particularly difficult. Environmental and social conditions that are distributed unequally and make certain categories of the population particularly vulnerable result more often in this evaluation of vulnerability and attendant feelings of ill-being. In the literature, risk perception is closely associated with race and gender. In public opinion surveys in the US, for example, white men as a group have been shown to have a low perception of environmental health risks, compared to women and nonwhite men. This has led some social scientists to posit the "white male" effect (Saterfield et al. 2004; Finucane et al. 2000), an objective social condition implying forms of basic vulnerability. Some authors have investigated the relevance of more subjective factors, such as the evaluation of personal vulnerability, as well as socio-political evaluations concerning environmental injustice (Saterfield et al. 2004). The results of the work of Saterfield et al. (2004) show that these four variables (race, gender, perceived vulnerability, and evaluation of environmental injustice) predict risk perception. Gender and the evaluation of social vulnerability appear to be the strongest predictors, followed by environmental injustice and race. Thus, the "white male" effect is always considered fundamental in explaining risk perception, but it is related to other variables such as age, education, income, political leanings, and religious affiliation. Likewise, it has been found that the perception of risks related to nanotechnology is strongly influenced by the evaluation of one's own vulnerability, and by an impression of environmental injustice (Conti et al. 2011).

Exposure to risks remains the most salient factor in the evaluation of vulnerability. In France, the *Observatoire des Zones Urbaines Sensibles* (ZUS) (the Observatory of Sensitive Urban Areas) has shown that towns containing such areas (i.e. ZUS) are twice as exposed as others to industrial risk as defined by the SEVESO directive (OZUS 2004). Over 40 % of people living in ZUS are exposed to industrial risks, that is, twice as many as in other neighborhoods (Champion et al. 2004). In addition, studies carried out on a local scale, particularly in French Overseas territories, the Nord-Pas-de-Calais *Région* and the Seine-Saint-Denis *Département*, show that poor populations are proportionally more often located close to mining sites, chemical plants, and areas with polluted ground (IFEN 2006; Faburel 2010). On the scale of the whole of the Ile-de-France (i.e. Paris and surrounding areas), although most of the towns exposed to a SEVESO risk have an average rather than a low socio-urban profile (47 % and 41 % of cases, respectively), the latter are characterized by a clear over-representation of this risk (Gueymard 2009). In Belgium, the populations exposed to industrial pollution by nitrogen oxides and fine particles generally live in poorer than average towns (Dozzi et al. 2008). Likewise, Walker et al. (2005) have shown an unequal distribution of industrial sites in England, with more of them located in deprived areas. In the US, studies have shown how Blacks, Hispanics and low-income communities live mostly in industrial and hazardous areas (Szasz and Meuser 1997). In Latin American countries, as well, environmental issues related to the constant threat of natural and industrial risks are linked to very marked social inequalities and/or vulnerabilities. Due to growing and often poorly controlled urbanization, and an unfavorable geodynamic and climatic context, Latin American countries are faced with many risks. Because of the common occurrence of natural phenomena (earthquakes, tsunamis, hurricanes, etc.) and the degradation of the urban surroundings of a large proportion of the populations in the big Latin American cities, evaluating risks and populations' adaptability remains a subject of fundamental importance. These environmental issues are further complicated by the presence of strong social, economic and political inequalities (poverty, discrimination, marginality) leading to multiple vulnerabilities. The causes of these vulnerabilities include: difficulties in accessing health and education services, increasingly wide social disparities, political corruption, spatial segregation and historical social conflicts (Navarro 2011; D'Ercole et al. 2009). These environmental inequalities appear in the form of geographical and social inequalities linked to history, land ownership and access to productive capital, or in the form of social conflicts over access to and use of vital resources such as water (Navarro 2008).

From a health perspective, environmental vulnerability goes hand in hand with social vulnerability. This is specifically the case of exposure to atmospheric pollution, whether of industrial origin or not (Catalan 2006). The poorest populations are most exposed to these sources of environmental risk. Residents close to industrial sites that pollute and carry technological risk are subject to several sources of vulnerability: environmental vulnerability linked to where they live; social vulnerability as essentially disadvantaged and economically fragile people are found within these same populations; and health vulnerability since these populations are subjected to health risks induced by daily exposure to industrial pollutants. Faced with these risks every day in their living environment, individuals may engage in adaptive strategies in order to limit the damaging effects on their psychological well-being. These may involve methods to assess the risks induced by their living environment (denial or acceptance of risk) but they may also lead to resignation or, conversely, the individual may engage in actions to resolve the problem, by joining a local advocacy group for example.

In short, vulnerability faced with environmental risks is directly correlated with inequality in relation to risks and pollution and, basically, with various forms of exposure. Research on the risks related to industrial activity (Roussel 2009) shows that twice as many people from disadvantaged groups live near a polluting industrial site, compared to other groups. Moreover, for the same level of exposure, sanitary effects and QOL can differ according to the vulnerability of the exposed people. For example, Evans (2004) has reported that poor children, compared to children from privileged backgrounds, are more exposed to family problems, violence, separation and instability at home. They have less social support and access to culture, and their parents are more authoritarian and less involved in school activities. The air and water that poor children consume are more polluted, their homes are more cluttered and noisy, and their neighborhoods more dangerous. The accumulation of environmental and social risks, in addition to exposure, is a particularly important aspect of poor children's vulnerability (Evans 2004), and constitutes a new approach to risk evaluation.

27.2.1 Vulnerability and Perception of Air Pollution

For many decades, anthropogenic environmental problems have generated increasing scientific, political and social interest, particularly in terms of their potential impact on people's quality of life and health. In fact, the quality and integrity of the environment and natural ecosystems are essential to the health and quality of life of human communities. Thus, the ecological disturbances caused by human activities unquestionably change the conditions of health and psychological well-being of these communities. Air pollution, in particular, is characterized by wide spatial diversity, including proximity to facilities or to polluting activities at a global level (Roy 2006). Air pollution is considered a strong determinant of health but very unequal, due to the variability of exposure to contaminants (Charles et al. 2007). The poorest urban populations are also often those who live in degraded environments and suffer most from environmental risks (Theys 2002). The fact of being exposed on a daily basis to toxins and the ecological degradation they cause, whether it is apparent or not, is in itself a criterion of vulnerability. Research also suggests that the evaluation of health risks, linked to exposure to different types of pollution, may be a significant explanatory factor of perceived health and subjective well-being (Lima 2004; Peek et al. 2009; Stenlund et al. 2009; Wind et al. 2004).

Air pollution is considered a major public health problem, given the significant technological and methodological advances that have shown not only a relationship between the gradients and time of exposure to air pollutants and the attendant problems of morbidity and mortality, but also that health risks increase in certain vulnerable, socially disadvantaged populations living in certain geographic areas (WHO Regional Office for Europe 2006). Moreover, in October 2013, the World Health Organization classified air pollution in the category of "some carcinogenic". Susceptibility to risk depends not only on environmental inequality but also on the capacity to resist (Morin 2006), as it involves a system of judgments and expectations that are socially constructed and expressed, and that can influence social groups' risk behaviors associated with collective welfare. The psychosocial approach to sanitary risks suggests that people's perception of air quality and its effects on health is determined not only by the objective conditions of exposure but also by knowledge that is socially constructed, through the media, and is marked by the history of social relations (Joffe 2003; Roussel 2007; Rouquette 2007). Studies have stressed the role of practical everyday experiences in how people perceive air quality. Research has shown that people largely perceive cars as the major source of pollution (Jacobi 1994; Howel et al. 2003). Industry is also often identified as a significant pollutant (Howel et al. 2003; Bickerstaff and Walker 2001). The visual perception of air pollution appears to increase with higher concentrations of suspended particles within certain size ranges (Schusky 1966) and with greater dust fall. In addition to vision, other forms of sensory awareness are important, such as olfactory evidence and the experience of physiological or psychological affects (Wakefield et al. 2001; Clarenburg 1973). Research has also found that the exposure level does not directly influence annoyance or other symptoms, and that these are instead mediated by perceived pollution and health risks. In fact, these perceptions play important roles in understanding and predicting environmentally-induced annoyance and health symptoms for non-toxic levels of exposure (Claeson et al. 2013). This was also identified in a study that showed that the recognition of health risks from air pollution during work-related commuting was high, but that awareness did not differ with the travel mode (Badland and Duncan 2009). Moreover, there has recently been a new understanding of the fact that people become aware of the extent of the risks associated with air pollution when their health is affected. However, it has been found that, here too, people's perceptions tend to be influenced less by scientifically-derived information than by local and personal experiences (Howel et al. 2003; Elliot et al. 1999).

In France, the environmental health barometer of the National Institute for Prevention and Health Education (INPES) examined the differences in the representation of atmospheric pollution among the population, through a telephone survey on a sample of 6,007 individuals aged from 18 to 75, and representative of the French population (Ménard et al. 2008). Overall, the results show that the vast majority of the population (70 %) feels well-informed about air pollution and its possible effects on health, but this sense of knowledge is associated with a pessimistic view of environmental problems and their evolution, with nine out of ten French individuals believing that air pollution has worsened. This concern stems from the perception of a serious health risk, since 85 % of the population believes that air pollution poses a high risk to health. These studies also show that, in France, there are differences in terms of representations of air pollution. For example, the most disadvantaged social classes have a more negative view of air pollution problems and perceive a higher health risk than the general population (Roussel et al. 2009). According to recent data, the air pollution phenomenon needs to be seen as an important social problem, which also arises from the conceptual framework of social risk management (Holzmann and Jørgensen 2001). In Germany (Kohlhuber et al. 2006), a study highlighted the fact that exposure to noise and air pollution is influenced by socio-economic status. Researchers used data from a large-scale population survey (7,275 adults, 40 % of whom were women) to show that environmental exposure is unevenly distributed and that economic differences play a fundamental role in the feeling of vulnerability.

A study on perceptions of air pollution since 1990 was undertaken in Latin America (in Chile, Brazil, Mexico and Colombia). Research in Mexico, in particular, showed the importance of including the population's perceptions in sanitary risk evaluation, in order to build indicators that match their conditions and expectations (Catalán 2006). For example, Catalán et al. (2001) studied Mexico City's adult population's perceptions of air pollution. In a sample of 394 subjects, 84 % perceived air pollution levels as being high or very high, and 93 % considered that the health risk was very high. In another study on an adolescent population (N = 680) (Catalán et al. 2009), over 80 % considered the pollution level to be very high and 60.9 % considered it to be the city's second biggest problem, after crime. Pollution was seen as a major health risk by 75 % of the respondents, while 71.6 % considered it to be a cause of death. In general, one of the characteristics of this type of risk is the accumulated effect of factors of ecological and socio-economic vulnerability. For groups exposed to a polluted environment, and who are also used to being in a socio-economically precarious situation, perceived levels of QOL are reduced (Fleury-Bahi et al. 2013). In other words, there is a formal, direct correlation between the evaluation of QOL and socio-economic level (Fleury-Bahi and Annabi-Attia 2014).

27.2.2 Vulnerability and Perception of Climate Change

Vulnerability with regard to climate change is considered to be the extent to which a system is likely to be able to deal with the harmful effects of climate change, or not. It depends not only on the physical characteristics of the events caused by the climate change to which the system is exposed, but also on that system's sensitivity and adaptive capacity (IPCC 2007). This definition suggests that certain individuals or groups may be more sensitive or susceptible than others, and therefore more likely to suffer more from the negative consequences of climate change. Likewise, it appears that the ability to adapt in the face of these events is not the same, due to differences in people's or group's access to resources and their ability to deal with a threat (what Safi et al. (2012) identify as coping, adjustment, adaptation, attenuation and survival). In the literature, the two components of vulnerability (sensitivity and ability to act) are generally defined in socio-economic terms (Adger 1999; Cutter et al. 2003), without taking into account other cultural factors (e.g. beliefs) or even psychological ones (cognitions, evaluations, and emotions).

Several studies that investigated the relationship between vulnerability and risk perception have shown its complexity, with varied results to explain it. For example, Brody et al. (2008) highlighted the fact that people living in areas where lives are lost due to environmental disasters, perceive the risk of climate change as more serious than do people living in areas where deaths from these phenomena are less common. Whitmarsh (2008) concludes that people suffering from air pollution have a keener perception of climate change than those who do not suffer from it.

As regards flood risks, the link with climate change has not always been clearly established. Some studies show that people living in areas where flooding has occurred over the centuries perceive the risk of climate change as less serious than do those who live in areas where flooding is less likely (Brody et al. 2008). Other researchers have found that people who have experienced a flood in the past 5 years perceive climate change in the same way as do the other respondents: the experience has had no effect on their perception of climate change and, in fact, they see the two risks as two different things (Whitmarsh 2008). By contrast, Spence et al. (2011) found that people who have experienced a flood are likely to have a more acute perception of climate change and feel more concerned.

A study that examined the correlation between vulnerability and perception of climate change risks (Safi et al. 2012) showed no correlations between the respondent's age, farming activity, and perception of risk related to climate change. Likewise, physical vulnerability or the fact of living in a high-risk area does not influence the perception of climate change risks. On the other hand, a gender effect was found insofar as women appear to be more sensitive to this risk than men. Thus, not only objective or even physical variables of vulnerability determine environmental risk perception; there seem to be other more subjective factors explaining it. For example, a strong correlation has been found between the belief in an anthropogenic cause of climate change and its perception as a risk. Hence, political orientation and beliefs regarding the environment are significant determinants of the perception of climate risks (Safi et al. 2012). Finally, more than physical vulnerability, it seems to be an individual's sensitivity and capacity to adapt that are likely to increase their perception of climate change risks. This clearly indicates the need to explore these more social and psychological dimensions of vulnerability.

27.2.3 Vulnerability and Perception of Flood Risks

Natural risks are a particular category of environmental risks, because direct responsibility cannot easily be attributed to human action (Navarro and Michel-Guillou 2014; see Michel-Guillou and Meur-Ferec in this section). These are the only risks for which there is no possibility of reconciling utility and acceptability, as in the case of other types of risk, especially technological or medical ones. Natural risks are often classified according to their size, and they indicate an event of major proportions that exceeds the coping capacities of a region or a country and its institutions. Benyakar (2003) explains that a disaster happens when an organization, society or individual loses their ability to deal with it. As noted above, for a disaster to happen, there have to be conditions of vulnerability, that is, a population likely to suffer damages or to have difficulties recovering. On this issue, and even though global risks, especially climate change, suggest a sort of equal exposure to environmental risks (Bourg et al. 2013), it has been found that the catastrophic potential of events remains greater among the most socially and economically vulnerable populations. A UN report (UNDP 2004) has even challenged the term "natural disasters", since they are due, above all, to social inequalities and vulnerabilities. For example, losses due to disasters in less developed countries (in Africa, for example) are primarily human (53 % of deaths are caused by natural disasters, against 2 % in developed countries such as Europe). Losses for developed countries are primarily economic (for every million dollars of losses in Africa, there are 77 million in Europe) (UNDP 2004).

Research on the evaluation of flood risks, which are the most common natural risks, shows an absence of consensus on the role of socio-economic factors in risk perception. Some contend that the inhabitants of flood-prone areas underestimate or even deny the risk, while others argue the contrary (Villa and Bélanger 2012). Likewise, the few studies that have examined the influence of socio-demographic factors on flood risk perception present contradictory results. Some indicate that risk perception is greater among the elderly (Kellens et al. 2011), while others report the opposite (Botzen et al. 2009). This same contradiction is found in the results of research analyzing socio-economic conditions. Some authors show a positive correlation between high income together with a high level of education, and an underestimation of risk (Botzen et al. op.cit.; Lindell and Hwang 2008). By contrast, others maintain that high-income, highly educated individuals have a keen awareness of the risk (Burningham et al. 2008; Tapsell and Tunstall 2008). Nevertheless, all these authors agree that the perception of flood risks is strongly determined by previous experience, which provides individuals with cognitive frameworks to define, understand and cope with risk (Weber et al. 2000).

Research on floods in France, for example, has identified two social representations among people living in flood-prone areas: one sees risk as part of the area and of daily life, while the other projects the responsibility for the event on the State, thus externalizing responsibility for its management (Castro et al. 2010; Weiss et al. 2006). Hence, common-sense knowledge developed around flood risks depends on available sources of information and on the perception of individual vulnerability. It is related to a precise form of attribution of responsibilities in crisis management and disaster-prevention policies. These forms of common-sense knowledge are also related to existing practices modulated by personal involvement. Such institutionalized practices, as well as access to information, show a form of inequality since they are determined by the social, political and institutional context.

Finally, it is important to note that exposure and proximity to flood risks have a major impact on the perception of that risk, and on the resulting strategies to cope with it. Research on people living in flood-risk areas shows that the closer their home is to a river, the higher their perception of the danger (Botzen et al. 2009; Burningham et al. 2008; Zhang et al. 2010). Likewise, former experience of floods seems to be a decisive factor in increasing risk perception (Villa and Bélanger 2012) because this perception helps in judging the probability of the same type of event occurring in the future (Botzen et al. 2009; Correia et al. 1998; Terpstra 2009; Zhang et al. 2010). However, it is also possible that people who have experienced a minor flood tend to underestimate the probability of the impact of future floods (Carroll et al. 2010; Ruin et al. 2007). This is due to familiarity with certain risks, that is, a sort of trivialization of certain low-intensity events (Bernardo 2013).

27.3 Proximity, Experience and Perception of Environmental Risks

Exposure to environmental risks is the first condition of vulnerability and the sign of the existence of a form of socio-environmental inequality (Charles et al. 2007). This exposure appears in the form of geographic proximity as well as the individual's experience of the threat. These two aspects correspond to what Sjoberg (2002) calls distal variables in risk perception: a context variable (proximity) and a personal variable (experience). Perceived vulnerability to risk is associated with direct or indirect experience of crisis situations. In the 1960s, for example, American researchers studied how people reacted to natural disasters (Burton and Kates 1964; Kates 1976). They looked at the impact of previous experiences on the adoption of preventive measures for future events. Their findings showed that residents' personal experiences determined their behavior with regard to natural disasters, whereas people who had not experienced the same type of event disregarded the possible consequences of these dangers. It also seems that people prefer to accept their losses rather than trying to reduce them. The majority of respondents preferred to pay the costs incurred due to a disaster rather than changing their lifestyle or even moving home. The same studies also showed that people living in high-risk areas tended to underestimate the danger more than those living in areas located far away. This was confirmed by a study on the perception of technological risks in Australia (Maderthaner et al. 1978), in which people living near a nuclear reactor were less afraid than those living further away. This is explained by the halo of proximity effect (Nisbett and Wilson 1977) or, more precisely, the Neighborhood Halo Effect (Bickerstaff 2004; Catalan et al. 2009), that is, individuals' tendency to perceive less risk in their close environment than further away. These results also emphasize the link between perceived vulnerability and physical proximity.

To illustrate the role of proximity of the threat in the feeling of vulnerability and risk evaluation, the results of research by the IFEN (2006) on the perception of natural risks among French people showed that, in general, people consider that they are not exposed to risks in their home. However, when the territorial scale is broadened, this feeling of being safe from danger decreases in proportion to the increase in the size of the area under consideration (Roy 2005). Thus, the perception of a threat and the feeling of vulnerability increase in direct proportion to the size of the territory (Uzzell 2000), to the collective nature of the risk, or to the number of people affected. In other words, as the threat becomes more collective, it is considered more disastrous.

27.4 Social Inequalities and Adaptability to Risks

Slovic (1987) argues that the perception of risk significantly influences people's decisions and actions – especially decisions marked by negative feelings triggered by a threat and regulated by the possibility of dealing with it. Associated emotions, such as fear or concern, are a factor that may influence the interpretation of a risk or the behaviors to adopt in particular situations. The perception of the risk changes according to the individual's impression of possible direct harm to their health or possessions. This is why risk perception is an important variable, which influences both stress and responses to the situation (López-Vázquez and Marván 2003, 2004). Thus, coping consists of all the strategies used to deal with a stressful situation. It can be defined as the set of "cognitive and behavioral efforts [through which the subject] manage[s] specific internal and/or external demands that are appraised as taxing or exceeding the resources of the person" (Lazarus and Folkman 1984: 141). This process accounts for the individual's actions and thoughts when faced with an unusual situation. Coping is a stabilizing factor that enables the subject to maintain psychosocial adaptation during long crisis periods (López-Vázquez and Marván 2012). It is the result of an evaluation of his/her intellectual and affective competences (Sordes et al. 1997). Coping also depends on the person's experience, existing practices and social and/or institutional ties. The most classic typology in the literature proposes the use of active or passive coping strategies (Moos and Billings 1986): the former to solve the problem and the latter to manage emotions.

This ability to deal with and adapt to circumstances is a crucial characteristic of social inequality. As noted above, social and environmental inequalities also relate to the capacity of populations to act in response to risks or threatening situations. It is generally agreed that the most disadvantaged populations more readily accept risky situations (Faburel 2008; Roussel 2009). They find it more difficult to react when confronted with public or economic authorities, also because they are not as well informed – and a sort of endless cycle sets in. It is "because social acceptance of degraded environments is greater among socially vulnerable populations that high-risk infrastructures or facilities can be built or can continue to pollute" (Roussel 2009: 7). Moreover, it was this awareness of the marginalized categories in the US that led to the environmental justice movement (Charles et al. 2007). In France, on the other hand, this awareness of inequalities has not always translated into legal proceedings, as environmental justice is not a popular advocacy movement; it is connected more to social inequalities concerning health. The struggle against health inequalities has given rise to movements in France such as the community health movement (Roussel 2009) in the 1960s and 1970s. This movement was formed to raise workers' awareness of their sanitary conditions and of the need to improve their habitat and habitus as regards health. It brings to mind the "psychology of communities" movement, well-known in North and South America, which stemmed from psychologists' concerns and sought to promote health. The psychology of communities is a politically engaged psychology that embraces a critical approach to the field of study (Moser 2006). It seeks to understand in order to transform, and its methods derive from the humanities and social sciences in general. Its target is therefore the community with its territory and living conditions.

As Beck (2001) has pointed out, the ability of populations to defend themselves in the face of risks is unequal. The implementation and effectiveness of strategies, both individual (cognitive or emotional, of avoidance or self-protection) and collective (political mobilization, popular demands) depend on conditions that are always unequal. A quest for equality or environmental justice seeks, fundamentally, to give populations and public authorities the means to improve their environment and their living conditions.

analyze social and economic living conditions, as well as the sanitary and environmental conditions that spawn or express inequalities, insofar as environment, health and QOL are closely interrelated. Environmental psychology affords a new perspective of this complexity by putting the relationship that people and populations establish with the environment and with their immediate surroundings at the center of the analysis. Research on environmental health shows that individuals' evaluation of their living environment is a better predictor of perceived QOL and health than are the environment's actual properties (Fleury-Bahi and Annabi-Attia 2014; Ellaway et al. 2001).

On the other hand, in the analysis of vulnerability, the recent emphasis on the capacity of populations to adapt represents a real change. This dimension relates to the involvement of stakeholders or actors, the formulation of responses and strategies, and the integration of the adaptive measures of the population – officials included – based on what exists, that is, on material, social and psychological resources. This change goes hand in hand with an evolution of quantitative approaches towards a more qualitative evaluation of vulnerability. A participative and qualitative approach to the analysis of vulnerability is considered to have great potential for scientific production and political actions (Rossignol et al. 2015). A participative research-action approach seems to offer an interesting way of working with marginalized populations (Downs et al. 2011). It requires the active engagement of the populations and public and private partners if they are to understand and reduce inequality gaps and improve their QOL.

27.5 Conclusion

Vulnerability has thus become a cross-cutting issue, which makes it possible to situate inequalities better in the field of environmental risk evaluation. Whether vulnerabilities are social, environmental or historically constructed, they always relate to inequalities. The objective of environmental psychology will be to identify and

References

Adger, W. N. (1999). Social vulnerability to climate change and extremes in coastal Vietnam. *World Development, 27*(2), 249–269.

Badland, H. M., & Duncan, M. J. (2009). Perceptions of air pollution during the work-related commute by adults in Queensland, Australia. *Atmospheric Environment, 43*, 5791–5795.

Beck, U. (2001). *La société du risque. Sur la voie d'une autre modernité*. Paris: Aubier.

Benyakar, M. (2003). *Lo disruptivo: amenazas individuales y colectivas: el psiquismo ante guerras, terrorismos y catastrofes sociales*. Argentina: Biblos.

Bernardo, F. (2013). Impact of place attachment on risk perception: Exploring the multidimensionality of risk and its magnitude. *Studies in Psychology, 34*(3), 323–329. doi:10.1174/021093913808349253.

Bickerstaff, K. (2004). Risk perception research: Sociocultural perspectives on the public experience of air pollution. *Environment International, 30*(6), 827–840.

Bickerstaff, K. J., & Walker, G. P. (2001). Public understandings of air pollution: The 'localisation' of environmental risk. *Global Environmental Change, 11*, 133–145.

Botzen, W. J. W., Aerts, J. C. J. H., & Van den Bergh, J. C. J. M. (2009). Dependence of flood risk perceptions on socioeconomic and objective risk factors. *Water Resources Research, 45*, 1–15.

Bourg, D., Joly, P.-B., & Kaufmann, A. (2013). *Du risque à la menace. Penser la catastrophe*. Paris: PUF.

Brody, S. D., Zahran, S., Vedlitz, A., & Grover, H. (2008). Examining the relationship between P.V. and public perception of global climate change in the United States. *Environment and Behavior, 40*(1), 72–95.

Brooks, N. (2003). *Vulnerability, risk and adaptation: A conceptual framework* (Tyndall Centre working paper No. 38). http://www.tyndall.ac.uk

Brulle, R. J., & Pellow, D. N. (2006). Environmental justice: Human health and environmental inequalities. *Annual Review of Public Health, 27*, 103–124. doi:10.1146/annurev.publhealth.27.021405.102124.

Bullard, R. (1994). Environmental justice for all: It's the right thing to do. *University of Oregon Journal of Environmental Law and Litigation, 9*, 281–308.

Bullard, R., & Johnson, G. (2000). Environmental justice: Grassroots activism and its impacts on public policy decision making. *Journal of Social Issues, 56*(3), 555–578.

Burningham, K., Fielding, J., & Thrush, D. (2008). "It'll never happen to me": Understanding public awareness of local flood risk. *Disasters, 32*(2), 216–238. doi:10.1111/j.1467-7717.2007.01036.x.

Burton, I., & Kates, R. W. (1964). The perception of natural hazards in resource management. *Natural Resources Journal, 3*, 412–441.

Carroll, B., Balogh, R., Morbey, H., & Araoz, G. (2010). Health and social impacts of a flood disaster: Responding to needs and implications for practice. *Disasters, 34*(4), 1045–1063. doi:10.1111/j.1467-7717.2010.01182.x.

Castro, P., Batel, S., Devine-Wright, H., Kronberger, N., Mouro, C., Weiss, K., & Wagner, W. (2010). Redesigning nature and managing risk: Social representation, change and resistance. In A. Abdel-Hadi, M. Tolman, & S. Soliman (Eds.), *Environment, health and sustainable development* (pp. 227–241). Göttingen: Hogrefe & Huber.

Catalán, M. (2006). Estudio de la percepción pública de la contaminación del aire y sus riesgos para la salud: perspectivas teóricas y metodológicas. *Revista Instituto Nacional de Enfermedades Respiratorias de México, 19*, 28–37.

Catalán, M., Rojas, R. M., & Pérez, N. J. (2001). La percepción que tiene la población adulta del Distrito Federal sobre la contaminación del aire. Estudio descriptivo. *Revista del Instituto Nacional de Enfermedades Respiratorias Mexico, 14*, 220–223.

Catalán, M., Riojas, H., Jarillo, E., & Delgadillo, H. (2009). Percepción de riesgo a la salud por contaminación del aire en adolescentes de la ciudad de México. *Salud Pública de México, 51*, 148–154.

Champion, J. B., Choffel, P., Dupont, E., et al. (2004). Les nuisances et les risques environnementaux. In *Rapport 2004 de l'Observatoire national des zones urbaines sensibles* (pp. 124–131). Paris: Ed. DIV.

Charles, L., Emelianoff, C., Ghorra-Gobin, C., Roussel, I., Roussel, F. -X., & Scarcwell, H. -J. (2007). Les multiples facettes des inégalités écologiques. *Développement durable et territoire, 9*. http://developpementdurable.revues.org.

Chaumel, M., & La Branche, S. (2008). Inégalités écologiques: vers quelle définition? *Espace Populations Sociétés, 1*, 101–110.

Chauvin, B., & Hermand, D. (2008). Contribution du paradigme psychométrique à l'étude de la perception des risques: une revue de littérature de 1978 à 2005. *L'Année Psychologique, 108*, 343–386.

Claeson, A. S., Lidén, E., Nordin, M., & Nordin, S. (2013). The role of perceived pollution and health risk perception in annoyance and health symptoms: A population-based study of odorous air pollution. *International Archives of Occupational and Environmental Health, 86*(3), 367–374.

Clarenburg, L. A. (1973). Penalization of the environment due to stench: A study of the perception of odorous air pollution by the population. *Atmospheric Environment, 7*(3), 333–342.

Conti, J., Satterfield, T., & Harthorn, B. H. (2011). Vulnerability and social justice as factors in emergent U.S. nanotechnology risk perceptions. *Risk Analysis, 31*(11), 1734–1748. doi:10.1111/j.1539-6924.2011.01608.x.

Correia, F. N., Fordham, M., Saraiva, M. D. G., & Bernardo, F. (1998). Flood hazard assessment and management: Interface with the public. *Water Resources Management, 12*, 209–227.

Cutter, S. L., Boruff, B. J., & Shirley, W. L. (2003). Social vulnerability to environmental hazards. *Social Science Quarterly, 84*(2), 242–261.

D'Ercole, R. (1994). Les vulnérabilités des sociétés et des espaces urbanisés: concepts, typologie, modes d'analyse. *Revue de Géographie Alpine, Croissance urbaine et risques naturels dans les montagnes des pays en développement, 4*, 87–96.

D'Ercole, R., Gluski, P., Hardy, S., & Sierra, A. (2009). Vulnérabilités urbaines dans les pays du Sud. *Cybergeo, European Journal of Geography*. doi:10.4000/cybergeo.22151.

Downs, T. J., Ross, L., Goble, R., Subedi, R., Greenberg, S., & Taylor, O. (2011). Vulnerability, risk perception, and health profile of marginalized people exposed to multiple built-environment. Stressors in Worcester, Massachusetts: A pilot project. *Risk Analysis, 31*(4), 609–628. doi:10.1111/j.1539-6924.2010.01548.x.

Dozzi, J., Lennert, M., & Wallenborn, G. (2008). Inégalités écologiques: analyse spatiale des impacts générés et subis par les ménages belges. *Espace Populations Sociétés, 1*, 127–143.

Ellaway, A., Macintyre, S., & Kearns, A. (2001). Perceptions of place and health in socially contrasting neighbourhoods. *Urban Studies, 38*(12), 2299–2316.

Elliot, S. J., Cole, K. C., Krueger, P., Voorberg, N., & Wakefield, S. (1999). The power of perception: Health risk attributed to air pollution in an urban industrial neighborhood. *Risk Analysis, 19*, 615–628.

Emelianoff, C. (2008). La problématique des inégalités écologiques, un nouveau paysage conceptual. *Ecologie et politique, 1*, 19–31.

Evans, G. W. (2004). The environment of childhood poverty. *American Psychologist, 59*(2), 77–92. doi:10.1037/0003-066X.59.2.77.

Evans, G. W., & Kantrowitz, E. (2002). Socioeconomic status and health: The potential role of environmental risk exposure. *Annual Review of Public Health, 23*, 303–331. doi:10.1146/annurev.publhealth.23.112001.112349.

Faburel, G. (2008). Les inégalités environnementales comme inégalités de moyens des habitants et des acteurs territoriaux. Pour que l'environnement soit réellement un facteur de cohésion urbaine. *Espace, Populations, Sociétés, 1*, 111–126.

Faburel, G. (2010). Current debates on environmental inequities: Greening our urban spaces. *Justice spatiale – spatial justice, 2*. http://www.jssj.org

Finucane, M. L., Slovic, P., Mertz, C. K., Flynn, J., & Satterfield, T. A. (2000). Gender, race, and perceived risk: The 'white male' effect. *Healthy Risk & Society, 2*(2), 159–172.

Fleury-Bahi, G. (2010). *Psychologie et environnement: des concepts aux applications*. Brussels: Editions De Boeck.

Fleury-Bahi, G., & Annabi-Attia, T. (2014). Psychologie environnementale et épidémiologie: de la mise en évidence d'objectifs scientifiques communs à leur opérationnalisation dans le champ de la santé environnementale. In L. Colbeau-Justin, D. Marchand, S. Depeau & K. Weiss (Eds.), *Psychologie environnementale: héritage et perspectives*. Paris: In Press.

Fleury-Bahi, G., Préau, M., Annabi-Attia, T., Marcouyeux, A., & Wittenberg, I. (2013). Perceived health and quality of life: The effect of exposure to atmospheric pollution. *Journal of Risk Research, 18(2)*. doi:10.1080/13669877.2013.841728

Fol, S., & Pflieger, G. (2010). Environmental justice in the US: Construction and uses of a flexible category: An application to transportation policies in the San Francisco area. *Justice Spatiale – Spatial Justice, 2*. http://www.jssj.org

Gilbert, C. (2009). La vulnérabilité, une notion vulnérable ? In S. Becerra & A. Peltier (Eds.), *Risques et environnement: recherches interdisciplinaires sur la vulnérabilité des sociétés* (pp. 23–40). Paris: L'Harmattan.

Gleyze, J.-F., & Reghezza, M. (2007). La vulnérabilité structurelle comme outil de compréhension des mécanismes d'endommagement. *Géocarrefour, 82*, 1–2. doi:10.4000/geocarrefour.1411.

Gueymard, S. (2009). *Inégalités environnementales en région Ile-de-France: répartition socio-spatiale des ressources, des handicaps et de satisfaction environnementale des habitants, thèse d'aménagement, urbanisme, politiques urbaines*. Thèse en urbanisme, Institut d'Urbanisme de Paris.

Henslin, J. M. (1967). Craps and magic. *American Journal of Sociology, 73*, 316–330.

Holzmann, R., & Jørgensen, S. (2001). Social risk management: A new conceptual framework for social protection, and beyond. *International Tax and Public Finance, 8*(4), 529–556.

Howel, D., Moffat, S., Bush, J., Dunn, C. E., & Prince, H. (2003). Public views on the links between air pollution and health in Northeast England. *Environmental Research, 91*, 163–171.

Institut Français de l'Environnement – IFEN. (2006). Les inégalités environnementales. In *L'Environnement en France* (pp. 419–430). Paris: La documentation française.

Intergovernmental Panel on Climate Change – IPCC. (2007). In Core Writing Team, R. K. Pachauri, & A. Reisinger (Eds.), *Climate change 2007: Synthesis report*. Geneva: IPCC.

Jacobi, P. R. (1994). Household and environment in the city of Sao Paulo: Problems, perceptions and solutions. *Environment and Urbanization, 6*(2), 87–110.

Joffe, H. (2003). Risk: From perception to representation. *British Journal of Social Psychology, 42*, 55–73.

Kates, R. W. (1976). Experiencing the environment as hazard. In H. M. Proshansky, W. H. Ittelson, & L. G. Rivlin (Eds.), *Environment psychology: People and their physical settings* (pp. 401–418). New York: Holt, Rinehart and Winston.

Kellens, W., Zaalberg, R., Neutens, T., Vanneuville, W., & De Maeyer, P. (2011). An analysis of the public perception of flood risk on the Belgian coast. *Risk Analysis, 31*(7), 1055–1068. doi:10.1111/j.1539-6924.2010.01571.x.

Kohlhuber, M., Mielckc, A., Weilanda, S. K., & Bolte, G. (2006). Social inequality in perceived environmental exposures in relation to housing conditions in Germany. *Environmental Research, 101*, 246–255.

Lazarus, R. S., & Folkman, S. (1984). *Stress, appraisal and coping*. New York: Springer Publishing Company.

Lazarus, R. S., & Launier, R. (1978). Stress-related transactions between person and environment. In L. A. Pervin & M. Lewis (Eds.), *Perspectives in interactional psychology* (pp. 287–327). New York: Plenum.

Leplat, J. (2006). Risque et perception du risque dans l'activité. In D. R. Kouabenan, B. Cadet, D. Hermand, & M. T. Muños Sastre (Eds.), *Psychologie du risque. Identifier, évaluer, prévenir* (pp. 19–33). Brussels: De Boeck & Larcier.

Lima, M. L. (2004). On the influence of risk perception on mental health: Living near an incinerator. *Journal of Environmental Psychology, 24*, 71–84.

Lima, M. L. (2008). Percepção de riscos e desigualdades sociais. In J. Pinto & P. Virgílio (Eds.), *Desigualdades, Desregulação e Riscos nas Sociedades Contemporâneas* (pp. 267–290). Porto: Afrontamento.

Lima, M. L., Barnett, J., & Vala, J. (2005). Risk perception and technological development at a societal level. *Risk Analysis, 25*(5), 1229–1239.

Lindell, M. K., & Hwang, S. N. (2008). Households' perceived personal risk and responses in a multihazard environment. *Risk Analysis, 28*(2), 539–556. doi:10.1111/j.1539-6924.2008.01032.x.

López-Vázquez, E., & Marván, M. L. (2003). Risk perception, stress and coping strategies in two catastrophe risk situations. *Social Behavior and Personality, 31*(1), 61–70.

López-Vázquez, E., & Marván, M. L. (2004). Validación de una escala de afrontamiento frente a riesgos extremos. *Salud Pública de México, 46*(3), 216–221.

López-Vázquez, E., & Marván, M. L. (2012). Volcanic risk perception, locus of control, stress and coping responses of people living near the Popocatépetl volcano in Mexico. *Journal of Risk Analysis and Crisis Response, 2*, 3–12. doi:10.2991/jracr.2012.2.1.1.

Maderthaner, R., Guttman, G., Swaton, E., & Otway, H. J. (1978). Effect of distance on risk perception. *Journal of Applied Psychology, 63*(3), 380–382.

Ménard, C., Girard, D., Léon, C., & Beck, F. (Eds.). (2008). *Baromètre santé environnement 2007*. St Denis: INPES.

Metzger, P., & D'Ercole, R. (2011). Les risques en milieu urbain: éléments de réflexion. *EchoGéo, 18*. doi:10.4000/echogeo.12640

Metzger, P., Couret, D., & Collectif Urbi. (2010). Vulnérabilité et pauvreté en milieu urbain: réflexions à partir des villes du Sud. In O. Coutard & J.-P. Levy (Eds.), *Écologies urbaines* (pp. 239–257). Paris: Economica.

Mitchell, R., & Popham, F. (2008). Effect of exposure to natural environment on health inequalities: An observational population study. *The Lancet, 372*(9650), 655–1660.

Moos, R. H., & Billings, A. G. (1986). Conceptualizing and measuring coping resource and processes. In L. Goldberger & S. Breznitz (Eds.), *Handbook of stress: Theoretical and clinical aspects* (pp. 212–230). New York: Free Press.

Morin, M. (2006). Pour une approche psycho-socio-environnementale des risques sanitaires. In K. Weiss & D. Marchand (Eds.), *Psychologie sociale de l'environnement* (pp. 165–177). Rennes: PUR.

Moser, G. (1998). Attribution causale et sentiment d'insécurité de victimes de différents types de vols. *Les Cahiers Internationaux de Psychologie Sociale, 39*, 43–52.

Moser, G. (2006). Psychologies sociales. Psychologie sociale, application de la psychologie sociale et psychologie sociale appliquée. *Les Cahiers Internationaux de Psychologie Sociale, 70*(2), 89–95. doi:10.3917/cips.070.0089.

Moser, G. (2009). *Psychologie environnementale. Les relations homme-environnement*. Brussels: De Boeck.

Navarro, O. (2008). L'eau comme enjeu: territoire, identité et conflits d'usage. In T. Kirat & A. Torre (Eds.), *Territoires de conflits. Analyses des mutations de l'occupation de l'espace*. Paris: L'Harmattan.

Navarro, O. (2011). Les enjeux socio-environnementaux du développement durable en Amérique du Sud. Considérations à partir du cas colombien. *Développement durable et territoires, 2*(3).

Navarro, O., & Michel-Guillou, E. (2014). Analyse des risques et menaces environnementales : un regard psycho-socio-environnementale. In D. Marchand, S. Depeau, & K. Weiss (Eds.), *L'individu au risque de l'environnement. Regards croisés de la psychologie environnementale* (pp. 271–298). Paris: Editions In Press.

Nisbett, R. E., & Wilson, T. D. (1977). The halo effect: Evidence for unconscious alteration of judgments. *Journal of Personality and Social Psychology, 35*, 250–256.

Nuissier, J. (1994). Le contrôle perçu et son rôle dans les transactions entre individus et événements stressants. In M. Bruchon-Schweitzer & R. Dantzer (Eds.), *Introduction à la psychologie de la santé* (pp. 67–97). Paris: PUF.

Observatoire des Zones Urbaines Sensibles. (2004). *Rapport 2004*. Paris: Editions de la DIV.

Peek, M., Cutchin, M., Freeman, D., Stowe, R., & Goodwin, J. (2009). Environmental hazards and stress: Evidence from the Texas city stress and health study. *Journal of Epidemiology Community Health, 63*(12), 985–990.

Pilisuk, M., & Acredolo, C. (1988). Fear of technological hazards: One concern or many? *Social Behaviour, 3*(1), 17–24.

Rossignol, N., Delvenne, P., & Turcanu, C. (2015). Rethinking vulnerability analysis and governance with emphasis on a participatory approach. *Risk Analysis, 35*(1), 129–141. doi:10.1111/risa.12233.

Rouquette, M.-L. (2007). Le rôle de l'implication personnelle dans la réception des campagnes d'information et de prévention. In L. Charles, P. Ebner, I. Roussel, & A. Weill (Eds.), *Evaluation et perception de l'exposition à la pollution atmosphérique* (pp. 115–120). Paris: La Documentation française.

Roussel, I. (2007). Information géographique, climat et pollution atmosphérique. In P. Carrega (Ed.), *Information géographique et climatologie* (pp. 129–191). Paris: Hermès Lavoisier.

Roussel, I. (2009). Les inégalités environnementales. *Air Pur, 76*, 5–12.

Roussel, I., Gailhard-Rocher, I., Lelievre, F., Lefranc, A., Tallec, A., Menard, C., & Beck, F. (2009). Diversité des perceptions de la pollution de l'air extérieur, disparités sociales et territoriales. Comment construire une politique égalitaire? *Air Pur, 76*, 30–35.

Roy, A. (2005). *La perception sociale des risques naturels* (Les données de l'environnement, 99). Orléans: Institut Français de l'Environnement (IFEN).

Roy, A. (2006). Les inégalités environnementales. In *L'environnement en France* (pp. 419–430). Orléans: Institut français de l'environnement (IFEN).

Ruin, I., Gaillard, J.-C., & Lutoff, C. (2007). How to get there? Assessing motorists' flash flood risk perception on daily itineraries. *Environmental Hazards, 7*, 235–244.

Safi, A. S., Smith, W. J., & Liu, Z. (2012). Rural Nevada and climate change: Vulnerability, beliefs, and risk perception. *Risk Analysis, 32*(6), 1041–1059.

Saterfield, T. A., Mertz, C. K., & Slovic, P. (2004). Discrimination, vulnerability and justice in the face of risk. *Risk Analysis, 24*(1), 115–129.

Savage, I. (1993). Demographic influences on risk perceptions. *Risk Analysis, 13*(4), 413–420.

Schusky, J. (1966). Public awareness and concern with air pollution in the St. Louis Metropolitan Area. *Journal of the Air Pollution Control Association, 16*(2), 72–76.

Seligman, M. E. P. (1975). *Helplessness: On depression, development, and death*. San Francisco: W. H. Freeman.

Sjöberg, L. (2002). Are received risk perception models alive and well? *Risk Analysis, 22*(4), 665–670.

Slovic, P. (1987). Perception of risk. *Science, 236*, 280–285.

Sordes-Ader, F., Esparbes-Pistre, S., & Tap, P. (1997). Adaptation et stratégies de coping à l'adolescence. *SPIRALE – Revue de Recherches en Éducation, 20*, 131–154.

Spence, A., Poortinga, W., Butler, C., & Pidgeon, N. F. (2011). Perceptions of climate change and willingness to save energy related to flood experience. *Nature Climate Change, 1*(1), 46–49.

Stenlund, T., Liden, E., Andresson, K., Garvill, J., & Nordin, S. (2009). Annoyance and health symptoms ant their influencing factors: A population-based air pollution intervention study. *Public Health, 123*, 339–345.

Stephens, C., Willis, R., & Walker, G. (2007). *Addressing environmental inequalities: Cumulative environmental impacts. Report*. Bristol: Environment Agency UK.

Szasz, A., & Meuser, M. (1997). Environmental inequalities: Literature review and proposals for new directions in research and theory. *Current Sociology, 45*(3), 100–120.

Tapsell, S. M., & Tunstall, S. M. (2008). I wish I'd never heard of Banbury: The relationship between 'place' and the health impacts from flooding. *Health & Place, 14*, 133–154.

Terpstra, T. (2009). *Flood preparedness: Thoughts, feelings and intentions of the Dutch public*. Thesis, University of Twente, Twente.

Theys, J. (2002). L'approche territoriale du « développement durable », condition d'une prise en compte de sa dimension sociale. *Développement durable et territoire, Dossier 1: Approches territoriales du Développement Durable*.

United Nations Development Programme – UNDP. (2004). *Reducing disaster risk: A challenge for development. A global report*. New York: Bureau for Crisis Prevention and Recovery.

Uzzell, D. (2000). The psycho-spatial dimension of global environmental problems. *Journal of Environmental Psychology, 20*, 307–318.

Vaughan, E. (1995). The socioeconomic context of exposure and response to environmental risk. *Environment and Behavior, 27*(4), 454–489.

Villa, J., & Bélanger, D. (2012). *Perception du risque d'inondation dans un contexte de changements climatiques: recension systématique des articles scientifiques sur sa mesure (1990–2011)*. Québec: Institut national de santé publique du Québec.

Wakefield, S., Elliot, S., Cole, D., & Eyles, J. (2001). Environmental risk and (re)action: Air quality, health and civic involvement in an urban industrial neighbourhood. *Health and Place, 7*, 163–177.

Walker, G., Mitchell, G., Fairburn, J., & Smith, G. (2005). Industrial pollution and social deprivation: Evidence and complexity in evaluating and responding to environmental inequality. *Local Environment, 10*(4), 361–377.

Ward, W. C., & Jenkins, H. M. (1965). The display of information and the judgment of contingency. *Canadian Journal of Psychology, 19*, 231–241.

Weber, J., Hair, J., & Fowler, C. (2000). Developing a measure of perceived environmental risk. *The Journal of Environmental Education, 32*(1), 28–35.

Weiss, K., Colbeau-Justin, L., & Marchand, D. (2006). Entre connaissance, mémoire et oublis représentations de l'environnement et réactions face à une catastrophe naturelle. In K. Weiss & D. Marchand (Eds.), *Psychologie sociale de l'environnement* (pp. 145–156). Rennes: PUR.

Whitmarsh, L. (2008). Are flood victims more concerned about climate change than other people? The role of direct experience in risk perception and behavioral response. *Journal of Risk Research, 11*(3), 351–374.

WHO Regional Office for Europe. (2006). *Air quality guidelines. Global update 2005*. Copenhagen: WHO.

WHO Regional Office for Europe. (2010). *Environment and health risks: A review of the influence and effects of social inequalities*. Copenhagen: WHO.

Wind, S., Van Sickle, D., & Wright, A. L. (2004). Health, place and childhood asthma in southwest Alaska. *Social Science & Medicine, 58*(1), 75–88.

Zhang, Y., Hwang, S. N., & Lindell, M. K. (2010). Hazard proximity or risk perception? Evaluating effects of natural and technological hazards on housing values. *Environment and Behavior, 42*(5), 597–624. doi:10.1177/0013916509334564.

28. Living in Industrial Areas: Social Impacts, Adaptation and Mitigation

Maria Luísa Lima and Sibila Marques

"From an individual point of view, a high quality of life in the context of sustainability might be best characterized by a people-environment congruity"

(Moser 2009, p. 351)

Sustainable development has been traditionally considered an important pillar of quality of life (QOL) definitions (Uzzell and Moser 2006). The rationale behind this is that the sustainable development of society can only be attained with the achievement of good environmental quality. QOL usually includes both objective and subjective indicators. The "European Common Indicators" proposed by the European Commission and the European Environment Agency in 1999 are a typical example, composed of a set of objective (e.g., availability of public areas and services, noise pollution) and subjective (e.g., citizen's satisfaction with their own community) assessment measures of QOL. In fact, according to Moser (2009), QOL can only be fully attained through a congruent person-environment situation based on the interrelation of two major factors: (a) the objective characteristics of the environment, and (b) the reported satisfaction with different aspects of the environment considering their interactive effects. Rather than looking at just the specific effect of a certain variable in isolation, such as noise or air pollution, it is extremely important to understand the way in which contextual interrelated residential *objective* and *subjective* factors affect the lives of individuals who live in that environment.

In this chapter, we are interested in exploring the impacts on the QOL of individuals living in a particular type of residential context: industrial neighborhoods. Based on previous findings (Lima 2004, 2006; Lima and Marques 2005; Marques and Lima 2011), our goal is to understand the main social impacts of this type of experience, considering the possibilities of adaptation to such an often threatening environment.

28.1 Industrial Contexts: Effects on Health of Objective and Subjective Factors

Most of the studies exploring the impacts of living in different settings have been conducted under a sociological umbrella, trying especially to uncover the role that certain social and economic aspects of neighborhoods play in an individual's overall adaptation and well-being (Wilson 1987). However, some studies have recently begun to investigate the impacts that more *physical* characteristics may have on individuals. This research has focused on the effects of urbanization, showing how this factor does indeed have significant influences on people's health. For instance, Haynes and Gale (1999) showed clear differences in mortality and deprivation in health among rural and urban residents in England;

M.L. Lima (✉) • S. Marques
Instituto Universitário de Lisboa (ISCTE-IUL), Cis-IUL, Lisbon, Portugal
e-mail: luisa.lima@iscte.pt; Sibila.Marques@iscte.pt

mortality and morbidity rates were higher in Inner London and other metropolitan cities than national average values. On the contrary, rural wards presented better health values. In fact, there is some evidence that higher urbanization levels are related to environment-related morbidity in both advanced (von Schirnding 2002) and low-income countries (Sclar et al. 2005).

These studies seem to show important effects on people's health depending on the degree of urbanization. More recently, and in a similar vein, some studies have presented compelling evidence that the level of *industrialization* (and not just urbanization) is related to poorer health. A good example of research into the effect of industrial contexts on health is the large-scale study conducted by Boardman and colleagues, which showed a significant association between living close to industrial activities and increased stress levels (Boardman et al. 2008).

One of the most fundamental concerns regarding the exploration of industrial contexts and health is that there is unfairness in the distribution of these impacts within the population. In fact, it seems that industrial neighborhoods are more occupied by the lower status groups in society, the poor and minorities, who end up being the effective targets of these types of hindering impacts (Lima 2008). Understanding the specific effects that living in these types of places has on these groups is also a matter of political and social justice.

Most of the studies exploring industrial neighborhood effects focus on the *objective* physical place characteristics. Many of them involve a one-by-one exploration of specific factors usually present in industrial areas, such as noise, air pollution and a lack of vegetation and restorative areas. These studies tend to show separately how living in industrial settings may pose real threats to residents' health and well-being.

Exposure to noise is considered a major threat to an individual's health. Studies show that 20 % of people living in Europe are consistently exposed to noise levels judged unacceptable by health experts (European Commission 1996). This has severe consequences for their health. In fact, many studies have shown the effects of noise on fundamental health outcomes, such as cardiovascular diseases (Jarup et al. 2008; Van Kempen and Babisch 2012) and sleep (Basner et al. 2010; Elmenhorst et al. 2012; Guifford 2014). The classic studies by Cohen on the impact of airport noise on children's blood pressure and academic performance (Cohen et al. 1980) were pioneering in this domain. At present, there is no doubt in the literature that exposure to noise leads to much more than hearing loss. In a recent WHO study, Babisch and Kim (2011) estimated the environmental burden of disease caused by traffic noise in Europe. Their research, based on a review of the evidence available for cardiovascular effects, shows that there are 4.8 myocardial infarcts and 30.1 ischemic heart disease cases per 100,000 population caused by traffic noise.

In another line of research, industrial activity has been associated with an increase in air pollutants in nearby areas with significant impacts on the health of residents (Kampa and Castanas 2008). Typical pollutants related to industrial activity are various gases (e.g., SO_2, NO_X, CO, ozone, and Volatile Organic Components), persistent organic pollutants (e.g., dioxins), heavy metals (e.g., mercury, lead) and particulate matter. Exposure to these types of air pollutants has been associated with an increase in hospitalization and increased mortality (Brunekreef and Holgate 2002; Sumpter and Chandramohan 2013), and poses a special danger to the cardiovascular and respiratory system (Kampa and Castanas 2008; Shah et al. 2013) and to mental health (e.g., Perera et al. 2012; Stansfeld et al. 2005).

Another important example of industrial neighborhood impacts is the lack of exposure to green areas, with significant negative effects on individuals' health. In fact, there is evidence that less exposure to natural environments is associated with both decreased physical (De Vries et al. 2003; Pretty et al. 2005) and mental well-being (Foundation 2000). In this vein, several studies have shown that exposure to natural environments favors positive emotions, greater focus and attention, a higher sense of efficacy and lower fatigue (Gatersleben 2008; Hartig 2008; Sequeira and Silva 2002). Specifically, regarding mental health effects, Mace and colleagues (Mace et al. 1999) reviewed over 100 studies suggesting that natural environments play an important role in facilitating recovery from stress. Hence, the lack of vegetation in industrial areas may pose a real threat to an individual's health and QOL.

Although there are more studies exploring the effects of *objective* characteristics of industrial contexts on health, there is also an important body of work showing the role played by more *subjective* environmental evaluations. In fact, some studies show that the way people perceive the environment is an important determinant of their levels of psychological health and well-being. For instance, research has consistently shown that people are annoyed by the level of noise and air quality, which are often inevitable effects of industrial activity. Environmental annoyance is generally defined as "a feeling of dissatisfaction associated with any agent or condition that is believed to affect individuals in an adverse way" (Steinheider and Winneke 1993, p. 353) and has often been associated with increased feelings of stress. Specifically, annoyance with noise has usually been identified as a source of low psychological and physical well-being in more general terms (e.g., Ouis 2001; Staples 1996) while annoyance with air quality has also been associated with harmful effects on psychological health (e.g., Cavalini et al. 1991; Chattopadhyay and Mukhopadhyay 1995).

Another variable that seems particularly important in the analysis of the consequences of exposure to more industrial contexts is risk perception. In fact, there is evidence showing that people who live near hazardous facilities have higher levels of general concern (Van der Pligt et al. 1986). This situation may be considered a source of chronic stress (Lazarus and Folkman 1984), which is usually associated with several types of psychological symptoms such as stress, anxiety and depression (Baum 1987; Baum and Fleming 1993; Baum et al. 1983).

All these *objective* and *subjective* factors interact, having a cumulative effect on the well-being of individuals living in industrial areas. Achieving congruency between objective and subjective evaluation seems to be an important factor affecting an individual's QOL (Moser 2009). Over the years, our team has collaborated in several multidisciplinary Environmental Impact Assessment Studies (EIAs) aiming to evaluate the effects on individuals of living near different types of hazardous facilities (such as incinerators, airports, and dams) (Lima 2004, 2006; Lima and Marques 2005; Marques and Lima 2011; Lima et al. 2012). Our goal is to use psychosocial theories and scientific methodologies (surveys, interviews, observational studies) to intervene in a rigorous and predictive manner in community participation processes and to describe and evaluate the impacts associated with living in potentially hazardous neighborhoods. As social psychologists, we are interested in understanding the interplay between social and cognitive factors affecting individual attitudes, choices and behaviors (Smith and Mackie 1995). In these studies, we focus on understanding how important psychosocial moderators and mediators – such as objective and perceived distance, place identity and perceived justice – influence significantly individuals' health and QOL.

Below, we present a case study where this evaluation procedure is described in more detail with the example of a solid waste incinerator.

28.2 Case Study: Living Near a Solid Waste Incinerator

The building of a waste incinerator is always controversial. Environmental movements stress the alternatives for waste management, local investors worry about the fall in house prices, local authorities manage the difficult decisions about locating the facility, and local residents try to create their own opinion, engaging more or less actively in the public participation processes that usually precede construction. They are typically concerned about health problems, a decrease in environmental quality (pollution, dust, noise), and an increased risk of technological accidents (Petts 1994).

In Portugal, our team followed the process of decision-making in the construction of a waste incinerator (Lima 2000, 2006) as well as the monitoring of the actual building and functioning process over 14 years (Lima 2004; Lima and Marques 2005; Marques and Lima 2011). In this part of the chapter, we describe the adaptation process of the residents that we observed during this period of time.

The decision to construct the waste incinerator in Oporto, as part of the waste management strategy of Lipor (the local waste management company), was received with moderate opposition by the residents. In a survey conducted in 1995 among 298 people who lived close to the site (100 living up to 2 km from the construction site – Lima 2006) the attitude towards the construction was 3.66 on a 1–5 Likert scale, with less positive attitudes in the closer subsample ($M = 3.42$, $SD = 1.00$; 20 % against). At that time, the incinerator was located in a rural area in the suburbs of Oporto, and the population living close to the site had low levels of education (73 % with less than 5 years at school). In this survey, only 33 % of the sample (50 % in the closer subsample) had heard about the construction of the incinerator, which was a very important result as it challenged the informed participation process that should have taken place. According to Lima (2006), the attitude towards the construction (among those informed) was predicted by the distance to the site (the closer, the less positive the attitude), the risk perception (the higher, the less positive the attitude), expectations (the more benefits expected, the more positive the attitude) and trust in experts (greater trust being associated with more positive attitudes). This pattern of results is interesting to analyze as it corresponds to an example of sense-making for a new and unfamiliar issue. At the time, there was no incinerator working in Portugal, and the attitude towards this new technology had to be built, based on previous knowledge (trust in experts) and beliefs about the future (expectations and risk perception).

Some years later, the Ministry of the Environment took the decision to allow the construction of the incinerator, conditioned by a monitoring process, which included, among other dimensions (such as public health and air pollution), monitoring the attitudes and mental health of the communities living nearby, in order to ensure that their quality of life was not threatened. From that time on, we followed this process and witnessed the adaptation of those residents and communities to the new site. According to our theoretical perspective, which stresses not only the impacts of the physical environment, but also the interpretation of these effects, the changes that required adaptation in this case study were:

1. *Changes in the environmental quality of the area.* These included noise and dust during the construction period, and increased air pollution and odors during the functioning of the incinerator. As seen above, these are typical sources of annoyance, environmental stress and ill-health.
2. *Changes in the meaning of the environment.* Living close to an unknown technology, associated with dreaded consequences that have the potential to affect people's quality of life, is certainly a situation perceived as highly risky (Slovic 1987). The public and media debates about the risks of the incineration make them salient, and this debate can raise worry and concern among residents (Matthies et al. 2000; Van der Pligt et al. 1986). Although it is difficult to prove the impact of risk perception on health because it co-occurs with other changes in the environment, it was clear to us that these concerns changed the meaning of the environment for the residents.
3. *Changes in the local identity.* Incinerators have a series of attributes that have been identified by Gregory et al. (1995) as typical of technologies that tend to stigmatize places (dreaded consequences, involuntary exposure; inequitable distribution of the impacts; uncertainty about the magnitude or persistence of the effects). Besides, risk perception in media coverage is an important factor in the stigmatization of places, with relevant consequences for those living there (Slovic et al. 1994). These consequences have been mainly analyzed in economic terms (decrease in the value of properties), but the theoretical framework used to define stigma stresses the importance of associated suffering. From this perspective, it is then also conceivable that this new incineration

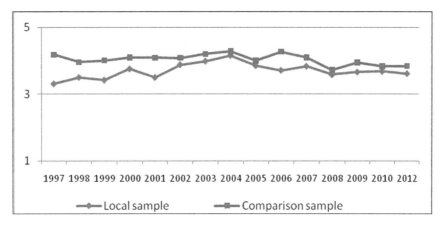

Fig. 28.1 Evolution of the attitude towards the incinerator from 1997 to 2012

can be seen as a threat to the local identity of the residents (Breakwell 1986, 1992; Devine-Wright and Lyons 1997; Twigger-Ross and Uzzell 1996), both in instrumental (e.g., economic aspects) and symbolic terms (namely positive distinctiveness, continuity, self-esteem, self-efficacy, see also Devine-Wright 2009).

4. *Changes in the uses of the spaces.* In some situations, environmental changes can introduce modifications in the structure (e.g., new highways) or the ambiance (e.g., smells) of the locality, so that they reduce the use of public spaces (Carr et al. 1992) and socialization practices. These interactions are vital to maintain social networks (Manzo 2003) and a sense of community (McMillan and Chavis 1986; Dines et al. 2006). For this reason, the costs of these environmental changes can also be associated with a diminished social support, an important variable to reduce and buffer stress (Cohen 2004).

Regular surveys have been conducted since 1997; first every 6 months, then on an annual basis and, more recently, with a 2-year interval. To date, 23 surveys have been carried out, with the evaluation of perceived risk and attitudes towards the incinerator, environmental annoyance and perceived quality, local identity and sense of community, mental health (stress, anxiety and depression) and well-being.

The results show a complex pattern that mixes the burden of adaptation to a threatening environment, familiarization with the threat and, in the end, no direct links to the local identity:

- The attitude towards the incinerator (Fig. 28.1) has always been more critical in the close than in the comparison sample (Lima 2006), but the difference between the groups has attenuated (Lima 2004) due to a diminished level of risk perception that occurred especially in the closer sample (Lima 2004). This familiarization with the threat (Lima 2004), associated with social interaction at a local level, probably contributed to a process of sense-making about the issue, and the integration of this facility into the local community.
- Annoyance due to noise, smell and dust increased in the closer sample when the incinerator started working but, in some cases, in a way not related to this facility. For example, in the July 2005 survey, the area suffered from nearby forest fires that increased environmental annoyance due to smell and dust. In the last six measurements, however, no differences were found with the comparison sample.
- Local attachment, evaluated by local identity and sense of community (Fig. 28.2), has decreased over the years in the local sample. However, the pattern is very similar in the comparison sample, and thus the differences

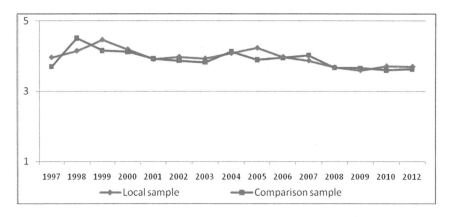

Fig. 28.2 Evolution of local attachment from 1997 to 2012

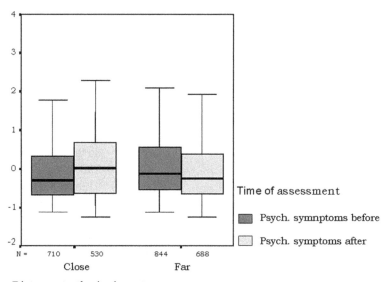

Fig. 28.3 Psychological symptoms before and after the incinerator started working (Lima 2004)

cannot be attributed to the installation of this incinerator in the neighborhood. These are localities that have grown significantly during these 15 years, and are now part of the metropolitan area of Oporto, losing their more rural/small town attributes.

- Psychological symptoms in the two samples have been extensively analyzed in Lima (2004, Fig. 28.3), comparing the periods before and after the incinerator started working. Following the habituation trend identified before, stress, anxiety and depression decreased in the comparison sample but not in the closer one. In the latter, although attitudes tended to improve and risk perception to decrease, psychological symptoms increased and this was interpreted as a special cognitive effort to minimize the perceived threat – a result that is compatible with the predictions from cognitive adaptation theory (Taylor and Brown 1988): the more threatened groups are those that need a greater change in the cognitive representation of the situation. Moreover, our results show that this adaptation process has a psychological cost. In this group, risk perception amplifies the effects of annoyance, as it introduces a sense of danger to environmental changes in the neighborhood. Residents become more attentive to these changes and they interpret them as hazardous and they have to deal with that. The normalization of the psychological symptoms occurred in the closer sample some years later, as in the last assessments there were no significant differences between the two groups. Besides, in a comparative study with other industrial areas performed some years later (Marques and Lima 2011), residents living close to the incinerator were not significantly different from a comparison non-industrial sample in terms of anxiety, depression and psychiatric comorbidity.

This study thus allowed us to follow, from a social psychological and health perspective, the dynamic process of community adjustment to a novel and intrusive facility in the neighborhood.

28.3 Adapting to Industrial Environments: The Role Played by Social Factors

Besides the descriptive analysis reported above, the literature identifies some psychosocial factors that might moderate the impacts of this environmental change on the quality of life of residents. In particular, when important environmental and landscape changes are at stake, an inclusive approach to the decision process is considered more favorable to promote a positive community adjustment (Lima et al. 2012). In fact, the participation of the residents in the planning phases of a large facility in the neighborhood is a way of both increasing the quality of the final project (as it includes local knowledge and interests) and facilitating the adaptation to the new facility. In terms of the psychosocial processes involved, perceived control, perceived justice and local identity are key variables, as they are linked to well-being and the quality of life of residents.

The idea that individuals' judgments and behaviors are shaped by their conception of what is just has been explored within the justice literature (e.g., Tyler et al. 1997; Tyler 2000) and applied to environmental psychology (Lima 2006; Van der Pligt 1992; Vlek and Stallen 1981; Zoellner et al. 2008). Research reveals a systematic positive association between perceived justice and the individual's attitudes, behaviors and acceptance of new local projects. For instance, in a recent study about the choice of location of a nuclear waste facility, Krutli et al. (2012) emphasized the need to take more account of issues related not only to distributive fairness (the way the costs and benefits are distributed) but also to procedural fairness (transparency in decision-making, opportunities to hear and include local voices in the project, and treating residents with respect). In the same manner, Lima (2006) showed that procedural justice plays a fundamental role in the prediction of attitudes towards new solid waste facilities. According to our research, procedural justice is particularly important in the initial phases of a change project, as it promotes the perceived control over the process, a fundamental variable in quality of life.

Within environmental psychology, several studies have extensively demonstrated the role of strong local identity in coping with negative aspects of the environment (Bonaiuto et al. 1996; Duarte and Lima 2005; Lima and Marques 2005). The moderator role of group identification has been highlighted both within environmental psychology research (Bonaiuto et al. 2008; Lima and Marques 2005) and, more broadly, within social psychology (O'Brien and Hummert 2006; Marques 2009). In a general manner, these studies explore the way high and low identifiers react when they are faced with a threat to their identity. Bonaiuto et al. (1996) showed that residents with higher levels of local or national identity perceived lower levels of pollution on local or national beaches. Duarte and Lima (2005) demonstrated that residents with higher levels of local identity perceived higher levels of environmental quality in their neighborhoods, maximizing its positive characteristics and minimizing the negative ones. Results from both studies suggested that individuals were using social creativity strategies to cope with potential threats to a positive social identity (Tajfel and Turner 1986), but they were not directly associated with better psychological health. However, Lima and Marques (2005) showed that local identity moderated the impacts of environmental annoyance on stress:

for those with higher levels of identification, environmental annoyance was not a predictor of stress as it was for those with lower levels of local identity. These results are important, as they go beyond the traditional social capital result that associates social cohesion with improved mental health (Petrou and Kupek 2008; Putnam 2000) to show a protective role for local identity. Besides, our research shows that local identity can be an important moderator of the impacts of perceived justice. Local issues are more relevant to highly identified residents, and the effects of procedural justice on attitudes towards new facilities are particularly strong in those people. It seems that, when faced with a possible threatening change to their environment, those who place a higher value on their place of residence are more affected by the way they perceive the process to be just and fair (Marques et al. 2015).

28.4 Conclusions

QOL is a multidimensional concept, depending upon the perceived congruency between objective and subjective factors of environmental contexts (Moser 2009). When thinking specifically about the effects of industrial places, one must consider not only the objective effects of typical stressors, such as noise, air pollution or limited exposure to green areas, but also the way associated environmental perceptions and evaluations may influence individuals' well-being.

By presenting the case of the impact assessment of a solid waste facility, we have shown how psychosocial factors may play a fundamental role and influence the adaptation process of individuals living in these neighborhoods. Overall, the results show more significant impacts on residents living near the solid waste facility when it was first installed, with higher risk perception and annoyance levels than comparison samples further away. However, the impacts on both these groups seem to converge over time, due to an adaptation process adopted by the nearby community. Hence, over time, local residents' perceptions toward the facility tend to become more positive, with clear benefits for their overall QOL. Our results also show that fundamental psychosocial factors, such as local identity and perceived justice, may have important moderator effects on the way individuals perceive the threat posed by these types of industrial facility.

This broad vision of individuals' QOL, considering both the convergence of objective and subjective factors, has proved to be very effective in monitoring the adaptation of communities to the installation of a hazardous facility – such as a solid waste incinerator – in their communities. Recognized by the public authorities in this domain, the proponents and the communities, this type of intervention proves the value of adopting a psychosocial perspective in environmental assessment procedures in order to guarantee a broader evaluation of individuals' QOL.

References

Babisch, W., & Kim, R. (2011). Traffic noise exposure and ischaemic heart disease. In M. Braubach, D. E. Jacobs, & D. Ormandy (Eds.), *Environmental burden of disease associated with inadequate housing. Methods for quantifying health impacts of selected housing risks in the WHO European Region* (pp. 97–111). Copenhagen: WHO Regional Office for Europe.

Basner, M., Griefahn, B., & Berg, M. (2010). Aircraft noise effects on sleep: Mechanisms, mitigation and research needs. *Noise and Health, 12*, 95–109.

Baum, A. (1987). Toxins, technology and natural disasters. In G. R. VandenBos & B. Bryant (Eds.), *Cataclysms, crises and catastrophes: Psychology in action* (pp. 5–54). Washington, DC: American Psychological Association.

Baum, A., & Fleming, I. (1993). Implications of psychological research on stress and technological accidents. *American Psychologist, 48*(6), 665–672.

Baum, A., Fleming, I., & Singer, J. E. (1983). Coping with victimization by technological disaster. *Journal of Social Issues, 39*, 117–138.

Boardman, J. D., Downey, L., Jackson, J. S., Merrill, J. B., Saint Onge, J. M., & Williams, D. R. (2008). Proximate industrial activity and psychological distress. *Population and Environment, 30*, 3–25. doi: http://dx.doi.org/10.1007/s11111-008-0075-8.

Bonaiuto, M., Breakwell, G. M., & Cano, I. (1996). Identity processes and environmental threat: The effects of nationalism and local identity upon perception of beach pollution. *Journal of Community & Applied Social Psychology, 6*(3), 157–175.

Bonaiuto, M., Bonnes, M., Ceccarelli, M., & Martorella, H. (2008). Local identity and the role of individual differences in the use of natural resources: The case of water consumption. *Journal of Applied Social Psychology, 4*, 947–967.

Breakwell, G. M. (1986). *Coping with threatened identity*. London: Methuen.

Breakwell, G. M. (1992). Processes of self-evaluation: Efficacy and estrangement. In G. M. Breakwell (Ed.), *Social psychology of identity and the self-concept* (pp. 35–55). Surrey: Surrey University Press.

Brunekreef, B., & Holgate, S. T. (2002). Air pollution and health. *Lancet, 360*, 1233.

Carr, S., Francis, M., Rivlin, L. G., & Stone, A. (1992). *Public space*. Cambridge: Cambridge University Press.

Cavalini, P. M., Koeter-Kemmerling, L. G., & Pulles, M. P. (1991). Coping with odour annoyance and odour concentrations: Three field studies. *Journal of Environmental Psychology, 11*, 123–142.

Chattopadhyay, P. K., & Mukhopadhyay, P. (1995). Air pollution and health hazards in human subjects: Physiological and self-report indices. *Journal of Environmental Psychology, 15*, 327–351.

Cohen, S. (2004). Social relationships and health. *American Psychologist, 59*(8), 676–684.

Cohen, S., Evans, G. W., Krantz, D. S., & Stokols, D. (1980). Physiological, motivational and cognitive effects of aircraft on children: Moving from the laboratory to the field. *American Psychologist, 35*, 231–243. doi: http://dx.doi.org/10.1037//0003-066X.35.3.231.

De Vries, S., Verheij, R., Groenewegen, P., & Spreeuwenberg, P. (2003). Natural environments-healthy environments? An exploratory analysis of the relationship between greenspace and health. *Environment and Planning A, 35*, 1717–1731.

Devine-Wright, P. (2009). Rethinking NIMBYism: The role of place attachment and place identity in explaining place-protective action. *Journal of Community & Applied Social Psychology, 19*(6), 426–441.

Devine-Wright, P., & Lyons, E. (1997). Remembering pasts and representing places: The construction of national identities in Ireland. *Journal of Environmental Psychology, 17*, 33–45.

Dines, N., Cattell, V., Gesler, W., & Curtis, S. (2006). *Public spaces, social relations and well-being in East London*. Bristol: Policy Press.

Duarte, A. P., & Lima, M. L. (2005). Análise dos conteúdos da identidade associada ao lugar [Analysis of place identity contents]. *Psicologia, 19*(1–2), 193–226.

Elmenhorst, E. M., Pennig, S., Rolny, V., Quehl, J., Mueller, U., Maaß, H., et al. (2012). Examining nocturnal railway noise and aircraft noise in the field: Sleep, psychomotor performance, and annoyance. *Science of the Total Environment, 424*, 48–56.

European Commission. (1996). Future noisy police. European Commission Green Paper, COM (96) 540 final.

Foundation, M. H. (2000). *Strategies for the living: Report of user led research into people's strategies for living with mental distress*. London: MHF.

Gatersleben, B. (2008). Humans and nature: Ten useful findings from environmental psychology research. *Counselling Psychology Review, 23*, 24–34.

Gregory, R., Flynn, J., & Slovic, P. (1995). Technological stigma. *American Scientist, 83*, 220–223.

Guifford, R. (2014). Environmental psychology matters. *Annual Review of Psychology, 65*, 541–579. doi: http://dx.doi.org/10.1146/annurev-psych-010213-115048.

Hartig, T. (2008). Green space, psychological restoration, and health inequality. *Lancet, 372*, 1614–1615.

Haynes, R., & Gale, S. (1999). Mortality, long-term illness and deprivation in rural and metropolitan wards in England and Wales. *Health & Place, 5*, 301–312.

Jarup, L., Babisch, W., Houthuijs, D., Pershagen, G., Katsouyanni, K., Cadum, E., et al. (2008). Hypertension and exposure to noise near airports: The HYENA study. *Environmental Health Perspectives, 116*, 329–333.

Kampa, M., & Castanas, E. (2008). Human health effects of air pollution. *Environmental Pollution, 151*, 362–367.

Krutli, P., Stauffacher, M., Pedolin, D., Moser, C., & Scholz, R. W. (2012). The process matters: Fairness in repository siting for nuclear waste. *Social Justice Research, 25*, 79–101.

Lazarus, R. S., & Folkman, S. (1984). *Stress appraisal and coping*. New York: Springer.

Lima, M. L. (2000). As controvérsias públicas nos Estudos de Impacte Ambiental. In E. Gonçalves (Ed.), *Ciência, Cultura Científica e Participação Pública* (pp. 139–151). Oeiras: Celta Editora.

Lima, M. L. (2004). On the influence of risk perception on mental health: Living near an incinerator. *Journal of Environmental Psychology, 24*(1), 71–84. doi: http://dx.doi.org/10.1016/S0272-4944(03)00026-4.

Lima, M. L. (2006). Predictors of attitudes towards the construction of a waste incinerator: Two case studies. *Journal of Applied Social Psychology, 36*, 441–466.

Lima, M. L. (2008). Percepção de riscos e desigualdades sociais. In J. Madureira Pinto & V. Borges Pereira (Eds.), *Desigualdades, Desregulação e Riscos nas Sociedades Contemporâneas* (pp. 267–290). Porto: Afrontamento.

Lima, M. L., & Marques, S. (2005). Towards successful psychosocial follow-up: A case study of a solid waste incinerator in the North of Portugal. *Impact Assessment and Project Appraisal, 23*(2), 227–233.

Lima, M. L., Moreira, S., & Marques, S. (2012). Participatory community involvement in planning processes of building project: A social psychological approach. *Umweltpsychologie, 16*, 68–87.

Mace, B., Bell, P., & Loomis, R. (1999). Aesthetic, affective, and cognitive effects of noise on natural landscape assessment. *Society and Natural Resources, 12*(3), 225–242.

Manzo, L. (2003). Beyond house and haven: Toward a revisioning of place attachment. *Journal of Environmental Psychology, 23*(1), 47–61.

Marques, S. (2009). *Is it age . . . or society? Aging stereotypes and older people's use of comparative optimism towards health*. Unpublished doctoral thesis. ISCTE-IUL, Lisbon.

Marques, S., Lima, M. L., Moreira, S., & Reis, J. (2015). Local identity as an amplifier: Testing its moderating role in the relationship between perceived justice and attitudes towards newdams projects. *Journal of Environmental Psychology, 44*(4), 63–73.

Marques, S., & Lima, M. L. (2011). Living in grey areas: Industrial activity and psychological health. *Journal of Environmental Psychology, 31*(4), 314–322. doi: http://dx.doi.org/10.1016/j.jenvp.2010.12.002.

Matthies, E., Hoeger, R., & Guski, R. (2000). Living on polluted soil: Determinants of stress symptoms. *Environment and Behavior, 32*, 270–286.

McMillan, D. W., & Chavis, D. (1986). Sense of community: A definition and theory. *Journal of Community Psychology, 14*, 6–23.

Moser, G. (2009). Quality of life and sustainability: Toward person-environment congruity. *Journal of Environmental Psychology, 29*, 351–357. doi: http://dx.doi.org/10.1016/j.jenvp.2009.02.002.

O'Brien, L. T., & Hummert, M. L. (2006). Memory performance of late middle-aged adults: Contrasting self-stereotyping and stereotype threat accounts of assimilation to age stereotypes. *Social Cognition, 24*, 338–358.

Ouis, D. (2001). Annoyance from road traffic noise: A review. *Journal of Environmental Psychology, 21*, 101–120.

Perera, F. P., Tang, D., Wang, S., Vishnevetsky, S., Zhang, B., Diaz, D., Camann, D., & Rauh, V. (2012). Prenatal polycyclic aromatic hydrocarbon (PAH) exposure and child behavior at age 6–7 years. *Environmental Health Perspectives, 120*(6), 921–926.

Petrou, S., & Kupek, E. (2008). Social capital and its relationship with measures of health status: Evidence from the health survey for England 2003. *Health Economics, 17*(1), 127–143.

Petts, J. (1994). Effective waste management: Understanding and dealing with public concerns. *Waste Management & Research, 12*(3), 207–222.

Pretty, J., Peacock, J., Sellens, M., & Griffin, M. (2005). The mental and physical health outcomes of green exercise. *International Journal of Environmental Health Research, 15*(5), 319–337.

Putnam, R. D. (2000). *Bowling alone: The collapse and revival of American community*. New York: Simon & Schuster.

Sclar, E. D., Garau, P., & Carolini, G. (2005). The 21st century health challenges of slums and cities. *The Lancet, 365*(9462), 901–903.

Sequeira, A., & Silva, M. N. (2002). O bem estar da pessoa idosa em meio rural. *Análise Psicológica, 20*(3), 505–516.

Shah, A. S. V., Langrish, J. P., Nair, N., McAllister, D. A., Hunter, A. L., Donaldson, K., Newby, D. E., & Mills, N. L. (2013). Global association of air pollution and heart failure: A systematic review and meta-analysis. *The Lancet, 382*(9897), 1039–1048. doi:10.1016/S0140-6736(13)60898-3.

Slovic, P. (1987). Perception of risk. *Science, 236*, 280–285.

Slovic, P., Flynn, J., & Gregory, R. (1994). Stigma happens: Social problems in the siting of nuclear waste facilities. *Risk Analysis, 14*, 773–777.

Smith, E. R., & Mackie, D. M. (1995). *Social psychology*. New York: Worth Publishers.

Stansfeld, S. A., Berglund, B., Clark, C., Lopez-Barrio, I., Fischer, P., Öhrström, E., Haines, M. M., Head, J., Hygge, S., Van Kamp, I., & Berry, B. F. (2005). Aircraft and road traffic noise and children's cognition and health: A cross-national study. *The Lancet, 365*, 1942–1949.

Staples, S. L. (1996). Human response to environmental noise: Psychological research and public policy. *American Psychologist, 51*(2), 143–150.

Steinheider, B., & Winneke, G. (1993). Industrial odours as environmental stressors: Exposure-annoyance associations and their modification by coping, age and perceived health. *Journal of Environmental Psychology, 13*, 353–363.

Sumpter, C., & Chandramohan, D. (2013). Systematic review and meta-analysis of the associations between indoor air pollution and tuberculosis. *Tropical Medicine and International Health, 18*(1), 101–108. doi:10.1111/tmi.12013.

Tajfel, H., & Turner, J. C. (1986). The social identity theory of intergroup behavior. In *Psychology of intergroup relations* (pp. 7–24). Chicago: Nelson-Hall.

Taylor, S. E., & Brown, J. D. (1988). Illusion and well being: A social cognitive perspective on mental health. *Psychological Bulletin, 106*, 231–248.

Twigger-Ross, C., & Uzzell, D. (1996). Place identity and place attachment. *Journal of Environmental Psychology, 16*(2), 205–220.

Tyler, T. R. (2000). Social justice: Outcome and procedure. *International Journal of Psychology, 35*(2), 117–125.

Tyler, T. R., Boeckmann, R., Smith, H., & Hou, Y. (1997). *Social justice in a diverse society*. Boulder: Westview Press.

Uzzell, D., & Moser, G. (2006). On the quality of life of environments. *European Review of Applied Psychology, 56*, 1–4.

Van der Pligt, J. (1992). *Nuclear energy and the public*. Oxford: Blackwell.

Van der Pligt, J., Eiser, J. R., & Spears, R. (1986). Attitudes towards nuclear energy: Familiarity and salience. *Environment and Behavior, 18*, 75–93.

Van Kempen, E., & Babisch, W. (2012). The quantitative relationship between road traffic noise and hypertension: A meta-analysis. *Journal of Hypertension, 30*, 1075–1086.

Vlek, C., & Stallen, P. (1981). Judging risks and benefits in the small and in the large. *Organizational Behavior & Human Performance, 28*(2), 235–271.

von Schirnding, Y. (2002). Health and sustainable development: Can we rise to the challenge? *The Lancet, 360*, 632–637.

Wilson, W. J. (1987). *The truly disadvantaged: The inner city, the underclass, and public policy*. Chicago: University of Chicago Press.

Zoellner, J., Schweizerries, P., & Wemheuer, C. (2008). Public acceptance of renewable energies: Results from case studies in Germany. *Energy Policy, 36*(11), 4136–4141.

Part VIII

Quality of Life and Environmental Threats – *Global Change, Energy and Emerging Risks*

Emerging Risks and Quality of Life: Towards New Dimensions of Well-Being?

29

Dorothée Marchand, Karine Weiss, and Bouchra Zouhri

Today, individual and collective preoccupations lead to new environmental issues. The development of Western societies, which seeks mainly to contribute to making the world more secure and to improving quality of life, paradoxically generates new risks of similar proportions to the activities that cause them. We are thus faced with poorly managed forms of pollution and change whose long-term consequences are unknown. As our environment develops, it becomes more complex by establishing links between multiple domains on different scales. This development contributes to the advent of new, unprecedented phenomena: climate change, pollution, health crises, etc. For example, despite the existence of widely recognized food risks, our society is experiencing significant changes in this sector, some of which are controversial and associated with high levels of uncertainty. Genetically modified organisms (GMOs) are a particularly interesting example: are they dangerous or do they represent progress? GMOs lead us to question some fundamental aspects of food – nature versus the monstrous and artificial (Wagner and Kronberger 2006), the visible versus the invisible, safety versus danger – and to deal with contested scientific sources (Spiroux de Vendômois et al. 2010).

In 2003, a report by the Organization for Economic Co-operation and Development (OECD 2003) on emerging risks in the twenty-first century was based on the idea that various major disasters could become increasingly likely. Risks that are considered conventional (such as storms and floods) will evolve to take on new forms, while new hazards will emerge, characterized by strong uncertainty and the potential to cause widespread, perhaps irreversible, damage. Although risk issues are generally well documented by the social sciences, this is not the case for emerging risks, which do not appear to be addressed as a specific object. However, the term "emerging" clearly places the understanding of new or unknown risks in a context of uncertainty and foresight, hence creating difficulties in information management and knowledge acquisition, which puts humans at the center of ever more important preoccupations in terms of health and quality of life. Thus, *emerging* risks are linked to the arrival of new dangers, exposures, behaviors and, more recently, increasing regulatory and collective awareness. They are the result of "a newly identified danger to which exposure is possible, or of exposure and/or a new or increased susceptibility to a known danger" and hence constitute new schemes to be integrated into our thought processes. They are linked to innovation and new

---

D. Marchand (✉)
CSTB, Centre Scientifique et Technique du Bâtiment, Paris, France
e-mail: dorothee.marchand@cstb.fr

K. Weiss • B. Zouhri
University of Nîmes, Nîmes, France
e-mail: karine.weiss@unimes.fr; b.zouhri@gmail.com

technologies, and can be concurrent with their use. The insufficiency (sometimes inexistence) or controversies of scientific knowledge about these risks make it difficult for scientists to describe them, and cause great concern for populations, who form their representations using eclectic and, more often than not, irrational elements. The emergence of these risks is experienced as a threat, which has a growing impact on quality of life in terms of health and well-being, mainly because of a lack of knowledge about and perception of their short-term effects, their systemic functioning, and their potential long-term effects on health. Unlike for better known, conventional risks (major risks, industrial risks), estimating emerging risks is often accompanied by marked uncertainty in terms of evaluation and the level of acceptability by the competent authorities and the exposed populations. Hence, the issue is to determine which actions to undertake when faced with risks that are characterized by such high uncertainty. The degree of uncertainty refers to three types of risk management: (1) foresight, when the risk is certain and identified and it is possible to take alternative action; (2) prevention, when the risk is identified, the probability of each consequence is known, and uncertainty refers to the predictability of the hazard; and (3) precaution, when the risk is assumed and its consequences are difficult to define (Erné-Heintz 2010). In the latter case, it is possible to evaluate all the consequences of a choice, whether in terms of governance, evaluation, decision-making and/or support. Nanoparticles, electromagnetic waves, radiofrequencies, micropollutants, nuclear sites and waste, climate change, GMOs, pesticides etc., are all examples for which a lack of knowledge associated with problems in terms of evaluation methods prevent ruling or legislating about the level of risk for populations. Furthermore, the representation of risk in society is tainted by erroneous communication, which is not realistic and often biased (non-contradictory), and can lead to an increase in litigation (conflict, jurisprudence, etc.) and unreasonable or even irrational behavior. Moreover, the lack of clarity and decisions in terms of risk management is a source of suspicion and controversy, which feeds the demands and fears of populations. The resulting distrust in public policy (Galand 2007) and experts is fuelled by lines of research and operationalization that focus more on the acceptability of risks than their uncertainty.

29.1 From Risk Perception to Emerging Risk Perception

The definition of risk is typically based on notions such as hazard, stakes and vulnerability. For example, Leplat (2006) defined risk as the product of an undesirable event's probability and the seriousness of its damage. Hence, this type of definition includes some uncertainty, which is associated not only with probability, but also with weak predictability of future events in terms of temporality, location or effects. For the social sciences, especially environmental social psychology, the notion of uncertainty is the basis of risk perception and risk representation literature (see Slovic 1987, 2000; Kouabenan et al. 2006; Weiss and Marchand 2006; Marchand et al. 2014; Chauvin 2014).

Beyond this probabilistic approach, risk is formed on the basis of choice (of a model, a measurement, or an index) and judgment (Breakwell 2007; Leplat 2006). Slovic's psychometric paradigm (1987, 2000) underlines how experts and laypeople evaluate risk differently. Thus, a study by Kraus et al. (1992) on the perception of risks linked with pesticides and food additives revealed significant differences between experts and laypeople in their judgment of chemical substances. The authors analyzed these differences through the scope of ideas, hypotheses and inferences related to these substances. The study was prolonged by Slovic (1995) in Canada, who showed that toxicologists perceive less risk than laypeople with regard to these substances. Slovic (1987) reported that when faced with such differences in judgment, experts dismiss laypeople's judgments, which they consider irrational. Peretti-Watel (2000) and later Chauvin (2014) highlighted a judgment bias: experts approach risk in rigid, quantitative terms, whereas laypeople's understanding of risk is more qualitative and submitted to psychological, social and cultural factors. These gaps in judgment are particularly interesting to analyze in the field of emerging risks, because they refer to another level of uncertainty by questioning scientific knowledge about its lack of perspective on recent issues, and the failure to take a politico-legal position allowing clear decision-making, or even the implementation of recommendations at the institutional level. Uncertainties in terms of an event's occurrence or its consequences on health reinforce these gaps and revive the view of public contestation as irrational. Experts tend to consider new forms of rejection and contestation as irrational, such as idiopathic environmental intolerance (Environmental Illness, IPCS/WHO 1996). This refers to symptoms associated with environmental factors, which are tolerated by most people and cannot be explained by known psychiatric or medical disorders (Sparks 2000). Examples include sick building syndrome, electrosensitivity, chemical odor intolerance syndrome, etc. (Marchand et al. in press). Reactions of rejection, or even pathologies related to these syndromes, produce expert explanations in terms of insanity or psychology, which are often confused. This can be observed in controversies about radiofrequencies, where expert and lay opinions continue to polarize.

29.2 Emerging Risks Faced with Socio-Political and Cultural Fundamentals

"The moment there is disagreement or controversy, that is to say, when someone says a risk is unacceptable, the question *ipso facto* becomes political" (Douglas and Wildavsky 1983, p. 65). In the case of emerging risks, social positioning depends on political attitudes and regulatory history, which change from one country to another and evolve slowly. The example of GMOs is a perfect illustration of differences in terms of the understanding of risks by the authorities. The European Union, including France, favors the *precautionary principle*, which appeared in legal texts in France in the 1980s (Peretti-Watel 2000). Thus, one directive (2001/18, article 4) deals with voluntary dissemination of GMOs and a law (178/2002, article 7) sets out the general requirements in terms of food safety. Both texts conform to the precautionary principle and aim "to take all appropriate action to avoid negative effects on human health or the environment that could result in the voluntary dissemination or marketing of GMOs" (2001/8, article 4) (Erné-Heintz 2010, pp. 17–18). If in doubt, "each European state can take proportionate and provisional measures (from a moratorium, while waiting for scientific studies, to field testing) on the condition that they are scientifically justified!" (Erné-Heintz 2010, p. 18). To evaluate a GMO-related risk, it should be compared to its conventional counterpart in order to determine the importance of nutritional, health and environmental differences, as well as any possible uncertainty.

Contrary to the French approach, authorities in the United States rely on the *equivalence principle* to evaluate and manage GMO-related risks, that is to say, GMOs are considered equivalent to their conventional counterpart in terms of risk. As such, GMOs are not a priori considered more risky than non-modified organisms (Erné-Heintz 2010). Thus, although both approaches are based on the equivalence principle for risk assessment, i.e. they both determine if a plant or genetically modified organism and a conventional organism are equivalent, they differ in terms of risk management: in Europe, the precautionary principle is used whilst taking these differences into account, whereas in the United States, when a GMO and its counterpart are found to be equivalent, they are no longer treated differently. This difference in terms of risk management strongly influences decision-making when it comes to GMO regulations in both cultures. For example, the United States possess no GMO-specific regulations, whereas such regulations are in place in Europe, as the precautionary principle prevails for legal risk management. These differences in risk management are reflected not only in regulations but also in society, in terms of general attitudes and understanding of risk. As there is no specific legislation in the United States, there are no labeling requirements, hence no traceability or oversight of GMOs on the market. In contrast, labeling is required in France, with the possibility of tracing GMOs. Yet, a study conducted on 142 French and American students (Crawford and Weiss 2013) showed that, in both samples, approximately half the respondents did not think that labeling indicated the presence of GMOs. In the same way, GMO-containing food was described as hard to identify by three quarters of the French and Americans.

Furthermore, authorities in the United States view GMOs as factors of progress whose problems can be managed; in Europe, where representational conflicts between nature and technology are rooted in the founding myths of its civilization (Wagner and Kronberger 2006), GMOs are seen more as an irreversible break with the natural order. This idea can be observed in popular representations, because American students share a positive representation of GMOs, which contribute to sustainable development, increase productivity and are created for the community, whereas French students reject these aspects and consider GMOs against nature, harmful, toxic, chemical, artificial and false. In the same way, more than half of the French consider that GMO-related risks are unacceptable and that there are reasons to be concerned about them, whereas most Americans accept the risk and do not view GMOs as a source of concern (Crawford and Weiss 2013). Nevertheless, the current state of scientific knowledge is not considered sufficient to eliminate all uncertainty about GMO-related risks. According to Löfstedt et al. (2002), the main health risks are resistance to antibiotics and the risk of allergens while the main environmental risks are hybridization, collateral damage of non-targeted species, and disturbance of ecosystems. It is because research has not yet provided conclusive results that the public is often fearful and has a strong sense of uncertainty. This uncertainty and its emotional consequences are partly why decisions in France are based on the precautionary principle and involve decision-makers' responsibility (Peretti-Watel 2000). The scientific research is highly contested; according to Spiroux de Vendômois et al. (2010), the debate about GMO-related health risks is based not only on theoretical considerations, but also on knowledge gained from scientific experiments conducted on mammals fed with GMOs. However, these experiments are insufficient because they are not legally required and are conducted over the short-term. Hence, their results cannot be generalized and cannot definitively confirm that GMOs have no negative effects.

Uncertainty in terms of scientific knowledge about health and environmental risks makes risk management choices highly political and moral (Löfstedt et al. 2002). In Europe, "unmanageable" uncertainty reigns, which results in the precautionary principle. However, for its opponents, this principle is an assault on science, scientific reasoning and commerce (Löfstedt et al. 2002). So who to believe? The issue of the public's trust in institutions is paramount for understanding social representations and risk management. Thus, American legislation is such that Americans may have unknowingly been eating GMO foods since the 1990s (Ackerman 2002).

Beyond cultural differences, attitudes towards emerging risks can vary within one country, hence revealing contradictory interests, which are themselves caused by controversies. For around 10 years, in France and abroad, the development of radiofrequency-emitting technology, and so of exposure to electromagnetic fields, has raised many questions and concerns in terms of risks to environmental health. A study by Marchand et al. (in press) on risks related to exposure to sources of electromagnetic fields revealed how a lack of scientific certainty about the health consequences of exposure to radiofrequencies is a source of concern and demands in civil society. For the development of mobile phones, industrial and political players rely on their strong economic benefits and the fact that medical and environmental sciences have not revealed any danger. Even though certain opinions can fall in-between, two different understandings of the precautionary principle continue to polarize the French debate. For some, it should be applied immediately to protect people from a real and perceived threat. For others, the precautionary principle is based on the plausibility of scientific evidence, which is not sufficiently developed to reach a decision. The safety culture, which is legally sanctioned by the precautionary principle, reveals a relationship to risk that does not tolerate risk. Zero-risk is impossible to attain in the case of emerging risks, namely because of their uncertainty. Nevertheless, the difficulty of its implementation nuances the boundaries of risk in a complex legal and semantic game, which ends in principles of restraint, moderation of exposure, etc.

29.3 A Psycho-Environmental Approach to Emerging Risks

Beyond risk perception and socio-political issues, situations involving emerging risks directly concern environmental psychology for two reasons. On one hand, the strong uncertainty and controversies involved in emerging risks indicate how they challenge the human-environment relationship. The lack of certainty, or even knowledge, about an identified source of risk can create a particularly insecure context. On the other hand, the historically integrated relationship with risk is modified. The idea of risk was first introduced at the end of the nineteenth century to replace the notion of "fault" to qualify industrial accidents. An increasingly secure relationship with risk was hence discussed in society: to what extent can a risk with many expected benefits be acceptable (both from an economic point of view and in terms of comfort aspirations)? The concept of risk evolves rapidly, both in terms of research and public policy. It has to incorporate new types of risk and produce new evaluation methods to gain new knowledge and new choices of governance. To meet these requirements, the socio-cognitive approach offers conceptual and methodological tools to analyze emerging risk issues in terms of the human-environment relationship, relying on risk perception on one hand, and social representations on the other.

Clearly, the context of the individual plays an important role in his/her risk perception. Thus, in the case of pesticide use, geo-climatic realities seem to justify farmers' choices, as they can reduce their use of phytosanitary products below the threshold authorized in conventional agriculture, as is the case in sunny regions in Southern France. Hence, they view themselves as "sustainable" or "organic" farmers in the absence of certification, and tend to minimize the risks associated with pesticide use. Their responses are very different from those of farmers in less agriculturally favorable regions, such as Brittany and Martinique, who mostly mention the negative aspects of pesticide use (pollution, health and environmental risks, danger), as well as the importance of using pesticides to maintain a crop and ensure a good yield (Zouhri et al. in press).

In risk perception, the human-environment relationship also reflects the dissemination of a threat in a space that is difficult to control. For example, among sources of radiofrequency emissions, mobile phone antennas are the subject of much controversy and civil rejection. These antennas are installed in public spaces. Other sources, such as new generation meters, induction hobs, wifi routers, etc., are omnipresent in private spaces but less controversial than spatially remote antennas. Analysis models by Hall (1966), Bronfenbrenner (1977) and Moser (2009) show that the relationship with the world can be understood in a complex socio-environmental system, on macro, global or distant environmental levels, or micro, intimate or physical levels. Mobile telephone (MT) use sometimes fits into the microsystem as a continuity of the body it affects (Hafetz et al. 2010) through handling, contact with the ear (upon which it emits heat) etc. When there is almost a physical continuity between the object and the individual, civil rejection is rarely expressed, yet it is precisely in this area that experts claim the risk is most threatening (ANSES 2013). Augner and Hacker (2009) questioned the psychological and psychobiological consequences of the perceived distance between a cell tower and the home. Their study showed that the estimation of distance is inversely proportional to recorded stress levels. The closer a cell tower is perceived, the more psychological and psychobiological stress levels increase. Hence, the authors put forward a psychosomatic hypothesis based on the fear of cell towers. However, they refer to cognitive dissonance to explain the low concern about symptoms described by individuals claiming to live close to antennas. Furthermore, Kristiansen and Elstein (2009) demonstrated that concern about MTs and cell towers increases with their use, but does not inhibit their use. An analysis of the literature on radiofrequency-related risk perception (ANSES 2013) revealed that preoccupations and concerns about exposure to electromagnetic fields vary according to the distance between the source and the home, but that family structure, gender, professional status and level of education do not affect MT use. The notion of *control* appears discriminant and explains why representations of risk vary depending on the context (Augner and Hacker 2009; Kristiansen and Elstein 2009; Van Kleef et al. 2010; Van Dongen et al. 2011). Thus, in domestic contexts, given that sources are considered controllable, the risk is perceived as less than for environmental sources upon which no control can be exerted (Van Dongen et al. 2011). One might hypothesize that denial of risk by MT users is linked to a lack of perceived control of cell towers. Sjöberg (2000) developed a hypothesis of control as an explicative factor of risk denial. The results of a survey conducted on 500 Bengalis revealed a consensus on the benefits of MTs (economic benefits, facilitation of commercial transactions, simplification of communication with friends and family and the ability to warn them in case of imminent danger, namely natural disasters such as cyclones and floods). Van Kleef et al. (2010) demonstrated the impact of cultural contexts on risk representation in a developing country frequently exposed to natural disasters, in order to show that the benefit gained from the deployment of mobile telephones hampers any health risk perception. Health risks are reduced and even perceived as inexistent by some. Cell towers are perceived positively as an opportunity for the country to develop and communicate. These results corroborate those obtained in research derived from cultural theory (Douglas and Wildavsky 1983; Thomson et al. 1990) and, more recently, in research on risk perception at a societal level (see Chauvin 2014, for a review).

29.4 Social Representations of Emerging Risks

Research showing the impact of context and distance on individual evaluations of risk raises questions about social representations of risk to explain the rejection of objects such as cell towers or GMOs. Combined with uncertainty, the absence of a recognized risk is a source of beliefs and rumors, which are in turn sources of conflict and controversies. This is why we are particularly interested in how these new threats are experienced and represented, and how they impact quality of life. The psychosocial approach especially seeks to understand the potential gap between knowledge, attitudes and behavior, in this case, preventive and protective behavior. The case of pesticide use shows that even though farmers know very well how to protect themselves when applying phytosanitary products, they generally do not have an accurate representation of the corresponding health risks, which partly explains the emergence of inappropriate behaviors (Martinez et al. 2004; Perry et al. 2002; Snipes et al. 2009). This non-recognition of health risks by farmers and their immediate social environment seems to be consensual (Baer and Penzell 1993; Quandt et al. 1998, 2006; Arcury et al. 2002). People only feel exposed in the presence of sensory indicators (smell, for example) (Elmore and Arcury 2001). Nonetheless, farmers using pesticides do not seem to adopt any protective behavior (Perry et al. 2002). The social representations approach (Moscovici 1961) enables the identification of social cognitions and any evolution in relation to the environment. In his seminal work, Moscovici (1961) seeks to show how a new scientific or political theory is diffused in a given culture, how it is transformed throughout this process, and how it changes the vision people have of themselves and of the world they live in. In other words, the individual simplifies, transforms and appropriates the multitude of information he/she is faced with.

Social representations are developed collectively, and can be defined as a set of shared beliefs, knowledge, opinions and judgments within a group regarding a social object. According to the structural approach (Abric 1994, 2003), social representations are defined as a socio-cognitive structure containing a regulating central system, which is linked to a peripheral system. Central core elements are fundamental as they determine the meaning and organization of the social representation, whereas peripheral elements change and are less stable. Identifying the central elements of a representation reveals the building blocks of the object, that is to say, what characterizes it most in the eyes of the subject. Central elements can be identified as those most often associated with the object and also considered the most important by individuals. The peripheral system, through its composing elements, can incorporate personal experiences. Thus, peripheral elements mediate between the central core and the real situation by fulfilling three essential roles: embodiment, regulation and protection of the central core, acting as a "bumper" (Flament 1989; Moliner 1994). Identifying this structure helps to understand how different, more or less involved groups relate to an object of representation.

In addition, some topics are harder to talk about than others, especially through the scope of the socio-representational model. It is difficult to express oneself spontaneously about "sensitive" objects and thus to identify the social representation of these sensitive objects. This is the case of farmers' social representation of pesticides, particularly regarding their practices: farmers do not easily talk about certain aspects of their profession that can be judged negatively (Weiss et al. 2006). Thus, when asked about their representation of the environment, they do not spontaneously mention pollution. Furthermore, they deny any liability that could be related to their practices by operating social comparisons that are favorable to them. This could be discussed in terms of denial, but it seems that it refers more to a "mute zone" of the social representation, i.e. counter-normative or dissonant elements that cannot be verbalized (Guimelli and Deschamps 2000): farmers' avoidance of health and environmental risks; the desire not to be seen as responsible for poorly defined pollution. This potentially marks a deviation from the in-group, hence the idea of counter-normativity. Thus, in order to identify the elements of the mute zone of a social representation, it is necessary to reduce normative pressure to facilitate verbalization. This normative pressure is both individual and social in the in-group.

The hierarchized association technique updates implicit or latent elements that could be lost or masked in discursive productions. Two other techniques are possible to reveal masked elements. The first is a substitution technique (Guimelli and Deschamps 2000) that reduces individual involvement by asking the subject to answer as would other members of the reference group. The second places the individual in a context that is distant from the reference group, enabling him/her to express his/her thoughts more freely by reducing risks of negative judgment by the interlocutor (Abric 2005).

The joint use of these techniques not only reveals the latent dimensions that structure the semantic field specific to the studied representations (De Rosa 1988), but also defines their structure and provides insight into which elements underlie decisions in terms of risk behavior and protection. In fact, two identical contents can correspond to two different representations, because the same elements can be structured differently and hence have different meanings. As a result, any approach that only takes into account the content and not the structure of the representation could lead to interpretation errors.

29.4.1 Representation of Health Risks with Regard to Radiofrequencies

A socio-cognitive approach to risk was the object of a study by Marchand et al. (2015). This French-Canadian team explores social representations of health risks linked to the environment, new technologies and wave exposure. Their hypothesis is based on the impact of social cognition on cultural expressions of health controversies in Europe and Quebec. A preliminary exploration of a more general perception of environmental health risks showed that radiofrequencies were spontaneously mentioned. The study (on 73 subjects, 37 in France and 36 in Quebec) was conducted with the hierarchized association technique. The results revealed a major cultural impact on the social representations of all three objects. A methodological limitation of the study prevented conclusions about the status of waves in the Quebec sample.[1] They did, however, have a strong structuring role in the representation of the environment in the French sample. Two common elements were observed between the French and Quebec central cores: health problems and physical illness, as well as psychological disorders. They differ with regard to a specific risk that is highly rooted in culture: electromagnetic fields characterize the relationship to risk in the French sample, whereas food is prominent when the Quebec sample talks about risks. Technological risks mobilize less consensual representations. Physical health problems are expressed in both samples. Addictions and dependency structure the French representation, whereas the Quebec representation is organized around isolation and psychological disorders. With regard to the representation of radiofrequency-related health risks, physical health problems are a common factor in both cultural groups. The French representation is more specifically structured by functional disorders, physical illness and electromagnetic fields, whereas the Quebec representation is structured by psychological disorders.

Navarro and Michel-Guillou (2014) highlight how much risk perception and even vulnerability are shaped by lay-knowledge, or social representations of risk. They are affected by normative constraints specific to a society or group, at a given time and in relation to its values, history and collective memory. Radiofrequencies, waves and electromagnetic fields are structuring factors of the French relationship to the environment. These elements appeared both in the test about the environment in general, and about radiofrequencies in particular. Yet, they are not structuring factors of the Quebec representation. The survey was conducted in March 2014, when French controversies were particularly vivid and highly publicized after the publication of the ANSES report on radiofrequency exposure and the adoption of a draft law by the National Assembly to limit exposure. In Quebec, the media continue to relay controversy about food irradiation, which has been made worse in Canada by technological changes since the 1980s (Gauthier 2008). Social representations of risk reflect a radicalization of rejection by the population. These representations suggest a link between individuals and environmental risks that is built around an emerging threat and the vulnerability it causes. The role of the media in risk perception has been discussed since the 1970s and is today considered one of the major hypotheses to explain cultural differences and similarities in terms of risk judgment (Nyland 1993; Neto et al. 2006; Chauvin 2014). As the object of ever more sensational publicity, emerging risks feed fear and feelings of vulnerability.

29.4.2 Social Representations of Pesticide-Related Risks

When focusing specifically on representations of risk associated with a particular object, the structural approach to social representations reveals differences that merit discussion, especially since they show specific relationships to emerging risks. It was with this in mind that 213 farmers were questioned in three areas of France (Southern France, Brittany and Martinique) about their representation of pesticides, using the hierarchized association technique and comparing standard and substitutive contexts (Zouhri et al. in press). On one hand, this study revealed regional differences that reflect the geo-climatic characteristics at the origin of agricultural practice choices regarding the use of pesticides. On the other hand, self-other substitution revealed elements that suggest the normative characteristics of the mute zones. Thus, when answering in the first person, thus activating their own representational system, farmers from Southern France structure their representation around central elements that favor operational aspects of pesticide use: treatments with phytosanitary products enable the

---

[1] The Quebec ethics committee requires subjects to be informed about the object of study in a survey. It was only included in the Quebec sample, not the French one.

completion of the crop cycle. This result is even more interesting when compared to the representational system of Breton farmers: the central core of their social representation includes the negative effects of pesticide use, for example pollution, environmental and health risks, danger, etc. Unlike for Southern France farmers, these elements are not efficiency goals but consequences in terms of risk and danger. Furthermore, no items refer to the "technical" benefits of pesticides for farmers. The same dimensions were found among Martinique farmers who, apart from the dangers of pesticide use, also mention elements about the numerous rules and norms that regulate this use.

The sample of farmers from Southern France who answered in a substitutive context, i.e. in the name of farmers in general, revealed a social representation of pesticides structured around their negative consequences for the environment and public health. Therefore, by distancing themselves from the object and answering what they think other farmers think, they feel free to express the negative aspects of pesticide use. This result is in line with the hypothesis of a mute zone (Guimelli and Rateau 2003) in the social representation of pesticides. The substitutive answers of Martinique and Breton farmers revealed a representation structured by the same elements as in the standard context, with the exception of an item regarding the direct action of pesticides for farmers, namely, "harvest a crop". The role of this item, only revealed by self-other substitution, leads us to question the role of pesticides in discourse and practice: Breton and Martinique farmers seem to have integrated the prevention discourse related to pesticide use, which they readily disclose in the standard context. Self-other substitution seems to provide a "free" space in which to express the necessity of pesticides for their profession. This finding is supported by Breton farmers who also mention, through self-other substitution, the stigma of agriculture or the negative image of their profession in the general population.

These results, which differ depending on the context and the methods used to identify the social representations, demonstrate the pertinence of the structural approach for understanding the cognitive phenomena that underlie risk perception, as well as the relevant protection and prevention behaviors.

29.5 Conclusion: Emerging Risks, Well-Being and Quality of Life

The ambivalent relationship between progress and risks is obvious in the examples developed in this chapter. Whether it is pesticide use or the development of GMOs in relation to health and environmental risks, or the effects of electromagnetic waves on human health, emerging risks perfectly demonstrate the paradox of modern times: coping with increasing demands in terms of comfort, technology and "sustainable well-being" (Weiss and Moser 2013, p. 574), while incorporating safety for people, property, and the environment. These elements all refer to the question of quality of life, which was a priority announced by the Brundtland report (1987) on sustainable development. The World Health Organization defines quality of life as "an individual's perception of their position in life in the context of the culture and value systems in which they live and in relation to their goals, expectations, standards and concerns. It is a broad ranging concept affected in a complex way by the person's physical health, psychological state, level of independence, social relationships, personal beliefs and their relationship to salient features of their environment" (WHOQOL 1994). Thus, quality of life is associated with health on one hand, and perceived environmental quality on the other: "a positive relationship to a dwelling is essential for individual well-being (…). Beyond residential requirements, such as home comfort, access to services, green spaces, criminality, and neighboring noise levels, other more general requirements contribute to quality of life, such as access to transport, the preservation of local natural areas, and water and air quality. This demonstrates how quality of life involves all environmental aspects that are dealt with by environmental psychology" (Moser 2009, p. 236). In fact, individual quality of life and health can be threatened not only by environmental conditions, but also by the individual's relationship to these same conditions (Uzzell and Moser 2006), hence the importance of the individual's perception of these conditions and their corresponding representations of risk. These risks, and emerging risks in particular, are susceptible to increasing stress levels in the relationship to an environment perceived as potentially degraded or threatening. Today, the challenge for society faced with progress and the requirements of sustainable development is to respond to increasing demands in terms of wellbeing and quality of life, while at the same time taking into account increasing calls for safety. In this context, the social sciences are and will be increasingly mobilized, no longer just to study factors of acceptability of new technologies or potentially polluting and disturbing infrastructures, but also to understand better the fragile balance between the desire for progress and concerns about the unknown results of this progress. "Environmental psychology may become more and more concerned about helping societies to develop sustainable environments. The application of this research ranges from one extreme, focusing on changes in the quality and quantity of demand (the need to change people's lifestyles), to another extreme, focusing on changes in the production process to make it more sustainable" (Gärling and Hartig 2000, p. 31). This sustainability can only be achieved by managing health and environmental risks. This is probably the reason why they constitute a field of increasing interest for populations.

References

Abric, J. C. (1994). *Pratiques sociales et représentations*. Paris: PUF.

Abric, J. C. (2003). L'approche structurale des représentations sociales: développements récents. *Psychologie et Société, 4*, 81–103.

Abric, J. C. (2005). A zona muda des representações sociais. In D. Oliveira & P. Campos (Eds.), *Representataçoes sociais uma teoria sem fronteiras* (p. 143). Rio: Editora Museu da Republica de Rio de Janeiro.

Ackerman, J. (2002). Food: How safe? How altered? *National Geographic, 201*(5), 2–51.

ANSES (2013). *Radiofréquences et santé. Mise à jour de l'expertise, Avis de l'Anses*. Rapport d'expertise collective.

Arcury, T. A., Quandt, S. A., & Russell, G. B. (2002). Pesticide safety among farmworkers: Perceived risk and perceived control as factors reflecting environmental justice. *Environmental Health Perspectives, 110*(2), 233–240.

Augner, C., & Hacker, G. W. (2009). Are people living next to mobile phone base stations more strained? Relationship of health concerns, self-estimated distance to base station, and psychological parameters. *Indian Journal of Occupational and Environmental Medicine, 13*(3), 141–145.

Baer, R. D., & Penzell, D. (1993). Susto and pesticide poisoning among Florida farmworkers. *Culture, Medicine and Psychiatry, 17*(3), 321–327.

Breakwell, G. M. (2007). *The psychology of risk*. Cambridge: Cambridge University Press.

Bronfenbrenner, U. (1977). Toward an experimental ecology of human development. *American Psychologist, 32*(7), 513–531.

Brundtland, G. H. (1987). *Our common future*. Oxford: Oxford University Press.

Chauvin, B. (2014). *La perception des risques. Apport de la psychologie à l'identification des déterminants du risque perçu*. Brussels: De Boeck.

Crawford, C., & Weiss, K. (2013). Risques et représentations sociales des OGM en France et aux USA. 7ème congrès de l'AFPSA, Lille, France, 17–19 December.

De Rosa, A. S. (1988). Sur l'usage des associations libres dans l'étude des représentations sociales de la maladie mentale. *Connexions, 51*, 27–50.

Douglas, M., & Wildavsky, A. (1983). *Risk and culture*. Berkley: UC Press.

Elmore, R. C., & Arcury, T. A. (2001). Pesticide exposure beliefs among Latino farmworkers in North Carolina's Christmas tree industry. *American Journal of Industrial Medicine, 40*, 153–160.

Erné-Heintz, V. (2010). *Les risques: subir ou prévenir ?* Paris: Ellipses.

Flament, C. (1989). Structure et dynamique des représentations sociales. In D. Jodelet (Ed.), *Les représentations sociales* (pp. 204–219). Paris: PUF.

Galand, J. P. (2007). Evaluer les risques et mieux prévenir les crises modernes. *Regard sur l'actualité L'Etat face aux risques, 328*, 5–12.

Gärling, T., & Hartig, T. (2000). Environmental psychology's relationship to the environmental (design) professions. *Newsletter of the International Association of Applied Psychology, 12*(1), 30–32.

Gauthier, E. (2008). *Les représentations sociales du risque dans le débat public sur l'irradiation des aliments au Canada*. PhD. Manuscript. Montreal: University of Quebec, Montreal.

Guimelli, C., & Deschamps, J. C. (2000). Effets de contexte sur la production d'associations verbales: le cas des représentations sociales des Gitans. *Les Cahiers Internationaux de Psychologie Sociale, 47*, 44–54.

Guimelli, C., & Rateau, P. (2003). Mise en évidence de la structure et du contenu d'une représentation sociale à partir du modèle des schèmes cognitifs de base (SCB): la représentation des études. *Nouvelle Revue de Psychologie Sociale, 2*(2), 251–262.

Hafetz, J. S., Jacobsohn, L. S., García-España, J. F., Curry, A. E., & Winston, F. K. (2010). Adolescent drivers' perceptions of the advantages and disadvantages of abstention from in-vehicle cell phone use. *Accident Analysis and Prevention, 42*(6), 1570–1576.

Hall, E. T. (1966). *The hidden dimension*. New York: Doubleday & Co.

IPCS/WHO. *Ministerial Meeting of the North Atlantic Council (NAC)/North Atlantic Cooperation Council (NACC)*. Berlin, Germany, 3–4 July 1996.

Kouabenan, D. R., Cadet, B., Hermand, D., & Munoz Sastre, M. T. (Eds.). (2006). *Psychologie du risque : identifier, évaluer, prévenir*. Brussels: De Boeck.

Kraus, N., Malmfors, T., & Slovic, P. (1992). Intuitive toxicology: Expert and lay judgments of chemical risks. *Risk Analysis, 12*, 215–232.

Kristiansen, I. S., & Elstein, A. S. (2009). Radiation from mobile phone systems: Is it perceived as a threat to people's health? *Bioelectromagnetics, 30*(5), 393–401.

Leplat, J. (2006). Risque et perception du risque dans l'activité. In D. R. Kouabenan, B. Cadet, D. Hermand, M. T. Munoz Sastre (Eds.), *Psychologie du risque: identifier, évaluer, prévenir* (ed. 2007, pp. 19–33). Paris: De Boeck.

Löfstedt, R. E., Fischhoff, B., & Fischhoff, I. R. (2002). Precautionary principles: General definitions and specific applications to Genetically Modified Organisms. *Journal of Policy Analysis and Public Management, 21*(3), 381–407.

Marchand, D., Depeau, S., & Weiss, K. (Eds.) (2014). *L'individu au risque de l'environnement : regards croisés de la psychologie environnementale*. Paris: In press.

Marchand, D., Brisson, G., Plante, S., Gauthier, M., Gauvin, D., & Zbinden, A. (2015). *Représentation du risque et coproduction de savoirs experts et profanes dans la résolution des controverses liées aux radiofréquences en France et au Québec*. CSTB/ANSES. Rapport intermédiaire.

Marchand, D., Weiss, K., Laffitte, J. D., Ramalho, O., Chaventré, F., & Collignan, B. (in press). L'incertitude, un facteur explicatif de l'évolution des crises environnementales. *Bulletin de Psychologie*.

Martinez, R., Gratton, T. B., Coggin, C., René, A., & Waller, W. (2004). A study of pesticides safety and health perceptions among pesticides applicators in Tarrant County, Texas. *Journal of Environmental Health, 66*(6), 34–37.

Moliner, P. (1994). Les méthodes de repérage et d'identification du noyau des représentations sociales. In C. Guimelli (Ed.), *Structures et transformation des représentations sociales* (pp. 199–232). Neuchâtel: Delachaux & Niestlé.

Moscovici, S. (1961). *La psychanalyse, son image, son public*. Paris: PUF.

Moser, G. (2009). *Psychologie environnementale*. Paris: De Boeck.

Navarro, O., & Michel-Guillou, E. (2014). Analyse des risques et menaces environnementales. Un regard psycho-socio-environnemental. In D. Marchand, S. Depeau, & K. Weiss (Eds.), *L'individu au risque de l'environnement* (pp. 271–297). Paris: In press.

Neto, F., Lazreg, C., & Mullet, E. (2006). Perception des risques et couverture médiatique. In D. R. Kouabenan, B. Cadet, D. Hermand, & M. T. Munoz Sastre (Eds.), *Psychologie du risque* (pp. 85–97). Brussels: De Boeck.

Nyland, L. G. (1993). *Risk perception in Brazil and Sweden* (Center for Risk Research Report n°15). Stockholm School of Economics.

OECD. (2003). *Les risques émergents au XXIème siècle. Vers un programme d'action*. Paris: OECD Publishing.

Peretti-Watel, P. (2000). *Sociologie du risque*. Paris: Armand Colin.

Perry, M. J., Marbella, A., & Layde, P. M. (2002). Compliance with required pesticide-specific protective equipment use. *American Journal of Industrial Medicine, 41*(1), 70–73.

Quandt, S. A., Arcury, T. A., Austin, C. K., & Saavedra, R. M. (1998). Farmworker and farmer perceptions of farmworker agricultural chemical exposure in North Carolina. *Human Organization, 57*(3), 359–368.

Quandt, S. A., Hernandez-Valero, M. A., Grzywacz, J. G., Hovey, J. D., Gonzales, M., & Arcury, T. A. (2006). Workplace, household, and personal predictors of pesticide exposure for farmworkers. *Environmental Health Perspectives, 114*(6), 943–952.

Sjöberg, L. (2000). Factors in risk perception. *Risk Analysis, 20*, 1–11.

Slovic, P. (1987). Perception of risk. *Science, 236*, 280–285.

Slovic, P. (1995). The construction of preference. *American Psychologist, 50*(5), 364–371.

Slovic, P. (2000). *The perception of risk*. London: Earthscan.

Snipes, S. A., Thompson, B., O'Connor, K., Shell-Duncan, B., King, D., Herrera, A. P., & Navarro, B. (2009). Pesticides protect the fruit, but not the people: Using community-based ethnography to understand farmworker pesticide-exposure risks. *American Journal of Public Health, 99*(S3), 616–621.

Sparks, P. J. (2000). Idiopathic environmental intolerances: Overview. *Occupational Medicine, 15*(3), 497–510.

Spiroux de Vendômois, J., Cellier, D., Vélot, C., Clair, E., Mesnage, R., & Séralini, G. E. (2010). Debate on GMOs health risks after statistical findings in regulatory tests. *International Journal of Biological Sciences, 6*(6), 590–598.

Thomson, P. B., Ellis, R., & Wildavsky, A. (1990). *Cultural theory*. Boulder: Westview Press.

Uzzell, D., & Moser, G. (2006). On the quality of life of environments. *European Review of Applied Psychology, 56*(1), 1–4.

Van Dongen, D., Smid, T., & Timmermans, D. R. M. (2011). Perception of health risks of electromagnetic fields by MRI radiographers and airport security officers compared to the general Dutch working population: A cross sectional analysis. *Environmental Health, 10*, 95.

Van Kleef, E., Fischer, A. R. H., Khan, M., & Frewer, L. J. (2010). Risk and benefit perceptions of mobile phone and base station technology in Bangladesh. *Risk Analysis, 30*(6), 1002–1015.

Wagner, W., & Kronberger, N. (2006). Le naturel et l'artificiel dans le nouveau monde du génie génétique. In K. Weiss & D. Marchand (Eds.), *Psychologie sociale de l'environnement*. Rennes: PUR.

Weiss, K., & Marchand, D. (Eds.). (2006). *Psychologie sociale de l'environnement*. Rennes: PUR.

Weiss, K., & Moser, G. (2013). Environnement. In L. Begue & O. Desrichard (Eds.), *Traité de Psychologie Sociale: la Science des interactions humaines* (pp. 567–584). Brussels: De Boeck.

Weiss, K., Moser, G., & Germann, C. (2006). Perception de l'environnement, conceptions du métier et pratiques culturales des agriculteurs dans le cadre du développement durable. *Revue Européenne de Psychologie Appliquée, 56*, 73–81.

WHOQOL. (1994). Development of the WHOQOL: Rationale and current status. *International Journal of Mental Health, 23*, 24–56.

Zouhri, B., Feliot-Rippeault, M., Weiss, K., Michel-Guillou, E. (in press). Social representation of pesticides in French farmers. *Environmental Science and Pollution Research*.

Energy Issues: Psychological Aspects

30

Rafaella Lenoir-Improta, Patrick Devine-Wright, José Q. Pinheiro, and Petra Schweizer-Ries

30.1 Introduction

Human beings have always needed energy in order to survive. In early civilizations, firewood provided the fuel for fires. Then animals began to be worked, bringing innovations such as the use of rotational machines. In the metal ages, machines were created to shape tools and metallurgy was developed.

Prior to the industrial revolution, renewable energy was viewed as a reliable way to keep machines running. Wind power was the driving force for sailing ships whilst water kept the mills turning. The use of firewood was still fundamental as a source of energy, especially for heating and cooking.

The development of the steam engine and further technological advances made it possible for humans to produce goods on a large scale. In order to do this, steam engines required vast quantities of fossil fuels, such as coal and oil-based fuels, giving rise to the industrial revolution (Sørensen 2011).

The attractions of the cities quickly drew people to them with the offer of work and prosperity. This led to a demand for services such as transport and heating. This period also coincided with the development of electric network grids which, through the use of transformers and alternating current, enabled electricity to be transported over long distances and made its use widespread. The consumer society was born and it had high energetic needs (Blas and Aragonés 1986; Sørensen 1991).

The greatest drawback of using fossil fuels (coal, natural gas and oil) is that, unlike other energy resources used previously, they are highly polluting throughout their extraction, processing and use. They also deplete when used, and for this reason, they are labeled as a non-renewable energy source (Bermann 2001).

The consumer society has continued to grow, but it faced a first and then a second oil crisis in the 1970s that challenged the belief in an infinite supply of oil. Energy dependence, as well as high oil prices due to the fear of running out of oil,

R. Lenoir-Improta (✉)
Department of Social Psychology,
University of Barcelona, Barcelona, Spain
e-mail: rafaellalenoir@gmail.com

P. Devine-Wright
Geography, College of Life and Environmental Sciences,
University of Exeter, Exeter, England
e-mail: P.G.Devine-Wright@exeter.ac.uk

J.Q. Pinheiro
Coordinator of People-Environment Research Group,
Federal University of Rio Grande do Norte, Natal, Brazil
e-mail: pinheiro@cchla.ufrn.br

P. Schweizer-Ries
Department of Teaching and Research Lab Sustainable Development, Integrated Institute for Sustainable Development, University of Applied Sciences Bochum, Bochum, Germany
e-mail: petra.schweizer-ries@fg-upsy.com

meant that governments across the world began to make fuel efficiency plans. Energy planning as a government responsibility was developed in many countries. Departments of Energy were set up to establish government authority and gain the expertise required for this new type of planning. In this period, governments also passed laws in order to make people save energy and use it more sensibly (Blas and Aragonés 1986; Sørensen 1991). Nevertheless, the fuel crisis not only brought home the cost of fossil fuel energy use, but also raised public awareness of the indirect costs associated with the use of fossil fuels to the environment (Bermann 2001).

However, the implementation of these measures has often neglected the social aspects inherent in this process, which has repeatedly led to a lack of public support for such measures and their rejection (Stern 1992).

The economic crisis and increased public awareness caused many countries to promote the use of renewable energy technologies as a way of overcoming such problems. At this time, renewable energy technology had advanced sufficiently to be considered commercially viable for electricity production, in particular via the use of wind turbines. This meant that, once again, renewable energies regained some of the ground lost to oil (Sørensen 1991). This period also saw the push towards the use of nuclear power. Nuclear energy faced, and still faces today, strong public resistance as public concern grew over the safe disposal of nuclear waste (Bechtel 1997; Sørensen 1991).

In the 1990s, renewable energies gained an established market space as a result of two factors. The first was the economic and political problems faced in the 1970s while the second was the increased public awareness of environmental issues derived from using fossil fuels. Consequently, international meetings were set up to discuss people's use of the environment and find ways of being less disruptive towards natural resources. Within the topics of discussion were the ongoing issue of fossil fuel use and its impact on the greenhouse effect and climate change (Bermann 2001; Winter 1996).

The first significant meeting dealing with climate change was the UN climate convention in Rio de Janeiro in 1992. This was followed by the Kyoto conference in 1998 where many countries signed the Kyoto protocol. The signatory countries agreed to decrease their 1990 carbon emissions by 5.2 % via the use of renewable energies so as to reduce their effect on climate change (United Nations 1998).

The current economic development system continues to rely heavily on the use of fossil fuels. This situation has worsened due to disorganized and accelerating population growth, as well as the emergence of industrialized countries, which have increased energy demand. Despite this system being unviable and unsustainable for our environment, it continues to grow (Bermann 2001; Clark 1995; Winter 1996).

Nowadays, many government campaigns promote and encourage a rational use of energy. Smart grids are becoming popular and the development of the Smart City concept is increasing. Such concepts are often seen as very technical and seek to raise public awareness and involve citizens in the rational use of energy, reducing and shifting energy consumption, as well as encouraging the production of environmentally clean energy by promoting the use of renewable energy technologies.

What is the role of psychology research in this area? As presented below, concern about energy use can be divided into stages, considering the different elements that have been incorporated into the discussion of this issue. Following this trend, the contribution of psychology can be separated into two broad areas of research: one-dimensional and multi-dimensional (Lenoir-Improta and Pinheiro 2011).

The first stage occurred primarily in the 1970s and 1980s. The research of that period can be categorized as one-dimensional as it was based solely on the economic factors derived from the wide use of fossil fuels. It focused on individual behavior, particularly in domestic energy use. This stage has been criticized for the technical, simplistic and neutral approach of its research, looking at energy use on an individual basis and ignoring social issues (Blas and Aragonés 1986; Bechtel 1997; Stern 1992).

The 1990s gave rise to a second, multi-dimensional, stage that remains in place today.

This period saw the popularization of the sustainable development concept and climate change. The related psychological research has mainly covered the broad scope of energy use (Stern 1992). As well as the economic and individual aspects that were developed in the one-dimensional perspective, social, cultural and environmental issues are now also taken into account. This "holistic view" has followed the trend that characterizes the research about sustainability/environmental conservation concern in environmental psychology during this period (Bonnes and Bonaiuto 2002; Pol 1993).

Unfortunately, in many cases, the social aspects of energy use have been limited or never considered in public policies, which has frequently produced acceptance problems. Some researchers have highlighted and criticized this lack (Devine-Wright 2005; Pol et al. 2006; Wolsink 2007, 2011), proposing new conceptual and practical perspectives (e.g. Lenoir-Improta, Di Masso and Pol 2015; Schweizer-Ries 2008). This and other subjects are explored in this chapter.

The purpose of this chapter is to provide an overview on what is the current state of the art on research about psychological aspects concerning the use of energy. In the next three sections, we explore the main contemporary issues concerning: (1) public acceptance of energy infrastructures, (2) energy consumption and conservation and, finally, (3) energy sufficiency.

30.2 Public Acceptance of Energy Production and Distribution Systems

There has been a considerable increase in empirical research on the acceptance or rejection of energy production and distribution systems over the past decade, stimulated by policy interest in the transition of energy systems to low carbon (renewable or nuclear) energy sources. Overarching these studies are two important concepts that have been the focus of much research: NIMBY (Not In My Back Yard) and social acceptance (Wüstenhagen et al. 2007).

NIMBYism is a way of understanding social conflicts around the siting of energy technologies, focusing upon the role of individuals who object and attributing opposition to spatial proximity and the personal characteristics of objectors (Burningham 2000). Despite its prevalence in media discourses, a consensus has emerged amongst energy researchers that it is not useful either for describing or for explaining public objections (Devine-Wright 2005; Wolsink 2006). NIMBYism narrowly conceives members of the public as technology objectors, skewing research towards only one mode of public response to energy production or distribution systems. It overlooks the multiple roles that the public can and does play in the low-carbon transition – as voters, technology adopters, investors, community organizers, opponents etc. (Walker and Cass 2007). It also presumes that public objections are based on ignorance, selfishness and emotionality (Devine-Wright 2005; Bell et al. 2013), presumptions that have been argued as being inappropriate and shown to be inaccurate (e.g., Wolsink 2006; Devine-Wright 2011).

An alternative to NIMBY (mis)understandings has been the concept of "social acceptance" (Ekins 2004). Wüstenhagen et al. (2007) developed a multi-dimensional approach to social acceptance encompassing societal-political, market and local community levels of analysis, each of which were considered inter-dependent and dynamic. This framework has been influential and frequently cited, reflecting the complexity of societal responses to novel energy technologies at multiple levels. However, the concept has also been criticized for maintaining an instrumental interest in public responses to energy technologies, with an altered label but retaining many of the suppositions inherent in the NIMBY concept. Empirical research has also pointed to shortcomings in existing studies, notably heterogeneity in empirical measures of "acceptance" and the need to distinguish between "acceptance" and "support" or "active acceptance" (Batel et al. 2013; Schweizer-Ries 2008). Related to the notion of social acceptance is the concept of the "social gap", which was initially proposed by Bell et al. (2005). This accounted for the gap between societal support for renewable energy (as suggested by opinion poll findings) and local objections (as revealed by NIMBY conflicts). One of the responses to the social gap was the idea of conditional acceptance (Bell et al. 2005, 2013) reflecting a series of issues or "conditions" under which general support for a given energy source can be translated into local objections. Empirical studies have revealed the complexity of factors (or "conditions") underlying local responses to siting proposals (both supportive and objecting), including place attachment (Devine-Wright 2009), environmental justice (both procedural and distributional, Cowell et al. 2012; Rau et al. 2012), and trust (Walker et al. 2010). These are now discussed in turn.

An important strand of research has focused upon spatial aspects of public responses to energy technologies, elaborating upon the "back yard" aspects of the NIMBY concept. Devine-Wright (2009) proposed that local objections might be founded upon concerns about changes to places with which residents self-identified, showing the relevance of concepts from environmental psychology, such as place attachment and place identity, to the understanding of siting conflicts. In a series of empirical studies on diverse technologies, such as offshore wind energy, tidal energy and high-voltage power lines (Devine-Wright and Howes 2010; Devine-Wright 2011, 2013), significant links were shown between person-place bonds and levels of community acceptance/objection. Interestingly, the nature of the relationship was shown to vary. In contexts where an energy project was perceived to threaten the character of a place, there was a negative relationship between the strength of place attachment and levels of acceptance (e.g., Devine-Wright and Howes 2010); in contexts where an energy project was perceived to enhance the character of a place, there was a positive relationship (e.g., Devine-Wright 2011; Lenoir-Improta and Pinheiro 2011).

These explanations point to the importance of capturing not only place attachments and identities in energy research, but also perceptions of the "fit" between place and technology, requiring a focus upon the meanings that are socially constructed concerning the nature of the technology and of the place and the degree to which each can or cannot fit well together (Devine-Wright 2009). These issues have also been shown to be important for the social acceptance of nuclear power. Venables et al. (2012) demonstrated that the acceptance of nuclear power stations in the UK was based upon widespread perceptions amongst local residents that existing nuclear power stations were a "normal" part of the place or landscape in which people lived, unremarkable and taken for granted. Studies have also shown how objections to energy infrastructures – from wind farms to high-voltage power lines – that are situated in rural landscapes often take the form of conflicting meanings associated with the countryside, seen as tranquil, beautiful and natural, and those associated with energy projects, seen as urban, industrial and technological (e.g., McLachlan 2009; Devine-Wright and Howes 2010).

Matters of justice have also been shown to play an important role in explaining public objections. These have multiple dimensions, and much energy research has focused upon procedural justice, including how decisions are taken and the nature of consultation activities between development organizations and local residents.

Gross (2007) revealed how fairness and equity were key to public responses to wind energy proposals in Australia, while "decide-announce-defend" institutional practices that close down opportunities for public engagement have been strongly criticized (Wolsink 2007; Ellis et al. 2007). A comparative analysis of renewable energy projects in Germany indicated that procedural justice was a key element in explaining levels of public support (Zoellner et al. 2008) and similar findings were indicated by a UK study of local responses to a proposed high-voltage power line (Devine-Wright 2013). There has also been interest in aspects of distributional justice, specifically the ways in which costs and benefits associated with energy projects are allocated across local, national and international scales, and specific actors such as local residents and company shareholders. Haggett (2008) has argued that public objections to wind farms arise from benefits at the global but not at the local scale and this has prompted a range of policy responses that attempt to address perceived imbalances in benefit/cost distribution (e.g., the payment of a community benefits package to communities affected by onshore wind farms in the UK).

There is an extensive literature on the payment of compensation to those affected by facilities siting and recent research has investigated these issues in the context of renewable energy technologies. Cass et al. (2010) observed that the payment of such benefits was motivated by diverse rationales, and the instrumentality of such measures has been criticized by researchers for overlooking fairness and justice (Cowell et al. 2012) as well as for creating potential problems in implementation that may not yield the expected levels of community acceptance (Aitken 2010). Cass et al. (2010) found that accusations of bribery could lead to benefit provision proving counter-productive in promoting social acceptance. This was the subject of a recent experimental study that compared public responses to a hypothetical wind farm proposal, showing that framing community benefits in ways that referred to controversy over bribery led to lower levels of public support than a purely benefit framing (Walker et al. 2014). The overlap between distributional and procedural justice was indicated by research that showed how the impact of compensation payments upon social acceptance was highest in scenarios where the instigating company had already engaged with the local community (Terwel et al. 2014) and lower in scenarios where no engagement had taken place.

The payment of benefits is not the only way to address perceptions of distributional injustice. In European countries such as Denmark and Germany, it is state policy to encourage citizen ownership of energy projects (e.g., all new wind farms have to be at least 20 % citizen-owned in Denmark). There is some evidence that public support for renewable energy projects is higher when projects such as wind farms are owned by the local community rather than by a local company (Warren and MacFadyan 2010). However, this topic needs further research, particularly in contexts where projects are jointly owned by local communities and private companies.

Trust has been found to play a key role in influencing public engagement and acceptance (Midden and Huijts 2009). Trust in developers – whether a private company or local landowner community – has been shown to be strongly associated with project support (Walker et al. 2010) while trust in local action groups has been associated with project objections, and particularly whether the character of the place is represented as under threat from "inappropriate" development (Devine-Wright and Howes 2010). Future studies could explore trust in multiple dimensions – for example, Poortinga and Pidgeon (2003) considered the role of factors that contribute to building trust in risk regulation, such as competence, objectivity, fairness, consistency, faith, honesty, accountability and responsibility. These factors could also be explored in research into the social acceptance of energy technologies.

Despite these advances, the literature has a number of weaknesses. First, it is predominantly based on case studies of single-technology projects, and is therefore poorly positioned to explain how novel energy technologies may be relationally understood through comparisons with more familiar technologies. Second, research is skewed towards a single-technology type – onshore wind. Far fewer studies have been conducted on projects such as large-scale solar farms or offshore wind energy. Third, many studies have adopted an individualistic and cognitivist perspective, researching "public perceptions" using quantitative survey methods. Such research overlooks how public beliefs and responses are shaped by the ways that societal actors engage with the public for a given technology (Batel and Devine-Wright 2014). One solution is to apply conceptual frameworks from social psychology (e.g., the theory of social representations, Moscovici 2000) to account for the ways that public understandings of particular technologies develop over time, are communicated and contested. Finally, there is a need for research that adopts a more systemic approach to energy production and distribution facilities. One important recent study applied a deliberative method to engage with citizens regarding energy *as a whole* and how it might need to change to achieve large cuts in carbon emissions (Parkhill et al. 2013). The research not only produced significant findings about public engagement with national energy policies, rather than just local energy projects (Pidgeon et al. 2014), but also revealed the importance of social values, which were identified as underlying how individuals perceived and evaluated different generation, distribution and consumption options.

30.3 Psychological Research on Energy Consumption and Conservation

Cities continue to grow, the world population keeps escalating, and humankind today counts on a huge amount and variety of services, not to mention electronic equipment that is part of the daily scenery. All this implies an increasing energy consumption, a tendency that may be reverted, or at least attenuated, by *energy efficiency*, and changes in our consumption habits and life style. Energy efficiency means being able to perform more while using less energy.

Due to the oil crisis at the beginning of the 1970s, the high cost of petroleum, which was relatively low during the previous decades, changed the outlook of the industrial and transportation sectors of the economy. By the 1970s and 1980s, supporters of the notion of energy efficiency were arguing in favor of its potential benefits: energy savings (reduction in energy consumption); decrease in pollutants, improving the environment; reduction of energy costs, both for final users and for utilities; less dependence of a country on imported energy sources; and a more competitive economy. Such an effort towards the optimization of energy use made a big difference. If energy efficiency had not been introduced during the 1970s, Europe and the USA would today be consuming about 50 % more energy than they do.

Research in behavioral sciences saw the implications of these movements. The bibliographic database PsycINFO (American Psychological Association) records a very clear increase in studies related to the consumption and conservation of energy, as shown in Table 30.1. In the 1970s, there was a jump from 23 articles in the previous decade to 89; however, the jump was even higher in the 1980s, when the database registered almost 300 articles.

During the 1990s, and particularly in the first decade of the twenty-first century, energy policies also emphasized the importance of energy efficiency for sustainable development and as a strategic form of climate change mitigation. Businesses and governments were gradually convinced of the need for change in patterns of energy consumption and started bringing a transformation into the economic scenario, despite the financial difficulties involved in the replacement of outdated technologies. Similar efforts were also tried with regard to the final users of energy, in sectors such as residential maintenance and personal transportation alternatives, by means of policies that were not always successful. This need for modifications in energy behavior in all economic sectors required more scientific research. An amplification of the scientific and social role of energy studies for society at large was again reflected in changes in psychologically oriented research. In the same bibliographic database (see Table 30.1), the 1990s seemed rather dormant, before an enormous leap to more than a thousand articles occurred in the first decade of the new century. Although textbooks about energy issues were scarce in the early decades of Environmental Psychology, with very few exceptions (e.g., Baum and Singer 1981; Stern and Aronson 1984), psychological research is now published in multidisciplinary periodicals

Table 30.1 Frequency of articles about the consumption or conservation of energy in the PsycINFO database for the decades indicated

Decades	Articles on energy consumption or energy conservation	Cumulative percentage
1951–1960	19	0
1961–1970	23	21.1
1971–1980	89	287.0
1981–1990	299	236.0
1991–2000	287	−4.0
2001–2010	1,131	294.1

such as *Energy Policy*, *Global Environmental Change*, and *Local Environment*, in addition to the traditional titles of the field, such as *Journal of Environmental Psychology* and *Environment and Behavior*.

The term *energy efficiency* usually elicits in people's minds an idea of technological devices or forms of energy that are capable of a better measurable performance than their predecessors and/or existing ones, justifying the replacement of earlier versions by newer and more efficient counterparts. Such a view of energy efficiency holds true not only for lay people, but also for specialized professionals, as can be appreciated in the definition of energy efficiency as "the ratio between service output or result and the energy input required to provide it" (Pérez-Lombard et al. 2013, p. 252). Energy efficiency indexes are available in stores next to the price tags of domestic appliances to guide the buyer when purchasing a refrigerator, for example.

Despite its apparently manageable precision, the concept of energy efficiency is elusive and rather ill-defined, at least from the point of view of the final user. If a householder wants to monitor their domestic electricity consumption, for instance, the (service provider) bill only gives the monthly total, and in some cases only the amount of money to be paid. Even when the total expenditure of electricity is reported on the bill, the householder does not know how to obtain a better performance in energy efficiency. This is comparable to a situation in which you want to save money at the grocery store but cannot find the price tags on the products and only receive a bill with the total to be paid to the cashier. You would not know how to reduce the costs; for instance, which products not to buy, the quantities of some items that could be reduced, and so forth (Stern and Aronson 1984).

The important point is that energy efficiency may present different meanings to different audiences (Hall et al. 2013). For example, the electrician who comes to the house to fix the lighting in the living room may recommend buying the latest generation of LED lamps because this will significantly reduce the electricity bill (as they consume much less electricity to provide the same level of illumination), even though they cost more than other types of lamp. The homeowner, however, may think that LED lamps take too long to repay the initial costs while his/her daughter may decide that the brightness generated by these lamps is inadequate for her studies and strains her eyes. People, groups and organizations rarely act as rational economic agents; they may act upon their values, dreams and social needs, instead of careful calculations, which takes the question of energy efficiency in this context into the realm of social and behavioral research. Maybe the point of view of McClelland and Canter (1981) is still applicable; they considered the scientific knowledge already investigated about the promotion of energy conservation to be the tip of an iceberg, whose complexity steadily challenges easy solutions.

Things may become even more complicated when this scenario of the multi-determination of practices about energy is extended to the level of policies intended to discipline the sector. In this context, an analogy seems pertinent. We may ask the same two questions proposed by environmental psychologist Robert Sommer (1972). First, a value question: "*which* goals are to be achieved?" and then a political question: "*whose* values are to be served?" (p. 85).

In their pioneering effort to analyze the human dimensions of energy use, the members of the Committee on Behavioral and Social Aspects of Energy Consumption and Production of the National Research Council/USA (Stern and Aronson 1984) identified four views of energy: as a commodity, an ecological resource, a social necessity, and strategic material. They also described the social, economic and political specifications and implications of each of the related policies. Making clear that "the way a society thinks about energy affects the way society makes decisions about energy" (p. 14), the Committee argued that the prevalent and omnipresent view of energy as a set of commodities unavoidably defies the diverse points of view of the final users.

When defining energy efficiency from the viewpoint of energy as a commodity – which is typically the case – the final user's satisfaction, quality of life and similar dimensions of energy use are left aside. As pointed out by Sachs (2007):

> None of the energy transitions of the past was performed due to the physical depletion of an energy source. Humankind history may be synthesized as the history of production and allocation of the economic surplus, cadenced by successive energy revolutions. All of them happened due to the identification of a new source of energy with superior qualities and inferior costs. So was the transition of energy of biomass to coal and from this to petroleum and natural gas. (p. 42)

History shows us that obtaining the scientific and technological fit is not always enough to ensure new energy technologies are adopted or utilized as intended; and the same applies for making them affordable. They may fail because designers do not adequately consider the final users' needs, practices and preferences. There may be different types of barriers (regulations, institutional structures, misplaced incentives, lack of decision-relevant information, among others), creating a large energy efficiency gap (Webler and Tuler 2010).

The residential sector is where such a gap is more noticeable, given the margin of control and options available to the users. At the same time, however, household actions guided by the adequate integration of existing technology and behavioral change may achieve significant reductions in energy consumption in the short term, without drastically altering life style (Dietz et al. 2009). Why, then, is there this gap in implementation and negative "side effects" that spoil energy efficiency initiatives?

A reasonable answer would tap into several distinct dimensions, including ethical, ideological and political aspects (Hall et al. 2013). However, in the remaining paragraphs of this section, we concentrate on people's knowledge of what energy is. Not the knowledge from some shallow definition, but a deep knowledge, addressing both production and consumption aspects of energy, comprising individual and community levels of representation and understanding, and intertwined with the problem of the (in)visibility of energy.

Several generations ago, our ancestors had to deal directly with energy questions if they wanted to have dinner ready a few hours later. Chopping wood and bringing the pieces to the stove, as well as carrying buckets of water from a nearby well, were daily routines that, besides assuring survival, afforded a clear cognitive (and affective) representation of what energy is. A similar experience would apply to planning and undertaking a 3-day journey on horseback between two villages, and many other daily routines demanding human expenditure of physical effort (energy). One is tempted to imagine how our relatives produced, distributed, mentally represented, negotiated, and saved energy back then. Whatever their practices were, one thing is certain: energy was "visible" to them.

Nowadays, our civilization is built upon the age of information and the great majority of our citizens do not have a clue about the forms of energy behind the technological wonders they are used to handling, minute by minute. Energy has become invisible, and its only form of visibility is as a commodity (Stern and Aronson 1984). Therefore, psychological research (and also applied interventions) faces the formidable challenge of making energy visible to lay people, so that it can be appropriately used, saved and administered by suitable policies. Ecological approaches (e.g., Stokols et al. 2009; Winkel et al. 2009) and proper environmental management strategies (e.g., Pol 2002) must be considered by psychological research when dealing with energy issues, otherwise the examined topic would only be people's relationships to services and applications of energy (the commodities), and not energy itself.

30.4 Energy Sufficiency

After studying the provision of energy services and their efficient use, we now concentrate on the demand side and the question of how to decide what is enough and how many energy services we really need. This topic is called energy sufficiency strategy within sustainability and the psychological question is: what makes people and societies live in an energy-sufficient way? This is often included and subsumed in the energy efficiency topic where it is named curtailment and often negatively related to "energy saving" by "cut back on amenities or comfort" e.g., by "turning off lights, unplugging appliances, or reducing appliance usage", associated with constraint and painful reduction (Karlin et al. 2014, p. 428 and 441). Regarding curtailment, Black et al. (1985) distinguished between regulating home temperature (e.g., setting one's thermostat) and minor curtailments, which they refer to as "energy services that might not be perceived as sacrifice" (p. 9). Curtailment is higher when people are older, less educated and have a lower income (Karlin et al. 2014, p. 432). Psychological predictors of efficiency behaviors include perceived personal benefits (Black et al. 1985), financial motivation (Cialdini and Schultz 2003), perceived cost (Nair et al. 2010), the importance of reducing energy use (Nair et al. 2010), environmental protection (Cialdini and Schultz 2003) and social and descriptive norms (Cialdini and Schultz 2003).

Karlin et al. (2014) identifies "two groups of behaviors" (p. 441): "(a) low-cost behaviors (curtailment) and (b) infrequent, high-cost behaviors (efficiency)". We see a more holistic understanding of the sufficiency strategy, which we highlight here. It is oriented towards taking only the energy services needed and saving the rest for later or for others. This is the mindful use of energy services, including the idea of sharing products, spaces and technologies. Darby (2007, p. 112) makes clear that "Energy Services are valued not just for themselves (heat, light), but for the activities and social relationships that they make possible: they raise questions of justice and emotional ties as well as practicalities of supply and demand". Sufficiency, according to Darby (2007), can be defined in two ways: as a quality when it is enough, meaning sufficiently available to fulfill the needs, knowing that needs and demands depend very much on the culture, the technologies available and subjective evaluations. The quantity gives the baseline, when something is sufficient and "ceilings" when it is "too much for safety or welfare in the short and long term" (Darby 2007, p. 111). Absolute needs are seen as the "need for clean water, daily food and basic shelter" (Darby 2007, p. 111) as they are also defined by Maslow (1987) and Max-Neef et al. (1986). The basic demand for energy is calculated to be 2000 W and our societies in the Global North use much more whereas many societies in the Global South still use much less, although targeting another life style. Knowing that 20 % of the global population (in the Global North) already uses 80 % of the natural resources, it is easy to see that this trend cannot be followed without risking conflicts about the fair distribution of natural reserves (see e.g., Trainer 2007).

The basic demand for energy services was worked out by Norgard (1991, p. 54) who summarized the following for one "standard-person": lighting for 6 h per day with 60 W lights with 1000 lm on average, refrigeration of a 200-l volume ($+5$ °C) and a 100-l freezer (-18 °C), 200 laundry washes per year, each 4 kg with warm water provided by non-electric energy, if needed, the use of some electronic devices including several hours of TV per day, listening to the radio and using a computer as well as other minor uses of electronics, ventilation for fresh air in high-rise buildings and some unspecified ventilation for cooling, and other uses as long as they are highly efficient electric devices. This would be the "minimum human right level of energy service per household and could sum up to about 1500 W per Person". This is less than the 2000 W-Society asked for in sustainability movements e.g., in Switzerland (Fischer 2009), and was increased to 4000 W-Society to be more realistic for the style of living in the Global North, which uses more than "only" household energy, when thinking of travelling and "gray" energy included in the products themselves (Spreng 1989). Citizens in North America are already using much more than 1500 W per Person, as are Western Europeans, whereas Indians, Brazilians and Chinese, for example, have used much less on average until now (Norgard 1991, p. 57). However, when life styles are changing as they are at the moment towards a more Western U.S. American style, it will be difficult to supply enough energy for all the demand and impossible with only oil-based energy production. In this calculation, there is no other solution than changing our energy supply and use system.

The sufficiency strategy requires life styles in the Global North to change for a fair and equal distribution of energy to all citizens in the world. This should also include preventing the Global South from adopting the energy-intensive life styles of the North. The analysis by economists supports "the observation that with thoughtful restructuring, highly developed countries could use a fraction of their current energy without any measurable loss in human development" (Steinberger and Roberts 2010, p. 432) and "truly sustainable social and environmental progress is only possible if the industrialized nations, which are currently using far more energy and emitting far more carbon dioxide per capita than they need for high standards of living, substantially reduce their consumption and emissions". As Morin (2010) names it, "the world citizenship" takes responsibility not only for future generations but also for those already living. In the psychological words of Kohlberg, it is vital to reach the moral stage of post-convention (Kohlberg 1995) in order to understand that it is important to share with others and not reserve and consume the resources for ourselves.

Alongside this knowledge and on this moral stage of post-convention, a social movement has started to develop in industrialized nations all over the world, called "voluntary simplicity" (see e.g. Alexander and Ussher 2011). Inside these communities, people enjoy living with less in a happier and wealthier way. They are voluntarily changing their way of life, including their energy consumption patterns This movement and the Transition Town Movement (Hopkins 2008) are motivated partly by the understanding that the peak oil will appear soon, if it has not already occurred (see e.g., Alexander 2011a, p. 6), and from then on, the reserve oil will become increasingly expensive. Therefore, it is not only a moral decision but also unwise not to change energy supply and use. On the use side, we have already referred to energy efficiency including the rebound effect due to human behavioral decisions on how to use energy-efficient technical devices.

In this section, we concentrate on the findings of what motivates people from a psychological point of view to live with fewer energy services. We do not refer to this as curtailment (see above) but as a new way of living with fewer demands, more relationships with oneself and others (relatedness) and healthier for humans and ecology. In sustainability, this is called "the good life" (Kaufmann-Hayoz 2006) or in the Andean culture "Sumakkawsay" (Lozano Castro 2013; García Álvarez 2011). The main question is: how can we live better in our world with less environmental and resource destruction and more social justice?

Here we look at the findings from psychology and bordering disciplines on what supports this new way of thinking and living in terms of the sufficiency strategy. As Ostrom, Gardner and Walker (1994) puts it: how to change the mental and behavioral habits of our societies towards a more sustainable living, here concentrating on energy sufficiency in thinking of a better life in the sense of post-consumerism investigated broadly by sociological investigations (Schor 1998; Pierce 2000; Kasser 2002; Craig-Lees and Hill 2002; Grisby 2004; Brown and Kasser 2005; Hamilton and Dennis 2005). In this culture, it will be agreed that "the human community must find a way to raise the material standards of living of the world's poorest people – who surely have a right to develop their economic capacities in some form – while at the same time reducing humanity's overall ecological footprint" (Meadows et al. 2004, p. 15) with the consequence, like Mahatma Gandhi (1997, pp. 306–307) called for, of "human beings to live simply so that others may simply live" (cited according to Alexander and Ussher (2011, p. 1) from the Australian "Voluntary Simplicity Movement".

Three main approaches can be seen for how to encourage these different ways of thinking and acting in the world: the individual, the community and the cultural approach. Although all this is based on psychology, other disciplines are currently more active in exploring this field of sufficiency and should also be taken into account. This is a huge task, which overlaps the boundaries of the disciplines (Riemer and Schweizer-Ries 2012; Lang et al. 2012).

As Kasser states (2002, according to Alexander and Ussher 2011, p. 5), "there is in fact a mounting body of sociological and psychological evidence indicating that lives orientated around achieving high levels of consumption often result in such things as time poverty, stress, physical and mental illness, wasteful status competition, loss of community, disconnection from nature, a sense of meaninglessness or alienation in life, and general unhappiness (not to mention ecological degradation)." Health and "the good life" seem therefore to be the motivation of people all over the world to join these social movements.

Alexander and Ussher (2011) conducted an internet survey with 50 closed questions on life style, behavior, values people "find best about living simply, what challenges they face in doing so, and what steps they think government could take to better support simple living" (Alexander and Ussher 2011, p. 6). Of the 2,268 participants, 970 were from North America, 871 from Australia, 147 from the UK, 108 from Western Europe without the UK, 77 from New Zealand, 4 from Japan and the rest from other parts of the world. 28 % lived in large cities, 18 % in medium-sized and 16 % in small cities, 17 % in small towns, and 21 % in rural (non-urban or farm) areas. "67 % acknowledged that they had reduced their incomes from what they had been in the past" (Alexander and Ussher 2011, p. 7). "38 % changed jobs or careers; 48 % reduced working hours; 16 % moved to the city or suburbs; 21 % moved rurally; and 22 % sold or changed their car. Furthermore, when asked whether they took steps to reduce household energy consumption, 46 % said they did so 'at every opportunity', 41 % did so 'often' and 12 % did so 'sometimes' while less than 1 % said they did 'not often' do so" (Alexander and Ussher 2011, p. 8). Concerning motivations for living simply, more than 80 % mentioned "environmental concerns"; around 70 % "to be healthier", "self-reliance/self-sufficiency", "decluttering life/minimalism"; about 60 % answered "to save money" and to live more spiritually or mindfully; about 50 % agreed with "more time with family", "more time for oneself"; around 40 % mentioned "more time for community involvement" and "humanitarian or social justice concerns" (Alexander and Ussher 2011, p. 9). Clearly, this social movement cannot come only from citizens on the ground, but also needs to be supported by politicians, technology developers, market designers and community developers. It is not an isolated movement of individuals but a wider change in a world community.

Transition towns are growing all over the world and can build energy-sustainable communities (Schweizer-Ries 2011). From psychological investigations, especially from Positive Psychology and Gestalt psychology, there are motivating factors other than the consumption of physical goods that support well-being. Hunnecke recently developed a theory on the psychological aspects of a post-growth society, including self-acceptance, self-sufficiency, mindfulness, sense-construction and solidarity (Hunnecke 2013).

30.5 Conclusions

This chapter provides a selective review of the current understanding of the social aspects of the energy issue. More specifically, we have focused on the social aspects of the use and production of electricity.

Energy overuse is a major problem and overlooking the social aspects associated with it might negatively impact the already difficult balance between the production and use of natural resources.

As observed throughout this chapter, many studies about the social aspects of the energy issue are being developed by different approaches. We have tried to cover some of the most significant ones, such as those related to the social acceptance of energy production and distribution systems, energy efficiency and sufficiency. In each of these approaches, researchers are focusing on different aspects, by using varied theoretical and methodological perspectives, and contributing overall to strong and robust progress toward a corpus of understanding of the social aspects of energy use.

This topic opens a rich field of research in Environmental Psychology. Not only well-established perspectives and concepts in Social Psychology are useful here, but also powerful alternative methods and theoretical frameworks such as Discursive Psychology (Billig 1991; Lenoir-Improta et al. 2015; Potter and Wetherell 1987). This wide variety of approaches to energy issues contributes to enriching, broadening and strengthening the field of Environmental Psychology.

When applied to daily life, the greatest contribution of these studies is that they present the different social implications of the use and production of energy in modern societies, some of which have limited or no public consciousness. Thus this raises, directly or indirectly, an awareness of the importance of this problem.

However, even with so much research and so many advances, there are still very few countries that really take into account this social aspect. If this does not happen more widely, it will be impossible to reach the much desired balance between human development and the use of natural resources.

References

Aitken, M. (2010). Wind power and community benefits – challenges and opportunities. *Energy Policy, 38*, 6066–6075.

Alexander, S. (2011). Property beyond growth: Toward a politics of voluntary simplicity. Doctoral Thesis. Melbourne Law School, University of Melbourne. http://www.simplicityinstitute.org/publications. Accessed 5 May 2011.

Alexander, S., & Ussher, S. (2011). *The voluntary simplicity movement: A multi-national survey analysis in theoretical context. Simplicity institute report.* Melbourne: Simplicity Institute.

Batel, S., & Devine-Wright, P. (2014). Towards a better understanding of people's responses to renewable energy technologies: Insights from social representations theory. *Public Understanding of Science.* doi:10.1177/0963662513514165.

Batel, S., Devine-Wright, P., & Tangeland, T. (2013). Beyond the social acceptance of renewable energy innovation: A discussion about acceptance and support. *Energy Policy, 58*, 1–5.

Baum, A., & Singer, J. E. (Eds.). (1981). *Advances in environmental psychology. Vol. 3. Energy: Psychological perspectives.* Hillsdale: Erlbaum.

Bechtel, R. (1997). Energy: A missed opportunity. In R. Bechtel (Ed.), *Environment and behavior: An introduction* (pp. 265–286). California: Sage.

Bell, D., Gray, T., & Haggett, C. (2005). The "social gap" in wind farm siting decisions: Explanations and policy responses. *Environmental Politics, 14*, 460–477.

Bell, D., Gray, T., Haggett, C., & Swaffield, J. (2013). Revisiting the "social gap": Public opinion and relations of power in the local politics of wind energy. *Environmental Politics, 22*, 115–135.

Bermann, C. (2001). *Energia no Brasil: Para quê? Para quem? – crise e alternativas para um país sustentável.* São Paulo: Livraria da Física.

Billig, M. (1991). *Ideology and opinions: Studies in rhetorical psychology.* London: Sage.

Black, J. S., Stern, P. C., & Elworth, J. T. (1985). Personal and contextual influences on household energy adaptations. *Journal of Applied Psychology, 70*, 3–21.

Blas, F. A., & Aragonés, J. I. (1986). Conducta ecológica responsable: la conservación de la energía. In J. F. Burillo & J. I. Aragonés (Eds.), *Introducción a la psicología ambiental* (pp. 303–329). Madrid: Alianza.

Bonnes, M., & Bonaiuto, M. (2002). Environmental psychology: From spatial-physical environment to sustainable development. In R. B. Bechtel & A. Churchman (Eds.), *Handbook of environmental psychology* (2nd ed., pp. 28–54). New York: Wiley.

Brown, K., & Kasser, T. (2005). Are psychological and ecological wellbeing compatible? The role of values, mindfulness, and lifestyle. *Social Indicators Research, 74*, 349–368.

Burningham, K. (2000). Using the language of NIMBY: A topic for research, not an activity for researchers. *Local Environment: The International Journal of Justice and Sustainability, 5*, 55–67.

Cass, N., Walker, G., & Devine-Wright, P. (2010). Good neighbours, public relations and bribes: The politics and perceptions of community benefit provision in renewable energy development in the UK. *Journal of Environmental Policy and Planning, 12*(3), 255–275.

Cialdini, R. B., & Schultz, W. (2003). *Understanding and motivating energy conservation via social norms (Tech. Rep.).* Menlo Park: William and Flora Hewlett Foundation.

Clark, M. E. (1995). Changes in Euro-American values needed for sustainability. *Journal of Social Issues, 51*(4), 63–82.

Cowell, R., Bristow, G., & Munday, M. (2012). *Wind energy and justice for disadvantaged communities: What role can community benefits play?* (Viewpoint for Joseph Rowntree Foundation). York: JRF.

Craig Lees, M., & Hill, C. (2002). Understanding voluntary simplifiers. *Psychology and Marketing, 19*(2), 187–210.

Darby, S. (2007). *The effectiveness of feedback on energy consumption. A review for Defra of the literature on metering, billing and direct displays.* http://www.eci.ox.ac.uk/research/energy/electric-metering.php

Devine-Wright, P. (2005). Beyond NIMBYism: Towards an integrated framework for understanding public perceptions of wind energy. *Wind Energy, 8*(2), 125–139.

Devine-Wright, P. (2009). Rethinking Nimbyism: The role of place attachment and place identity in explaining place protective action. *Journal of Community and Applied Social Psychology, 19*(6), 426–441.

Devine-Wright, P. (2011). Public engagement with large-scale renewable energy: Breaking the NIMBY cycle. *Wiley Interdisciplinary Reviews: Climate Change, 2*, 19–26.

Devine-Wright, P. (2013). Explaining "NIMBY" objections to a power line: The role of personal, place attachment and project-related factors. *Environment and Behavior, 45*, 761–781.

Devine-Wright, P., & Howes, Y. (2010). Disruption to place attachment and the protection of restorative environments: A wind energy case study. *Journal of Environmental Psychology, 30*, 271–280.

Dietz, T., Gardner, G. T., Gilligan, J., Stern, P. C., & Vandenbergh, M. P. (2009). Household actions can provide a behavioral wedge to rapidly reduce U.S. carbon emissions. *Proceedings of the National Academies of Sciences, 106*(44), 18452–18456.

Ekins, P. (2004). Step changes for decarbonising the energy system: Research needs for renewables, energy efficiency and nuclear power. *Energy Policy, 32*, 1891–1904.

Ellis, G., Barry, J., & Robinson, C. (2007). Many ways to say no, different ways to say yes: Applying Q-Methodology to understand public acceptance of wind farm proposals. *Journal of Environmental Planning and Management, 50*(4), 517–551.

Fischer, S. (2009). 2000-Watt Society – The Swiss vision for the creation of sustainable low energy communities. 2000-Watt Society, 45th ISOCAP Congress. http://www.isocarp.net/Data/case_studies/1379.pdf. Derived June 2015.

Gandhi, M. (1997). My quest for simplicity. In M. Rahnema & V. Bawtree (Eds.), *The post development reader*. London: Zed Books.

García Álvarez, S. (2011). El sumak kawsay y la política económica del gobierno. La Tendencia. *Revista de análisis politico, 12*, 82–86.

Grigsby, M. (2004). *Buying time and getting by: The voluntary simplicity movement*. Albany: State University of New York Press.

Gross, C. (2007). Community perspectives of wind energy in Australia: The application of a justice and community fairness framework to increase social acceptance. *Energy Policy, 35*(5), 2727–2736.

Haggett, C. (2008). Over the sea and far away? A consideration of the planning, politics, and public perceptions of offshore wind farms. *Journal of Environmental Policy and Planning, 10*(3), 289–306.

Hall, S. M., Hards, S., & Bulkeley, H. (2013). New approaches to energy: Equity, justice and vulnerability. Introduction to the special issue. *Local Environment, 18*(4), 413–421.

Hamilton, C., & Denniss, R. (2005). *Affluenza: When too much is never enough*. Crows Nest: Allen & Unwin.

Hopkins, R. (2008). *The transition handbook: From oil dependency to local resilience (transition guides)*. London: Green Books.

Hunnecke, M. (2013). *Psychological resources for sustainable development*. Bonn: Foundation for Cultural Renewal.

Karlin, B., Davis, N., Sanguinetti, A., Gamble, K., Kirkby, D., & Stokols, D. (2014). Dimensions of conservation: Exploring differences among energy behaviors. *Environment and Behavior, 46*(4), 423–452.

Kasser, T. (2002). *The high price of materialism*. Cambridge: MIT Press.

Kaufmann-Hayoz, R. (2006). Human action in context. A model framework for interdisciplinary studies in view of sustainable development. *Umweltpsychologie, 10*(1), 154–177.

Kohlberg, L. (1995). *Die Psychologie der Moralentwicklung*. Frankfurt/M: Suhrkamp.

Lang, D., Wieck, A., Bergmann, M., Stauffacher, M., Martens, P., Moll, P., Swilling, M., & Thomas, C. J. (2012). Transdisciplinary research in sustainability science – practice, principles, and challenges. *Sustainability Science, 7*(Supplement I, S.), 25–43.

Lenoir-Improta, R., Di Masso, A., Pol, E. (2015). Stakeholders' views of a new wind farm: Exploring acceptance and rejection as discursive accomplishments. Proceedings from Royal Geographical Society-IBG annual international conference, Exeter, England.

Lenoir-Improta, R., & Pinheiro, J. Q. (2011). Socio-environmental impacts of Brazil's first large-scale wind farm. In P. Devine-Wright (Ed.), *Renewable energy and the public; From NIMBY to participation* (pp. 219–231). London: Earthscan.

Lozano Castro, A. (2013). Runa Yachachiy: orenamiento territorial y buen vivir – Sumak Kawsay. *Revista electrónica digital*, I Semestre, Quito, Ecuador.

Maslow, A. (1987). *Motivation and personality*. New York: Harper and Row.

Max-Neef, M., Elisalde, A., & Hopenhayn, M. (1986). *Desarrollo a escala humana: una opción para el futuro*. Santiago: Fundación Dag Hammerskjöld.

McClelland, L., & Canter, R. J. (1981). Psychological research on energy conservation: Context, approaches, methods. In A. Baum & J. E. Singer (Eds.), *Advances in environmental psychology, Vol. 3. Energy: Psychological perspectives* (pp. 1–25). Hillsdale: Erlbaum.

McLachlan, C. (2009). "You don't do a chemistry experiment in your best china": Symbolic interpretations of place and technology in a wave energy case. *Energy Policy, 37*, 5342–5350.

Meadows, D., Randers, J., & Meadows, D. (2004). *Limits to growth: The 30-year update*. White River Junction: Chelsea Green Publish Company.

Midden, C. J. H., & Huijts, N. (2009). The role of trust in the affective evaluation of novel risks: The case of CO2 storage. *Risk Analysis, 29*, 743–751.

Morin, E. (2010). *Die sieben Fundamente des Wissens für eine Erziehung der Zukunft*. Hamburg: Krämer.

Moscovici, S. (2000). *Social representations: Explorations in social psychology*. London: Polity Press.

Nair, G., Gustavsson, L., & Mahapatra, K. (2010). Factors influencing energy efficiency investments in existing Swedish residential buildings. *Energy Policy, 38*, 2956–2963.

Norgard, J. S. (1991). Energy conservation through efficiency and sufficiency. In *Conference proceedings "Global Collaboration on a Sustainable Energy Development"*. Physics Department, Technical University of Denmark, DK-2800 Lyngby, Denmark.

Ostrom, E., Gardner, R., & Walker, J. (1994). *Rules, games, and common-pool resources*. Ann Arbor: University of Michigan Press.

Parkhill, K., Demski, C., Butler, C., Spence, A., & Pidgeon, N. (2013). *Transforming the UK energy system: Public values, attitudes and acceptability – synthesis report*. London: UKERC.

Pérez-Lombard, L., Ortiz, J., & Velázquez, D. (2013). Revisiting energy efficiency fundamentals. *Energy Efficiency, 6*, 239–254.

Pidgeon, N., et al. (2014). Creating a national citizen engagement process for energy policy. *Proceedings of the National Academy of Sciences, 111*(4), 13606–13613.

Pierce, L. B. (2000). *Choosing simplicity: Real people finding peace and fulfillment in a complex world*. Carmel: Gallagher Press.

Pol, E. (1993). *Environmental psychology in Europe: From architectural psychology to green psychology*. Aldershot: Avebury.

Pol, E. (2002). Environmental management: A perspective from environmental psychology. In R. B. Bechtel & A. Churchman (Eds.), *Handbook of environmental psychology* (2nd ed., pp. 55–84). New York: Wiley.

Pol, E., Di Masso, A., Castrechini, A., Bonet, M. R., & Vidal, T. (2006). Psychological parameters to understand and manage the NIMBY effect. *Revue Européenne de Psychologie Appliquée, 56*, 43–51.

Poortinga, W., & Pidgeon, N. (2003). Exploring the dimensionality of trust in risk regulation. *Risk Analysis, 23*, 961–973.

Potter, J., & Wetherell, M. (1987). *Discourse and social psychology: Beyond attitudes and behaviour*. London: Sage.

Rau, I., Schweizer-Ries, P., & Hildebrand, J. (2012). Participation strategies – the silver bullet for public acceptance? In S. Kabisch, A. Kunath, P. Schweizer-Ries, & A. Steinführer (Eds.), *Vulnerability, risk and complexity: Impacts of global change on human habitats* (pp. 177–192). Leipzig: Hogrefe.

Riemer, M., & Schweizer-Ries, P. (2012). Complexity, normativity, and transdisciplinarity: Is psychology ready to meet the sustainability challenges? *Umweltpsychologie, 16*(1), 143–166.

Sachs, I. (2007). A revolução energética do século XXI [The energy revolution of the 21st century]. *Estudos Avançados, 21*(59), 21–38.

Schor, J. (1998). *The overspent American: Upscaling, downshifting, and the new consumer*. New York: Basic Books.

Schweizer-Ries, P. (2008). Energy sustainable communities: Environmental psychological investigations. *Energy Policy, 36*, 4126–4135.

Schweizer-Ries, P. (2011). Socio-environmental research on energy sustainable communities: Participation experiences of two decades. In P. Devine-Wright (Ed.), *Public engagement with renewable energy: From Nimby to participation* (pp. 187–202). London: Earthscan.

Sommer, R. (1972). *Design awareness*. San Francisco: Holt, Rinehart & Winston.

Sørensen, B. (1991). A history of renewable energy technology. *Energy Policy, 19*(1), 8–12.

Sørensen, B. (2011). *Renewable energy: Physics, engineering, environmental impacts, economics & planning*. Burlington: Academic Press-Elsevier.

Spreng, D. (1989). *Wieviel Energie braucht die Energie? Energiebilanzen von Energiesystemen*. Zurich: vdf – Verlag der Fachvereine.

Steinberger, J. K., & Roberts, J. T. (2010). From constraint to sufficiency: The decoupling of energy and carbon from human needs, 1975–2005. *Ecological Economics, 70*, 425–433.

Stern, P. (1992). What psychology knows about energy conservation. *American Psychologist, 47*(10), 1224–1232.

Stern, P. C., & Aronson, E. (Eds.). (1984). *Energy use: The human dimension [National Research Council, Committee on Behavioral and Social Aspects of Energy Consumption and Production]*. New York: Freeman.

Stokols, D., Misra, S., Runnerstrom, M. G., & Hipp, J. A. (2009). Psychology in an age of ecological crisis – from personal angst to collective action. *American Psychologist, 64*(3), 181–193.

Terwel, B. W., Koudenberg, F. A., & Ter Mors, E. (2014). Public responses to community compensation: The importance of prior consultations with local residents. *Journal of Community and Applied Social Psychology, 24*, 479–490.

Trainer, T. (2007). *Renewable energy cannot sustain a consumer society*. Dordrecht: Springer.

United Nations. (1998). Kyoto Protocol to the United Nations framework convention on climate change. http://unfccc.int/resource/docs/convkp/kpeng.pdf. Accessed 8 Oct 2014.

Venables, D., Pidgeon, N., Parkhill, K. A., Henwood, K., & Simmons, P. (2012). Living with nuclear power: Sense of place, proximity, and risk perceptions in local host communities. *Journal of Environmental Psychology, 32*(4), 371–383.

Walker, G. P., & Cass, N. (2007). Carbon reduction, "the public" and renewable energy: Engaging with socio-technical configurations. *Area, 39*, 458–469.

Walker, G., Devine-Wright, P., Hunter, S., High, H., & Evans, B. (2010). Trust and community: Exploring the meanings, contexts and dynamics of community renewable energy. *Energy Policy, 38*, 2655–2633.

Walker, B., Wiersma, B., & Bailey, E. (2014). Community benefits, framing and the social acceptance of offshore wind farms: An experimental study in England. *Energy Research & Social Science, 3*, 46–54.

Warren, C., & MacFadyan, M. (2010). Does community ownership affect public attitudes to wind energy? A case study from south-west Scotland. *Land Use Policy, 27*, 204–213.

Webler, T., & Tuler, S. P. (2010). Getting the engineering right is not always enough: Researching the human dimensions of the new energy technologies. *Energy Policy, 38*, 2690–2691.

Winkel, G., Saegert, S., & Evans, G. W. (2009). An ecological perspective on theory, methods, and analysis in environmental psychology: Advances and challenges. *Journal of Environmental Psychology, 29*(3), 318–328.

Winter, D. D. (1996). *Ecological psychology: Healing the split between planet and self.* New York: Harper Collins.

Wolsink, M. (2006). Comment – invalid theory impedes our understanding: A critique on the persistence of the language of NIMBY. *Transactions. Institute of British Geographers, 31*, 85–91.

Wolsink, M. (2007). Wind power implementation: The nature of public attitudes: Equity and fairness instead of "backyard motives". *Renewable and Sustainable Energy Reviews, 11*(6), 1188–1207.

Wolsink, M. (2011). Discourses on the implementation of wind power: Stakeholder views on public engagement. In P. Devine-Wright (Ed.), *Renewable energy and the public; from NIMBY to participation* (pp. 75–87). London: Earthscan.

Wustenhagen, R., Wolsink, M., & Burer, M. J. (2007). Social acceptance of renewable energy innovation: An introduction to the concept. *Energy Policy, 35*, 2683–2691.

Zoellner, J., Schweizer-Ries, P., & Wemheuer, C. (2008). Public acceptance of renewable energies: Results from case studies in Germany. *Journal of Energy Policy, 36*(11), 4136–4141.

Global Challenges for Environmental Psychology: The Place of Labor and Production

31

David Uzzell, Nora Räthzel, Ricardo García-Mira, and Adina Dumitru

It is some 45 years since Proshansky et al. (1970) published their seminal reader *Environmental psychology: man and his physical setting*, which Enric Pol referred to as "... the most emblematic text" (Pol 2007). The book emerged out of a growing realization by the three researchers that, although they were principally concerned with the impacts of psychiatric wards on patients and vice versa, generalizations could be made about the role of the environment on people's behavior. For some 20 years after its publication, the interests that dominated environmental psychology changed little, focusing largely on the built environment and what would become core issues in EnvPsy101 courses (e.g., crowding, privacy, territoriality, user satisfaction and post-occupancy evaluation). Since 1990, global warming, sustainable development and what we now refer to as climate change have taken center stage. This has been a significant shift and arguably has done much to prevent the area "fading away" as Proshansky warned in 1987. He argued that if environmental psychology was to survive it needed to "strengthen itself as a social institution" (Proshansky 1987: 1486). Environmental psychology has achieved this in no small part, contributing to both our understanding of the policies and practices towards mitigating and adapting to climate change.

31.1 Some Pitfalls of Environmental Psychology Research on Climate Change

This major shift in environmental psychology has entailed a focus on studying how individual consumers contribute to climate change, as households in the European Union are responsible for 25 % of the total emissions resulting from fossil fuels, according to a report released by the European Environment Agency (2011). It is now known from studies of the environmental impact of different human activities that the categories of shelter, mobility, food and the consumption of manufactured products are responsible for the majority of direct and indirect greenhouse gas (GHG) emissions in the European Union (Hertwich 2011). The approach in environmental psychology has been to focus largely on internal factors that might determine discrete, specific behaviors, and to develop models for predicting these behaviors. Internal factors, such as values,

D. Uzzell (✉)
University of Surrey, Surrey, UK
e-mail: d.uzzell@surrey.ac.uk

N. Räthzel
University of Umeå, Umeå, Sweden
e-mail: nora.rathzel@soc.umu.se

R. García-Mira • A. Dumitru
University of A Coruna, A Coruña, Spain
e-mail: ricardo.garcia.mira@udc.es;
adina.dumitru@udc.es

beliefs, worldviews, personal norms, goals, identities, and perceptions of control over behavior, have been among those most studied and this body of research has yielded some interesting results on what determines pro-environmental behavior. Nevertheless, psychology has been more successful concentrating on environmentally convenient rather than environmentally significant actions (Stern 2000), or more systemic changes toward sustainable lifestyles. This has implied a concept of sustainable lifestyles that is predominantly additive (e.g., the more sustainable behaviors a person performs, the more sustainable the lifestyle), but ignores the contradictions that lead to specific lock-in situations, even in those cases where a large number of pro-environmental behaviors are performed (Jackson and Papathanasopoulou 2008).

In terms of social factors, social norms have been among the most studied, and research has found evidence that they have a greater weight than many of the individual factors (Carrus et al. 2009; Fornara et al. 2011; García-Mira et al. 2003; Schultz et al. 2008). This is not unexpected in environmental psychology, as social psychology shows that human behavior must be understood within its societal context. Looking more closely, the other individual factors for which research has found evidence of impact are social in nature. Identity is the most obvious one, but values, beliefs and worldviews also result from processes of social construction. Besides being a result of socialization processes, identity formation is fundamentally a relational process and the moment-to moment experience of identity is lived relationally. As social identity theory contends, we define our identities in relation to others to whom we are bound by establishing similarities and differences and defining in-groups and out-groups (Tajfel and Turner 1986). Our very definition of ourselves is thus essentially as being in relationship with and part of a collective.

On a second level, focusing on discrete behaviors might play a role in the usually modest percentages of variance that environmental psychology research can explain. Besides the oft-mentioned idea that behavior is also determined by factors outside the individual, such as conditions pertaining to the physical structures or social structures in which we live out our lives, individual behaviors are interrelated in a complex web of human activities and patterns of time use (Dumitru et al. 2014) that are undertaken for different reasons and have associated consumption and resulting emissions. For example, by studying the factors affecting private car use to travel between work and home, one can list a catalogue of individual, social and structural factors that might influence this behavior, yet such explanations would ignore the fact that mobility behaviors are related to how we conceive of our lives in the city, the associated disparate requirements of having to do so much in a given period of time, and how our perceptions of a good and safe life have changed. For example, we now take children to school by car due to our perceptions of insecurity (García-Mira and Goluboff 2005). Focusing on how to diminish individual car use by looking for effective alternatives or internal factors can only take us so far, if we do not approach changes in lifestyles as an organic reality that needs a systemic approach to change. This has started to be signaled by both researchers and policy-makers as climate change, and its associated problems have been conceived as "wicked" problems, that is, large-scale and constituting long-term policy dilemmas in which multiple risks and uncertainties combine with different public views of them (Balint et al. 2011).

The notion used to conceptualize a systemic approach to climate change is 'sustainable development', in which the economic, the social, and the environmental are three separate entities that overlap in certain moments/spaces. Elsewhere, we have suggested that this is a limited model and we need instead to think of relationships of production and consumption, and political relationships as being closely intertwined, feeding off, shaping and limiting each other. These relationships are created not least through the fact that all individuals are active in all of these fields at different moments (Räthzel and Uzzell 2009b; Uzzell and Räthzel 2009).

31.2 Environmental Psychology and the Need for Transformative Collective Action

In order to overcome some of the limitations of environmental psychology, the predominantly individualistic approaches in the discipline and its major focus on consumption need to be addressed. People live, work, and act in cooperation with others and this cooperation is shaped by and shapes individual and collective identities and actions. Environmental problems are often social or collective and will only be addressed through social theories and action. Such work has, however, invariably received less attention as governments have sought to explain climate change through the errant consumption behaviors of individuals and address it through behavior change strategies. Collectivities and collective action need to be placed on the agenda of psychology in general and of environmental psychology specifically.

When engaging with other disciplines to contribute to solving "wicked" problems, we notice that environmental psychology is more geared to explaining stability rather than change. It is better equipped to explain why people act in a certain way and how they will react to a certain environment or policy, rather than how people start environmental initiatives, maintain motivation over time, or influence others to join them. It is not so much that environmental psychology does not have the tools for this, but rather that the disciplinary culture is still more rooted in a cognitive-behavioral paradigm and its reactive approach. To explain how people exert what sociologists call "agency" requires a change in our basic approach. While it is important to understand how people react to their physical and social environments, the nature and dimension of the problems of climate change also require that we understand how people are enabled to mobilize collectively to transform their lifestyles, communities and economies in a sustainable direction. Mobilizing agency and exercising it in a transformative direction goes beyond making decisions not to use so much energy or to start using public transport. It requires people to engage in creative action with others to transform their ways of living and propose and explore alternatives, and for social scientists in conjunction with policy-makers and practitioners to test and roll them out into larger sectors of society. External structural transformations have to be accompanied by self-transformations such that the individual understands what those transformations entail, and how participation in different collectives and collective action is beneficial (and thus becomes an intrinsic motivation).

Furthermore, it is now commonplace to say that transitions to a green economy will require a shift in our systems of production and consumption supported by alternative economic frameworks and institutions. Technological fixes focused on improving resource efficiency are recognized as insufficient, due to rebound effects (Hertwich 2005). A change in our understanding of 'environmentally virtuous' lifestyles and well-being is necessary, together with an understanding of the structures that would support such new concepts and their achievement in practice. These transformations require coordinated and collective human action, not only knowledge of behavioral determinants and change. The problem with a behavioral perspective is that it takes power and agency out of the equation (in psychology, the closest would be the concept of autonomy in self-determination theory (Deci and Ryan 2000)). Thus, we need to understand what drives people to engage in transformation/transformative efforts and to connect with others to do so. For example, the processes of interaction in communities and groups provide powerful explanations for the initiation of social movements.

Any discipline is a lens through which a section of the world is interpreted. By adopting the lens of individual behavior and of the factors being processed through perception and leading to a behavior, environmental psychology has sometimes shied away from explaining the complex processes of group creativity and the complex relational nature of the ways in which individuals act. This is not to say that environmental psychology does not have the potential to do this. A great deal of thorough theorizing (e.g., on social influence, group creativity, etc.) and methodological innovation (e.g., moment-to-moment measurements, experimental approaches) in psychology can be brought to bear upon these issues, if a change in focus is adopted. Innovation continues to be part of our responsibility as scientists (García-Mira and Dumitru 2014b).

Nudging, or what has been marketed as 'behavioral insight', is the latest solution proffered to steer individual or consumer behavior (Thaler and Sunstein 2008). It refers to structuring the choice architecture that influences and enables individuals to make choices in desirable directions. Of course, whose desire is the question, and Thaler and Sunstein refer to nudging as "libertarian paternalism". Employing nudging as a behavior change strategy has been spurred by at least two key factors: the first is cognitive research into the limitations of human decision-making capacities due to limited cognitive resources and the bounded rationality nature of our functioning (Kahneman and Tversky 1979). Due to information processing limitations, people rely more on habits in their everyday life so that initiating habit change through making more information about options available requires considerable mental resources (Verplanken et al. 1998). Second, the urgency of climate change requires immediate action and it is claimed that nudging can act as a necessary shortcut for the rapid reduction in emissions. However, many scholars argue that nudging is limited as it only produces short-lived results and not lasting changes in people's lifestyles. Moreover, and more importantly, we contend that such choice reduction leaves the human potential for transformation untapped, as it does little to mobilize the ability for bottom-up sustainable innovation.

In order to take the measures needed to avoid the negative consequences of climate change, social sciences need to focus attention on how to mobilize the creative potential of individuals in collectives, and to understand how they can be supported in identifying and implementing innovative and acceptable solutions for changes in lifestyles that go beyond token actions for the environment. Moreover, the mobilization of such potential can provide a way out of the lock-in of opposing interests (Unruh 2000). Industries are sometimes slow in implementing changes as they consider that they (and their shareholders) have much to lose in the short term. The same is true of politicians who are reticent to act against climate change as they think the public might punish them if they legislate drastic measures. Individuals organized in collective structures are already initiating and promoting change both through social innovation initiatives for alternative lifestyles and economies, and through more traditional forms of collective organization such as trade unions. One of the central places where people act collectively is the workplace. This includes the production process itself, which is based on cooperation as well as collective organizing in trade unions. It is thus surprising that environmental psychologists have undertaken comparatively little work in the workplace.

Where are the workers? In 2009, Giuliani and Scopelliti published a comprehensive empirical analysis of how research interests in environmental psychology have changed over the years. They employed a systematic and sophisticated statistical analysis of papers published in the *Journal of Environmental Psychology* and *Environment and Behavior* since 2005. The papers "were classified in relation to the following criteria: mode of human-environment transaction, research topic, type of setting and function of places, sociodemographic characteristics and environmental role of people, mode of presentation of the setting, sampling procedure, and source of data." (Giuliani and Scopelliti 2009: 375).

Using Multiple Correspondence Analysis and Hierarchical Cluster Analysis, the authors identified six clusters of research activity. Work environments comprised the fifth cluster with only 7 % of published papers, and were seen to focus largely on "the responsive mode of person-environment transaction[s]" (*ibid*: 380). Most of this research involved post-occupancy evaluation studies and concentrated on reactions to different environmental conditions (e.g., lighting, temperature, and windows) and their effects on work performance, well-being and satisfaction (Rioux et al. 2013). Reviewing publications over the last few years within and beyond these journals suggests that the proportion of papers on work environments and the topics investigated have not changed much, although there has been a growth in recent years on the greening of the office (Bringslimark et al. 2009; Uzzell 2013) and changing sustainable behaviors in the workplace (Young et al. 2013). While this research is important, the office and the service sector are not ubiquitous working environments. Industrial employment accounts for 23 % of global employment, while just under 32 % of the world's workforce were employed in the agricultural sector in 2013 (International Labour Office 2014). There is little environmental psychology research in these settings. In all countries, the majority of the population spends most of their waking hours going to, at, or coming home from work. Yet the attention given to this 'paramount reality' for most people is very limited compared with other areas of environmental activity and sectors.

In addition, the current emphasis remains on reactive rather than active modes of transaction with the environment (Stokols 1978). More research on active modes that focus on 'bottom-up' solutions to sustainability problems and employee participation by those who work at the interface of production processes might lead to a more creative and effective impact by environmental psychologists than has hitherto been achieved. Conducting research outside the office in industrial and agricultural settings would also extend the potential impact of environmental psychology as well as the validity of its findings as it attempts to search for more generic understandings of people-environment relationships. If the aim is to create a society that provides 'quality of life' as envisaged in the first part of this handbook, then research cannot only focus on working spaces and working conditions, as important as these might be; it is the worker and her/his environment-transforming practices that need to come to the forefront of research, such as in a recent book on workers' practices at work and at home (Bolzan de Campos 2012).

31.3 Example One: Constructing Futures with Workers

Concerns about workplace well-being have emphasized autonomy as one important contributory factor (Moreau and Mageau 2012; Trépanier et al. 2013), which has also been proposed as a fundamental human need (Deci and Ryan 2000). Organizational research has occasionally discussed worker well-being or satisfaction and tried to make the business case for taking up these issues in terms of productivity. A transformative perspective, however, seeks to go further by questioning whether constantly increasing productivity should even be a goal in organizations. This goal has often led to the introduction of technologies requiring fewer people, broadening the gap between the skilled and the unskilled, and contributing to more stress and associated illnesses and thus a poorer quality of life. This is not to say that we should not improve productivity or end drudge jobs, but it does raise questions at a societal level concerning the way we think about and value work (Schor 2010).

Our own research on organizations has shown that creating contexts for autonomy contributes to workers being more pro-active in suggesting ways to transform everyday practices in the workplace. The issue of autonomy is a complex one when it comes to environmental practices. Organizational constraints, such as one's job description, organizational infrastructure or technology, can edit individual choices, in some cases making them almost automatic and thus unquestionable and unchallenged. 'Hard' strategies focusing on technological fixes and ecological modernization (Mol et al. 2009) can clearly play a role, along with 'soft' approaches such as green defaulting (Pichert and Katsikopoulos 2008), but they all have the disadvantage that they do little to address the underlying causes of environmental degradation (Foster 2001). Moreover, it has been found that automated technology may impair personal responsibility and undermine even simple actions to support sustainability (Murtagh et al. 2015).

Research has shown that pro-environmental self-identity is a determinant of pro-environmental behavior (García-Mira and Dumitru 2014b) and could potentially, in some circumstances, contribute to the transference of practices from one life domain to another (e.g., between home and work). One of the conditions of the development of pro-environmental self-identity is carrying out behaviors that are pro-environmental and being aware of it. Initiating these behaviors is more powerful in terms of identity than just passively carrying them out. Moreover, those workers who are frontrunners in the development of pro-environmental practices can actually see their initiatives stifled by a too tightly constrained environment (García-Mira and Dumitru 2014a). Furthermore, if all autonomy is restricted by replacing choice with automation (e.g., lights switching on and off automatically), sustainable behaviors will be limited to a given life domain and no transference of practices is possible to other life domains such as the home, or between behavioral dimensions (Thøgersen and Ölander 2003, 2006; Whitmarsh and O'Neill 2010), so that consensus about end-goals for sustainable transformations becomes unlikely.

One of the organizations we studied in the LOCAW project (García-Mira et al. 2014) is a public university, which is a type of institution with a relatively high level of worker autonomy. Participatory bodies are an integral part of decision-making processes, and thus there is considerable room for the sustainable initiatives of workers to be taken up by management if these are perceived as having worker support. These initiatives could then be translated by management into specific measures and policies for sustainable everyday practices in the areas of energy consumption, waste generation and management, and work-related mobility. However, we found that, in spite of such high levels of autonomy and the existence of a wealth of effective suggestions to promote pro-environmental everyday practices by workers, the management of the university perceived that there was a low level of demand for pro-environmental options to be made available (e.g., in products in cafeterias, etc.). Middle-level decision-makers in each university center complained about the lack of spaces and contexts in which to share experience with other colleagues confronting the same issues, in order to come up with creative solutions to problems that have an environmental sustainability dimension. Thus, it can be seen how, even in organizations with a high degree of worker autonomy, contexts of peer-exchange are not necessarily present, nor are there contexts where innovative solutions can be found.

Involving workers in transforming organizations should go beyond solving everyday problems to encouraging strategic sustainability initiatives for which spaces are made available for their development, testing and implementation. Such contexts are likely to produce consensus about the final goals and outcomes of organizational change as well as successful pathways for transformation in organizations. In our own research, we explored the potential for worker participation in the design of sustainable future visions and solutions for the organization. We undertook a series of scenario development workshops, using a backcasting method, to obtain a worker-led perspective of what transition to a sustainable organization with a significantly lower level of CO_2 emissions would mean.

Backcasting scenarios constitute a relatively new methodology in the field of sustainability and climate change (García-Mira et al. 2012). Despite its appearance and theorization in the 1970s, it has only recently become widely used as an instrument in helping decision-making processes in policy-making. The backcasting scenario methodology appeared in response to the discontent with traditional methods of trend extrapolation in energy forecasting, where it was assumed that energy demand would increase gradually and renewable energy technologies and energy conservation efforts were ignored (Vergragt and Quist 2011).

In sustainability studies, backcasting scenarios allow us to envisage and analyze different types of sustainable futures and develop agendas, strategies and pathways to reach them (Vergragt and Quist 2011). The method has a strong normative component, as it starts from desirable future states or a set of objectives and then analyzes the steps and policies that are needed to get there, in order to design agendas that normally require cooperation and communication among different types of actors in complex socio-economic and political environments. It is considered a useful tool in moving towards alternative climate futures (Giddens 2011).

Backcasting methodologies seem to perform better when taking into account the systemic nature and high degree of uncertainty associated with the environmental problems we are now facing. Generally, they also assume that systemic changes in society are needed in order to reach the normative objectives established. Also called third-generation scenarios, they have characteristics that make them especially suitable for facilitating transitions to sustainability: they constitute a systemic approach; they are comprehensive and rely on the participation of relevant stakeholders; they acknowledge uncertainty and complexity as key characteristics of the analysis of the future; and they establish a normative stance in mapping the future.

An important debate in the field of backcasting for sustainability is centered on the question of who should develop the future visions. Some argue that they should be created by experts, while others are strong supporters of involving stakeholders in defining both the future visions and the strategic measures needed to get there (Robinson 1990; Robinson et al. 2011), as this produces higher-order learning (Brown and Vergragt 2008; Quist et al. 2011), a greater attachment to the goals, and stronger feelings of empowerment. When taking on a sustainability agenda in organizations, it seems rather obvious that it is necessary to involve stakeholders in the creation of the vision, as well as in the definition of the complex pathways to make it possible: participation in the establishment of goals is fundamental to personal identification with those goals and thus an important determinant of the willingness to put them into practice.

We used a backcasting scenario approach that is process-oriented, participative and iterative. Scenarios were narrative, as these are easier to handle by stakeholders than abstract representations about the future. An important novelty in our research was the use of backcasting scenarios with organizational stakeholders in order to envisage future sustainable visions of the organization within a sustainable regional and European context. This is scarce in backcasting research, as most studies have been developed around future visions of a region or city, in order to help policy-making for local, regional or national governments. Almost no studies have been undertaken to support transformation and sustainable changes in private and public organizations.

We used a multi-method approach, using focus groups to develop the scenarios, inspired in part by that used by Svenfelt et al. (2011) in their study on decreasing energy use in buildings but significantly adapted to fit the objectives of LOCAW. We combined this with the stepwise approach of Kok et al. (2011) to orient the process and help stakeholders disengage with the present and create truly innovative visions of the future, one of the hardest aspects of backcasting scenarios with both stakeholders and experts (Svenfelt et al. 2011).

The two backcasting exercises followed a different structure, as their objectives were different. The first exercise dealt with scenario development, which was achieved in two workshops: one dealing with the creation of visions for the future, and the second with defining the strategic pathways to reach them and the social actors who should be involved. The second exercise focused on providing feedback to participants on how policy measures would function in a simulated environment and asking the participants to suggest corrections to their initial proposals and the model design.

The backcasting exercises of the university were interesting due to the nature of the organization. As a public university, it is concerned both with being at the forefront of sustainability efforts and with maintaining the values associated with a high-quality education, such as collaborative face-to-face interaction. This limits the preferences for some sustainable options such as flexible working. The value-laden nature of the institution was visible in, for example, the long discussions related to the nature and purpose of education and the philosophy that should drive it. One could also observe a certain conservative approach, which was overcome after a while by the interventions of information technology specialists who provided a historical perspective of the development of technology in recent decades.

Several policy tracks and interventions were then developed from these scenarios to be implemented in the organizational simulations, using agent-based modeling (Matthews et al. 2007; Sánchez-Maroño et al. 2012). These policies were tested in different combinations to see their effects on the performing of certain behaviors and related emission levels, with highly informative conclusions on the types of policies that were likely to be most effective. Policies were derived targeting changes in three areas of environmental practice: the consumption of materials and energy, waste generation and management, and work-related mobility. More attention was paid to those practices responsible for higher emissions in each organization under study.

Besides the useful policy recommendations that such an exercise provides, the value of this approach lies in its potential use of organizational contexts to produce innovative worker-led sustainability proposals for transforming both production processes and everyday practices in the workplace. Trade unions are especially suitable groups for enabling this as they act at the intersection of societal concerns for jobs, worker well-being, and environmental sustainability. One could imagine trade unions from the North and South coming together to create alternative futures for transnational corporations, which would transform existing patterns of resource depletion, worker exploitation and conflicting power relations.

Finally, both workers and managers often need to learn about how to transform organizations and make them more sustainable, or how to find solutions to the intricate connections between maintaining jobs, a good quality of life for the workers and keeping within the healthy environmental boundaries of our planet. As these problems are complex, finding solutions requires contexts of social learning, exploration and testing of options, as well as a sense of shared responsibility between workers and managers. The workplace can be seen as a community of practice, in which individuals learn and construct their identities (Wenger 1998). The 'Communities of Practice' approach stresses the importance of creating adequate conditions to link experiences, reflection, and experimentation between individuals and groups (Reed et al. 2010).

Communities of practice are important for the functioning of any organization, but they become crucial for those that recognize knowledge as a key asset. Knowledge is created, shared, organized, revised, and communicated within and between these communities. Communities of practice fulfill a number of functions with respect to the creation, accumulation, and diffusion of knowledge in an organization because they are nodes for the *exchange and interpretation of information*; they can retain knowledge and *steward competencies* to keep the organization at the cutting edge and, finally, they provide *homes for identities* (Wenger 2000).

Communities of practice structure an organization's learning potential in two ways: through the knowledge they develop at their *core* and through interactions at their *boundaries*. Like any asset, these communities can become liabilities if their own expertise becomes insular. It is therefore important to pay as much attention to the boundaries of communities of practice as to their core, and to make sure that there is enough activity at these boundaries to renew learning. For while the core is the center of expertise, radically new insights often arise at the boundary between communities. Communities of practice truly become organizational assets when their core and their boundaries are active in complementary ways.

31.4 Example Two: Trade Unions as Environmental Actors

The second study focuses on the efforts of workers and trade unions to contribute to environmentally sustainable production and curb the damaging effects of climate change (García-Mira et al. 2014; Räthzel and Uzzell 2011, 2013). We chose trade unions as an example for collective action because they are potentially a major force that can address global issues like climate change. Transnational corporations (TNCs) are the largest carbon emitters (nearly two thirds of all the carbon emitted between 1751 and 2010 can be traced back to 90 companies (Heede 2014)) and are organized on a global scale (e.g., International Chamber of Commerce, World Economic Forum). Trade unions are the only organization in production that can potentially counter the power of TNCs. Trade unions are 'glocal' organizations in that they are organized at local, regional, and global levels, simultaneously close to local specificities and global processes. They are the largest international democratically-elected body of members representing all sectors of the economy. One of our questions was how and whether this potential is realized in trade union environmental policies.

As expected, the major issue workers and trade unions are struggling with is what they experience as the contradiction between the protection of jobs and the protection of the environment (Räthzel and Uzzell 2011). Numerous reports, papers and analyses have been published by trade unions, the International Labour Organization (ILO) and research institutes (e.g., Poschen 2012; ILO Regional Office for Asia and the Pacific and ILO Employment-Intensive Investment Programme 2011) showing that in a 'green economy', more jobs can be created than are lost when strong carbon-emitting production processes are closed down and replaced by low carbon production. As some unionists, especially in countries of the global South, have argued such future perspectives are no consolation to those who fear the loss of their jobs here and now. Thus, environmentally engaged unionists use a number of strategies to overcome the apparent conflict between jobs and the environment. These range from a belief in technological fixes through to the connection of immediate workers' interests with ecological transformations, an emphasis on connecting technological change with social change, and appealing to workers' social interests as workers *and* citizens. Especially important for environmental psychology are the ways in which workers' identities are tied to specific occupations and professions. For example, a truck driver cannot easily be trained to become an office worker, since for him (the vast majority of drivers are still male), such work is just 'paper pushing' (Räthzel and Uzzell 2013; Uzzell 2010). Our research demonstrates the necessity of a contextual analysis of individuals' practices. For example, even strong trade unions are limited in their possibilities to pursue an ambitious environmental agenda by the economic calamities that determine workers' lives. In the countries of the global South, the needs of immediate survival can result in global warming sinking to the bottom of the trade union agenda (as a unionist quoted workers in India saying: 'I will die sooner of poverty than from climate change'), while in the global North, unions like the Comisiones Obreras (CCOO) in Spain, which had created a nationwide network of environmental scientists to advise workers and unions across regional and sectorial spectra, felt they had to cut back their work severely with the arrival of the 2008 financial crisis.

Another perspective derives from our research on international trade unions and unions situated in the global South. International solidarity has been a defining characteristic of trade unions since the nineteenth century (Waterman and Timms 2005), but it has usually meant unions supporting each other in their local struggles. To incorporate a global phenomenon like climate change into the trade union agenda requires unions and their members to investigate the global effects of local processes and, if not form global alliances, at least develop empathy. As our research shows, perhaps the most serious obstacle for a global trade union strategy is the divide between unions of the global North and those of the global South. The history of colonialism continues to be reflected in North-South relationships between unions. While there are also multiple differences *between* Southern unions and *between* Northern unions, these are cross-cut by what Southern unionists experience as domination by Northern unions (Uzzell and Räthzel 2013). As one unionist from South Africa said: 'The Northern unions created the international unions a hundred years ago, they have the biggest resources, they own them. When the big boys want to do something it happens, no matter how much resistance there is from unions in the South'. This was not an exceptional comment. Talking to unionists in Brazil, South Africa, and India, we heard similar descriptions of North-South relationships within international unions from all our 13 interview partners in these countries, who belonged to nine different unions. Some unionists in the North self-critically confirmed these perceptions (Räthzel and Uzzell 2013). What this demonstrates is that a simple coming together of Northern and Southern actors does not solve the problem of existing power relations. Overcoming distance – psychological as well as physical – is not simply an issue of goodwill, since the barrier to collective actions was the structure and organization of international trade union bodies. Southern unions are under-represented in the leading bodies of these organizations, though not as much as Southern academics are under-represented in the decision-making bodies of international journals (see below). Some representatives of Northern unions whom we interviewed recognized this and some international unions have tried to bring about change, like the International Trade Union Confederation (ITUC), which uses some of its resources to fund the participation of Southern unions in international meetings. At the same time, exposure to the problems of the South has had a significant impact on Northern unions' environmental policies, which reveals that sharing experiences and listening to each other is *one* way of finding common ground, even if it is not a sufficient condition for effective solidarity.

Thus, another lesson to be learned for future research in environmental psychology is that institutions are not only 'environments' in the sense of their physicality, but that collective action is immersed within their structures and may be hampered or facilitated by them, as well as reproducing them. A research question for environmental psychology would be to examine the institutional structures that would favor people's collective practices. This would include a historical analysis of present-day institutions. To treat currently observable institutions and the patterns of practices within them as indicators of how people 'are' ignores the ways in which such practices came into being. Through a historical perspective that transcends the taken-for-granted here and now and sees organizations as a result of specific social practices within specific societal and institutions conditions, it is possible to envisage alternative practices enabled by alternative societal and institutional conditions.

31.5 Environmental Psychology and North-South Relationships

Climate change is recognized as a global problem, yet the concept of 'global' in psychological terms has come in for little scrutiny or definition such that it is almost an empty signifier. What is the global and where is it? Is it just somewhere else? Or is it everywhere because we are all globalized now? And how can psychologists tackle something that seems to be at the opposite end of the spectrum from the realm of the individual, their normal focus?

The global is relational, like space in general (Massey 2005). It can only be defined from a specific position, and from any specific position the global is elsewhere. One might argue then that there is no such thing as a 'global level'. Every place is local, and the global is the collective noun for the millions of places that are local where millions of actions occur that create global environmental change. Indeed, when it comes to climate change, this seems to be the appropriate definition. Wherever in the world emissions are produced, they add to global warming, which then has its effect on all other parts of the world. Yet, to describe global processes in this way is already to simplify them. Emissions are not produced evenly in all parts of the world and the detrimental consequences of carbon emissions for environmental degradation and health are not experienced equally across the globe (Patz et al. 2007). The highly industrialized countries in the global North produce the largest amount of emissions per capita,[1] while the countries most affected by global warming are situated in the global South. Some argue that globalization is a process that victimizes the local. Massey reminds us, however, that this is a one-sided view. The relationship between global and local *forces* has to be introduced into the analysis: local places like the City of London, she argues, "are the places in and through which globalization is *produced*: the moments through which the global is constituted, invented, coordinated" (Massey 2005: 101). In other words, societal power relationships within and across nation-states, regions and continents determine the ways in which climate change develops and affects people.

While the global challenge of climate change has become an important area of investigation and teaching in environmental psychology, the 'global' is also the guest who is rarely invited to dinner: there is an under-representation of global South perspectives on environmental issues in the most prominent journals of the global North. This is not only a result of where research is undertaken, and by whom, it also reflects a failure to recognize the non-universal context of theories and paradigms for understanding our world and the nature of environmental well-being.

Twenty years ago, Stokols (1995) suggested five areas of future research that would reflect society's concerns: "(a) toxic contamination of environments and rapid changes in the global ecosystem, (b) the spread of violence at regional and international levels, (c) the pervasive impact of information technologies on work and family life, (d) escalating costs of health care delivery and the growing importance of disease prevention and health promotion strategies, and (e) processes of societal aging in the United States and other regions of the world" (Stokols 1995: 828). Prediction is always a risky endeavor, and while these areas are reflected in journal publications as well as the programs of national research funding agencies, it would be difficult to make the case that they have assumed the significance anticipated in the environmental psychology research agenda. An exception is the first issue if it is reformulated in terms of pro-environmental behavior change and sustainable development. Stokols generously and self-critically reflects that his own 'geographic and cultural frame of reference on the environment and behavior field' influenced his priorities (p. 832), and subsequent history suggests that his frame and priorities were and remain congruent with the majority view in environmental psychology: Euro/US-centric research priorities have largely become normative and taken-for-granted by and for the rest of the world.

The issues of the global South only received a passing mention at the very end of Stokols' paper in a footnote: 'Additional topics that are likely to receive greater research attention in the future are the design of environments for living and working in outer space .... and the formulation of effective policies for reducing conflicts among industrialized and developing countries related to the contamination of shared environments and the depletion of natural resources' (Stokols 1995: 832).

It would be wrong to characterize environmental psychology as being completely dominated by the North. The two leading environmental psychology journals (i.e., *Journal of Environmental Psychology* and *Environment and Behavior*) attract papers from scholars across the globe; the editor of the former reports that he received submissions from over 40 countries in 2014 (Gifford 2015). But there are many parts of the world – whole continents – that are under-represented, even where there are active researchers. The problem is largely a structural one. Of course, one cannot publish environmental psychology research if there are no researchers doing such research, and it is not being suggested that where such research taking place it is purposely being excluded. Moreover, the dominance of the English language (as the international scientific language) is clearly an

---

[1] According to the World Bank, China produced 8,286,892 thousand metric tons of carbon dioxide emissions in 2010, while its per capita production was 6.2 metric tons. The USA produced 5,433,057 thousand metric tons in the same year while its per capita production was 17.6 metric tons. World Bank, World Development Indicators. http://wdi.worldbank.org/table/3.8# Accessed November 8, 2014.

impediment. It is a characteristic of work in the social sciences (as distinct from the natural sciences) to be context specific, and analyses do not necessarily travel well across borders. Is it not then imperative in the social sciences that when we are dealing with globalized problems, we do all we can to have a global understanding of them? The challenge for our sub-discipline is to try and find ways by which we can begin to counter some of these structural barriers. This might start with an examination of the editorial boards of 'international' academic journals.

The editorial board is meant to be the touchstone of a journal, ensuring standards, identifying trends and developments, advising on publication policy, attracting new authors and subject matter and providing expertise. Perhaps most importantly, editorial boards act as gatekeepers to a discipline. If we take the two aforementioned environmental psychology journals, we find that combined they have 128 scholars on their editorial boards. Of these, only two are from a country of the global South.

If an editorial board largely comprises scholars whose knowledge and interests are situated in the global North, this might accentuate 'othering'. How can editorial boards fulfill their brief of "identifying trends and developments ... attracting new authors and subject matter and providing expertise" if it is difficult for them, in terms of languages and the contexts they come from to engage with the majority of the world's academic community?

Unequal North-South relationships are also reflected in the ways in which research is conducted. Psychologists, for example, are required to operate under ethical principles and codes of conduct. If we take an example from the UK to illustrate the issue, a general statement prefaces the detailed ethics' code of the British Psychological Society: '*In all circumstances, investigators must consider the ethical implications and psychological consequences for the participants in their research. The essential principle is that the investigation should be considered from the standpoint of all participants; foreseeable threats to their psychological well-being, health, values or dignity should be eliminated*' (British Psychological Society 2009). Are the implications of research always considered from the standpoint of all participants? It may well be that environmental psychology is in a unique position here, especially if it is argued that the 'local' and the 'global' are inseparable. If research is conducted among a British population into attitudes towards biofuels, are farmers in Borneo, for instance, who provide the raw material for the fuel, interviewed about *their* attitudes too so that a more balanced evaluation of the desirability, implications and benefits of such a development can be made? In addition to the local/global issue, it might be argued that because psychology has focused on the individual's psychology, less attention has been paid to the wider community and ethical implications of people-environment relationships.

A consideration of North-South relationships opens up a field of investigation about the ways in which these are experienced and conceptualized in public discourses and in the everyday. In a study conducted among students in Sweden and the UK (Räthzel and Uzzell 2009a), we found that respondents saw the South as producing what they saw as global environmental degradation. Students identified poverty in developing countries, the industrialization of developing countries, and overpopulation as being the most significant causes of environmental degradation. These answers need to be seen in the context of constructions of the 'Other' through which the West/the global North defines itself. The complex and multidimensional ways in which this happens have been analyzed by many scholars, notably Edward Said (1979). Said argued that the East, in contrast to the evaluation highlighted above, could be the object of desire and admiration when it stands for the exotic, and the adventures and emotions that are forbidden within Western cultures. In '*Risk and the Other*', Joffe (1999) discusses this mechanism of projecting internal conflicts onto a threatening 'Other' for group relationships. Joffe argues that in periods of crisis, the out-group moves from being a vague threat and challenge to society's way of life, to being seen as the 'purveyor of chaos' (Joffe 1999: 23), thus creating a righteous 'us' and a disruptive and transgressive 'them' (Douglas 2005). In the context of environmental issues, this is highly pertinent: media reports over the last decade of 'tiger economies', 'China building two power stations every week' and 'the rise of the Indian middle classes demanding middle class consumer lifestyles' only seem to confirm that 'they' are responsible for bringing the [natural] system 'out of control' and into a state of chaos – *sine qua non* for 'otherness'. The denigration and fear of 'the Other' is also invoked in scenarios of the consequences of climate change, such as increased migration from the South to the North, which is then automatically associated with conflict.

What is not taken into consideration in these images of the South is that industrial development and consequent environmental degradation in the South and their wider impact on carbon emissions and climate change for the whole world are fuelling consumerism in the North *at the invitation of the North*. Many of our consumer goods, including designer and sports' clothes and electrical goods, are produced in China for Western consumption. At least 20 % of industrial investments in China come from Western corporations and smaller companies. The Chinese and Western economies are also interdependent in terms of trade and finance (Das Argument 2006; Arrighi 2005). For instance, in 2009, China was the number one trading partner of the EU in terms of imports. The EU imported products from China worth 214.8bn Euros, while it exported goods to China worth roughly a third of this, 81.6bn Euros (Eurostat 2010:16).

One of the much-lauded innovations to reduce carbon emissions has been the research and development of biofuels to replace fossil fuels. However, this has been at the expense of countries and livelihoods in the South. The US usage of corn for the production of ethanol significantly escalated the price of corn in Mexico leading to the so-called Tortilla crisis in 2007 (McMichael 2009). Another example is the cutting down of trees in large areas of South-East Asian forests to cultivate palm trees for biodiesel, leading to the drying out of the wetlands. Ironically, because this releases carbon, this has a greater detrimental effect than the positive effect intended by using it to replace fossil fuels (Henseling 2008: 830).

These 'power geometries', to use Massey's (2005) concept, can lead to processes of displacement in the global South, which can be understood as the mirror image of the global North. For example, in our research on trade unions and their environmental policies in India, South Africa, Brazil, Spain, the UK, and Sweden (Räthzel and Uzzell 2013), we have come across opinions in India that describe climate change as a conspiracy of the North to prevent the global South from developing. In South Africa, a union representative warned that: '... *for us in the poor South, we fear that some of the roots of climate change politics may be an attempt to monitor how we live our lives and to curb our consumption in the global South. We find implicit in some of the demands made by environmentalists in the West which seek to place limits on growth in the developing world, nothing but eco-imperialism*' (Jim 2009: 2).

The competition between workers of the global South and the global North for jobs has its effects on the ways in which climate change measures are judged and experienced in both hemispheres. In their concern to defend the means for their survival, workers in the global North and the global South overlook that they share vital interests, namely the need to protect the nature that nurtures them and to resist the economic system that sets them in competition with each other and endangers the very basis of their and the earth's well-being.

Why should any of this be of interest to environmental psychologists, either in the global North or in the global South? If environmental psychologists want to contribute to an understanding of the relationship between people and their environment from the point of view of the individual, they cannot focus on individuals alone as we have argued. Individuals do not act independently of the societal contexts in which they live, are socialized, develop their knowledge, worldviews and values, and create and are confronted with affordances for action. To understand individuals in their social-spatial contexts requires concepts and theories that are informed by and have been developed in relationship to the specific contexts they seek to understand

and interpret. In this we follow Raewyn Connell (2007), who refutes the idea that a universal set of concepts and theories can be used to explain social life in any corner of the world. Since the unequal process of globalization increases the interdependency of people across the planet and has corresponding effects on virtually all individuals, although in very different ways, it is essential for environmental psychologists, in collaboration with other social scientists, to enter into a global process of communication with a willingness to learn from each other. One of the touchstones of environmental psychology has been its recognition of the importance of the social and environmental context of human action. While the problems we face may be experienced in our homes and workplaces, neighborhoods and cities, their ontology is global. A new challenge for environmental psychology is to recognize that the 'production' of environmental psychology should be global and diverse.

Acknowledgement We would like to express our gratitude to the European Union 7th Framework Programme (Grant Agreement no 265155), the Swedish Research Council (DNR 421-2010-1990) and the Swedish Research Council for Health, Working Life and Welfare, (DNR: 2007-1491) for supporting some of the research discussed in this chapter.

References

Arrighi, G. (2005). Hegemony unravelling, part I. *New Left Review, 32*, 23–80.

Balint, P. J., Stewart, R. E., Desai, A., & Walters, L. C. (2011). *Wicked environmental problems: Managing uncertainty and conflict*. Washington: Island Press.

Bolzan de Campos, C. (2012). *Gestión ambiental y comportamiento proambiental de trabajadores: Aproximación de una muestra brasileña*. Saarbrücken: Editorial Académica Española.

Bringslimark, T., Hartig, T., & Patil, G. G. (2009). The psychological benefits of indoor plants: A critical review of the experimental literature. *Journal of Environmental Psychology, 29*(4), 422–433.

British Psychological Society. (2009). *Code of ethics and conduct*. The British Psychological Society. http://www.bps.org.uk/system/files/documents/code_of_ethics_and_conduct.pdf

Brown, H. S., & Vergragt, P. J. (2008). Bounded sociotechnical experiments as agents of systemic change: The case of a zero-energy residential building. *Technological Forecasting and Social Change, 75*(1), 107–130.

Carrus, G., Nenci, A. M., & Caddeo, P. (2009). The role of ethnic identity and perceived ethnic norms in the purchase of ethnical food products. *Appetite, 52*(1), 65–71.

Connell, R. (2007). *Southern theory: The global dynamics of knowledge in social science*. Cambridge/Malden: Polity Press.

Das Argument. (2006). *Grosser Widerspruch China. Das Argument 268 (Heft 5/6)*. Österr: Ökologie-Inst.

Deci, E. L., & Ryan, R. M. (2000). The "what" and "why" of goal pursuits: Human needs and the self-determination of behavior. *Psychological Inquiry, 11*(4), 227–268.

Douglas, M. (2005). *Purity and danger: An analysis of concept of pollution and taboo*. London/New York: Routledge.

Dumitru, A., García Mira, R., Craig, A., Omann, I., Hernandez, B., & Rauschmeyer, F. (2014). *A spatial and temporal approach to lifestyles*. Presented at the 1st pressure cooker symposium. DRIFT, 21st November 2014, Rotterdam.

European Environment Agency. (2011). *End-user GHG emissions from energy: Reallocation of emissions from energy industries to end users 2005–2009* (No. 19/2011). http://www.eea.europa.eu/publications/end-use-energy-emissions

Eurostat. (2010). *External and intra-European Union trade data 2004-09*. Luxembourg: Publications Office of the European Union. http://ec.europa.eu/eurostat

Fornara, F., Carrus, G., Passafaro, P., & Bonnes, M. (2011). Distinguishing the sources of normative influence on proenvironmental behaviors: The role of local norms in household waste recycling. *Group Processes & Intergroup Relations, 14*(5), 623–635.

Foster, J. B. (2001). Ecology against capitalism. *Monthly Review, 53*(5), 2–3.

García-Mira, R., & Dumitru, A. (2014a). Final report: LOCAW: low carbon at work: Modelling agents and organisations to achieve transition to a low carbon Europe. Report to EU, Brussels, FP7-CP-FP.

García-Mira, R., & Dumitru, A. (2014b). *Urban sustainability: Innovative spaces, vulnerabilities and opportunities*. A Coruña: Provincial de A Coruña and IEIP.

García-Mira, R., & Goluboff, M. (2005). Perception of urban space from two experiences: Pedestrian and automobile passengers. In R. García Mira, D. Uzzell, J. E. Real Deus, & J. Romay (Eds.), *Housing, space and quality of life* (pp. 7–16). Aldershot: Ashgate Publishing.

García-Mira, R., Real Deus, J. E., Durán, M., & Romay, J. (2003). Predicting environmental attitudes and behavior. In G. Moser (Ed.), *People, places and sustainability* (pp. 302–311). Göttingen: Hogrefe & Huber.

García-Mira, R., Dumitru, A., Vega-Marcote, P., Alonso-Betanzos, A. (2012). Patterns of sustainable production and consumption in large-scale organizations: Multi-method approaches to the study of workplace practices. In W. Sieber & M. Schweighofer (Eds.),

SPC meets industry. *Proceedings of the 15th European roundtable on sustainable consumption and production.* Bregenz: Ökologie-Inst.

García-Mira, R., Bonnes, M., Craig, A., Dumitru, A., Ilin, C., Räthzel, N.,...Uzzell, D. (2014). *LOCAW: Low carbon at work: Modelling agents and organisations to achieve transition to a low carbon Europe* (No. EU Grant Agreement No 265155). A Coruña: Institute of Psychosocial Studies and Research "Xoan Vicente Viqueira". http://www.locaw-fp7.com/index.php?pagina=reports

Giddens, A. (2011). *The politics of climate change.* Cambridge/Malden: Polity Press.

Gifford, R. (3 Jun 2015). Personal communication.

Giuliani, M. V., & Scopelliti, M. (2009). Empirical research in environmental psychology: Past, present, and future. *Journal of Environmental Psychology, 29*(3), 375–386.

Heede, R. (2014). Tracing anthropogenic carbon dioxide and methane emissions to fossil fuel and cement producers, 1854–2010. *Climatic Change, 122*(1–2), 229–241.

Henseling, K. O. (2008). Die Große Transformation. Strategischer Umbau der Stoff- und Energiewirtschaft. *Das Argument, 279*(Heft 6), 827–838.

Hertwich, E. G. (2005). Consumption and the rebound effect: An industrial ecology perspective. *Journal of Industrial Ecology, 9*(1-2), 85–98.

Hertwich, E. G. (2011). The life cycle environmental impacts of consumption. *Economic Systems Research, 23*(1), 27–47.

ILO Regional Office for Asia & the Pacific., & ILO Employment-Intensive Investment Programme. (2011). *Local investments for climate change adaptation: Green jobs through green works.* Bangkok: ILO.

International Labour Office. (2014). *Global employment trends 2014 risk of a jobless recovery?.* Geneva: International Labour Office. http://public.eblib.com/choice/publicfullrecord.aspx?p=1641543

Jackson, T., & Papathanasopoulou, E. (2008). Luxury or "lock-in"? An exploration of unsustainable consumption in the UK: 1968 to 2000. *Ecological Economics, 68*(1), 80–95.

Jim, I. (2009). *Global capitalism and the challenge of climate change.* Presented at the Climate Change Meeting IMF, Germany, 14–15 October 2009.

Joffe, H. (1999). *Risk and "the Other".* Cambridge: Cambridge University Press.

Kahneman, D., & Tversky, A. (1979). Prospect theory: An analysis of decision under risk. *Econometrica: Journal of the Econometric Society, 47*(2), 263–291.

Kok, K., van Vliet, M., Bärlund, I., Dubel, A., & Sendzimir, J. (2011). Combining participative backcasting and exploratory scenario development: Experiences from the SCENES project. *Technological Forecasting and Social Change, 78*(5), 835–851.

Massey, D. B. (2005). *For space.* London/Thousand Oaks: Sage.

Matthews, R. B., Gilbert, N. G., Roach, A., Polhill, J. G., & Gotts, N. M. (2007). Agent-based land-use models: A review of applications. *Landscape Ecology, 22*(10), 1447–1459.

McMichael, P. (2009). A food regime analysis of the "world food crisis". *Agriculture and Human Values, 26*(4), 281–295.

Mol, A. P. J., Sonnenfeld, D. A., & Spaargaren, G. (Eds.). (2009). *The ecological modernisation reader: Environmental reform in theory and practice.* London/New York: Routledge.

Moreau, E., & Mageau, G. A. (2012). The importance of perceived autonomy support for the psychological health and work satisfaction of health professionals: Not only supervisors count, colleagues too! *Motivation and Emotion, 36*(3), 268–286.

Murtagh, N., Gatersleben, B., & Uzzell, D. (2015). Does perception of automation undermine pro-environmental behaviour? Findings from three everyday settings. *Journal of Environmental Psychology, 42*, 139–148.

Patz, J. A., Gibbs, H. K., Foley, J. A., Rogers, J. V., & Smith, K. R. (2007). Climate change and global health: Quantifying a growing ethical crisis. *EcoHealth, 4*(4), 397–405.

Pichert, D., & Katsikopoulos, K. V. (2008). Green defaults: Information presentation and pro-environmental behaviour. *Journal of Environmental Psychology, 28*(1), 63–73.

Pol, E. (2007). Blueprints for a history of environmental psychology (II): From architectural psychology to the challenge of sustainability. *Medio Ambiente Y Comportamiento Humano, 8*(1/2), 1–28.

Poschen, P. (2012). *Working towards sustainable development opportunities for decent work and social inclusion in a green economy.* Geneva: International Labour Office. http://site.ebrary.com/id/10583481

Proshansky, H. M. (1987). The field of environmental psychology: Securing its future. *Handbook of Environmental Psychology, 2,* 1467–1488.

Proshansky, H. M., Ittelson, W. H., & Rivlin, L. G. (1970). *Environmental psychology: Man and his physical setting.* New York: Holt, Rinehart and Winston.

Quist, J., Thissen, W., & Vergragt, P. J. (2011). The impact and spin-off of participatory backcasting: From vision to niche. *Technological Forecasting and Social Change, 78*(5), 883–897.

Räthzel, N., & Uzzell, D. (2009a). Changing relations in global environmental change. *Global Environmental Change, 19*(3), 326–335.

Räthzel, N., & Uzzell, D. (2009b). Transformative environmental education: A collective rehearsal for reality. *Environmental Education Research, 15*(3), 263–277.

Räthzel, N., & Uzzell, D. (2011). Trade unions and climate change: The jobs versus environment dilemma. *Global Environmental Change, 21*(4), 1215–1223.

Räthzel, N., & Uzzell, D. (2013). *Trade unions in the green economy: Working for the environment.* New York: Routledge.

Reed, M., Evely, A. C., Cundill, G., Fazey, I. R. A., Glass, J., Laing, A., ...Raymond, C. (2010). What is social learning? *Ecology and Society*, *15*(4). http://www.ecologyandsociety.org/vol15/iss4/resp1/

Rioux, L., Le Roy, J., Rubens, L., & Le Conte, J. (2013). *Le confort au travail: que nous apprend la psychologie environnementale ?* Québec: Presses de l'Université Laval.

Robinson, J. B. (1990). Futures under glass: A recipe for people who hate to predict. *Futures*, *22*(8), 820–842.

Robinson, J., Burch, S., Talwar, S., O'Shea, M., & Walsh, M. (2011). Envisioning sustainability: Recent progress in the use of participatory backcasting approaches for sustainability research. *Technological Forecasting and Social Change*, *78*(5), 756–768.

Said, E. W. (1979). *Orientalism*. New York: Vintage Books.

Sánchez-Maroño, N., Alonso-Betanzos, A., Fontenla-Romero, O., Bolón-Canedo, V., Gotts, N. M., Polhill, J. G., ...García-Mira, R. (2012). An agent-based prototype for enhancing sustainability behavior at an academic environment. In *Highlights on practical applications of agents and multi-agent systems* (pp. 257–264). New York: Springer.

Schor, J. (2010). *Plenitude: The new economics of true wealth*. New York: Penguin Press.

Schultz, W. P., Khazian, A. M., & Zaleski, A. C. (2008). Using normative social influence to promote conservation among hotel guests. *Social Influence*, *3*(1), 4–23.

Stern, P. C. (2000). Toward a coherent theory of environmentally significant behavior. *Journal of Social Issues*, *56*(3), 407–424.

Stokols, D. (1978). Environmental psychology. *Annual Review of Psychology*, *29*, 252–295.

Stokols, D. (1995). The paradox of environmental psychology. *American Psychologist*, *50*(10), 821.

Svenfelt, Å., Engström, R., & Svane, Ö. (2011). Decreasing energy use in buildings by 50 % by 2050: A backcasting study using stakeholder groups. *Technological Forecasting and Social Change*, *78*(5), 785–796.

Tajfel, H., & Turner, J. C. (1986). The social identity theory of intergroup behavior. In S. Worschel & W. G. Austin (Eds.), *Psychology of intergroup relations* (pp. 7–24). Chicago: Nelson-Hall.

Thaler, R. H., & Sunstein, C. R. (2008). *Nudge: Improving decisions about health, wealth, and happiness*. New Haven: Yale University Press.

Thøgersen, J., & Ölander, F. (2003). Spillover of environment-friendly consumer behaviour. *Journal of Environmental Psychology*, *23*(3), 225–236.

Thøgersen, J., & Ölander, F. (2006). To what degree are environmentally beneficial choices reflective of a general conservation stance? *Environment and Behavior*, *38*(4), 550–569.

Trépanier, S.-G., Fernet, C., & Austin, S. (2013). The moderating role of autonomous motivation in the job demands-strain relation: A two sample study. *Motivation and Emotion*, *37*(1), 93–105.

Unruh, G. C. (2000). Understanding carbon lock-in. *Energy Policy*, *28*(12), 817–830.

Uzzell, D. (2010). Psychology and climate change: Collective solutions to a global problem. *British Academy Review*, *16*, 15–16.

Uzzell, D. (2013). Greening the office and job satisfaction. In L. Rioux, J. Le Roy, L. Rubens, & J. Le Conte (Eds.), *Le confort au travail: que nous apprend la psychologie environnementale ?* (pp. 61–81). Québec: Presses de l'Université Laval.

Uzzell, D., & Räthzel, N. (2009). Transforming environmental psychology. *Journal of Environmental Psychology*, *29*(3), 340–350.

Uzzell, D., & Räthzel, N. (2013). Local place and global space: Solidarity across borders and the question of the environment. In N. Räthzel & D. Uzzell (Eds.), *Trade unions in the green economy: Working for the environment* (pp. 241–256). Oxford: Routledge.

Vergragt, P. J., & Quist, J. (2011). Backcasting for sustainability: Introduction to the special issue. *Technological Forecasting and Social Change*, *78*(5), 747–755.

Verplanken, B., Aarts, H., Knippenberg, A., & Moonen, A. (1998). Habit versus planned behaviour: A field experiment. *British Journal of Social Psychology*, *37*(1), 111–128.

Waterman, P., & Timms, J. (2005). Trade union internationalis and global civil society in the making. In H. Anheier, M. Kaldor, & M. Glasius (Eds.), *Global civil society 2004/5* (pp. 175–202). London: Sage.

Wenger, E. (1998). *Communities of practice: Learning, meaning, and identity*. Cambridge: Cambridge University Press.

Wenger, E. (2000). Communities of practice and social learning systems. *Organization*, *7*(2), 225–246.

Whitmarsh, L., & O'Neill, S. (2010). Green identity, green living? The role of pro-environmental self-identity in determining consistency across diverse pro-environmental behaviours. *Journal of Environmental Psychology*, *30*(3), 305–314.

Young, W., Davis, M., McNeill, I. M., Malhotra, B., Russell, S., Unsworth, K., & Clegg, C. W. (2013). Changing behaviour: Successful environmental programmes in the workplace. *Business Strategy and the Environment*. http://doi.org/10.1002/bse.1836.

CPSIA information can be obtained
at www.ICGtesting.com
Printed in the USA
LVHW10*2345170918
590499LV00007B/216/P